Microbial Metal and Metalloid Metabolism

ADVANCES
AND
APPLICATIONS

Microbial Metal and Metalloid Metabolism

ADVANCES AND APPLICATIONS

Edited by
John F. Stolz and
Ronald S. Oremland

ASM PRESS

Washington, DC

Cover images: From the top to the bottom: images one and three are electron micrographs of elemental Se precipitates produced by *Bacillus* sp. strain NS3, and image two is a TEM image of nanospheres produced by *B. selenitireducens* (Chapter 16, Fig. 1 and 2). Image four is a high-resolution TEM image of biogenic nanoparticles of UO_2 obtained from *Shewanella oneidensis* MR-1 under nongrowth conditions (Chapter 8, Fig. 1). Some images have been colorized.

Copyright © 2011 ASM Press. ASM Press is a registered trademark of the American Society for Microbiology. All rights reserved. No part of this publication may be reproduced or transmitted in whole or in part or reused in any form or by any means, electronic or mechanical, including photocopying and recording, or by any information storage and retrieval system, without permission in writing from the publisher.

Disclaimer: To the best of the publisher's knowledge, this publication provides information concerning the subject matter covered that is accurate as of the date of publication. The publisher is not providing legal, medical, or other professional services. Any reference herein to any specific commercial products, procedures, or services by trade name, trademark, manufacturer, or otherwise does not constitute or imply endorsement, recommendation, or favored status by the American Society for Microbiology (ASM). The views and opinions of the author(s) expressed in this publication do not necessarily state or reflect those of ASM, and they shall not be used to advertise or endorse any product.

Library of Congress Cataloging-in-Publication Data

Microbial metal and metalloid metabolism: advances and applications / edited by John F. Stolz and Ronald S. Oremland.
    p. ; cm.
  Includes bibliographical references and index.
  ISBN-13: 978-1-55581-536-3 (hardcover: alk. paper)
  ISBN-10: 1-55581-536-7 (hardcover: alk. paper) 1. Microbial metabolism.   2. Microorganisms—Composition. 3. Organometallic compounds.   I. Stolz, John F. II. Oremland, R. S. (Ronald S.)

QR88.M528 2011
572′.429—dc22

2011002134

10 9 8 7 6 5 4 3 2 1

Printed in the United States of America.

Address editorial correspondence to ASM Press, 1752 N St., N.W., Washington, DC 20036-2904, USA.

Send orders to ASM Press, P.O. Box 605, Herndon, VA 20172, USA.
Phone: 800-546-2416; 703-661-1593. Fax: 703-661-1501.
E-mail: books@asmusa.org
Online: http://estore.asm.org

DEDICATION

This volume begins with a tribute to Dr. Terry Beveridge as we felt it very appropriate to dedicate this volume to him (Chapter 1). Terry, who passed away in 2007, was a major contributor to the field of microbe/metal interactions and one who touched many of our lives, either directly or through his seminal work on microbe/mineral interactions.

CONTENTS

Contributors ix
Preface xiii

1. Geochemical Reactivity of Bacterial Surfaces: a Tribute to T. J. Beveridge
   F. Grant Ferris
   1

I.   ENVIRONMENTS   11

2. From Geocycles to Genomes and Back
   Arpita Bose, Sebastian Kopf, and Dianne K. Newman
   13

3. Hyperthermophile-Metal Interactions in Hydrothermal Environments
   James F. Holden, Angeli Lal Menon, and Michael W. W. Adams
   39

4. Microbe-Metal Interactions on Seafloor Basalts
   Jason B. Sylvan, Amanda G. Turner, and Katrina J. Edwards
   65

5. Microbial Transformations of Arsenic in the Subsurface
   Jonathan R. Lloyd, Andrew G. Gault, Marina Héry, and Jean D. MacRae
   77

II.   PROCESSES   91

6. Mineralogical Controls on Microbial Reduction of Fe(III) (Hydr)oxides
   Colleen M. Hansel and Christopher J. Lentini
   93

7. **Microorganisms and Processes Linked to Uranium Reduction and Immobilization**
Joel E. Kostka and Stefan J. Green
117

8. **Direct and Indirect Processes Leading to Uranium(IV) Oxidation**
Rizlan Bernier-Latmani and Bradley M. Tebo
139

9. **Anaerobic Respiratory Iron(II) Oxidation**
J. Cameron Thrash, Sarir Ahmadi, and John D. Coates
157

10. **Accentuate the Positive: Dissimilatory Iron Reduction by Gram-Positive Bacteria**
Kelly C. Wrighton, Anna E. Engelbrektson, Iain C. Clark, Ryan A. Melnyk, and John D. Coates
173

11. **Regulation of Arsenic Metabolic Pathways in Prokaryotes**
Chad W. Saltikov
195

III. NEW TECHNOLOGIES 211

12. **Transcriptome Analysis of Metal-Reducing Bacteria**
Dwayne A. Elias and Matthew W. Fields
213

13. **Application of Proteomics in Bioremediation**
Peter Chovanec, Partha Basu, and John F. Stolz
247

14. **Monitoring Microbial Activity with GeoChip**
Joy D. Van Nostrand, Sanghoon Kang, Ye Deng, Yuting Liang, Zhili He, and Jizhong Zhou
261

15. **Methods for Detection of Arsenate-Respiring Bacteria: Advances, Cautions, and Caveats**
John F. Stolz, Mahmoud M. Berekaa, Edward Fisher, Ganna Polshyna, Mirunalni Thangavelu, Rishu Dheer, Antonio Garcia Moyano, Samy El Assar, and Partha Basu
283

16. **Nanoparticles Formed from Microbial Oxyanion Reduction of Toxic Group 15 and Group 16 Metalloids**
Carolyn I. Pearce, Shaun M. Baesman, Jodi Switzer Blum, Jonathan W. Fellowes, and Ronald S. Oremland
297

17. **Microbial Respiration of Anodes and Cathodes in Electrochemical Cells**
Kelvin B. Gregory and Dawn E. Holmes
321

Index 361

CONTRIBUTORS

Michael W. W. Adams
Department of Biochemistry and Molecular Biology, University of Georgia,
Athens, GA 30606

Sarir Ahmadi
Department of Plant and Microbial Biology, University of California,
Berkeley, CA 94720

Shaun M. Baesman
U.S. Geological Survey, Water Resources Division, Menlo Park, CA 94025

Partha Basu
Department of Chemistry and Biochemistry, Duquesne University, Pittsburgh, PA 15282

Mahmoud M. Berekaa
Environmental Sciences Department, Alexandria University, Alexandria, Egypt

Rizlan Bernier-Latmani
Environmental Microbiology Laboratory, Ecole Polytechnique Fédérale de Lausanne,
CH 1015 Lausanne, Switzerland

Arpita Bose
Department of Organismic and Evolutionary Biology, Harvard University,
Cambridge, MA 02138

Peter Chovanec
Department of Biological Sciences, Duquesne University, Pittsburgh, PA 15282

Iain C. Clark
Department of Plant and Microbial Biology, University of California,
Berkeley, CA 94720

John D. Coates
Department of Plant and Microbial Biology, University of California,
and Earth Sciences Division, Ernest Orlando Lawrence Berkeley National Laboratory,
Berkeley, CA 94720

Ye Deng
Institute for Environmental Genomics, Department of Botany and Microbiology,
University of Oklahoma, Norman, OK 73019

Rishu Dheer
Department of Biological Sciences, Duquesne University, Pittsburgh, PA 15282

Katrina J. Edwards
Geomicrobiology Group, Department of Biological Sciences,
Marine Environmental Biology Section, and Departments of Biological Sciences and
Earth Sciences, University of Southern California, Los Angeles, CA 90089

Samy El Assar
Botany Department, Alexandria University, Alexandria, Egypt

Dwayne A. Elias
Biosciences Division, Oak Ridge National Laboratory, Oak Ridge, TN 37831-6036

Anna E. Engelbrektson
Department of Plant and Microbial Biology, University of California,
Berkeley, CA 94720

Jonathan W. Fellowes
School of Earth, Atmospheric & Environmental Sciences, University of Manchester,
Manchester, M13 9PL United Kingdom

F. Grant Ferris
Department of Geology, University of Toronto, 22 Russell St.,
Toronto, ON M5S 3B1, Canada

Matthew W. Fields
Department of Microbiology, Center for Biofilm Engineering, Montana State University,
Bozeman, MT 59717

Edward Fisher
Department of Biological Sciences, Duquesne University, Pittsburgh, PA 15282

Andrew G. Gault
Department of Geological Sciences and Geological Engineering, Queen's University,
Kingston ON K7L 3N6, Canada

Stefan J. Green
Georgia Institute of Technology, School of Biology and School of Earth and
Atmospheric Sciences, Atlanta, GA 30332-0230

Kelvin B. Gregory
Department of Civil and Environmental Engineering, Carnegie Mellon University,
119 Porter Hall, Pittsburgh, PA 15213

Colleen M. Hansel
School of Engineering and Applied Sciences, Department of Earth and Planetary Sciences,
Pierce Hall, Room 118, 29 Oxford Street, Harvard University, Cambridge, MA 02138

Zhili He
Institute for Environmental Genomics, Department of Botany and Microbiology,
University of Oklahoma, Norman, OK 73019

Marina Héry
HydroSciences UMR 5569 CNRS—Universités Montpellier I and II—IRD, Place Eugene Bataillon, CC MSE 34095 Montpellier cedex 5, France

James F. Holden
Department of Microbiology, University of Massachusetts, Amherst, MA 01003

Dawn E. Holmes
Physical and Biological Sciences, S305, Western New England College, Springfield, MA 01119

Sanghoon Kang
School of Science and Computer Engineering, University of Houston—Clear Lake, Houston, TX 77058

Sebastian Kopf
Division of Geological and Planetary Sciences, Caltech, Pasadena, CA 91125

Joel E. Kostka
Georgia Institute of Technology, School of Biology and School of Earth and Atmospheric Sciences, Atlanta, GA 30332-0230

Christopher J. Lentini
School of Engineering and Applied Sciences, Engineering Sciences Laboratory, 58 Oxford Street, Room 305, Harvard University, Cambridge, MA 02138

Yuting Liang
Institute for Environmental Genomics, Department of Botany and Microbiology, University of Oklahoma, Norman, OK 73019

Jonathan R. Lloyd
School of Earth, Atmospheric and Environmental Sciences and Williamson Research Centre for Molecular Environmental Science, University of Manchester, Manchester M13 9PL, United Kingdom

Jean D. MacRae
Department of Civil and Environmental Engineering, University of Maine, Orono, ME 04469-5711

Ryan A. Melnyk
Department of Plant and Microbial Biology, University of California, Berkeley, CA 94720

Angeli Lal Menon
Department of Biochemistry and Molecular Biology, University of Georgia, Athens, GA 30606

Antonio Garcia Moyano
Centro de Biología Molecular, Universidad Autonoma de Madrid, Madrid, Spain

Dianne K. Newman
Divisions of Biology and Geological and Planetary Sciences and Howard Hughes Medical Institute, Caltech, Pasadena, CA 91125

Ronald S. Oremland
U.S. Geological Survey, Water Resources Division, Menlo Park, CA 94025

Carolyn I. Pearce
Pacific Northwest National Laboratory, 902 Battelle Boulevard, P.O. Box 999,
MSIN K8-96, Richland, WA 99352

Ganna Polshyna
Department of Chemistry and Biochemistry, Duquesne University,
Pittsburgh, PA 15282

Chad W. Saltikov
Department of Microbiology and Environmental Toxicology,
University of California, Santa Cruz, CA 95064

John F. Stolz
Department of Biological Sciences, Duquesne University, Pittsburgh, PA 15282

Jodi Switzer Blum
U.S. Geological Survey, Water Resources Division, Menlo Park, CA 94025

Jason B. Sylvan
Geomicrobiology Group, Department of Biological Sciences,
Marine Environmental Biology Section, University of Southern California,
Los Angeles, CA 90089

Bradley M. Tebo
Division of Environmental and Biomolecular Systems, Oregon Health &
Science University, 20000 NW Walker Rd., Beaverton, OR 97006

Mirunalni Thangavelu
Department of Biological Sciences, Duquesne University, Pittsburgh, PA 15282

J. Cameron Thrash
Department of Plant and Microbial Biology, University of California,
Berkeley, CA 94720

Amanda G. Turner
Department of Earth Sciences, University of Southern California,
Los Angeles, CA 90089

Joy D. Van Nostrand
Institute for Environmental Genomics, Department of Botany and
Microbiology, University of Oklahoma, Norman, OK 73019

Kelly C. Wrighton
Department of Plant and Microbial Biology, University of California,
Berkeley, CA 94720

Jizhong Zhou
Institute for Environmental Genomics, Department of Botany and
Microbiology, University of Oklahoma, Norman, OK 73019

PREFACE

The impact of microbial activity in shaping the composition of the soil, water, and atmosphere of the Earth over the millennia is now well recognized. Moreover, in a remarkable feat of biological evolution, over a third of the elements in the periodic table have found some use in the chemistry of life. Of the 114 elements typically listed in the table, fewer than half are inert or have no known biological function (Wackett et al., 2004). Over 60 elements (in elemental form or as compounds) are involved in some form of microbial structure or activity (Color Plate 1). Those who grow organisms in culture recognize the major (i.e., carbon, hydrogen, nitrogen, oxygen, phosphorus, and sulfur), minor (calcium, potassium, magnesium, sodium, and chlorine), and trace (cobalt, copper, iron, manganese, molybdenum, nickel, selenium, tungsten, vanadium, and zinc) elements as typical constituents of a defined medium. But there are other elements that are necessary for cellular processes, such as quorum sensing (e.g., boron), and in the active sites of metalloenzymes. Even toxic metals and metalloids can be involved in energy generation and carbon fixation. For example, tellurate can serve as an electron acceptor in anaerobic respiration (Chapter 16) and arsenite can serve as an electron donor in photoautotrophy (Chapter 11). The latter speaks directly to the fact that photosynthesis can be distilled down to the microbe having sufficient quanta of light of the appropriate wavelength for which its photosystem has been tuned (e.g., light harvesting and reaction center pigments), a source of electrons [e.g., H_2O, H_2S, $Fe(II)$, or $As(III)$], and carbon.

The recent discovery that arsenate may replace some of the cellular phosphate to allow for growth under phosphate-limited conditions suggests that crafty organisms may substitute for other essential elements (Wolfe-Simon et al., 2010), such as selenium substituting for sulfur. For those elements that do not have a direct biological function (e.g., uranium), their chemical form, toxicity, and mobility can nonetheless be influenced by microbes because they may be bound to the cell, transported out of the cell, oxidized, reduced, or methylated (Chapters 7 and 8). Iron continues to provide surprises with anaerobic iron oxidation (Chapter 9) and iron reduction by gram-positive bacteria (Chapter 10). Microbe-metal interactions are regulated by mineralogical (Chapter 6) and genetic (Chapter 12) controls. The net results of these activities and interactions are robust biogeochemical cycles.

A more practical ramification of microbe-mineral interactions is the realization that one cannot assume that an element will remain in a given chemical form or oxidation state once it has been released into the environment and exposed to microbial activity. Despite this revelation, we continue to introduce new chemicals and compounds into the environment with only rudimentary toxicology studies. Case in point, the organoarsenical roxarsone (3-nitro-4-hydroxy-benzene arsonic acid) has been used extensively in the production of broiler chickens in the United States as a prophylactic against coccidiosis and as a growth stimulant (Chapter 15). The compound passes through the chicken unmetabolized and into the litter. The litter is then applied to fields as fertilizer, whereupon soil microbes and those associated with the chicken frass degrade the roxarsone, releasing inorganic arsenic. Given that a sizeable percentage of the 9 billion chickens raised each year are given the feed additive and each chicken releases about 150 mg of roxarsone in its lifetime, poultry farming can be a significant non-point source of arsenic.

Many major discoveries and innovations have been made since the publication of the first volume of *Environmental Microbe-Mineral Interactions* ten years ago (Lovley, 2000). Recent advances in microbial ecology have been the result of employing both reductionist and holistic methodologies in an integrated approach, "from genes to geocycles" (Chapter 2). No longer dependent on culture, the geomicrobiologist has been free to investigate extreme environments where the microbial transformation of metals and metalloids is essential for survival, such as hydrothermal systems (Chapter 3), seafloor basalts (Chapter 4), and the subsurface (Chapter 5). Advances in technology such as pyrosequencing, microarrays, and mass spectrometry have resulted in a proliferation of the "–omics": genomics, metagenomics, transcriptomics, proteomics, metabolomics, and metallomics. The number of annotated microbial genomes is now over 1200 with more than four times as many in production. The latter has facilitated advanced proteomic, genomic, and metagenomic studies of microbe-metal interactions in pure culture and natural environments using transcriptomics (Chapter 12), proteomics (Chapter 13), and geochip (Chapter 14) as well as functional biochemical and molecular probes (Chapter 15). The discovery of pili that function as "nanowires" and the development of microbial fuel cells (Chapter 17) are particularly noteworthy, because they have resulted in a fundamental advancement in our understanding of microbial energy generation. Metal-reducing bacteria, such as *Geobacter sulfurreducens*, immobilized on an electrode can literally generate free electrons and be used to power small electronics.

The purpose of this volume is to provide an overview of the current state of the field as well as some prognostication for future directions. It covers a wide range of topics and approaches with contributions by both established leaders in the field and by up-and-coming new investigators. There are both general introductory chapters and more focused contributions; thus, this volume should be appropriate for advanced students, researchers, and professionals. We see its primary use as a reference, but it could be adopted for a graduate-level course (e.g., Applied and Environmental Microbiology, Biotechnology and Bioremediation). While we regret that we may be missing a few things, it does leave the door open for the next edition.

REFERENCES

Lovley, D. R. 2000. *Environmental Microbe-Metal Interactions*. ASM Press, Washington, DC.

Wackett, L. P., A. G. Dodge, and L. B. M. Ellis. 2004. Microbial genomics and the periodic table. *Appl. Environ. Microbiol.* **70:**647–655.

Wolfe-Simon, F., J. Switzer Blum, T. R. Kulp, G. W. Gordon, S. E. Hoeft, J. Pett-Ridge, J. F. Stolz, S. M. Webb, P. K. Weber, P. C. W. Davies, A. D. Anbar, and R. S. Oremland. 2010. A bacterium that can grow by using arsenic instead of phosphorus. *Science* doi: 10.1126/science.1197258.

JOHN F. STOLZ
RONALD S. OREMLAND

GEOCHEMICAL REACTIVITY OF BACTERIAL SURFACES: A TRIBUTE TO T. J. BEVERIDGE

F. Grant Ferris

An essential lesson to learn as a graduate student with Terry Beveridge was "Nature never jests," a quote I attribute to Swiss physiologist and poet Victor Albrecht von Haller (1708–1777; *Réflexions sur le systême de la génération de M. de Buffon*). This attitude is perhaps best reflected in Terry's pioneering studies on the molecular basis of bacterial cell wall reactivity, sorption of metals, as well as mineral precipitation and dissolution reactions. The essential link between these areas of research is the fundamental view that design and construction impart functionality.

Before I venture too much further, it is important to note that this tribute to Terry and his accomplishments is based largely on my own memories and experiences. If anything, my recollections are far too narrow in scope to fully capture the true breadth and scope of what Terry achieved as a scientist. Indeed, Terry's contributions to medical microbiology and his work on the development of new techniques in electron microscopy are not considered. That being said, it is important to recognize that Terry was a true and fearless natural philosopher who was comfortable with interdisciplinary research. Moreover, his distinguished mentorship is something that everyone who worked in the Beveridge laboratory retains as a deep and fond memory.

At this point, the best way forward for me will be to revisit and expand upon the importance of distinguished mentorship. The next direction to take will be to explore some of the early work on bacterial cell envelopes that paved the way for Terry and his studies on the geochemical reactivity of bacterial cell surfaces. Then, with the stage set, I will explore how the field of bacterial-metal-mineral interactions was and continues to be advanced by Terry's tutelage.

INSPIRATION FROM EXTRAORDINARY PEOPLE

I could not be more proud or inspired by my scientific family; Ph.D. supervisor Terry Beveridge (Fellow of the Royal Society of Canada, F.R.S.C.), Terry's and my postdoctoral supervisor W. S. Fyfe (Fellow of the Royal Society, F.R.S.C., Order of Canada), and Terry's Ph.D. supervisor R. G. E. Murray (Past President of the American Society for Microbiology, F.R.S.C., Order of Canada).

The first time I became acquainted with Terry was as a third-year undergraduate student, and I remember him as a demanding

---

F. Grant Ferris, Department of Geology, University of Toronto, Toronto, ON M5S 3B1, Canada.

instructor. Even then, he inspired curiosity and a desire to learn. This prompted me to complete my fourth-year undergraduate thesis under Terry's supervision, which thankfully, as a newly minted graduate in microbiology, landed me a job as a technician in his laboratory. A year later, I found myself in graduate school with Terry as my supervisor. There was never a question too silly, an idea too malformed, or criticism for taking a different line of experimentation. Just enduring encouragement to explore and be fearless.

Within a few years Terry introduced me first to R. G. E. Murray and then W. S. Fyfe. Little did I know at the time that the two of them were scientific giants in their disciplines of microbiology and geochemistry, respectively; instead, I came to know them as supportive individuals who were keen on the progress of my Ph.D. and postdoctoral research. I owe them much, but I owe even more to Terry for bringing me into the fold.

BREAKING THE RESOLUTION BARRIER OF LIGHT MICROSCOPY

The wavelength of visible light limits the resolution of light microscopes to about 0.2 µm. Higher resolution can be achieved by several means (e.g., atomic force microscopy), but with electromagnetic radiation shorter wavelengths are needed (Holt and Beveridge, 1982). This is the main advantage of scanning and transmission electron microscopes that boast electron beams with much smaller wavelengths than visible light. Above all, electron microscopy was the workhorse for much of Terry's research; certainly, a great deal can be learned about the microscopic world when it is brought into sharp focus.

The first practical electron microscope was developed by E. F. Burton in the Department of Physics at the University of Toronto in the late 1930s (Murray, 1988). Curiously, the initial purpose of the electron microscope was to study colloids. At present, these are perhaps better, or more infamously, known as nanoparticles.

In the early 1950s, using light and electron microscopy (at Cambridge, United Kingdom), R. G. E. Murray and C. F. Robinow demonstrated that, after ether-induced cytolysis, *Bacillus cereus* retained its shape owing to some kind of an enveloping cell wall (Robinow and Murray, 1953). It was not until 1954 that R. G. E. Murray acquired his own electron microscope at the University of Western Ontario to begin his more detailed investigations on the ultrastructure of bacterial cell envelopes (Murray, 1988).

A critical development in the years immediately after R. G. E. Murray and C. F. Robinow's exposition of the existence of a cell wall among gram-positive bacteria (e.g., the *Firmicutes*) was E. Kellenberger's use of osmium tetroxide and uranyl acetate to fix bacteria for thin-sectioning and electron microscopy (Kellenberger and Ryter, 1958). This technique not only improved contrast for electron microscopic studies, but also revealed the fine ultrastructural detail of bacterial cells for the first time, as well as the intracellular development of bacteriophage.

Thus began the undressing of bacterial cells, or as one might call it, the truly revealing *haute couture* period of bacteriology. Specifically, throughout the late 1950s and 1960s, heavy metal staining and chemical fixation techniques improved, as well as embedding resins for thin-sectioning. Moreover, structured surface arrays on bacterial cells were discovered, prompting R. G. E. Murray (1988) to observe, "It was obvious that these arrays, or S-layers as they are now termed, were a common feature of bacteria in nature . . . They needed serious study" (Fig. 1).

The enigma of S-layers persisted until the mid-1970s when two of R. G. E. Murray's graduate students managed to isolate S-layer proteins from *Aquaspirillum serpens* (F. Buckmire) and *A. putridiconchylium* (T. J. Beveridge). A key discovery was that these proteins would self-assemble on isolated outer membranes in the presence of Ca^{2+}, but were stripped off by Na^+ (Buckmire and Murray, 1976; Beveridge and Murray, 1976a). Herein begins the real

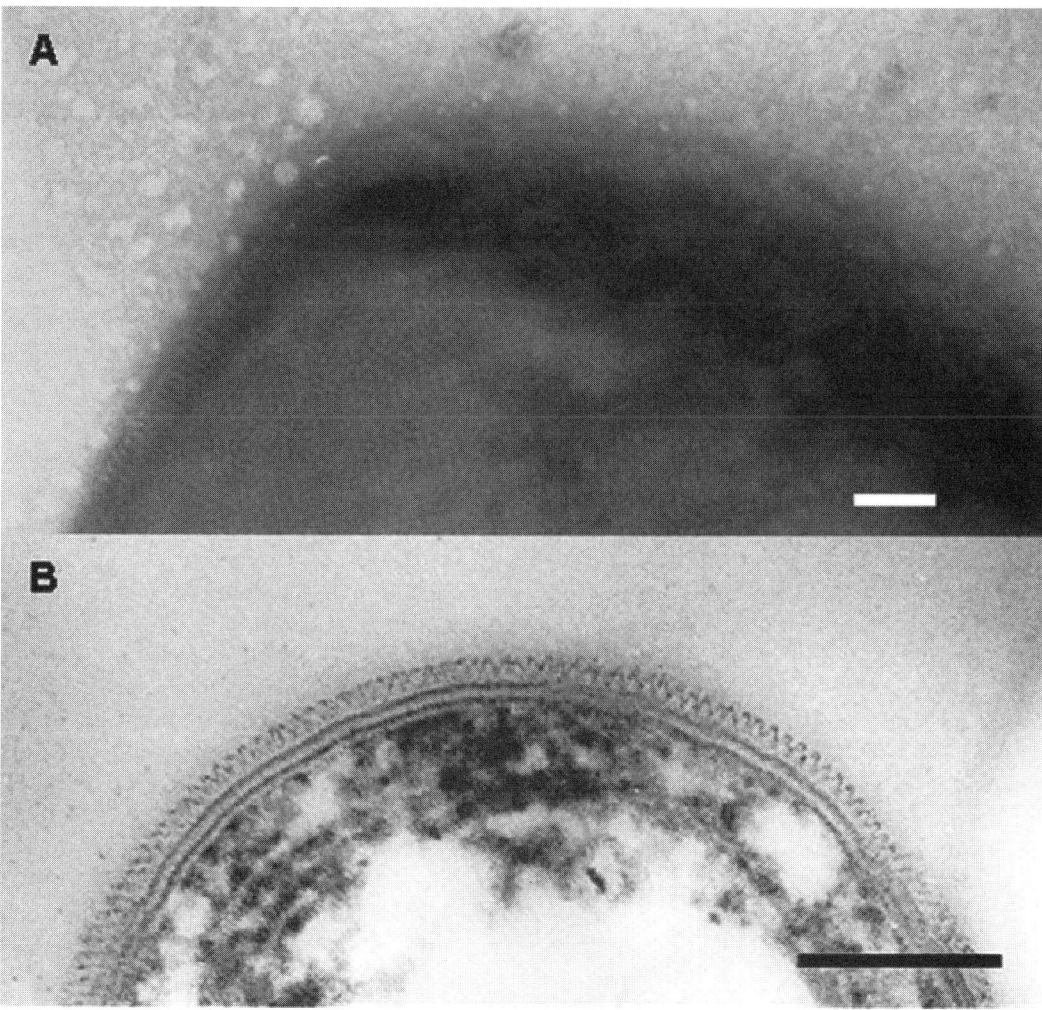

FIGURE 1 (A) Negative stain of a regularly structured surface array (S-layer) on an unidentified bacterium isolated from Green Dragon Spring in Yellowstone National Park, Wyoming (bar = 250 nm). (B) Thin section showing an S-layer on a bacterium in a microbial mat from Terrace Spring in Yellowstone National Park, Wyoming (bar = 250 nm).

story of bacterium-metal-mineral interactions and T. J. Beveridge.

THE CURIOUS MATH OF TERRY BEVERIDGE

As an astute graduate student of R. G. E. Murray, Terry recognized that the enhanced electron contrast provided by heavy metals in thin sections and whole mounts of bacteria had something to do with how the metals interacted with the macromolecular constituents of the cells. Furthermore, it was clear from his Ph.D. work that certain metals like Ca^{2+} were important for the maintenance of bacterial cell envelope structure and integrity. These two observations gave rise to four hypotheses, which have since curiously proven that one plus one equals four.

The four hypotheses were (i) metal ions bind to specific macromolecular constituents in the bacterial cell envelope, (ii) the amount of metal that is bound can be quantified, (iii) sites of metal ion binding can be identified, and (iv) the binding of different metals can be distinguished by transmission electron microscopy owing to atomic number contrast. There is a subtle elegance to these hypotheses, and working with gram-positive *Bacillus subtilis*, Terry compiled a substantial body of evidence to support each and every one (Beveridge and Murray, 1976b, 1980).

An obvious challenge to address after unraveling the gram-positive cell wall story was the issue of metal ion interactions with the cell envelope of gram-negative bacteria. Initial work with isolated whole cell envelopes from *Escherichia coli* K-12 confirmed that metal binding occurred, but not to the same extent as with *B. subtilis* cell walls (Beveridge and Koval, 1981). Then the peptidoglycan sacculus and outer membrane were implicated as separate entities capable of binding metallic ions. Finally, as part of my Ph.D. thesis work with Terry, the phosphate groups of lipopolysaccharide were identified as major sites for metal binding by using ^{31}P-nuclear magnetic resonance and paramagnetic metal cation probes (Ferris and Beveridge, 1985, 1986).

SORPTION TO MINERAL PRECIPITATION

As R. G. E. Murray (1988) observed:

> When T.J. Beveridge was working with me on his doctorate . . . we had many occasions to discuss revealing substructure by staining with heavy metal salts . . . But our geologist colleague, Professor W.S. Fyfe, was more interested in why many ore bodies have a high percentage of organic residues and particular selections of metals.

The result of these ruminations was an especially important paper, at least in my opinion, "Diagenesis of metals chemically complexed to bacteria: laboratory formation of metal phosphate, sulfide, and organic condensates in artificial sediments." These experiments demonstrated clearly that bacterial cells can have a profound impact on the kinds of minerals that develop in sediments under different geochemical conditions. They certainly set the stage for later laboratory and field experiments.

At some point in time during my Ph.D. thesis work, Terry dropped a copy of the diagenesis paper on my desk. I was perplexed because I did not know the meaning of either diagenesis or authigenesis, so I looked them up.

> Diagenesis—all physical, chemical, and biological modifications undergone by a sediment after its initial deposition.
>
> Authigenesis—the process in which new mineral phases are crystallized in sediments during diagenesis.

The definitions of these processes were intriguing and somehow managed to get stuck in the back of my mind. Not too long afterward, it was postdoc time!

To paraphrase R. G. E. Murray: when I was working with Terry on my doctorate, we had many an occasion to discuss bacterial interactions with metallic ions ... but a geologist colleague of Terry's and R. G. E. Murray, W. S. Fyfe, was more interested in bacterial mineral formation. The bait was taken hook, line, and sinker. Best of all, postdoctoral studies turned into an avalanche of revelations.

The idea that bacteria might influence mineral precipitation in natural sediments resulted in some curious undertakings on my part, all spurred on by Terry. First, for transmission electron microscopy, I decided to treat sediments as biological specimens. Second, my chemically fixed sediment samples were embedded in resin and thin-sectioned as normal for transmission electron microscopy. These initiatives had two outcomes of opposite degrees of happiness; thin-sectioning of the embedded sediment samples not only caused damage to diamond knives (in trouble, not happy), but also yielded thin sections for transmission electron microscopy that revealed an exceptionally intimate association between bacteria and minerals in natural sediments (forgiven, exceptionally happy) (Ferris et al., 1986, 1987a, 1987b, 1989b).

Inspired by the wonderful examples of bacteria-mineral interactions observed in hot spring (Fig. 2A) and lake sediments (Fig. 2B), I started to experiment to reproduce what was occurring naturally. These efforts were especially successful in terms of microbial silicification and preservation of bacterial microfossils (Ferris et al., 1988). An even greater advance was the idea to conduct experiments in the field to quantify microbial sorption of metals under acidic and neutral pH conditions (Ferris et al., 1989a). The field experimentation worked incredibly well, and is a true testament to the faith and understanding of both Terry and W. S. Fyfe.

These heady times were accompanied by substantial growth in the Beveridge Laboratory; more graduate students, more technicians, more postdocs, new equipment, a refurbished laboratory! Terry also insisted that I attend the International Symposium on Environmental Biogeochemistry (ISEB) in Nancy, France. This was my first scientific meeting outside of North America, and to this day, I could argue that it was one of the best things that Terry facilitated for me. It was at this meeting that I was introduced to the international geomicrobiological community among which I note especially H. L. Ehrlich, W. C. Ghiorse, R. Hallberg, W. E. Krumbein, and J. Berthelin among others; however, I also became acquainted with J. B. Thompson and the world of microbial carbonate precipitation (Thompson et al., 1990; Thompson and Ferris 1990).

At the same time, this was the beginning of the end for me as a postdoc with W. S. Fyfe and Terry. J. B. Thompson soon moved into the Beveridge laboratory as a postdoc to promote further studies on microbial carbonate precipitation, especially those of Terry's Ph.D. student S. Douglas who examined the role of S-layers in carbonate mineral precipitation induced by cyanobacteria. For me, it was off to the oil patch to deal with microbial corrosion of steel and secondary oil recovery.

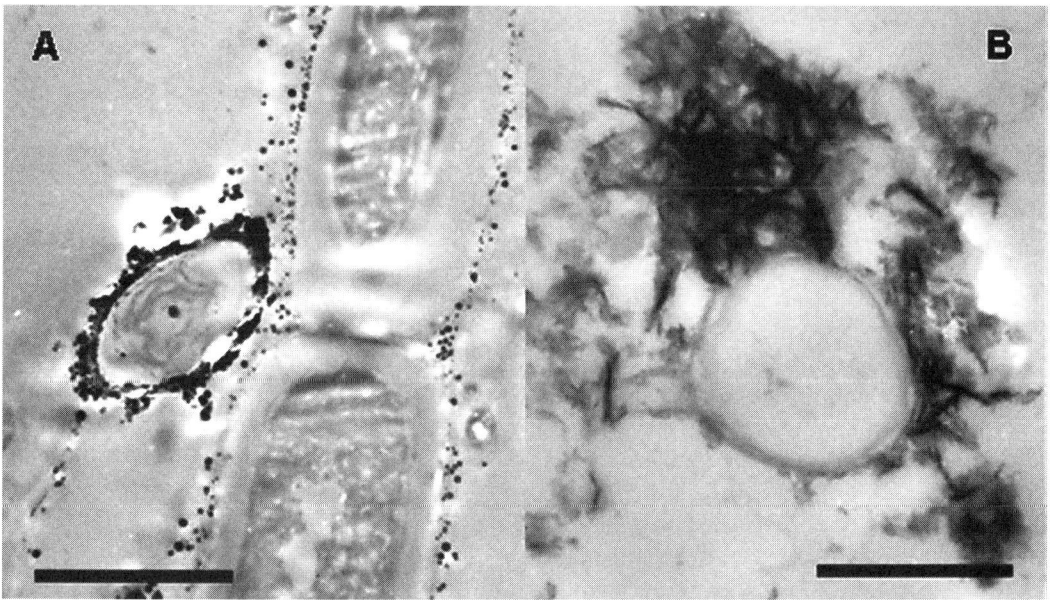

FIGURE 2 Unstained thin-section electron micrographs of cyanobacteria undergoing silicification in an Icelandic hot spring (bar = 2.0 µm) (A), and iron-silicate mineral precipitates accreting on the surface of a bacterial cell in sediment from Moose Lake, Ontario, Canada (B). The lake is impacted by acid mine drainage (bar = 1.0 µm).

CARBONATES AND SILICA REDUX

After a very rewarding time in the private sector, I returned to the academic world at the University of Toronto, my mom's alma mater, not to mention Terry's as well. It did not take long to reconnect with Terry. Shortly thereafter, working again with W. S. Fyfe, we forged an interesting connection between microbial carbonate precipitation, silicate weathering, and atmospheric carbon dioxide (Ferris et al., 1994). This has grown into a hot topic that neither I nor Terry might have anticipated when the study was first published.

I also caught up with J. B. Thompson and S. Douglas (now a Ph.D. graduate of Terry's and postdoc with me), which landed us in Central British Columbia, Canada, to study microbial carbonate precipitation in saline alkaline lakes and groundwater discharge zones (Ferris et al., 1995). A curious thing is that, in 1997, we described modern laminated freshwater microbialites in Kelly Lake (Ferris et al., 1997). Since then, Kelly Lake has been overshadowed by neighboring Pavilion Lake 14 km to the south, because it too has microbialites that have garnered attention as a much publicized Earth analogue study site for astrobiologists (Lim et al., 2009).

In a similar vein, microbial silicification was revisited with several studies in Iceland involving S. Douglas and K. O. Konhauser (Ph.D. graduate student of W. S. Fyfe, and postdoc with Terry and me). Fueled in our minds by the idea that such examples were good analogues for how microbial fossils were formed billions of years ago on Earth (Schultze-Lam et al., 1985), it was perhaps good fortune that the astrobiology community seized upon silicification as a good candidate for mineral preservation of microorganisms on extraterrestrial bodies. Certainly, excellent studies on microbial silicification continue to be conducted because this is by far the best known mechanism for the preservation of structurally intact microbial fossils, whether it be on Earth or arguably elsewhere (Ferris and Magalhaes, 2008).

SYMPHONY IN BACTERIAL CELL SURFACE REACTIVITY

There is little doubt that we live in a small world, and the world continues to get smaller for many reasons that need not be discussed here; however, Terry was a never ending source of contacts (remember that distinguished mentorship thing). From very early on (actually, as a graduate student) I came to know Y. A. Gorby (visiting Terry's laboratory), then later E. E. Rodin (through a postdoc of Terry's), and K. Pedersen (introduced by W. S. Fyfe, my postdoc supervisor via Terry and R. G. E. Murray). This interesting mix, coupled with some new developments on quantitative approaches to metal binding by bacteria, prompted a venture into the interfacial surface chemistry of bacterial cell-metal ion-mineral interactions.

Some of the more important outcomes of these connections were that iron oxide precipitation on bacterial cells is thermodynamically favored compared with precipitation in solution (Warren and Ferris, 1998); natural and synthetic composite mixtures of bacterial cells and iron oxides are important natural sorbents for metal ions (Ferris et al., 1999, 2000; Small et al., 1999); individual metal binding sites can be quantified in terms of total metal binding capacity (Martinez and Ferris, 2001; Smith and Ferris, 2001; Martinez et al., 2004); bacterial cell surface iron oxide precipitates interact chemically with specific surface functional groups and reduce the metal binding capacity of the composite relative to the individual end members (Smith and Ferris, 2003); and bacterial cell surfaces are characterized by a heterogeneous surface charge distribution (Cox et al., 1999; Sokolov et al., 2001). Moreover, surface charge distribution is remarkably consistent between individual species of a single genus implying that this physicochemical trait is at once highly conserved and critical to the survival of the species (Phoenix et al., 2007). This was my last publication with Terry.

As a final movement to the symphony, I find some solace in knowing that the last visit of a graduate student of mine to Terry's laboratory turned out to be a rondo of sorts that harkened back to the work of E. Kellenberger on bacteriophage and heavy metal staining of bacteria. I refer here to Jennifer Kyle who was engaged in exploring the geochemical reactivity of bacteriophage and captured some of the most compelling electron microscopic images of natural bacteriophage that have ever been obtained (Kyle et al., 2008a, 2008b). I was humbled as much as Terry would have approved.

SOME PARTING COMMENTS

I cannot end without emphasizing some additional aspects concerning the geochemical reactivity of bacterial cell surfaces that were emphasized by Terry. To begin with, bacteria (and bacteriophage) are subject to diffusion and viscous shear rather than advection in turbulent eddies (i.e., a Kolmogorov scale larger than the size range of bacteria and bacteriophage, even at high rates of fluid shear). This stresses the importance of diffusion-limited reactions, in particular, in terms of bacterial design strategies; for example, maximize surface area to volume, minimize diffusion distances, steepen diffusion gradients.

Next, it is important to recognize that all bacteria are obligated to grow from the inside out, and necessarily shed older cell wall fragments (and sometimes bacteriophage) into their surroundings. This means that new metal binding sites will be generated as bacterial cells grow, so absorbed as well as precipitated metals will slough off and be removed from the cell. In this sense, metal binding and precipitation by bacteria will be a continual source of particulate metals in natural systems.

Finally, one should never forget economy and efficiency of design. Specifically, energy is expended to synthesize and assemble all cell wall components. In this context, given the antiquity of prokaryotes, one must conclude that cell wall reactivity is representative of an early and highly conserved evolutionary adaptation.

NATURE NEVER JESTS!

Dedicated to my teacher, supervisor, mentor, colleague, and friend, TJB

The microbe is so very small
You cannot make him out at all.

But many sanguine people hope
To see him down a microscope.

Of lovely pink and purple spots
Composed of forty separate . . . dots

His eyebrows of a tender green
All these have never yet been seen.

But scientists who ought to know
Assure us that they must be so.

Oh! Let us never doubt
What nobody is sure about.

Hilaire Belloc
More Beasts for Worse Children (1897)

REFERENCES

Beveridge, T. J., and S. F. Koval. 1981. Binding of metals to cell envelopes of *Escherichia coli* K-12. *Appl. Environ. Microbiol.* **42:**325–335.

Beveridge, T. J., J. D. Meloche, W. S. Fyfe, and R. G. E. Murray. 1983. Diagenesis of metals chemically complexed to bacteria: laboratory formation of metal phosphate, sulfide, and organic condensates in artificial sediments. *Appl. Environ. Microbiol.* **45:**1094–1108.

Beveridge, T. J., and R. G. E. Murray. 1976a. Reassembly in vitro of superficial cell-wall components of *Spirillum putridiconchylium*. *J. Ultrastruct. Res.* **55:**105–118.

Beveridge, T. J., and R. G. E. Murray. 1976b. Uptake and retention of metals by cell walls of *Bacillus subtilis*. *J. Bacteriol.* **127:**1502–1518.

Beveridge, T. J., and R. G. E. Murray. 1980. Sites of metal deposition in the cell wall of *Bacillus subtilis*. *J. Bacteriol.* **141:**876–887.

Buckmire, R., and R. G. E. Murray. 1976. Substructure and in vitro assembly of outer structured layer of *Spirillum serpens*. *J. Bacteriol.* **125:**290–299.

Cox, J. S., D. S. Smith, L. A. Warren, and F. G. Ferris. 1999. Characterizing heterogeneous bacterial surface functional groups using discrete affinity spectra for proton binding. *Environ. Sci. Technol.* **33:**4514–4521.

Ferris, F. G., and T. J. Beveridge. 1985. Binding of a paramagnetic metal cation to *Escherichia coli*

K-12 outer membrane vesicles. *FEMS Microbiol. Lett.* **24**:43–46.

Ferris, F. G., and T. J. Beveridge. 1986. Site specificity of metallic ion binding in *Escherichia coli* lipopolysaccharide. *Can. J. Microbiol.* **32**:52–55.

Ferris, F. G., T. J. Beveridge, and W. S. Fyfe. 1986. Iron-silica crystallite nucleation by bacteria in a geothermal sediment. *Nature* **320**:609–611.

Ferris, F. G., C. M. Fratton, J. P Gerits, S. Schultze-Lam, and B. Sherwood Lollar. 1995. Microbial precipitation of a strontium carbonate phase at a groundwater discharge zone near Rock Creek, British Columbia, Canada. *Geomicrobiol. J.* **13**:57–67.

Ferris, F. G., W. S. Fyfe, and T. J. Beveridge. 1987a. Bacteria as nucleation sites for authigenic minerals in a metal contaminated lake sediment. *Chem. Geol.* **63**:225–232.

Ferris, F. G., W. S. Fyfe, and T. J. Beveridge. 1987b. Manganese-oxide deposition in a hot spring microbial mat. *Geomicrobiol. J.* **5**:33–42.

Ferris, F. G., W. S. Fyfe, and T. J. Beveridge. 1988. Metallic ion binding by *Bacillus subtilis*: implications for the fossilization of microorganisms. *Geology* **16**:149–152.

Ferris, F. G., R. O. Hallberg, B. Lyven, and K. Pedersen. 2000. Retention of strontium, cesium, lead, and uranium by bacterial iron oxides from a subterranean environment. *Appl. Geochem.* **15**:1035–1042.

Ferris, F. G., K. O. Konhauser, B. Lyven, and K. Pedersen. 1999. Accumulation of metals by bacteriogenic iron oxides in a subterranean environment. *Geomicrobiol. J.* **16**:181–192.

Ferris, F. G., and E. Magalhaes. 2008. Interfacial energetics of bacterial silicification. *Geomicrobiol. J.* **25**:333–337.

Ferris, F. G., S. Schultze, T. C. Witten, W. S. Fyfe, and T. J. Beveridge. 1989a. Metal interactions with microbial biofilms in acidic and neutral pH environments. *Appl. Environ. Microbiol.* **55**:1249–1257.

Ferris, F. G., K. Tazaki, and W. S. Fyfe. 1989b. Iron oxides in acid mine drainage environments and their association with bacteria. *Chem. Geol.* **74**:321–330.

Ferris, F. G., J. B. Thompson, and T. J. Beveridge. 1997. Modern freshwater stromatolites and thrombolites from Kelly Lake, British Columbia. *PALIOS* **12**:213–219.

Ferris, F. G., R. G. Wiese, and W. S. Fyfe. 1994. Precipitation of carbonate minerals by microorganisms: implications for silicate weathering and the global carbon dioxide budget. *Geomicrobiol. J.* **12**:1–13.

Holt, S. C., and T. J. Beveridge. 1982. Electron microscopy: its development and application to microbiology. *Can. J. Microbiol.* **28**:1–53.

Kellenberger, E., and A. Ryter. 1958. Cell wall and cytoplasmic membrane of *Escherichia coli*. *J. Biophys. Biochem. Cytol.* **4**:323–326.

Kyle, J. E., H. S. C. Eydal, F. G. Ferris, and K. Pedersen. 2008a. Viruses in granitic groundwater from 69 to 450 m depth of the Aspo hard rock laboratory, Sweden. *ISME J.* **2**:571–574.

Kyle, J. E., K. Pedersen, and F. G. Ferris. 2008b. Virus mineralization at low pH in the Rio Tinto, Spain. *Geomicrobiol. J.* **25**:338–345.

Lim, D. S. S., B. E. Laval, G. Slater, D. Antoniades, A. Forrest, W. Pike, R. Pieters, M. Saffari, D. Reid, D. Andersen, and C. P. McKay. 2009. Limnology of Pavilion Lake B.C.—Characterization of a microbialite forming environment. *Fundam. Appl. Limnol.* **173**:329–351S.

Martinez, R. E., and F. G. Ferris. 2001. Chemical equilibrium modeling techniques for the analysis of high resolution bacterial metal sorption data. *J. Colloid Interface Sci.* **243**:73–80.

Martinez, R. E., K. Pedersen, and F. G. Ferris. 2004. Cadmium complexation by bacteriogenic iron oxides from a subterranean environment. *J. Colloid Interface Sci.* **275**:82–89.

Murray, R. G. E. 1988. A structured life. *Annu. Rev. Microbiol.* **42**:1–34.

Phoenix, V. R., A. A. Korenevsky, F. G. Ferris, Y. A. Gorby, and T. J. Beveridge. 2007. Influence of lipopolysaccharide on the surface proton-binding behavior of *Shewanella* spp. *Curr. Microbiol.* **55**:152–157.

Robinow, C. F., and R. G. E. Murray. 1953. The differentiation of cell wall, cytoplasmic membrane and cytoplasm of Gram positive bacteria by selective staining. *Exp. Cell Res.* **4**:390–407.

Schultze-Lam, S., F. G. Ferris, K. O. Konhauser, and R. G. Wiese. 1995. In situ silicification of an Icelandic hot spring microbial mat: implications for microfossil formation. *Can. J. Earth Sci.* **32**:2021–2026.

Small, T. D., L. A. Warren, E. E. Roden, and F. G. Ferris. 1999. Sorption of strontium by bacteria, Fe(III) oxide and bacteria-Fe(III) oxide composites. *Environ. Sci. Technol.* **33**:4465–4470.

Smith, D. S., and F. G. Ferris. 2001. Computational and experimental approaches to studying metal interactions with microbial biofilms. *Methods Enzymol.* **337**:225–242.

Smith, D. S., and F. G. Ferris. 2003. Specific surface chemical interactions between hydrous ferric oxide and iron reducing bacteria determined using pK spectra. *J. Colloid Interface Sci.* **266**:60–67.

Sokolov, I., D. S. Smith, G. S. Henderson, Y. A. Gorby, and F. G. Ferris. 2001. Cell surface electrochemical heterogeneity of the Fe(III)-reducing bacteria *Shewanella putrefaciens*. *Environ. Sci. Technol.* **35:**341–347.

Thompson, J. B., and F. G. Ferris. 1990. Cyanobacterial precipitation of gypsum, calcite, and magnesite from natural alkaline lake water. *Geology* **18:**995–998.

Thompson, J. B., F. G. Ferris, and D. Smith. 1990. Geomicrobiology and sedimentology of the mixolimnion and chemocline in Fayetteville Green Lake, New York. *PALIOS* **5:**52–75.

Warren, L. A., and F. G. Ferris. 1998. Continuum between sorption and precipitation of Fe(III) on bacterial cell surfaces. *Environ. Sci. Technol.* **32:**2331–2337.

ENVIRONMENTS

I

FROM GEOCYCLES TO GENOMES AND BACK

Arpita Bose, Sebastian Kopf, and Dianne K. Newman

2

INTRODUCTION

A holy grail for environmental microbiologists is being able to predict the effects of any given microbial community on a particular environment. In an era of increasingly dramatic changes in global climate, this goal is becoming evermore important. It is now well accepted that microorganisms have had and continue to have a profound influence on shaping the chemistry of the Earth. It would thus be both intellectually satisfying and practically useful if we could enumerate the microbial players in a specific locale, and, knowing their metabolic potential and how they regulate their various metabolisms, make predictions about how their presence would shape the geochemistry of that locale as it evolves in time.

Despite significant progress that has been made in developing tools that would aid in this effort in the past decade, we are still very far from being able to accomplish this. This is due to many factors, including our limited understanding of how microorganisms catalyze reactions that have a geochemical impact. Although we have become good at identifying which organisms inhabit a particular site, we seldom have a complete grasp of their metabolic potential and how their metabolisms are regulated, and we have an even poorer understanding of the stability and catalytic rates of their biogeochemically relevant enzymes. While impressive efforts have been made in describing the metabolic potential of specific microbial communities through large sequencing projects (e.g., metagenomic and metaproteomic reconstruction of microbial communities in acid-mine drainage systems [Wilmes et al., 2009; Allen and Banfield, 2005]), large gaps in our understanding remain.

In this chapter, we discuss ways a budding geomicrobiologist might embark on the quest to understand how microbial communities affect their environment and to predict how they will respond in the face of environmental change. To this end, we first introduce various methods geomicrobiologists have at their disposal to achieve this goal, including both traditional (nonmolecular) and molecular methods. Because this book concerns microbial interactions with metals, we have chosen to focus our discussion on iron—one of the most ubiquitous and biogeochemically relevant metals in

---

Arpita Bose, Department of Organismic and Evolutionary Biology, Harvard University, Cambridge, MA 02138. *Sebastian Kopf*, Division of Geological and Planetary Sciences, Caltech, Pasadena, CA 91125. *Dianne K. Newman*, Divisions of Biology and Geological and Planetary Sciences and Howard Hughes Medical Institute, Caltech, Pasadena, CA 91125.

the environment. We provide a brief review of the (bio)geochemistry of this element before concluding the final portion of this chapter with a description of Lake Matano, an iron-rich environment that is geochemically fascinating with respect to metal cycling. We use Lake Matano as a case study to illustrate how the approaches described in the first section might be applied to gain insight into the complex interplay between microorganisms and geochemistry in a specific context.

METHODS AVAILABLE TO STUDY MICROBIAL COMMUNITIES

Today, geomicrobiologists can take advantage of numerous approaches to ask questions about the roles microorganisms play in any given place. Although lately there has been great enthusiasm for molecular techniques, there is still great value in using traditional methods to characterize the contributions of microorganisms to a system. We begin with a description of the latter, because nonmolecular methods provide a foundation upon which to perform molecular studies.

Nonmolecular Approaches

Nonmolecular methods include techniques that involve studying microbial processes in the field (in situ) and those that involve studying a microbial process in the laboratory in microcosms or as isolated reactions (in vitro). Each approach has its advantages and disadvantages, as we will discuss.

IN SITU METHODS

The success of in situ methods depends heavily on having information about a given environment. For most commonly studied environments, such as lakes, rivers, wetlands, soils, and sediments, this would include geochemical parameters such as pH, redox potential (E_h), solute composition, carbon sources, and the availability of terminal electron acceptors in addition to geophysical properties relevant to the respective site (temperature, water depth and stratification, soil horizons, seasonality, wind regimen, etc.). This information sets the stage for use of the various in situ approaches described hereafter. One has to be aware, however, that in situ approaches in general have a few caveats. When studying a natural system, the role of microbes cannot always be based on "guilt by association," i.e., the presence of an organism at a site does not necessarily imply that it is mediating the geomicrobial process under question. In addition, microbes almost never occur in isolation. Either they form stable associations called "consortia" or they form transient associations where only one or a few members of an association are of interest and the rest just hitchhike their way into one's data. In addition, microbes undergo processes of dispersal, where they might be present in a sample only because they were in transit through the area being sampled.

To simplify in situ studies, workers have divided microbes found in a particular sample into (i) indigenous microbes, which are part of the normal microflora of the sample site and most likely are the geomicrobial agents being sought; (ii) adventitious microbes, which are transient passerby organisms that came into the sample site during dispersal; and (iii) contaminants, which are organisms introduced during sampling (Ehrlich and Newman, 2009b). Using some assumptions, we can classify the organisms present in a particular sample. Indigenous microbes will likely represent the numerically dominant species. The adventitious microbes, in many cases, might be incapable of growth under the prevailing conditions in the sample site, while contaminants might be recognized as organisms that are unlikely to be present in the setting where they were found (Ehrlich and Newman, 2009b).

In Situ Microscopy. Use of microscopy to study the role of geomicrobial organisms is a very simple means of studying microbial diversity at a given site. Visual examination followed by light microscopy is a traditional tool for observing microbes when they occur in abundance (Brock, 1978). Samples can either be directly visualized using light microscopy or after acquiring the microbes of interest, such as by the buried slide method. In this method a slide buried at a location of interest, for example, in the sediments of a lake or river bed, is retrieved after a few days of incubation and is washed and stained appropriately (Lawrence et al., 1997). Photosynthetic organisms are autofluorescent and thus can be visualized using fluorescence microscopy (Lawrence et al., 1997). Capillaries have also been used to draw up microbes, which were then studied under the microscope (Perfil'ev and Gabe, 1969). However, various forms of electron microscopy are also used in the field for less abundant organisms. These include transmission electron microscopy (TEM), scanning electron microscopy (SEM), and environmental SEM (Baker and Banfield, 1998; Ghiorse and Balkwill, 1983; Edwards, et al., 1999; Jannasch and Wirsen, 1981; Sieburth, 1975) (Fig. 1).

In Situ Study of Geomicrobial Activity. Radioisotopes can be used to study the biogeochemistry of certain substances to determine ongoing geomicrobial processes. They are especially useful because of the high sensitivity of their measurement, which allows the experimentalist to add very little radioactivity to a sample even for the detection of slow geomicrobial processes. For example, the globally relevant contribution of microbial communities to the sulfur cycle in the form of sulfate reduction and sulfide oxidation could be studied using either $^{35}SO_4^{2-}$ or $H^{35}S^-$. A small, predetermined amount of these chemicals could be added to a closed vessel at the depth from which the sample was originally taken (Ivanov, 1968). After incubation and the likely action of the geomicrobial organisms present in the sample, $^{35}S^{2-}$ or $^{35}SO_4^{2-}$ could be separated and quantified. This technique is also applicable for the investigation of microbial interactions with metals. In the case of manganese, for example, it can be used to determine the existence and extent of biological manganese oxidation. There are two approaches: the first method assumes that reduced Mn is insoluble and that the change in dissolved $^{54}Mn^{2+}$ after removal of the precipitated form is a measure of microbial $^{54}Mn^{2+}$ oxidation; the second method measures microbially assimilated ^{54}Mn using a filter-based assay (Burdige and Kepkay,

A.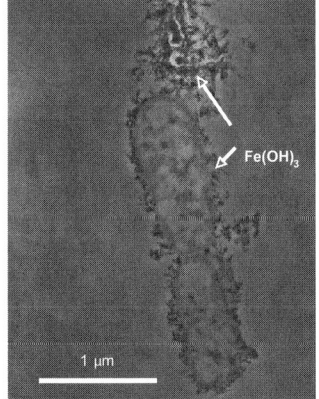

B.

FIGURE 1 TEM and SEM of *Rhodobacter* strain SW2 grown photoferrotrophically. (A) TEM image of *Rhodobacter* strain SW2 grown photoferrotrophically for 5 days. Arrows indicate Fe(III) precipitates. Image previously published as Figure 4B in Adamczyk et al. (2003). (B) SEM of *Rhodobacter* strain SW2 grown photoferrotrophically for 4 weeks showing the crystalline, regularly shaped Fe(III) precipitates.

1983). One could envision similar experiments being performed with radioisotopes of iron to quantify the rates of iron transformations, both oxidative and reductive, mediated by microorganisms under various conditions (e.g., in the presence or absence of nitrate; in the presence or absence of light).

On a different scale, voltammetric methods can be used to measure changes in concentrations of metals over short distances. Sensitive voltammetric microelectrodes exist for measuring oxygen, Mn^{2+}, Fe^{2+}, and HS^- (Luther et al., 1998, 2001). These techniques are particularly amenable to application in sediment or soil environments as well as microbial biofilms, where the density of microorganisms is higher than in the water column, and geochemical gradients can be expected to change over much smaller scales. However, for environments with slow chemical turnover and/or relatively fast diffusion, we would not expect the concentrations of such chemicals to change substantially and it would be necessary to use in vitro methods to better understand the steady-state in situ observations.

IN VITRO METHODS

In vitro methods can help support conclusions drawn regarding whether any given microbial process that is observed by use of in situ methods plays a role in a particular environment. However, such methods work best for simple microbial communities, and their success relies heavily on the accurate reproduction of the environment of the sampling site. In many cases, measurements made in the field represent steady-state concentrations of chemicals. However, microorganisms mediating various geomicrobial processes might require higher concentrations of electron donors and acceptors in nonflowthrough systems (e.g., batch culture), so care must be taken in the design of relevant in vitro experiments. In vitro laboratory reconstruction typically provides more energetically favorable conditions that can enable the selection of organisms that mediate the geobiological process of interest or help stably maintain a pure culture or consortium. In this regard, in vitro methods commonly deviate from the actual environment that they strive to reproduce. With this caveat in mind, geomicrobiologists have devised a number of methods to study microbial communities or pure cultures. These are divided into batch cultures and chemostats (White, 2000). Even though neither of the two methods exactly simulates the natural environment, they permit changes in microbial physiology along with changes in the chemical composition of the culture vessel to be monitored.

Batch Cultures and Chemostats. The batch culture method is especially popular for studying microbial processes mediated by pure cultures of microbes and is a closed-system approach. It consists of growing cells with a predetermined amount of carbon source, electron donor, and electron acceptor (either a rich/undefined medium made in a consistent manner or a defined medium can be used for such experiments). The growth of the microbial population leads to continuous changes in the medium; substrates deplete, products accumulate, and many inhibitory compounds that are products of metabolism result in defined phases of growth (White, 2000; Ehrlich and Newman, 2009b). These include a lag, exponential, stationary, and death phase (for details, read Novick [1955] and Ehrlich and Newman [2009b]). Batch culture experiments can establish the rates at which particular microorganisms catalyze a given reaction under well-defined conditions.

A chemostat is an open-system method where the total volume of medium remains constant. However, unlike a batch culture, the spent medium is replaced by fresh medium at a predetermined rate called the flow rate (for details, read Ehrlich and Newman [2009b], White [2000], and Herbert et al. [1956]). A steady-state situation allows the microbial population to grow at a maximum rate called the growth rate. However, if the populations were limited for any nutrient, the growth rate would change and adjust until a new steady state is reached. Using such concepts, the geomicrobiologist can determine the concentration of a limiting nutrient for a microbial population. Depending on how accurate the laboratory simulation is to the natural environment, chemostat experiments can be used to assess what nutrient the population might be limited for in its natural setting (Jannasch, 1967, 1969).

Culturing Techniques and the Enrichment Method. In vitro experiments usually rely on the utilization of pure strains or enrichment cultures. The pure-culture technique has classical microbiological roots and has served the field of medical microbiology very well (Schlegel, 1993). However, for the environmental microbiologist, this technique has not been as rewarding. This is based on the realization of the "great plate anomaly," i.e., that we can culture only 0.1 to 1% of the microbes from any given sample (Staley and Konopka, 1985). Thus, any inference that one makes about a community based on pure culture techniques is always incomplete. Nonetheless, to move from finding the microbe present in an environment to understanding its real role, one has to try to obtain pure cultures. To do so, we must realize that organisms like to grow under the conditions where they are found and that they might prefer growing in stable microbial communities. The role of geochemists in helping microbiologists understand the chemical composition of the environment is becoming apparent. Careful design of medium composition, microencapsulation followed by flow cytometry, and optical tweezer methods have helped obtain many species in pure culture (Kaeberlein et al., 2002; Zengler et al., 2002; Rappe et al., 2002; Svensson et al., 2004). However, the ability to sequence partially pure cultures to give rise to whole genome sequences for "*Candidatus*" strains helps minimize the anguish caused by the inability to obtain pure cultures (Duan et al., 2009; Tran-Nguyen et al., 2008; Pelletier et al., 2008; Jargeat et al., 2004; Wernegreen et al., 2002).

Together, these traditional approaches can be used to begin to describe the microbial contribution to geochemical processes in any environment, by trying to isolate and enrich for some of the strains involved in the geochemical reactions of interest, and by exploring their physiology and metabolic potential in batch cultures and chemostats. To follow up on such studies and better understand the variables that control the microbial community in situ, to study unculturable organisms, and to assess the actual function and importance of isolated strains in a particular environment, work on the molecular level is necessary.

Molecular Approaches

In the past few decades, environmental microbiology has been influenced heavily by advances in molecular biology. The ease and affordability of DNA sequencing has aided the incorporation of many molecular tools into the geomicrobiologist's tool kit. Molecular methods combined with classical microbial genetics and biochemistry can help us appreciate how microbes perform many geochemical processes. Three simple questions one can answer with molecular tools are: Who is there? What are they doing? How are they doing it? Ultimately, all three questions must be understood to predict the influence of a microbial population on a given environment.

WHO IS THERE?

Culture-independent approaches have taken environmental microbiology to a whole new level. The need for such methods arose with the realization that most microbes from a given environment are recalcitrant to being isolated in pure culture (Eilers et al., 2000). A failure to reproduce the natural conditions in which the organisms reside often contributes to this problem. In addition, microbes rarely exist in isolation and thus trying to purify them may disrupt important cell-cell interactions that are integral to their survival. One way around this is to extract lipids, DNA, and RNA from natural samples, and use them to infer the identity of the organisms present in that sample. 16S rDNA, i.e., the DNA sequence that encodes the 16S rRNA, has become the molecule of choice for phylogenetic identification (Stahl, 1997; Ward et al., 1990). This is because ribosomal RNA undergoes only minor sequence changes since it is part of the ribosome. Translation is an essential information pathway and extreme changes in the key machinery are rarely tolerated. This realization led Woese and coworkers to show that 16S rRNA sequences from diverse organisms can be used for phylogenetic analysis (Olsen and Woese, 1993).

The 16S rRNA molecule has been tested for its rigor at assigning an unknown microbe a phylogenetic identity for nearly four decades, and has been proven robust. Many variations have been developed over the years to exploit 16S rRNA. However, it should be noted that, while 16S rRNA/DNA-based approaches might help assign an unknown organism an identity, in most cases this does not predict its metabolism (McArthur, 2006). This is because microorganisms are capable of diverse metabolisms, some of which have been moved around through horizontal gene transfer over evolutionary history, giving their genomes a fluidity that restricts the utility of 16S rRNA/DNA to identification and phylogeny (Doolittle, 2000). In addition, other genes that are key players in other information pathways, such as transcription and DNA replication, can also add robustness to a phylogenetic assignment based on 16S rRNA/DNA (Olsen and Woese, 1993). Some organisms have multiple divergent *rrn* operons, i.e., ribosomal RNA encoding genes, which complicates identification of an unknown organism (Klappenbach et al., 2001). With these caveats in mind, 16S rRNA/DNA still stands as one of the most commonly used biomolecules for identification and phylogenetic assignment.

In the following sections, we briefly summarize various culture-independent techniques that have become widely used in the past few decades. Such molecular methods can be applied to any environment of interest.

PCR-Based Methods. PCR-based methods allow identification of known and unknown organisms exploiting the ability of degenerate primers targeting 16S rDNA to amplify divergent sequences (Guyer and Koshland, 1989). In practice, DNA is purified using commercial kits from any new isolate or complex microbial community. PCR and degenerate primers are used to amplify the 16S rDNA, and sequenced directly for a pure culture or prepared into a PCR clone library followed by sequencing individual clones. This sequence is then compared with public databases such as GenBank (http://www.ncbi.nlm.nih.gov/Genbank/) or Ribosome Database Project (http://rdp.cme.msu.edu/). Because the 16S rDNA sequence for any new isolate or microbial community studied is submitted to these public databases, a quick comparison of the unknown sequence with this database allows geomicrobiologists to find the closest phylogenetic relative of the microbe under study (Color Plate 2). However, the level of sequence identity required to assign a microorganism at the species level is debated. Recent studies show that closely related 16S rDNA sequences can be binned into clusters that represent bacterial taxa and are called an operational taxonomic unit (OTU). OTUs are defined as clusters with up to 2.5% sequence divergence in 16S rRNA (Hughes et al., 2001). This is based on the observed divergence seen within populations of known species (Stackebrandt and Goebel, 1994). Other workers have considered each 16S rRNA sequence type as a distinct OTU. However, no sequence-based OTU corresponds to the fundamental units of bacterial ecology; as pointed out earlier, 16S rDNA and other single-gene-based approaches, in most cases, are limited to identification (Cohan, 2002; Staley, 2003).

An application of the PCR-based approach that gives insight into community composition is denaturing gradient gel electrophoresis. This method separates PCR products of similar size but differing sequence (Color Plate 2). 16S rDNA can be amplified from a community and then hybridized with probes specific for a particular species being sought. In addition, individual bands of interest can be cut and eluted for reamplification followed by DNA sequencing to identify the associated microbe (Muyzer et al., 1993; Stahl, 1997; Ward et al., 1992, 1998). Another technique especially useful for comparing communities is terminal restriction fragment length polymorphism. One of the PCR primers used has a fluorescent label. DNA is amplified using PCR and digested with restriction enzymes. The DNA bands resulting from this digestion are separated according to size, and each fragment is detected by use of a laser detector that can detect the fluorescent label. Although a good measure of community structure, this method does not help in identification of individual organisms or OTUs (Clement et al., 1998; Dunbar et al., 2001; Liu et al., 1997). Finally, the use of DNA microarrays to catalogue microbial diversity is becoming increasingly popular. This method involves hybridization techniques using amplified DNA probed against a library of spotted DNA molecules on a glass slide that correspond to known organisms (Wu et al., 2006; Gentry et al., 2006).

FISH Methods. The large diversity of microbial 16S rDNA sequences available allow us to design methods to use these known sequences to identify close relatives in uncharacterized communities. The approach that has exploited growing 16S rDNA sequences most is fluorescence in situ hybridization (FISH), although autoradiography-based approaches have also been used (Giovannoni et al., 1988; Amann et al., 1990, 1995). rRNA is an ideal biomolecule for identification of microbes for the following reasons: (i) Ribosomes are essential to survival of all forms of life including microbes. Under most physiological conditions microbes have thousands of ribosomal particles, which results in natural amplification of a signal. (ii) 16S rDNA is conserved evolutionarily as discussed above. In addition, RNA molecules do not undergo evolution of the third (Wobble) base of each codon as happens in protein-coding genes. Thus, probes that span much larger regions of RNA can be designed, and probes can be constructed that detect much larger taxonomic units of *Bacteria* and *Archaea*. In practice, performing FISH is fairly straightforward (Color Plate 2); however, the success of FISH depends on many variables, which will be discussed later. First, the microbial cells in a given sample are fixed to stabilize cell morphology and permeabilize cells for later hybridization. Then, the cells are incubated with a labeled probe, at which point the probe enters the cell and hybridizes with the rRNA sequence; excess probe is washed off to reduce background. The sample can then be visualized by epifluorescence microscopy or the cells of interest can be sorted using flow cytometry. A database of successful and tested FISH probes is available on ProbeBase (http://www.microbial-ecology.net/probebase/) (Amann and Fuchs, 2008). One can also design a new FISH probe based on phylogenetic analysis from databases like SILVA (http://www.arb-silva.de/fish-probes/probe-design/) (Amann and Fuchs, 2008). However, success of the new probe entails iterative testing and optimization, making it a tedious, time-consuming process. It also requires that a pure culture of the organism be tested, although techniques such as Clone-FISH, which allows probes to be optimized without the need for pure cultures, can circumvent this issue (Amann and Fuchs, 2008). Various modifications of the basic FISH procedure have been developed to counter issues of failure to observe any/weak signal due to low ribosomal content or lack of active transcription. This includes catalyzed reported deposition FISH, a method that uses horseradish peroxidase-labeled oligonucleotide probes. The horseradish peroxidase catalyzes the deposition of tyramine molecules, which results in signal amplification (Pernthaler et al., 2002).

Whole-Genome Sequence Approaches (i.e., Phylogenomics). The caveats of using single-gene approaches to understand phylogeny ultimately affect our ability to identify organisms with certainty. Whole-genome sequencing is now affordable and routine in microbiology owing to the relatively small size of microbial genomes. This has led to the development of the concept of phylogenomics that uses all the genes in an organism to determine its phylogeny (for details, read Delsuc et al. [2005]). Availability of genomic data helps counter the effects due to too small sample size that phylogenetics faces by expanding the number of characters that can be used in phylogenetic analysis by orders of magnitude. Phylogenomics unlike phylogenetics involves the development of tools to analyze large sets of genomic data and makes phylogenetic inferences from it. What emerge from these studies are not only more robust phylogenetic trees, but also new species-like characters that are based on genome structure, such as rare genomic changes (Philippe and Laurent, 1998; Rokas and Holland, 2000).

Caveats of Using DNA-Based Methods. All the above-mentioned methods are DNA based; therefore, they all suffer from some basic problems. These include (i) the inability to access microbial cells that adhere to particles, (ii) the inability to lyse cells and isolate DNA, (iii) copurification of PCR inhibitors that affect downstream applications, (iv) shearing of DNA, (v) PCR bias, and (vi) PCR-based errors (Ehrlich and Newman, 2009a). Although the recognition of these problems leads to the development of better technology, another way to circumvent such issues is to use other approaches like those mentioned below. Information gained from each approach can then converge into a unified and reliable data set.

WHAT ARE THEY DOING?

When analyzing a particular environment for microbial activity, the first step is often to determine which organisms are present in that niche ("Who is there?"). However, owing to the concepts of indigenous, adventitious, and contaminant organisms introduced earlier, the presence of an organism does not always mean it is playing a direct role in a specific geomicrobial process of interest. The nonmolecular in situ and in vitro methods described earlier can shed light on the specific role of microbes present in a given environment; in addition, a number of molecular methods can help answer the "What are they doing?" question with finer resolution. Such methods can be applied to single cells or microbial communities.

Single-Cell Approaches. The general principle of single-cell methods combines identification of single cells with separation of the desired population of these cells, and then determining their particular characteristics (Color Plate 3).

FISH-MAR. This was one of the first single-cell approaches developed to understand microbial communities. It involves using radio-labeled substrates and monitoring their incorporation into macromolecules of a desired set of organisms that are identified in parallel with FISH approaches. The use of radio-labeled substrates makes FISH-microautoradiography (FISH-MAR) very sensitive such that short incubations suffice. However, a disadvantage is that radio-labeled compounds not incorporated into macromolecules are lost during sample preparation for FISH analysis. Thus, the nature of the metabolic products is never obvious (Lee et al., 1999; Ouverney and Fuhrman, 1999). FISH-MAR also has a number of other limitations. (i) It requires active growth of the microbial population under study, (ii) it requires prior knowledge about the kind of organisms that are present in a given sample, (iii) only a limited number of populations can be visualized simultaneously using FISH owing to the requirement for distinct fluorophores, (iv) FISH analysis might not be possible for some microbes, and (v) the desired radio-labeled substrate is unavailable (Ehrlich and Newman, 2009a). The techniques described below help circumvent some of these issues.

FISH-SIMS. This method combines FISH-based identification of an organism followed by analysis of the stable isotopic composition of the desired cells by use of an ion microprobe. The advantage of secondary ion mass spectrometry (SIMS)-based approaches is the ability to analyze the surface of a microbe or microbial assembly to resolve the spatial distribution of small isotopic differences. This powerful technique thus allows spatial tracing of isotopic signals (whether from naturally occurring signatures such as isotopically light methane or labeled ^{13}C and ^{15}N incubations) as they are assimilated, incorporated, and propagated by the different microbial populations fluorescently tagged with FISH. FISH-SIMS has been, for example, very successfully used in the past decade to investigate consortia of methane-oxidizing archaea with sulfate-reducing bacteria responsible for anaerobic methane oxidation in numerous natural environments (Orphan et al., 2001; Pernthaler et al., 2008).

Magneto-FISH. Another modification of FISH makes use of magnetic beads attached to antibodies specific for the fluorophore used in the catalyzed reported deposition FISH technique. This allows immunoseparation of a desired phylogenetic group of microbes. The separated cells can then be used to isolate DNA for metagenomic analysis. It is especially useful in isolating microbial consortia (Pernthaler et al., 2008).

Single-Cell PCR. Recent advances in single-cell PCR approaches have enhanced our ability to assess single cells of microorganisms while simultaneously linking these capabilities to their phylogeny. Microfluidic digital PCR allows amplification of 16S rDNA sequences along with other genes that serve as markers for a specific metabolic capability. This technique is therefore unbiased with respect to transcriptional levels and protein content, which are major limitations of FISH-based approaches (Ottesen et al., 2006). Single-cell PCR, combined with single-cell whole-genome sequencing, is a powerful tool and paves the road to single-cell phylogenomics.

Community Approaches. Although single-cell approaches are useful to understand the metabolic capabilities of individual cells, how these metabolisms combine to give rise to an observed geochemical profile entails studying a whole community. A few of such community-based methods are discussed below. However, newer methods that combine the single-cell approaches with the community-based approaches are becoming increasingly popular.

SIP. Stable isotope probing (SIP) involves tracking stable isotopes from particular substrates into components of microbial cells that provide phylogenetic information (biomarkers). This process has primarily been used to identify the microorganisms involved in specific biogeochemical transformations that are important in global elemental cycling. The first instance when SIP was used was to study incorporation of ^{13}C into polar lipid-derived fatty acids (PLFAs) (Petsch et al., 2003). However, the phylogenetic resolution offered by PLFAs is much lower in comparison with 16S rRNA/DNA-based methods. Thus, DNA-SIP and RNA-SIP were devised.

DNA labeled with stable isotopes can be isolated from mixed microbial communities, based on the increase in buoyant density associated with isotopic enrichment. Density centrifugation in CsCl gradients can then be used to separate labeled from unlabeled DNA, and 16S rDNA clone libraries constructed from labeled DNA can be sequenced to obtain the identity of organisms assimilating the labeled substrate. Although DNA-SIP offers phylogenetic resolution superior to PLFA-SIP, it requires a high level of isotopic enrichment. For instance, DNA must contain at least 15 to 20% ^{13}C before it can be isolated on the basis of buoyant density (Radajewski et al., 2000). DNA synthesis is related to replication, but bacterial replication in most environments is slow. Therefore, the incorporation of stable isotopes into DNA may not be efficient for DNA-SIP to be applied. The use of RNA as a biomarker in SIP helps circumvent this drawback of DNA-SIP. Transcription occurs with a much higher turnover rate and so the incorporation of the stable isotope is much higher in RNA-SIP. Labeled RNA can be isolated by density centrifugation, on cesium trifluoroacetate gradients. Followed by reverse transcription,

PCR and sequencing then provide the phylogenetic information desired (Dumont and Murrell, 2005; Whiteley et al., 2006). The advantage of SIP methods is the ability to identify hitherto unknown organisms involved in biogeochemical processes.

Isotope Array. Isotope array methods combine the utility of DNA microarrays with the benefits of stable isotope probing. RNA-SIP is performed, and the isolated RNA is then labeled with a fluorescent dye. This labeled RNA is then used to probe a 16S rRNA microarray of predetermined microbes. Isotope arrays unfortunately are limited in application because of the requirement to have prior knowledge about the organisms being sought (Hesselsoe et al., 2009; Adamczyk et al., 2003).

Metagenomics, Metatranscriptomics, and Metaproteomics. Metagenomic analysis ensues the direct isolation of genomic DNA from an environment and thus circumvents culturing the organisms under study. Subsequently cloning this DNA into a cultured organism (such as *Escherichia coli*) confines it for study and preservation (Riesenfeld et al., 2004). Metagenomics has seen numerous advances such that it is even possible to reconstruct whole genomes of uncultured organisms with some certainty (Woyke et al., 2009). It has also provided new genes that have later been pursued using classical genetic and biochemical tools and shown to encode novel enzymes (Hoff et al., 2008). However, metagenomics alone does not directly tell us the metabolic role a particular organism (or gene) might play in a given environment, because DNA only stores information. The transcription, translation, and regulation of the gene products allow organisms to affect their surroundings, resulting in a complex geomicrobial process. Despite its limitations, metagenomics can offer powerful initial insights into the possible microbes and microbial processes that might occur in a given environment.

Metatranscriptomics helps understand which genes in a community are transcribed at any given time. Thus, unlike metagenomics, direct inferences can be drawn about which genes are important under the condition being assessed. Like metagenomics, metatranscriptomics (or environmental transcriptomics) involves random sequencing of microbial community mRNA. Because no primers or probes are required for direct sequencing, there is no need to anticipate important genes a priori and transcripts from microbial assemblages are sequenced without bias. Furthermore, highly similar sequences, which might cross-hybridize on a microarray, can be distinguished by having a unique sequence (Warnecke and Hess, 2009; Shi et al., 2009). Experimental metatranscriptomics involves assessing changes in transcription in a particular environment and is a powerful tool for understanding the timing and regulation of complex microbial processes within communities and consortia, as well as microbial dexterity in response to changing conditions (Warnecke and Hess, 2009). Community expression profiling via direct sequencing saves data obtained from individual metatranscriptomic studies and added to community databases. These include CAMERA (http://camera.calit2.net/), MG-RAST (http://metagenomics.nmpdr.org/), or IMG/M (http://img.jgi.doe.gov/cgibin/m/main.cgi).

Metaproteomics, the identification of proteins isolated from a given environment is a recent development in environmental microbiology due to advances in sophisticated protein isolation and mass spectrometric techniques (for recent reviews, see Wilmes et al. [2008], Maron et al. [2007], Klaassens et al. [2007], and Wilmes and Bond [2006]). Metaproteomics has a great advantage over metagenomics and metatranscriptomics because it measures the actual molecular catalysts (i.e., enzymes) that mediate geochemical transformations. Thus, it circumvents the concern that gene expression is not an accurate predictor of the activity of a given gene product in any given environment.

HOW ARE THEY DOING IT?

Knowledge about which organisms are present in an environment, along with information about the metabolic reactions they perform can prompt questions such as: How do these organisms do what they do? What are the genes responsible? Do these genes resemble other known genes? If these genes encode enzymes, what is their mechanism and how are their rates of catalytic activity affected by relevant environmental variables? Where are the products of these genes located in the cell? What other molecular factors are required for their assembly? How are these genes or gene products regulated? Only by answering these questions can geomicrobiologists achieve a deeper level of understanding over the variables that control microbial activities. Such depth of knowledge, ultimately, is necessary for being able to predict how microbial communities will respond to and, in turn, modify an environment as it changes over time. To acquire this information, geomicrobiologists can draw on a wealth of tools from the fields of classical genetics, biochemistry, and cell biology.

Genetics. Classical genetics is extremely powerful and techniques in microbial genetics have been honed immensely over decades. Genetics has also kept abreast with advances in molecular biology such that molecular genetics is now commonplace. Through tools of both random and directed mutagenesis, scientists can identify genes that are linked to a particular metabolic capability (Maloy and Cronan, 1994). The prerequisites for performing random mutagenesis on an organism are (i) the existence of a pure or partially pure culture, (ii) the ability of the organism to form colonies on solidified media, (iii) the ability to transfer DNA into the isolated organism, (iv) the ability to assay the phenotype being sought, and (v) the ability of one or more plasmids to stably replicate in the native host for complementation experiments. If all these requirements are met, random mutagenesis can be performed using various established protocols (Color Plate 4) (Salyers et al., 2000). The use of targeted gene deletion requires, in addition to the requirements for random mutagenesis, the knowledge of the genome sequence or at least the region surrounding the locus to be deleted. Two methods, both based on homologous recombination, are generally used: (i) insertional inactivation and (ii) markerless deletion (Color Plate 4) (Rother and Metcalf, 2005; Guss et al., 2005; Pritchett et al., 2004).

In case an organism does not fulfill the requirements listed above, the method of heterologous complementation can be applied using a closely related genetically tractable organism. DNA isolated from the organism under study is cloned into plasmids that replicate in the host organism. The host organism is then tested for gain of the activity that is being sought (Color Plate 5). Many examples exist where this approach has been successfully applied (Croal et al., 2007; Beja et al., 2000; Martinez et al., 2007).

Bioinformatics and Subsequent Studies. The identification of a genetic locus that confers a certain metabolic capability to an organism leads to the question of what these genes encode. Standard bioinformatics tools provided as public online user interfaces are available on the World Wide Web. NCBI blast (http://blast.ncbi.nlm.nih.gov/Blast.cgi), a basic alignment search tool, helps find the closest homologs of an unknown gene from available sequences submitted to GenBank. Other websites such as Expert Protein Analysis System (http://www.expasy.ch/) and EMBL-EBI servers (http://www.ebi.ac.uk/Tools/) provide numerous tools that allow users to determine the likely protein or RNA domains and motifs present in a DNA sequence. Although bioinformatics points to the putative role of a protein or RNA, the actual role of a biomolecule needs to be determined using experimental methods; for instance, the biochemical analysis of proteins or the study of function of small RNAs. Localization of proteins can also be studied using methods developed by cell biologists involving the use of sophisticated microscopic tools in conjunction with fluorescent reporters (green fluorescent protein and its derivatives) (Valdivia et al., 2006). Gene regulation studies can also be performed using quantitative reverse transcription-PCR and classical reporter gene fusions (Seeber and Boothroyd, 1996; Miller and Hershberger, 1984).

THE (BIO)GEOCHEMISTRY OF IRON

Iron at the Earth's Surface

Iron is the most abundant element on Earth, contributing 32% by weight to the bulk composition of our planet (Morgan and Anders, 1980). A large portion of the iron is concentrated in the inner core, but the element still constitutes a remarkable 5% of the Earth's crust on average, exceeded in crustal abundance only by oxygen, silicon, and aluminum (Rankama and Georg Sahama, 1950). The chemical reactivity of iron and its role as one of the most important reduction-oxidation (redox) active elements implicate iron in a myriad of highly relevant geochemical as well as biochemical transformations.

Igneous rocks rich in iron-bearing mafic minerals, such as olivine and pyroxene, are the primary source of iron to the Earth's crust and surface environment. The (bio)geochemical cycling of iron at the Earth's surface starts with the mobilization of iron during erosion and weathering of exposed igneous rock, as well as during hydrothermal circulation of seawater at oceanic spreading centers (the sources). Subduction of iron-bearing oceanic crust and seafloor sediments ultimately returns iron to the mantle (the sink) where the rocks can melt, form a magma, and reenter the igneous rock cycle. Once in a mobile phase as a particulate mineral or dissolved ion, many physicochemical processes can occur, altering properties such as solubility and chemical speciation (Ussher et al., 2004). Iron is distributed throughout the hydrosphere by riverine and marine fluxes, transported in the atmosphere in aerosols and dust particles, and spread throughout the lithosphere by sedimentary processes, mineral precipitation, and diagenesis, making up a large proportion of soils and sedimentary rocks from where it can be remobilized. Its complex biogeochemical cycle affects not only its own distribution in the lithosphere, hydrosphere, atmosphere, and biosphere, but also the distribution and processes of other important element cycles such as those of carbon, nitrogen, and sulfur.

Iron is most commonly observed in its ferrous (oxidation state +2) or ferric (oxidation state +3) form. Reducing conditions in the mantle and deep crust make the ferrous form dominant in magma and fresh igneous rock, but ferric iron is stable at present oxic atmospheric conditions, making +3 the most common oxidation state in surface environments. However, the dynamic and variable chemistry of many surface environments sometimes allows for considerable amounts of ferrous iron and co-occurrence of both oxidation states. The following chemical reactions exemplify two typical redox transformations of iron: (I) the oxidation of Fe^{2+} by molecular oxygen to form a simple, insoluble Fe^{3+} hydroxide and (II) the reduction of Fe^{3+} by excess hydrogen sulfide, partially precipitating as an iron sulfide mineral (pyrite). Both reactions are relevant redox pathways in natural environments.

$$4\ Fe^{2+} + O_2 + 10\ H_2O \rightarrow 4\ Fe(OH)_3\ (s) + 8\ H^+ \quad (I)$$

$$2Fe^{3+} + 2\ H_2S \rightarrow 2\ FeS_2\ (s) + Fe^{2+} + 4\ H^+ \quad (II)$$

The solubility and chemical behavior of iron in aqueous environments is heavily influenced by its speciation, that is, the actual form in which it is present in solution. Besides the oxidation state of the element, speciation depends largely on acidity/alkalinity of the environment (pH) and the availability of ligands. Ferric iron is generally much less soluble than ferrous iron, especially at circumneutral pH, although strong ligands can improve its stability in solution. Ferric iron in marine environments, for example, has been found to be >99% complexed by strong organic ligands (Ussher et al., 2004). The solubility of the ions is ultimately limited by the stability of various minerals, for example, siderite and pyrite for Fe^{2+}, and (oxy)hydroxides and oxides like ferrihydrite, goethite, and hematite for Fe^{3+}.

The biogeochemical cycling of iron strongly influences other element cycles by means of mineral formation and dissolution, coupling of redox cycles, as well as enzymatic catalysis. The distribution and transformation of iron in the Earth's surface environment have a particularly noteworthy impact on the previously mentioned, globally important cycling of nitrogen, sulfur, and carbon (illustrated schematically in Fig. 2). While iron and nitrogen redox cycling is coupled directly in the microbial process of anaerobic respiration of ferrous iron with nitrate, iron impacts the nitrogen cycle most pervasively in its role in enzymatic catalysis (Weber et al., 2006). A wide variety of metalloproteins and cofactors incorporate iron in their active sites, and virtually all processes of the microbially mediated cycling of nitrogen (nitrogen assimilation, fixation, denitrification, nitrification) require small amounts of iron for this purpose. Many surface environments are rarely limited by iron availability, but the low concentrations of iron in ocean waters can crucially affect the productivity and ecology of the oceans by limiting nitrogen fixation (Falkowski et al., 1998; Morel, 2003).

The sulfur and iron cycles are closely linked through the mutually controlled formation of iron-sulfur minerals, such as the aforementioned pyrite, and through the reductive interaction of sulfide species with ferric iron (Canfield et al., 1992). The two cycles are, however, most intimately intertwined in their prominent role as iron-sulfur (FeS) clusters in a variety of metalloproteins and cofactors involved in electron transfer reactions that impact a variety of other element cycles.

The biogeochemical cycling of iron is linked to the carbon cycle in various direct and indirect ways. Direct links include the formation

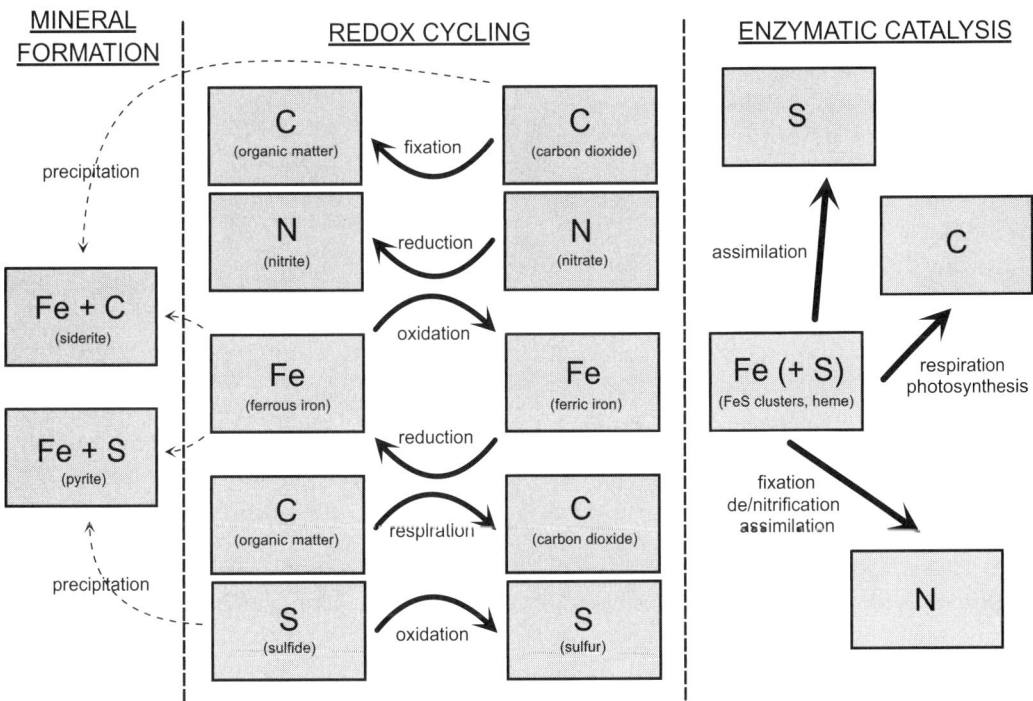

FIGURE 2 Simplified illustration of the coupling of the biogeochemical cycling of iron with the carbon, nitrogen, and sulfur cycles. Shown are interactions of the element cycles during mineral formation and dissolution (left column: arrows indicate precipitation, parentheses provide examples of mineral species), redox cycling (center column: arrows illustrate redox transformations coupled to iron oxidation/reduction, parentheses provide examples of the reduced and oxidized species involved), and enzymatic catalysis (right column: arrows indicate examples of iron-dependent metabolic processes for sulfur, carbon, and nitrogen, respectively).

of the iron-carbonate mineral siderite as well as the coupling of redox cycling in phototrophic iron oxidation (where ferrous iron is used as a source of electrons to fix carbon) and dissimilatory iron reduction (where organic substrates are respired using ferric iron as the terminal electron acceptor). In its catalytic function in enzymes, iron is central to redox metabolism (Fraústo da Silva and Williams, 2001). It plays an active role, for example, in many enzymes involved in the electron transfer reactions of both respiration and photosynthesis, and furthermore influences the carbon cycle indirectly through its effects on the nitrogen cycle.

Finally, the interaction of iron with the carbon and sulfur cycle has played an important role in the accumulation of atmospheric oxygen throughout Earth's history. The accumulation of oxygen at the Earth's surface is largely controlled by the direct burial of organic carbon and the indirect burial of the reducing power from organic carbon in the form of pyrite. The direct burial of organic carbon prevents some of the biomass produced during oxygenic photosynthesis to be respired again with the produced oxygen, and thus frees some of the oxygen to accumulate. In indirect burial, the organic matter produced by oxygenic photosynthesis drives sulfate reduction and iron reduction, producing pyrite and oxidizing the organic matter to CO_2. Burial of pyrite through the couplings of the carbon, sulfur, and iron cycle thus liberates oxygen to the atmosphere (Garrels and Lerman, 1981; Berner and Maasch, 1996; Canfield et al., 2005).

Iron and the Biosphere

The biosphere has a tremendous impact on the distribution of iron at the Earth's surface. Microorganisms, in particular, heavily influence the geochemical cycling of iron in various ways. Most organisms assimilate small amounts of iron to satisfy their nutritional demands for the element. In the process, they heavily influence the mobilization of iron from rocks, soils, and sediments and alter the speciation of iron in aqueous environments. This is a result of both active and passive microbial activities through the deliberate production of molecules for iron acquisition and by action of common metabolic products such as organic acids, respectively. Many microorganisms can catalyze the oxidation or reduction of iron both actively by metabolizing iron to generate energy for growth and passively, e.g., by influencing the chemistry of their immediate surroundings or providing catalytic surfaces for mineral formation.

IRON AS A NUTRIENT

Iron is an essential nutrient for the survival, growth, and reproduction of almost all forms of life. As a key catalyst in several enzymatic electron transfer reactions, iron is required by most organisms for vital cellular processes ranging from respiration to photosynthesis, nitrogen fixation, and the oxidative stress response (Fraústo da Silva and Williams, 2001). Many of the iron-bearing enzymes involved in these processes probably evolved at a time in Earth's history when ferrous iron was abundant in the world's oceans because the Earth's surface environment was still anoxic and the much less soluble ferric form of iron could not yet stabilize (Anbar and Knoll, 2002; Saito et al., 2003; Glass et al., 2009). With the evolutionary invention of oxygenic photosynthesis and the rise of atmospheric oxygen around 2.4 billion years ago, the eventual oxidation of the Earth's surface and oceans brought about drastic changes in atmospheric and marine chemistry including a severe limitation in the bioavailability of iron. Despite being the most abundant transition metal in the biosphere, most iron is bound in insoluble oxides and hydroxides at near-neutral pH at present atmospheric oxygen levels. It is thus hardly surprising that the biosphere has evolved many elegant ways to influence the cycling of iron to allow for assimilation. In response to the scarcity of bioavailable iron in many natural environments, many microorganisms have, for example, evolved the ability to synthesize potent chelating agents to aid in iron acquisition. The fascinating and complex process of mobilization and acquisition of iron by these iron-specific chelators, commonly known as siderophores, is discussed in detail by Kraemer (2004) and Kraemer et al. (2005).

IRON AS AN ENERGY SOURCE

All nonphototrophic organisms take advantage of concentrations of oxidizable and reducible species that are far from those demanded by thermodynamic redox equilibria and kinetically too slow to proceed on their own. This is what happens in respiration, where organic matter (highly reduced) and molecular oxygen (highly oxidized) are out of thermodynamic equilibrium, but the kinetics of abiotic combustion at standard conditions are so slow that they do not just recombine spontaneously. Microorganisms gain energy by catalyzing such redox reactions in a controlled manner to restore thermodynamic equilibrium. Microbially mediated transformations of iron are often much faster than their abiotic equivalents and are ubiquitous in natural environments. A wide variety of physiologically different prokaryotes can use iron for energy generation, either by using ferric iron in dissimilatory iron reduction as an oxidant to respire more reduced substrates, or by metabolizing ferrous iron in dissimilatory iron oxidation with a stronger oxidant. These microbially mediated redox transformations often occur in combination and can contribute to extensive (re-)cycling of iron in the environment. Canfield et al. (1993) estimated repeated oxidation and reduction in coastal sediments between 100 and 300 times before ultimate burial into the sediment. A good review of microbially mediated iron redox processes is provided by Kappler and Straub (2005).

To gain energy from the oxidation of ferrous iron with molecular oxygen in aerobic environments, microorganisms have to compete with abiotic oxidation. At circumneutral pH, the chemical reaction proceeds within minutes and neutrophilic iron oxidizers have to catalyze iron oxidation extremely fast (Stumm and Morgan, 1995). Emerson and Weiss (2004) provide a detailed description of aerobic iron oxidation at neutral pH. At more acidic conditions (pH < 5), the stability of ferrous iron increases and chemical oxidation proceeds much more slowly, allowing microorganisms to compete more easily. However, the energy gained from oxidation decreases with pH, and much larger quantities of iron need to be oxidized to meet cellular energy requirements. A variety of acidophilic iron-oxidizing bacteria from all across the proteobacteria (most prominently *Acidithiobacillus ferrooxidans*), nitrospira, firmicutes, and acidobacteria, as well as several archaeal strains from lineages from the *Thermoplasmatales* and *Sulfolobales* (e.g., Edwards et al. [2000]) have been identified, in particular, in acid mine drainage environments where acidic conditions prevail and sufficient quantities of ferrous iron, mostly in the form of pyrite, are available (see Baker and Banfield [2003] for a review). In addition to aerobic iron oxidizers, several microorganisms have been discovered that can oxidize ferrous iron anaerobically by respiring nitrate at neutral pH (see Weber et al. [2006]). Although considered a possible form of anoxic phototrophic metabolism for a long time, the so-called photoferrotrophs, bacteria that grow photosynthetically with ferrous iron as their sole source of reducing power, were only discovered in the early 1990s by Widdel et al. (1993). In the past decade, several cultures of iron-oxidizing phototrophic bacteria from all three major phylogenetic lineages of anoxygenic phototrophs (purple sulfur bacteria, purple nonsulfur bacteria, and green sulfur bacteria) have been isolated from such diverse environments as iron-rich springs, freshwater marshes, and marine sediments (Kappler and Straub, 2005).

Instead of harnessing energy from existing thermodynamic disequilibria, fixation of solar energy by photosynthesis actually creates new thermodynamic gradients that can be harnessed to support cell growth. As such, photoferrotrophy is fundamentally different from all other microbial interactions with the iron cycle. It is, in fact, the only microbial process that can change the redox state of iron species in natural environments against the thermodynamic resting state. As such, photoferrotrophy holds the power to drive redox cycling on the Earth's surface independent of oxygenic photosynthesis. Although such an isolated occurrence of photoferrotrophy, completely decoupled from oxygenic photosynthesis, is unlikely to occur today owing to the ubiquity and evolutionary success of oxygenic photosynthesis, it might still constitute an important contribution to primary productivity and the generation of oxidants in shallow sediments and iron-rich microenvironments. Furthermore, assuming an early evolution of anoxygenic photosynthesis preceding the evolution of cyanobacteria and oxygenic photosynthesis, photoferrotrophs could have been the dominant primary producers in a ferrous iron-rich Archean ecosystem.

This last assertion is of particular interest in the context of Earth's evolution because of the occurrence of massive iron-rich sedimentary deposits known as banded iron formations (BIFs) from a time in Earth's history when dissolved ferrous iron was abundant in the Earth's oceans but molecular oxygen was still extremely scarce. Photochemical oxidation of ferrous iron by UV radiation was considered a possible mechanism, but recent findings have cast doubt on the efficiency of such a process for the deposition of BIF, and it is generally believed that chemical or microbially mediated oxidation of ferrous iron by molecular oxygen from early oxygenic photosynthesis played the main role in BIF formation (Konhauser et al., 2007). More recently, however, the possibility of photoferrotrophic iron oxidation has been advanced as an alternative or complementary mechanism for banded iron formation (e.g., Kappler et al. [2005] and Posth et al. [2008]). A good review on the extent and significance of several major banded iron formations is provided by Beukes and Gutzmer (2008).

The ability to use the product of ferrous iron oxidation, ferric iron, as a terminal electron acceptor is widespread; numerous bacteria and archaea capable of dissimilatory ferric iron reduction at circumneutral and acidic pH have been isolated in the past decade. The environmentally most important inorganic reductant for ferric iron is hydrogen sulfide, which constitutes a kinetically competitive abiotic pathway and commonly contributes to iron cycling, especially when produced in large quantities during microbial sulfate reduction. A variety of organic and inorganic substrates can be respired and some organic substrates fermented with ferric iron depending on the strain and mineral form of the iron available, although not all iron-reducing bacteria can grow with iron as the sole electron sink. It is unclear whether this occurs because of an inability to gain energy from the process or whether the energy gained is insufficient to support growth but could still be important for survival. The ubiquity of iron minerals in sediments and breadth of substrate flexibility makes dissimilatory iron reduction an important metabolic pathway for the anaerobic mineralization of organic matter, especially in environments where sulfate and nitrate are unavailable terminal electron acceptors. However, a challenge of this metabolism is that organisms have to use an insoluble electron acceptor and cope with the difficulty of either solubilizing the iron mineral or transferring electrons from the cell to the mineral surface. Three different strategies seem to have evolved in response to this problem, although their distribution in natural environments remains inconclusive: (i) physical contact with the mineral surface for direct electron transfer, (ii) synthesis of iron chelators to solubilize ferric iron, and (iii) synthesis of redox-active molecules to shuttle electrons to the mineral surface. See Lovley et al. (2004) for a comprehensive review.

Because the majority of ferrous iron-oxidizing and ferric iron-reducing prokaryotes were isolated during the past decade, it is hardly surprising that our knowledge of these microorganisms, their metabolism, and especially their contribution to the biogeochemical cycling of iron is still in its infancy (Kappler and Straub, 2005). This is particularly true for the most recently discovered group of anoxygenic phototrophic iron oxidizers whose metabolic potential and place in microbial ecosystems has only recently started to be investigated and fully appreciated.

FROM GEOCYCLES TO GENOMES AND BACK: LAKE MATANO AS A CASE STUDY

A recent study of the iron-rich Lake Matano in Indonesia by Crowe et al. (2008a) provides an interesting case study for a joint geochemical and microbiological effort to investigate the roles that microorganisms play in shaping the geochemistry of this environment. In this final section, we describe what is known about Lake Matano and discuss how the traditional and molecular microbiological approaches described above may be used to gain insight into how microorganisms affect the biogeochemical cycling of iron and other elements in this environment.

Lake Matano, a part of the Malili Lakes system of Indonesia, is among the ten deepest lakes on Earth. It is estimated to be between one and four million years old (Brooks, 1950). It has relatively stable physical characteristics leading to species endemism, which has been studied in this lake by numerous workers (Sabo, 2006; Myers et al., 2000). Its unique iron geochemistry makes the lake a particularly interesting analogue for the chemistry of the oceans on early Earth (Crowe et al., 2008a), which differed markedly from modern environments. As discussed earlier, the lack of oxygen, presence of low concentrations of sulfide, along with the abundance of ferrous iron have implicated microbially mediated Fe(II) oxidation as a possible mechanism for the formation of extensive BIFs in the Archean ocean (Canfield et al., 2000, 2005; Isley and Abbott, 1999). Because Lake Matano's geochemistry may resemble the composition of ancient oceans (Crowe et al., 2008a), it provides an opportunity to study the closely linked biogeochemical cycles of iron, carbon, and sulfur in an environment that can be characterized both geochemically and microbiologically (Crowe et al., 2008b). As discussed in detail below, numerous interesting observations have emerged from the study of the geochemistry of this lake and point to the likelihood of novel microbial metabolisms working within it that actively shape the geochemistry of the lake (Crowe, 2008; Crowe et al., 2008b).

Lake Matano is a tropical lake and therefore shows many typical characteristics of lake systems at low latitudes, such as higher annual irradiance, lack of seasonal variation, and high amounts of Fe and Mn hydroxides supplied from extensive weathering of iron-rich country rock in the drainage basin (Crowe et al., 2008b). The lack of large temperature fluctuations can facilitate poor mixing of the different layers of water, often leading to seasonal stratification, i.e., a separation of the water column into stable layers of differing water densities because of temperature and salinity differences. In the case of Lake Matano, the great depth and relatively small surface area of the lake allows this commonly seasonal phenomenon to persist, leading to a permanently stratified water column.

Stratification of lakes can lead to subsequent chemical stratification of redox-sensitive elements because of the redox activity of microorganisms. Measurements of physical parameters in Lake Matano reveal that it has a permanent pycnocline at ~100 m depth stably separating the mixolimnion, the upper mixed water layer in contact with the atmosphere, from the monimolimnion, the deep anoxic waters. The redox or chemocline, marking the gradual chemical transition from completely oxygenated surface waters to increasingly reducing deep waters, extends from ~100 m to ~220 m depth (Crowe et al., 2008b).

The unique geochemistry of Lake Matano arises owing to the interplay between the geophysical characteristics of the environment and the activity of resident microbes in the lake. Spatiochemical stratification of electron acceptors is observed in the lake as follows: oxygen concentrations drop until they reach undetectable levels at ~100 m depth. At this depth the dissolved Mn concentration increases to 10 µM and of Fe(II) to 140 µM. Both Fe(II) and dissolved Mn concentrations do not drop dramatically from this concentration as the depth increases to 300 m (Crowe et al., 2008b). At the same time, the sulfate concentrations in Lake Matano remain very low (<20 µM). Modeling studies predict that microbially mediated sulfate reduction is not expected to occur in the water column but rather in the sediment, whereas sulfide accumulation in the water column is limited to below detection by the low solubility of iron sulfide minerals and the high concentration of Fe(II) (Crowe et al., 2008b).

Spatiochemical stratification of various electron donors is intimately linked to microbial oxidation of organic matter. The autochthonous organic carbon from primary productivity in the lake is very low, and it is the degradation of allochthonous organic matter, accounting for most of the dissolved organic carbon, that likely leads to spatiochemical stratification. The dissolved organic carbon is mineralized completely between 100 and 200 m depth concomitant with the increase in Fe(II) concentrations, suggesting microbial Fe(III) reduction as a possible means of substrate oxidation (Crowe et al., 2008b). It is likely that an iron cycle operates across the chemocline with some flux of iron from the sediment to the water column. Modeling studies suggest that a combination of various phenomena might lead to the steady-state concentrations of Fe in this region. These include (i) the descent of insoluble Fe(III) hydroxides, (ii) the appearance of Fe(II) due to microbially mediated Fe(III) reduction, (iii) the diffusion of Fe(II) from

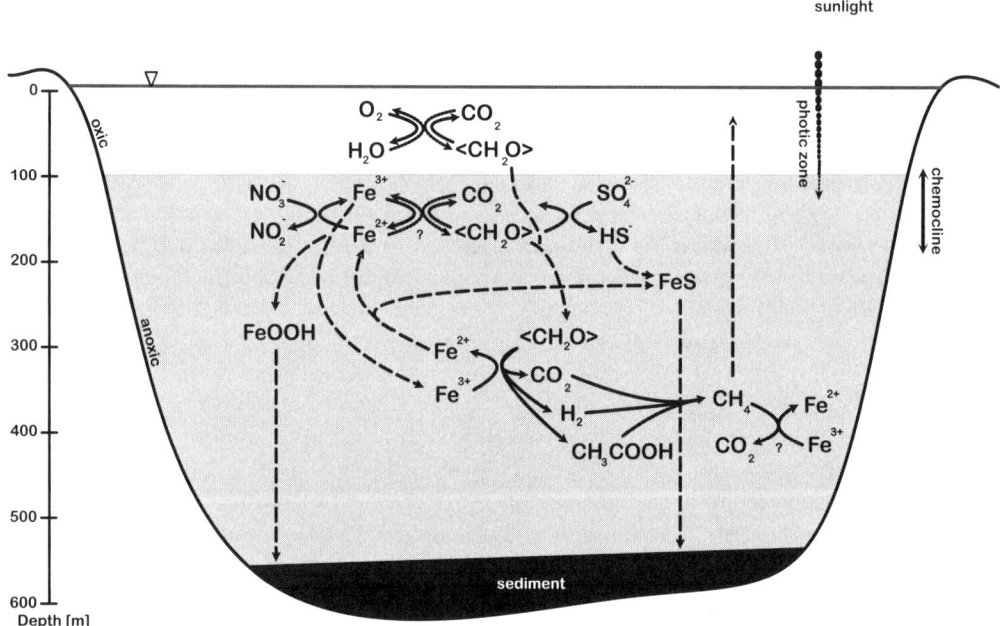

FIGURE 3 Simplified schematic illustrating the interactions of the iron cycle with other element cycles in Lake Matano. The oxic surface waters are shown with a white background, the anoxic monimolimion is shown in light grey. Penetration of sunlight and phototrophic transformations are indicated with dotted lines, nonphototrophic microbial transformations are indicated with solid lines, and precipitation and diffusion are shown by dashed lines. Coupled arrows, such as organic matter oxidation to carbon dioxide (<CH_2O> to CO_2) with sulfate reduction to hydrogen sulfide (SO_4 to HS^-), illustrate closely linked redox transformations. Several hypothesized, but still insufficiently investigated potential processes in Lake Matano, such as photoferrotrophy and iron-dependent anaerobic methane oxidation, are highlighted with a question mark.

the monimolimnion, (iv) the regeneration of Fe(III) by Fe(II) oxidation at the redox boundary, and (v) the upward flux of soluble Fe(II) from the sediment to monimolimnion (Fig. 3) (Crowe et al., 2008b).

In the anoxic deep waters, once anaerobic microorganisms have utilized the more favorable electron acceptors, methanogenesis and anaerobic methane oxidation (AOM) can contribute to the carbon cycle. Sulfate reduction is more energetically favorable than methanogenesis, and sulfate-reducing bacteria outcompete methanogens for acetate as a carbon source (Capone and Kiene, 1988; Reeburgh, 2007). The absence of sulfate in the Lake Matano water column precludes this competition (Crowe et al., 2008b). However, the abundance of Fe and Mn introduces the possibility of microbes that reduce these minerals competing with methanogens for acetate. In addition, the absence of sulfate and nitrate in Lake Matano, the two well-established electron acceptors that are coupled to AOM, raises the possibility that Fe and Mn are more important players in AOM in this environment (Crowe, 2008; Boetius et al., 2000; Raghoebarsing et al., 2006). The recent demonstration that marine sediments can couple anaerobic methane oxidation to Fe and Mn reduction shows that such reactions are feasible (Beal et al., 2009). Methane production occurs primarily in the sediment despite the presence of considerable amounts of Fe and Mn. Disappearance of methane below the pycnocline between 100 and 200 m suggests likelihood of AOM in this zone (Crowe, 2008).

The concentrations of nitrogen and phosphorus are below detection limits in the mixolimnion. Ammonium is detected in the monimolimnion as the predominant N species, while soluble phosphate is the predominant P species across the chemocline (Crowe et al., 2008b). The appearance of soluble phosphate increases with the Fe(II) concentration suggesting that the P and Fe cycles are linked in this environment, probably involving sorption and removal of phosphate from the mixolimnion by the particulate Fe(III) species. The lack of essential nutrients along with the high concentration of chromium [Cr(VI)] might account for the low primary productivity in the mixolimnion of Lake Matano (Crowe et al., 2008b).

Given the many spatiochemical gradients of common substrates for microbial metabolisms in Lake Matano, the likelihood of finding organisms that actively contribute to shaping these gradients is high. For example, photosynthetic pigment measurements in the lake have revealed that chlorophyll *a* is present at low levels in the surface oxic layer where cyanobacteria would be commonly expected to contribute significantly to primary productivity in nutrient-richer environments. The low levels of light-absorbing pigments, such as chlorophyll *a*, in the surface layer allow light to penetrate to the deeper layers where bacteriochlorophyll *e* predominates. This pigment is characteristic of anoxygenic photosynthetic bacteria that belong to the *Chlorobiaceae* that thrive under lower-light conditions. Consistently with this, 16S rDNA studies have indicated the presence of various members of this family in the redoxcline of Lake Matano (Crowe et al., 2008a). The high concentration of Fe(II) in this layer suggests that it might be serving as an electron donor for anoxygenic photosynthesis for these organisms and that their activities might influence iron cycling in the lake. However, whether this is, in fact, the case, or whether other Fe(II)-oxidizing and Fe(III)-reducing organisms make greater contributions to the iron cycle is not clear. Indeed, it is possible that different groups of microorganisms control the iron biogeochemistry of the lake at different times, and that these microbial communities change altogether from year to year. Such depth of knowledge, ultimately, is necessary for being able to predict how microbial communities will respond to and, in turn, modify an environment like Lake Matano as it changes over time. How can we resolve this?

As described at the beginning of this chapter, a variety of traditional and molecular microbiological methods could be used to provide some insight and inform our understanding of the microbial communities and biogeochemical dynamics in this environment. Fortunately, recent work on Lake Matano has provided us with detailed information on the geochemistry of the lake (Crowe, 2008; Crowe et al., 2008a, 2008b), paving the road for some of the in situ techniques. Traditional light microscopy could be used to study microbial populations in the lake sediments and their spatial variability close to the chemocline (e.g., by the buried slide method). Because of low concentrations of microorganisms in the water column itself, however, more powerful techniques such as TEM and SEM could be necessary, which would provide the additional benefit of allowing investigation of the physical association of microorganisms with freshly formed minerals, such as iron (hydro)oxides from microbial iron oxidation. A reliable assessment of in situ geomicrobial activity via isotope labeling at the chemocline, in the case of phototrophic organisms, could be foiled by slow growth in the relatively low light at this depth; however, chemotrophic organisms could be operating on faster time scales if external carbon sources to the lake provide sufficient substrate for growth. While limited by slow growth, advanced isotope techniques such as RNA-SIP are of particular interest for an environment like Lake Matano, where a high degree of species endemism suggests the possible occurrence and likely importance of novel microorganisms and unique metabolic pathways that could be identified by this technique.

In vitro batch culture and chemostat studies of microorganisms from this environment are highly desirable to assess their metabolic potential but depend on enrichment and isolation of organisms from the water column. Slow growth might be a limiting factor in the ultimate success of this approach. Culture-independent techniques, however, could be used regardless, and FISH could be used, for example, to study the small-scale spatial distribution, abundance, and physical association of various bacterial and archaea groups identified by 16S fingerprinting to be present in the lake sediments. Modern extensions of the approach, such as FISH-SIMS, could be similarly successful in investigating microbial populations in the lake, possibly aiding identification and study of a so-far elusive but suspected syntrophic community that combines anaerobic methane oxidation with metal reduction in this unique environment. Although technically and computationally challenging, the combination of metagenomic, -transcriptomic, and -proteomic techniques could significantly advance our understanding of the genetic metabolic potential, actively expressed metabolic pathways, and geochemically active enzymes the microbial communities in Lake Matano command. Detailed genetic studies of the organisms, however, would require reasonably fast-growing pure cultures, whose metabolic machinery for iron oxidation or reduction, for example, could then be assessed by random mutagenesis (for genetically tractable organisms) or heterologous complementation (for intractable strains with suitable tractable strains that are closely related). Once whole genomes of novel organisms from this environment would become available, bioinformatics provides powerful additional tools to search for metabolic key components that contribute to the biogeochemical cycling of iron and other elements in this environment, and whose identification is crucial for our understanding of the biogeochemical dynamics of this system.

CONCLUSIONS

We began this chapter with the assertion that a holy grail for environmental microbiologists is to understand the biogeochemistry of an environment sufficiently well to predict its behavior. This is a tall order and requires tremendous effort on the part of many groups to achieve for any given environment. One might reasonably ask if the complexity of real-world systems is so vast as to make this impossible to achieve? Perhaps, but we can hope that first-order predictions about the behavior of a given biogeochemical system are attainable, provided the dominant pathways are known and the controlling variables are well defined.

In the case of the example we discussed in detail in this chapter, Lake Matano, although we know basic aspects about its geochemistry and microbial communities, much remains to be learned. We do not have a good appreciation for how the structure of its microbial community changes over time, or how changes to this structure affect the geochemical profiles of the lake. We do not understand which environmental variables control the success of particular members of the community, or how functionally redundant it is (e.g., particular organisms may come and go, but the geochemical reactions they catalyze might be similarly catalyzed by a different group of organisms). The better able we are to characterize these aspects, and to point to underlying molecular catalysts (e.g., metabolic enzymes), their rates, and the variables that regulate them, the better able we will be to make predictions about how the lake might respond to environmental perturbations.

Despite these knowledge gaps, the current state of research in microbe-metal interactions provides a fascinating outlook. From years of investigations in geochemistry and microbiology, we can appreciate the tremendous scope of possible pathways. We are working in a historically opportune moment, when enough is known about specific microbial processes to allow us to venture into assessing their relative contributions and importance to the complex cycling of iron and other elements in situ. The coevolution of microbial life and the environment is much akin to a terrific puzzle where we know enough of the pieces to get a first blurry glimpse of the magnificent full picture, but we do not know yet where each one goes nor how the pieces fit together. It will be satisfying to see the details of this picture emerge and sharpen over the coming years in a variety of systems.

REFERENCES

Adamczyk, J., M. Hesselsoe, N. Iversen, M. Horn, A. Lehner, P. H. Nielsen, M. Schloter, P. Roslev, and M. Wagner. 2003. The isotope array, a new tool that employs substrate-mediated labeling of rRNA for determination of microbial community structure and function. *Appl. Environ. Microbiol.* **69:**6875–6887.

Allen, E. E., and J. F. Banfield. 2005. Community genomics in microbial ecology and evolution. *Nat. Rev. Microbiol.* **3:**489–498.

Amann, R., and B. M. Fuchs. 2008. Single-cell identification in microbial communities by improved fluorescence *in situ* hybridization techniques. *Nat. Rev. Microbiol.* **6:**339–348.

Amann, R. I., L. Krumholz, and D. A. Stahl. 1990. Fluorescent-oligonucleotide probing of whole cells for determinative, phylogenetic, and environmental studies in microbiology. *J. Bacteriol.* **172:**762–770.

Amann, R. I., W. Ludwig, and K. H. Schleifer. 1995. Phylogenetic identification and *in situ* detection of individual microbial cells without cultivation. *Microbiol. Mol. Biol. Rev.* **59:**143–169.

Anbar, A. D., and A. H. Knoll. 2002. Proterozoic ocean chemistry and evolution: a bioinorganic bridge? *Science* **297:**1137–1142.

Baker, B. J., and J. F. Banfield. 2003. Microbial communities in acid mine drainage. *FEMS Microbiol. Ecol.* **44:**139–152.

Baker, W. W., and J. F. Banfield. 1998. Zones of chemical and physical interaction at interfaces between microbial communities and minerals: a model. *Geomicrobiol. J.* **15:**223–244.

Beal, E. J., C. H. House, and V. J. Orphan. 2009. Manganese- and iron-dependent marine methane oxidation. *Science* **325:**184–187.

Beja, O., L. Aravind, E. V. Koonin, M. T. Suzuki, A. Hadd, L. P. Nguyen, S. B. Jovanovich, C. M. Gates, R. A. Feldman, J. L. Spudich, E. N. Spudich, and E. F. DeLong. 2000. Bacterial rhodopsin: evidence for a new type of phototrophy in the sea. *Science* **289:**1902–1906.

Berner, R. A., and K. A. Maasch. 1996. Chemical weathering and controls on atmospheric O_2 and CO_2: fundamental principles were enunciated by J. J. Ebelmen in 1845. *Geochim. Cosmochim. Acta* **60:**1633–1637.

Beukes, N. J., and J. Gutzmer. 2008. Origin and paleoenvironmental significance of major iron formations at the Archean-Paleoproterozoic boundary, p. 5–47. *In* S. Hagemann, C. Rosiere, J. Gutzmer, and N. J. Beukes (ed.), *Reviews in Economic Geology*, vol. 15. *Banded Iron Formation-Related High-Grade Iron Ore.* Society of Economic Geologists, Denver, CO.

Boetius, A., K. Ravenschlag, C. J. Schubert, D. Rickert, F. Widdel, A. Gieseke, R. Amann, B. B. Jorgensen, U. Witte, and O. Pfannkuche. 2000. A marine microbial consortium apparently mediating anaerobic oxidation of methane. *Nature* **407:**623–626.

Brock, T. D. 1978. *Thermophilic Microorganisms and Life at High Temperatures*. Springer-Verlag, New York, NY.

Brooks, J. L. 1950. Speciation in ancient lakes. *Q. Rev. Biol.* **25:**131–176.

Burdige, D. J., and P. E. Kepkay. 1983. Determination of bacterial manganese oxidation rates in sediments using an *in situ* dialysis technique. I. Laboratory studies. *Geochim. Cosmochim. Acta* **47:**1907–1916.

Canfield, D. E., K. S. Habicht and B. Thamdrup. 2000. The Archean sulfur cycle and the early history of atmospheric oxygen. *Science* **288:**658–661.

Canfield, D. E., B. Thamdrup, and J. W. Hansen. 1993. The anaerobic degradation of organic matter in Danish coastal sediments: iron reduction, manganese reduction, and sulfate reduction. *Geochim. Cosmochim. Acta* **57:**3867–3883.

Canfield, D. E., E. Kristensen, and B. Thamdrup. 2005. *Aquatic Geomicrobiology*. Elsevier, San Diego, CA.

Canfield, D. E., R. Raiswell, and S. Bottrell. 1992. The reactivity of sedimentary iron minerals toward sulfide. *Am. J. Sci.* **292:**659–683.

Capone, D. G., and R. P. Kiene. 1988. Comparison of microbial dynamics in marine and fresh water sediments: contrasts in anaerobic carbon catabolism. *Limnol. Oceanogr.* **33:**725–749.

Clement, B. G., L. E. Kehl, K. L. DeBord, and C. L. Kitts. 1998. Terminal restriction fragment patterns (TRFPs), a rapid, PCR-based method for the comparison of complex bacterial communities. *J. Microbiol. Methods* **31:**135–142.

Cohan, F. M. 2002. What are bacterial species? *Annu. Rev. Microbiol.* **56:**457–487.

Croal, L. R., Y. Jiao, and D. K. Newman. 2007. The *fox* operon from *Rhodobacter* strain SW2 promotes phototrophic Fe(II) oxidation in *Rhodobacter capsulatus* SB1003. *J. Bacteriol.* **189:**1774–1782.

Crowe, S. A., S. Katsev, K. Leslie, A. Sturm, C. Magen, S. Nomosatryo, M. A. Pack, J. D. Kessler, W. S. Reeburgh, J. A. Roberts, L. Gonzáles, G. D. Haffner, A. Mucci, B. Sundby, and D. A. Fowle. 2010. The methane cycle in ferruginous Lake Matano. *Geobiology* **9**(1):61–78. http://dx.doi.org/10.1111/j.1472-4669.2010.00257.x.

Crowe, S. A., C. Jones, S. Katsev, C. Magen, A. H. O'Neill, A. Sturm, D. E. Canfield, G. D. Haffner, A. Mucci, B. Sundby, and D. A. Fowle. 2008a. Photoferrotrophs thrive in an Archean Ocean analogue. *Proc. Natl. Acad. Sci. USA* **105:**15938–15943.

Crowe, S. A., A. H. O'Neill, S. Katsev, P. Hehanussa, G. D. Haffner, B. Sundby, A. Mucci, and D. A. Fowle. 2008b. The biogeochemistry of tropical lakes: a case study from Lake Matano, Indonesia. *Limnol. Oceanogr.* **53:**319–331.

Delsuc, F., H. Brinkmann, and H. Philippe. 2005. Phylogenomics and the reconstruction of the tree of life. *Nat. Rev. Genet.* **6:**361–375.

Doolittle, W. F. 2000. Uprooting the tree of life. *Sci. Am.* **282:**90–95.

Duan, Y., L. Zhou, D. G. Hall, W. Li, H. Doddapaneni, H. Lin, L. Liu, C. M. Vahling, D. W. Gabriel, K. P. Williams, A. Dickerman, Y. Sun, and T. Gottwald. 2009. Complete genome sequence of citrus huanglongbing bacterium, 'Candidatus *Liberibacter asiaticus*' obtained through metagenomics. *Mol. Plant-Microbe Interact.* **22:**1011–1020.

Dumont, M. G., and J. C. Murrell. 2005. Stable isotope probing—linking microbial identity to function. *Nat. Rev. Microbiol.* **3:**499–504.

Dunbar, J., L. O. Ticknor, and C. R. Kuske. 2001. Phylogenetic specificity and reproducibility and new method for analysis of terminal restriction fragment profiles of 16S rRNA genes from bacterial communities. *Appl. Environ. Microbiol.* **67:**190–197.

Edwards, K. J., P. L. Bond, T. M. Gihring, and J. F. Banfield. 2000. An archaeal iron-oxidizing extreme acidophile important in acid mine drainage. *Science* **287:**1796–1799.

Edwards, K. J., B. M. Goebel, T. M. Rodgers, M. O. Schrenk, T. M. Gihring, M. M. Cardona, B. Hu, M. M. McGuire, R. J. Hamers, N. R. Pace, and J. F. Banfield. 1999. Geomicrobiology of pyrite (FeS2) dissolution: case study at Iron Mountain, California. *Geomicrobiol. J.* **16:**155–179.

Ehrlich, H. L., and D. K. Newman. 2009a. Molecular methods in geomicrobiology, p. 139–156. *In* H. L. Ehrlich and D. K. Newman (ed.), *Geomicrobiology*. CRC Press, Boca Raton, FL.

Ehrlich, H. L., and D. K. Newman. 2009b. Nonmolecular methods in geomicrobiology, p. 117–138. *In* H. L. Ehrlich and D. K. Newman (ed.), *Geomicrobiology*. CRC Press, Boca Raton, FL.

Eilers, H., J. Pernthaler, F. O. Glockner, and R. Amann. 2000. Culturability and *in situ* abundance of pelagic bacteria from the North Sea. *Appl. Environ. Microbiol.* **66:**3044–3051.

Emerson, D., and J. V. Weiss. 2004. Bacterial iron oxidation in circumneutral freshwater habitats: findings from the field and the laboratory. *Geomicrobiol. J.* **21:**405–414.

Falkowski, P. G., R. T. Barber, and V. Smetacek. 1998. Biogeochemical controls and feedbacks on ocean primary production. *Science* **281:**200–206.

Fraústo da Silva, J. J. R., and R. J. P. Williams. 2001. *The Biological Chemistry of the Elements: the Inorganic Chemistry of Life*. Oxford University Press, New York, NY.

Garrels, R. M., and A. Lerman. 1981. Phanerozoic cycles of sedimentary carbon and sulfur. *Proc. Natl. Acad. Sci. USA* **78:**4652–4656.

Gentry, T. J., G. S. Wickham, C. W. Schadt, Z. He, and J. Zhou. 2006. Microarray applications in microbial ecology research. *Microb. Ecol.* **52:**159–175.

Ghiorse, W. C., and D. L. Balkwill. 1983. Enumeration and morphological characterization of bacteria indigenous to subsurface environments. *Dev. Ind. Microbiol.* **24:**213–224.

Giovannoni, S. J., E. F. Delong, G. J. Olsen, and N. R. Pace. 1988. Phylogenetic group-specific oligonucleotide probes for identification of single microbial cells. *J. Bacteriol.* **170:**720–726.

Glass, J. B., F. Wolfe-Simon, and A. D. Anbar. 2009. Coevolution of metal availability and nitrogen assimilation in cyanobacteria and algae. *Geobiology* **7:**100–123.

Guss, A. M., B. Mukhopadhyay, J. K. Zhang, and W. W. Metcalf. 2005. Genetic analysis of mch mutants in two *Methanosarcina* species demonstrates multiple roles for the methanopterin-dependent C-1 oxidation/reduction pathway and differences in H_2 metabolism between closely related species. *Mol. Microbiol.* **55:**1671–1680.

Guyer, R. L., and D. E. Koshland, Jr. 1989. The molecule of the year. *Science* **246:**1543–1546.

Herbert, D., R. Elsworth, and R. C. Telling. 1956. The continuous culture of bacteria; a theoretical and experimental study. *J. Gen. Microbiol.* **14:**601–622.

Hesselsoe, M., S. Fureder, M. Schloter, L. Bodrossy, N. Iversen, P. Roslev, P. H. Nielsen, M. Wagner, and A. Loy. 2009. Isotope array analysis of *Rhodocyclales* uncovers functional redundancy and versatility in an activated sludge. *ISME J.* **3:**1349–1364.

Hoff, K. J., M. Tech, T. Lingner, R. Daniel, B. Morgenstern, and P. Meinicke. 2008. Gene prediction in metagenomic fragments: a large scale machine learning approach. *BMC Bioinf.* **9:**217–231.

Hughes, J. B., J. J. Hellmann, T. H. Ricketts, and B. J. Bohannan. 2001. Counting the uncountable: statistical approaches to estimating microbial diversity. *Appl. Environ. Microbiol.* **67:**4399–4406.

Isley, A. E., and D. H. Abbott. 1999. Plume-related mafic volcanism and the deposition of banded iron formations. *J. Geophys. Res. Solid Earth* **104:**15461–15477.

Ivanov, M. V. 1968. *Microbiological Processes in the Formation of Sulfur Deposits.* U.S. Department of Agriculture and the National Science Foundation (Israel Program for Scientific Translations), Washington, DC.

Jannasch, H. W. 1967. Growth of marine bacteria at limiting concentration of organic carbon in seawater. *Limnol. Oceanogr.* **12:**264–271.

Jannasch, H. W. 1969. Estimations of bacterial growth in natural waters. *J. Bacteriol.* **99:**156–160.

Jannasch, H. W., and C. O. Wirsen. 1981. Morphological survey of microbial mats near deep-sea thermal vents. *Appl. Environ. Microbiol.* **41:**528–538.

Jargeat, P., C. Cosseau, B. Ola'h, A. Jauneau, P. Bonfante, J. Batut, and G. Becard. 2004. Isolation, free-living capacities, and genome structure of "Candidatus *Glomeribacter gigasporarum*," the endocellular bacterium of the mycorrhizal fungus *Gigaspora margarita*. *J. Bacteriol.* **186:**6876–6884.

Kaeberlein, T., K. Lewis, and S. S. Epstein. 2002. Isolating "uncultivable" microorganisms in pure culture in a simulated natural environment. *Science* **296:**1127–1129.

Kappler, A., C. Pasquero, K. O. Konhauser, and D. K. Newman. 2005. Deposition of banded iron formations by anoxygenic phototrophic Fe(II)-oxidizing bacteria. *Geology* **33:**865–868.

Kappler, A., and K. L. Straub. 2005. Geomicrobiological cycling of iron. *Rev. Mineral. Geochem.* **59:**85–108.

Klaassens, E. S., W. M. de Vos, and E. E. Vaughan. 2007. Metaproteomics approach to study the functionality of the microbiota in the human infant gastrointestinal tract. *Appl. Environ. Microbiol.* **73:**1388–1392.

Klappenbach, J. A., P. R. Saxman, J. R. Cole, and T. M. Schmidt. 2001. rrndb: the ribosomal RNA operon copy number database. *Nucleic Acids Res.* **29:**181–184.

Konhauser, K. O., L. Amskold, S. V. Lalonde, N. R. Posth, A. Kappler, and A. D. Anbar. 2007. Decoupling photochemical Fe(II) oxidation from shallow-water BIF deposition. *Earth Planet. Sci. Lett.* **258:**87–100.

Kraemer, S. 2004. Iron oxide dissolution and solubility in the presence of siderophores. *Aquat. Sci.* **66:**3–18.

Kraemer, S. M., A. Butler, P. Borer, and J. Cervini-Silva. 2005. Siderophores and the dissolution of iron-bearing minerals in marine systems. *Rev. Mineral. Geochem.* **59:**53–84.

Lawrence, J. R., D. R. Korber, G. M. Wolfaardt, and D. E. Caldwell. 1997. Analytical imaging and microscopy techniques. *In* C. J. Hurst, G. R. Knudsen, M. J. McInerney, L. D. Stetzenbach, and M. V. Walter (ed.), *Manual of Environmental Microbiology*. ASM Press, Washington, DC.

Lee, N., P. H. Nielsen, K. H. Andreasen, S. Juretschko, J. L. Nielsen, K. H. Schleifer, and M. Wagner. 1999. Combination of fluorescent *in situ* hybridization and microautoradiography-a new tool for structure-function analyses in microbial ecology. *Appl. Environ. Microbiol.* **65:**1289–1297.

Liu, W. T., T. L. Marsh, H. Cheng, and L. J. Forney. 1997. Characterization of microbial diversity by determining terminal restriction fragment length polymorphisms of genes encoding 16S rRNA. *Appl. Environ. Microbiol.* **63:**4516–4522.

Lovley, D. R., D. E. Holmes, and K. P. Nevin. 2004. Dissimilatory Fe(III) and Mn(IV) reduction. *Adv. Microb. Physiol.* **49:**219–286.

Luther, G. W., P. J. Brendel, B. L. Lewis, B. Sundby, L. Lefrancois, N. Silverberg, and D. B. Nuzzio. 1998. Simultaneous measurement of O_2, Mn, Fe, I-, and S(-II) in marine pore waters with a solid-state voltammetric microelectrode. *Limnol. Oceanogr.* **43:**325–333.

Luther, G. W., B. T. Glazer, L. Hohmann, J. I. Popp, M. Taillefert, T. F. Rozan, P. J. Brendel, S. M. Theberge, and D. B. Nuzzio. 2001. Sulfur speciation monitored *in situ* with solid state gold amalgam voltammetric microelectrodes: polysulfides as a special case in sediments, microbial mats and hydrothermal vent waters. *J. Environ. Monit.* **3:**61–66.

Maloy, S. R., and J. E. Cronan. 1994. *Microbial Genetics.* Jones and Bartlett Publishers, Sudbury, MA.

Maron, P. A., L. Ranjard, C. Mougel, and P. Lemanceau. 2007. Metaproteomics: a new approach for studying functional microbial ecology. *Microb. Ecol.* **53:**486–493.

Martinez, A., A. S. Bradley, J. R. Waldbauer, R. E. Summons, and E. F. DeLong. 2007. Proteorhodopsin photosystem gene expression enables photophosphorylation in a heterologous host. *Proc. Natl. Acad. Sci. USA* **104:**5590–5595.

McArthur, J. V. 2006. *Microbial Ecology: An Evolutionary Approach.* Academic Press, San Diego, CA.

Miller, F. D., and C. L. Hershberger. 1984. A quantitative beta-galactosidase alpha-complementation assay for fusion proteins containing human insulin B-chain peptides. *Gene* **29:**247–250.

Morel, F. M. M. 2003. The biogeochemical cycles of trace metals in the oceans. *Science* **300:**944–947.

Morgan, J. W., and E. Anders. 1980. Chemical composition of Earth, Venus, and Mercury. *Proc. Natl. Acad. Sci. USA* **77:**6973–6977.

Muyzer, G., E. C. de Waal, and A. G. Uitterlinden. 1993. Profiling of complex microbial populations by denaturing gradient gel electrophoresis analysis of polymerase chain reaction-amplified genes coding for 16S rRNA. *Appl. Environ. Microbiol.* **59:**695–700.

Myers, N., R. A. Mittermeier, C. G. Mittermeier, G. A. da Fonseca, and J. Kent. 2000. Biodiversity hotspots for conservation priorities. *Nature* **403:**853–858.

Novick, A. 1955. Growth of bacteria. *Annu. Rev. Microbiol.* **9:**97–110.

Olsen, G. J., and C. R. Woese. 1993. Ribosomal RNA: a key to phylogeny. *FASEB J.* **7:**113–123.

Orphan, V. J., C. H. House, K. U. Hinrichs, K. D. McKeegan, and E. F. DeLong. 2001. Methane-consuming archaea revealed by directly coupled isotopic and phylogenetic analysis. *Science* **293:**484–487.

Ottesen, E. A., J. W. Hong, S. R. Quake, and J. R. Leadbetter. 2006. Microfluidic digital PCR enables multigene analysis of individual environmental bacteria. *Science* **314:**1464–1467.

Ouverney, C. C., and J. A. Fuhrman. 1999. Combined microautoradiography-16S rRNA probe technique for determination of radio-isotope uptake by specific microbial cell types in situ. *Appl. Environ. Microbiol.* **65:**1746–1752.

Pelletier, E., A. Kreimeyer, S. Bocs, Z. Rouy, G. Gyapay, R. Chouari, D. Riviere, A. Ganesan, P. Daegelen, A. Sghir, G. N. Cohen, C. Medigue, J. Weissenbach, and D. Le Paslier. 2008. "Candidatus *Cloacamonas acidaminovorans*": genome sequence reconstruction provides a first glimpse of a new bacterial division. *J. Bacteriol.* **190:**2572–2579.

Perfil'ev, B. V., and D. R. Gabe. 1969. *Capillary Methods for Studying Microorganisms.* University of Toronto Press, Toronto, Canada.

Pernthaler, A., A. E. Dekas, C. T. Brown, S. K. Goffredi, T. Embaye, and V. J. Orphan. 2008. Diverse syntrophic partnerships from deep-sea methane vents revealed by direct cell capture and metagenomics. *Proc. Natl. Acad. Sci. USA* **105:**7052–7057.

Pernthaler, A., J. Pernthaler, and R. Amann. 2002. Fluorescence *in situ* hybridization and catalyzed reporter deposition for the identification of marine bacteria. *Appl. Environ. Microbiol.* **68:**3094–3101.

Petsch, S. T., K. J. Edwards, and T. I. Eglinton. 2003. Abundance, distribution and delta C-13 analysis of microbial phospholipid-derived fatty acids in a black shale weathering profile. *Org. Geochem.* **34:**731–743.

Philippe, H., and J. Laurent. 1998. How good are deep phylogenetic trees? *Curr. Opin. Genet. Dev.* **8:**616–623.

Posth, N. R., F. Hegler, K. O. Konhauser, and A. Kappler. 2008. Alternating Si and Fe deposition caused by temperature fluctuations in Precambrian oceans. *Nat. Geosci.* **1:**703–708.

Pritchett, M. A., J. K. Zhang, and W. W. Metcalf. 2004. Development of a markerless genetic exchange method for *Methanosarcina acetivorans* C2A and its use in construction of new genetic tools for methanogenic archaea. *Appl. Environ. Microbiol.* **70:**1425–1433.

Radajewski, S., P. Ineson, N. R. Parekh, and J. C. Murrell. 2000. Stable-isotope probing as a tool in microbial ecology. *Nature* **403:**646–649.

Raghoebarsing, A. A., A. Pol, K. T. van de Pas-Schoonen, A. J. P. Smolders, K. F. Ettwig, W. I. C. Rijpstra, S. Schouten, J. S. S. Damste, H. J. M. Op den Camp, M. S. M. Jetten, and M. Strous. 2006. A microbial consortium couples anaerobic methane oxidation to denitrification. *Nature* **440:**918–921.

Rankama, K., and T. Georg Sahama. 1950. *Geochemistry*. University of Chicago Press, Chicago, IL.

Rappe, M. S., S. A. Connon, K. L. Vergin, and S. J. Giovannoni. 2002. Cultivation of the ubiquitous SAR11 marine bacterioplankton clade. *Nature* **418:**630–633.

Reeburgh, W. S. 2007. Oceanic methane biogeochemistry. *Chem. Rev.* **107:**486–513.

Riesenfeld, C. S., P. D. Schloss, and J. Handelsman. 2004. Metagenomics: genomic analysis of microbial communities. *Annu. Rev. Genet.* **38:**525–552.

Rokas, A., and P. W. Holland. 2000. Rare genomic changes as a tool for phylogenetics. *Trends Ecol. Evol.* **15:**454–459.

Rother, M., and W. W. Metcalf. 2005. Genetic technologies for Archaea. *Curr. Opin. Microbiol.* **8:**745–751.

Sabo, E., D. Roy, P. B. Hamilton, P. E. Hehanussa, R. McNeely, and G. D. Haffner. 2008. The plankton community of Lake Matano: factors regulating plankton composition and relative abundance in an ancient, tropical lake of Indonesia. *Hydrobiologia* **615:**225–235. http://dx.doi.org/10.1007/s10750-008-9560-4.

Saito, M. A., D. M. Sigman, and F. M. M. Morel. 2003. The bioinorganic chemistry of the ancient ocean: the co-evolution of cyanobacterial metal requirements and biogeochemical cycles at the Archean-Proterozoic boundary? *Inorg. Chim. Acta* **356:**308–318.

Salyers, A. A., G. Bonheyo, and N. B. Shoemaker. 2000. Starting a new genetic system: lessons from bacteroides. *Methods* **20:**35–46.

Schlegel, H. G. 1993. *General Microbiology*. Cambridge University Press, New York City, NY.

Seeber, F., and J. C. Boothroyd. 1996. *Escherichia coli* beta-galactosidase as an in vitro and in vivo reporter enzyme and stable transfection marker in the intracellular protozoan parasite *Toxoplasma gondii*. *Gene* **169:**39–45.

Shi, Y., G. W. Tyson, and E. F. DeLong. 2009. Metatranscriptomics reveals unique microbial small RNAs in the ocean's water column. *Nature* **459:**266–269.

Sieburth, J. M. 1975. *Microbial Seascapes. A Pictorial Essay on Marine Microorganism and Their Environments*. University Park Press, Baltimore, MD.

Stackebrandt, E., and B. M. Goebel. 1994. A place for DNA : DNA reassociation and 16S rRNA sequence analysis in the present species definition in bacteriology. *Int. J. Syst. Bacteriol.* **44:**846–849.

Stahl, D. A. 1997. Molecular approaches for the measurement of density, diversity and phylogeny, p. 102–114. *In* C. J. Hurst, G. R. Knudsen, M. J. McInerney, L. D. Stetzenbach, and M. V. Walter (ed.), *Manual of Environmental Microbiology*. ASM Press, Washington, DC.

Staley, J. T. 2003. Speciation and bacterial phylospecies, p. 40–48. *In* A. T. Bull (ed.), *Microbial Diversity and Bioprospecting*. ASM Press, Washington, DC.

Staley, J. T., and A. Konopka. 1985. Measurement of *in situ* activities of nonphotosynthetic microorganisms in aquatic and terrestrial habitats. *Annu. Rev. Microbiol.* **39:**321–346.

Stumm, W., and J. J. Morgan. 1995. *Aquatic Chemistry: Chemical Equilibria and Rates in Natural Waters*. Wiley, New York, NY.

Svensson, E., A. Skoog, and J. P. Amend. 2004. Concentration and distribution of dissolved amino acids in a shallow hydrothermal system, Vulcano Island (Italy). *Org. Geochem.* **35:**1001–1014.

Tran-Nguyen, L. T., M. Kube, B. Schneider, R. Reinhardt, and K. S. Gibb. 2008. Comparative genome analysis of "Candidatus Phytoplasma australiense" (subgroup tuf-Australia I; rp-A) and "Ca. Phytoplasma asteris" strains OY-M and AY-WB. *J. Bacteriol.* **190:**3979–3991.

Ussher, S. J., E. P. Achterberg, and P. J. Worsfold. 2004. Marine biogeochemistry of iron. *Environ. Chem.* **1:**67–80.

Valdivia, R. H., B. P. Cormack, and S. Falkow. 2006. The uses of green fluorescent protein in prokaryotes. *Methods Biochem. Anal.* **47:**163–178.

Ward, D. M., M. M. Bateson, R. Weller, and A. L. Ruffroberts. 1992. Ribosomal-RNA analysis of microorganisms as they occur in nature. *Adv. Microb. Ecol.* **12:**219–286.

Ward, D. M., M. J. Ferris, S. C. Nold, and M. M. Bateson. 1998. A natural view of microbial biodiversity within hot spring cyanobacterial mat communities. *Microbiol. Mol. Biol. Rev.* **62:**1353–1370.

Ward, D. M., R. Weller, and M. M. Bateson. 1990. 16S rRNA sequences reveal numerous uncultured microorganisms in a natural community. *Nature* **345:**63–65.

Warnecke, F., and M. Hess. 2009. A perspective: metatranscriptomics as a tool for the discovery of novel biocatalysts. *J. Biotechnol.* **142:**91–95.

Weber, K. A., J. Pollock, K. A. Cole, S. M. O'Connor, L. A. Achenbach, and J. D. Coates. 2006. Anaerobic nitrate-dependent iron(II) bio-oxidation by a novel lithoautotrophic betaproteobacterium, strain 2002. *Appl. Environ. Microbiol.* **72:**686–694.

Wernegreen, J. J., A. B. Lazarus, and P. H. Degnan. 2002. Small genome of Candidatus *Blochmannia*, the bacterial endosymbiont of *Camponotus*, implies irreversible specialization to an intracellular lifestyle. *Microbiology* **148:**2551–2556.

White, D. 2000. *The Physiology and Biochemistry of Prokaryotes.* Oxford Unversity Press, New York, NY.

Whiteley, A. S., M. Manefield, and T. Lueders. 2006. Unlocking the 'microbial black box' using RNA-based stable isotope probing technologies. *Curr. Opin. Biotechnol.* **17:**67–71.

Widdel, F., S. Schnell, S. Heising, A. Ehrenreich, B. Assmus, and B. Schink. 1993. Ferrous iron oxidation by anoxygenic phototrophic bacteria. *Nature* **362:**834–836.

Wilmes, P., and P. L. Bond. 2006. Metaproteomics: studying functional gene expression in microbial ecosystems. *Trends Microbiol.* **14:**92–97.

Wilmes, P., S. L. Simmons, V. J. Denef, and J. F. Banfield. 2009. The dynamic genetic repertoire of microbial communities. *FEMS Microbiol. Rev.* **33:**109–132.

Wilmes, P., M. Wexler, and P. L. Bond. 2008. Metaproteomics provides functional insight into activated sludge wastewater treatment. *PLoS One* **3:**e1778.

Woyke, T., G. Xie, A. Copeland, J. M. Gonzalez, C. Han, H. Kiss, J. H. Saw, P. Senin, C. Yang, S. Chatterji, J. F. Cheng, J. A. Eisen, M. E. Sieracki, and R. Stepanauskas. 2009. Assembling the marine metagenome, one cell at a time. *PLoS One* **4:**e5299.

Wu, L. Y., X. Liu, C. W. Schadt, and J. Z. Zhou. 2006. Microarray-based analysis of subnanogram quantities of microbial community DNAs by using whole-community genome amplification. *Appl. Environ. Microbiol.* **72:**4931–4941.

Zengler, K., G. Toledo, M. Rappe, J. Elkins, E. J. Mathur, J. M. Short, and M. Keller. 2002. Cultivating the uncultured. *Proc. Natl. Acad. Sci. USA* **99:**15681–15686.

HYPERTHERMOPHILE-METAL INTERACTIONS IN HYDROTHERMAL ENVIRONMENTS

James F. Holden, Angeli Lal Menon, and Michael W. W. Adams

3

Marine hydrothermal environments are found where seawater within the crust circulates and reacts with a magmatic heat source. During this reaction, volatile compounds, such as H_2, H_2S, and CO_2, and heavy metals are leached from geothermally heated rock into the circulating fluids. Hot buoyant hydrothermal fluid then rises through cracks in the upper layers of ocean crust until it reaches the seafloor at temperatures up to 400°C. Upon contact with seawater, the dissolved metals and sulfides in the hydrothermal fluid precipitate forming the "black smoker" plumes typically associated with these environments. Cold seawater that is in direct contact with hot hydrothermal fluid above the seafloor is rapidly heated causing the precipitation of sulfate minerals from seawater (e.g., anhydrite [$CaSO_4$]) that forms a barrier between the exiting hydrothermal fluids and the surrounding seawater (Haymon, 1983). The resulting conduit that forms serves as a scaffold for the accumulation of precipitated metal sulfides and the subsequent formation of what is known as a chimney (Color Plate 6).

An active hydrothermal sulfide deposit can reach tens of meters in height and across the base. A temperature gradient forms across the wall of the chimney largely because of the mixture of seawater and hydrothermal fluid in the chimney wall, although some conductive heat transfer also occurs. Oftentimes, hydrothermal fluids mix with seawater within the crust and exit the seafloor at temperatures too low (<150°C) to cause sulfate precipitation. In these cases, so-called diffuse vents form where 2 to 150°C fluids flow directly out of cracks in the seafloor basalt without any associated metal-sulfide deposits (Color Plate 6).

The chemical disequilibrium formed in the interior of black smoker chimneys and in diffuse vent fluids nourishes microbes that inhabit their pore spaces and outer surfaces (see reviews by Huber and Holden, 2008; Schrenk et al., 2008). The hottest of these microorganisms are hyperthermophiles, which are defined as organisms with an optimal growth temperature above 80°C (Stetter, 1999). Most hyperthermophiles are anaerobes that require metalloenzymes with low reduction potentials to catalyze essential metabolic reactions. Many proteins from aerobic hyperthermophiles also contain metal cofactors. A few hyperthermophiles are also capable of using metals in a dissimilatory manner as terminal electron

---

James F. Holden, Department of Microbiology, University of Massachusetts, Amherst, MA 01003. *Angeli Lal Menon and Michael W. W. Adams*, Department of Biochemistry and Molecular Biology, University of Georgia, Athens, GA 30606.

Microbial Metal and Metalloid Metabolism: Advances and Applications
Edited by John F. Stolz and Ronald S. Oremland © 2011 ASM Press, Washington, DC

acceptors during respiration. The goal of this chapter is to explore the metal chemistry of different types of marine hydrothermal environments, the metal requirements of hyperthermophiles, the nature and constraints of their interactions with metals, and the biogeochemical implications of hyperthermophile-metal interactions.

METAL CHEMISTRY OF HYDROTHERMAL FLUIDS

The metal composition of hydrothermal fluids is controlled by temperature, chlorinity, reduction potential, pH, dissolved sulfide, and source rock composition (Metz and Trefry, 2000). The primary metal sulfide minerals formed in hydrothermal vent deposits are chalcopyrite [$(Fe,Cu)S_2$], sphalerite, wurtzite [both $(Fe,Zn)S_2$], pyrite and marcasite (both FeS_2), and pyrrhotite (FeS). Temperature is the primary factor controlling the solubility and distribution of (Fe,Cu)- and (Fe,Zn)-sulfides as well as many trace metals in hydrothermal systems. The solubility of $(Fe,Cu)S_2$ decreases rapidly below 350°C and leads to chalcopyrite formation, while $(Fe,Zn)S_2$ solubility decreases below 200°C and leads to sphalerite and wurtzite formation (Haymon, 1983). As a result, the central hydrothermal fluid conduit within black smoker chimneys is commonly lined with chalcopyrite with outlying layers of wurtzite, sphalerite, and marcasite in the cooler regions of a chimney. Elements such as molybdenum, cobalt, and selenium are commonly associated with $(Fe,Cu)S_2$ and tend to be removed from solution with copper at higher temperatures (Metz and Trefry, 2000). Molybdenum concentrations in hydrothermal fluids are generally lower than molybdenum concentrations in seawater (Table 1); therefore, the primary source of molybdenum for metalloenzymes is seawater.

Elevated chloride concentrations in hydrothermal fluids also increase the solubility of metals, especially those that form chloride complexes with neutral or negative charge (e.g., $FeCl_2$, $ZnCl_2$) (Metz and Trefry, 2000). There is a strong linear relationship between iron and chloride concentrations in hydrothermal fluids demonstrating the importance of brines in controlling the major element content of these fluids (Butterfield and Massoth, 1994). Unlike copper, zinc concentrations in hydrothermal fluids above 200°C are better correlated with chloride concentrations (Metz and Trefry, 2000). Hydrothermal fluid in subseafloor reaction zones can phase separate into two fluid phases at very high reaction temperatures forming a low-chlorinity, low-metal, vapor-rich fluid that discharges from hydrothermal environments prior to the release of a low-vapor, high-metal, high-chlorinity fluid (Butterfield et al., 1997). Therefore, metal concentrations in hydrothermal fluids depend in part on which fluid phase is emitted at that time. Low pH and more oxidizing conditions also enhance metal levels in hydrothermal fluids. Copper concentrations were 2 to 6 times higher in a more oxidizing synthetic hydrothermal fluid ($fO_2 \approx 10^{-24}$ at 400°C) than in more reducing fluid ($fO_2 \approx 10^{-26}$ at 400°C) (Seyfried and Ding, 1993, 1995).

The fluid chemistry of hydrothermal environments also varies significantly with the type of rock that hosts the circulating fluids. Most deep-sea studies are focused on one of three types of host-rock hydrothermal systems: mafic rock, ultramafic rock, and andesitic rock. Mafic rock hydrothermal environments are commonly found along mid-ocean spreading centers where new seafloor is formed by periodic magmatic eruptions as two tectonic plates spread apart from one another. Ultramafic rock hydrothermal environments are also found in zones of plate spreading where tectonic forces, not volcanism, drive plate movement. Both are igneous basalts and gabbros with high concentrations of MgO and FeO, but they differ in their silica content with ultramafic rocks having concentrations less than 45%, while mafic rocks have concentrations above 45%. The Earth's mantle is composed primarily of ultramafic rock, and occasionally blocks of this material are tectonically displaced to the surface or shallow subsurface where they are exposed to circulating seawater. Serpentinization is the hydrous altera-

TABLE 1 Chemistry of hydrothermal fluids from various sites and host-rock environments

Location	pH	H_2 (mM)	H_2S (mM)	Fe (mM)	Zn (µM)	Ni (µM)	Co (µM)	W (nM)	Mo (nM)
Mafic rock									
Juan de Fuca Ridge[a]	2.8–4.5	0.16–0.53	1.4–8.1	0.01–16.4	2–600	ND	0.05–1.4	ND	0–33
East Pacific Rise[b]	3.1–3.8	0.14–1.7	2.9–12.2	0.4–12.6	2–106	ND	ND	ND	ND
Mid-Atlantic Ridge[c]	3.1–5.0	0.37–1.03	0.6–6.0	0.02–5.2	0.2–83	< 2	< 2	ND	3–11
Ultramafic rock									
Rainbow[d]	2.8	16	1.2	24.0	160	3.0	13	ND	2
Logatchev[e]	3.3	12–19	0.8–3.6	2.4–2.5	29–38.3	< 2	0.33–1.01	ND	4.1–18.5
Kairei[f]	3.4–3.5	2.5–8.2	4.0	3.5–6.0	67–90	ND	ND	0.21	2
Volcanic arcs									
Lau Basin[g]	2.4–2.7	0.05–0.10	7.0–9.0	1.2–12.8	1,200–3,100	ND	ND	ND	ND
Izu-Bonin-Mariana Arc[h]	3.5–5.3	0.01	2.4–14.6	0.46	7	ND	ND	15	5
Manus Basin[i]	2.1–2.5	ND	5.3–9.7	0.01–4.4	ND	ND	ND	ND	ND
Seawater[j]	7.8	0.0004	0	0.000061	0.076	0.112	0.007	0.5	104

[a]Butterfield and Massoth, 1994; Butterfield et al., 1990, 1994; Lilley et al., 1993; Metz and Trefry, 2000. ND, not determined.
[b]Von Damm, 1995; Douville et al., 1999; Lilley et al., 1993.
[c]Douville et al., 1999.
[d]Charlou et al., 2002; Douville et al., 2002.
[e]Charlou et al., 1998; Douville et al., 2002; Schmidt et al., 2007.
[f]Kishida et al., 2004; Takai et al., 2004; Gamo et al., 2001; Gallant and Von Damm, 2006.
[g]Takai et al., 2008; Douville et al., 1999; Fouquet et al., 1991.
[h]Kishida et al., 2004; Nakagawa et al., 2006; Gamo et al., 2004.
[i]Douville et al., 1999; Gamo et al., 1997.
[j]Turekian, 1968.

tion of ultramafic rocks (primarily olivine) into serpentine and magnetite (Fe_3O_4) with abundant concomitant H_2 production (McCollom and Bach, 2009). The low silica activity of the rocks results in the formation of minerals that largely exclude Fe(II) leading to the formation of magnetite and H_2. In mafic basalts, a greater proportion of Fe(II) is sequestered in silicate alteration minerals rather than being converted to Fe(III), resulting in less H_2 production (McCollom and Bach, 2009). As a result, hydrothermal fluids that interact with ultramafic rocks have H_2 concentrations that are significantly higher than those that have interacted with hot mafic rocks (Table 1). This in turn increases the reduction potential of the hydrothermal fluids and the solubility of metals in them.

In contrast to mid-ocean spreading centers, volcanic arcs form at convergent plate boundaries and can host mafic, ultramafic, and andesitic rock types. As the subducted plate melts, it supplies magma for a nearby volcanism and submarine hydrothermal circulation that runs along the subduction zone. Hydrous minerals in the subducted slab, which formed during hydrothermal circulation at the mid-ocean ridge, are dehydrated and changed to more stable anhydrous forms releasing water and soluble elements. Unlike mid-ocean spreading centers, andesitic rocks tend to be acidic. As a result, the hydrothermal fluids that interact with these rocks tend to have a lower pH (Table 1). The trace elements in andesite-hosted hydrothermal fluids are strongly influenced by fluid-acid rock interactions (Douville et al., 1999).

With the exception of molybdenum, all of the major metals that are assimilated into metalloenzymes in hyperthermophiles (i.e., iron, zinc, cobalt, nickel, and tungsten) are present in hydrothermal fluids at concentrations that are significantly higher than those in seawater (Table 1). Furthermore, they are also present within the mineral deposits formed where hydrothermal fluids exit the seafloor and interact with seawater (Kristall et al., 2006).

METALS USED IN HYPERTHERMOPHILE METABOLISM

Among hyperthermophiles, 20 archaeal genera and two bacterial genera are commonly

TABLE 2 Genera of hyperthermophiles found in marine hydrothermal environments

Taxonomic group	Genera	Electron donors	Electron acceptors
Aquificaceae (family)	*Aquifex*	H_2, S^0, $S_2O_3^{2-}$	O_2, S^0
Archaeoglobaceae (family)	*Archaeoglobus, Ferroglobus, Geoglobus*	H_2, $S_2O_3^{2-}$, Fe(II), formate, acetate, fatty acids, peptides, sugars	H^+, SO_4^{2-}, SO_3^{2-}, $S_2O_3^{2-}$, $HFeO_2$, NO_3-
Desulfurococcaceae (family)	*Aeropyrum, Desulfurococcus, Ignicoccus, Staphylothermus, Stetteria, Thermodiscus*	H_2, peptides, sugars	H^+, S^0, $S_2O_3^{2-}$, O_2
Korarchaeota (phylum)	*Korarchaeum*	Peptides	H^+
Methanocaldococcaceae (family)	*Methanocaldococcus, Methanotorris*	H_2	CO_2
Methanopyraceae (family)	*Methanopyrus*	H_2	CO_2
Nanoarchaeota (phylum)	*Nanoarchaeum*	Parasitic	
Pyrodictiaceae (family)	*Hyperthermus, Pyrodictium, Pyrolobus*	H_2, peptides	S^0, $S_2O_3^{2-}$, SO_3^{2-}, $HFeO_2$, NO_3-, O_2
Thermococcaceae (family)	*Thermococcus, Paleococcus, Pyrococcus*	Peptides, sugars	H^+, S^0, Fe_2S_3, O_2
Thermotogaceae (family)	*Thermotoga*	Peptides, sugars	H^+, S^0, $HFeO_2$

associated with marine hydrothermal environments (Table 2) (see review by Holden, 2009). They all require metal-containing enzymes for CO_2 assimilation, catabolism of organic compounds, and respiration. Hyperthermophilic autotrophs in hydrothermal environments are ecologically important for their ability to assimilate CO_2 in the absence of both sunlight and oxygen and serve as primary producers. They include three genera of methanogens (*Methanocaldococcus*, *Methanotorris*, and *Methanopyrus*) that are obligate autotrophs, use H_2 to reduce CO_2 to CH_4 and H_2O, and assimilate CO_2 using the acetyl coenzyme A (acetyl-CoA) pathway (Ferry, 1999). The key metalloenzymes for hyperthermophilic methanogenesis are the tungsten-dependent enzyme formylmethanofuran dehydrogenase (Vorholt et al., 1997b), the cobalt-dependent N^5-methyltetrahydromethanopterin: coenzyme M (CoM) methyltransferase (Kengen et al., 1992), the nickel-dependent enzymes methyl-CoM reductase (Rospert et al., 1991) and carbon monoxide dehydrogenase/acetyl-CoA synthase (Gong et al., 2008), and iron-containing electron carriers such as ferredoxin. Members of the *Archaeoglobaceae* similarly assimilate CO_2 using the acetyl-CoA pathway when grown autotrophically (Vorholt et al., 1995, 1997a) but lack methyl-CoM reductase that is needed for methane formation (Klenk et al., 1997). Other autotrophs include genera from the *Crenarchaeota*: *Hyperthermus*, *Pyrodictium*, *Pyrolobus*, *Ignicoccus*, and *Stetteria*. *Ignicoccus* and *Aquifex* assimilate CO_2 using the dicarboxylate/4-hydroxybutyrate cycle and the reductive citric acid cycle, respectively (Beh et al., 1993; Huber et al., 2008). All of these autotrophs use H_2 as their electron donor, and the hydrogenases used to oxidize H_2 typically contain iron and nickel (Vignais et al., 2001).

Heterotrophy is also common among hyperthermophiles with peptides serving as the most widely used organic carbon source (Table 2). Peptides are catabolized by deamination of amino acids to form 2-ketoacids (Robb et al., 1992), which are then decarboxylated to acyl-CoAs and subsequently used to generate ATP and carboxylic acids using acyl-CoA synthetases (Mai and Adams, 1996a). There are four decarboxylases that are used and each is a ferredoxin-dependent oxidoreductase that contains 4Fe-4S clusters (Blamey and Adams, 1993; Heider et al., 1996; Mai and Adams, 1994, 1996b). Many hyperthermophilic heterotrophs also catabolize carbohydrates using a modified Embden-Meyerhof pathway that relies on ADP-dependent kinases (Kengen et al., 1994), a tungsten-dependent glyceraldehyde-3-phosphate oxidoreductase (Mukund and Adams, 1995), and ferredoxin as the electron carrier. Other tungsten-dependent enzymes in hyperthermophilic heterotrophs include aldehyde oxidoreductase and formaldehyde oxidoreductase (Mukund and Adams, 1991; Roy et al., 1999).

The majority of marine hyperthermophiles are strict anaerobes; the most commonly used terminal electron acceptors are sulfur compounds (Table 2). *Pyrodictium abyssi* produces a membrane-bound H_2:sulfur oxidoreductase complex that contains iron, nickel, and copper (Dirmeier et al., 1998). In contrast, *Pyrococcus furiosus* appears to use a soluble NAD(P)H- and coenzyme A-dependent sulfur reductase during sulfur reduction (Schut et al., 2007). *Archaeoglobus* species are the only archaea known to grow on sulfate and sequentially use ATP sulfurylase, adenosine-5′-phosphosulfate reductase, and dissimilatory sulfite reductase in a manner similar to that found in sulfate-reducing bacteria (Dahl et al., 1993; Speich et al., 1994; Sperling et al., 1998). Adenosine-5′-phosphosulfate reductase and dissimilatory sulfite reductase are cytoplasmic proteins that contain two 4Fe-4S clusters and a siroheme-4Fe-4S cluster, respectively (Dahl et al., 1993; Speich et al., 1994; Schiffer et al., 2008).

Many heterotrophic hyperthermophiles, such as *P. furiosus*, produce H_2 from H^+ using a membrane-bound hydrogenase coupled with the generation of a sodium ion motive force (Sapra et al., 2003) and soluble NADH-dependent hydrogenases (Bryant and Adams, 1989; Ma et al., 2000). These all contain iron and nickel cofactors. *Pyrolobus fumarii*, *Ferroglobus placidus*, and *Pyrobaculum aerophilum* grow by denitrification (Blöchl et al., 1997; Hafenbradl et al., 1996; Völkl et al., 1993). In contrast with gram-negative bacteria, all four denitrification enzymes in *P. aerophilum* are membrane bound and use menaquinol as the electron donor (de Vries and Schröder, 2002). Its dissimilatory nitrate reductase contains molybdenum and several iron sulfur clusters (Afshar et al., 2001), while its nitric oxide contains heme and nonheme iron (de Vries et al., 2003). Aerobic respiration in *Pyrobaculum* utilizes iron-sulfur clusters and iron-containing hemes in a bc_1-like complex (Henninger et al., 1999) and a copper-containing SoxM-type terminal oxidase (Nunoura et al., 2005). Facultatively aerobic *Pyrobaculum* species also possess superoxide dismutase and catalase that contain manganese and iron (Whittaker and Whittaker, 2000; Amo et al., 2002, 2003).

METAL ASSIMILATION INTO METALLOENZYMES

It is estimated, based on bioinformatic analyses, that close to 50% of the members of the six known classes of enzymes require or contain metals (Andreini et al., 2008). Metalloenzymes form a major part of metalloproteomes and reflect the "bioavailability" of metals within a given environment that are (or were at some point during their evolution) the most accessible (Dupont et al., 2006). As mentioned above, trace metal bioavailability in marine hydrothermal vent environments can be dramatically influenced by fluctuations in temperature, salinity, sulfide concentrations, and dissolved oxygen. To deal with these changes and ensure survival, hyperthermophiles occupying this unique niche must have mechanisms to import/export metals, sense metal availability, distinguish metal types, and regulate metal uptake (see review by Waldron et al. [2009]).

Iron

Iron is essential for the growth of almost all organisms and is predicted to be one of the most abundant metals in hyperthermophilic archaea (Andreini et al., 2007, 2009; Dupont et al., 2006). The first reported example of hyperthermophilic iron assimilation involves a novel assimilatory ferric reductase from the anaerobic archaeon *Archaeoglobus fulgidus* (Chiu et al., 2001; Vadas et al., 1999). This cytoplasmic ferric reductase, a 40-kDa homodimer covalently linked by a disulfide bridge, catalyzes the flavin-mediated reduction of Fe^{3+}-EDTA using NAD(P)H as the electron donor. The high cellular abundance and specific activity of the enzyme also suggests a possible role as a terminal electron acceptor in a ferric-based respiratory pathway (Vadas et al., 1999). In addition, the *A. fulgidus* homolog of the bacterial diphtheria toxin regulator (DtxR), a metal-dependent repressor, was shown to be dramatically upregulated in response to decreased metal availability in the growth medium (Bell et al., 1999). DtxR is known to be a global iron-sensitive regulatory element in *Corynebacterium diphtheriae* and may serve the same function in *A. fulgidus* (Tao et al., 1994).

The first attempt to understand the iron regulon in archaea utilized the hyperthermophilic archaeon *Thermococcus kodakaraensis*. This organism requires iron for growth and has homologs of known iron-uptake systems, namely two FeoB family proteins similar to those involved in high-affinity Fe(II) uptake in bacterial systems and homologs of the major bacterial transcriptional factors involved in iron utilization and homeostasis, DtxR and the ferric uptake repressor Fur (Cartron et al., 2006; Louvel et al., 2009). Unexpectedly, Fur expression and activity in *T. kodakaraensis* were not regulated by iron. Microarray data analyses suggested that the DtxR homolog in this organism may control expression of iron-related genes because a mutant strain defective in DtxR resulted in derepressed expression of the high-affinity FeoAB-like transporters (Louvel et al., 2009). Interestingly, while a putative Fur homolog was found in the genome of the closely related archaeon *P. furiosus*, no homologs were found in two other *Pyrococcus* species (*P. abyssi* and *P. horikoshii*) raising the question as to its role in iron assimilation in hyperthermophilic archaea. In contrast, at least one homolog of the DtxR was found in each of these hyperthermophiles (Louvel et al., 2009).

The FeoB protein family is widely distributed among prokaryotes; however, the exact role of FeoB in high-affinity Fe(II) uptake is still unclear (i.e., whether it is directly involved in Fe(II) uptake or acts as a sensor of intracellular or external Fe concentrations) (Cartron et al., 2006). It is the only known membrane protein with an N-terminal GTPase domain similar to that found in small GTPase (G) proteins known to play an important role in the regulation of cellular processes by switching between active GTP-binding and inactive GDP-binding conformational states (Caldon and March, 2003). The high-resolution crystal structures of the soluble N-terminal GTPase of FeoB from *Methanocaldococcus jannaschii* (Köster et al., 2009) and the GTPase and GDP dissociation inhibitor-like spacer domain of the FeoB-like protein from *Thermotoga maritima* (Hattori et al., 2009), while showing distinct differences, strongly suggest that these domains could function in the regulation and transport of iron in these hyperthermophiles.

Molybdenum and Tungsten

The assimilation of molybdenum and tungsten will be addressed together because these two elements have much in common. Both are rare in nature, ranking 53rd and 54th in abundance, respectively (Greenwood and Earnshaw, 1984). They most likely exist as stable, highly soluble tetrahedral oxoanions, molybdate (MoO_4^{2-}) and tungstate (WO_4^{2-}), in aqueous environments, making these the most likely bioavailable forms for cellular uptake. The biological distribution of molybdenum and tungsten is not uniform and organisms can be divided into four categories based on usage: organisms that are strictly molybdenum dependent, organisms that are strictly tungsten dependent, organisms that use either metal depending on availability, and organisms that simultaneously use both metals for distinct functions (Bevers et al., 2009). While molybdenum utilization occurs in all domains of life, tungsten utilization appears to be restricted to (facultatively) anaerobic prokaryotes. Many anaerobic archaea and some bacteria appear to be molybdenum independent and require tungsten for growth. In fact, most of these are hyperthermophilic anaerobes that are found in hydrothermal vent systems with abundant tungstate present and most closely represent the environment of primitive Earth (Johnson et al., 1996).

All known enzymes that utilize molybdenum or tungsten catalyze redox reactions where the redox chemistry of the metal is controlled by the metallocofactor and enzyme environment (Hille, 2002). With the exception of bacterial nitrogenase, which has an iron-molybdenum cofactor (FeMoCo) in its active site, all molybdenum and tungsten cofactors are indirectly bound to proteins via variants of a common organic cofactor, the metal binding-pterin cofactor (MPT) (Bevers et al., 2009; Schwarz et al., 2009). The molybdenum or tungsten in the active sites of these enzymes are coordinated by two or four dithiolene sulfur atoms from one or two pterin molecules that make up the cofactor. MPT enzymes have been classified into four families based on cofactor composition and sequence homology. The three molybdenum families are exemplified by sulfite oxidase, xanthine oxidase, and dimethyl sulfoxide reductase while the two tungsten families consist of the formate dehydrogenases (a subset of the dimethyl sulfoxide reductase family) and aldehyde oxidoreductases. The presence of tungsten- and/or molybdenum-containing enzymes within cells is dictated by their ability to specifically transport the correct metals into the cell, synthesize the MPT cofactors, and regulate both uptake and metalloenzyme biosynthesis.

The transport of molybdenum via high-affinity molybdate ABC transporters has been well characterized in mesophilic bacterial systems and the molybdate-binding protein (ModA) was shown to bind both molybdate and tungstate equally well (K_D, 3 and 7 μM, respectively) (Rech et al., 1996). The molybdate ABC transporter system is found in over

90% of molybdenum-utilizing bacteria but only in ~40% of sequenced archaeal genomes, including *T. kodakaraensis* (Zhang and Gladyshev, 2008; Zhang et al., 2009). The first tungsten-specific transporter, TupABC, was identified in the obligate tungstate-utilizing mesophilic bacterium *Eubacterium acidaminophilum*. The substrate-binding protein (TupA) was shown to have a higher binding affinity for tungsten over molybdenum by several orders of magnitude (K_D, 0.5 μM) (Makdessi et al., 2001, 2004). Homologs of the TupABC transporter are equally distributed in bacteria (26%) and archaea (33%), including the hyperthermophilic *Pyrobaculum* species (Zhang and Gladyshev, 2008). The second tungsten-specific transport system (WtpABC) was discovered in *P. furiosus* (Bevers et al., 2006), a marine archaeon with a strict requirement for tungsten and a tungsten proteome that predominantly consists of five previously characterized aldehyde oxidoreductases (Mukund and Adams, 1996; Sevcenco et al., 2009). Its genome also encodes two putative formate dehydrogenase family proteins (Robb et al., 2001). The Wtp ABC transporter appears to be the most highly selective for tungstate over molybdate, which could be due to the uniquely distorted octahedral substrate coordination that was revealed from the high-resolution crystal structure of tungsten-bound WtpA from *A. fulgidus* and four additional archaeal species (Hollenstein et al., 2009). The periplasmic W-transport protein (WtpA) from this transporter has a binding affinity for tungstate in the picomolar range ($K_D \sim$ 17 pM) and molybdate in the nanomolar range ($K_D \sim$ 11 nM) (Bevers et al., 2006). WtpABC is the most common transporter in archaea (64%) and was found in only 3.5% of sequenced bacteria (Zhang and Gladyshev, 2008).

Nickel and Cobalt

The transition metals nickel and cobalt are essential cofactors for a number of prokaryotic enzymes, including some that are found in marine hyperthermophiles. There are eight known nickel enzymes, and seven of these catalyze the use and/or production of gases that play important roles in biological carbon, nitrogen, and oxygen cycles (see reviews by Fontecilla-Camps et al., 2009; Ragsdale, 2009). As described above, marine hydrothermal vents are rich in H_2 and CO_2 and provide methanogenic hyperthermophiles with the substrates required for the production of methane (Childress and Fisher, 1992). Key enzymes in methanogenesis (and anaerobic methane oxidation) are (i) the nickel-containing methyl-CoM reductase that generates methane, (ii) the nickel-containing carbon monoxide dehydrogenase/acetyl-CoA synthase (CODH/ACS) that catalyzes the reduction of CO_2 to a carbon monoxyl group followed by its ligation to a methyl group and coenzyme A to form acetyl-CoA, and (iii) the [NiFe]-hydrogenases that oxidize H_2 and enable a wide range of autotrophs, including methanogens, to use this gas as a source of energy and electrons (Fontecilla-Camps et al., 2009; Ragsdale, 2009).

Cobalt is found in two types of prokaryotic protein. It is an essential component of several enzymes such as metalloproteases, like methionine aminopeptidase and prolidase in *Pyrococcus* sp., and is present as a divalent cation. However, although these enzymes show a strong preference for cobalt and this metal is found in the purified proteins, they are not absolutely cobalt-specific and in vitro are active with other metal cations such as zinc or iron (Ghosh et al., 1998; Theriot et al., 2009; Tsunasawa et al., 1997). In contrast, in the other type of cobalt-containing enzyme, cobalt is absolutely required and cannot be substituted because it is present as an organometallic vitamin B_{12}-based cofactor termed cobalamin. Such cobalt cofactors are found in three classes of prokaryotic enzyme (Banerjee and Ragsdale, 2003): adenosylcobalamin-dependent isomerases (e.g., the class II ribonucleotide reductase from *P. furiosus* [Riera et al., 1997]), methylcobalamin-dependent methyltransferases (e.g., the methionine synthase from *T. maritima* [Huang et al., 2007]), and B_{12}-dependent reductive halogenases.

Comparative genomics of vitamin B_{12} metabolism and regulation in prokaryotes using characterized B_{12}-dependent enzymes, cofactor biosynthesis pathways, and uptake systems has established that the only known transport system in prokaryotes is BtuFCD (Rodionov et al., 2003), an ABC transporter that has been shown to actively transport vitamin B_{12} and other corrinoids in enteric bacteria (Borths et al., 2002). Among the hyperthermophiles whose genome sequences are available, *Aquifex aeolicus* and *Ignicoccus hospitalis* contain no known B_{12}-dependent enzymes, cofactor biosynthesis pathways, or uptake systems (Rodionov et al., 2003; Zhang et al., 2009). Only three hyperthermophiles containing known B_{12}-dependent enzymes are able to synthesize cobalamin de novo (*A. fulgidus*, *M. jannaschii*, and *Methanopyrus kandleri*) and two of these also have the BtuFCD transporter (*A. fulgidus* and *M. jannaschii*). The remaining hyperthermophiles with B_{12}-containing enzymes (*T. maritima*, *Pyrococcus* species, and *T. kodakaraensis*) do not possess the (complete) B_{12} biosynthetic pathway, but they do contain the BtuFCD uptake system, suggesting that they can acquire B_{12} and/or salvage B_{12} precursors from their external environment (Escalante-Semerena, 2007; Rodionov et al., 2003; Zhang et al., 2009).

In silico comparative genomic analyses have identified a novel group of ABC transporters, the homologous CbiMNQO and NikMNQO transport systems, as the most widespread prokaryotic transporters for cobalt and nickel ions, respectively, and include most marine hyperthermophiles (Rodionov et al., 2006, 2009). The metal specificity (cobalt or nickel) of these hypothetical transporters was predicted based on the identification and presence of specific regulatory elements, namely, riboswitches (B_{12} elements), binding sites for the nickel-dependent repressor (NikR), and colocalization with genes encoding B_{12}-dependent enzymes involved in B_{12} biosynthesis or nickel-dependent enzymes. These transporters encode a cytoplasmic ATP-binding protein (CbiO/NikO) and a transmembrane protein (CbiQ/NikQ), but lack an extracytoplasmic solute-binding protein, an essential component in uptake ABC transporters, suggesting a different transport mechanism (Rodionov et al., 2009). The CbiN/NikN proteins both have two transmembrane domains with an extracytoplasmic loop; however, they exhibit no significant sequence similarity and are of unknown function. The CbiM/NikM proteins are a unique family of integral membrane proteins with two subfamilies based on the presence of distinct sets of conserved residues within each group that could confer a preference for cobalt or nickel, respectively. Experimental analysis validated both metal uptake activity and metal specificity (Rodionov et al., 2006).

Among the marine hyperthermophiles, only *A. aeolicus* and *I. hospitalis* lack cobalt utilization. In addition, while both encode NiFe hydrogenases, only *A. aeolicus* has a known Ni-uptake system, the HupE/UreJ secondary transporter (Eitinger et al., 2005), and neither contains the bacterial nickel-dependent repressor, NikR (Dosanjh and Michel, 2006). The reverse is true for *T. maritima* and *Aeropyrum pernix*, which have no known nickel proteins or transporters but do contain B_{12} proteins and the B_{12} transporter, BtuFCD. The remaining hyperthermophiles (*A. fulgidus*, *M. jannaschii*, *M. kandleri*, *T. kodakaraensis*, and *Pyrococcus* species) utilize both nickel and cobalt. The CbiMNQO and NikMNQO transport systems were identified in *A. fulgidus* and *M. jannaschii*, whereas *M. kandleri*, *T. kodakaraensis*, and *P. furiosus* have the NikMNQO system only and *P. horikoshii* and *P. abyssi* have neither, indicating that as yet unidentified nickel transporters may be present in these archaea (Zhang et al., 2009). All the nickel and cobalt users have orthologs of the bacterial nickel-dependent repressor, NikR, in their genomes. The *Thermococcales* contain a phylogenetically distinct group of NikRs, and while two members (*T. kodakaraensis* and *P. furiosus*) appear to have a unique predicted NikR-binding signal, candidate NikR-binding sites were not identified in *P. horikoshii* and *P. abyssi* (Rodionov et al., 2006). However, the crystal structures of NikR from *Escherichia coli* and *P. horikoshii* reveal a plausible mechanism for nickel-dependent promoter recognition (Chivers and Tahirov, 2005; Schreiter et al., 2003).

DISSIMILATORY METAL REDUCTION

The Chemistry of Metal Respiration

All organisms gain energy mainly by catalyzing redox reactions at rates faster than the same thermodynamically favorable abiotic reaction (Amend and Shock, 2001). With use of H_2 as the common electron donor, the free energy of various terminal electron-accepting processes at 100°C can be predicted and compared (Table 3). The free energy of iron reduction is very dependent upon the standard free energy of iron mineral formation for each mineral and the amount of water produced in the reduction reaction.

At near-neutral pH, iron oxides are the most thermodynamically favorable and the most environmentally relevant form of iron for microbial respiration (Lovley, 1991; Straub et al., 2001). Several forms of FeOOH with varying free energies of formation, such as poorly crystalline iron oxide ($\Delta G° = -427.8$ kJ mol^{-1}), lepidocrocite (-477.7 kJ mol^{-1}), goethite (-488 kJ mol^{-1}), and akaganéite (-752.7 kJ mol^{-1}), can be reduced to magnetite ($-1,027.2$ kJ mol^{-1}) with the free energies of the reaction dependent entirely upon the inherent mineral stability of each substrate (Table 3) (Schwertmann and Cornell, 2000). Poorly crystalline iron oxide behaves like an aqueous species despite its insolubility owing to its amorphous state and large surface area-to-volume ratio, and the large free energy yielded upon its reduction to magnetite makes this a very favorable terminal electron acceptor.

Ferrihydrite [Fe(OH)$_3$] is the most bioavailable mineral form of iron oxide due to its large surface area. While akaganéite and ferrihydrite ($\Delta G° = -699.0$ kJ mol^{-1}) have similar free energies of formation, the reduction of ferrihydrite to magnetite is far more energetically favorable than the reduction of akaganéite to magnetite primarily because of the 10 moles of water that are formed per H_2 consumed

TABLE 3 Values of $\Delta G°$ (kJ per mol H_2) for various chemolithoautotrophic reactions using 1 mol of H_2 as the electron donor at 100°C and saturation pressures for H_2O (calculated from Amend and Shock, 2001; Schwertmann and Cornell, 2000)

Growth reaction	$\Delta G^0_{100°C}$ (kJ per mol H_2)
Iron reduction	
$H_{2(aq)} + 6\ FeOOH_{(aq)}$ (poorly crystalline) $\rightarrow$ 2 Fe_3O_4 (magnetite) + 4 $H_2O_{(l)}$	-472.24
$H_{2(aq)} + 6\ Fe(OH)_3$ (ferrihydrite) $\rightarrow$ 2 Fe_3O_4 + 10 $H_2O_{(l)}$	-303.34
$H_{2(aq)} + 6$ -FeOOH (lepidocrocite) $\rightarrow$ 2 Fe_3O_4 + 4 $H_2O_{(l)}$	-172.66
$H_{2(aq)} + 6$ -FeOOH (goethite) $\rightarrow$ 2 Fe_3O_4 + 4 $H_2O_{(l)}$	-107.26
$H_{2(aq)} + 3\ Fe_2O_3$ (hematite) $\rightarrow$ 2 Fe_3O_4 + $H_2O_{(l)}$	-50.86
$H_{2(aq)} + 6$ -FeOOH (akaganéite) $\rightarrow$ 2 Fe_3O_4 + 4 $H_2O_{(l)}$	$1,477.34$
$H_{2(aq)} + FeS_2$ (pyrite) $\rightarrow$ FeS (pyrrhotite) + $H_2S_{(aq)}$	8.12
$H_{2(aq)} + (Fe,Cu)S_2$ (chalcopyrite) $\rightarrow$ (Fe,Cu)S + $H_2S_{(aq)}$	41.87
$H_{2(aq)} + (Fe,Zn)S_2$ (sphalerite) $\rightarrow$ (Fe,Zn)S + $H_2S_{(aq)}$	48.78
Other respiratory reactions	
$H_{2(aq)} + 0.5\ O_{2(aq)} \rightarrow H_2O_{(l)}$	-258.44
$H_{2(aq)} + 0.4\ NO_3^- + 0.4\ H^+ \rightarrow 0.2\ N_{2(aq)} + 1.2\ H_2O_{(l)}$	-253.34
$H_{2(aq)} + 0.25\ SO_4^{2-} + 0.5\ H^+ \rightarrow 0.25\ H_2S_{(aq)} + H_2O_{(l)}$	-78.87
$H_{2(aq)} + 0.25\ CO_{2(aq)} \rightarrow 0.25\ CH_4 + 0.5\ H_2O_{(l)}$	-45.25
$H_{2(aq)} + S_4^{2-} \rightarrow H_2S_{(aq)} + S_3^{2-}$	-43.22

during the reduction of ferrihydrite that pulls the reaction in a thermodynamically favorable direction (Table 3). It is often found as a coating on iron-bearing minerals, suspended in groundwater, and as a water column precipitate at the oxic-anoxic interface in many natural waters (Zachara et al., 2002). In laboratory studies, dissimilatory iron-reducing microorganisms were shown to convert ferrihydrite into a suite of more stable, secondary iron oxides such as hematite, goethite, magnetite, green rust (mixed valence hydroxide), vivianite [$Fe_3(PO_4)_2$], and siderite ($FeCO_3$) (Zachara et al., 2002). The type of microbial iron reduction depended upon factors such as the crystalline structure of the iron oxides, the electron donor-to-acceptor ratio, solution conditions, pH, the partial pressure of CO_2, and coprecipitated ions. Furthermore, dissimilatory iron reduction need not be limited to just the reduction of iron oxides but may also expand to iron sulfides. In highly sulfidic environments, the abiotic formation of pyritic compounds such as marcasite, sphalerite, wurtzite, and pyrrhotite will be favored over iron oxides. While the reduction of FeS_2 compounds to FeS and H_2S is not thermodynamically favorable under standard conditions (Table 3), it may become favorable in highly reduced and sulfidic environments where the formation of pyrrhotite occurs naturally.

Description of Hyperthermophilic Iron Reducers

It has been suggested that many, if not all, hyperthermophiles are capable of some degree of enzymatic dissimilatory iron reduction (Vargas et al., 1998). However, dissimilatory iron reduction has been studied extensively in just four groups of hyperthermophilic archaea and one hyperthermophilic bacterium. *Pyrobaculum islandicum*, which was isolated on elemental sulfur, was the first hyperthermophile shown to grow in a dissimilatory manner on iron (Kashefi and Lovley, 2000; Vargas et al., 1998). It reduced both soluble iron [Fe(III) citrate] and poorly crystalline Fe(III) oxide as well as Mn(IV) oxide and other toxic metals in a growth-dependent manner (Kashefi and Lovley, 2000). Dissimilatory iron reduction has since been demonstrated in other *Pyrobaculum* species such as *P. aerophilum*, *P. arsenaticum*, and *P. calidifontis* (Feinberg and Holden, 2006; Feinberg et al., 2008). Members of this genus are highly versatile metabolically and are capable of reducing a wide range of other compounds such as NO_3^-, $S°$, $S_2O_3^{2-}$, various heavy metals, and low levels of O_2.

Geoglobus ahangari and *Geothermobacterium ferrireducens* were the first hyperthermophilic, obligately iron-reducing organisms isolated from marine and terrestrial environments, respectively (Kashefi et al., 2002a, 2002b). *G. ferrireducens* had previously only been detected in hot springs from Yellowstone National Park using 16S rRNA analysis (Kashefi et al., 2002a). *Ferroglobus placidus*, which was originally characterized as an Fe(II)-oxidizing nitrate reducer (Hafenbradl et al., 1996), is another marine hyperthermophile that was shown to reduce Fe(III) oxide and is a close relative of *G. ahangari* (Tor et al., 2001). It and *G. ahangari* can use acetate as an electron donor and carbon source during growth on iron. Other deep-sea hyperthermophilic iron reducers found are *Pyrodictium* strain Su06 and *Hyperthermus* strain Ro04 that were isolated from the interiors of black smoker chimneys. These organisms are autotrophs and were present in much larger abundances than hyperthermophilic methanogens in all samples analyzed from a site in the northeastern Pacific Ocean (Ver Eecke et al., 2009). One of the hottest known organisms in culture is an obligate iron reducer belonging to the family *Pyrodictiaceae* (strain 121) that is capable of growth at temperatures up to 121°C (Kashefi and Lovley, 2003).

Electron Transfer across the Cell Wall

The respiration of insoluble metals can be considered a two-step process: (i) electron transfer from the cytoplasm to the exterior surface of the cell and (ii) transfer from the cell surface to the insoluble metal. Both

Geobacter and *Shewanella* require polyheme (2 to 12 hemes per protein) *c*-type cytochromes to bridge the transfer of electrons from the cytoplasmic membrane, across the periplasm, and to the outer surface of the outer cell wall membrane (Beliaev et al., 2001; Butler et al., 2004; Leang et al., 2003; Lloyd et al., 2003; Myers and Myers, 1997, 2000; Pitts et al., 2003). Based on their genome sequences, *P. aerophilum*, *P. islandicum*, and *P. arsenaticum* lack open reading frames with more than one CXXCH motif (each motif coordinates a *c*-type cytochrome heme) showing that they do not use polyheme *c*-type cytochromes (Feinberg and Holden, 2006; Feinberg et al., 2008). Only *P. calidifontis* contains even a single putative polyheme *c*-type cytochrome (Feinberg et al., 2008). *P. aerophilum* has two monoheme *c*-type cytochromes that are believed to be part of a bc_1 respiration complex and nitrite reductase. When *P. aerophilum* was grown on Fe(III) citrate and nitrate, it produced one and two *c*-type cytochrome-containing proteins, respectively, which matches the expected outcome based on the annotation (Feinberg and Holden, 2006). *P. islandicum* produced one *c*-type cytochrome only when it was grown on thiosulfate, and there were no *c*-type cytochromes produced when cultures were grown on either Fe(III) citrate or elemental sulfur (Feinberg et al., 2008). Therefore, even monoheme cytochromes appear to have little to no role in dissimilatory iron reduction in these organisms. Both organisms appear to produce membrane-bound NADH oxidases and ATP synthases and presumably respire when grown on iron, but the mechanism of electron transfer and proton translocation in these organisms has not been explored.

DNA microarray analyses of *P. aerophilum* before and after a shift from aerobic growth to growth on Fe(III) citrate did not show any significantly upregulated genes in response to growth on iron (Cozen et al., 2009). Only 17 open reading frames were upregulated on iron and these were typically weak and transient. These encode succinate dehydrogenase, thiosulfate sulfurtransferase, some redox-related proteins (e.g., ferredoxin), and eight hypothetical proteins. Similarly, proteomic data suggest that there is little to no increase in new membrane protein expression in *P. aerophilum* and *P. islandicum* when grown on iron relative to growth on either nitrate (*P. aerophilum*) or thiosulfate (*P. islandicum*) (Feinberg and Holden, 2006; Feinberg et al., 2008). Instead, both species show increased soluble NADH-dependent ferric reductase activity when grown on iron relative to the other electron acceptors (Feinberg and Holden, 2006; Feinberg et al., 2008). The lack of known assimilatory ferric reductase homologs (e.g., FeR from *A. fulgidus*, Fre from *E. coli* and *Azotobacter vinelandii*) in both genomes indicates that these reductases are novel. This ferric reductase may have the capacity to operate in both a dissimilatory and an assimilatory manner. How this would be linked to energy conservation is unknown.

Electron Transfer to Insoluble Metals

Respiration couples electron transfer through the cytoplasmic membrane to an external terminal electron acceptor with the generation of an electrochemical proton gradient across the membrane resulting in H^+-dependent energy formation across the cytoplasmic membrane. Electron carriers such as cytochromes and membrane-soluble quinones shuttle electrons between enzymes within the cell wall. While the physiological mechanisms for O_2, NO_3^-, and SO_4^{2-} reduction are very well understood, little is known about the physiology and biochemistry of dissimilatory iron reduction. Most of what is known comes from the study of mesophilic iron reducers. In general, dissimilatory iron reduction involves three mechanisms (Fig. 1): direct cell contact with iron; the reduction and release of extracellular electron shuttles for nonenzymatic, extracellular iron reduction; and the secretion of iron chelators that solubilize iron and return it to the cell for reduction. Each of these will be discussed based on what is known from mesophiles and then applied to our understanding of hyperthermophilic dissimilatory iron reduction.

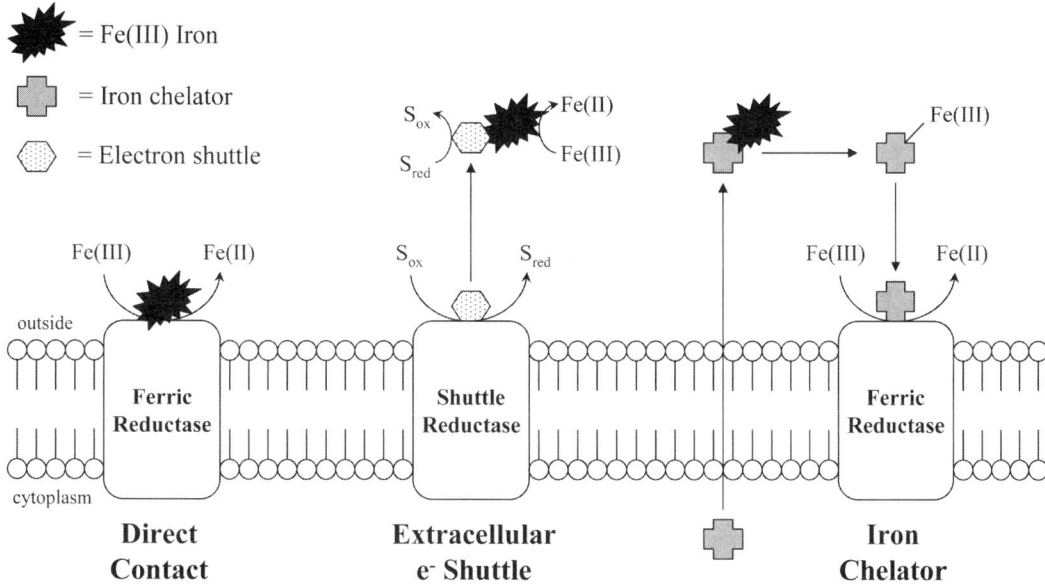

FIGURE 1 Models for the mechanisms of dissimilatory iron reduction.

Direct Contact with Iron

Determination of whether dissimilatory metal-reducing microorganisms require direct contact with insoluble iron oxides involves the sequestration of the iron in a semipermeable membrane from the cells. These barriers consist of dialysis tubing (12 to 300 kDa pore sizes), microporous alginate beads (12 kDa pore size), and agar (>300 kDa). The mesophile *Geobacter metallireducens* did not reduce insoluble ferric oxides when separated from it by a barrier (Nevin and Lovley, 2000). Instead, it and *Geobacter sulfurreducens* produce pili only when grown on insoluble Fe(III) oxide and Mn(IV) oxide, and these pili are highly electrically conductive (Childers et al., 2002; Reguera et al., 2005). The pili serve as biological "nanowires" that transfer electrons from the cell surface to the surface of Fe(III) oxides. The pili also play a structural role and are essential for biofilm formation, and efficient electron transfer through the biofilm requires the presence of electrically conductive pili (Reguera et al., 2006, 2007). *Shewanella oneidensis* MR-1 also produces electrically conductive pili in direct response to electron-acceptor limitation (Gorby et al., 2006). Thin-section electron microscopy showed that nanocrystalline magnetite formed linear features along presumptive nanowires when cultures were grown on silica ferrihydrite (Gorby et al., 2006). Scanning electron microscopy revealed a network of nanowires linking *S. oneidensis* cells to each other and to mineral surfaces throughout an artificial saturated sand column suggesting that nanowires serve as conduits of electron transfer throughout large portions of sedimentary environments (Ntarlagiannis et al., 2007).

Four *Pyrobaculum* species (*P. aerophilum, P. islandicum, P. arsenaticum,* and *P. calidifontis*) were tested for their ability to grow on Fe(III) oxide when separated from it by a dialysis membrane barrier (12,000 to 14,000 Da pore size) (Feinberg and Holden, 2006; Feinberg et al., 2008). The growth and Fe(II) production rates of *P. aerophilum* were the same with and without the presence of the barrier suggesting that direct contact with Fe(III) oxide is not necessary (Feinberg and Holden, 2006). *P. aerophilum* forms a biofilm when grown on insoluble iron and aggregates with the Fe(III) oxide; however, the cells are only loosely associated with the iron and dissociate from it after gravitational settling of the iron (J. F. Holden et al., unpublished data). In contrast, *P. islandicum* and *P. calidifontis* both required direct contact with Fe(III) oxide for growth, while *P. arsenaticum* did not require direct contact but its growth was retarded relative to growth without the barrier (Feinberg et al., 2008). This difference may be reflected in the end products of Fe(III) oxide reduction where *P. islandicum* produces magnetite (Kashefi et al., 2008; J. F. Holden, personal observation) while *P. aerophilum* produces soluble Fe(II) (Holden, personal observation).

Extracellular Electron Shuttles

Redox-active compounds such as flavins, phenazines, soluble quinones, melanin, thiol-containing molecules, and humic compounds have all demonstrated the ability to shuttle electrons outside the cell (Hernandez and Newman, 2001). Although *S. oneidensis* appears to produce conductive pili for Fe(III) oxide reduction, it is also capable of growth on insoluble iron without direct contact (Lies et al., 2005). It was shown that *S. oneidensis* and other *Shewanella* species secrete flavins within a biofilm that adsorb quickly to Fe(III) and Mn(IV) oxide surfaces and mediate extracellular electron transfer (Marsili et al., 2008; von Canstein et al., 2008). In *Pseudomonas* species, exogenous redox-active molecules known as phenazines nonenzymatically shuttle electrons to poorly crystalline ferrihydrite and highly crystalline hematite (Hernandez et al., 2004; Wang and Newman, 2008). Reduction rates decreased with increasing pH between pH 5 and 8, and the rate was higher for poorly crystalline ferrihydrite than from hematite (Wang and Newman, 2008).

As mentioned, the hyperthermophile *P. aerophilum* does not require direct contact with insoluble iron for growth whereas *P. islandicum* does (Feinberg and Holden, 2006; Feinberg et al., 2008). The cell-free spent supernatant of *P. aerophilum* and *P. islandicum* grown on Fe(III) oxide, Fe(III) citrate, nitrate, thiosulfate, and elemental sulfur was analyzed using high-performance liquid chromatography, and three peaks were found only in *P. aerophilum* supernatant primarily when cultures were grown on Fe(III) oxide. By use of tandem mass spectrometry, the compounds in two of the three peaks were identified as monoribonucleosides, primarily adenosine (the compound in the third peak did not ionize) (Feinberg et al., 2008). These may be minor components of a biofilm because nucleotides have recently been shown to be important structural components of biofilms (Steinberger and Holden, 2005; Whitchurch et al., 2002).

Indeed, *P. aerophilum* produces a biofilm when grown on Fe(III) oxide that consists primarily of glucose and mannose (Holden et al., unpublished). Both glucose and the ribose in nucleosides are reducing sugars. With increasing temperature, these cyclic molecules linearize, exposing a highly reactive aldehyde group that readily reduces metals. The resulting carboxyl group on the oxidized sugar may then also chelate and transport additional Fe^{3+} for further reduction by the cells. Ribose and glucose abiotically reduced Fe(III) oxide in *P. aerophilum* growth medium at temperatures above 55 and 75°C, respectively (Holden et al., unpublished). Neither the nonreducing sugar sucrose nor adenosine reduced Fe(III) oxide at 95°C; however, at 100°C ribonucleosides depurinate and depyrimidinate and release ribose at rates that are 1,000-fold higher than at 25°C (Marguet and Forterre, 1994). Therefore, the biofilms of hyperthermophiles may serve a dual function by keeping Fe(III) oxide in close proximity to the cells and as a mediator of extracellular electron transfer between the cells and the metal.

Iron Chelators

Most organisms produce iron chelators called siderophores to solubilize Fe(III) for assimilatory iron reduction. Magnetotactic bacteria, which produce intracellular magnetosomes for orientation within a magnetic field, release siderophores to scavenge iron under low-nutrient conditions (Calugay et al., 2003). These iron chelators are low molecular weight (500 to 1,000 Da) to transport iron across the cell membrane for subsequent reduction and incorporation into biomass (Vartivarian and Cowart, 1999). Soluble Fe^{3+} concentrations increased with the growth of the mesophilic dissimilatory iron reducers *Geothrix fermentans* and *Shewanella algae* (Nevin and Lovley, 2002a, 2002b) on insoluble iron suggesting that they release extracellular iron chelators. The addition of the nonphysiological chelator nitriloacetic acid to the same cultures increased their rates of dissimilatory iron reduction. *Pyrobaculum* species and *G. ahangari* are the only known hyperthermophiles that reduce soluble Fe(III) chelated to citrate (Kashefi and Lovley, 2000, 2003; Feinberg and Holden, 2006; Feinberg et al., 2008) suggesting that they may have a mechanism for iron reduction not found in other hyperthermophilic iron-reducing genera that will only reduce insoluble forms of iron.

Environmental Constraints on Iron Reduction

Ferrous iron is stable and soluble under anoxic and acidic conditions but readily oxidizes to ferric iron in oxic environments above pH 5. This Fe(III) forms precipitates with hydroxides, oxides, phosphates, and sulfates with very low solubility (10^{-18} M) (Schröder et al., 2003). The solubility of Fe(III) increases in the presence of organic compounds either found in the environment or produced by the organism in the form of a chelator. Ferrihydrite reduction depends upon Fe(II) concentrations and is a coupled biotic-abiotic process (Hansel et al., 2003). An initially low Fe(II) concentration (<0.3 mM) results in the rapid precipitation of goethite and an inhibition of magnetite nucleation. At higher initial Fe(II) concentrations (>0.3 mM), both goethite and magnetite precipitate. Goethite precipitates more quickly than magnetite, but magnetite will accumulate over time eventually at the expense of goethite formation.

The reduction potential and pH of an environment also appear to affect the growth rates of hyperthermophiles on iron. *P. islandicum* growth on elemental sulfur and thiosulfate and *P. aerophilum* growth on nitrate were optimal at pH 6, pH 6, and pH 7, respectively (Feinberg et al., 2008). However, for both organisms growth on Fe(III) oxide and Fe(III) citrate was optimal at pH 7 to 8 and extended to pH 9, which increased the known pH range for growth of both organisms. Similarly, *P. islandicum* growth on sulfur and thiosulfate increased with increasing reduction potential down to −571 mV, but growth on Fe(III) citrate and Fe(III) oxide was more favorable under only mildly reducing conditions (>−210 mV) (Feinberg et al., 2008). *P. aerophilum* growth on Fe(III) oxide, Fe(III) citrate, and nitrate only occurred when reduction potentials were −210 mV or lower. Therefore, these *Pyrobaculum* species appear to have a preference for mildly reducing and neutral to mildly basic conditions.

BIOGEOCHEMISTRY

With regard to microbe-metal biogeochemistry and geomicrobiology, questions remain on the factors that determine the distribution patterns of hyperthermophiles in hydrothermal environments and whether these environments possess any type of biogenic signature that can be used as a biomarker. The use of tracers in various extant environments and in the geologic record is perhaps the least explored area related to microbe-metal interactions and is a significant area for future work.

Magnetite is often the end product of hyperthermophilic iron respiration (Kashefi et al., 2002, 2008) and may be useful as a biomarker of this metabolism. Magnetite precipitated by magnetotactic bacteria has been proposed to be a biomarker in the sedimentary record (Chang and Kirschvink, 1989; Kirschvink, 1982). However, *G. metallireducens* produced 5,000-fold more magnetite than an equivalent biomass of magnetotactic bacteria (Lovley et al., 1987), and under low pCO_2 and high ferric oxide conditions, this magnetite could potentially be the dominant source of magnetization in sediments (Vali et al., 2004). The discovery of magnetite in the deep subsurface, presumably of microbial origin, led in part to the proposal of the deep hot biosphere (Gold, 1992). The presence of magnetite has been used in part as evidence for life in the Martian meteorite ALH84001 (McKay et al., 1996; Thomas-Keprta et al., 2001). The magnetite crystals found within the meteorite showed several features that were originally suggested to be unique to biogenesis by magnetotactic bacteria. However, these crystals were later suggested to more closely resemble those formed abiotically as a thermal decomposition product of Fe-rich carbonate produced by inorganic hydrothermal precipitation in laboratory experiments (Golden et al., 2004). Although the latter results are inconclusive, the potential for using magnetite as a biomarker of iron respiration has considerable potential.

The fractionation of stable metal isotopes is another potential biomarker. To date, most studies have focused on iron fractionation. There are four naturally occurring stable iron isotopes (^{54}Fe [5.84%], ^{56}Fe [91.76%], ^{57}Fe [2.12%], and ^{58}Fe [0.28%]), and typically a sample's ^{56}Fe-to-^{54}Fe ratio is measured relative to a standard (δ^{56}Fe (‰) = [^{56}Fe/^{54}Fe$_{sample}$/^{56}Fe/^{54}Fe$_{Ig\,Rxs}$ − 1] × 1,000) (Johnson et al., 2008). Most igneous rocks, rock alterations, and weathering products have ^{56}Fe values near zero; therefore, nonzero values for minerals and rocks must be produced by interactions with iron-bearing fluids. Some of the largest fractionations in the isotopic composition of iron occur between oxidized and reduced forms. Therefore, biochemistry that involves redox state changes of iron has been the focus for tracing biogeochemical phenomena. Reductive dissolution of iron oxide minerals by the dissimilatory iron-reducing bacteria *Geobacter* and *Shewanella* produces Fe(II)$_{aq}$ and Fe^{2+}-bearing minerals such as siderite and magnetite. The iron isotope composition of Fe(II)$_{aq}$ that is produced by dissimilatory iron reduction (δ^{56}Fe = −2.5‰ to −1.0‰) and sulfide interactions during dissimilatory sulfate reduction (δ^{56}Fe = 0 to +0.5‰) are different from each other and the host rock, and thus may be a means for identifying these processes in nature (Johnson et al., 2005). Iron isotope fractionations produced by *G. sulfurreducens* and *Shewanella putrefaciens* during reduction of hematite and goethite under identical conditions were similar, suggesting a common mechanism (Crosby et al., 2007). Modern mid-ocean ridge hydrothermal fluids differ somewhat by having δ^{56}Fe values between −0.8‰ and −0.1‰ that could potentially complicate detecting microbial signals in hydrothermal environments. To date, iron isotope fractionation has not been examined in hyperthermophiles.

The fractionation of other transition metals used in biology may serve a more specific tracer of certain forms of activity. For example, it was suggested that nickel fractionation (^{60}Ni/^{58}Ni) is indicative of methanogens since three of the eight known nickel enzymes are found in these organisms (Cameron et al., 2009). Abiotic mantle and crust samples showed very little variation in Ni isotope composition (δ^{60}Ni = −0.1‰ to 0.4‰) (Cameron et al., 2009). In contrast, nickel assimilation during the growth of methanogens, including the marine hyperthermophile *M. jannaschii*, led to cells that were isotopically light in nickel relative to the starting media ($\Delta\delta^{60}$Ni = −0.4‰ to −1.5‰). While this method shows significant potential as a general biomarker, the δ^{60}Ni values resulting from the synthesis of other nonmethanogenic nickel-containing enzymes (e.g., hydrogenase) need to be determined to establish whether nickel fractionation can be specific for methanogens. The in vitro and in vivo fractionation of copper and zinc during incorporation into proteins similarly showed that organisms incorporate lighter isotopes of transition metals preferentially and that transition metal fractionation occurs stepwise along their pathways within biological systems during their uptake (Zhu et al., 2002).

CONCLUSIONS

It is clear that there is extensive interaction between metals and hyperthermophilic microbes from hydrothermal environments, and yet our study of the nature of these interactions is in its infancy. Hyperthermophiles contain many metalloproteins, but we have yet to define the full metalloproteome of any organism or determine the full function of these proteins or their mechanisms of metal uptake and incorporation. The respiration of some metals is highly energetically favorable and may lead to a competitive advantage over other forms of respiration; however, dissimilatory metal reduction may be limited by the organism's ability to transfer electrons from its cytoplasm to the typically insoluble metal acceptor. Therefore, it is necessary to determine these respiratory mechanisms and how they lead to energy conservation, especially because the physiological mechanisms for metal reduction in hyperthermophiles differ from those found in mesophilic bacteria based on the lack of polyheme c-type cytochromes, a periplasm, and an outer cell wall membrane. We are also unaware of the environmental constraints on microbe-metal interactions in nature, and our ability to detect hyperthermophile-metal interactions in the rock record is only at the beginning of its development. However, given the ancient nature of hyperthermophile metabolisms and the importance of hydrothermal environments in the modern subsurface biosphere, the creation of habitable space during the Hadean, Archean, and Proterozoic periods, and potentially on moons and other planets, the further study of hyperthermophile-metal interactions in hydrothermal environments will create a new understanding of the basic principles that govern a broad array of metabolic processes and a significant portion of the Earth's biosphere.

ACKNOWLEDGMENTS

This work was supported in part by grants from the National Science Foundation (OCE-0732611 to J.F.H.) and from the Department of Energy (DE-FG0207ER64326 and DE-FG05-95ER20175 to M.W.W.A.).

REFERENCES

Afshar, S., E. Johnson, S. de Vries, and I. Schröder. 2001. Properties of a thermostable nitrate reductase from the hyperthermophilic archaeon *Pyrobaculum aerophilum*. *J. Bacteriol.* **183:** 5491–5495.

Amend, J. P., and E. L. Shock. 2001. Energetics of overall metabolic reactions of thermophilic and hyperthermophilic Archaea and Bacteria. *FEMS Microbiol. Rev.* **25:**175–243.

Amo, T., H. Atomi, and T. Imanaka. 2002. Unique presence of a manganese catalase in a hyperthermophilic archaeon, *Pyrobaculum calidifontis* VA1. *J. Bacteriol.* **184:**3305–3312.

Amo, T., H. Atomi, and T. Imanaka. 2003. Biochemical properties and regulated gene expression of the superoxide dismutase from the facultatively aerobic hyperthermophile *Pyrobaculum calidifontis*. *J. Bacteriol.* **185:**6340–6347.

Andreini, C., L. Banci, I. Bertini, S. Elmi, and A. Rosato. 2007. Non-heme iron through the three domains of life. *Proteins* **67:**317–324.

Andreini, C., I. Bertini, G. Cavallaro, G. L. Holliday, and J. M. Thornton. 2008. Metal ions in biological catalysis: from enzyme databases to general principles. *J. Biol. Inorg. Chem.* **13:**1205–1218.

Andreini, C., I. Bertini, and A. Rosato. 2009. Metalloproteomes: a bioinformatic approach. *Acc. Chem. Res.* **42:**1471–1479.

Banerjee, R., and S. W. Ragsdale. 2003. The many faces of vitamin B_{12}: catalysis by cobalamin-dependent enzymes. *Annu. Rev. Biochem.* **72:**209–247.

Beh, M., G. Strauss, R. Huber, K. O. Stetter, and G. Fuchs. 1993. Enzymes of the reductive citric acid cycle in the autotrophic eubacterium *Aquifex pyrophilus* and in the archaebacterium *Thermoproteus neutrophilus*. *Arch. Microbiol.* **160:**306–311.

Beliaev, A. S., D. A. Saffarini, J. L. McLaughlin, and D. Hunnicutt. 2001. MtrC, and outer membrane decahaem c cytochrome required for metal reduction in *Shewanella putrefaciens* MR-1. *Mol. Microbiol.* **39:**722–730.

Bell, S. D., S. S. Cairns, R. L. Robson, and S. P. Jackson. 1999. Transcriptional regulation of an archaeal operon in vivo and in vitro. *Mol. Cell* **4:**971–982.

Bevers, L. E., P. L. Hagedoorn, G. C. Krijger, and W. R. Hagen. 2006. Tungsten transport protein A (WtpA) in *Pyrococcus furiosus*: the first member of a new class of tungstate and molybdate transporters. *J. Bacteriol.* **188:**6498–6505.

Bevers, L. E., P. L. Hagedoorn, and W. R. Hagen. 2009. The bioinorganic chemistry of tungsten. *Coordination Chem. Rev.* **253:**269–290.

Blamey, J. M., and M. W. W. Adams. 1993. Purification and characterization of pyruvate ferredoxin oxidoreductase from the hyperthermophilic archaeon *Pyrococcus furiosus*. *Biochim. Biophys. Acta* **1161:**19–27.

Blöchl, E., R. Rachel, S. Burggraf, D. Hafenbradl, H. W. Jannasch, and K. O. Stetter. 1997. *Pyrolobus fumarii*, gen. and sp. nov., represents a novel group of archaea, extending the upper temperature limit for life to 113°C. *Extremophiles* **1:**14–21.

Borths, E. L., K. P. Locher, A. T. Lee, and D. C. Rees. 2002. The structure of *Escherichia coli* BtuF and binding to its cognate ATP binding cassette transporter. *Proc. Natl. Acad. Sci. USA* **99:**16642–16647.

Bryant, F. O., and M. W. W. Adams. 1989. Characterization of hydrogenase from the hyperthermophilic archaebacterium, *Pyrococcus furiosus*. *J. Biol. Chem.* **264:**5070–5079.

Butler, J. E., F. Kaufmann, M. V. Coppi, C. Núñez, and D. R. Lovley. 2004. MacA, a diheme *c*-type cytochrome involved in Fe(III) reduction by *Geobacter sulfurreducens*. *J. Bacteriol.* **186:**4042–4045.

Butterfield, D. A., I. R. Jonasson, G. J. Massoth, R. A. Feely, K. K. Roe, R. E. Embley, J. F. Holden, R. E. McDuff, M. D. Lilley, and J. R. Delaney. 1997. Seafloor eruptions and evolution of hydrothermal fluid chemistry. *Phil. Trans. R. Soc. Lond. A* **355:**369–386.

Butterfield, D. A., and G. J. Massoth. 1994. Geochemistry of north Cleft segment vent fluids: temporal changes in chlorinity and their possible relation to recent volcanism. *J. Geophys. Res.* **99:**4951–4968.

Butterfield, D. A., G. J. Massoth, R. E. McDuff, J. E. Lupton, and M. D. Lilley. 1990. Geochemistry of hydrothermal fluids from Axial Seamount Hydrothermal Emissions Study vent field, Juan de Fuca Ridge: subseafloor boiling and subsequent fluid-rock interaction. *J. Geophys. Res.* **95:**12895–12921.

Butterfield, D. A., R. E. McDuff, M. J. Mottl, M. D. Lilley, J. E. Lupton, and G. J. Massoth. 1994. Gradients in the composition of hydrothermal fluids from the Endeavour segment vent field: phase separation and brine loss. *J. Geophys. Res.* **99:**9561–9583.

Caldon, C. E., and P. E. March. 2003. Function of the universally conserved bacterial GTPases. *Curr. Opin. Microbiol.* **6:**135–139.

Calugay, R. J., H. Miyashita, Y. Okamura, and T. Matsunaga. 2003. Siderophore production by the magnetic bacterium *Magnetospirillum magneticum* AMB-1. *FEMS Microbiol. Lett.* **218:**371–375.

Cameron, V., D. Vance, C. Archer, and C. H. House. 2009. A biomarker based on the stable isotopes of nickel. *Proc. Natl. Acad. Sci. USA* **106:**10944–10948.

Cartron, M. L., S. Maddocks, P. Gillingham, C. J. Craven, and S. C. Andrews. 2006. Feo-transport of ferrous iron into bacteria. *Biometals* **19:**143–157.

Chang, S. R., and J. L. Kirschvink. 1989. Magnetofossils, the magnetization of sediments, and the evolution of magnetite biomineralization. *Ann. Rev. Earth Planet. Sci.* **17:**169–195.

Charlou, J. L., J. P. Donval, Y. Fouquet, P. Jean-Baptiste, and N. Holm. 2002. Geochemistry of high H_2 and CH_4 vent fluids issuing from ultramafic rocks at the Rainbow hydrothermal field (36°14'N, MAR). *Chem. Geol.* **191:**345–359.

Charlou, J. L., Y. Fouquet, H. Bougault, J. P. Donval, J. Etoubleau, P. Jean-Baptiste, A. Dapoigny, P. Appriou, and P. A. Rona. 1998. Intense CH_4 plumes generated by serpentinization of ultramafic rocks at the intersection of the 15° 20'N fracture zone and the Mid-Atlantic Ridge. *Geochim. Cosmochim. Acta* **62:**2323–2333.

Childers, S. E., S. Ciufo, and D. R. Lovley. 2002. *Geobacter metallireducens* accesses insoluble Fe(III) oxide by chemotaxis. *Nature* **416:**767–769.

Childress, J. J., and C. R. Fisher. 1992. The biology of hydrothermal vent animals: physiology, biochemistry, and autotrophic symbioses. *Oceanogr. Mar. Biol. Annu. Rev.* **30:**337–441.

Chiu, H. J., E. Johnson, I. Schröder, and D. C. Rees. 2001. Crystal structures of a novel ferric reductase from the hyperthermophilic archaeon *Archaeoglobus fulgidus* and its complex with $NADP^+$. *Structure* **9:**311–319.

Chivers, P. T., and T. H. Tahirov. 2005. Structure of *Pyrococcus horikoshii* NikR: nickel sensing and implications for the regulation of DNA recognition. *J. Mol. Biol.* **348:**597–607.

Cozen, A. E., M. T. Weirauch, K. S. Pollard, D. L. Bernick, J. M. Stuart, and T. M. Lowe. 2009. Transcriptional map of respiratory versatility in the hyperthermophilic crenarchaeon *Pyrobaculum aerophilum*. *J. Bacteriol.* **191:**782–794.

Crosby, H. A., E. E. Roden, C. M. Johnson, and B. L. Beard. 2007. The mechanisms of iron isotope fractionation produced during dissimilatory Fe(III) reduction by *Shewanella putrefaciens* and *Geobacter sulfurreducens*. *Geobiology* **5:**169–189.

Dahl, C., N. M. Kredich, R. Deutzmann, and H. G. Trüper. 1993. Dissimilatory sulfite reductase from *Archaeoglobus fulgidus*: physicochemical properties of the enzyme and cloning, sequencing and analysis of the reductase genes. *J. Gen. Microbiol.* **139:**1817–1828.

de Vries, S., and I. Schröder. 2002. Comparison between the nitric oxide reductase family and its aerobic relatives, the cytochrome oxidases. *Biochem. Soc. Trans.* **30:**662–667.

de Vries, S., M. J. F. Strampraad, S. Lu, P. Moënne-Loccoz, and I. Schröder. 2003. Purification and characterization of the MQH$_2$: NO oxidoreductase from the hyperthermophilic archaeon *Pyrobaculum aerophilum*. *J. Biol. Chem.* **278:**35861–35868.

Dirmeier, R., M. Keller, G. Frey, H. Huber, and K. O. Stetter. 1998. Purification and properties of an extremely thermostable membrane-bound sulfur-reducing complex from the hyperthermophilic *Pyrodictium abyssi*. *Eur. J. Biochem.* **252:**486–491.

Dosanjh, N. S., and S. L. J. Michel. 2006. Microbial nickel metalloregulation: NikRs for nickel ions. *Curr. Opin. Chem. Biol.* **10:**123–130.

Douville, É., J.-L. Charlou, J.-P. Donval, D. Hureau, and P. Appriou. 1999. As and Sb behavior in fluids from various deep-sea hydrothermal systems. *Earth Planet. Sci.* **328:**97–104.

Douville, É., J. L. Charlou, E. H. Oelkers, P. Bienvenu, C. F. Jove Colon, J. P. Donval, Y. Fouquet, D. Prieur, and P. Appriou. 2002. Trace metals in hot acidic fluids from a deep-sea hydrothermal system in an ultra-mafic environment: rainbow vent field (36°14'N MAR). *Chem. Geol.* **184:**37–48.

Dupont, C. L., S. Yang, B. Palenik, and P. E. Bourne. 2006. Modern proteomes contain putative imprints of ancient shifts in trace metal geochemistry. *Proc. Natl. Acad. Sci. USA* **103:**17822–17827.

Eitinger, T., J. Suhr, J. Moore, and J. A. C. Smith. 2005. Secondary transporters for nickel and cobalt ions: theme and variations. *BioMetals* **18:**399–405.

Escalante-Semerena, J. C. 2007. Conversion of cobamide into adenosylcobamide in bacteria and archaea. *J. Bacteriol.* **189:**4555–4560.

Feinberg, L. F., and J. F. Holden. 2006. Characterization of dissimilatory Fe(III) versus NO$_3^-$ reduction in the hyperthermophilic archaeon *Pyrobaculum aerophilum*. *J. Bacteriol.* **188:**525–531.

Feinberg, L. F., R. Srikanth, R. W. Vachet, and J. F. Holden. 2008. Constraints on anaerobic respiration in the hyperthermophilic archaea *Pyrobaculum islandicum* and *Pyrobaculum aerophilum*. *Appl. Environ. Microbiol.* **74:**396–402.

Ferry, J. G. 1999. Enzymology of one-carbon metabolism in methanogenic pathways. *FEMS Microbiol. Rev.* **23:**13–38.

Fontecilla-Camps, J. C., P. Amara, C. Cavazza, Y. Nicolet, and A. Volbeda. 2009. Structure-function relationships of anaerobic gas-processing metalloenzymes. *Nature* **460:**814–822.

Fouquet, Y., U. von Stackelberg, J. L. Charlou, J. P. Donval, J. P. Foucher, J. Erzinger, P. Herzig, R. Mühe, M. Wiedicke, S. Soakai, and H. Whitechurch. 1991. Hydrothermal activity in the Lau back-arc basin: sulfides and water chemistry. *Geology* **19:**303–306.

Gallant, R. M., and K. L. Von Damm. 2006. Geochemical controls on hydrothermal fluids from the Kairei and Edmond Vent Fields, 23°-25°S, Central Indian Ridge. *Geochem. Geophys. Geosyst.* **7:**Q06018.

Gamo, T., H. Chiba, T. Yamanaka, T. Okudaira, J. Hashimoto, S. Tsuchida, J. Ishibashi, S. Kataoka, U. Tsunogai, K. Okamura, Y. Sano, and R. Shinjo. 2001. Chemical characteristics of newly discovered black smoker fluids and associated hydrothermal plumes at the Rodriquez Triple Junction, Central Indian Ridge. *Earth Planet. Sci. Lett.* **193:**371–379.

Gamo, T., H. Masuda, T. Yamanaka, K. Okamura, J. Ishibashi, E. Nakayama, H. Obata, K. Shitashima, Y. Nishio, H. Hasumoto, M. Watanabe, K. Mitsuzawa, N. Seama, U. Tsunogai, F. Kouzuma, and Y. Sano. 2004. Discovery of a new hydrothermal venting site in the southernmost Mariana Arc: Al-rich hydrothermal plumes and white smoker activity associated with biogenic methane. *Geochem. J.* **38:**527–534.

Gamo, T., K. Okamura, J. L. Charlou, T. Urabe, J. M. Auzende, J. Ishibashi, K. Shitashima, H. Chiba, and shipboard scientific party of the ManusFlux cruise. 1997. Acidic and sulfate-rich hydrothermal fluids from the Manus back-arc basin, Papua New Guinea. *Geology* **25:**139–142.

Ghosh, M., A. M. Grunden, D. M. Dunn, R. Weiss, and M. W. W. Adams. 1998. Characterization of native and recombinant forms of an unusual cobalt-dependent proline dipeptidase (prolidase) from the hyperthermophilic archaeon *Pyrococcus furiosus*. *J. Bacteriol.* **180:**4781–4789.

Gold, T. 1992. The deep, hot biosphere. *Proc. Natl. Acad. Sci. USA* **89:**6045–6049.

Golden, D. C., D. W. Ming, R. V. Morris, A. Brearley, H. V. Lauer, A. H. Treiman, M. E. Zolensky, C. S. Schwandt, G. E. Lofgren, and G. A. McKay. 2004. Evidence for exclusively inorganic formation of magnetite in Martian meteorite ALH84001. *Amer. Mineral.* **89:**681–695.

Gong, W., B. Hao, Z. Wei, D. J. Ferguson, T. Tallant, J. A. Krzycki, and M. K. Chan. 2008. Structure of the $\alpha_2\varepsilon_2$ Ni-dependent CO dehydrogenase component of the *Methanosarcina barkeri* acetyl-CoA decarbonylase/synthase complex. *Proc. Natl. Acad. Sci. USA* **105:**9558–9563.

Gorby, Y. A., S. Yanina, J. S. McLean, K. M. Rosso, D. Moyles, A. Dohnalkova, T. J. Beveridge, I. S. Chang, B. H. Kim, K. S. Kim, D. E. Culley, S. B. Reed, M. F. Romine, D. A. Saffarini, E. A. Hill, L. Shi, D. A. Elias,

D. W. Kennedy, G. Pinchuk, K. Watanabe, S. Ishii, B. Logan, K. H. Nealson, and J. K. Fredrickson. 2006. Electrically conductive bacterial nanowires produced by *Shewanella oneidensis* strain MR-1 and other microorganisms. *Proc. Natl. Acad. Sci. USA* **103:**11358–11363.

Greenwood, N. N., and A. Earnshaw. 1984. *Chemistry of the Elements*, p. 1167–1168. Pergamon Press, Oxford, United Kingdom.

Hafenbradl, D., M. Keller, R. Dirmeier, R. Rachel, P. Roßnagel, S. Burggraf, H. Huber, and K. O. Stetter. 1996. *Ferroglobus placidus* gen. nov., sp. nov., a novel hyperthermophilic archaeum that oxidizes Fe^{2+} at neutral pH under anoxic conditions. *Arch. Microbiol.* **166:**308–314.

Hansel, C. M., S. G. Benner, J. Neiss, A. Dohnalkova, R. K. Kukkadapu, and S. Fendorf. 2003. Secondary mineralization pathways induced by dissimilatory iron reduction of ferrihydrite under advective flow. *Geochim. Cosmochim. Acta* **67:**2977–2992.

Hattori, M., Y. Jin, H. Nishimasu, Y. Tanaka, M. Mochizuki, T. Uchiumi, R. Ishitani, K. Ito, and O. Nureki. 2009. Structural basis of novel interactions between the small-GTPase and GDI-like domains in prokaryotic FeoB iron transporter. *Structure* **17:**1345–1355.

Haymon, R. M. 1983. Growth history of hydrothermal black smoker chimneys. *Nature* **301:**695–698.

Heider, J., X. Mai, and M. W. W. Adams. 1996. Characterization of 2-ketoisovalerate ferredoxin oxidoreductase, a new and reversible coenzyme A-dependent enzyme involved in peptide fermentation by hyperthermophilic archaea. *J. Bacteriol.* **178:**780–787.

Henninger, T., S. Anemüller, S. Fitz-Gibbon, J. H. Miller, G. Schäfer, and C. L. Schmidt. 1999. A novel Rieske iron-sulfur protein from the hyperthermophilic crenarchaeon *Pyrobaculum aerophilum*: sequencing of the gene, expression in *E. coli* and characterization of the protein. *J. Bioenerg. Biomembr.* **31:**119–128.

Hernandez, M. E., A. Kappler, and D. K. Newman. 2004. Phenazines and other redox-active antibiotics promote microbial mineral reduction. *Appl. Environ. Microbiol.* **70:**921–928.

Hernandez, M. E., and D. K. Newman. 2001. Extracellular electron transfer. *Cell. Mol. Life Sci.* **58:**1562–1571.

Hille, R. 2002. Molybdenum and tungsten in biology. *Trends Biochem. Sci.* **27:**360–367.

Holden, J. F. 2009. Extremophiles: hot environments, p. 127–146. *In* M. Schaechter (ed.), *Encyclopedia of Microbiology*. Elsevier Press, Oxford, United Kingdom.

Hollenstein, K., M. Comellas-Bigler, L. E. Bevers, M. C. Feiters, W. Meyer-Klaucke, P. L. Hagedoorn, and K. P. Locher. 2009. Distorted octahedral coordination of tungstate in a subfamily of specific binding proteins. *J. Biol. Inorg. Chem.* **14:**663–672.

Huang, S., G. Romanchuk, K. Pattridge, S. A. Lesley, I. A. Wilson, R. G. Matthews, and M. Ludwig. 2007. Reactivation of methionine synthase from *Thermotoga maritima* (TM0268) requires the downstream gene product TM0269. *Protein Sci.* **16:**1588–1595.

Huber, H., M. Gallenberger, U. Jahn, E. Eylert, I. A. Berg, D. Kockelkorn, W. Eisenreich, and G. Fuchs. 2008. A dicarboxylate/4-hydroxybutyrate autotrophic carbon assimilation cycle in the hyperthermophilic Archaeum *Ignicoccus hospitalis*. *Proc. Natl. Acad. Sci. USA* **105:**7851–7856.

Huber, J. A., and J. F. Holden. 2008. Modeling the impact of diffuse vent microorganisms along mid-ocean ridges and flanks, p. 215–231. *In* R. P. Lowell, J. S. Seewald, A. Metaxas, and M. R. Perfit (ed.), *Magma to Microbe: Modeling Hydrothermal Processes at Ocean Spreading Centers*. Geophysical Monograph 178. AGU Press, Washington, DC.

Johnson, C. M., B. L. Beard, and E. E. Roden. 2008. The iron isotope fingerprints of redox and biogeochemical cycling in modern and ancient Earth. *Annu. Rev. Earth Planet. Sci.* **36:**457–493.

Johnson, C. M., E. E. Roden, S. A. Welch, and B. L. Beard. 2005. Experimental constraints on Fe isotope fractionation during magnetite and Fe carbonate formation coupled to dissimilatory hydrous ferric oxide reduction. *Geochim. Cosmochim. Acta* **69:**963–993.

Johnson, M. K., D. C. Rees, and M. W. W. Adams. 1996. Tungstoenzymes. *Chem. Rev.* **96:**2817–2840.

Kashefi, K., D. E. Holmes, A.-L. Reysenbach, and D. R. Lovley. 2002a. Use of Fe(III) as an electron acceptor to recover previously uncultured hyperthermophiles: isolation and characterization of *Geothermobacterium ferrireducens* gen. nov., sp. nov. *Appl. Environ. Microbiol.* **68:**1735–1742.

Kashefi, K., and D. R. Lovley. 2000. Reduction of Fe(III), Mn(IV), and toxic metals at 100°C by *Pyrobaculum islandicum*. *Appl. Environ. Microbiol.* **66:**1050–1056.

Kashefi, K., and D. R. Lovley. 2003. Extending the upper temperature limit for life. *Science* **301:**934.

Kashefi, K., B. M. Moskowitz, and D. R. Lovley. 2008. Characterization of extracellular minerals produced during dissimilatory Fe(III) and U(VI) reduction at 100°C by *Pyrobaculum islandicum*. *Geobiology* **6:**147–154.

Kashefi, K., J. M. Tor, D. E. Holmes, C. V. Gaw Van Praagh, A.-L. Reysenbach, and D. R. Lovley. 2002b. *Geoglobus ahangari* gen.

nov., sp. nov., a novel hyperthermophilic archaeon capable of oxidizing organic acids and growing autotrophically on hydrogen with Fe(III) serving as the sole electron acceptor. *Int. J. Syst. Evol. Microbiol.* **52:**719–728.

Kengen, S. W. M., P. J. H. Daas, E. F. G. Duits, J. T. Keltjens, C. Vanderdrift, and G. D. Vogels. 1992. Isolation of a 5-hydroxybenzimidazolyl cobamide-containing enzyme involved in the methyltetrahydromethanopterin-coenzyme M methyltransferase reaction in *Methanobacterium thermoautotrophicum*. *Biochim. Biophys. Acta* **1118:**249–260.

Kengen, S. W. M., F. A. M. de Bok, N. D. Vanloo, C. Dijkema, A. J. M. Stams, and W. M. de Vos. 1994. Evidence for the operation of a novel Embden-Meyerhof pathway that involves ADP-dependent kinases during sugar fermentation by *Pyrococcus furiosus*. *J. Biol. Chem.* **269:**17537–17541.

Kirschvink, J. L. 1982. Paleomagnetic evidence for fossil biogenic magnetite in western Crete. *Earth Planet. Sci. Lett.* **59:**388–392.

Kishida, K., Y. Sohrin, K. Okamura, and J. Ishibachi. 2004. Tungsten enriched in submarine hydrothermal fluids. *Earth Planet. Sci. Lett.* **222:**819–827.

Klenk, H.-P., R. A. Clayton, J.-F. Tomb, O. White, K. E. Nelson, K. A. Ketchum, R. J. Dodson, M. Gwinn, E. K. Hickey, J. D. Peterson, D. L. Richardson, A. R. Kerlavage, D. E. Graham, N. C. Kyrpides, R. D. Fleischmann, J. Quackenbush, N. H. Lee, G. G. Sutton, S. Gill, E. F. Kirkness, B. A. Dougherty, K. McKenney, M. D. Adams, B. Loftus, S. Peterson, C. I. Reich, L. K. McNeil, J. H. Badger, A. Glodek, L. Zhou, R. Overbeek, J. D. Gocayne, J. F. Weidman, L. McDonald, T. Utterback, M. D. Cotton, T. Spriggs, P. Artiach, B. P. Kaine, S. M. Sykes, P. W. Sadow, K. P. D'Andrea, C. Bowman, C. Fujii, S. A. Garland, T. M. Mason, G. J. Olsen, C. M. Fraser, H. O. Smith, C. R. Woese, and J. C. Venter. 1997. The complete genome sequence of the hyperthermophilic, sulphate-reducing archaeon *Archaeoglobus fulgidus*. *Nature* **390:**364–370.

Köster, S., M. Wehner, C. Herrmann, W. Kühlbrandt, and O. Yildiz. 2009. Structure and function of the FeoB G-domain from *Methanococcus jannaschii*. *J. Mol. Biol.* **392:**405–419.

Kristall, B., D. S. Kelley, M. D. Hannington, and J. R. Delaney. 2006. Growth history of a diffusely venting sulfide structure from the Juan de Fuca Ridge: a petrological and geochemical study. *Geochem. Geophys. Geosyst.* **7:**Q07001.

Leang, C., M. V. Coppi, and D. R. Lovley. 2003. OmcB, a *c*-type polyheme cytochrome, involved in Fe(III) reduction in *Geobacter sulfurreducens*. *J. Bacteriol.* **185:**2096–2103.

Lies, D. P., M. E. Hernandez, A. Kappler, R. E. Mielke, J. A. Gralnick, and D. K. Newman. 2005. *Shewanella oneidensis* MR-1 uses overlapping pathways for iron reduction at a distance and by direct contact under conditions relevant for biofilms. *Appl. Environ. Microbiol.* **71:**4414–4426.

Lilley, M. D., D. A. Butterfield, E. J. Olson, J. E. Lupton, S. A. Macko, and R. E. McDuff. 1993. Anomalous CH_4 and NH_4^+ concentrations at an unsedimented mid-ocean-ridge hydrothermal system. *Nature* **364:**45–47.

Lloyd, J. R., C. Leang, A. L. Hodges-Myersen, M. V. Coppi, S. Cuifo, B. Methé, S. J. Sandler, and D. R. Lovley. 2003. Biochemical and genetic characterization of PpcA, a periplasmic *c*-type cytochrome in *Geobacter sulfurreducens*. *Biochem. J.* **369:**153–161

Louvel, H., T. Kanai, H. Atomi, and J. N. Reeve. 2009. The Fur iron regulator-like protein is cryptic in the hyperthermophilic archaeon *Thermococcus kodakaraensis*. *FEMS Microbiol. Lett.* **295:**117–128.

Lovley, D. R. 1991. Dissimilatory Fe(III) and Mn(IV) reduction. *Microbiol. Rev.* **55:**259–287.

Lovley, D. R., J. F. Stolz, G. L. Nord, Jr., and E. J. P. Phillips. 1987. Anaerobic production of magnetite by a dissimilatory iron-reducing microorganism. *Nature* **330:**252–254.

Ma, K., R. Weiss, and M. W. W. Adams. 2000. Characterization of hydrogenase II from the hyperthermophilic archaeon *Pyrococcus furiosus* and assessment of its role in sulfur reduction. *J. Bacteriol.* **182:**1864–1871.

Mai, X., and M. W. W. Adams. 1994. Indolepyruvate ferredoxin oxidoreductase from the hyperthermophilic archaeon *Pyrococcus furiosus*. *J. Biol. Chem.* **269:**16726–16732.

Mai, X., and M. W. W. Adams. 1996a. Purification and characterization of two reversible acyl-CoA synthetases (ADP-forming) from the hyperthermophilic archaeon *Pyrococcus furiosus*. *J. Bacteriol.* **178:**5897–5903.

Mai, X., and M. W. W. Adams. 1996b. Characterization of a fourth type of 2-keto acid-oxidizing enzyme from a hyperthermophilic archaeon: 2-ketoglutarate ferredoxin oxidoreductase from *Thermococcus litoralis*. *J. Bacteriol.* **178:**5890–5896.

Makdessi, K., J. R. Andreesen, and A. Pich. 2001. Tungstate uptake by a highly specific ABC transporter in *Eubacterium acidaminophilum*. *J. Biol. Chem.* **276:**24557–24564.

Makdessi, K., K. Fritsche, A. Pich, and J. R. Andreesen. 2004. Identification and characterization of the cytoplasmic tungstate/molybdate-binding protein (Mop) from *Eubacterium acidaminophilum*. *Arch. Microbiol.* **181:**45–51.

Marguet, E., and P. Forterre. 1994. DNA stability at temperatures typical for hyperthermophiles. *Nucleic Acids Res.* **22:**1681–1686.

Marsili, E., D. B. Baron, I. D. Shikhare, D. Coursolle, J. A. Gralnick, and D. R. Bond. 2008. *Shewanella* secretes flavins that mediate extracellular electron transfer. *Proc. Natl. Acad. Sci. USA* **105:**3968–3973.

McCollom, T. M., and W. Bach. 2009. Thermodynamic constraints on hydrogen generation during serpentinization of ultramafic rocks. *Geochim. Cosmochim. Acta* **73:**856–875.

McKay, D. S., E. K. Gibson, K. L. Thomas-Keprta, H. Vali, C. S. Romanek, S. J. Clemett, X. D. F. Chillier, C. R. Maechling, and R. N. Zare. 1996. Search for past life on Mars: possible relic biogenic activity in Martian meteorite ALH84001. *Science* **273:**924–930.

Metz, S., and J. H. Trefry. 2000. Chemical and mineralogical influences on concentrations of trace metals in hydrothermal fluids. *Geochim. Cosmochim. Acta* **64:**2267–2279.

Mukund, S., and M. W. W. Adams. 1991. The novel tungsten-iron-sulfur protein of the hyperthermophilic archaebacterium, *Pyrococcus furiosus*, is an aldehyde ferredoxin oxidoreductase: evidence for its participation in a unique glycolytic pathway. *J. Biol. Chem.* **266:**14208–14216.

Mukund, S., and M. W. W. Adams. 1995. Glyceraldehyde-3-phosphate ferredoxin oxidoreductase, a novel tungsten-containing enzyme with a potential glycolytic role in the hyperthermophilic archaeon *Pyrococcus furiosus*. *J. Biol. Chem.* **270:**8389–8392.

Mukund, S., and M. W. W. Adams. 1996. Molybdenum and vanadium do not replace tungsten in the catalytically active forms of the three tungstoenzymes in the hyperthermophilic archaeon *Pyrococcus furiosus*. *J. Bacteriol.* **178:**163–167.

Myers, C. R., and J. M. Myers. 1997. Cloning and sequencing of *cymA*, a gene encoding a tetraheme cytochrome *c* required for reduction of iron(III), fumarate, and nitrate by *Shewanella putrefaciens* strain MR-1. *J. Bacteriol.* **179:**1143–1152.

Myers, J. M., and C. R. Myers. 2000. Role of the tetraheme cytochrome CymA in anaerobic electron transport in cells of *Shewanella putrefaciens* MR-1 with normal levels of menaquinone. *J. Bacteriol.* **182:**67–75.

Nakagawa, T., K. Takai, Y. Suzuki, H. Hirayama, U. Konno, U. Tsunogai, and K. Horikoshi. 2006. Geomicrobiological exploration and characterization of a novel deep-sea hydrothermal system at the TOTO caldera in the Mariana Volcanic Arc. *Environ. Microbiol.* **8:**37–49.

Nevin, K. P., and D. R. Lovley. 2000. Lack of production of electron-shuttling compounds or solubilization of Fe(III) during reduction of insoluble Fe(III) oxide by *Geobacter metallireducens*. *Appl. Environ. Microbiol.* **6:**2248–2251.

Nevin, K. P., and D. R. Lovley. 2002a. Mechanisms for accessing insoluble Fe(III) oxide during dissimilatory Fe(III) reduction by *Geothrix fermentans*. *Appl. Environ. Microbiol.* **68:**2294–2299.

Nevin, K. P., and D. R. Lovley. 2002b. Mechanisms for Fe(III) oxide reduction in sedimentary environments. *Geomicrobiol. J.* **19:**141–159.

Ntarlagiannis, D., E. A. Atekwana, E. A. Hill, and Y. Gorby. 2007. Microbial nanowires: is the subsurface "hardwired"? *Geophys. Res. Lett.* **34:**L17305.

Nunoura, T., Y. Sako, T. Wakagi, and A. Uchida. 2005. Cytochrome aa_3 in facultatively aerobic and hyperthermophilic archaeon *Pyrobaculum oguniense*. *Can. J. Microbiol.* **51:**621–627.

Pitts, K. E., P. S. Dobbin, F. Reyes-Ramirez, A. J. Thomson, D. J. Richardson, and H. E. Seward. 2003. Characterization of the *Shewanella oneidensis* MR-1 decaheme cytochrome MtrA. *J. Biol. Chem.* **278:**27758–27765.

Ragsdale, S. W. 2009. Nickel-based enzyme systems. *J. Biol. Chem.* **284:**18571–18575.

Rech, S., C. Wolin, and R. P. Gunsalus. 1996. Properties of the periplasmic ModA molybdate-binding protein of *Escherichia coli*. *J. Biol. Chem.* **271:**2557–2562.

Reguera, G., K. D. McCarthy, T. Mehta, J. S. Nicoll, M. T. Tuominen, and D. R. Lovley. 2005. Extracellular electron transfer via microbial nanowires. *Nature* **435:**1098–1101.

Reguera, G., K. P. Nevin, J. S. Nicoll, S. F. Covalla, T. L. Woodard, and D. R. Lovley. 2006. Biofilm and nanowire production leads to increased current in *Geobacter sulfurreducens* fuel cells. *Appl. Environ. Microbiol.* **72:**7345–7348.

Reguera, G., R. B. Pollina, J. S. Nicoll, and D. R. Lovley. 2007. Possible nonconductive role of *Geobacter sulfurreducens* pilus nanowires in biofilm formation. *J. Bacteriol.* **189:**2125–2127.

Riera, J., F. T. Robb, R. Weiss, and M. Fontecave. 1997. Ribonucleotide reductase in the archaeon *Pyrococcus furiosus*: a critical enzyme in the evolution of DNA genomes? *Proc. Natl. Acad. Sci. USA* **94:**75–78.

Robb, F. T., J.-B. Park, and M. W. W. Adams. 1992. Characterization of an extremely thermostable glutamate dehydrogenase: a key enzyme in the primary metabolism of the hyperthermophilic archaebacterium, *Pyrococcus furiosus*. *Biochim. Biophys. Acta* **1120:**267–272.

Robb, F. T., D. L. Maeder, J. R. Brown, J. DiRuggiero, M. D. Stump, R. K. Yeh, R. B. Weiss, and D. M. Dunn. 2001. Genomic sequence of hyperthermophile, *Pyrococcus furiosus*: implications for physiology and enzymology. *Methods Enzymol.* **330:**134–157.

Rodionov, D. A., P. Hebbeln, A. Eudes, J. ter Beek, I. A. Rodionova, G. B. Erkens, D. J. Slotboom, M. S. Gelfand, A. L. Osterman, A. D. Hanson, and T. Eitinger. 2009. A novel class of modular transporters for vitamins in prokaryotes. *J. Bacteriol.* **191:**42–51.

Rodionov, D. A., P. Hebbeln, M. S. Gelfand, and T. Eitinger. 2006. Comparative and functional genomic analysis of prokaryotic nickel and cobalt uptake transporters: evidence for a novel group of ATP-binding cassette transporters. *J. Bacteriol.* **188:**317–327.

Rodionov, D. A., A. G. Vitreschak, A. A. Mironov, and M. S. Gelfand. 2003. Comparative genomics of the vitamin B12 metabolism and regulation in prokaryotes. *J. Biol. Chem.* **278:**41148–41159.

Rospert, S., J. Breitung, K. Ma, B. Schworer, C. Zirnbibl, R. K. Thauer, D. Linder, R. Huber, and K. O. Stetter. 1991. Methyl-coenzyme M reductase and other enzymes involved in methanogenesis from CO_2 and H_2 in the extreme thermophile *Methanopyrus kandleri*. *Arch. Microbiol.* **156:**49–55.

Roy, R., S. Mukund, G. J. Schut, D. M. Dunn, R. Weiss, and M. W. W. Adams. 1999. Purification and molecular characterization of the tungsten-containing formaldehyde ferredoxin oxidoreductase from the hyperthermophilic archaeon *Pyrococcus furiosus*: the third of a putative five-member tungstoenzyme family. *J. Bacteriol.* **181:**1171–1180.

Sapra, R., K. Bagramyan, and M. W. W. Adams. 2003. A simple energy-conserving system: proton reduction coupled to proton translocation. *Proc. Natl. Acad. Sci. USA* **100:**7545–7550.

Schiffer, A., K. Parey, E. Warkentin, K. Diederichs, H. Huber, K. O. Stetter, P. M. H. Kroneck, and U. Ermler. 2008. Structure of the dissimilatory sulfite reductase from the hyperthermophilic archaeon *Archaeoglobus fulgidus*. *J. Mol. Biol.* **379:**1063–1074.

Schmidt, K., A. Koschinsky, D. Garbe-Schönberg, L. M. de Carvalho, and R. Seifert. 2007. Geochemistry of hydrothermal fluids from the ultramafic-hosted Logatchev hydrothermal field, 15°N on the Mid-Atlantic Ridge: temporal and spatial investigation. *Chem. Geol.* **242:**1–21.

Schreiter, E. R., M. D. Sintchak, Y. Guo, P. T. Chivers, R. T. Sauer, and C. L. Drennan. 2003. Crystal structure of the nickel-responsive transcription factor NikR. *Nat. Struct. Biol.* **10:**794–799.

Schrenk, M. O., J. F. Holden, and J. A. Baross. 2008. Magma-to-microbe networks in the context of sulfide hosted microbial ecosystems, p. 233–258. *In* R. P. Lowell, J. S. Seewald, A. Metaxas, and M. R. Perfit (ed.), *Magma to Microbe: Modeling Hydrothermal Processes at Ocean Spreading Centers*. Geophysical Monograph 178. AGU Press, Washington, DC.

Schröder, I., E. Johnson, and S. de Vries. 2003. Microbial ferric iron reductases. *FEMS Microbiol. Rev.* **27:**427–447.

Schut, G. J., S. L. Bridger, and M. W. W. Adams. 2007. Insights into the metabolism of elemental sulfur by the hyperthermophilic archaeon *Pyrococcus furiosus*: characterization of a coenzyme A-dependent NAD(P)H sulfur oxidoreductase. *J. Bacteriol.* **189:**4431–4441.

Schwarz, G., R. R. Mendel, and M. W. Ribbe. 2009. Molybdenum cofactors, enzymes and pathways. *Nature* **460:**839–847.

Schwertmann, U., and R. M. Cornell. 2000. *Iron Oxides in the Laboratory: Preparation and Characterization*, 2nd ed. Wiley-VCH Verlag, Weinheim, Germany.

Sevcenco, A. M., M. W. H. Pinkse, E. Bol, G. C. Krijger, H. T. Wolterbeek, P. D. E. M. Verhaert, P. L. Hagedoorn, and W. R. Hagen. 2009. The tungsten metallome of *Pyrococcus furiosus*. *Metallomics* **1:**395–402.

Seyfried, W. E., Jr., and K. Ding. 1993. The effect of redox on the relative solubilities of copper and iron in Cl-bearing aqueous fluids at elevated temperatures and pressures: an experimental study with application to subseafloor hydrothermal systems. *Geochim. Cosmochim. Acta* **57:**1905–1917.

Seyfried, W. E., Jr., and K. Ding. 1995. Phase equilibria in subseafloor hydrothermal systems: a review of the role of redox, temperature, pH and dissolved Cl on the chemistry of hot springs at mid-ocean ridges, p. 248–272. *In* S. E. Humphris, R. A. Zierenberg, L. S. Mullineaux, and R. E. Thomson (ed.), *Seafloor Hydrothermal Systems. Physical, Chemical, Biological, and Geological Interactions*. Geophysical Monograph 91. AGU Press, Washington, DC.

Speich, N., C. Dahl, P. Heisig, A. Klein, F. Lottspeich, K. O. Stetter, and H. G. Trüper. 1994. Adenylsulfate reductase from the sulfate-reducing archaeon *Archaeoglobus fulgidus*: cloning and characterization of the genes and comparison of the enzyme with other iron-sulfur flavoproteins. *Microbiology-UK* **140:**1273–1284.

Sperling, D., U. Kappler, A. Wynen, C. Dahl, and H. G. Trüper. 1998. Dissimilatory ATP sulfurylase from the hyperthermophilic sulfate reducer *Archaeoglobus fulgidus* belongs to the group of homo-oligomeric ATP sulfurylases. *FEMS Microbiol. Lett.* **162:**257–264.

Steinberger, R. E., and P. A. Holden. 2005. Extracellular DNA in single- and multiple-species unsaturated biofilms. *Appl. Environ. Microbiol.* **71:**5404–5410.

Stetter, K. O. 1999. Extremophiles and their adaptation to hot environments. *FEBS Lett.* **452:**22–25.

Straub, K. L., M. Benz, and B. Schink. 2001. Iron metabolism in anoxic environments at near neutral pH. *FEMS Microbiol. Ecol.* **34:**181–186.

Takai, K., T. Gamo, U. Tsunogai, N. Nakayama, H. Hirayama, K. H. Nealson, and K. Horikoshi. 2004. Geochemical and microbiological evidence for a hydrogen-based, hyperthermophilic subsurface lithoautotrophic microbial ecosystem (HyperSLiME) beneath an active deep-sea hydrothermal field. *Extremophiles* **8:**269–282.

Takai, K., T. Nunoura, J. Ishibashi, J. Lupton, R. Suzuki, H. Hamasaki, Y. Ueno, S. Kawagucci, T. Gamo, Y. Suzuki, H. Hirayama, and K. Horikoshi. 2008. Variability in the microbial communities and hydrothermal fluid chemistry at the newly discovered Mariner hydrothermal field, southern Lau Basin. *J. Geophys. Res.* **113:**G02031.

Tao, X., N. Schiering, H. Y. Zeng, D. Ringe, and J. R. Murphy. 1994. Iron, DtxR, and the regulation of diphtheria toxin expression. *Mol. Microbiol.* **14:**191–197.

Theriot, C. M., S. R. Tove, and A. M. Grunden. 2009. Characterization of two proline dipeptidases (prolidases) from the hyperthermophilic archaeon *Pyrococcus horikoshii*. *Appl. Microbiol. Biotechnol.* doi:10.1007/s00253-009-2235-x.

Thomas-Keprta, K. L., S. J. Clemett, D. A. Bazylinski, J. L. Kirschvink, D. S. McKay, S. J. Wentworth, H. Vali, E. K. Gibson, Jr., M. F. McKay, and C. S. Romanek. 2001. Truncated hexa-octahedral magnetite crystals in ALH84001: presumptive biosignatures. *Proc. Natl. Acad. Sci. USA* **98:**2164–2169.

Tor, J. M., K. Kashefi, and D. R. Lovley. 2001. Acetate oxidation coupled to Fe(III) reduction in hyperthermophilic microorganisms. *Appl. Environ. Microbiol.* **67:**1363–1365.

Tsunasawa, S., Y. Izu, M. Miyagi, and I. Kato. 1997. Methionine aminopeptidase from the hyperthermophilic archaeon *Pyrococcus furiosus*: molecular cloning and overexpression in *Escherichia coli* of the gene, and characteristics of the enzyme. *J. Biochem.* **122:**843–850.

Turekian, K. K. 1968. *Oceans*. Prentice-Hall, Englewood Cliffs, NJ.

Vadas, A., H. G. Monbouquette, E. Johnson, and I. Schröder. 1999. Identification and characterization of a novel ferric reductase from the hyperthermophilic archaeon *Archaeoglobus fulgidus*. *J. Biol. Chem.* **274:**36715–36721.

Vali, H., B. Weiss, Y.-L. Li, S. K. Sears, S. S. Kim, J. L. Kirschvink, and C. L. Zhang. 2004. Formation of tabular single-domain magnetite induced by *Geobacter metallireducens* GS-15. *Proc. Natl. Acad. Sci. USA* **101:**16121–16126.

Vargas, M., K. Kashefi, E. L. Blunt-Harris, and D. R. Lovley. 1998. Microbiological evidence for Fe(III) reduction on early Earth. *Nature* **395:**65–67.

Vartivarian, S. E., and R. E. Cowart. 1999. Extracellular iron reductases: identification of a new class of enzymes by siderophore-producing microorganisms. *Arch. Biochem. Biophys.* **364:**75–82.

Ver Eecke, H. C., D. S. Kelley, and J. F. Holden. 2009. Abundances of hyperthermophilic autotrophic Fe(III) oxide reducers and heterotrophs in hydrothermal sulfide chimneys of the northeastern Pacific Ocean. *Appl. Environ. Microbiol.* **75:**242–245.

Vignais, P. M., B. Billoud, and J. Meyer. 2001. Classification and phylogeny of hydrogenases. *FEMS Microbiol. Rev.* **25:**455–501.

Völkl, P., R. Huber, E. Drobner, R. Rachel, S. Burggraf, A. Trincone, and K. O. Stetter. 1993. *Pyrobaculum aerophilum* sp. nov., a novel nitrate-reducing hyperthermophilic archaeum. *Appl. Environ. Microbiol.* **59:**2918–2926.

von Canstein, H., J. Ogawa, S. Shimizu, and J. R. Lloyd. 2008. Secretion of flavins by *Shewanella* species and their role in extracellular electron transfer. *Appl. Environ. Microbiol.* **74:**615–623.

Von Damm, K. L. 1995. Controls in the chemistry and temporal variability of seafloor hydrothermal fluids, p. 222-247. *In* S. E. Humphris, R. A. Zierenberg, L. S. Mullineaux, and R. E. Thomson (ed.), *Seafloor Hydrothermal Systems: Physical, Chemical, Biological, and Geological Interactions*. Geophysical Monograph 91. American Geophysical Union, Washington, DC.

Vorholt, J., J. Kunow, K. O. Stetter, and R. K. Thauer. 1995. Enzymes and coenzymes of the carbon monoxide dehydrogenase pathway for autotrophic CO_2 fixation in *Archaeoglobus lithotrophicus* and the lack of carbon monoxide dehydrogenase in the heterotrophic *A. profundus*. *Arch. Microbiol.* **163:**112–118.

Vorholt, J. A., D. Hafenbradl, K. O. Stetter, and R. K. Thauer. 1997a. Pathways of autotrophic CO_2 fixation and of dissimilatory nitrate reduction to N_2O in *Ferroglobus placidus*. *Arch. Microbiol.* **167:**19–23.

Vorholt, J. A., M. Vaupel, and R. K. Thauer. 1997b. A selenium-dependent and a selenium-independent formylmethanofuran dehydrogenase and their transcriptional regulation in the hyperthermophilic *Methanopyrus kandleri*. *Mol. Microbiol.* **23:**1033–1042.

Waldron, K. J., J. C. Rutherford, D. Ford, and N. J. Robinson. 2009. Metalloproteins and metal sensing. *Nature* **460:**823–830.

Wang, Y., and D. K. Newman. 2008. Redox reactions of phenazine antibiotics with ferric (hydr)oxides and molecular oxygen. *Environ. Sci. Technol.* **42:**2380–2386.

Whitchurch, C. B., T. Tolker-Nielsen, P. C. Ragas, and J. S. Mattick. 2002. Extracellular DNA required for bacterial biofilm formation. *Science* **295:**1487.

Whittaker, M. M., and J. W. Whittaker. 2000. Recombinant superoxide dismutase from a hyperthermophilic archaeon, *Pyrobaculum aerophilum*. *J. Biol. Inorg. Chem.* **5:**402–408.

Zachara, J. M., R. K. Kukkadapu, J. K. Fredrickson, Y. A. Gorby, and S. C. Smith. 2002. Biomineralization of poorly crystalline Fe(III) oxides by dissimilatory metal reducing bacteria (DMRB). *Geomicrobiol. J.* **19:**179–207.

Zhang, Y., and V. N. Gladyshev. 2008. Molybdoproteomes and evolution of molybdenum utilization. *J. Mol. Biol.* **379:**881–899.

Zhang, Y., D. A. Rodionov, M. S. Gelfand, and V. N. Gladyshev. 2009. Comparative genomic analyses of nickel, cobalt and vitamin B_{12} utilization. *BMC Genomics* **10:**78–103.

Zhu, X. K., Y. Guo, R. J. P. Williams, R. K. O'Nions, A. Matthews, N. S. Belshaw, G. W. Canters, E. C. de Waal, U. Weser, B. K. Burgess, and B. Salvato. 2002. Mass fractionation processes of transition metal isotopes. *Earth Planet. Sci. Lett.* **200:**47–62.

MICROBE-METAL INTERACTIONS ON SEAFLOOR BASALTS

Jason B. Sylvan, Amanda G. Turner, and Katrina J. Edwards

4

INTRODUCTION

More than 70% of planet Earth is covered by ocean, the bottom of which constitutes one of the most poorly studied environments on this planet, the ocean crust. Lithospheric ocean crust is largely composed of volcanic extrusive basalt, which outcrops at mid-ocean ridge spreading centers and seamounts. Nearly 200,000 km² of metal-rich basalt outcrops at the ocean floor, where it interacts with deep seawater, rendering this the largest continuous potential endolithic microbial rock habitat on Earth. Seawater also circulates through the porous and permeable upper ocean basement (upper 500 m below sediment), carrying seawater solutes and microbes into the subseafloor, where they may take up residence in the deep biosphere. This chapter broadly considers mineral-microbe interactions at and below the ocean floor, the epilithic and endolithic microorganisms harbored in rock habitats in the dark ocean, and metal and redox cycling at rock and mineral surfaces. We discuss empirical studies on materials retrieved from the seafloor and subseafloor and discuss new research frontiers using novel in situ microbial observatories for time-series experiments in these environments.

Microbes have been known to interact with minerals for most of scientific oceanographic exploration. This includes marine sediments, minerals in the water column (e.g., dust, marine snow), and more recently, rocks and minerals in the deep sea (e.g., basalt, hydrothermal sulfides). In the past 15 years, our understanding of the role that microbes play in alteration of basaltic igneous crust both at and below the seafloor has grown enormously, and it is apparent that at least some of these mineral-microbe interactions are closely linked with oceanic alteration and weathering cycles in the deep sea. Weathering and alteration of rocks and minerals in the ocean plays an important role in balancing a variety of biogeochemical elemental budgets and cycles, including the carbon cycle (Edmond et al., 1979; Staudigel and Hart, 1983; Brady and Gislason, 1997; Wolff-Boenisch et al., 2004). Laboratory and field studies have also indicated that alteration of rocks and minerals in the deep sea is connected to carbon fixation, and thereby dark ocean primary production (Edwards et al., 2005). Furthermore, seawater circulation in the subseafloor

---

Jason B. Sylvan, Geomicrobiology Group, Department of Biological Sciences, Marine Environmental Biology Section, University of Southern California, Los Angeles, CA 90089. *Amanda G. Turner*, Department of Earth Sciences, University of Southern California, Los Angeles, CA 90089. *Katrina J. Edwards*, Geomicrobiology Group, Department of Biological Sciences, Marine Environmental Biology Section, and Department of Earth Sciences, University of Southern California, Los Angeles, CA 90089.

basalt-hosted aquifer system below the ocean floor suggests that these interactions and reactions may propagate deep into the subsurface, supporting a biosphere below the seafloor with far-ranging implications for the biogeochemical function of the Earth system.

On a global basis, approximately 200,000 km^2 of basalt is exposed or thinly sedimented at the bottom of the ocean at mid-ocean ridge spreading centers, ridge flanks, and seamounts (Edwards et al., 2005). The scale of this basalt biome renders it potentially the largest continuous endolithic rock habitat on Earth. In this chapter, we review recent insights into dark ocean mineral-mineral microbe interactions, principally as associated with this potentially largest endolithic surface on Earth: seafloor and subseafloor basalt. We also discuss experimental lines of research on alteration and weathering of the igneous ocean crust conducted in recent years both at and below the seafloor, and offer insights into new means for directly assessing the role of microbes in metal cycling and rock alteration.

ROCK ALTERATION AT AND BELOW THE SEAFLOOR

Production and Alteration of Seafloor Basalts at Mid-Ocean Ridges, Flanks, and Seamounts

Magma is extruded during volcanic eruptions at the ocean floor in the form of pillow lavas and sheet flows at the rate of ~2.6 km^2 year^{-1} (Cogné and Humler, 2004). This volcanic extrusive package typically makes up the upper ~700 m of basaltic ocean crust, which is underlain by a sheeted dike complex (~1.5 km) and slowly cooled, crystalline gabbroic rock (~4 km). The depth to which microbe-mineral interactions occur associated with the oceanic crust is not known, but it is likely that the dominant processes occur where fluid and pore space connect the oceanic fluids to these solids, i.e., exposed basaltic lava at the seafloor, and the highly permeable extrusive upper basement (200 to 500 m) through which seawater circulates hydrothermally (Fisher, 1998; Fisher and Becker, 2000). Basalt is composed of glass, olivine, sulfides, and, to a lesser extent, plagioclase and clinopyroxene, which collectively have a composition of 50 weight % SiO_2 and 4 to 15% each of MgO, FeO, CaO, and Al_2O_3. Basalt is rich in reduced chemical species and is susceptible to changes in chemical speciation as a result of the redox gradients that exist at the interface between this reduced solid and oxidizing seawater (Edwards et al., 2005).

Over time, basalt alters to form a complex mixture of reaction products (clays, carbonates, and amorphous oxyhydroxides), which are collectively referred to as "palagonite" (Color Plate 7). Alteration reactions that result in palagonization occur in a series of stages. Basaltic rock begins to chemically weather on exposure to oxidizing seawater at the ocean floor, resulting in dissolution and secondary precipitation reactions (e.g., Zhou and Fyfe, 1989; Jercinovic and Ewing, 1992; Stroncik and Schmincke, 2001), or by incongruent dissolution and leaching of cations from glass (Thorseth et al., 1991; Stroncik and Schmincke, 2002). At the seafloor, exposed basalts also interact with hydrothermal fluids—to varying degrees depending on their proximity to venting fluids and vent fluid chemistry. In addition, minerals that form in the plume environment rain out onto the underlying basalt that is exposed at the seafloor. These seafloor processes collectively result in crusts of oxidized Mn, Fe, and Al minerals, which accumulate and thicken with age until the basalts are repaved by fresh eruptions or become buried by sediment accumulation.

Seafloor alteration of basalt involves significant changes in the redox state of chemical constituents in basalt and hydrothermal fluids (Bach and Edwards, 2003). These reactions are energy yielding (Bach et al., 2006) and, hence, have been a topic of significant interest to microbiologists interested in metal cycling and links between metal cycling, seafloor alteration of basalt, and chemolithoautotrophy as a result of these processes in the dark ocean.

Fluid Flow and the Potential for Microbial Life below the Bottom of the Ocean

Weathering occurs both at and below the seafloor. The upper 200 to 500 m of the basaltic ocean crust is characterized by very high permeabilities (10^{-12} to 10^{-15} m^2) and rampant circulation of hydrothermal fluids (Fisher, 1998; Fisher and Becker, 2000). Basalt outcrops (e.g., exposures of basalt near mid-ocean ridges, on ridge flanks, and at seamounts) serve as conduits for fluid flow either into or out of the basaltic aquifer. The volume equivalent of all the seawater that resides in the ocean basins is circulated through the subseafloor aquifer system approximately every 200,000 years (Johnson and Priuis, 2003). This subseafloor oceanic aquifer system serves as a geochemical processing zone, where biochemical reactions and elemental cycling and exchange occur with globally significant ramifications. For example, the global hydrothermal iron input to the oceans is approximately equivalent to that which is delivered from global riverine transport (Wheat and Mottl, 2000). The size of this habitat, the potential for microbes to be involved in energy-yielding reactions involving metals, and the resultant biogeochemical influences indicate that the subseafloor basalt biome may have consequence for the balance of many elemental cycles at global scales.

Evidence of Fe(II) oxidation over the first 20 million years (My) after basalt formation indicates the potential for microbes to participate in oxidation of Fe(II) and S in seafloor basalts (Bach and Edwards, 2003). These measurements are based only on oxidation reactions within seafloor basalts; other metabolisms such as Fe reduction and nitrate reduction are also likely to fuel microbial populations in this subseafloor aquifer system. Evidence for sulfate reduction up to 50 My after basalt formation pushes the extent of life in the deep subsurface to even older crust (Rouxel et al., 2008), while similar evidence from δ^{34}S isotopes in sulfides from basaltic ophiolites on Macquarie Island shows that sulfate reduction occurs in younger basement as well, in this case, ~10 My (Alt et al., 2003). It is also possible that the autotrophic growth supported by oxidation of Fe and S, as well as H$_2$-consuming reactions such as methanogenesis and sulfate reduction, provide as much organic carbon production in basalts as the degradation of organic matter deposited from the water column in marine sediments (Bach and Edwards, 2003). These revelations have provided fresh impetus for studying microbial life on and in basalts in an effort to better constrain the carbon cycle and understand the life that exists within this enormous biome.

BASALT MICROBIOLOGY

Abundance and Diversity of Basalt-Hosted Prokaryotes

Microbes were first detected on basalts collected by the Ocean Drilling Program from the eastern Equatorial Pacific Ocean (Giovannoni et al., 1996). Cells that stained with DNA binding dyes and the presence of pitting at the interface between fresh basalt and secondary alteration products were the first indications that basalts host an active microbial community. These channels were later shown to be areas of anomalously high C, N, and P concentrations, further proof of microbial life (Fisk et al., 1998). The discovery of microbial life hosted by subseafloor basalts, which are a challenge to collect, spurred on studies of seafloor basalts, which are easier to collect and analogs for their subseafloor counterparts. Indeed, all subseafloor basalts were once exposed at the seafloor and carried with them to the subsurface whatever microbial population was present as the rock was slowly covered by sediment.

Seafloor basalts harbor prokaryotic cell counts of 6×10^5 to 1×10^9 cells per gram of rock, 3 to 4 orders of magnitude greater than the overlying seawater (Einen et al., 2008; Santelli et al., 2008). This number appears to be stable with rock age and water depth of rock collection (Einen et al., 2008), although only a few studies exist that quantify basalt microbes deep in the marine subsurface. Estimates indicate that 6×10^7 cells per gram of rock may be supported through alteration reactions on and in basalts (Santelli et al., 2008). Studies on sulfide minerals indicate that the population size of the endolithic chemoautolithotroph community appears to be influenced by the rate at which the rocks release reduced chemicals that may support their growth (Edwards et al., 2003a; Rogers et al., 2003) and it is likely that the same factors control population size on basalts.

The greatest effort of the basalt microbiology studies to date has been defining their microbial community. Bacterial communities on basalts are extremely diverse, more so than any other marine habitat (Santelli et al., 2008). They are also far more diverse than the archaeal communities on the same basalts (Thorseth et al., 2001; Lysnes et al., 2004; Mason et al., 2007). Basalt microbiological diversity studies still number in the few, but studies from diverse environments are beginning to coalesce into a picture of what the basalt microbial community looks like.

On basalts, *Bacteria* represent approximately 90% or more of the prokaryotic population (Santelli et al., 2008). The role of *Archaea* in deep-subsurface marine sediments is currently debated; some workers argue for a dominant role by *Archaea* (Biddle et al., 2006, 2008; Lipp et al., 2008), while others argue that *Bacteria* are more important (Schippers et al., 2005), but the story is much clearer on basalts. *Archaea* on basalts comprise a minor but well-defined component of the prokaryotic community. Marine Group I (MGI) *Crenarchaeota* dominate basalt clone libraries (Thorseth et al., 2001; Lysnes et al., 2004; Mason et al., 2007, 2009). Only one study reports the presence of any other group of *Archaea*, the Marine Benthic Group B (Mason et al., 2009).

While the basalt rocks themselves do not appear to be ideal substrates for most archaea, they are common in habitats intimately associated with basalts. For example, flocculent material sampled from basalt surfaces on 0.5 My ridge flank basalts at 9°27′N East Pacific Rise (EPR) contained thermophilic archaea within the *Desulfurococcales* and *Thermoproteales* orders of *Crenarchaeota*, within the *Methanopyrus* and *Thermococcus* genera of the *Euryarchaeota*, and the *Korarchaeota* (Ehrhardt et al., 2007). Thermophilic archaea are also common to subsurface hydrothermal fluids on the basalt-hosted Juan de Fuca Ridge (Cowen et al., 2003; Huber et al., 2006). It appears that archaea are carried within fluids in the subsurface basalt aquifer, but they do not colonize the basalts themselves. Little is known about what ecological role the archaea that are present on basalts carry out; future efforts focused on culturing should aim to resolve this basic question.

Bacterial populations and diversity on basalts are far more complex than their archaeal counterparts. Analysis of sequences amplified from basalts in the public databases indicates that there may be 11 monophyletic clades originating from ocean crust (which includes sediment, basalt, and gabbro): 7 bacterial and 4 archaeal (Mason et al., 2007). These Ocean Crust Clades (OCCs) comprise bacterial clades within the α-(*Sulfitobacter*) and γ-proteobacteria, *Acidobacteria*, *Actinobacteria*, and *Verrucomicrobia*. All of the archaeal OCCs are from the MGI *Crenarchaeota*, supporting the above discussion of low archaeal diversity on basalts. Among these 4 archaeal groups, α-MGI OCC IX is the only archaeal OCC composed entirely of sequences cloned from basalts (they are not found in seawater samples). Geographic distribution of the OCC groups is cosmopolitan, indicating that similar microbial communities exist on geographically distant basalts (Mason et al., 2007). Other authors (Santelli et al., 2008) have confirmed this.

Basalts from the Arctic harbor *Actinobacteria*, *Bacteroidetes*, *Chloroflexi*, *Firmicutes*, and α-, δ-, ε- and γ-proteobacteria (Thorseth et al., 2001; Lysnes et al., 2004). Basalts from 9°N on the EPR are dominated by γ-proteobacteria, which were present on all basalts collected and represent 25 to 33% of clones recovered. γ-Proteobacteria dominated a basalt collected on the Vailulu'u Seamount, west of Samoa, and proteobacteria in the α-, δ- γ-, and ε-subdivisions comprised 80% of the clones recovered (Sudek et al., 2009). δ-Proteobacteria, *Planctomycetes*, and *Actinobacteria* were present on all samples from the EPR in one study (Santelli et al., 2009), and also in basalts collected from the Juan de Fuca Ridge (Mason et al., 2009), indicating their likely importance on basalts. β-Proteobacteria are rarely recovered from basalts. Sequences from methane oxidizers, sulfide oxidizers, and some psychrophiles were also present at the EPR (Santelli et al., 2009). Many of the EPR sequences were related to *Comamonadaceae*, which are facultative chemolithoautotrophs capable of hydrogen oxidation. Hydrogen is likely produced within basalts abiotically; therefore, the substrate necessary for hydrogen oxidation is endemic to basalts (Templeton et al., 2005) and explains the presence of hydrogen oxidizers on the rocks. *Gemmatimonadetes* sequences have been recovered only from rocks on the Juan de Fuca Ridge (Mason et al., 2009). Unclassifiable clones represent a significant proportion of total clones on basalts, up to 13% in some samples, indicating an endemic population that is not known (Lysnes et al., 2004; Santelli et al., 2009; Sudek et al., 2009).

The bacterial communities on basalts are not only diverse, but they are also unique. Comparisons of microbial communities from basalt habitats with communities from background seawater indicate that the microbes on basalts are different, even if they share similar phylogenetic makeup at the class level (Mason et al., 2009; Santelli et al., 2009). For example, the γ-proteobacteria found in the seawater overlying basalts rocks are not the same as those found on the rocks. Rarefaction curves for 16S rRNA clone libraries from basalt samples at Loihi Seamount, Hawaii, 9°N EPR, and Vailulu'u Seamount are nearly identical, indicating some level of similarity between these samples (Santelli et al., 2008; Sudek et al., 2009).

Bacterial communities on basalts are correlated to the age of the rock (Lysnes et al., 2004; Mason et al., 2009; Santelli et al., 2009), which is likely related to the substrate state in terms of amount of oxidation of the rock and electron donors and acceptors present to allow for microbial metabolism. This has been found at sampling sites from the Arctic, EPR, and Juan de Fuca Ridge. *Actinobacteria* were more prevalent on older rocks (~104 years to 2 My) than young rocks (20 years) in the Arctic (Lysnes et al., 2004). The same study found *Bacteroidetes* to be distributed across all rock ages, while later work indicated they were not present on young rocks of similar age to those collected from the Arctic (Santelli et al., 2009). Furthermore, Santelli et al. (2009) showed that, for environmental samples collected at EPR, the most diverse communities were found on the oldest rocks in the sample suite. Young Arctic basalts with no sediment cover and fresh glass harbor branching and twisted stalks representative of the iron oxidizers *Gallionella*, but the older rocks do not (Thorseth et al., 2001). In hindsight, it is likely that the twisted stalks indicate the presence of *Mariprofundus ferrooxidans* gen. nov. sp. nov., an Fe-oxidizing ζ-proteobacterium isolated from Loihi seamount, off Hawaii (Emerson et al., 2007). *M. ferrooxidans* produces Fe stalks that bear an appearance remarkably similar to *Gallionella* stalks.

Sequences of ζ-proteobacteria are increasingly being recovered from marine environments where Fe oxidation and deposition are occurring (Kato et al., 2009; Rassa et al., 2009; Sudek et al., 2009) suggesting that, in the marine realm, ζ-proteobacteria related to "*Mariprofundus*" likely dominate over β-proteobacteria related to *Gallionella*, which was not detected by Thorseth and colleagues (2001), and is generally not detected in clone libraries from the marine realm. In addition, microfossils on basalts that can confidently be dated to ~3 billion years ago and are enriched in C, N, and P bear physical characteristics remarkably similar to the stalks made by *M. ferrooxidans*, indicating that these or similar organisms were likely present in the archaean subsurface biosphere (Banerjee et al., 2007). Readers interested in further information about ancient microfossils on basalt are encouraged to refer to a recent review on the topic (Staudigel et al., 2008).

The difference in community composition by rock age is supported by mineralogy—young rocks contain fewer Mn oxides and are found to host fewer Mn oxidizers (Templeton et al., 2005). The current hypothesis is that newly formed basalts are colonized initially by chemolithoautotrophs such as Fe(II) oxidizers and methanogens, which produce organic carbon that can be utilized by the heterotrophs that colonize later.

Basalt Microbial Physiology

Knowledge about what bacteria are present on basalts is growing steadily, while data about archaeal assemblages is still lacking. Despite this, few physiological data still are available about basalt-hosted microbes. One study used a functional gene array called GeoChip (He et al., 2007) to analyze functional genes on a basalt from Juan de Fuca Ridge (Mason et al., 2009) and detected genes for methanogenesis (*mcr* genes), nitrogen fixation, anammox, denitrification, iron reduction, and dissimilatory sulfate reduction. Based on the presence of these genes, it appears that an active N-cycle may be active on basalts. Other authors have also detected sequences of annamox bacteria (Santelli et al., 2008), but there is as yet no physiological proof of the annamox reaction occurring on basalts. In addition, this is the first incidence of sequences from methanogens being detected, although any 16S rRNA genes are still undetected. However, methanogens have been detected via the production of methane in enrichment cultures with collected basalts, but no sequences matching known methanogens were retrieved from those enrichments (Lysnes et al., 2004). It seems that methanogens are very likely present on and in basalts, but substantial proof of their presence is still lacking. Their activity would certainly have an impact on global carbon cycling and is an avenue for further research.

While studies on direct physiology of in situ basalt microbes are lacking, focusing on culturing microbes from basalt rocks and characterizing the isolates has been successful. Studies focusing on enrichment cultures are necessarily limited to which organisms can be detected, but they allow one to identify physiological traits of the microbes isolated. Culturing efforts revealed that Fe redox cycling is important on basalts. Fe(III)-reducers, including *Shewanella frigidimarina*, were isolated from basalts (Lysnes et al., 2004). While chemolithoautotrophic Fe(II) oxidizers have not yet been isolated directly from basalts, strains isolated from seafloor environments are able to grow with basalt as the only Fe(II) source (Edwards et al., 2003b; Bailey et al., 2009). It is interesting that the ability to oxidize Fe(II) is shared across multiple proteobacterial lineages, including representatives from α-, γ- and ζ-proteobacteria, possibly indicating horizontal gene transfer of this ability, or perhaps hinting at the possibility that iron oxidation is an ancestral trait within the *Proteobacteria*.

Culturing efforts using basalts from the Ascension Fracture Zone along the Mid-Atlantic Ridge yielded isolates of α-, β- and γ-proteobacteria, *Actinobacteria*, and bacilli (Rathsack et al., 2009). Cultured γ-proteobacteria included *Marinobacter* and *Halomonas*, two genera common to hydrothermal systems (Kaye and Baross, 2000). Rathsack et al. (2009) discuss the diversity of cultivates from different substrates, but it should be noted that diversity of cultivates cannot be considered in the same way as when probing with universal primers for PCR because most of the diversity represented in the environment cannot be recovered by cultivation methods. Therefore, any discussion of diversity based on isolates is biased by the specific culturing methods used. However, cultivates can be used for laboratory physiology studies, which provide valuable information that cannot be obtained by molecular surveys alone, as discussed below.

A diversity of Mn(II)-oxidizing bacteria are also present on basalts (Templeton et al., 2005). They are evident on young rocks (8 to 128 years) in association with Mn oxides, indicating that Mn(II) oxidizers colonize basalts soon after rock formation. Isolated Mn(II) oxidizers include α- and γ-proteobacteria, *Actinobacteria*, and gram-positive bacteria. The most common Mn(II)-oxidizers cultured are *Pseudoalteromonas* (Thorseth et al., 2001; Templeton et al., 2005) and *Sulfitobacter* (Alpha OCC I in Mason et al. [2007]). Like Fe(II) oxidation, it appears from this work that the ability to oxidize Mn(II) is widespread throughout bacterial lineages. Unlike Fe(II) oxidizers, none of the Mn(II) oxidizers currently isolated are autotrophs. It seems likely that the Mn oxidizers colonize basalts following the initial colonization by autotrophs, including chemolithoautotrophic Fe(II) oxidizers.

Fe oxidation on basalts is clearly an important biogeochemical process there, but it is still unclear why some Fe-oxidizing species appear on basalts and others do not. Fe-oxidizing bacteria that are capable of growing on basalts appear to require a lower but sustainable flux of Fe. It is possible there are two different groups of Fe oxidizers: those that metabolize sulfides (high Fe flux demand) and those that metabolize silicates (low Fe flux demand). Currently, this answer remains elusive because there are few solid data on the physiology of Fe oxidizers. Fe-oxidation rates of up to 0.249 min^{-1} and accounting for up to 71% of Fe oxidation in the environment have been measured in natural freshwater systems (Rentz et al., 2007), but nearly nothing is known of the rates or physiological processes of Fe oxidation in the deep sea. Molecular diagnostics are needed for investigating pathways of Fe oxidation on both sulfides and on silicates.

Culture studies show that bacteria growing on basalts participate in significant release of Si, Fe, and Mn from the rocks (Daughney et al., 2004; Edwards et al., 2004). This shows that the endolithic microbes accelerate mobilization of these elements, and possibly others, from the rocks. In addition, basalt substrate composition (e.g., amount of oxidation) influences what microbes can grow on basalt (Bailey et al., 2009). This result from cultivation studies supports the field work showing different communities on younger versus older rocks, as discussed above (Lysnes et al., 2004; Mason et al., 2009; Santelli et al., 2009).

IN SITU ALTERATION AND WEATHERING EXPERIMENTS: INCUBATION STUDIES

Seafloor Incubation Studies

The rock microbial biotopes that are harbored at and below the seafloor present an enormous challenge for scientists to study. Oceanic crustal rock represents the largest potential epi- and endolithic microbial habitat on the planet, and yet we are only just beginning to understand its microbial ecology, its functional role in global biogeochemical cycles such as C, Fe, Mn, S, and more. To overcome the challenges that are intrinsic to working at and below the bottom of the ocean, scientists have increasingly moved toward in situ experimentation as a means of conducting time-series studies of mineral-microbe interactions. In situ experimentation at and below the seafloor provides powerful means of getting at mechanisms, processes, and functions of microorganisms and their interactions with rocks and minerals. In situ experimentation has become increasingly sophisticated and multifaceted in scope, design, and means of interfacing with relevant downstream data analysis pipelines.

A standard methodology for these types of experiments is to adhere mineral substrates in the form of rock chips to glass slides that are placed at the bottom of the ocean for some time and collected and analyzed later. This allows for multiple lines of analysis (as shown in Color Plate 8). Uses for in situ incubations include devolution of the different pathways of basalt alteration—what the influences of interaction with hydrothermal plumes at the seafloor are, what the influences of differing plume chemistries are, for example, high Fe (e.g., Loihi seamount) versus high S (e.g., EPR 9°N) plume fields, and what the influence of basalt-seawater interaction is. Exposure experiments and incubation studies also allow direct testing of the colonization potential of different mineral substrates.

Surface-colonizing microbes are generally referred to as epilithic, whereas microbes that inhabit pore spaces in rocks are considered endoliths. Examination of natural rocks and minerals at the seafloor considers the entirety of three-dimensional rock-colonizing communities—epi- and endolithic. Studies that have conducted time-series evaluations of colonization of rock and minerals at the seafloor have begun to reveal some important common threads and suggest that microbes colonizing rock at the seafloor have a very strong tendency to select for an endolithic lifestyle, if they can find a suitable feature to colonize. For example, Edwards et al. (2003a) found that microbes preferentially colonized pits and pores on the sulfur and sulfide minerals reacted at the seafloor when they were present. They hypothesized that one of the prominent colonizing microbial groups, lithotrophic Fe-oxidizing bacteria, colonized pits specifically because these locations would allow them, via respiration, to draw down local oxygen concentrations most rapidly, enabling them to establish the conditions needed for their metabolism. Colonization densities of bacteria on the incubated chips of sulfur and sulfide minerals increase as the number of dissolution pits increase and secondary alteration products accumulate (Edwards et al., 2003a). By examining alteration products on different mineralogies deployed within the same incubation vessel, Edwards et al. were able to determine that the majority of alteration products observed had resulted from oxidative dissolution in situ. Colonization densities also appear to correlate with substrate reactivity, suggesting not only that microbes selectively colonize materials conducive to their growth and colonization, but also that abiotic kinetics may play a role in determining this selectivity (Edwards et al., 2003a; Toner et al., 2009).

Further observations demonstrate that preferential colonization of pits is a very common feature of seafloor microbes among diverse taxa. Colonizing microbes on mineral chips incubated at the seafloor show a correlation between microbial colonization and surface features such as polishing grooves and pits, suggesting that microbial attachment is strongly influenced by surface topography (Edwards and Rutenburg, 2001). The size and, more importantly, the shape of microtopographic features on the surface of a mineral can significantly alter the propensity for and specificity of bacterial attachment. Smaller nanometer-size surface features may be more habitable for microbes than larger ones because they increase the area for contact and the binding potential for attachment (Edwards and Rutenburg, 2001). This is highlighted in Color Plate 9, which shows confocal microscopy of stained/hybridized microbes on several polished basalt panels following reaction at the seafloor for 6 to 12 months at Loihi Seamount, Hawaii. Pits and pores are preferentially colonized in all cases. Even small surface features such as polishing scratches are preferentially colonized over planar features on the surfaces. The occurrence of microbial cells on the surfaces also corresponds to localized regions of basalt dissolution and secondary mineral formation (Color Plate 9), suggesting an important role for microbes in alteration processes. Similar observations of a localized feedback between surface features such as scratches and pits and microbial colonization have previously been made based on laboratory studies (Edwards and Rutenburg, 2001) and suggested based on field studies (Taunton et al., 2000; Rogers and Bennett, 2004), but have generally been difficult to directly observe in a spatially controlled fashion.

Subseafloor Incubation Studies

Although incubation studies on seafloor exposed basalts yield an understanding of microbial growth preferences, the majority of basalts are located in the subseafloor, currently making access to them limited to seafloor drilling. Efforts to study subseafloor biosphere basalts (and the sedimentary subseafloor as well) are still in their infancy, but observatories created within drill boreholes that are then sealed off show potential for yielding critical information about the subseafloor environment. Once a casing is fit in a borehole to keep it open and a unit called a CORK (Davis et al., 1992a, 1992b) put in place, sealing off the borehole from overlying seawater, the boreholes can become subseafloor observatories. The first study of the crustal subseafloor biosphere used a sealed CORK observatory on the Juan de Fuca ridge flank to pump in situ crustal fluids to the seafloor, where they were collected for molecular and geochemical analysis (Cowen et al., 2003). This study revealed that the aging formation fluid from the crustal habitat was greatly enriched in NH_4^+, Si, and Mg and depleted in SO_4^{2-}, having a high degree of similarity to venting fluids sampled from a basaltic outcrop a few kilometers away on the presumed flow path (Wheat et al., 2000; Wheat and Mottl, 2000). Supportive of the observed chemistry, 16S rRNA clones for possible chemolithoautotrophic ammonia oxidizers, N_2 fixers (NH_4^+ is a product of N fixation), and SO_4^{2-} reducers were recovered. This first demonstration of the utility of borehole CORKed observatories for microbial studies in the ocean crust revealed the enormous potential that exists for new research in mineral-microbe studies below the bottom of the ocean.

The utility of using borehole subsurface observatories for microbiological studies has advanced in the years since the initial study by Cowen et al. (2003), with additional subseafloor experiments designed and deployed to replicate mineral-microbe studies conducted previously at the seafloor (Edwards et al., 2003a; Toner et al., 2009). In 2004, during an Integrated Ocean Drilling Program Expedition on the Juan de Fuca Ridge (Leg 301), novel in situ microbial colonization chambers were deployed in several boreholes (Fisher et al., 2005). The results of these experiments are still being recovered, but they provide some motivation for further design modification and a broadening of design possibilities. More recently, novel flowthrough colonization experiments, referred to as FLOCS (for FLow-through Osmo Colonization System; Color Plate 10) have been designed and tested (Orcutt et al., 2010). Future borehole observatories are slated to include FLOCS incubation chambers to collect chemical and microbiological samples below the seafloor at other CORKed boreholes in the Atlantic and Pacific Oceans.

CONCLUSION

The basaltic oceanic crust represents the largest and most poorly explored microbiological biome on Earth. The mineral-microbe interactions that occur at rock and mineral surfaces at the bottom of the ocean and below the ocean floor influence changes in mineralogy and speciation of metals and organic matter. Many of these reactions are energy yielding, but the degree to which they are harnessed by lithotrophic microbes is poorly understood.

The deep ocean is also very difficult to access for scientific study in comparison with nearly any other environment that may be of interest to microbiologists. Bottom access to rock samples, in particular, requires a submarine, remotely operated vehicle, or a drilling ship, which may present a daunting hurdle to scientists interested in mineral-microbe interactions on the ocean floor. However, it is important to note that, despite the challenging accessibility, it is fundamentally critical that we develop a more comprehensive understanding of the biogeochemical processes occurring in the seafloor and subseafloor realms, because their potential magnitude of influence may scale to the vast size of these biomes. Hence, despite the challenges presented to studying dark, deep-sea rocky provinces, their value in the Earth system as a whole warrants further consideration. Furthermore, we believe that the recently developed new strategies for conducting time-series studies and conducting active experimentation at and below the seafloor offer powerful means for studying these remote systems, and will offer broad and extensive research results in scientific advancement.

REFERENCES

Alt, J. C., G. J. Davidson, D. A. H. Teagle, and J. A. Karson. 2003. Isotopic composition of gypsum in the Macquarie Island ophiolite: implications for the sulfur cycle and the subsurface biosphere in oceanic crust. *Geology* **31**:549–552.

Bach, W., and K. J. Edwards. 2003. Iron and sulfide oxidation within the basaltic ocean crust: implications for chemolithoautotrophic microbial biomass production. *Geochim. Cosmochim. Acta* **67**:3871–3887.

Bach, W., K. J. Edwards, J. M. Hayes, J. Huber, S. Sievert, and M. Sogin. 2006. Energy in the dark: fuel for life in the deep ocean and beyond. *EOS Trans. AGU* **7**:73, 78.

Bailey, B., A. Templeton, H. Staudigel, and B. M. Tebo. 2009. Utilization of substrate components during basaltic glass colonization by Pseudomonas and Shewanella isolates. *Geomicrobiol. J.* **26**:648–656.

Banerjee, N. R., A. Simonetti, H. Furnes, K. Muehlenbachs, H. Staudigel, L. Heaman, and M. J. Van Kranendonk. 2007. Direct dating of Archean microbial ichnofossils. *Geology* **35**:487–490.

Biddle, J. F., S. Fitz-Gibbon, S. C. Schuster, J. E. Brenchley, and C. H. House. 2008. Metagenomic signatures of the Peru Margin subseafloor biopshere show a genetically distinct environment. *Proc. Natl. Acad. Sci. USA* **105**:10583–10588.

Biddle, J. F., J. S. Lipp, M. A. Lever, K. G. Lloyd, K. B. Sorensen, R. T. Anderson, H. F. Fredricks, M. Elvert, T. J. Kelly, D. P. Schrag, M. Sogin, J. E. Brenchley, A. Teske, C. H. House, and K.-U. Hinrichs. 2006. Heterotrophic archaea dominate sedimentary subsurface ecosystems of Peru. *Proc. Natl. Acad. Sci. USA* **103**:3846–3851.

Brady, P. V., and S. R. Gislason. 1997. Seafloor weathering controls on atmospheric CO_2 and global climate. *Geochim. Cosmochim. Acta* **61**:965–973.

Cogné, J.-P., and E. Humler. 2004. Temporal variation of oceanic spreading and crustal production rates during the last 180 My. *Earth Planet. Sci. Lett.* **227**:427–439.

Cowen, J. P., S. J. Giovannoni, F. Kenig, H. P. Johnson, D. Butterfield, M. S. Rappe, M. Hutnak, and P. Lam. 2003. Fluids from aging ocean crust that support microbial life. *Science* **299**:120–123.

Daughney, C. J., J. P. Rioux, D. Fortin, and T. Pichler. 2004. Laboratory investigation of the role of bacteria in the weathering of basalt near deep sea hydrothermal vents. *Geomicrobiol. J.* **21**:21–31.

Davis, E. E., K. Becker, T. L. Pettigrew, B. Carson, and R. Macdonald. 1992a. CORK: a hydrological sea and downhole observatory for deep-ocean boreholes. *Proc. ODP Init. Rep.* **139**:43–53.

Davis, E. E., D. S. Chapman, M. J. Mottl, W. J. Bentkowski, K. Dadey, C. B. Forster, R. Harris, S. Nagihara, K. Rohr, C. G. Wheat, and M. Whiticar. 1992b. FlankFlux: an experiment to study the nature of hydrothermal circulation in young oceanic crust. *Can. J. Earth Sci.* **29**:925–952.

Edmond, J. M., C. Measures, R. E. McDuff, L. H. Chan, and C. B. Grant. 1979. Ridge crest hydrothermal activity and the balances of the major and minor elements in the ocean: the Galapagos data. *Earth Planet. Sci. Lett.* **46**:1–18.

Edwards, K. J., W. Bach, and T. M. McCollom. 2005. Geomicrobiology in oceanography: microbe-mineral interactions at and below the seafloor. *Trends Microbiol.* **13**:449–456.

Edwards, K. J., W. Bach, T. M. McCollom, and D. R. Rogers. 2004. Neutrophilic iron-oxidizing bacteria in the ocean: their habitats, diversity, and roles in mineral deposition, rock alteration, and biomass production in the deep-sea. *Geomicrobiol. J.* **21**:393–404.

Edwards, K. J., T. M. McCollom, H. Konishi, and P. R. Buseck. 2003a. Seafloor bioalteration of sulfide minerals: results from in-situ incubation studies. *Geochim. Cosmochim. Acta* **67**:2843–2856.

Edwards, K. J., D. R. Rogers, C. O. Wirsen, and T. M. McCollom. 2003b. Isolation and characterization of novel psychrophilic, neutrophilic, Fe-oxidizing, chemolithoautotrophic alpha- and, gamma-Proteobacteria from the deep sea. *Appl. Environ. Microbiol.* **69**:2906–2913.

Edwards, K. J., and A. D. Rutenburg. 2001. Microbial response to surface microtopography: the role of metabolism in localized mineral dissolution. *Chem. Geol.* **180**:19–32.

Ehrhardt, C. J., R. M. Haymon, M. G. Lamontagne, and P. A. Holden. 2007. Evidence for hydrothermal Archaea within the basaltic flanks of the East Pacific Rise. *Environ. Microbiol.* **9**:900–912.

Einen, J., I. H. Thorseth, and L. Ovreas. 2008. Enumeration of Archaea and Bacteria in seafloor basalt using real-time quantitative PCR and fluorescence microscopy. *FEMS Microbiol. Lett.* **282**:182–187.

Emerson, D., J. A. Rentz, T. G. Liburn, R. E. Davis, H. Aldrich, C. Chan, and C. L. Moyer. 2007. A novel lineage of Proteobacteria involved in formation of marine Fe-oxidizing microbial mat communities. *PLoS ONE* **2**:e677.

Fisher, A. T. 1998. Permeability within basaltic oceanic crust. *Rev. Geophys.* **36**:143–182.

Fisher, A. T., and K. Becker. 2000. Channelized fluid flow in oceanic crust reconciles heat-flow and permeability data. *Nature* **403**:71–74.

Fisher, A. T., C. G. Wheat, K. Becker, E. E. Davis, H. W. Jannasch, D. Schroeder, R. Dixon, T. L. Pettigrew, R. Meldrum, R.

Macdonald, M. Nielsen, M. R. Fisk, J. P. Cowen, W. Bach, and K. J. Edwards. 2005. Scientific and technical design and deployment of long-term subseafloor observatories for hydrogeological and related experiments, IODP expedition 301, eastern flank of Juan de Fuca Ridge. In A. T. Fisher (ed.), *Proc. IODP*. doi:10.2204/iodp.proc.2301.2005. Integrated Ocean Drilling Program, College Station, TX.

Fisk, M. R., S. J. Giovannoni, and I. H. Thoreth. 1998. Alteration of oceanic volcanic glass: textural evidence of microbial activity. *Science* **281**:978–980.

Giovannoni, S. J., M. R. Fisk, T. D. Mullins, and H. Furnes. 1996. Genetic evidence for endolithic microbial life colonizing basaltic glass/seawater interfaces. *Proc. Ocean Drilling Program, Scientific Results* **148**:207–214.

He, Z. L., T. J. Gentry, C. W. Schadt, L. Y. Wu, J. Liebich, S. C. Chong, Z. J. Huang, W. M. Wu, B. H. Gu, P. Jardine, C. Criddle, and J. Zhou. 2007. GeoChip: a comprehensive microarray for investigating biogeochemical, ecological, and environmental processes. *ISME J.* **1**:67–77.

Huber, J. A., H. P. Johnson, D. A. Butterfield, and J. A. Baross. 2006. Microbial life in ridge flank crustal fluids. *Environ. Microbiol.* **8**:88–99.

Jercinovic, M. J., and R. C. Ewing. 1992. Corrosion of geological and archaeological glasses, p. 330–371. In D. E. Clark and B. K. Zoitos (ed.), *Corrosion of Glass, Ceramics, and Ceramic Superconductors*. Noyes Publications, Park Ridge, NJ.

Johnson, H. P., and M. J. Priuis. 2003. Fluxes of fluid and heat from the oceanic crustal reservoir. *Earth Planet. Sci. Lett.* **216**:565–574.

Kato, S., K. Yanagawa, M. Sunamura, Y. Takano, J. Ishibashi, T. Kakegawa, M. Utsumi, T. Yamanaka, T. Toki, T. Noguchi, K. Kobayashi, A. Moroi, H. Kimura, Y. Kawarabayasi, K. Marumo, T. Urabe, and A. Yamagishi. 2009. Abundance of Zetaproteobacteria within crustal fluids in back-arc hydrothermal fields of the Southern Mariana Trough. *Environ. Microbiol.* **11**:3210–3222.

Kaye, J. Z., and J. A. Baross. 2000. High incidence of halotolerant bacteria in Pacific hydrothermal-vent and pelagic environments. *FEMS Microbiol. Ecol.* **32**:249–260.

Lipp, J. S., Y. Morono, H. Inagaki, and K.-U. Hinrichs. 2008. Significant contribution of Archaea to extant biomass in marine subsurface sediments. *Nature* **454**:991–994.

Lysnes, K., I. H. Thorseth, B. O. Steinsbu, L. Ovreas, T. Torsvik, and R. B. Pedersen. 2004. Microbial community diversity in seafloor basalt from the Arctic spreading ridges. *FEMS Microbiol. Ecol.* **50**:213–230.

Mason, O. U., C. A. Di Meo-Savoie, J. D. Van Nostrand, J. Z. Zhou, M. R. Fisk, and S. J. Giovannoni. 2009. Prokaryotic diversity, distribution, and insights into their role in biogeochemical cycling in marine basalts. *ISME J.* **3**:231–242.

Mason, O. U., U. Stingl, L. J. Wilhelm, M. M. Moeseneder, C. A. Di Meo-Savoie, M. R. Fisk, and S. J. Giovannoni. 2007. The phylogeny of endolithic microbes associated with marine basalts. *Environ. Microbiol.* **9**:2539–2550.

Orcutt, B., C. G. Wheat, and K. J. Edwards. 2010. Subseafloor ocean crust microbial observatories: development of FLOCS (FLow-through Osmo Colonization System) and evaluation of borehole construction materials. *Geomicrobiol. J.* **27**:143–157.

Rassa, A. C., S. M. McAllister, S. A. Safran, and C. L. Moyer. 2009. Zeta-Proteobacteria dominate the colonization and formation of microbial mats in low-temperature hydrothermal vents at Loihi Seamount, Hawaii. *Geomicrobiol. J.* **26**:623–638.

Rathsack, K., E. Stackebrandt, J. Reitner, and G. Schumann. 2009. Microorganisms isolated from deep sea low-temperature influenced oceanic crust basalts and sediment samples collected along the Mid-Atlantic Ridge. *Geomicrobiol. J.* **26**:264–274.

Rentz, J. A., C. Kraiya, G. W. Luther, and D. Emerson. 2007. Control of ferrous iron oxidation within circumneutral microbial iron mats by cellular activity and autocatalysis. *Environ. Sci. Technol.* **41**:6084–6089.

Rogers, D. R., C. M. Santelli, and K. J. Edwards. 2003. Geomicrobiology of deep-sea deposits: estimating community diversity from low-temperature seafloor rocks and minerals. *Geobiology* **1**:109–117.

Rogers, J. R., and P. C. Bennett. 2004. Mineral stimulation of subsurface microorganisms: release of limiting nutrients from silicates. *Chem. Geol.* **203**:91–108.

Rouxel, O., S. H. Ono, J. Alt, D. Rumble, and J. Ludden. 2008. Sulfur isotope evidence for microbial sulfate reduction in altered oceanic basalts at ODP Site 801. *Earth Planet. Sci. Lett.* **268**:110–123.

Santelli, C. M., V. P. Edgcomb, W. Bach, and K. J. Edwards. 2009. The diversity and abundance of bacteria inhabiting seafloor lavas positively correlate with rock alteration. *Environ. Microbiol.* **11**:86–98.

Santelli, C. M., B. N. Orcutt, E. Banning, W. Bach, C. L. Moyer, M. L. Sogin, H. Staudigel, and K. J. Edwards. 2008. Abundance and diversity of microbial life in ocean crust. *Nature* **453**:653–656.

Schippers, A., L. N. Neretin, J. Kallmeyer, T. G. Ferdelman, B. A. Cragg, R. J. Parkes, and B. B. Jorgensen. 2005. Prokaryotic cells of the deep sub-seafloor biosphere identified as living bacteria. *Nature* **433**:861–865.

Staudigel, H., H. Furnes, N. McLoughlin, N. R. Banerjee, L. B. Connel, and A. Templeton. 2008. 3.5 billion years of glass bioalteration: volcanic rocks as a basis for microbial life? *Earth-Sci. Rev.* **89**:156–176.

Staudigel, H., and S. R. Hart. 1983. Alteration of basaltic glass: mechanisms and significance for the oceanic crust-seawater budget. *Geochim. Cosmochim. Acta* **47**:337–350.

Stroncik, N. A., and H. U. Schmincke. 2001. Evolution of palagonite: crystallization, chemical changes, and element budget. *Geochem. Geophy. Geosys.* **2**:2000GC000102.

Stroncik, N. A., and H. U. Schmincke. 2002. Palagonite—a review. *Int. J. Earth Sci.* **91**:680–697.

Sudek, L. A., A. S. Templeton, B. M. Tebo, and H. Staudigel. 2009. Microbial ecology of Fe (hydr)oxide mats and basaltic rock from Vailulu'u Seamount, American Samoa. *Geomicrobiol. J.* **26**:581–596.

Taunton, A. E., S. A. Welch, and J. F. Banfield. 2000. Microbial controls on phosphate and lanthanide distributions during granite weathering and soil formation. *Chem. Geol.* **169**:371–382.

Templeton, A. S., H. Staudigel, and B. M. Tebo. 2005. Diverse Mn(II)-oxidizing bacteria isolated from submarine basalts at Loihi Seamount. *Geomicrobiol. J.* **22**:127–139.

Thorseth, I. H., H. Furnes, and O. Tumyr. 1991. A textural and chemical study of Icelandic palagonite of varied composition and its bearing on the mechanisms of the glass-palagonite transformation. *Geochim. Cosmochim. Acta* **55**:731–749.

Thorseth, I. H., T. Torsvik, V. Torsvik, F. L. Daae, R. B. Pedersen, and the Keldysh-98 Scientific Party. 2001. Diversity of life in ocean floor basalt. *Earth Planet. Sci. Lett.* **194**:31–37.

Toner, B. M., S. C. Fakra, S. J. Manganini, C. M. Santelli, M. A. Marcus, J. W. Moffett, O. Rouxel, C. R. German, and K. J. Edwards. 2009. Preservation of iron(II) at hydrothermal vents within carbon-rich matrices. *Nat. Geosci.* doi: 10.1038/NGEO1433.

Wheat, C. G., H. W. Jannasch, J. N. Plant, C. L. Moyer, F. J. Sansone, and G. M. McMurtry. 2000. Continuous sampling of hydrothermal fluids from Loihi Seamount after the 1996 event. *J. Geophys. Res.-Solid Earth* **105**:19353–19367.

Wheat, C. G., and M. J. Mottl. 2000. Composition of pore and spring waters from Baby Bare: global implications of geochemical fluxes from a ridge flank hydrothermal system. *Geochim. Cosmochim. Acta* **64**:629–642.

Wolff-Boenisch, D., S. R. Gislason, E. H. Oelkers, and C. V. Putnis. 2004. The dissolution rates of natural glasses as a function of their composition at pH 4 and 10.6 and temperatures from 25 to 74°C. *Geochim. Cosmochim. Acta* **68**:4843–4858.

Zhou, Z., and W. S. Fyfe. 1989. Palagonitization of basaltic glass from DSDP site 335, Leg 37: textures, chemical composition, and mechanism of formation. *Am. Mineralogist* **74**:1045–1053.

MICROBIAL TRANSFORMATIONS OF ARSENIC IN THE SUBSURFACE

*Jonathan R. Lloyd, Andrew G. Gault,
Marina Héry, and Jean D. MacRae*

5

INTRODUCTION

Arsenic is a highly toxic element that supports a surprising range of biogeochemical transformations mediated by prokaryotes and eukaryotes. The biochemical basis of these microbial-metalloid interactions is described, with an emphasis on energy-yielding redox biotransformations that cycle between the As(V) and As(III) oxidation states. A particular focus is the microbial reduction of As(V) that is sorbed onto subsurface sediments, and the subsequent mobilization of As(III) into water that is abstracted for drinking and irrigation. The health of tens of millions is threatened worldwide through the widespread consumption of these arsenic-contaminated groundwaters abstracted from aquifers, most notably in Southeast Asia. The organisms implicated in these subsurface processes are described, alongside the mechanisms involved and other microbial processes that may be able to help attenuate the release of As into groundwaters.

MICROBIAL CYCLING OF ARSENIC

Although arsenic is highly toxic to humans, and indeed most other forms of life, some microorganisms have evolved to tolerate relatively high concentrations of the metalloid. Some specialist prokaryotes can even thrive on the element, using it as a source of energy for growth. For example, in Californian alkali soda lakes such as Mono and Searles Lakes, the combination of alkali pH (pH 9.8), high carbonate/bicarbonate concentrations, and naturally elevated concentrations of the metalloid from hydrothermal waters can result in millimolar (high parts per million) concentrations of dissolved arsenic that sustain a diversity of specialist microorganisms adapted to using the arsenic oxyanions in energy-yielding redox reactions (Lloyd and Oremland, 2006; Oremland et al., 2004). However, even at the very low concentrations of arsenic found in most other marine and freshwater environments, microorganisms can transform or even accumulate arsenic to concentrations many times those encountered in the environment they inhabit. Together with inorganic and physical processes, these biological processes constitute

---

Jonathan R. Lloyd, School of Earth, Atmospheric and Environmental Sciences and Williamson Research Centre for Molecular Environmental Science, University of Manchester, Manchester M13 9PL, United Kingdom. *Andrew G. Gault*, Department of Geological Sciences and Geological Engineering, Queen's University, Kingston, ON K7L 3N6, Canada. *Marina Héry*, HydroSciences UMR 5569 CNRS—Universités Montpellier I and II—IRD, Place Eugene Bataillon, CC MSE, 34095 Montpellier cedex 5, France. *Jean D. MacRae*, Department of Civil and Environmental Engineering, University of Maine, Orono, ME 04469-5711.

the global arsenic cycle described, for example, in Mukhopadhyay et al. (2002). In particular, the redox chemistry of the metalloid makes it a useful energy source to sustain prokaryotic life. Therein lies the potential for microorganisms to play a critical (and potentially lethal) role in controlling the solubility of arsenic, which is closely linked to its oxidation state in the subsurface and leads to the widespread contamination of aquifers that provide drinking and irrigation water to tens of millions of people worldwide.

There are four oxidation states of arsenic: −III, 0, +III, and +V. However, the predominant forms of inorganic arsenic are +V (arsenate; $H_2AsO_4^-$ and $HAsO_4^{2-}$), which sorbs efficiently to a range of subsurface minerals, and +III (arsenite; $H_3AsO_3^0$ and $H_2AsO_3^-$), which is comparatively more mobile. Under anaerobic conditions, specialist "dissimilatory arsenate-reducing prokaryotes" are able to respire arsenate as the electron acceptor in place of oxygen, reducing it to As(III). Although pH and concentration dependent, the oxidation/reduction potential of As(V)/As(III) is between +60 and +135 mV (Oremland and Stolz, 2003), sufficient to support growth when organic matter or sulfide is supplied as the electron donor. As(III) is also a suitable electron donor for microbial respiratory processes, with electrons from the oxidation of As(III) to As(V) passed to a suitable electron acceptor such as oxygen or nitrate under anaerobic conditions. Finally, a wide diversity of microorganisms have also evolved an energy-requiring detoxification process catalyzed by the *ars* operon, linked to the intracellular reduction of As(V) by the ArsC protein and its efflux as As(III). A synopsis of these microbial processes is illustrated in Fig. 1, with the underlying biochemistry shown and described in more detail in the next section of this chapter.

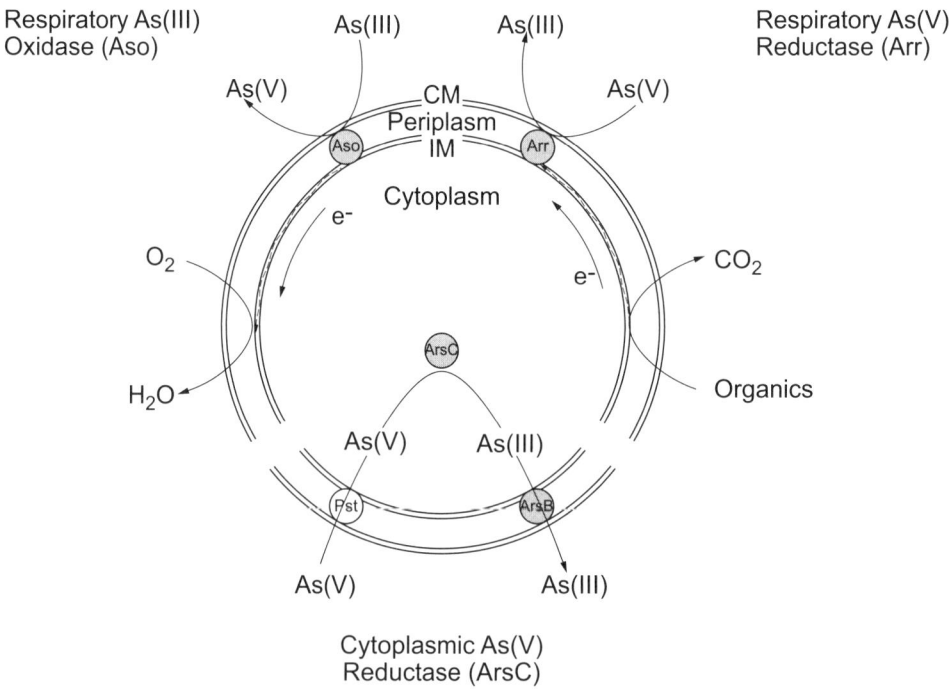

FIGURE 1 Biochemical transformations of arsenic oxyanions by microbial cells.

THE BIOCHEMISTRY OF MICROBIAL ARSENIC TRANSFORMATIONS

Microbial Resistance to As(V) via the Arsenic Operon

Arsenate is a chemical surrogate for phosphate and can therefore enter the microbial cell via transporters meant for the uptake of this essential nutrient. Once within the cell it can interfere with phosphate-based energy-generating processes, inhibiting, for example, oxidative phosphorylation. Arsenite, on the other hand, enters via a different route (aquaglycerolporins) and targets a broader range of cellular processes, binding to the thiol groups in important cellular proteins such as pyruvate dehydrogenase and 2-oxo-glutatarate dehydrogenase. Microorganisms have evolved multiple strategies to protect themselves from toxic arsenic. For example, fungi can use methylation as a detoxification strategy, producing monomethylarsonic acid or dimethylarsinic acid. Prokaryotes (including representatives from both the *Archaea* and *Bacteria*) can also produce volatile methylated arsines, removing arsenic from the local environment. An alternative strategy used by bacteria and yeast (e.g., *Saccharomyces cerevisiae*) is based upon the "ArsC" arsenate reductase protein. The gene encoding this enzyme, along with those for other proteins required for arsenic detoxification, is often located on plasmids, making them highly amenable to genetic study and subsequent protein crystallography studies, e.g., for ArsC proteins (Mukhopadhyay et al., 2002). This type of resistance determinant is very common in the prokaryotic world; more than 100 *ars* operons have been sequenced (Mukhopadhyay et al., 2002), and this number is growing steadily with the increase in the number of genomic sequences deposited in genetic databases. ArsC is a small (13 to 16 kDa) protein that is found in the cytoplasm of the microbial cell and mediates the reduction of arsenate to arsenite, with glutaredoxin, glutathione, or thioredoxin supplying the reducing power (Silver and Phung, 2005). The toxic arsenite is then excreted by an energy-requiring ATP-dependent efflux pump ArsB (Fig. 1). However, despite its wide distribution throughout the prokaryotic world, at present, there is little evidence linking ArsC activity directly to redox transformations of arsenic in the subsurface, especially the reduction and mobilization of the metalloid into groundwaters.

Gaining Energy from Arsenic: the Dissimilatory Reduction of As(V) under Anaerobic Conditions

Although the ArsC-mediated detoxification system requires energy in the form of ATP to function, some specialist bacteria are able to obtain energy for growth through the "dissimilatory" reduction of As(V). Two closely related representatives of the ε-proteobacterial phylogenetic group (*Sulfurospirillum arsenophilum* and *S. barnesii*) were the first dissimilatory arsenate-reducing bacteria identified in the mid-1990s, and representatives have since been identified in other phylogenetic groups, including the γ- and δ-proteobacteria, the low GC "gram-positive" bacteria, thermophilic *Eubacteria* and *Crenarchaeota* (Oremland and Stolz, 2003). Of particular relevance to the mobilization of arsenic in aquifer sediments is the observation that organisms that can respire soluble As(V) are also able to reduce the pentavalent metalloid when sorbed to a range of mineral phases. This in turn results in significant mobilization of As(III), which sorbs less strongly to surfaces. Although sharing the common ability to obtain energy for growth through the reduction of sorbed or soluble As(V), the overall metabolic diversity of these dissimilatory arsenate-respiring organisms is surprisingly wide. Certainly, given the low concentrations of As(V) in most environments, compared with other electron acceptors, a high level of respiratory diversity may be crucial to the survival of these organisms. Indeed, to date, only one obligate arsenate-reducing prokaryote has been identified (strain MLMS-1 isolated from Mono Lake [Hoeft et al., 2004]). Other electron acceptors used by these organisms are strain specific, but include selenate, nitrate, nitrite, fumarate, Fe(III), thiosulfate, elemental sulfur, dimethyl sulfoxide, and trimethylamine oxide. They can also use a range of electron donors, including hydrogen, sulfide, acetate, formate, pyruvate, butyrate, citrate, succinate, fumarate, glucose, and aromatics such as benzoate.

Despite the environmental importance of the process, the mechanism of As(V) reduction by dissimilatory arsenate-reducing bacteria has not been studied in the same depth as the ArsC detoxification system described above. The first "respiratory arsenate reductase" characterized was from the gram-negative bacterium *Chrysiogenes arsenatis*, isolated from gold mine wastewater. The protein consisted of 87- and 29-kDa subunits and is related to the dimethyl sulfoxide family of mononuclear molybdenum-containing enzymes (Krafft and Macy, 1998). Similar proteins (and the corresponding genes designated *arrA* and *arrB*) were subsequently characterized in *Bacillus selenitireducens* (Afkar et al., 2003) and *Shewanella* sp. strain ANA-3 (Saltikov and Newman, 2003). This poses a considerable conundrum because all Arr proteins described to date reside in the periplasm and are unlikely to access sorbed As(V) directly. Hence, there is a need for more detailed research into the precise mechanism of reduction of sorbed arsenate in the subsurface; see Fig. 2 for an overview of the potential mechanisms. However, genetic studies using *Shewanella* sp. strain ANA-3 have made it possible to study the interplay between the ArsC detoxification (an intracellular process) and ArrA respiratory systems in laboratory cultures, which both potentially result in As(V) reduction to As(III) (Saltikov et al., 2003). Strains that were able to synthesize the ArrA protein were able to respire both soluble As(V) and also As(V) sorbed onto hy-

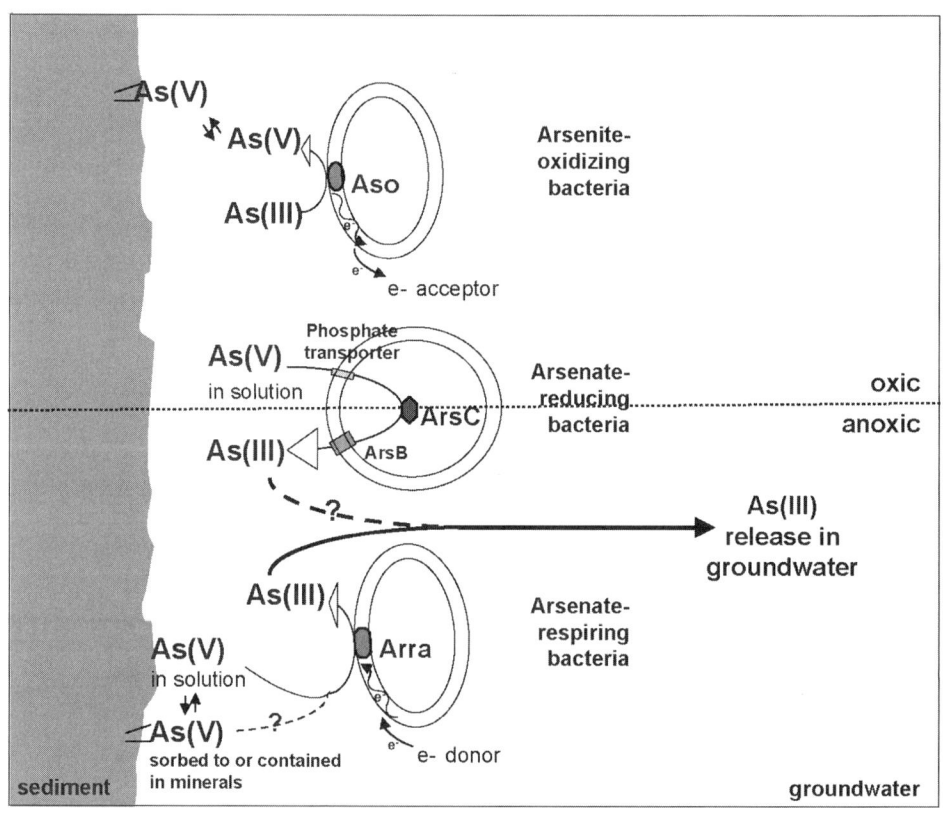

FIGURE 2 Microbial interactions with arsenic in subsurface sediments.

drous ferric oxide (Malasarn et al., 2004), confirming the earlier work discussed above by Zobrist and Oremland (Zobrist et al., 2000). In contrast, *Shewanella* sp. CN-8, a detoxifying arsenate reducer carrying the *ars* operon that cannot respire arsenate because it lacks the *arrA* and *arrB* genes, reduced soluble As(V) but not arsenate sorbed onto Fe(III) oxides (Langner and Inskeep, 2000), supporting the hypothesis of Arr-mediated (but not ArsC-mediated) reduction and mobilization of sorbed arsenate in sediments. Studies with wild type and mutants of *Shewanella* sp. ANA-3 have been extended recently to examine As desorption from ferrihydrite- and goethite-coated sands presorbed with As(V), providing additional evidence for a more important role for Arr in arsenic mobilization compared with Fe(III) reduction (Tufano et al., 2008). Finally, PCR primers have also been developed to target *arrA* genes in environmental samples, e.g., Malasarn et al. (2004) and Song et al. (2009), making it possible to retrieve *arrA* genes from species present in the environment, but not currently available in culture. For example, this approach has been used successfully to investigate the diversity of arsenate-respiring bacteria in Mono Lake, California (Kulp et al., 2006, 2007) and Cambodian sediments (Lear et al., 2007) (see "Microbially driven mobilization of arsenic in aquifers" below). Such molecular tools also offer the potential to quantify active As(V)-reducing bacteria through real-time quantitative PCR techniques, which may give a relatively rapid indication of the rates of arsenic reduction (and potentially mobilization) in field samples.

Closing the Arsenic Cycle: the Oxidation of As(III)

Finally, arsenite-oxidizing bacteria are also known to couple the oxidation of As(III) to the reduction of either molecular oxygen or nitrate, the latter under anaerobic conditions. As(III) oxidation is also catalyzed by a wide range of microorganisms; more than 30 strains from at least nine different genera have been identified so far (Oremland and Stolz, 2003), including chemolithoautotrophic examples that conserve energy for growth via arsenite oxidation, using some of this energy to fix CO_2 for the synthesis of cellular material. A comparatively well-studied example is strain NT-26, which was isolated from a gold mine in the Northern Territory of Australia and shown to be related to soil bacteria in the *Agrobacterium/Rhizobium* branch of the α-proteobacteria (Santini et al., 2000). A fairly recent review has also given an excellent overview of the genetic basis of arsenite oxidation by *Alcaligenes faecalis*, based on the genome sequence for this soil bacterium (Silver and Phung, 2005). The first organism (*Marinobacter santoriniensis*) shown unequivocally to gain energy from either the aerobic oxidation of As(III) or the anaerobic reduction of As(V) has also been described recently (Handley et al., 2009). It was isolated from hydrothermal sediments of the Greek island of Santorini, and although it contains the genes required for the arsenite oxidase (*asoA* and *asoB)*, the genes thought necessary for the dissimilatory reduction of As(V) are absent from the sequenced genome, suggesting the utilization of novel genes and corresponding proteins for As(V) respiration (J. R. Lloyd and K. M. Handley, unpublished data). There are indeed surprising similarities between the respiratory arsenate reductases of dissimilatory arsenate-reducing prokaryotes and respiratory arsenite oxidases; both are molybdoproteins comprising two subunits. Thus, it is conceivable that, in *M. santoriniensis* and potentially other organisms, a single biochemical system could catalyze both oxidative and reductive transformations of arsenic oxyanions. Indeed, the genome of *Alkalilimnicola ehrlichii*, an organism capable of anaerobic chemolithoautotrophic growth by coupling the oxidation of As(III) to the reduction of nitrate and carbon dioxide, lacks the As(III) oxidase Aox, but does possess the operons for two putative respiratory arsenate reductases (Arr) (Richey et al., 2009). Only one of the homologs is expressed under chemolithoautotrophic conditions and exhibits both arsenite oxidase and arsenate reductase activity. In addition, the same study showed that the respiratory arsenate reductases of *Alkaliphilus oremlandii* and *Shewanella* sp. strain ANA-3 can also function as As(III) oxidases in vitro (Richey et al., 2009).

MICROBIALLY DRIVEN MOBILIZATION OF ARSENIC IN AQUIFERS

Contamination of groundwater from arsenic naturally occurring in the subsurface poses a global public health crisis in countries including Mexico, China, Hungary, Argentina, Chile, Cambodia, India, and Bangladesh (Smith et al., 2000). For example, 28 to 62% of the 125 million inhabitants of Bangladesh are at risk of arsenic poisoning from water abstracted from the subsurface and used for drinking, cooking, and irrigation. Long-term exposure to arsenic can lead to scaling of the skin, circulatory and nervous system disorders, and skin, lung, and bladder cancers (Smedley and Kinniburgh, 2002). In West Bengal and Bangladesh where the problem has received the most attention, the aquifer sediments are derived from weathered materials from the Himalayas. Arsenic typically occurs in concentrations of 2 to 100 ppm in these sediments, much of it sorbed onto a variety of mineralogical hosts including hydrated ferric oxides, phyllosilicates, and sulfides (Nickson et al., 2000; Smedley and Kinniburgh, 2002). The mechanism of arsenic release from these sediments has been a topic of intense debate, and microbial processes such as those described above (and shown in Fig. 2) as well as complementary/competing chemical processes have been invoked (Akai et al., 2004; Chowdhury et al., 1999; Das et al., 1996; Nickson et al., 1998; Oremland and Stolz, 2003). The oxidation of arsenic-rich pyrite has been proposed as one possible mechanism (Chowdhury et al., 1999; Das et al., 1996), while other studies have suggested that the reductive dissolution of arsenic-rich Fe(III) oxyhydroxides deeper in the aquifer may lead to the release of arsenic into the groundwater (Harvey et al., 2002; Nickson et al., 1998, 2000; Smedley and Kinniburgh, 2002). Additional factors that may add further complication to potential arsenic-release mechanisms from sediments include the predicted mobilization of sorbed arsenic by phosphate generated from the intensive use of fertilizers (Acharyya et al., 1999), by carbonate (Appelo et al., 2002) produced via microbial metabolism (Harvey et al., 2002), or by changes in the sorptive capacity of ferric oxyhydroxides (Smedley and Kinniburgh, 2002).

However, microbially mediated reduction of assemblages comprising arsenic (most likely as arsenate) sorbed to ferric oxyhydroxides has gained consensus as the dominant mechanism for the mobilization of arsenic into these groundwaters (Akai et al., 2004; Islam et al., 2004; Nickson et al., 2000; Smedley and Kinniburgh, 2002; Van Geen et al., 2004), and the acceptance of this mechanism of arsenic release has paralleled advances in the microbiology of organisms that can respire sorbed As(V) described above. For example, an early microcosm-based study from our group in Manchester provided direct evidence for the role of indigenous metal-reducing bacteria in the formation of toxic, mobile As(III) in sediments from the Ganges Delta (Islam et al., 2004). This study showed that the addition of acetate to anaerobic sediments, as a proxy for organic matter and a potential electron donor for metal reduction, resulted in stimulation of microbial reduction of Fe(III) followed by As(V) reduction and the subsequent release of As(III), presumably by As(V)-respiring bacteria that were previously respiring Fe(III). Microbial communities responsible for metal reduction and As(III) mobilization in the stimulated anaerobic sediment were analyzed using molecular (PCR) and cultivation-dependent techniques. Both approaches confirmed an increase in numbers of metal-reducing bacteria, principally *Geobacter* species. However, subsequent studies suggested that *Geobacter* strains that were available in culture at that time did not possess the *arrA* genes required to support the reduction of sorbed As(V) and the mobilization of As(III). Indeed, in *Geobacter sulfurreducens*, a strain lacking the biochemical machinery for

As(V) reduction, Fe(II) minerals formed during respiration on Fe(III) have proved to be potent sorbents for arsenic, preventing mobilization of arsenic during active iron reduction (Islam et al., 2005). Incorporation of As(V) into the Fe(II)-bearing mineral phase is even possible, with synchrotron studies showing that the pentavalent metalloid can substitute into magnetite formed during respiration of Fe(III) oxyhydroxides (Coker et al., 2006). However, the genomes of at least two newly isolated *Geobacter* species (*G. uraniireducens* and *G. lovleyi*) do contain putative *arrA* genes; and, interestingly, genes closely related to the *G. uraniireducens* and *G. lovleyi* putative *arrA* gene sequences have been identified recently in Southeast Asian sediments stimulated for Fe(III) and As(V) reduction in several studies from our group (see below). Thus, some *Geobacter* species could play a role in arsenate reduction and release of As(III) from Southeast Asian sediments, perhaps alongside other well-known arsenate-reducing bacteria, including *Sulfurospirillum* species that have also been detected using molecular PCR-based techniques (Lear et al., 2007; Rowland et al., 2009). Recent advances in these areas will now be described.

Microbial Ecology of Arsenic-Impacted Aquifers: Hunting for the Organisms That Mobilize Arsenic

Although the biogeochemical conditions that promote microbial arsenic mobilization are becoming clearer, it remains a major challenge to identify the organisms that have the potential to cause the reduction and mobilization of the metalloid among the complex microbial communities that exist in the subsurface. One approach that has recently proved useful is the application of stable isotope probing, which can link the active fraction of a microbial community to a particular biogeochemical process. Here, sediments are supplemented with a ^{13}C-labeled substrate, and the components of the microbial community that assimilate the substrate are identified by PCR-based analysis of the "heavy" labeled DNA or RNA separated from unlabeled "light" nucleic acids using an ultracentrifuge (Radajewski et al., 2000). This technique has been used to identify the functional components of several microbial processes (Neufeld et al., 2007), and has been used recently to identify active As(V)-respiring bacteria in Cambodian aquifer sediments (Lear et al., 2007) implicated in the reductive mobilization of arsenic (Rowland et al., 2007). ^{13}C-Labeled acetate was used as a proxy for organic matter in these experiments, promoting the reduction of the As(V) present naturally in the sediments, concomitant with the detection of 16S rRNA genes affiliated with the known arsenate-respiring bacteria *Desulfotomaculum* sp. and *Desulfosporosinus* sp. In the presence of 10 mM added As(V), most of which was associated with the mineral phases in the microcosms, an organism closely related to the arsenate-reducing organism *Sulfurospirillum* strain NP4 (MacRae et al., 2007) was identified, which was also closely related to clones identified previously in West Bengal sediments associated with high arsenic concentrations. Functional gene analysis of sediments amended with ^{13}C-labeled acetate and As(V) targeted the As(V) respiratory reductase gene (*arrA*) using highly specific primers, and identified gene sequences most closely related to those found in *S. barnesii* and *G. uraniireducens* (see the phylogenetic tree in Fig. 3 that includes published and unpublished sequences retrieved by our group in a range of studies using Southeast Asian aquifer sediments). Neither *arrA* nor 16S rRNA genes affiliated with known arsenate-respiring bacteria could be detected in the initial sediment, before incubation of the microcosms, presumably because they were below the threshold of detection.

Although stable isotope-probing experiments have proved useful in identifying anaerobic organisms that are potentially involved in the reductive mobilization of arsenic, an obvious limitation of this approach is the need to manipulate the conditions in laboratory incubations through the addition of labeled substrates, potentially altering the geochemical matrix of the sediment. This contrasts with the approach used in a recent complementary study where sediments collected from depths

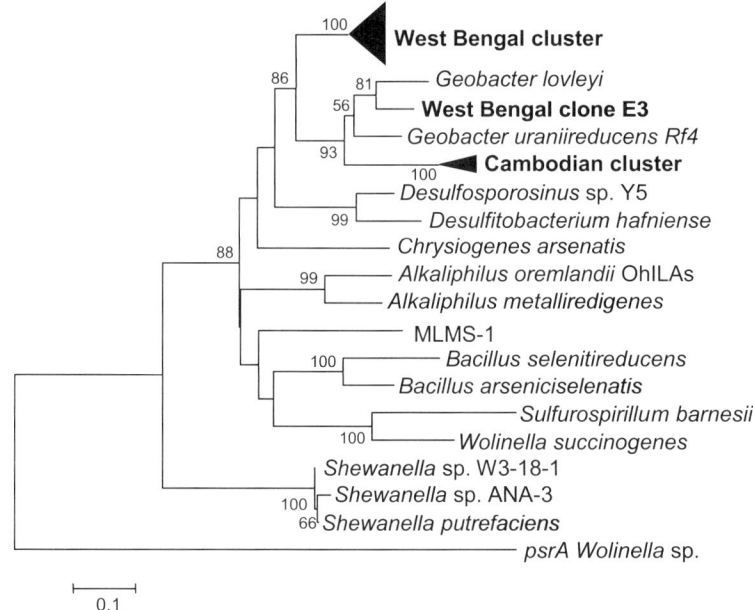

FIGURE 3 ArrA phylogenetic tree based on 203 amino acid sequences using the neighbor-joining method and the Dayhoff model. The "Cambodian cluster" contains sequences derived from *arrA* clones from stable isotope-probing experiments using ^{13}C-labeled acetate (Lear et al., 2007); the "West Bengal cluster" contains sequences derived from *arrA* clones from low organic carbon sediments supporting As(V) reduction and mobilization (Héry et al., 2010).

of 8 to 30 meters at an arsenic "hotspot" in the Nadia district in West Bengal (Bandyopadhyay, 2002) were incubated in the absence of an exogenous carbon source and added arsenate (Rowland et al., 2007). Analyses of the sediments showed the presence of indigenous organics, including petroleum compounds previously hypothesized to play a role in promoting the anaerobic metabolism of metals (Rowland et al., 2007, 2009). The rates and extent of As(III) release and Fe(III) reduction were quantified alongside changes in bacterial community structure. Again, 16S rRNA analyses suggested a potential role of species from the genera *Sulfurospirillum* and *Geobacter* in the respiration of Fe(III), the reductive mobilization of arsenic and the oxidation of organic matter, including in this case petroleum compounds (Rowland et al., 2009). Thus, there are converging lines of evidence that suggest organisms affiliated with *Sulfurospirillum* and *Geobacter* may play a role in the mobilization of arsenic through the activation of genes that encode respiratory arsenate-reducing enzyme systems. The ability of *Sulfurospirillum* species to respire As(V), leading to As(III) mobilization, has been well documented (Zobrist et al., 2000); however, *Geobacter* species have yet to be shown to catalyze this transformation despite accumulating molecular evidence. For this reason, experiments have been conducted recently to determine whether new *Geobacter* isolates that potentially encode As(V) respiratory reductases can indeed reduce and mobilize the metalloid in axenic culture.

Arsenate Reduction and Mobilization by Pure Cultures of *G. uraniireducens*: Implications for Aquifer Biogeochemistry

Although multiple lines of evidence described above point toward a potential role for *Geobacter* species in the mobilization of arsenic, perhaps by direct reduction of sorbed As(V) to the more mobile As(III), the reduction of arsenate by any *Geobacter* species has yet to be reported formally. Because *arrA* genes affiliated with the recently sequenced genome of *G. uraniireducens* were detected by stable isotope probing of sediments from an arsenic-rich aquifer in Cambodia that exhibited marked As(V) reduction (Lear et al., 2007), we have recently focused our attention on the ability of this organism to reduce As(V). *G. uraniireducens* Rf4 was obtained from the culture collection of D. R. Lovley (University of Massachusetts) and grown in the dark at 30°C in a standard modified freshwater medium (Caccavo et al., 1994) containing 20 mM acetate as the electron donor and 40 mM fumarate as the electron acceptor. Preliminary experiments were conducted to assess the tolerance of *G. uraniireducens* to 10 mM As(V) added to the minimal medium. Growth of the organism was clearly visible within a few days of inoculation, and ion chromatography-inductively coupled plasma mass spectrometry analysis (Islam et al., 2004) indicated that approximately 5% of the initial 10 mM As(V) was reduced to As(III) over 1 month of incubation. Thus, *G. uraniireducens,* in common with *G. sulfurreducens* (Islam et al., 2005), is able to tolerate relatively high concentrations of arsenate without relying on detoxification via a conventional *ars* operon that would result in the intracellular reduction of As(V) to As(III) and its subsequent efflux and detection at >5% of the total arsenic pool. Indeed, analysis of the *G. uraniireducens* genome suggested that the presence of a functional *ars* operon was unlikely, because the genome lacks a gene with significant homology to that of the intracellular arsenate reductase (ArsC) that forms an integral component of arsenate detoxification. Genes with homology to other components of *ars* operons, e.g., *arsR*, *arsA*, and *arsD*, were present, however, but clearly not part of a conventional *ars* operon. The mechanisms of resistance to millimolar concentrations of As(V) in both *G. uraniireducens* and *G. sulfurreducens* clearly warrant further investigation.

Having determined that *G. uraniireducens* was tolerant to 10 mM As(V) in defined medium, and able to reduce a small but significant fraction of the metalloid, experiments were conducted to determine whether the organism could respire As(V) added as the sole electron acceptor. Medium containing 0.1 to 5 mM As(V) as the sole electron acceptor and 20 mM acetate as the electron donor was inoculated with a fresh culture of cells grown on fumarate as the electron acceptor (10% vol/vol inoculum), and samples were removed periodically for arsenic speciation analysis (Islam et al., 2004). Reduction of As(V) to As(III) was observed at all concentrations tested [Fig. 4; only data for 0.5 mM As(V) shown] and no methylated (monomethylarsonic acid and dimethylarsinic acid) arsenic species were detected. Complete reduction was obtained in medium amended with 0.1 mM and 0.5 mM As(V). At higher arsenate concentrations, *G. uraniireducens* was able to reduce 73%, 22%, and 21% of dissolved As(V) initially present at levels of 1 mM, 2.5 mM, and 5 mM, respectively, over the 45-day incubation period. No As(V) reduction was noted in control cultures not inoculated with active biomass. The bulk (62 to 80%) of As(V) reduction occurred within the first week of all experiments. Cell numbers were also quantified at higher concentrations of arsenate by direct counting after staining using acridine orange, and showed an approximate twofold increase in cell numbers from $(3.0 \pm 0.8) \times 10^7$ cells ml^{-1} in As(V)-free controls to $(6.0 \pm 1.5) \times 10^7$ cells ml^{-1}, $(5.1 \pm 0.6) \times 10^7$ cells ml^{-1}, and $(6.1 \pm 2.1) \times 10^7$ cells ml^{-1} at 1, 2.5, and 5 mM As(V), respectively, suggesting that the organism is capable of conserving energy for growth from the reduction of As(V). Although the microcosms were not electron donor limited, the maximum absolute concentration of As(V) reduced in experiments containing the higher concentrations of dissolved arsenic did not exceed 1 mM, suggesting that a plateau in the growth conserved by *G. uraniireducens* from As(V) respiration is reached at 0.5 mM < As(V) < 1 mM. The reasons for this unexpected observation require further investigation.

To determine whether *G. uraniireducens* influences the solubility of arsenic sorbed to

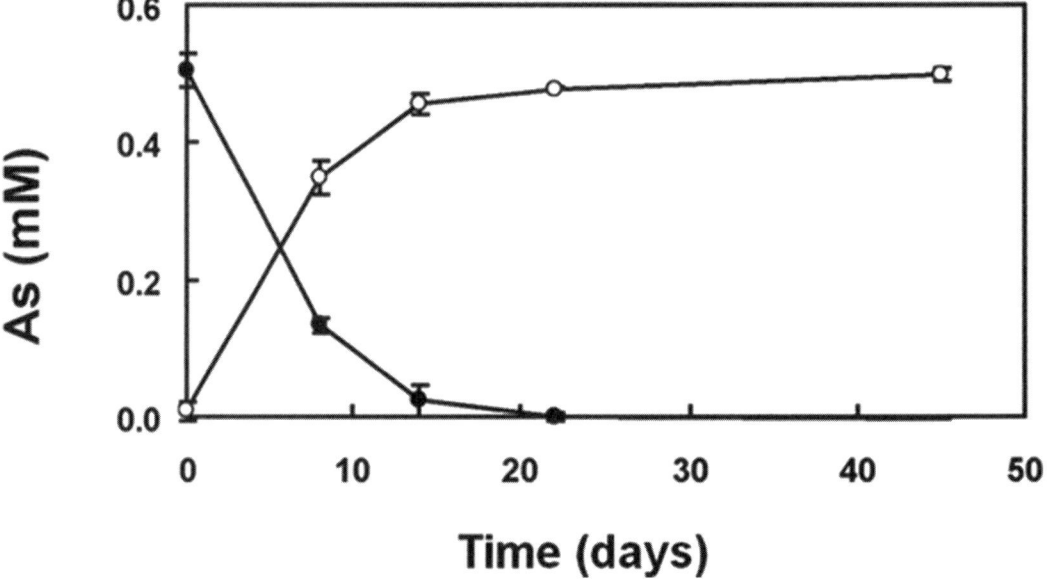

FIGURE 4 Change in dissolved arsenic speciation in cultures inoculated with *G. uraniireducens* amended with 0.5 mM As(V): As(V) (●) and As(III) (○). The mean and one standard deviation from experiments conducted in triplicate are displayed.

poorly soluble iron minerals, twice-washed resting cells were incubated with 10 mmol liter^{-1} ferrihydrite, mixed with 100 μM sodium arsenate in 30 mM sodium bicarbonate containing 20 mM sodium acetate (pH 7.1). *G. sulfurreducens* was tested as a comparator that does not reduce As(V) (Islam et al., 2005). Both species caused some iron to be solubilized, as measured by inductively coupled plasma-atomic emission spectrometry, but only *G. uraniireducens* caused the release of arsenic, with 64% of the added arsenic in solution after 7 days incubation at 30°C and 79% after 21 days. No soluble arsenic was detected in the *G. sulfurreducens* cultures (Fig. 5A). The speciation of arsenic released upon incubation with *G. uraniireducens* was almost completely in the form of As(III) (Fig. 5B).

The reduction of As(V) by *G. uraniireducens* clearly has a number of important environmental implications. This organism is the first member of the family *Geobacteraceae* to be shown to reduce As(V) and this activity is supported by the identification of genes potentially encoding a respiratory arsenate reductase (Lear et al., 2007). For example, the putative *arrA* gene sequence of *G. uraniireducens* potentially encodes a protein with 57% and 55% identity to ArrA from *Chrysiogenes arsenatis* and *Desulfitobacterium hafniense*, respectively. This putative protein contains an iron-sulfur cluster-coordinating motif (CQGCTXWCX$_{27}$C) recognized in ArrA from *Shewanella* sp. strain ANA-3, *C. arsenatis*, *D. hafniense*, and *Bacillus selenitireducens*, and an N-terminal sequence with a twin arginine transporter signal sequence (conserved motif RRXFLK), also present in these respiratory arsenate reductases. Further genetic studies are needed to confirm the role of these genes in arsenic metabolism in *G. uraniireducens*, and explore the diversity of As(V)-reducing *Geobacter* species and the ecophysiology of these organisms in arsenic-impacted subsurface sediments.

CONCLUSIONS AND FUTURE DIRECTIONS

Knowledge of aspects of the microbial arsenic cycle is improving in a range of environments, especially those with "extreme" geochemi-

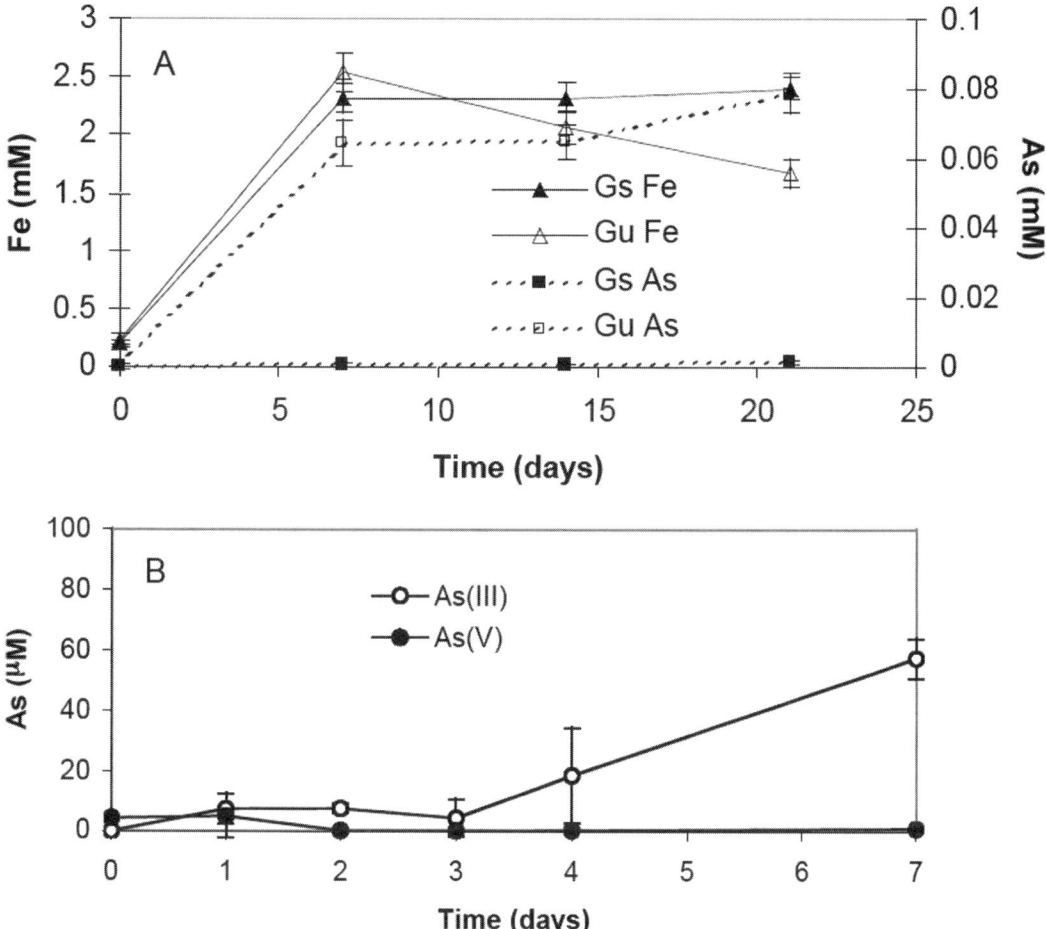

FIGURE 5 Release of dissolved arsenic species by resting cells incubated with 10 mM ferrihydrite and 100 μM sodium arsenate. (A) Dissolved arsenic and iron in *G. uraniireducens* (Gu) and *G. sulfurreducens* (Gs) cultures. (B) Dissolved As(III) (○) and As(V) (●) in *G. uraniireducens* cultures. All experiments were done in triplicate; error bars indicate one standard deviation.

cal conditions that lead to elevated soluble concentrations of the metalloid. In aquifer sediments, which have more complex geochemical and mineralogical backgrounds and support diverse microbial communities, the role of specific microorganisms in mobilizing low (typically parts per billion) but significant concentrations of arsenic is less clear. However, several recent studies suggest that we are approaching a consensus that metal-reducing bacteria, especially dissimilatory As(V)-respiring bacteria, may play an important role in this process. The microbiological focus of research in this area is shifting now toward the unequivocal identification of the organisms involved, which will in turn lead to studies on the underlying mechanisms at a molecular (genetic) level and the factors that result in the activation of these physiological processes in situ. Given the complex interplay between arsenic-metabolizing microorganisms, and the arsenic-bearing minerals (often Fe phases) that they interact with, there is a need for a better understanding at the nanoscale of the microbe-mineral interface. There is also an urgent need for improvements in the understanding of the mechanisms of delivery of key nutrients and electron donors that stimulate or sustain arsenate-reducing bacteria. These factors relate, in turn, to the hydrological conditions in situ (e.g., through drawdown of highly labile organic matter from surface waters through extensive water extractions) and/or the distribution of alternative extant electron donors in aquifer sediments. Other important research priorities include a better understanding of the in situ activation of alternative microbial processes that can reverse the mobilization of arsenic, e.g., sulfate reduction or Fe(II) oxidation that can lead to sulfide or Fe(III) minerals, respectively, that could sorb arsenic efficiently. For the latter process, the co-stimulation of microbial oxidation of soluble As(III) to As(V) (which sorbs preferentially to a broad range of minerals) also requires further investigation. It is important that very recent studies suggest that sulfate reduction can be supported even in very low carbon aquifer sediments that contain low concentrations of electron donor, resulting in removal of arsenic mobilized by dissimilatory metal reduction (Héry et al., 2010). Complementary field data suggest that microbial sulfate reduction may even be minimizing the impact of arsenic mobilization in Southeast Asian aquifers (van Geen, personal communication). These latter observations are an important step in translating our rapidly expanding knowledge of As-microbe interactions to field-scale models that will inform safer water use in the future.

ACKNOWLEDGMENTS

The authors gratefully acknowledge funding from the UK NERC (grant NE/D013291/1) and National Science Foundation (grant 0134054).

REFERENCES

Acharyya, S. K., P. Chakraborty, S. Lahiri, B. C. Raymahashay, S. Guha, and A. Bhowmik. 1999. Arsenic poisoning in the Ganges delta. *Nature* **401**:545.

Afkar, E., J. Lisak, C. Saltikov, P. Basu, R. Oremland, and J. Stolz. 2003. The respiratory arsenate reductase from *Bacillus selenitireducens* strain MLS10. *FEMS Microbiol. Lett.* **226**:107–112.

Akai, J., K. Izumi, H. Fukuhara, H. Masuda, S. Nakano, T. Yoshimura, H. Ohfuji, H. M. Anawar, and K. Akai. 2004. Mineralogical and geomicrobiological investigations on groundwater arsenic enrichment in Bangladesh. *Appl. Geochem.* **19**:215–230.

Appelo, C. A. J., M. J. J. Van der Weiden, C. Tournassat, and L. Charlet. 2002. Surface complexation of ferrous iron and carbonate on ferrihydrite and the mobilization of arsenic. *Environ. Sci. Technol.* **36**:3096–3103.

Bandyopadhyay, R. K. 2002. Hydrochemistry of arsenic in Nadia district, West Bengal. *J. Geol. Soc. India* **59**:33–46.

Caccavo, F., Jr., D. J. Lonergan, D. R. Lovley, M. Davis, J. F. Stolz, and M. J. McInerney. 1994. *Geobacter sulfurreducens* sp. nov., a hydrogen and acetate-oxidizing dissimilatory metal reducing microorganism. *Appl. Environ. Microbiol.* **60**:3752–3759.

Chowdhury, T. R., G. K. Kumar Basu, B. K. Mandal, B. K. Biswas, G. Samanta, U. K. Chowdhury, C. R. Chanda, D. Lodh, S. L. Roy, K. C. Saha, S. Roy, S. Kabir, Q. Quamruzzaman, and D. Chakraborti. 1999. Arsenic poisoning in the Ganges delta. *Nature* **401**:545–546.

Coker, V. S., A. G. Gault, C. I. Pearce, G. v. d. Laan, N. D. Telling, J. M. Charnock, D. A. Polya, and J. R. Lloyd. 2006. XAS and XMCD evidence for species-dependent partitioning of arsenic during microbial reduction of ferrihydrite to magnetite. *Environ. Sci. Technol.* **40**:7745–7750.

Das, D., G. Samanata, B. K. Mandal, T. R. Chowdhury, C. R. Chanda, P. P. Chowdhury, G. K. Basu, and D. Chakraborti. 1996. Arsenic in groundwater in six districts of West Bengal, India. *Environ. Geochem. Health* **18**:5–15.

Handley, K. M., M. Hery, and J. R. Lloyd. 2009. Redox cycling of arsenic by the hydrothermal marine bacterium *Marinobacter santoriniensis*. *Environ. Microbiol.* **11**:1601–1611.

Harvey, C. F., C. H. Swartz, A. B. M. Badruzzaman, N. Keon-Blute, W. Yu, M. Ashraf Ali, J. Jay, R. Beckie, V. Niedan, D. Brabander, P. M. Oates, K. N. Ashfaque, S. Islam, H. F. Hemond, and M. F. Ahmed. 2002. Arsenic mobility and groundwater extraction in Bangladesh. *Science* **298**:1602–1606.

Héry, M., B. E. v. Dongen, F. Gill, D. Mondal, D. J. Vaughan, R. D. Pancost, D. A. Polya, and J. R. Lloyd. 2010. Biogeochemical cycling of arsenic in low organic carbon aquifer sediments from West Bengal. *Geobiology* **8:**155–168.

Hoeft, S. E., T. R. Kulp, J. F. Stolz, J. T. Hollibaugh, and R. S. Oremland. 2004. Dissimilatory arsenate reduction with sulfide as electron donor: experiments with Mono Lake water and isolation of strain MLMS-1, a chemoautotrophic arsenate respirer. *Appl. Environ. Microbiol.* **70:**2741–2747.

Islam, F., A. G. Gault, C. Boothman, D. A. Polya, J. M. Charnock, D. Chatterjee, and J. R. Lloyd. 2004. Role of metal-reducing bacteria in arsenic release from Bengal Delta sediments. *Nature* **430:**68–71.

Islam, F. S., R. L. Pederick, A. G. Gault, L. K. Adams, D. A. Polya, J. M. Charnock, and J. R. Lloyd. 2005. Reduction of Fe(III) by *Geobacter sulfurreducens* and the capture of arsenic by biogenic Fe(II) minerals. *Appl. Environ. Microbiol.* **71:**8642–8648.

Krafft, T., and J. M. Macy. 1998. Purification and characterization of the respiratory arsenate reductase of *Chrysiogenes arsenatis*. *Eur. J. Biochem.* **255:**647–653.

Kulp, T. R., S. Han, C. W. Saltikov, B. D. Lanoil, K. Zargar, and R. S. Oremland. 2007. Effects of imposed salinity gradients on dissimilatory arsenate reduction, sulfate reduction, and other microbial processes in sediments from two California soda lakes. *Appl. Environ. Microbiol.* **73:**5130–5137.

Kulp, T. R., S. E. Hoeft, L. G. Miller, C. Saltikov, J. N. Murphy, S. Han, B. Lanoil, and R. S. Oremland. 2006. Dissimilatory arsenate and sulfate reduction in sediments of two hypersaline, arsenic-rich soda lakes: Mono and Searles lakes, California. *Appl. Environ. Microbiol.* **72:**6514–6526.

Langner, H. W., and W. P. Inskeep. 2000. Microbial reduction of arsenate in the presence of ferrihydrite. *Environ. Sci. Technol.* **34:**3131–3136.

Lear, G., B. Song, A. G. Gault, D. A. Polya, and J. R. Lloyd. 2007. Molecular analysis of arsenate-reducing bacteria within Cambodian sediments following amendment with acetate. *Appl. Environ. Microbiol.* **73:**1041–1048.

Lloyd, J. R., and R. S. Oremland. 2006. Microbial transformations of arsenic in the environment; from soda lakes to aquifers. *Elements* **2:**85–90.

MacRae, J. D., I. N. Lavine, K. A. McCaffery, and K. Ricupero. 2007. Isolation and characterization of NP4, arsenate-reducing *Sulfurospirillum*, from Maine groundwater. *J. Environ. Eng.-ASCE* **133:**81–88.

Malasarn, D., C. W. Saltikov, K. M. Campbell, J. M. Santini, J. G. Hering, and D. K. Newman. 2004. *arrA* is a reliable marker for As(V) respiration. *Science* **306:**455.

Mukhopadhyay, R., B. Rosen, L. Phung, and S. Silver. 2002. Microbial arsenic: from geocycles to genes and enzymes. *FEMS Microbiol. Rev.* **26:**311.

Neufeld, J. D., M. Wagner, and J. C. Murrell. 2007. Who eats what, where and when? Isotope-labelling experiments are coming of age. *ISME J.* **1:**103–110.

Nickson, R., J. McArthur, W. Burgess, K. M. Ahmed, P. Ravenscroft, and M. Rahman. 1998. Arsenic poisoning of Bangladesh groundwater. *Nature* **395:**338.

Nickson, R. T., J. M. McArthur, P. Ravenscroft, W. G. Burgess, and K. M. Ahmed. 2000. Mechanism of arsenic release to groundwater, Bangladesh and West Bengal. *Appl. Geochem.* **15:**403–413.

Oremland, R. S., and J. F. Stolz. 2003. The ecology of arsenic. *Science* **300:**939–944.

Oremland, R. S., J. F. Stolz, and J. T. Hollibaugh. 2004. The microbial arsenic cycle in Mono Lake, California. *FEMS Microbiol. Ecol.* **48:**15–27.

Radajewski, S., P. Ineson, N. R. Parekh, and J. C. Murrell. 2000. Stable-isotope probing as a tool in microbial ecology. *Nature* **403:**646–649.

Richey, C., P. Chovanec, S. E. Hoeft, R. S. Oremland, P. Basu, and J. F. Stolz. 2009. Respiratory arsenate reductase as a bidirectional enzyme. *Biochem. Biophys. Res. Commun.* **382:**298–302.

Rowland, H. A. L., C. Boothman, R. Pancost, A. G. Gault, D. A. Polya, and J. R. Lloyd. 2009. The role of indigenous microorganisms in the biodegradation of naturally occurring petroleum, the reduction of iron, and the mobilization of arsenite from West Bengal aquifer sediments. *J. Environ. Qual.* **38:**1598–1607.

Rowland, H. A. L., R. L. Pederick, D. A. Polya, R. D. Pancost, B. E. Van Dongen, A. G. Gault, D. J. Vaughan, C. Bryant, B. Anderson, and J. R. Lloyd. 2007. The control of organic matter on microbially mediated iron reduction and arsenic release in shallow alluvial aquifers, Cambodia. *Geobiology* **5:**281–292.

Saltikov, C. W., A. Cifuentes, K. Venkateswaran, and D. K. Newman. 2003. The ars detoxification system is advantageous but not required for As(V) respiration by the genetically tractable *Shewanella species* strain ANA-3. *Appl. Environ. Microbiol.* **69:**2800–2809.

Saltikov, C. W., and D. K. Newman. 2003. Genetic identification of a respiratory arsenate reductase. *Proc. Natl. Acad. Sci. USA* **100:**10983–10988.

Santini, J. M., L. I. Skly, R. D. Schnagll, and J. M. Macy. 2000. A new chemolithoautotrophic arsenite-oxidizing bacterium isolated from a gold mine: phylogenetic, physiological, and preliminary biochemical studies. *Appl. Environ. Microbiol.* **66:**92–97.

Silver, S., and L. T. Phung. 2005. Genes and enzymes involved in bacterial oxidation and reduction of inorganic arsenic. *Appl. Environ. Microbiol.* **71:**599–608.

Smedley, P. L., and D. G. Kinniburgh. 2002. A review of the source, behaviour and distribution of arsenic in natural waters. *Appl. Geochem.* **17:**517–568.

Smith, A. H., E. O. Lingas, and M. Rahman. 2000. Contamination of drinking-water by arsenic in Bangladesh: a public health emergency. *Bull. WHO* **78:**1093–1103.

Song, B., E. Chyun, P. R. Jaffe, and B. B. Ward. 2009. Molecular methods to detect and monitor dissimilatory arsenate-respiring bacteria (DARB) in sediments. *FEMS Microbiol. Ecol.* **68:**108–117.

Tufano, K. J., C. Reyes, C. W. Saltikov, and S. Fendorf. 2008. Reductive processes controlling arsenic retention: revealing the relative importance of iron and arsenic reduction. *Environ. Sci. Technol.* **42:**8283–8289.

Van Geen, A., J. Rose, S. Thoral, J. Garnier, Y. Zheng, and J. Bottero. 2004. Decoupling of As and Fe release to Bangladesh groundwater under reducing conditions. Part II: Evidence from sediment incubations. *Geochim. Cosmochim. Acta* **68:**3475–3486.

Zobrist, J., P. R. Dowdle, J. A. Davis, and R. S. Oremland. 2000. Mobilization of arsenite by dissimilatory reduction of adsorbed arsenate. *Environ. Sci. Technol.* **34:**4747–4753.

PROCESSES

MINERALOGICAL CONTROLS ON MICROBIAL REDUCTION OF Fe(III) (HYDR)OXIDES

Colleen M. Hansel and Christopher J. Lentini

6

INTRODUCTION

Iron oxides, oxyhydroxides, and hydroxides [hereinafter referred to collectively as (hydr)oxides] are ubiquitous in the environment with contents ranging from one to several hundred grams per kilogram in aerobic soils (Cornell and Schwertmann, 2003). The reduction of Fe(III) (hydr)oxide phases regulates the degradation of carbon, the flow of electrons within subsurface sediments and soils, and the fate and transport of nutrients and contaminants. In particular, Fe (hydr)oxides are considered one of the most important sinks for (in)organic contaminants and nutrients within soils, sediments, and waters. As a consequence of their high surface area and density of reactive surface sites, Fe(III) (hydr)oxides may sorb numerous organics (e.g., pesticides, p-aminohippuric acids), nutrients (e.g., phosphate), and metals (e.g., Pb, As, U) (Cornell and Schwertmann, 2003). Consequently, the dissolution of metal-laden Fe(III) (hydr)oxide phases poses a potential threat to water quality should anaerobiosis occur. In contrast, the reduction of Fe(III) (hydr)oxides may also be coupled to the degradation and/or sequestration of contaminants within natural and engineered systems. In fact, stimulated in situ microbial Fe(III) reduction has received considerable attention as a promising means of environmental remediation of a myriad of (in)organic contaminants. Targeted stimulation of dissimilatory metal-reducing bacteria (DMRB) poses a particularly attractive and flexible means of environmental in situ remediation (Lovley, 2001). Through both direct enzymatic and indirect metabolic reactions, DMRB can degrade, immobilize, and transform contaminants ranging from benzene to uranium (Anderson and Lovley, 2000; Lovley, 2001). These processes include hydrocarbon oxidation coupled to Fe(III) reduction, sequestration of soluble contaminants (e.g., As) within Fe secondary phases (e.g., magnetite), and/or immobilization of redox-active contaminants (e.g., Cr) by reaction with the reactive metabolite Fe(II). Thus, central to many proposed remedial approaches are Fe(III)-bearing minerals.

In studies in which treatment progression depends on reduction of Fe(III) phases, mineral phase reactivity and longevity currently represent a key limitation. Previous studies have revealed that microbial reduction of various Fe(III) (hydr)oxides is transient, having

---

Colleen M. Hansel, School of Engineering and Applied Sciences, Department of Earth and Planetary Sciences, Harvard University, Cambridge, MA 02138. *Christopher J. Lentini*, School of Engineering and Applied Sciences, Engineering Sciences Laboratory, Harvard University, Cambridge, MA 02138.

a diminished reductive capacity over time (Benner et al., 2002; Hansel et al., 2003a, 2004). Iron (hydr)oxides are mineralogically diverse, with the most studied and most utilized phase being ferrihydrite. While ferrihydrite is considered the most "bioavailable" Fe(III) (hydr)oxide for microbial reduction (Lovley and Phillips, 1986), under some conditions, more recalcitrant phases, such as goethite, can be more reactive (C. M. Hansel, unpublished data). Furthermore, ferrihydrite reactivity has been shown to rapidly decline over time (Hansel et al., 2003a). Differences in rates of reaction both between mineral phases and over time are significant, yet an understanding of the controlling factors on reduction rates is incomplete. Our lack of knowledge on what controls the short- and long-term reduction capacity of Fe(III) minerals currently limits our ability to convert this promising biogeochemical process into a successful remediation strategy. In this chapter, we will discuss the current state of knowledge regarding the geochemical controls on the initial and sustained reduction of Fe(III) minerals. We will introduce the major controlling factors for microbial Fe(III) reduction; yet, it is important to note that the properties of Fe minerals are not mutually exclusive and, thus, defining the individual role of these properties is difficult, if not impossible.

PHYSICOCHEMICAL PROPERTIES OF Fe MINERALS

The Fe(III)-Fe(II) redox couple is an important electron-transfer mediator for many reactions of significance to biogeochemical cycles, especially in soils and sediments. In the reduced state, Fe(II) exists as a hydrated ion in solution, adsorbs onto solid phases, or precipitates as reactive Fe(II) phases (e.g., siderite, vivianite). The oxidized Fe(III) species can exist as a number of primary and secondary minerals, including Fe-containing phyllosilicates and (hydr)oxides (including oxides, oxyhydroxides, and hydroxides). Above acidic pH values (pH > 4), low-solubility Fe(III) (hydr)oxides are the predominant form of Fe(III) in most soils and sediments (Cornell and Schwertmann, 2003). Fe(III) (hydr)oxides consist of an array of Fe^{3+} and O^{2-} or OH^- ions, and differ in how their basic structural units, $Fe(O/OH)_6$ or FeO_6, are arranged in space (Fig. 1). The term ferrihydrite encompasses a group of oxyhydroxide minerals that lack long-range order and have varying degrees of crystallinity. The two end members in the crystallinity continuum are referred to as two-line and six-line ferrihydrite, based on the presence of either two or six to eight reflections in X-ray diffraction patterns (Cornell and Schwertmann, 2003). Frequently, these phases are erroneously referred to as "amorphous" despite the

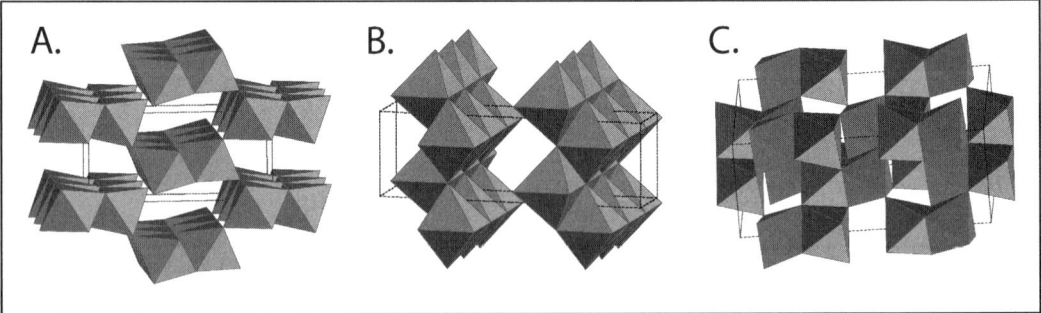

FIGURE 1 Structures of Fe(III) (hydr)oxides commonly used in microbial reduction experiments. (A) Goethite structure composed of octahedral double chains linked through corners. H atoms not shown. (B) Lepidocrocite structure composed of octahedral double chains in corrugated layers. The layers are cross-linked through edges. H atoms not shown. (C) Hematite structure composed of octahedra linked through edge- and corner-sharing as well as face-sharing along the c axis. Unit cell outlined in dashed black line.

presence of distinct short-range order. The local structure of ferrihydrite resembles those of well crystalline phases, including hematite, akageneite, or goethite, depending on growth conditions (Manceau and Drits, 1993). Interconversion of Fe (hydr)oxides also occurs whereby thermodynamically less stable phases (e.g., ferrihydrite, lepidocrocite) transform to more stable phases (e.g., goethite, hematite). This "ripening" of Fe(III) (hydr)oxides is catalyzed by reductants, including cysteine and Fe(II) (Cornell et al., 1991; Fredrickson et al., 2003; Hansel et al., 2005; Jang et al., 2003; Zachara et al., 2002).

Iron(III) (hydr)oxides exist in a spectrum of crystallinities and subsequent bioavailabilities; ranging from ferrihydrite (poorly crystalline/least thermodynamically stable) to hematite (well crystalline/most thermodynamically stable) (Cornell and Schwertmann, 2003). The thermodynamic stability of Fe (hydr)oxides is a function of crystal structure and particle size, which ultimately controls the solubility of the phases. Because solubility is a function of several properties (ionic strength, temperature, particle size, crystal defects), there is a considerable amount of uncertainty with the solubility product values obtained in the pH ranges of most environmental conditions. This has led to large discrepancies in the literature between calculated solubility products by different authors (Cornell and Schwertmann, 2003). However, the solubility of common, pure Fe(III) (hydr)oxides generally progresses in the order of two-line ferrihydrite > six-line ferrihydrite > lepidocrocite > goethite > hematite at circumneutral pH (Baes and Mesmer, 1976; Cornell and Schwertmann, 2003; Langmuir, 1969). The physicochemical properties of Fe(III) (hydr)oxides vary widely, including morphology, crystal size, crystal structure, surface area, density, magnetism, conductivity, solubility, and free energy of formation (Table 1) (Cornell and Schwertmann, 2003).

Trivalent and divalent metal cations may isomorphously substitute into Fe(III) (hydr)oxide structures. Within the environment, natural Fe (hydr)oxides always contain coprecipitated ions, frequently containing several moles percent of foreign ions (Singh and Gilkes, 1992; Trolard et al., 1995). Aluminum(III) is one of the most prevalent and effective coprecipitates within Fe(III) (hydr)oxides, with substitution reaching 33 mol% in goethite, for example (Cornell and Schwertmann, 2003). A number of trace metals may also substitute into Fe(III) (hydr)oxides, including Cd(II), Co(III), Cr(III), Cu(II), Mn(III), Ni(II), V(III), and Zn(II). While Mn(III) can occupy a significant fraction (e.g., approximately 47% in goethite) of Fe(III) sites within Fe(III) (hydr)oxides, the remaining trace metals typically substitute to levels below 10%. Substitution of Fe(III) within (hydr)oxide phases impacts crystallite size, morphology, surface area, solubility, surface chemistry, and rates of acid and reductive dissolution (Cornell and Schwertmann, 2003). For the most part, metal substitution reduces the solubility and interconversion of

TABLE 1 Physicochemical properties of Fe(III) (hydr)oxides commonly used in bioreduction experiments

Fe(III) (hydr)oxide	Parameter					
	Formula	Crystal system	Dominant morphology	Range of surface area (m^2/g)	Standard free energy (kJ/mol)	Solubility product (25°C) (log K_{SO})
Ferrihydrite	$Fe_5HO_8 \cdot 4H_2O$[a]	Hexagonal	Spheres	100–700	−699	−39
Lepidocrocite	γ-FeOOH	Orthorhombic	Laths	15–260	−477.7	−39.5
Goethite	α-FeOOH	Orthorhombic	Acicular	8–200	−488.6	−40.7
Hematite	α-Fe_2O_3	Hexagonal	Plates	5–200	−742.7	−42.75

[a]The exact formula for ferrihydrite is not established; others include $Fe(OH)_3$, $Fe(OH)_3 \cdot nH_2O$, and $5Fe_2O_3 \cdot 9H_2O$.

Fe (hydr)oxides relative to pure phases (Cornell and Schwertmann, 2003).

The reduction potential (E_h) of Fe is a function of the phase or complex (Fig. 2). The E_h for Fe(III/II) redox couples varies over 700 mV, while that of Fe(III) (hydr)oxide phases spans over 300 mV (Favre et al., 2006; Thamdrup et al., 2000). Remarkably, when corrected for typical environmental conditions, the E_h of more crystalline (stable) Fe(III) (hydr)oxides (e.g., goethite, hematite) falls below the potential for sulfate reduction. Based on a partial equilibrium model, the existence of overlapping Fe(III)- and sulfate-reducing zones within various sedimentary environments can be explained by this large range in Fe (hydr)oxide reduction potentials (Jakobsen and Postma, 1999; Postma and Jakobsen, 1996). Thus, it is predicted that the distribution and proximity of Fe(III)- and sulfate-reducing zones will vary depending on the composition and stability of the Fe(III) phases (Postma and Jakobsen, 1996).

MICROBIAL Fe(III) REDUCTION

The importance of microorganisms in the biogeochemical cycling of Fe has long been recognized (Starkey and Halvorson, 1927). The

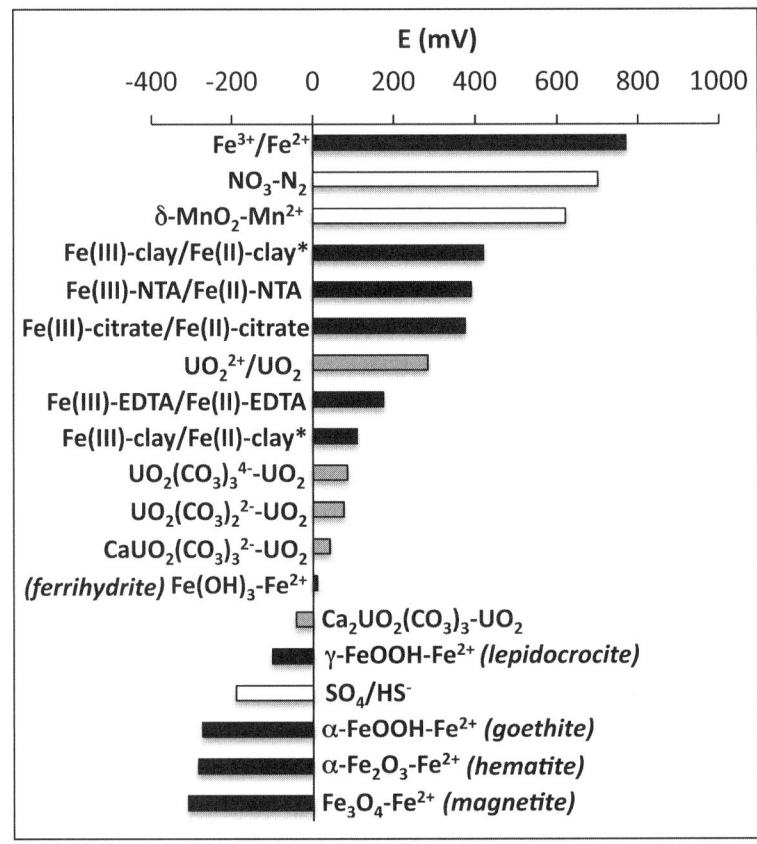

FIGURE 2 Redox potentials of various Fe couples compared with other couples of relevance in groundwater and contaminated systems (modified from Brooks et al. [2003] and Thamdrup et al. [2000]). A variety of U species are included to illustrate the role of complexation on reduction potential of soluble complexes. Temperature = 25°C, pH = 7, $[Fe^{2+}] = [Mn^{2+}] = [NO_3^-] = 10$ μM, $[U(VI)] = 50$ μM, $[Ca^{2+}] = 5$ mM, $[HCO_3^-] = 28.1$ to 28.7 mM, $[SO_4^{2-}] = 10$ mM, $[HS^-] = 1$ μM, $P_{N2} = 1$ atm. Fe(III) clay potentials presented for SWa-1 with $[Na^+] = 100$ μM and either $m_{rel} = 0.02$ ($E_h = 420$ mV) or $m_{rel} = 0.70$ (E = 110 mV) (Favre et al., 2006). NTA, nitrilotriacetic acid.

microbial reduction of Fe(III) may be a result of Fe uptake or incorporation into cellular components (assimilatory), or respiration whereby Fe(III) serves as a terminal electron acceptor (dissimilatory). Historically, geochemical and redox gradients within sediments hinted at the presence of microorganisms that link the oxidation of organic carbon to the reduction of Fe(III) (Froelich et al., 1979; Reeburgh, 1983). Yet, sedimentary Fe(III)-reducing zones were primarily attributed to fermentative bacteria (Jones et al., 1984), which are thought to reduce Fe(III) as a supplementary terminal electron acceptor (Bromfield, 1954a, 1954b; Lovley, 1987). The recent discovery and isolation of bacteria that couple growth to the reduction of Fe(III) (Lovley and Phillips, 1988; Myers and Nealson, 1988) prompted countless studies that unveiled the widespread importance of microbial respiration in the cycling of Fe(III). Fe(III)-reducing microbes are found in a wide range of environmental conditions (e.g., temperature, pH) and include various members of the domains *Bacteria* and *Archaea*, facultative and obligate anaerobes, and heterotrophs and autotrophs (Ehrlich and Newman, 2009; Lovley, 2000; Lovley et al., 2004; Weber et al., 2006). Because of the ubiquity, abundance, phylogenetic diversity, and metabolic versatility of Fe(III)-reducing microorganisms (Lovley, 1991; Lovley et al., 1991b; Nealson and Saffarini, 1994), it is proposed that metal-reducing bacteria catalyze most of the Fe(III) reduction occurring under nonsulfidogenic, anaerobic conditions (Lovley et al., 1991b). Furthermore, the metabolic versatility of these organisms includes the reduction of a number of metal(loid)s (e.g., Cr, As) and radionuclides (e.g., U, Tc) (Caccavo et al., 1992; Fredrickson et al., 2000; Lloyd et al., 2000, 2002; Lovley et al., 1991a; Saltikov and Newman, 2003) highlighting their potential for environmental remediation and biotechnology. Yet, it is becoming increasingly apparent that fermentative bacteria are also important players in Fe(III) reduction within pristine and contaminated sediments (Hansel et al., 2008; Petrie et al., 2003), and some fermenting organisms may, in fact, conserve energy through Fe(III) reduction (Dobbin et al., 1999).

Unlike other metabolisms where the terminal electron acceptor may diffuse into the cell, DMRB respire solid Fe minerals, presumably making cytoplasmic-bound terminal reductases inaccessible. Mechanisms by which bacteria may access Fe(III) from sparingly soluble Fe (hydr)oxides include (i) use of exogenous (e.g., humics) or endogenous soluble electron shuttles (e.g., phenazines) (Hernandez et al., 2004; Lovley et al., 1996), (ii) use of exogenous or endogenous Fe(III)-chelating compounds (e.g., siderophores) (Nevin and Lovley, 2002), and (iii) direct electron transfer through enzymes either embedded on the bacterial outer membrane (DiChristina et al., 2005; Myers and Myers, 1998, 2001) or on extracellular appendages (e.g., nanowires, pili) (El-Naggar et al., 2008; Gorby et al., 2006) (for reviews, see DiChristina et al. [2005], Gralnick and Newman [2007], and Kappler and Straub [2005]). It has been proposed that the mechanism of electron transfer may vary among microbial species (Nevin and Lovley, 2002). In particular, the need for direct microbial contact with the mineral surface is still ambiguous and may in fact vary among microbial species and environmental conditions (Lies et al., 2005; Nevin and Lovley, 2000, 2002). In fact, multiple electron transfer pathways may be operative within single species biofilms (Hernandez and Newman, 2001). The operative reduction pathway may be a consequence of the distribution and orientation of the microbe within redox and chemical gradients at the Fe(III) (hydr)oxide surface. For instance, within three-dimensional biofilm structures on Fe(III) (hydr)oxide surfaces, organisms proximal to the mineral surface may preferentially reduce Fe(III) using extracellular membrane-bound enzymes, while those embedded within the biofilm matrix may use shuttles or electrically conductive cellular appendages to transfer electrons. Interestingly, it has recently been demonstrated that the DMRB *Shewanella putrefaciens* strain 200 and *S. oneidensis* MR-1 produce soluble organic Fe(III) complexes during anaerobic respiration of various Fe(III) (hydr)oxides (Taillefert et al., 2007). The Fe(III)-solubilizing organic ligands, however, are not siderophores, and mutants impaired in their ability to produce soluble organic-Fe(III) and reduce Fe(III) retained wild-type siderophore production capability (Fennessey et al., 2010; Jones et al., 2010). Furthermore, increasing evidence indicates that a number of species may reduce Fe(III) (hydr)oxides indirectly via soluble electron shuttles (e.g., 9,10-anthraquinone-2,6-disulfonic acid) (Kappler and Straub, 2005) or via sulfide production (Haveman et al., 2008; Straub and Schink, 2004). Low sulfide levels favor sulfur cycle-mediated Fe(III) reduction by microbes and this metabolism supports growth in pure culture of species such as *Sulfurospirillum deleyianum*. Interestingly, in the presence of ferrihydrite, only catalytic amounts of thiosulfate (50 µM) are required to maintain growth, and sulfur was cycled up to 60 times in these cultures (Straub and Schink, 2004). Thus, direct enzymatic reduction may not be required for electron transfer to solid Fe(III) substrates, and, in fact, DMRB incapable of effectively transferring electrons to more crystalline Fe(III) phases may attempt to overcome this constraint by producing Fe(III)-solubilizing organic ligands and/or by utilizing exogenous electron shuttles. The binding and reduction efficiency of these inorganic and organic electron shuttles will undoubtedly be a function of the properties of the Fe(III) phases, in particular, the phase solubility.

MINERALOGICAL AND GEOCHEMICAL CONTROLS ON MICROBIAL Fe(III) REDUCTION

At circumneutral pH, Fe(III) (hydr)oxides are sparingly soluble. Despite this, it is clear that microorganisms have evolved the unique metabolic ability to respire on solid substrates including poorly crystalline Fe(III) (hydr)oxides (Phillips et al., 1993) and iron-containing clays (Kostka et al., 1996, 1999; Vorhies and Gaines, 2009). The ability of DMRB to directly transfer electrons to more crystalline forms of Fe(III) (hydr)oxides, however, is not resolved (Lovley et al., 2004). The inability of crystalline Fe(III) (hydr)oxides to serve as effective terminal electron acceptors was initially suspected by findings in natural environments in which Fe(III) reduction correlated well with poorly crystalline phases (Roden and Wetzel, 2002; Thamdrup, 2000). Nevertheless, the reduction of highly crystalline phases (goethite, hematite, magnetite) has been observed in the laboratory (Roden and Zachara, 1996; Zachara et al., 1998), within column flow experiments (Hansel et al., 2004; Roden et al., 2000), and in natural soils (Stucki et al., 2007). Yet, cultured model DMRB show diminished abilities to reduce these more crystalline phases, only reducing a small fraction of the potentially available Fe(III). Incomplete microbial reduction of Fe(III) (hydr)oxides is pervasive throughout the literature, regardless of nutrient conditions, Fe(III) (hydr)oxide structure, and hydrologic conditions (dynamic versus static). For instance, merely 25% of ferrihydrite, 5% goethite, and 1% hematite were reduced by *S. putrefaciens* strain CN32 in an artificial groundwater medium under flow conditions (Hansel et al., 2004)—a trend consistent with previous experiments conducted in batch systems (Fredrickson et al., 2003; Roden and Zachara, 1996). The cessation of microbial ferrihydrite reduction is easily attributed to the conversion of surface layers to more crystalline phases (e.g., goethite, magnetite) during Fe(II)-induced remineralization processes (Fredrickson et al., 1998; Hansel et al., 2003a; Zachara et al., 2002). In contrast, secondary mineralization does not occur following reduction of goethite and hematite (Hansel et al., 2004). While Fe(II) sorption to bacterial or mineral surface sites has been previously implicated in this loss of reduction capacity (Roden and Urrutia, 2002; Urrutia et al., 1999), mounting evidence reveals that the lifetime of Fe(II) on the surface of Fe(III) (hydr)oxides is extremely short. Instead, it has been shown both empirically and through ab initio modeling that, after Fe(II) sorption, electrons are injected into the (hydr)oxide structure and either delocalized within the conduction band (Larese-Casanova and Scherer, 2007; Williams and Scherer, 2004) or lost at distal defect sites (Grantham et al., 1997; Rosso et al., 2003b).

Numerous studies have attempted to tease out the underlying geochemical and mineralogical controls on Fe(III) reduction using Fe(III) phases of varying size, crystallinity, structure, composition, and solubility. However, the major factors controlling Fe(III) reduction remain unclear because of the intimate association these properties share. For instance, differences in solubility may arise from variations in specific surface area, crystallinity, and impurity content (Cornell and Schwertmann, 2003). In addition, difficulties occur when comparing the same mineral synthesized under different methods and/or different laboratory conditions. These factors have led to an ambiguous understanding of the factors controlling microbial Fe(III) reduction. The following sections will attempt to summarize our current understanding of the geochemical and mineralogical constraints on microbial Fe(III) (hydr)oxide reduction with the realization that the field is far from a resolution or consensus on what factors ultimately control the transfer of electrons to Fe(III) mineral surfaces.

Surface Area

The seminal paper exploring the relationship between Fe (hydr)oxide properties and microbial Fe(III) reduction identified a link between mineral surface area and reduction rates by *Shewanella algae* strain BrY (Roden and Zachara, 1996). While cell-normalized rates for pure ferrihydrite, lepidocrocite, goethite, and hematite ranged nearly two orders of magnitude [$\sim 2 \times 10^{-10}$ to 2×10^{-12} mmol Fe(II) h^{-1} cell^{-1}], a linear correlation was observed between initial rates of Fe(III) reduction and Fe(III) (hydr)oxide surface area. This correlation was less apparent at higher surface areas (Roden, 2003b, 2006). For instance, given an equivalent particle size and surface area, both chemical (ascorbate, pH 3) and microbial reduction of lepidocrocite exceeded that of goethite. A similar correlation was observed with a spectrum of synthetic Fe(III) (hydr)oxides and different microorganisms (*Geobacter sulfurreducens* and *S. putrefaciens*) (Roden, 2003b, 2006). These results suggested that surface area is the major factor controlling the initial rates of microbial Fe(III) (hydr)oxide reduction, independent of the type of Fe(III) (hydr)oxide present (Roden, 2006). In addition to these pure-culture studies, acetate-limited enrichment cultures of wetland sediments suggested that Fe(III)-reducing microorganisms respiring crystalline goethite and hematite in freshwater outcompeted methanogens only if the Fe(III) (hydr)oxides were provided in considerably larger amounts (Roden, 2003a). Hence, methanogenesis could be equally impeded by crystalline Fe(III) (hydr)oxides and "amorphous HFO" only when both were available at equivalent surface loadings. The investigators concluded that these results provided further support that surface area is a dominant control on microbial iron reduction.

Recent evidence brings to question the role of surface area in controlling microbial Fe(III) (hydr)oxide reduction rates. For instance, Bonneville et al. (2009) observed that maximum Fe(III) reduction rates (v_{max}) showed a stronger correlation with solubility ($r^2 = 0.90$ on a log-log scale) than surface area ($r^2 = 0.54$ on a log-log scale). Furthermore, a recent study by Cutting et al. (2009) convincingly demonstrated that initial surface area-normalized Fe(III) reduction rates of various Fe(III) (hydr)oxides vary by approximately 2 orders of magnitude (Fig. 3). These differences were attributed to subtle variations in the crystallinity and morphology of the synthesized Fe(III) phases, which were characterized in detail using a combination of X-ray diffraction, transmission electron microscopy, and X-ray photoelectron spectroscopy. These studies highlighted the need to thoroughly characterize solid-phase Fe(III) substrates using a suite of complimentary spectroscopic and microscopic techniques because of the limitations of using operationally defined chemical extraction techniques to assess mineral reactivity and reduction and the large variability in (hydr)oxide

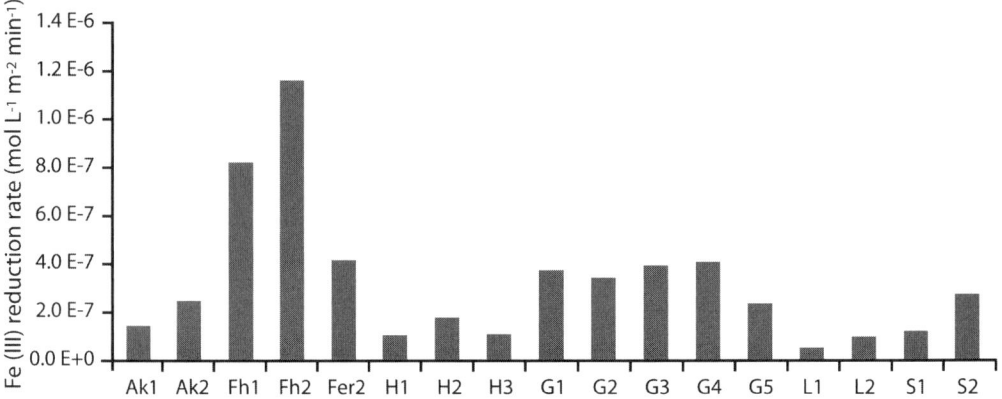

FIGURE 3 Surface area-normalized reduction rates (mol liter^{-1} m^{-2} min^{-1}) for various Fe(III) (hydr)oxides by *G. sulfurreducens* (Cutting et al., 2009). Hematite (H), goethite (G), lepidocrocite (L), feroxyhyte (Fh), akaganeite (Ak), Schwertmannite (S), and two-line ferrihydrite (Fer2) were formed using different synthesis procedures to generate phases varying in size, morphology, surface area, and crystallinity. Reprinted from Cutting et al. (2009) with permission from Elsevier.

properties with synthesis procedures. As detailed below, a number of other geochemical parameters, including solubility (Bonneville et al., 2004, 2009) and crystallinity (Cutting et al., 2009), have been implicated in controlling Fe(III) (hydr)oxide reduction kinetics and have been used to explain deviation from a single surface area-normalized initial reduction rate (see below).

Solubility

Recently, microbial Fe(III) reduction of various Fe(III) (hydr)oxide substrates was shown to have a Michaelis-Menten kinetic dependence with respect to the concentration of Fe(III) substrate (Bonneville et al., 2004). The authors implicitly demonstrate that maximum reduction rates per cell (v_{max}) correlate positively with the solubility of the Fe(III) phase. Because differences in solubility result from variations in synthesis method, Bonneville et al. (2004) determined the solubility of the Fe(III) (hydr)oxides used in their experiment through pε-pH titrations of Fe(III) phase/Fe^{2+}(aq) suspensions (Bonneville et al., 2004). However, between pH 4 and 7, this method was not suitable with crystalline and low-solubility phases. To overcome this issue, Bonneville et al. (2009) later employed a dialysis bag pε-pH titration method under acidic conditions (pH 1 to 2.5). With use of this improved methodology, the solubility and maximum cell-normalized reduction rates (v_{max}) for both the crystalline and poorly crystalline Fe(III) (hydr)oxides showed a positive linear relationship (log-log plot) (Fig. 4). Given the correlation between solubility and surface area ($r^2 = 0.8$), the role of surface area in microbial reduction was also considered. However, the correlation between v_{max} and solubility ($r^2 = 0.90$ on a log-log scale) was far better than for surface area ($r^2 = 0.54$ on a log-log scale) (Bonneville et al., 2009).

The solubility of Fe(III) (hydr)oxides is influenced by endogenous or exogenous organic complexes. As mentioned above, *S. putrefaciens* produces a soluble Fe(III)-organic complex before Fe(III) reduction of different Fe(III) phases, ranging from ferric citrate to hematite (Taillefert et al., 2007). Results from this study suggested that the rate of reduction is linearly dependent on the initial rate of production of this soluble Fe(III)-organic complex, suggesting that Fe(III) solubilization is requisite for Fe(III) reduction. Depending on the size of this organic complex [in relation to N$_2$ and Fe (hydr)oxide pores], the rates of reduction should depend on the amount of reactive surface sites available to complex,

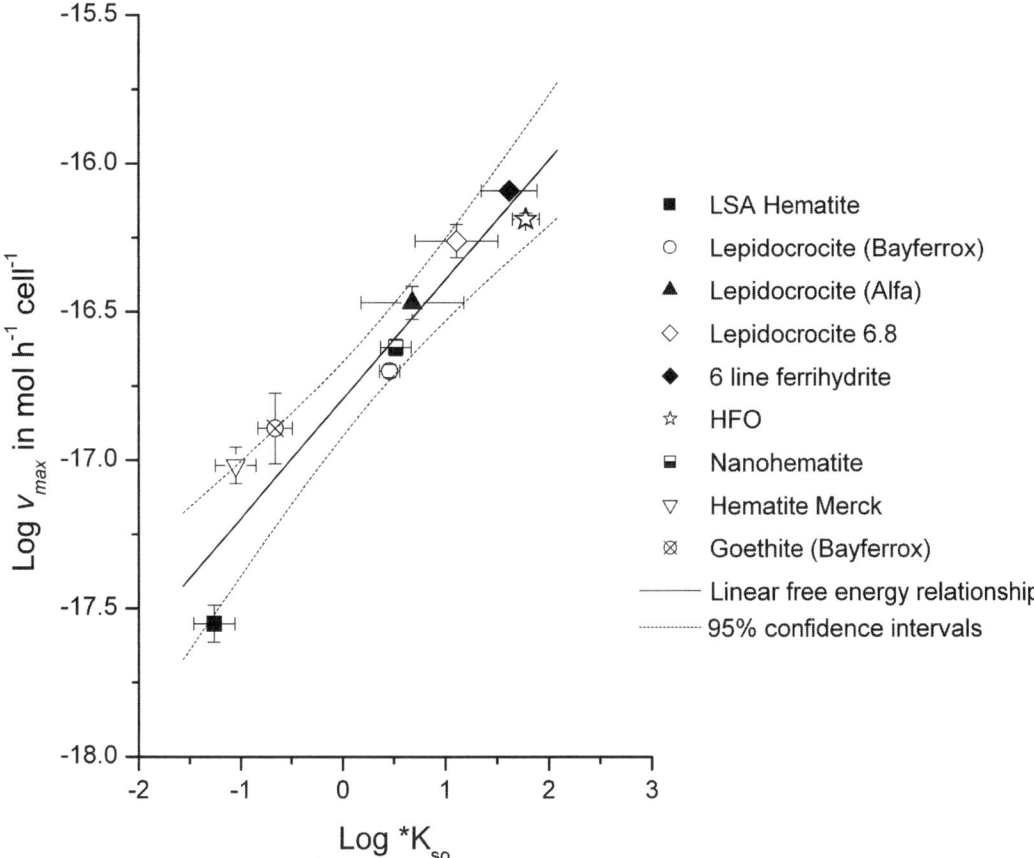

FIGURE 4 Relationship between the solubility product ($^*K_{so}$) of various Fe(III) (hydr)oxides and the maximum initial Fe(III) reduction rate per cell (v_{max}) of *S. putrefaciens* strain 200R (Bonneville et al., 2009). The solubility products were measured for each phase using a dialysis bag technique under acidic conditions (pH 1 to 2.5) at 25°C and defined as $^*K_{so} = a_{Fe3+} \cdot a_{H+}^{-n}$. Reprinted from Bonneville et al. (2009) with permission from Elsevier.

which could correlate well with surface area even in aggregates. Assuming that the same organic molecule (metabolite) is produced in the presence of various Fe(III) (hydr)oxides, the solubility of the Fe(III) phase will also exert a large control on the rates and extent of Fe(III) complexation.

Structural Environment

The role of Fe(III) (hydr)oxide crystallinity has recently been implicated in the rate of Fe(III) (hydr)oxide reduction (Cutting et al., 2009). By use of a variety of methods, a number of Fe(III) (hydr)oxide minerals were synthesized having inter- and intramineral variations in crystallinity, morphology, surface area, and crystallite size. When normalized to surface area, the rates of microbial Fe(III) reduction varied significantly and instead the rates of reduction correlated well with crystallinity, with the poorly crystalline minerals showing higher rates of reduction compared with their crystalline counterparts (Cutting et al., 2009) (Fig. 3). This trend was observed both within and among mineral types. This is particularly evident for the feroxyhyte samples, which have similar surface area, but the poorly crystalline sample (Fh2) was reduced more readily than the relatively crystalline sample (Fh1). These findings implicated crystallinity as the major controlling factor in microbial Fe(III) reduction.

In an attempt to measure the intrinsic susceptibility of Fe(III) (hydr)oxides to microbial reduction, Neal et al. (2003) accounted for surface area and crystallinity effects by employing three single crystal Fe(III) oxide surfaces. Cell accumulation of *S. oneidensis* strain MR-1, which was used as a proxy for cell activity, showed significant differences on the surfaces of the three oxides, with accumulation being greatest on the hematite (001) face compared with the magnetite (111) and (100) faces. Calculations using Marcus and ab initio density functional theory were used to model electron transfer rates from the outer membrane cytochrome to the mineral surface, which showed direct positive correlation between electron transfer rates and the amount of cell accumulation seen in the experiment (Neal et al., 2003).

Constraints on electron transfer to different Fe(III) (hydr)oxides may arise at the protein level. For instance, recent findings indicated that the putative Fe(III) reductase OmcA can bind with high affinity and transfer electrons directly to hematite but not to goethite (Xiong et al., 2006). Adhesion forces have been used to characterize the interactions and affinity between cells or proteins and different mineral surfaces or crystal faces. Lower et al. (2001) used biological force microscopy to measure the adhesion force between *S. oneidensis* MR-1 cells and the (010) surface of the isostructural minerals goethite (α-FeOOH) and diaspore (α-AlOOH). The regulation of a 150-kDa putative Fe(III) reductase to the outer membrane suggested that this protein may be specific for goethite reduction and implied molecular recognition of the Fe(III) (hydr)oxide surface relative to the Al substrate (Lower et al., 2001). Using a similar approach, Neal et al. (2005) investigated the fine scale effects of surface structures on cell adhesion by *S. oneidensis* to the hematite (001) and magnetite (111) and (100) crystal faces. Adhesive forces differed significantly for the three crystal faces, with the hematite (001) face showing statistically significant greater adhesion than both of the magnetite faces (Neal et al., 2005) (Fig. 5). Furthermore, two of the cell-attached cantilevers showed statistical differences between the (111) and (100) magnetite faces. These findings correlated well with the observed pH point of zero charge of the bulk minerals (considering a negative charge for the outer membrane of *Shewanella*) and also with the density of ferric sites. In summary, these findings suggest that microbes and proteins can discriminate between different mineral types and even crystal faces.

Defects in mineral surfaces can lead to an increase in highly reactive sites compared with mineral terraces. Defects (screw step, edge dislocation, kink, etc.) have long been considered important in both surface and bulk chemical reactions (Brown et al., 1999). The importance of defects on bacterial reduction of hematite crystals was investigated with *S. putrefaciens* strain CN32 (Rosso et al., 2003b). Reduction of hematite did not happen at the point of contact of the cell-mineral interface but rather "beyond the footprint of the organism." This was initially attributed to the production of quinone-like electron-shuttling compounds that attacked high-energy surface defects at a distance from the microbe-mineral interface. Ensuing calculations showed that electron conduction away from the initial site of electron transfer within the hematite crystal to surface defects was a (more) likely explanation (Kerisit and Rosso, 2006; Rosso et al., 2003a).

The limited ability of cultured DMRB to respire crystalline Fe(III) (hydr)oxides (Hansel et al., 2004) may be a consequence of ferrihydrite impurities or regions of enhanced disorder within the mineral structures. Upon consumption of these higher-energy (disordered) sites, reduction may cease leading to only minor amounts of Fe(III) reduction. By use of changes in ascorbate-catalyzed reductive dissolution rates, it was estimated that hematite, goethite, and high-surface area goethite contained 0.04, 0.10, and 0.32% "ferrihydrite-like" impurities (Roden, 2006). The residual disorder, however, had little impact on the Fe(III) reduction rates because the amount of Fe(III) reduction by *G. sulfurreducens* was

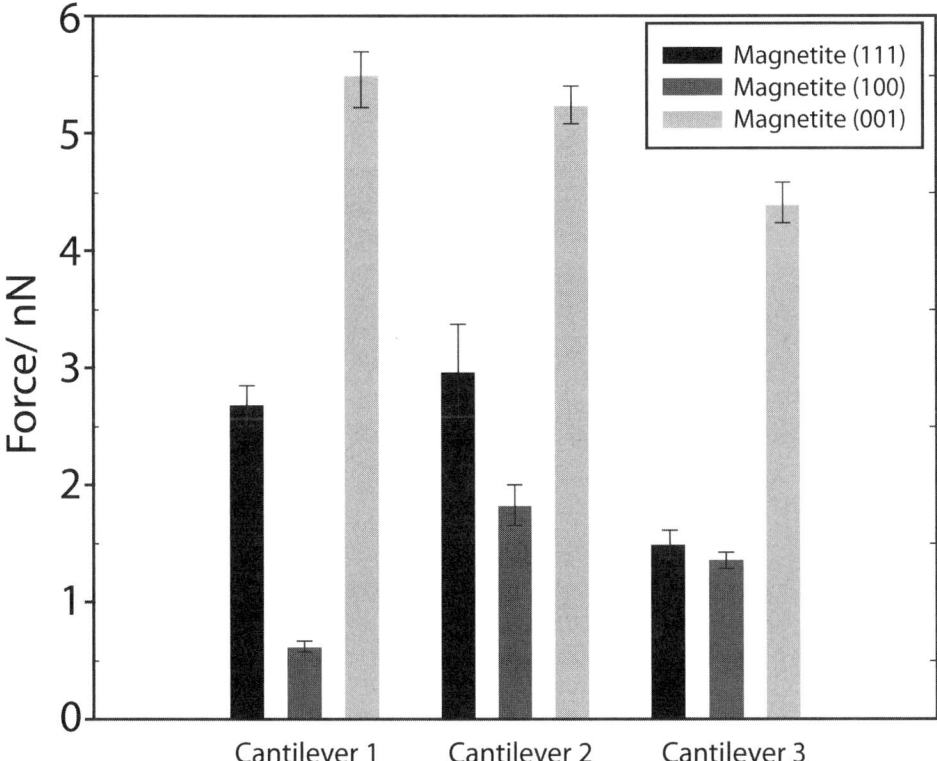

FIGURE 5 Mean force adhesion upon retraction of *S. oneidensis* MR-1 embedded on cantilevers and Fe(III) oxide single crystal faces (Neal et al., 2005). Pairwise comparisons between the means across the three cantilevers indicated significant differences between the two magnetite faces and between the magnetite and hematite faces. Figure reprinted from Neal et al. (2005) under the open access license agreement.

5- to 95-fold higher than the estimated abundance of the ferrihydrite impurities within the (hydr)oxide phases. Recently, we observed that synthesized Fe(III) (hydr)oxides contained a substantial initial degree of disorder, which can approach 25 mol% Fe, depending on the synthesis and washing procedures (Hansel, unpublished). These values are predicted based on the fractional abundance of a ferrihydrite spectral component required to reconstruct the Fe(III) (hydr)oxide extended X-ray absorption fine-structure spectra using linear combination fitting. We found that the amount of Fe(III) reduced within goethite and hematite by *S. putrefaciens* strain CN32 within flowthrough columns is, in fact, a function of the initial degree of disorder within these more stable phases, suggesting that the more crystalline component of the (hydr)oxides is not bioavailable for microbial reduction (Hansel et al., 2004). These contrasting findings further highlight the uncertainty of crystalline (hydr)oxide availability for microbial reduction, at least for presently cultured model DMRB. Observations of microbial reduction of more crystalline (e.g., goethite) Fe(III) (hydr)oxides within natural sediments (Stucki et al., 2007) raise the possibility that presently unrecognized groups of microbes are responsible for reducing more recalcitrant Fe phases within the environment (Hori et al., 2010; Lehours et al., 2009; Lin et al., 2007).

Size

Crystal size reflects the growth conditions under which the Fe(III) (hydr)oxide was formed and determines its surface area (Cornell and

Schwertmann, 2003). As a result, teasing out which property is controlling microbial reduction is difficult. For instance, because of a log-log relationship between surface area and average particle size, surface area and particle size both correlate strongly with the microbial reduction rates of Fe(III) (hydr)oxides (Roden and Zachara, 1996).

Using empirically derived evidence from nanohematite sorption onto *S. putrefaciens*, Bonneville et al. (2006) constructed a model that emphasized the role of microbe-mineral adhesion in the reduction of Fe(III) (hydr)oxide phases. The rate of reduction correlated well with the amount of Fe(III) (hydr)oxide surface coverage of the cell. Based on their model, the apparent half-saturation constant increased with particle size, probably because of a decrease in microbe-mineral interaction. This suggested that, at similar Fe(III) concentrations, cell surface coverage for larger particles will be smaller resulting in lower rates of reduction compared with their smaller counterparts. When applied to previously collected data (Bonneville et al., 2004), the model was able to reproduce initial reduction rates for five Fe(III) (hydr)oxides of varying sizes (Bonneville et al., 2006).

Two studies that explored the role of particle size on hematite nanoparticle reduction found contrasting trends depending on the normalization parameter. By use of a novel aerosol synthesis method that accurately controlled surface area, the reduction rates for three nanohematite (10, 30, and 50 nm) particles revealed a trend with mass-normalized rates decreasing with increasing particle size (Yan et al., 2008). Yet, when normalized to surface area, this trend was reversed. Similarly, surface area–normalized reduction rates of nanohematite ranging in size from 11 to 99 nm in diameter by *S. oneidensis* MR-1 were an order of magnitude higher for larger (99 nm) particles compared with the smallest particle (11 nm) (Bose et al., 2009). These discrepancies are likely attributed to the dependence of Fe(III) reduction rates on the aggregation state and subsequent microbial accessibility of reactive surface sites (see below).

Mineral Aggregation

Iron(III) (hydr)oxide nanometer- and micrometer-sized particles are metastable in the environment but, because of their high surface energy, they tend to aggregate and/or ripen with increased aging (Waychunas et al., 2005). The time frame within which this happens is still poorly understood, but a number of investigations have implicated aggregation of Fe(III) (hydr)oxides as a control on microbial Fe(III) reduction.

One noted exception to the linear trend between surface area of synthetic Fe(III) (hydr)oxides and microbial reduction rates was attributed to mineral aggregation. Aggregation resulted in two-line ferrihydrite and feroxyhyte having lower reduction rates even though they possessed larger surface areas compared with the other Fe(III) (hydr)oxides. These aggregated Fe (hydr)oxides deviated far from the linear trend line and could not be used in the regression analysis for these data (Roden, 2003a; Roden and Zachara, 1996). Since the Brunauer-Emmett-Teller (BET) surface area is measured through N_2 adsorption, it does not necessarily correlate well with reactive site density with respect to microbial accessibility. These data highlight the potential complexity of trying to relate total surface area, measured through the BET surface area, with microbial reduction rates. As has been noted, microbial reduction solely based on surface area may be an oversimplification, owing to specific limitations of microbial accessibility within aggregated minerals, which is not accounted for in chemical adsorption experiments (Zachara et al., 1998).

Attempts to tease out the role of particle size on microbial reduction by using nanometer-sized Fe(III) (hydr)oxides have proved problematic because of particle aggregation. As mentioned above, reduction rates for nanohematite particles showed an expected trend with rates decreasing with increasing particle size. However, when normalized for surface area and cell density, reduction rates were greatest for the large particles (30 and 99 nm) and lowest for small particles (10 and 11 nm) (Bose et al., 2009; Yan et al., 2008). A decrease in "effective" surface area for the small particles was likely due to aggregation and resulted in the relative decrease in surface area and cell-normalized reduction rates. The authors argue that aggregated particles correlated better with the area of cell-particle contact rather than total surface area because direct contact was required for reduction (Bose et al., 2009; Yan et al., 2008). Alternatively, Bose et al. (2009) argue that differences in the aggregation state of the Fe(III) (hydr)oxides may induce different microbial Fe(III) reduction mechanisms (i.e., indirect versus direct).

Surface Sorbates and Coprecipitates

Because of their ubiquity and reactive surface chemistry, Fe(III) (hydr)oxides serve as important sorbents and repositories for many nutrients (e.g., phosphate) and metals (e.g., Al, Pb) (Cornell and Schwertmann, 2003). Sorbed ions may impact surface site reactivity and/or accessibility for reductive dissolution. Furthermore, isomorphic substitution or coprecipitation of metal cations within Fe(III) (hydr)oxides impacts the physicochemical properties of the phase, in particular, solubility. Correspondingly, high concentrations of surface sorbed or substituted ions may diminish the "bioavailability" of these phases for microbial reduction. In fact, the extent of microbial reduction showed an inverse linear correlation with phosphate loading on two-line ferrihydrite within batch incubations (Borch et al., 2006). At full surface coverage, the production of Fe(II) decreased nearly 5-fold compared with pure two-line ferrihydrite. Similarly, arsenate sorption on the surface of ferrihydrite resulted in a decreased extent of microbial Fe(III) reduction (Kocar et al., 2006) compared with that observed previously for pure ferrihydrite (Hansel et al., 2003a) under equivalent flowthrough column conditions. More recently, the impact of arsenic sorption on Fe(III) (hydr)oxide reduction was found to be a function of the arsenate concentration (Chow and Taillefert, 2009). Interestingly, the presence of low As(V) concentrations (≤ 1 µM) stimulated Fe(III) reduction, while higher As(V) levels (>1 µM) inhibited Fe(III) (hydr)oxide reduction, and, instead, arsenate reduction to arsenite prevailed.

The impact of cation substitution on microbial Fe(III) oxide reduction, however, is not as straightforward as that for surface sorbates. For instance, cation substitution within goethite (Al, Co, Mn, Cr) and ferrihydrite (Ni, Pb) decreased the microbial reduction of the Fe(III) phases (Bousserrhine et al., 1999; Fredrickson et al., 2001; Parmar et al., 2001; Sturm et al., 2008), yet Ni substitution within goethite did not impact the susceptibility of goethite to bioreduction (Zachara et al., 2001). Furthermore, Si substitution (1 and 5 mol%) within ferrihydrite had little effect on the rate and extent of microbial Fe(III) reduction and biomineralization by *S. putrefaciens* CN32 (Kukkadapu et al., 2004). The presence of merely 5% structural Al within goethite dramatically decreased the rate and extent of bacterial Fe(III) reduction by the anaerobic fermenter *Clostridium butyricum* (Bousserrhine et al., 1999). Furthermore, both abiotic reductive and acid dissolution of Al-containing goethite decreased with increasing substitution (Bousserrhine et al., 1998, 1999; Schwertmann, 1984; Torrent et al., 1987). Yet, reduction of natural Al-substituted goethite by the common DMRB *S. putrefaciens* strain CN32 was found to be comparable to pure, synthetic goethite (Kukkadapu et al., 2001), which was attributed to enhanced disorder or microheterogeneities counteracting the stabilizing effect of Al within Fe(III) (hydr)oxides (Kukkadapu et al., 2001; Zachara et al., 1998). Thus, the impact of substituted ions on Fe(III) (hydr)oxide reduction is complex, varying with cation type, (hydr)oxide, microbial species, and microbial metabolism (e.g., fermentation versus respiration).

More recently, the impact of Al doping on the rate and extent of Fe(III) reduction by *S. putrefaciens* strain CN32 was shown to vary with Fe(III) (hydr)oxide (Ekstrom et al., in press). In brief, Al substitution dramatically decreased the extent and rate of Fe(III) reduction for ferrihydrite but did not significantly impact lepidocrocite and goethite. Interestingly, the extent of Fe(III) reduction consistently declined 1% for every mole percent increase in Al substitution in ferrihydrite. As Al substitution increased, the rates of Fe(III) reduction for ferrihydrite began to converge with those for lepidocrocite and goethite. Based on projected rates, at Al molar concentrations above 16%, the rate of ferrihydrite reduction would be less than that of both lepidocrocite and goethite (Fig. 6). Recent investigations reveal that the dominant Fe(III)-bearing phases utilized by indigenous metal-reducing microbial populations within uranium-contaminated subsurface soils are Al-substituted goethite and Fe-rich phyllosilicates (Stucki et al., 2007). Goethite, in particular, was readily reduced even though the oxides contained a high degree of Al substitution with mass percentages as high as 35% (Arnseth and Turner, 1988; Phillips et al., 2007; Stucki et al., 2007). This disparity of Al influence on Fe(III) (hydr)oxide bioavailability may explain the persistence of ferrihydrite within sedimentary environments, whereby the capacity for Al-substituted ferrihydrite to serve as a terminal electron acceptor is dramatically diminished, making more crystalline phases favorable.

Reduction Potential

Iron(III) reduction rates were previously shown to be similar over a large range of Fe(III) (hydr)oxide half-cell potentials (Roden, 2003b, 2006). Furthermore, reduction of Fe(III) (hydr)oxides within wetland soil incubations proceeded when thermodynamic calculations predicted they were less favorable than methanogenesis (Roden, 2003a). Thus, it was hypothesized that the thermodynamic properties of Fe(III) (hydr)oxides do not affect bacterial enzymatic electron transfer because microbial recognition is the same regardless of crystal structure (Roden, 2003a). In contrast, there is evidence to suggest that the thermodynamic gain of the redox couple may, in fact, affect microbial Fe(III) reduction (Liu et al., 2001), which will vary substantially given the large Eh range for the different Fe(III) (hydr)oxide phases (Fig. 2).

DMRB display an amazing respiratory versatility, which may be attributed to broad specificity of particular c-type cytochromes

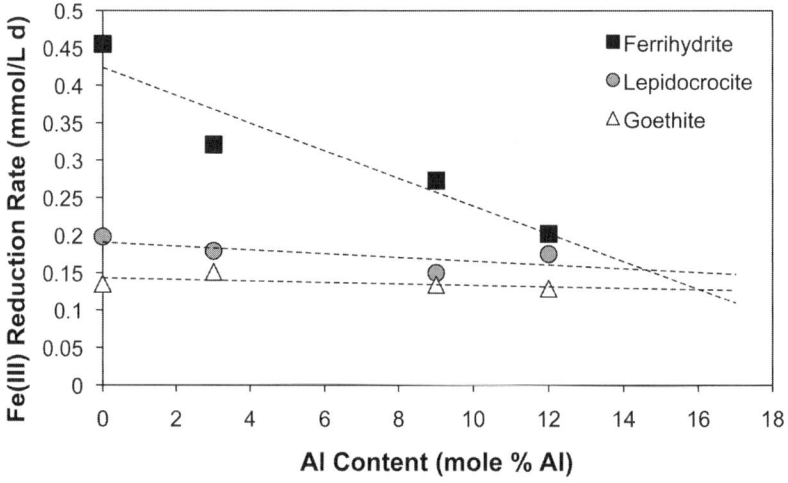

FIGURE 6 Maximum initial Fe(III) reduction rate (mmol liter^{-1} day^{-1}) as a function of Al substitution. Trend lines extended to project Fe(III) reduction rates at higher Al levels. Equivalent Fe(III) reduction trends and crossover point were obtained when Fe(III) (hydr)oxides were provided as a slurry or precipitated onto quartz sand. Modified from Ekstrom et al., 2010.

and/or a large set of cytochromes with widely varying midpoint redox potentials (Dale et al., 2007). For instance, the electron transfer properties of two decaheme cytochromes (OmcA and MtrC) that are considered terminal reductases in *Shewanella* sp. have different properties and mechanisms for binding and transferring electrons to hematite surfaces (Lower et al., 2007; Wigginton et al., 2007). Furthermore, OmcA does not appear to adhere to goethite, likely inhibiting electron transfer (Xiong et al., 2006). Thus, the individual role of the putative terminal reductases in transferring electrons to Fe(III) (hydr)oxides as well as their capacity to interact with phases spanning a large Eh range is unresolved and still under investigation.

Recent investigations have clearly illustrated differential expression of proteins involved in metal respiration (e.g., cytochromes) in the common DMRB *Geobacter* sp. under varying electron acceptor conditions (Ding et al., 2006, 2008). For instance, the physiological status of *G. sulfurreducens* varied in the presence of soluble [i.e., Fe(III) citrate or fumarate] versus insoluble (i.e., ferrihydrite) electron acceptors with a higher abundance of proteins involved in extracellular electron transfer, such as OmcS. Furthermore, it was recently demonstrated that protein expression of *Geobacter uraniireducens* was significantly different when grown in a defined culture medium versus subsurface sediments (Holmes et al., 2009). The difference in redox potential of the electron acceptors [e.g., Fe(III) citrate versus ferrihydrite, ΔE = ~375 mV] will likely have a substantial impact on DMRB physiology; yet, this influence will be masked by the broader signature due to the electron acceptors being in different phases (dissolved versus solid species).

Recently, voltammetric fingerprints of *G. sulfurreducens* adhered to graphite electrodes were found to vary depending on the applied potential on the anode (Busalmen et al., 2008). More energy was produced when higher potentials (600 mV, relative to Ag/AgCl⁻ electrode with E_{NHE} = 197 mV) were applied relative to lower potentials (100 mV), suggesting a greater electron transport capacity at higher E_h. Of particular relevance, however, are recent findings that the succession and composition of anode-hosted bacterial populations differed substantially at different anode potentials (300 versus 600 mV relative to Ag/AgCl⁻) (White et al., 2009). Microbial fuel cells inoculated with seawater contained four dominant bacterial groups, δ-, γ-, and ε-*Proteobacteria* and *Flavobacterium-Cytophaga-Bacteroides* (FCB). While *Geobacter* spp. dominated the bacterial community during peak power production on anodes at 600 mV, FCB phylotypes predominated at lower reduction potentials (300 mV). The reduction potential range for these experiments (corrected for reference to Ag/AgCl⁻) is approximately −100 to 400 mV spanning Fe(III) chelates [e.g., Fe(III) citrate] to lepidocrocite (γ-FeOOH, E = −275 mV) (see Fig. 2). While these systems did not attempt to mimic the potential of low E_h metal substrates (e.g., goethite), these novel studies revealed that the reduction potential of anodes impacts the phylogeny and physiology of anode-associated communities—findings that can be translated to mineral surfaces. The role of E_h in impacting the structure, activity, and physiology of Fe(III)-reducing populations remains an unexplored area and one that warrants investigation.

IMPACT OF MICROBIAL Fe(III) (HYDR)OXIDE REDUCTION ON THE FATE AND TRANSPORT OF METALS AND RADIONUCLIDES

As a consequence of their high surface area and density of reactive surface sites, Fe(III) (hydr)oxides serve as reactive templates in soils and sediments. They coprecipitate many metals and sorb numerous organics (e.g., pesticides, *p*-aminohippuric acids), nutrients (e.g., phosphate), and metals (e.g., Pb, As, U). Microbial Fe(III) reduction results in either mineral dissolution or conversion to more stable (e.g., more crystalline) secondary phases having a lower sorptive capacity, such as goethite (Hansel et al., 2005; Tufano and Fendorf, 2008; Tufano et al., 2008). Consequently, the reduction of metal-laden Fe(III) (hydr)oxide phases may release associated contaminants into solution. For instance, the release of sorbed and coprecipiated metals (e.g., As, Ni, Co) and nutrients (e.g., phosphate) during microbial reduction of pure Fe(III) (hydr)oxide phases (Fredrickson et al., 2001; Zachara et al., 2001) and Fe(III)-bearing sediments (Crowe et al., 2007; Cummings et al., 1999; Landa et al., 1991) has been frequently observed. The ultimate fate of the associated ions, however, will be a function of ion speciation, mineral transformations, and the aqueous chemical environment. In fact, the fate of many contaminants will be a function, in large part, of the mineralogy of the initial Fe(III) sorbent and products of microbial Fe(III) reduction.

A number of recent investigations have illustrated that reductive dissolution of Fe(III) (hydr)oxides may actually enhance contaminant attenuation, in particular, through sequestration within secondary precipitates. In particular, the formation of secondary magnetite following microbial reduction of Fe(III) (hydr)oxides in the presence of Zn(II) or Ni(II) results in immobilization of the cations, presumably within the magnetite structure (Cooper et al., 2000; Fredrickson et al., 2001). However, batch experiments examining Fe(III) reduction of synthetic and natural Fe(III) (hydr)oxides in the presence of Zn illustrate that the immobilization of Zn may be inhibited in the presence of complexing agents, including nitrilotriacetic acid, clays, and the humic acid analog 9,10-anthraquinone-2,6-disulfonic acid (Coby and Picardal, 2006). Furthermore, lanthanide (Nd, Gd, Tb, Ho, Er)- and transition metal (Cr, Mn, Co, Ni, and Zn)-substituted magnetites have been formed via microbial Fe(III) reduction of akaganeite (β-FeOOH) (Moon et al., 2007a, 2007b). Recent studies also illustrate that sequestration of arsenic under Fe(III)-reducing conditions may be a function of magnetite formation (Kocar et al., 2006; Tufano and Fendorf, 2008; Pedersen et al., 2006). Microbial reduction of As(V)-sorbed ferrihydrite under advective flow resulted in the enhanced retention of As, as both As(III) and As(V), relative to systems not undergoing magnetite precipitation (Kocar et al., 2006). Similarly, recent studies indicate that As(V) is strongly sorbed (Islam et al., 2005) or is incorporated into the magnetite structure (Coker et al., 2006) following microbial reduction of both ferrihydrite or soluble Fe(III). These findings all contradict the previous assumptions and laboratory observations (Cummings et al., 1999) that microbial reduction of Fe(III) (hydro)oxides, which are strong sorbents of metal(loid)s, would result in the enhanced mobilization of As. Similar behavior has recently been observed for uranium (Neiss et al., 2007). In the presence of Ca, the formation of ternary $Ca-UO_2-CO_3$ complexes results in the preferential reduction of Fe(III) (Brooks et al., 2003) and subsequent transformation of ferrihydrite to magnetite. The preferential reduction of Fe(III) and mineralization to magnetite results in the preservation and sequestration of the oxidized U(VI) species, which is found adsorbed onto or incorporated into secondary goethite and magnetite precipitates (Stewart et al., 2009; Nico et al., 2009). Thus, the potential for secondary magnetite formation to immobilize contaminants is more extensive than originally thought. While these findings hold promise for introducing additional pathways of contaminant immobilization within reducing environments, the stability of these magnetite complexes is not known, and they therefore pose a potential means of contaminant remobilization upon changing environmental conditions. In particular, while microbial reduction of pure magnetite is limited at circumneutral pH (Kostka and Nealson, 1995), distortion of the magnetite structure (Coker et al., 2006) through ion incorporation will likely impact the stability and bioavailability of magnetite for microbial reduction.

Microbial reduction of Fe(III) (hydr)oxides may also result in a cascade of subsequent redox and/or complexation reactions because of the reactivity of the metabolite Fe(II). Microbial Fe(III) reduction provides a strong reductant and complexing ligand (Fe^{II}). The microbial reduction of Fe(III) (hydr)oxides may therefore indirectly result in the immobilization of a number of anion or redox-active contaminants and nutrients. For instance, Fe(II)-stimulated removal of phosphate has received increasing attention as a possible means of attenuating phosphate from septic systems through the formation of the ferrous phosphate mineral vivianite (Robertson, 2000). Furthermore, ferrous Fe is a particularly strong reductant of many contaminants including metals (e.g., Cr) (Fendorf et al., 2000), radionuclides (e.g., U, Tc) (Fredrickson et al., 2004), and organic compounds (e.g., nitroamines) (Hofstetter et al., 1999). Consequently, the rate of contaminant reduction will be a function of Fe(III) reduction rates and the bioavailability of Fe(III), which will be a function of the geochemical and mineralogical factors discussed above. Interestingly, laboratory simulations indicate that effective reduction of redox-active metals (e.g., Cr) can be sustained on catalytic amounts of Fe(III) (Wielinga et al., 2001). For instance, biogenic Fe(II) abiotically reacts with Cr(VI), resulting in the formation of mixed Fe(III)-Cr(III) (hydr)oxide phases (Eary and Rai, 1989; Hansel et al., 2003b; Sass and Rai, 1987), resulting in the regeneration of microbially available Fe(III) (primarily as ferrihydrite) (Hansel et al., 2003b). Theoretically, this continued Fe-Cr redox cycling can result in sustained Cr(VI) reduction, considering minimal Fe loss from the system. In some cases, the rate, extent, and products of Fe(II)-induced contaminant reduction are a function of the bonding environment of the biogenic Fe(II). While aqueous Fe(II) is a potent reductant of chromate (Cr^{VI}), the rates of Tc(VI) and U(VI) reduction are dramatically enhanced via reaction with surface-bound (adsorbed) or coprecipitated Fe(II) (e.g., magnetite) (Fredrickson et al., 2004; Lloyd et al., 2000). For instance, while dissolved concentrations of uranyl (U^{VI}) are unchanged in the presence of aqueous Fe(II) over an 80-h reaction period, Fe(II) adsorbed on hematite results in complete removal of U(VI) via reduction within 60 h (Liger et al., 1999).

CONCLUSIONS

Understanding the geochemical controls on microbial Fe(III) reduction is crucial for defining the reducing capacity of soils and sediments, predicting the fate and transport of (in)organic contaminants, delineating the controls on carbon degradation and cycling, and forecasting electron flow and operative terminal electron accepting processes in subsurface environments. Variations in the rate and extent of microbial reduction of Fe(III) (hydr)oxides has been linked to surface area, solubility, crystallite size, thermodynamics, crystal structure, disorder, aggregation, surface sorbates, and cation substitution. Yet, as a consequence of the intimate association between these physicochemical properties and their evolution with reaction progression, isolating the contribution of various factors on constraining microbial Fe(III) reduction has proved a challenging task. We contend that, to further advance our knowledge, more sophisticated techniques and experimental approaches will need to be used, including surface-sensitive and time-resolved spectroscopy to accurately interrogate the molecular environment. We also anticipate that using poised anodes as surrogate minerals holds promise in addressing the role of reduction potential in electron transfer processes. Ultimately, unraveling the enigma of the microbe-mineral interface will require a multidisciplinary approach requiring an appreciation of the physics and chemistry of mineral surfaces, the enzymatic and nonenzymatic pathways responsible for electron transfer, and the ecology of metal-reducing microbes within a complex mineral framework.

REFERENCES

Anderson, R. T., and D. R. Lovley. 2000. Anaerobic bioremediation of benzene under sulfate-reducing conditions in a petroleum-contaminated aquifer. *Environ. Sci. Technol.* **34:**2261–2266.

Arnseth, R. W., and R. S. Turner. 1988. Sequential extraction of iron, manganese, aluminum, and silicaon in soils from 2 contrasting watersheds. *Soil Sci. Soc. Am. J.* **52:**1801–1807.

Baes, C. F., and R. E. Mesmer. 1976. *The Hydrolysis of Cations.* Wiley, New York, NY.

Benner, S. G., C. M. Hansel, B. W. Wielinga, T. M. Barber, and S. Fendorf. 2002. Reductive dissolution and biomineralization of iron hydroxide under dynamic flow conditions. *Environ. Sci. Technol.* **36:**1705–1711.

Bonneville, S., T. Behrends, and P. Van Cappellen. 2009. Solubility and dissimilatory reduction kinetics of iron(III) oxyhydroxides: a linear free energy relationship. *Geochim. Cosmochim. Acta* **17:**5273–5282.

Bonneville, S., T. Behrends, P. Van Cappellen, C. Hyacinthe, and W. F. M. Roling. 2006. Reduction of Fe(III) colloids by Shewanella putrefaciens: a kinetic model. *Geochim. Cosmochim. Acta* **70:**5842–5854.

Bonneville, S., P. Van Cappellen, and T. Behrends. 2004. Microbial reduction of iron(III) oxyhydroxides: effects of mineral solubility and availability. *Chem. Geol.* **212:**255–268.

Borch, T., Y. Masue, R. K. Kukkadapu, and S. Fendorf. 2006. Phosphate imposed limitations on biological reduction and alteration of ferrihydrite. *Environ. Sci. Technol.* **41:**166–172.

Bose, S., M. F. Hochella, Jr., Y. A. Gorby, D. W. Kennedy, D. E. McCready, A. S. Madden, and B. H. Lower. 2009. Bioreduction of hematite nanoparticles by the dissimilatory iron reducing bacterium Shewanella oneidensis MR-1. *Geochim. Cosmochim. Acta* **73:**962–976.

Bousserrhine, N., U. G. Gasser, E. Jeanroy, and J. Berthelin. 1998. Effect of aluminum substitution on ferri-reducing bacterial activity and dissolution of goethites. *C. R. Acad. Sci.* **326:**617–624.

Bousserrhine, N., U. G. Gasser, E. Jeanroy, and J. Berthelin. 1999. Bacterial and chemical reductive dissolution of Mn-, Co-, Cr-, and Al-substituted goethites. *Geomicrob. J.* **16:**245–258.

Bromfield, S. M. 1954a. The reduction of iron oxide by bacteria. *J. Soil. Sci.* **5:**129–139.

Bromfield, S. M. 1954b. Reduction of ferric compounds by soil bacteria. *J. Gen. Microbiol.* **11:**1–6.

Brooks, S. C., J. K. Fredrickson, S. L. Carroll, D. W. Kennedy, J. M. Zachara, A. E. Plymale, S. D. Kelly, K. M. Kemner, and S. Fendorf. 2003. Inhibition of bacterial U(VI) reduction by calcium. *Environ. Sci. Technol.* **37:**1850–1858.

Brown, G. E., Jr., V. E. Henrich, W. H. Casey, D. L. Clark, C. Eggleston, A. Felmy, D. W. Goodman, M. Gratzel, G. Maciel, M. I. McCarthy, K. H. Nealson, D. A. Sverjensky, M. F. Toney, and J. M. Zachara. 1999. Metal oxide surfaces and their interactions with aqueous solutions and microbial organisms. *Chem. Rev.* **99:**77–174.

Busalmen, J. P., A. Esteve-Nunez, and J. M. Feliu. 2008. Whole cell electrochemistry of electricity-producing microorganisms evidence an adaptation for optimal exocellular electron transport. *Environ. Sci. Technol.* **42:**2445–2450.

Caccavo, F., Jr., R. P. Blakemore, and D. R. Lovley. 1992. A hydrogen-oxidizing, Fe(III)-reducing microorganism from the Great Bay Estuary, New Hampshire. *Appl. Environ. Microbiol.* **58:**3211–3216.

Chow, S. S., and M. Taillefert. 2009. Effect of arsenic concentration on microbial iron reduction and arsenic speciation in an iron-rich freshwater sediment. *Geochim. Cosmochim. Acta* **73:**6008–6021.

Coby, A. J., and F. W. Picardal. 2006. Influence of sediment components on the immobilization of Zn during microbial Fe-(hydr)oxide reduction. *Environ. Sci. Technol.* **40:**3813–3818.

Coker, V. S., A. G. Gault, C. I. Pearce, G. Van der Laan, N. D. Telling, J. M. Charnock, D. A. Polya, and J. R. Lloyd. 2006. XAS and XMCD evidence for species-dependent partitioning of arsenic during microbial reduction of ferrihydrite to magnetite. *Environ. Sci. Technol.* **40:**7745–7750.

Cooper, D. C., F. Picardal, J. Rivera, and C. Talbot. 2000. Zinc immobilization and magnetite formation via ferric oxide reduction by Shewanella putrefaciens 200. *Environ. Sci. Technol.* **34:**100–106.

Cornell, R. M., W. Schneider, and R. Giovanoli. 1991. Phase transformations in the ferrihydrite/cysteine system. *Polyhedron* **8:**2829–2834.

Cornell, R. M., and U. Schwertmann. 2003. *The Iron Oxides: Structure, Properties, Reactions, Occurrence and Uses,* 2nd ed. VCH, Weinheim, Germany.

Crowe, S. A., J. A. Roberts, C. G. Weisener, and D. A. Fowle. 2007. Alteration of iron-rich lacustrine sediments by dissimilatory iron-reducing bacteria. *Geobiology* **5:**63–73.

Cummings, D. E., F. Caccavo, Jr., S. Fendorf, and R. F. Rosenzweig. 1999. Arsenic mobilization by the dissimilatory Fe(III)-reducing bacterium Shewanella alga BrY. *Environ. Sci. Technol.* **33:**723–729.

Cutting, R. S., V. S. Coker, J. W. Fellowes, J. R. Lloyd, and D. J. Vaughan. 2009. Mineralogical and morphological constraints on the reduction of Fe(III) minerals by Geobacter sulfurreducens. *Geochim. Cosmochim. Acta* **73:**4004–4022.

Dale, J. R., R. Wade, Jr., and T. J. DiChristina. 2007. A conserved histidine in cytochrome *c* maturation permease CcmB of Shewanella putrefaciens is required for anaerobic growth below a threshold standard redox potential. *J. Bacteriol.* **189:**1036–1043.

DiChristina, T. J., J. K. Fredrickson, and J. M. Zachara. 2005. Enzymology of electron transport: energy generation with geochemical consequences. *Rev. Mineral. Geochem.* **59:**27–52.

Ding, Y.-H. R., K. K. Hixson, M. A. Aklujkar, M. S. Lipton, R. D. Smith, D. R. Lovley, and T. Mester. 2008. Proteome of *Geobacter sulfurreducens* grown with Fe(III) oxide or Fe(III) citrate as the electron acceptor. *Biochim. Biophys. Acta* **1784**:1935–1941.

Ding, Y.-H. R., K. K. Hixson, C. S. Giometti, A. Stanley, A. Esteve-Nunez, T. Khare, S. L. Tollaksen, W. Zhu, J. N. Adkins, M. S. Lipton, R. D. Smith, T. Mester, and D. R. Lovley. 2006. The proteome of dissimilatory metal-reducing microorganism *Geobacter sulfurreducens* under various growth conditions. *Biochim. Biophys. Acta* **1764**:1198–1206.

Dobbin, P. S., J. P. Carter, C. Garcia-Salamanca San Juan, M. von Hobe, A. K. Powell, and D. J. Richardson. 1999. Dissimilatory Fe(III) reduction by *Clostridium beijerinckii* isolated from freshwater sediment using Fe(III) maltol enrichment. *FEMS Microbiol. Lett.* **176**:131–138.

Eary, L. E., and D. Rai. 1989. Kinetics of chromate reduction by ferrous-ions derived from hematite and biotite at 25-degrees-C. *Am. J. Sci.* **289**:180–213.

Ehrlich, H. L., and D. K. Newman. 2009. *Geomicrobiology*, 5th ed. CRC Press, Boca Raton, FL.

Ekstrom, E. B., D. R. Learman, A. S. Madden, and C. M. Hansel. 2010. Contrasting effects of Al substitution on microbial reduction of Fe(III) (hydr)oxides. *Geochim. Cosmochim. Acta* **74**:7086–7099.

El-Naggar, M. Y., Y. A. Gorby, W. Xia, and K. H. Nealson. 2008. The molecular density of states in bacterial nanowires. *Biophys. J.* **95**:L10–L12.

Favre, F., J. W. Stucki, and P. Boivin. 2006. Redox properties of structural Fe in ferruginous smectite. A discussion of the standard potential and its environmental implications. *Clays Clay Miner.* **54**:466–472.

Fendorf, S., B. W. Wielinga, and C. M. Hansel. 2000. Chromium transformations in natural environments: the role of biological and abiological processes in chromium(VI) reduction. *Int. Geol.* **42**:691–701.

Fennessey, C. M., M. E. Jones, M. Taillefert, and T. J. DiChristina. 2010. Siderophores are not involved in Fe(III) solubilization during anaerobic Fe(III) respiration by *Shewanella oneidensis* MR-1. *Appl. Environ. Microbiol.* **76**:2425–2432.

Fredrickson, J. K., R. K. Kukkadapu, C. K. Liu, and J. M. Zachara. 2003. Influence of electron donor/acceptor concentrations on hydrous ferric oxide (HFO) bioreduction. *Biodegradation* **14**:91–103.

Fredrickson, J. K., J. M. Zachara, D. W. Kennedy, H. Dong, T. C. Onstott, N. W. Hinman, and S.-M. Li. 1998. Biogenic iron mineralization accompanying the dissimilatory reduction of hydrous ferric oxide by a groundwater bacterium. *Geochim. Cosmochim. Acta* **62**:3239–3257.

Fredrickson, J. K., J. M. Zachara, D. W. Kennedy, M. C. Duff, Y. A. Gorby, S.-M. W. Li, and K. M. Krupka. 2000. Reduction of U(VI) in goethite (a-FeOOH) suspension by a dissimilatory metal-reducing bacterium. *Geochim. Cosmochim. Acta* **64**:3085–3098.

Fredrickson, J. K., J. M. Zachara, D. W. Kennedy, R. Kukkadapu, J. P. McKinley, S. M. Heald, C. Liu, and A. E. Plymale. 2004. Reduction of TcO4- by sediment-associated biogenic Fe(II). *Geochim. Cosmochim. Acta* **68**:3171–3187.

Fredrickson, J. K., J. M. Zachara, R. K. Kukkadapu, Y. A. Gorby, S. C. Smith, and C. F. Brown. 2001. Biotransformation of Ni-substituted hydrous ferric oxide by an Fe(III)-reducing bacterium. *Environ. Sci. Technol.* **35**:703–712.

Froelich, P. N., G. P. Klinkhammer, M. L. Bender, N. A. Luedtke, G. R. Heath, D. Cullen, P. Dauphin, D. Hammond, B. Hartman, and V. Maynard. 1979. Early oxidation of organic matter in pelagic sediments of the eastern equatorial Atlantic: suboxic diagenesis. *Geochim. Cosmochim. Acta* **43**:1075–1090.

Gorby, Y. A., S. Yanina, J. S. McLean, K. M. Rosso, D. Moyles, A. Dohnalkova, T. J. Beveridge, I. S. Chang, B.-H. Kim, K. S. Kim, D. E. Culley, S. B. Reed, M. F. Romine, D. Saffarini, E. A. Hill, L. Shi, D. Elias, D. W. Kennedy, G. Pinchuk, K. Watanabe, S. I. Ishii, B. Logan, K. H. Nealson, and J. K. Fredrickson. 2006. Electrically conductive bacterial nanowires produced by *Shewanella oneidensis* strain MR-1 and other microorganisms. *Proc. Natl. Acad. Sci. USA* **103**:11358–11363.

Gralnick, J. A., and D. K. Newman. 2007. Extracellular respiration. *Mol. Microbiol.* **65**:1–11.

Grantham, M. C., P. M. Dove, and T. J. DiChristina. 1997. Microbially catalyzed dissolution of iron and aluminum oxyhydroxide mineral surface coatings. *Geochim. Cosmochim. Acta* **61**:4467–4477.

Hansel, C. M., S. G. Benner, and S. Fendorf. 2005. Competing Fe(II)-induced mineralization pathways of ferrihydrite. *Environ. Sci. Technol.* **39**:7147–7153.

Hansel, C. M., S. G. Benner, J. Neiss, A. Dohnalkova, R. K. Kukkadapu, and S. Fendorf. 2003a. Secondary mineralization pathways induced by dissimilatory iron reduction of ferrihydrite under advective flow. *Geochim. Cosmochim. Acta* **67**:2977–2992.

Hansel, C. M., S. G. Benner, P. Nico, and S. Fendorf. 2004. Structural constraints of ferric (hydr)oxides on dissimilatory iron reduction and the fate of Fe(II). *Geochim. Cosmochim. Acta* **68:**3217–3229.

Hansel, C. M., S. Fendorf, P. M. Jardine, and C. A. Francis. 2008. Changes in bacterial and archaeal community structure and functional diversity along a geochemically variable soil profile. *Appl. Environ. Microbiol.* **74:**1620–1633.

Hansel, C. M., B. W. Wielinga, and S. Fendorf. 2003b. Structural and compositional evolution of Cr/Fe solids after indirect chromate reduction by dissimilatory iron-reducing bacteria. *Geochim. Cosmochim. Acta* **67:**401–412.

Haveman, S. A., R. J. DiDonato, L. Villanueva, E. S. Shelobolina, B. L. Postier, B. Xu, A. Liu, and D. R. Lovley. 2008. Genome-wide gene expression patterns and growth requirements suggest that *Pelobacter carbinolicus* reduces Fe(III) indirectly via sulfide production. *Appl. Environ. Microbiol.* **74:**4277–4284.

Hernandez, M. E., and D. Newman. 2001. Extracellular electron transfer: review. *Cell. Mol. Life Sci.* **58:**1562–1571.

Hernandez, M. E., A. Kappler, and D. K. Newman. 2004. Phenazines and other redox active antibiotics promote microbial mineral reduction. *Appl. Environ. Microbiol.* **70:**921–928.

Hofstetter, T. B., C. G. Heijman, S. B. Haderlein, C. Holliger, and R. P. Schwarzenbach. 1999. Complete reduction of TNT and other (poly)nitroaromatic compounds under iron-reducing subsurface conditions. *Environ. Sci. Technol.* **33:**1479–1487.

Holmes, D. E., R. A. O'Neil, M. A. Chavan, L. A. N'Guessan, H. A. Vrionis, L. A. Perpetua, M. J. Larrahondo, R. DiDonato, A. Liu, and D. R. Lovley. 2009. Transcriptome of *Geobacter uraniireducens* growing in uranium-contaminated subsurface sediments. *ISME J.* **3:**216–230.

Hori, T., A. Muller, Y. Igarashi, R. Conrad, and M. W. Friedrich. 2010. Identification of iron-reducing microorganisms in anoxic rice paddy soil by 13C-acetate probing. *ISME J.* **4:**267–278.

Islam, F. S., R. L. Pederick, A. G. Gault, L. K. Adams, D. A. Polya, J. M. Charnock, and J. R. Lloyd. 2005. Interactions between the Fe(III)-reducing bacterium Geobacter sulfurreducens and arsenate, and capture of the metalloid by biogenic Fe(II). *Appl. Environ. Microbiol.* **71:**8642–8648.

Jakobsen, R., and D. Postma. 1999. Redox zoning, rates of sulfate reduction and interactions with Fe-reduction and methanogenesis in a shallow sandy aquifer, Romo, Denmark. *Geochim. Cosmochim. Acta* **63:**137–151.

Jang, J.-H., B. A. Dempsey, G. L. Catchen, and W. D. Burgos. 2003. Effects of Zn(II), Cu(II), Mn(II), Fe(II), NO_3^-, or SO_4^{2-} at pH 6.5 and 8.5 on transformations of hydrous ferric oxide (HFO) as evidenced by Mossbauer spectroscopy. *Colloids Surfaces A Physicochem. Eng. Aspects* **221:**55–68.

Jones, J., S. Gardener, and B. M. Simon. 1984. Reduction of ferric iron by heterotrophic bacteria in lake sediments. *J. Gen. Microbiol.* **130:**45–51.

Jones, M. E., C. M. Fennessey, T. J. DiChristina, and M. Taillefert. 2010. Shewanella oneidensis MR-1 mutants selected for their inability to produce soluble organic-Fe(III) complexes are unable to respire Fe(III) as an electron acceptor. *Environ. Microbiol.* **12:**938–950.

Kappler, A., and K. L. Straub. 2005. Geomicrobiological cycling of iron. *Rev. Mineral. Geochem.* **59:**85–108.

Kerisit, S., and K. M. Rosso. 2006. Computer simulation of electron transfer at hematite surfaces. *Geochim. Cosmochim. Acta* **70:**1888–1903.

Kocar, B. D., M. J. Herbel, K. J. Tufano, and S. Fendorf. 2006. Contrasting effects of dissimilatory iron(III) and arsenic(V) reduction on arsenic retention and transport. *Environ. Sci. Technol.* **40:**6715–6721.

Kostka, J. E., and K. H. Nealson. 1995. Dissolution and reduction of magnetite by bacteria. *Environ. Sci. Technol.* **29:**2535–2540.

Kostka, J. E., J. W. Stucki, K. H. Nealson, and J. Wu. 1996. Reduction of structural Fe(III) in smectite by a pure culture of Shewanella putrefaciens strain MR-1. *Clays Clay Miner.* **44:**522–529.

Kostka, J. E., J. Wu, K. H. Nealson, and J. W. Stucki. 1999. The impact of structural Fe(III) reduction by bacteria on the surface chemistry of smectite clay minerals. *Geochim. Cosmochim. Acta* **63:**3705–3713.

Kukkadapu, R., J. M. Zachara, J. K. Fredrickson, and D. W. Kennedy. 2004. Biotransformation of two-line silica-ferrihydrite by a dissimilatory Fe(III)-reducing bacterium: formation of carbonate green rust in the presence of phosphate. *Geochim. Cosmochim. Acta* **68:**2799–2814.

Kukkadapu, R. K., J. M. Zachara, S. C. Smith, J. K. Fredrickson, and C. Liu. 2001. Dissimilatory bacterial reduction of Al-substituted goethite in subsurface sediments. *Geochim. Cosmochim. Acta* **65:**2913–2924.

Landa, E. R., E. J. P. Phillips, and D. R. Lovley. 1991. Release of Ra-226 from uranium mill tailings by microbial Fe(III) reduction. *Appl. Geochem.* **6:**647–652.

Langmuir, D. 1969. The Gibbs free energies of substrates in the system Fe-O_2-H_2O-CO_2 at 25C, p. B180–B184. U.S. Geol. Surv. Prof. Paper 650-B. U.S. Geological Survey, Reston, VA.

Larese-Casanova, P., and M. M. Scherer. 2007. Fe(II) sorption on hematite: new insights based on spectroscopic measurements. *Environ. Sci. Technol.* **41:**471–477.

Lehours, A.-C., I. Batisson, A. Guedon, G. Mailhot, and G. Fonty. 2009. Diversity of culturable bacteria from the anaerobic zone of the meromictic Lake Pavin, able to perform dissimilatory-iron reduction in different in vitro conditions. *Geomicrobiol. J.* **26:**212–223.

Lies, D. P., M. E. Hernandez, A. Kappler, R. E. Mielke, J. A. Gralnick, and D. K. Newman. 2005. Shewanella oneidensis MR-1 uses overlapping pathways for iron reduction at a distance and by direct contact under conditions relevant for biofilms. *Appl. Environ. Microbiol.* **71:**4414–4426.

Liger, E., L. Charlet, and P. Van Cappellen. 1999. Surface catalysis of uranium(VI) reduction by iron(II). *Geochim. Cosmochim. Acta* **63:**2939–2955.

Lin, B., C. Hyacinthe, S. Bonneville, M. Braster, P. Van Cappellen, and W. F. M. Roling. 2007. Phylogenetic and physiological diversity of dissimilatory ferric iron reducers in sediments of the polluted Scheldt estuary, Northwest Europe. *Environ. Microbiol.* **9:**1956–1968.

Liu, C. X., S. Kota, J. M. Zachara, J. K. Fredrickson, and C. K. Brinkman. 2001. Kinetic analysis of the bacterial reduction of goethite. *Environ. Sci. Technol.* **35:**2482–2490.

Lloyd, J. R., J. Chesnes, S. Glasauer, D. J. Bunker, F. R. Livens, and D. R. Lovley. 2002. Reduction of actinides and fission products by Fe(III)-reducing bacteria. *Geomicrobiol. J.* **19:**103–120.

Lloyd, J. R., V. A. Sole, C. V. G. Van Praagh, and D. R. Lovley. 2000. Direct and Fe(II)-mediated reduction of technetium by Fe(III)-reducing bacteria. *Appl. Environ. Microbiol.* **66:**3743–3749.

Lovley, D. R. 1987. Organic matter mineralization with the reduction of ferric iron: a review. *Geomicrobiol. J.* **5:**375–399.

Lovley, D. R. 1991. Dissimilatory Fe(III) and Mn(IV) reduction. *Microbiol. Rev.* **55:**259–287.

Lovley, D. R. 2000. Fe(III)- and Mn(IV)-reducing prokaryotes. *In* M. Dworkin, S. Falkow, E. Rosenberg, K. Schleifer, and E. Stackebrandt (ed.), *The Prokaryotes*. Springer-Verlag, New York, NY.

Lovley, D. R. 2001. Anaerobes to the rescue. *Science* **293:**1444–1446.

Lovley, D. R., J. D. Coates, E. L. Blunt-Harris, E. J. P. Phillips, and J. C. Woodward. 1996. Humic substances as electron acceptors for microbial respiration. *Nature* **382:**445–448.

Lovley, D. R., D. E. Holmes, and K. P. Nevin. 2004. Dissimilatory Fe(III) and Mn(IV) reduction. *Adv. Microb. Physiol.* **49:**219 286.

Lovley, D. R., and E. J. P. Phillips. 1986. Availability of ferric iron for microbial reduction in bottom sediments of the freshwater tidal Potomac River. *Appl. Environ. Microbiol.* **52:**751–757.

Lovley, D. R., and E. J. P. Phillips. 1988. Novel mode of microbial energy metabolism: organic carbon oxidation coupled to dissimilatory reduction of iron or manganese. *Appl. Environ. Microbiol.* **54:**1472–1480.

Lovley, D. R., E. J. P. Phillips, Y. A. Gorby, and E. R. Landa. 1991a. Microbial reduction of uranium. *Nature* **350:**413–416.

Lovley, D. R., E. J. P. Phillips, and D. J. Lonergan. 1991b. Enzymatic versus nonenzymatic mechanisms for Fe(III) reduction in aquatic sediments. *Environ. Sci. Technol.* **25:**1062–1067.

Lower, B. H., L. Shi, R. Yongsunthon, T. Droubay, D. E. McCready, and S. K. Lower. 2007. Specific bonds between iron oxide surface and outer membrane cytochromes MtrC and OmcA from *Shewanella oneidensis* MR-1. *J. Bacteriol.* **189:**4944–4952.

Lower, S. K., M. F. Hochella, and T. J. Beveridge. 2001. Bacterial recognition of mineral surfaces: nanoscale interactions between Shewanella and alpha-FeOOH. *Science* **292:**1360–1363.

Manceau, A., and V. A. Drits. 1993. Local structure of ferrihydrite and feroxyhite by EXAFS spectroscopy. *Clay Miner.* **28:**165–184.

Moon, J. W., Y. W. Roy, L. W. Yeary, R. J. Lauf, C. J. Rawn, L. J. Love, and T. J. Phelps. 2007a. Microbial formation of lanthanide-substituted magnetites by Thermoanaerobacter sp. TOR-39. *Extremophiles* **11:**859–867.

Moon, J. W., L. W. Yeary, A. J. Rondinone, C. J. Rawn, M. J. Kirkham, Y. Roh, L. J. Love, and T. J. Phelps. 2007b. Magnetic response of microbially synthesized transition metal- and lanthanide-substituted nano-sized magnetites. *J. Magn. Mater.* **313:**283–292.

Myers, C. R., and K. H. Nealson. 1988. Bacterial manganese reduction and growth with manganese oxide as the sole electron acceptor. *Science* **240:**1319–1321.

Myers, J. M., and C. R. Myers. 1998. Isolation and sequence of *omcA*, a gene encoding a decaheme outer membrane cytochrome *c* of *Shewanella putrefaciens* MR-1, and detection of *omcA* homologs in other strains of *S. putrefaciens*. *Biochim. Biophys. Acta* **1373:**237–251.

Myers, J. M., and C. R. Myers. 2001. Role of outer membrane cyctochromes OmcA and OmcB of *Shewanella putrefaciens* MR-1 in reduction of manganese dioxide. *Appl. Environ. Microbiol.* **67:**260–269.

Neal, A. L., T. L. Bank, M. F. Hochella, and K. M. Rosso. 2005. Cell adhesion of Shewanella oneidensis to iron oxide minerals: effect of different single crystal faces. *Geochem. Trans.* **6:**77–84.

Neal, A. L., K. M. Rosso, G. G. Geesey, Y. A. Gorby, and B. J. Little. 2003. Surface structure effects on direct reduction of iron oxides by *Shewanella oneidensis*. *Geochim. Cosmochim. Acta* **67**:4489–4503.

Nealson, K. H., and D. Saffarini. 1994. Iron and manganese in anaerobic respiration: environmental significance, physiology, and regulation. *Annu. Rev. Microbiol.* **48**:311–343.

Neiss, J., B. D. Stewart, P. S. Nico, and S. Fendorf. 2007. Speciation-dependent microbial reduction of uranium within iron-coated sands. *Environ. Sci. Technol.* **41**:7343–7348.

Nevin, K. P., and D. R. Lovley. 2000. Lack of production of electron-shuttling compounds or solubilization of Fe(III) during reduction of insoluble Fe(III) oxide by *Geobacter metallireducens*. *Appl. Environ. Microbiol.* **66**:2248–2251.

Nevin, K. P., and D. R. Lovley. 2002. Mechanisms of Fe(III) oxide reduction in sedimentary environments. *Geomicrobiol. J.* **19**:141–159.

Nico, P. S., B. D. Stewart, and S. Fendorf. 2009. Incorporation of oxidized uranium into Fe(hydr)oxides during Fe(II) catalyzed remineralization. *Environ. Sci. Technol.* **43**:7391–7396.

Parmar, N., Y. A. Gorby, T. J. Beveridge, and F. G. Ferris. 2001. Formation of green rust and immobilization of nickel in response to bacterial reduction of hydrous ferric oxide. *Geomicrobiol. J.* **18**:375–385.

Pedersen, H. D., D. Postma, and R. Jakobsen. 2006. Release of arsenic associated with the reduction and transformation of iron oxides. *Geochim. Cosmochim. Acta* **70**:4116–4129.

Petrie, L., N. N. North, S. L. Dollhopf, D. L. Balkwill, and J. E. Kostka. 2003. Enumeration and characterization of iron(III)-reducing microbial communities from acidic subsurface sediments contaminated with uranium(VI). *Appl. Environ. Microbiol.* **69**:7467–7479.

Phillips, D. H., D. B. Watson, and Y. Roh. 2007. Uranium deposition in a weathered fractured saprolite/shale. *Environ. Sci. Technol.* **41**:7653–7660.

Phillips, E. J. P., D. R. Lovley, and E. E. Roden. 1993. Composition of non-microbially reducible Fe(III) in aquatic sediments. *Appl. Environ. Microbiol.* **59**:2727–2729.

Postma, D., and R. Jakobsen. 1996. Redox zonation: equilibrium constraints on the Fe(III)/SO_4^- reduction interface. *Geochim. Cosmochim. Acta* **60**:3169–3175.

Reeburgh, W. S. 1983. Rates of biogeochemical processes in anoxic sediments. *Annu. Rev. Earth Planet. Sci.* **11**:269–298.

Robertson, W. D. 2000. Treatment of wastewater phophate by reductive dissolution of iron. *J. Environ. Qual.* **29**:1678–1685.

Roden, E. E. 2003a. Diversion of electron flow from methanogenesis to crystalline Fe(III) oxide reduction in carbon-limited cultures of wetland sediment microorganisms. *Appl. Environ. Microbiol.* **69**:5702–5706.

Roden, E. E. 2003b. Fe(III) oxide reactivity toward biological versus chemical reduction. *Environ. Sci. Technol.* **37**:1319–1324.

Roden, E. E. 2006. Geochemical and microbiological controls on dissimilatory iron reduction. *C. R. Geosci.* **338**:456–467.

Roden, E. E., and M. M. Urrutia. 2002. Influence of biogenic Fe(II) on bacterial crystalline Fe(III) oxide reduction. *Geomicrobiol. J.* **19**:209–251.

Roden, E. E., M. M. Urrutia, and C. J. Mann. 2000. Bacterial reductive dissolution of crystalline Fe(III) oxide in continuous-flow column reactors. *Appl. Environ. Microbiol.* **66**:1062–1065.

Roden, E. E., and R. G. Wetzel. 2002. Kinetics of microbial Fe(III) oxide reduction in freshwater wetland sediments. *Limnol. Ocean.* **41**:1733–1748.

Roden, E. E., and J. M. Zachara. 1996. Microbial reduction of crystalline iron(III) oxides: influence of oxide surface area and potential for cell growth. *Environ. Sci. Technol.* **30**:1618–1628.

Rosso, K. M., D. M. A. Smith, and M. Dupuis. 2003a. An ab initio model of electron transport in hematite (alpha-Fe2O3) basal planes. *J. Chem. Phys.* **118**:6455–6466.

Rosso, K. M., J. M. Zachara, J. K. Fredrickson, Y. A. Gorby, and S. C. Smith. 2003b. Nonlocal bacterial electron transfer to hematite surfaces. *Geochim. Cosmochim. Acta* **67**:1081–1087.

Saltikov, C. W., and D. K. Newman. 2003. Genetic identification of a respiratory arsenate reductase. *Proc. Natl. Acad. Sci. USA* **100**:10983–10988.

Sass, B. M., and D. Rai. 1987. Solubility of amorphous chromium(III)-iron(III) hydroxide solid solutions. *Inorg. Chem.* **26**:2228–2232.

Schwertmann, U. 1984. The influence of aluminum on Fe oxides IX. Dissolution of Al-goethites in 6 M HCl. *Clays Clay Miner.* **19**:9–19.

Singh, B., and R. J. Gilkes. 1992. Properties and distribution of iron oxides and their association with minor elements in the soils of south-western Australia. *J. Soil Sci.* **43**:77–98.

Starkey, R. L., and H. O. Halvorson. 1927. Studies on the transformations of iron in nature. II. Concerning the importance of microorganisms in the solution and precipitation of iron. *Soil Sci.* **24**:381–402.

Stewart, B. D., P. S. Nico, and S. Fendorf. 2009. Stability of uranium incorporated into Fe(Hydr)oxides under fluctuating redox conditions. *Environ. Sci. Technol.* **43**:4922–4927.

Straub, K. L., and B. Schink. 2004. Ferrihydrite-dependent growth of Sulfurospirillum deleyianum through electron transfer via sulfur cycling. *Appl. Environ. Microbiol.* **70**:5744–5749.

Stucki, J. W., K. Lee, B. A. Goodman, and J. E. Kostka. 2007. Effects of in situ biostimulation on iron mineral speciation in a sub-surface soil. *Geochim. Cosmochim. Acta* **71:**835–843.

Sturm, A., S. A. Crowe, and D. A. Fowle. 2008. Trace lead impacts biomineralization pathways during bacterial iron reduction. *Chem. Geol.* **249:**282–293.

Taillefert, M., J. S. Beckler, E. Carey, J. L. Burns, C. M. Fennessey, and T. J. DiChristina. 2007. Shewanella putrefaciens produces an Fe(III)-solubilizing organic ligand during anaerobic respiration on insoluble Fe(III) oxides. *J. Inorg. Biochem.* **101:**1760–1767.

Thamdrup, B. 2000. Bacterial manganese and iron reduction in aquatic sediments, p. 41–84. *Advances in Microbial Ecology*, vol. 16. Kluwer Academic/Plenum Publishing, New York, NY.

Thamdrup, B., R. Rossello-Mora, and R. Amann. 2000. Microbial manganese and sulfate reduction in Black Sea shelf sediments. *Appl. Environ. Microbiol.* **66:**2888–2897.

Torrent, J., U. Schwertmann, and V. Barron. 1987. The reductive dissolution of synthetic goethite and hematite in dithionite. *Clays Clay Miner.* **22:**329–337.

Trolard, F., G. Bourrie, E. Jeanroy, A. J. Herbillon, and H. Martin. 1995. Trace elements in natural iron oxides from laterites: a study using selective kinetic extraction. *Geochim. Cosmochim. Acta* **59:**1285–1297.

Tufano, K. J., and S. Fendorf. 2008. Confounding impacts of iron reduction on arsenic retention. *Environ. Sci. Technol.* **42:**4777–4783.

Tufano, K. J., C. Reyes, C. W. Saltikov, and S. Fendorf. 2008. Reductive processes controlling arsenic retention: revealing the relative importance of iron and arsenic reduction. *Environ. Sci. Technol.* **42:**8283–8289.

Urrutia, M. M., E. E. Roden, and J. M. Zachara. 1999. Influence of aqueous and solid-phase Fe(II) complexants on microbial reduction of crystalline iron(III) oxides. *Environ. Sci. Technol.* **33:**4022–4028.

Vorhies, J. S., and R. R. Gaines. 2009. Microbial dissolution of clay minerals as a source of iron and silica in marine sediments. *Nat. Geosci.* **2:**221–225.

Waychunas, G. A., C. S. Kim, and J. F. Banfield. 2005. Nanoparticulate oxide minerals in soils and sediments: unique properties and contaminant scavenging mechanisms. *J. Nanoparticle Res.* **7:**409–433.

Weber, K. A., L. A. Achenbach, and J. D. Coates. 2006. Microorganisms pumping iron: anaerobic microbial iron oxidation and reduction. *Nature* **4:**752–764.

White, H. K., C. E. Reimers, E. E. Cordes, G. Dilly, and P. R. Girguis. 2009. Quantitative population dynamics of microbial communities in plankton-fed microbial fuel cells. *ISME J.* **3:**635–646.

Wielinga, B., M. M. Mizuba, C. M. Hansel, and S. Fendorf. 2001. Iron promoted reduction of chromate by dissimilatory iron-reducing bacteria. *Environ. Sci. Technol.* **35:**522–527.

Wigginton, N. S., K. M. Rosso, and M. F. Hochella, Jr. 2007. Mechanisms of electron transfer in two decaheme cytochromes from a metal-reducing bacterium. *J. Phys. Chem B* **111:**12857–12864.

Williams, A. G. B., and M. M. Scherer. 2004. Spectroscopic evidence for Fe(II)-Fe(III) electron transfer at the iron oxide-water interface. *Environ. Sci. Technol.* **38:**4782–4790.

Xiong, Y., L. Shi, B. Chen, M. U. Mayer, B. H. Lower, Y. Londer, S. Bose, M. F. Hochella, Jr., J. K. Fredrickson, and T. C. Squier. 2006. High-affinity binding and direct electron transfer to solid metals by the *Shewanella oneidensis* MR-1 outer membrane *c*-type cytochrome OmcA. *J. Am. Chem. Soc.* **128:**13978–13979.

Yan, B. Z., B. A. Wrenn, S. Basak, P. Biswas, and D. E. Giammar. 2008. Microbial reduction of Fe(III) in hematite nanoparticles by Geobacter sulfurreducens. *Environ. Sci. Technol.* **42:**6526–6531.

Zachara, J. M., J. K. Fredrickson, S.-M. Li, D. W. Kennedy, S. C. Smith, and P. L. Gassman. 1998. Bacterial reduction of crystalline Fe^{3+} oxides in single phase suspensions and subsurface materials. *Am. Miner.* **83:**1426–1443.

Zachara, J. M., J. K. Fredrickson, S. C. Smith, and P. L. Gassman. 2001. Solubilization of Fe(III) oxide-bound trace metals by a dissimilatory Fe(III) reducing bacterium. *Geochim. Cosmochim. Acta* **65:**75–93.

Zachara, J. M., R. K. Kukkadapu, J. K. Fredrickson, Y. A. Gorby, and S. C. Smith. 2002. Biomineralization of poorly crystalline Fe(III) oxides by dissimilatory metal reducing bacteria (DMRB). *Geomicrobiol. J.* **19:**179–207.

MICROORGANISMS AND PROCESSES LINKED TO URANIUM REDUCTION AND IMMOBILIZATION

Joel E. Kostka and Stefan J. Green

7

INTRODUCTION

Metal contaminants, many of which are radioactive and/or toxic, are concentrated in subsurface aquifers and represent a global scale threat to groundwater that is used for drinking. In the United States alone the volume of radionuclide-contaminated subsurface materials is larger than that of the Great Lakes combined, and is spread over 120 sites in 36 states (NABIR, 2003; National Research Council, 1993). During the Cold War era, the extraction and processing of metal radionuclides for weapons production resulted in extensive subsurface contamination. Cleanup projects for these sites are projected to cost billions of dollars and to last many decades. Within the nuclear weapons complex managed by the U.S. Department of Energy (DOE), hexavalent uranium [U(VI)] is the most common radionuclide contaminant found in sediments and groundwaters (Riley and Zachara, 1992). Nitrate is often a co-contaminant with U(VI) because of the use of nitric acid in the processing of uranium and uranium-bearing waste. Because of their widespread significance as groundwater contaminants in subsurface aquifers, much of the research on metal contaminants has centered on the remediation of U(VI) and nitrate.

Current remediation practices favor reductive immobilization of U(VI) catalyzed by microorganisms in close proximity to the contaminant source zone, coupled with natural attenuation and monitoring elsewhere (Anderson et al., 2003; Istok et al., 2004; Wu et al., 2006b, 2006c; Anderson, 2006). Previous studies have indicated that no net U(VI) reduction occurs until nitrate and denitrification intermediates are removed. Once nitrate is depleted, U(VI) and Fe(III) are reduced concurrently (Finneran et al., 2002; Elias et al., 2003; Istok et al., 2004; Edwards et al., 2007). Conversely, reduced U can subsequently be oxidized and remobilized via biotic or abiotic reactions with oxygen, nitrate, and denitrification intermediates such as nitrite and iron oxide minerals (Senko et al., 2002, 2005; Finneran et al., 2002; Beller, 2005). An alternative remediation strategy, employing precipitation of U(VI) phosphate minerals via the introduction of phosphate minerals, polyphosphates, or organophosphates to the subsurface, has the advantage that it does not require anoxic conditions (Beazley et al., 2007; Martinez et al., 2007; Wellman et al., 2008a; 2008b). The biogeochemical cycles of C, N,

Joel E. Kostka and Stefan J. Green, Georgia Institute of Technology, School of Biology and School of Earth and Atmospheric Sciences, Atlanta, GA 30332-0230.

P, Fe, and U are intimately linked in subsurface environments and strongly impacted by microbial activity. Close consideration of these cycles and the microbial community is critical for the design and implementation of appropriate remediation strategies,

Remediation potential in the subsurface is largely dictated by the physiological requirements for the growth and metabolism of microorganisms, and these requirements are most likely manifested at the community level (Tiedje, 1993; Loeffler and Edwards, 2006; Lovley et al., 2008). Through cultivation-independent characterization of the microbial groups that catalyze the relevant biogeochemical reactions, the "potential" physiological mechanisms controlling radionuclide fate and transport may be predicted and site-specific remediation strategies may be developed. Major advances in DNA-sequencing capability have begun to provide us with an unprecedented view of the community composition and genetic potential of the subsurface microbial world. However, the physiology and metabolism of an organism cannot be extrapolated from its genome sequence alone. Furthermore, the small-subunit ribosomal RNA (rRNA) gene, which has been used most often to describe microbial community composition in the subsurface, lacks resolution at the species level (Zengler, 2008). In other words, different strains that are genotypically or phylogenetically defined as the same species can have very different phenotypes (Prakash et al., 2009). The metabolic "potential" gleaned from community sequencing must be coupled to the cultivation and physiological characterization of model microorganisms along with biogeochemical studies in order to effectively interrogate microbially mediated mechanisms of U transformation. The metabolism or function of subsurface microbial communities must be directly linked to their phylogenetic structure for the elucidation of biogeochemical mechanisms.

Despite recent advances, the preponderance of microorganisms in nature remain uncultivated, and genetic databases are largely composed of sequences for which few or no cultivated representatives are available in culture collections (Schmidt et al., 2008; Janssen, 2008). This is especially true for anaerobic or subsurface microorganisms of significance to the mobility and fate of metal radionuclides. In addition, studies of subsurface microorganisms have been limited primarily to the domain *Bacteria*. Even though eukaryotic microbes (fungi) and mesophilic members of the *Archaea* have been shown to play a key role in the biogeochemical cycles of soils (Gadd, 2007; Schmidt et al., 2008; Hayatsu et al., 2008), these domains remain relatively unexplored in the subsurface. New approaches have begun to change the paradigm of the "uncultured majority" (Zengler, 2008). Approaches include the development of high-throughput cultivation (Zengler et al., 2002), combining cultivation with culture-independent methods for quick identification of isolates, design and use of novel growth chambers, and application of single-cell separation procedures (Giovannoni et al., 2007; Zengler, 2008). For example, these new techniques have increased our ability to access novel microorganisms from the marine biosphere, resulting in the successful cultivation of obligate oligotrophs that are outcompeted in nutrient-rich media (Giovannoni et al., 2007). However, to our knowledge, similar innovative strategies have rarely been applied to the cultivation of anaerobes or to subsurface environments.

An extensive review of all microorganisms and processes that affect the fate and transport of U(VI) is beyond the scope of this chapter. We will focus on the microorganisms and electron transfer processes that are likely to impact the reductive immobilization of U(VI) in the contaminated terrestrial subsurface. We offer a perspective on how microbial eukaryotes may play a role in U(VI) biotransformation, and we end with a discussion of the nonreductive immobilization of U(VI) through microbially facilitated precipitation with phosphate.

THE CHEMISTRY OF URANIUM BIOIMMOBILIZATION

Uranium contamination originates from a variety of sources, including the nuclear power industry, the production of nuclear weapons, the mining of uranium, the combustion of coal, and the application of phosphate fertilizer (Lloyd and Macaskie, 2000; Markich, 2002; Wilkins et al., 2006). In the environment, uranium exists as primarily 3 isotopes (^{238}U, ^{235}U, ^{234}U), all of which are radioactive and chemically toxic. Uranium chemistry in terrestrial subsurface environments, where most contamination is concentrated, is highly complex and difficult to model (Fig. 1). The most common valence states of uranium in the environment are U(IV) and U(VI). In oxic groundwaters and sediments, hexavalent uranium is present mainly as the uranyl ion (UO_2^{2+}), which is highly soluble, stable, and mobile over a wide pH range (Murphy and Shock, 1999). Uranyl tends to form stable complexes with common groundwater constituents, including carbonate, phosphate, and calcium. Although carbonate exerts a strong influence on U(VI) speciation at circumneutral pH, it plays a diminished role at lower pH when it becomes protonated (Beazley, 2009).

The mobility of uranium in porous media is mainly controlled by complexation and redox reactions (Fig. 1) (Suzuki and Suko, 2005). Uranium speciation is highly sensitive to redox chemistry. U(VI) can be microbiologically or abiotically immobilized from water by its reduction from UO_2^{2+} to insoluble U(IV) oxides, such as uraninite (UO_2). The half-cell potential for this reaction is "intermediate," and at circumneutral pH, various chemical reductants [such as complexed iron, Fe(II), or hydrogen sulfide] or microorganisms will

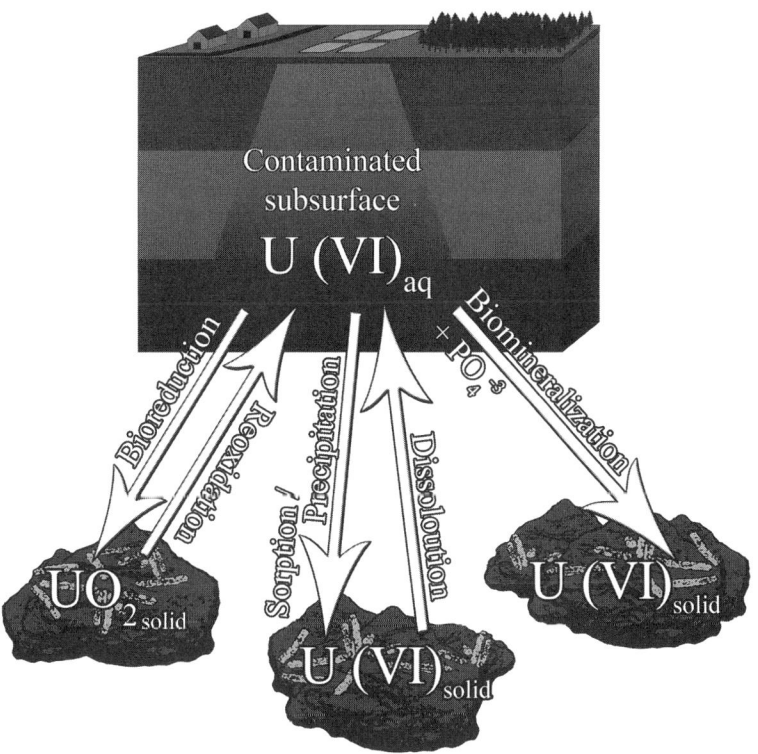

FIGURE 1 Schematic summarizing the predominant biogeochemical reactions and processes impacting U(VI) mobility in the contaminated subsurface. U(IV) in the solid phase is represented by uraninite (UO_2), but other mineral forms may be present.

mediate U(VI) reduction to U(IV). Microbial U(VI) reduction may be catalyzed by both direct (enzymatic) and indirect (abiotic) mechanisms. The products of microbial Fe(III) and sulfate reduction, Fe(II) and hydrogen sulfide, can react abiotically to reduce U(VI) (Liger et al., 1999; Mohagheghi et al., 1985; Scott et al., 2005; Hua et al., 2006). Although less information is available on the abiotic mechanisms, the general consensus is that microbial U(VI) reduction proceeds at faster rates than abiotic reduction under most environmental conditions (Suzuki and Suko, 2006; Wilkins et al., 2006).

A large portion of uranium is associated with the solid phase of porous media. Uranium geochemistry is strongly linked through sorption/precipitation processes to the surface chemistry of minerals, including oxyhydroxides, phyllosilicates, phosphates, carbonates, and sulfides (Duff et al., 2002; Wang et al., 2005; Kelly et al., 2008; Luo et al., 2009; Singer et al., 2009). Iron minerals have been extensively shown to be efficient sorbents of U(VI), both in the laboratory and in subsurface sediments (Catalano and Brown, 2005). Uranyl forms microprecipitates and sorbs in fractures and cavities within sediment grains (Liu et al., 2004). Sorption and precipitation of uranyl will largely impact its mobility (Ilton et al., 2008) and electron transfer reactions leading to its sequestration (Jeon et al., 2004; Liu et al., 2006; Kelly et al., 2008).

One of the more promising strategies for the in situ remediation of uranium waste involves the "biostimulation" of uranium immobilization. Biostimulation is defined as the addition of nutrients (e.g., carbon, nitrogen, phosphorus sources) that serve to increase the number or activity of indigenous microorganisms available for bioremediation activity. Biostimulation, which often results in significant changes in subsurface redox conditions, can lead to the creation of a permeable treatment zone in contaminated aquifers that removes radionuclides from the aqueous phase before they enter sensitive water supplies. Field experiments at DOE sites have indicated that biostimulation of U(VI) reduction by electron donor addition is a promising remediation strategy for uranium-contaminated groundwaters (Anderson et al., 2003; Istok et al., 2004; Wu et al., 2006b, 2006c, 2007).

METABOLIC AND PHYLOGENETIC DIVERSITY OF CULTIVATED U(VI)-REDUCING TAXA

A metabolically and phylogenetically diverse group of 48 microbial taxa has been demonstrated to reduce U(VI) in pure culture (Fig. 2). U(VI)-reducing taxa are spread throughout the domain *Bacteria* and a single representative of the *Archaea* was shown to reduce U(VI) (Kashefi and Lovley, 2000). The vast majority of known U(VI) reducers are within the phyla *Proteobacteria* and *Firmicutes* (Fig. 2). The U(VI)-reducing *Proteobacteria* primarily comprise *Geobacteraceae* members and mesophilic sulfate reducers within the *Deltaproteobacteria*, while the remainder are primarily members of the *Gammaproteobacteria* within the families *Pseudomonadaceae* and *Shewanellaceae*.

Unifying biochemical features of U(VI) reduction mechanisms and their relationship to energy-generating metabolism remain largely a mystery among cultivated U(VI) reducers. In addition, rigorous physiological testing across physicochemical conditions relevant to U(VI)-contaminated systems has generally not been performed. However, some shared phenotypes are recognized. All taxa are heterotrophic and grow anaerobically at a redox potential sufficiently low to support U(VI) reduction. The majority of uranium-reducing taxa fit into three categories according to their primary metabolism: (i) dissimilatory Fe(III)-reducing bacteria (FeRB), (ii) dissimilatory sulfate-reducing bacteria (SRB), and (iii) fermentative bacteria (Nealson et al., 2002; Lovley et al., 2004; Sani et al., 2004; DiChristina et al., 2005; Wu et al., 2006b). A substantial minority of U(VI) reducers are thermophilic, with the majority mesophilic.

To our knowledge, only seven of these species have been shown to conserve energy for growth with U(VI) as the sole electron acceptor (*Anaeromyxobacter dehalogenans, Geobacter lovleyi, G. metallireducens, G. sulfurreducens, Shewanella oneidensis, Thermoterrabacterium ferrireducens, Desulfotomaculum reducens*) (Fig. 2). All of these organisms are FeRB, and a single strain is also capable of sulfate respiration (*D. reducens*; Tebo and Obraztsova, 1998). Wall and Krumholz (2006) hypothesized that growth coupled to U(VI) reduction may be restricted to bacteria able to use insoluble minerals as terminal electron acceptors. All studies to date support this hypothesis (Wall and Krumholz, 2006; Sanford et al., 2007). However, few strains have been rigorously tested using more sensitive molecular methods for growth detection, and future studies may reveal a broader diversity of microbes capable of growth with U(VI) as the electron acceptor. As it stands now, the metabolic capabilities of microbial groups capable of U(VI) reduction cannot be distinguished from those capable of the reduction of Fe(III) minerals. All of the microorganisms capable of U(VI) reduction fit into two categories similar to Fe(III) reducers: (i) FeRB that grow with Fe(III) as the sole electron acceptor and (ii) SRB or fermenters that can shunt electrons to Fe(III) but cannot support growth from metal respiration. The large overlap between organisms capable of U(VI) and Fe(III) reduction is not surprising considering the substantial gaps in our knowledge of the physiology and biochemistry of electron transfer to oxidized metals (DiChristina et al., 2005; Fredrickson and Zachara, 2008). The complex speciation of U(VI) in groundwater results in varied redox potentials (e.g., Brooks et al., 2003), and thus far, no single "silver bullet" terminal reductase has been identified in microorganisms that respire U(VI). Most likely, the physiological mechanisms of metal reduction are as complex as metal redox chemistry, and include a diverse array of protein networks, organic ligands, and electron shuttles. Considering that Fe(III) minerals intimately impact the fate and transport of U(VI) through sorption and coprecipitation reactions, all FeRB are linked to U(VI) immobilization in one way or another.

ECOLOGY OF U(VI)-REDUCING TAXA IN THE TERRESTRIAL SUBSURFACE

A functional genetic marker for uranium reduction has not yet been established. Thus, our view of microbial U(VI) reduction in the environment is still limited by processes that we can verify in laboratory pure cultures, and the ecology of U(VI) reducers (including distribution, abundance, and in situ metabolism) remains largely unknown. We have compiled a list of microbial taxa detected in native and treated uranium-contaminated sediment and groundwater samples using cultivation-independent surveys of rRNA genes (Table 1). Not surprisingly, many of the dominant taxa detected in the subsurface overlap with the taxa that have been shown to reduce U(VI) in culture. This is because the sequences of known U(VI)-reducing organisms are selected for and annotated in genetic databases. Known lineages of denitrifying, sulfate-reducing, and iron-reducing bacteria within the *Proteobacteria* are most commonly detected, reflecting the key terminal electron-accepting processes in contaminated subsurface environments. *Deltaproteobacteria* are most frequently associated with U(VI)-reducing environments, and contain the major Fe(III)-reducing lineage *Geobacteraceae* and abundant SRB lineages such as the *Desulfovibrionaceae* (Anderson et al., 2003; North et al., 2004; Vrionis et al., 2005; Akob et al., 2008).

In soils and sediments of the terrestrial subsurface where most U(VI) accumulates, environmental conditions are often characterized by a lack of light, large fluctuations in redox conditions, and the low or intermittent availability of water, organic carbon, and nutrients. In short, many of these systems can be considered as "extreme" habitats for microbial growth and survival (Griebler and Lueders, 2009). In U(VI)-contaminated ecosystems, these extreme conditions are often exacerbated by high levels of acidity and high concentrations of toxic metals or organic compounds (Riley and Zachara, 1992; Brooks, 2001; Moon et al., 2005). The DOE is tasked with managing sites

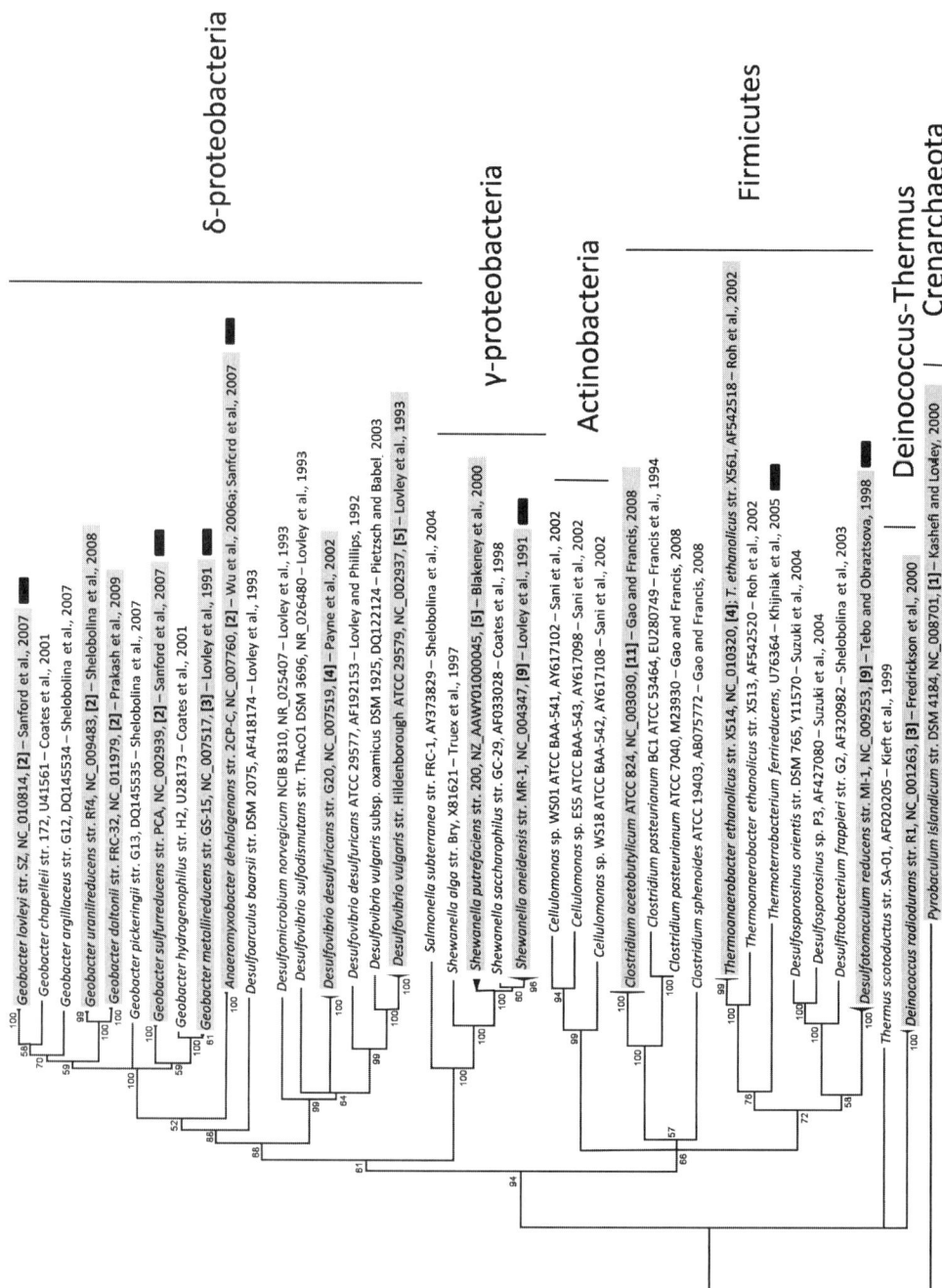

in the United States impacted by nuclear legacy waste. Much of the information available on microbial communities that impact the fate and transport of U(VI) has been collected under the auspices of the DOE's Subsurface Biogeochemical Research Program (http://esd.lbl.gov/research/projects/ersp/). Two sites, at Rifle, Colorado, and Oak Ridge, Tennessee, have been targeted, in particular. This section will focus on the state of the field with regard to the ecology of microbial communities that impact electron flow and may catalyze U(VI) reduction in the subsurface of these sites.

The chemistry of water and sediments in the subsurface is driven by microbial metabolism that is in turn controlled by the availability of electron donors and acceptors (Chapelle, 2000). Even though DOE sites can be highly contaminated with U(VI), it is important to consider that U is most often present in trace amounts relative to other redox-active constituents. Thus, the fate of U(VI) will primarily depend on the microbial transformation of other electron acceptors and donors. Reduction-oxidation activity in subsurface environments contaminated with metal radionuclides is most often limited by carbon or electron donor availability (Istok et al., 2004). However, autochthonous carbon sources are present, and contrary to popular belief, unamended groundwaters are often suboxic to anoxic (http://www.esd.ornl.gov/orifrc/). Furthermore, current remediation practices at DOE sites favor biostimulation by electron donor amendment. Thus, electron transfer processes should be considered in the context of a subsurface that is flooded with electron donor. Metal reduction in the environment is often thought to be primarily catalyzed by respiratory organisms that gain an advantage by coupling metal respiration to growth (Lovley et al., 2004; Canfield et al., 2005). However, when electron donor is not limiting, do respiratory organisms really outcompete those that shunt electrons to metals under fermentative conditions? In addition, microbial community composition, and therefore the mechanisms of U(VI) reduction, will be determined by the choice of electron donors amended to the subsurface during bioremediation (e.g., Akob et al., 2008; Madden et al., 2007). To appropriately design U(VI) bioremediation strategies, the potential function and phylogenetic structure of microbial populations that mediate electron flow must be understood under site-specific conditions.

As stated above, FeRB and SRB comprise the two best-studied groups of metal-reducing organisms that are capable of U(VI) reduction (Nealson et al., 2002; Lovley et al., 2004; Sani et al., 2004; DiChristina et al., 2005; Wall and Krumholz, 2006). Under in situ conditions in the terrestrial subsurface, FeRB are likely to outcompete SRB because Fe(III) is usually a much more abundant electron acceptor than sulfate in subsurface sediments. We note, however, that sulfate levels in the U(VI)-contaminated subsurface are variable and occasionally

FIGURE 2 Phylogenetic tree of U(VI)-reducing microorganisms. Bootstrapped neighbor-joining trees were generated using partial and full-length 16S rRNA gene sequences obtained from the National Center for Biotechnology Information and from the DOE's Joint Genome Institute. Sequences were aligned using the software package Greengenes (DeSantis et al., 2006) and analyzed within the phylogenetic software package MEGA (Kumar et al., 2008). Bootstrap values greater than 50% are indicated at each node, and polytomies indicate branching points that were not consistently supported by bootstrap analyses. Organisms for which genome sequences are available are highlighted in grey, with all the 16S rRNA genes present in the genome compressed into a single cluster. The number of 16S rRNA genes present in the genome is indicated in brackets adjacent to the accession number. Relevant references are indicated for each organism. Species shown to conserve energy using U(VI) as a sole electron acceptor are indicated with a black box. The scale bar represents 0.05 substitutions per nucleotide position. Missing from the tree are the short gene sequences from the genome of *Cellulomonas flavigena* ATCC 482 (NZ_ABTJ00000000), as well as several organisms for which no sequences were available, including *Pseudomonas denitrificans* ATCC 13867, *Pseudomonas* sp. CRB5, and *Veillonella alcalescens* (formerly *Micrococcus lactilyticus*).

TABLE 1 Taxonomic profiling of small-subunit rRNA gene sequences recovered from cultivation-independent analyses of uranium-contaminated environments[a]

Taxon	OTU	Sequences
Proteobacteria	1,217	5,461
Geobacteraceae (D)	264	1,220
Desulfovibrionaceae (D)	25	333
Desulfuromonaceae (D)	53	209
Desulfobacteraceae (D)	33	77
Desulfobulbaceae (D)	17	58
Rhodocyclaceae (B)	119	764
Oxalobacteraceae (B)	46	467
Comamonadaceae (B)	60	440
Hydrogenophilaceae (B)	30	199
Gallionellaceae (B)	12	67
Burkholderiaceae (B)	19	56
Alcaligenaceae (B)	13	39
Xanthomonadaceae (G)	60	293
Pseudomonadaceae (G)	26	185
Sphingomonadaceae (A)	33	99
Hyphomicrobiaceae (A)	38	69
Bradyrhizobiaceae (A)	24	66
Acidobacteria	315	825
Actinobacteria	116	538
Firmicutes	206	415
Peptococcaceae	42	141
Bacillaceae	24	50
Clostridiaceae	27	48
Thermoanaerobacteraceae	10	17
Bacteroidetes	132	321
Chloroflexi	41	87
OD1	28	70
Gemmatimonadetes	28	49
Verrucomicrobia	22	30
Planctomycetes	20	26

high, and allow the substantial development of SRB under some conditions (Anderson et al., 2003; Wu et al., 2007). As a soluble electron acceptor, sulfate may also be favored relative to iron oxide minerals in the subsurface. Furthermore, geochemical evidence from subsurface aquifers indicates that Fe(III) and sulfate reduction zones may overlap (Jakobsen and Postma, 1999; Vrionis et al., 2005; Wu et al., 2007), and physiological studies corroborate the geochemistry to suggest that these processes may be catalyzed by overlapping populations. Many SRB can reduce Fe(III) (Lovley et al., 2004), and a subset of these may conserve energy for growth from Fe(III) respiration (Tebo and Obraztsova, 1998; Holmes et al., 2004). Thus, both SRB and FeRB are thought to have a high bioremediation potential in U(VI)-contaminated subsurface environments. Viable counts from U(VI)-contaminated subsurface sediments and groundwater support this hypothesis. Although FeRB are far more abundant than SRB under unamended or in situ conditions (Petrie et al., 2003), both SRB and FeRB are abundant in subsurface materials that have undergone U(VI) bioremediation by biostimulation (Wu et al., 2006b, 2006c; Cardenas et al., 2008).

One of the earliest field demonstrations of the reductive immobilization of U(VI) via biostimulation was conducted at the DOE's Uranium Mill Tailings Remedial Action site, a former uranium-processing facility near Rifle, Colorado. The subsurface treatment zone of the Rifle site is characterized by alluvial permeable sandy sediments, neutral pH, moderate sulfate concentrations, low nitrate concentrations, and

[a]Clone sequences were retrieved from GenBank by searching for the keyword "uranium" and combined with unpublished clone library sequences from sediment and groundwater at the Oak Ridge Integrated Field Research Challenge (OR-IFRC) site. These sequences were aligned within the Ribosomal Database Project (RDP) Pyrosequencing pipeline (Cole et al., 2007), operational taxonomic units (OTU) were identified using the Complete Linkage Clustering method with 97% similarity used as a threshold, and putative taxonomic identification was performed using the RDP Classifier. The abundance of sequences within select families of these phyla is also shown. The number of estimated OTU within each taxon is overestimated because of the different lengths and regions of the small-subunit rRNA gene sequence used for these analyses.

a relatively predictable flow pathway. Acetate was injected as an electron donor into a gallery of wells placed perpendicular to groundwater flow. During multiple field manipulation experiments, groundwater geochemistry and microbiology were monitored over a 3- to 4-month period postinjection (Anderson et al., 2003; Vrionis et al., 2005; N'Guessan et al., 2008). Uranium(VI) concentrations decreased to below the prescribed treatment level of 0.18 µM, close to the U.S. Environmental Protection Agency maximum contaminant level for drinking water of 0.126 µM in many of the monitoring wells (Anderson et al., 2003). Uranium reduction was concurrent with Fe(III) reduction. After the initial metal reduction phase, sulfate reduction was detected and subsequently groundwater U(VI) concentrations increased. The authors hypothesized that this was due to weak U(VI)-reduction activity by acetate-oxidizing, sulfate-reducing microorganisms.

The microbial groups linked to electron flow during bioremediation at the Rifle site were investigated by comparison of rRNA gene sequence libraries generated from groundwater and sediment samples collected over space and time. Microbial communities in the background or control wells were dominated by *Alphaproteobacteria* and *Betaproteobacteria*, and the predominant taxa did not change substantially during the treatment period. In contrast, a large enrichment of bacteria from the family *Geobacteraceae* (*Deltaproteobacteria*) was observed in the treated wells during the metal reduction phase, and sequences of SRB were abundant in treatment wells near the point of injection during the sulfate reduction phase. As expected, acetate-utilizing FeRB (*Geobacter*) and SRB (*Desulfobacter*) were detected in abundance during field manipulation. However, gram-positive sulfate-reducing genera (e.g., *Desulfotomaculum* and *Desulfosporosinus*), which have not been demonstrated to utilize acetate, were also detected (Anderson et al., 2003; Chang et al., 2005). It was concluded that U(VI) reduction was primarily catalyzed by members of the *Geobacteraceae*. However, evidence for the involvement of SRB in U(VI) immobilization remains equivocal (Anderson et al., 2003; Chang et al., 2005).

Detailed investigations of U(VI) bioremediation have also been conducted at the DOE's Oak Ridge Integrated Field Research Challenge (OR-IFRC) site, Oak Ridge, Tennessee. This site was formerly referred to as the Oak Ridge Field Research Center. In contrast to the Rifle site, the OR-IFRC subsurface is highly contaminated and largely acidic; it contains a plethora of contaminants including radionuclides [U(VI) and technetium], toxic metals, chelating agents, chlorinated hydrocarbons, polychlorinated biphenyls, and fuel hydrocarbons (http://www.esd.ornl.gov/orifrc/; Brooks, 2001; Moon et al., 2005). The subsurface treatment zone of the OR-IFRC is characterized by clay-rich impermeable sediments, highly variable pH, high nitrate and sulfate concentrations, and a less predictable flow pathway along fracture zones (Moon et al., 2005). Current in situ U(VI) bioremediation and natural attenuation/monitoring experiments center on two sampling areas along the S-3 contaminant plume. Area 3 is close to the source zone and contains the highest contaminant levels (average groundwater chemistry: 50 to 500 µM U, 100 to 1,000 mM nitrate, 10 to 50 mM sulfate, pH 3 to 5), while area 2 contains more moderate levels of contamination (average groundwater chemistry: 5 to 10 µM U, 10 to 100 mM nitrate, 0 to 2 mM sulfate, pH 5 to 7). U(VI) and nitrate are major groundwater contaminants at DOE sites and therefore these contaminants drive remediation efforts at the OR-IFRC. Furthermore, the combination of heterogeneous mineralogy, a low pH, and mixed metal contamination in a subsurface environment is representative of legacy nuclear waste sites worldwide.

OR-IFRC field manipulations have centered on the use of ethanol or glucose as an electron donor to stimulate bioreduction of U(VI), and pH neutralization was used in the highly acidic contaminant source zone. During multiple field manipulation experiments, groundwater and sediment geochemistry revealed the stoichiometric reduction of nitrate, Fe(III), and sulfate that more or less follows with thermodynamic predictions in the presence of excess electron donor. Uranium(VI) concentrations decreased in a matter of weeks to below the maximum contaminant level (0.130 µM) in many of the monitoring wells during multiple biostimulation experiments (Istok et al., 2004; Wu et al., 2007). In corroboration with microcosm tests conducted with subsurface sediments under controlled conditions in the laboratory (Finneran et al., 2002; Edwards et al., 2007; Akob et al., 2008), U(VI) reduction did not proceed until nitrate was depleted (Istok et al., 2004; Wu et al., 2006b, 2006c, 2007). Remobilization/reoxidation of U was observed concurrently with denitrification and the accumulation of denitrification intermediates. Unlike the observations at the Rifle site, U(VI) reduction was associated with both Fe(III) and sulfate reduction phases of field manipulations at the OR-IFRC (Istok et al., 2004; Wu et al., 2006, 2007).

Extensive microbial community characterization has revealed a diverse assemblage of microbes encompassing all phyla within the domain *Bacteria* in the subsurface of the OR-IFRC (Petrie et al., 2003; Yan et al., 2003; North et al., 2004; Fields et al., 2005; Wan et al., 2005; Akob et al., 2007, 2008; Cardenas et al., 2008; Hwang et al., 2009). The metabolism of these communities is limited by labile carbon availability, acidic pH, and co-contaminants such as nitrate and toxic metals (Al, Ni) (Petrie et al., 2003; Istok et al., 2004; Edwards et al., 2007). Under native, unamended conditions, microbial communities were composed of *Betaproteobacteria, Alphaproteobacteria, Firmicutes, Actinobacteria*, and *Bacteroidetes*, with the *Proteobacteria* predominating. In general, biostimulation resulted in a substantial increase in both the abundance and diversity of microorganisms in subsurface sediments and groundwaters from the highly contaminated areas of the OR-IFRC (Istok et al., 2004; North et al., 2004; Peacock et al., 2004; Wu et al., 2006b; Cardenas et al., 2008; Hwang et al., 2009). However, other studies observed a decrease in diversity upon biostimulation near the source zone of contamination (Chang et al., 2005; Spain et al., 2007). It is likely that we have not completely covered the subsurface microbial community with conventional cloning and sequencing of 16S rRNA genes. Because of the range of environmental conditions at DOE sites and the potential for dramatic shifts during remediation, deep sequencing methods should be applied to fully characterize native microbial communities and account for rare community members.

In agreement with studies at the Rifle site, members of the *Deltaproteobacteria* were most often detected in conjunction with U(VI) reduction activity at the OR-IFRC. North et al. (2004) observed that members of the *Deltaproteobacteria* increased from 5% to almost 40% in 16S rRNA gene sequence libraries following biostimulation of U(VI) reduction. Quantitative PCR results further confirmed an increase in abundance of *Geobacteraceae* and *Desulfovibrionaceae* groups in parallel with U(VI) reduction upon biostimulation (North et al., 2004; Hwang et al., 2009). In particular, an extensive data set has been compiled to indicate that bacteria of the genus *Geobacter* are diverse, abundant, and metabolically active in OR-IFRC materials during biostimulation (North et al., 2004; Chang et al., 2005; Akob et al., 2008). However, although the members of the *Geobacteraceae* often make an important contribution, mounting evidence indicates that a much broader diversity of taxa participate in or even predominate over subsurface metal reduction at the OR-IFRC. Ample evidence indicates that bacteria from the genus *Desulfovibrio* and gram-positive spore-forming bacteria such as those from the genera *Desulfosporosinus* or *Desulfotomaculum* may mediate U(VI) reduction in the subsurface (Chang et al., 2001; Wu et al., 2007; Cardenas et al., 2008). Members of the genus *Anaeromyxobacter* (*Deltaproteobacteria*), for which U(VI) reduction supports growth (Sanford et al., 2007), have also been commonly detected in abundance in parallel with U(VI)-reducing activity (Petrie et al., 2003; North et al., 2004; Cardenas et al., 2008; Thomas et al.,

2009). In fact, sequences of the FeRB genera *Ferribacterium* (*Betaproteobacteria*) and *Geothrix* (*Acidobacteria*) were among the most abundant sequence types retrieved in recent field U(VI) bioremediation experiments (Wu et al., 2007; Cardenas et al., 2008; Hwang et al., 2009). However, to our knowledge, these FeRB have not yet been tested for their ability to reduce U(VI).

The choice of electron donor is likely to have a large effect on the biostimulation of U(VI) immobilization by impacting the structure-function relationships of U(VI)-reducing microorganisms. Glucose, acetate, and ethanol have been utilized most often in field bioremediation tests. Acetate is a nonfermentable substrate that stimulates a hierarchy of terminal electron-accepting pathways in the order in which microorganisms glean free energy from each available electron acceptor (O_2, NO_3^-, Fe(III), Mn(IV), SO_4^{2-}). In contrast, ethanol has the potential to stimulate a larger diversity of metabolisms including respiration coupled to complete ethanol oxidation, syntrophic fermentation, incomplete ethanol oxidation to acetate (e.g., some SRB), and subsequent respiration coupled to acetate or H_2 oxidation. Both acetate and ethanol stimulated the fairly rapid reduction of U(VI) at neutral pH in field tests as described above. However, acetate is present as an undissociated acid at acidic pH values frequently found at radionuclide-contaminated sites, and can inhibit microbial metabolism by passing through the cell membrane and uncoupling proton motive force (Baronofsky et al., 1984). Under acidic conditions, ethanol would be a more appropriate choice for biostimulation. However, when the bioremediation strategy employs pH neutralization in parallel with electron donor amendment, the inhibition of microbial metabolism by acetate may not be an issue.

Differences in the microbial populations stimulated by the two electron donors may directly impact U(VI) biotransformation. After extended acetate treatment at the Rifle site, the growth of acetate-utilizing SRB was stimulated near the injection point. U(VI) concentrations rebounded to initial levels in the groundwater, and it was hypothesized that the complete consumption of acetate under sulfate-reducing conditions prevented further U(VI) reduction downgradient from the injection gallery (Anderson et al., 2003). The predominant SRB detected near the injection gallery were members of the family *Desulfobacteraceae*, which have so far not been demonstrated to reduce U(VI) in pure culture. In contrast, SRB of the *Desulfovibrionaceae* and the gram-positive *Peptococcaceae* were shown to reduce U(VI) in culture (Fig. 1) (Lovley and Phillips, 1992; Lovley et al., 1993). Whereas the *Desulfobacteraceae* are known for acetate utilization and the complete oxidation of carbon substrates, bacteria from the genus *Desulfovibrio* and gram-positive SRB are known for incomplete carbon oxidation and the utilization of hydrogen or ethanol as an electron donor (Muyzer and Stams, 2008). Thus, it follows that sulfate reduction was closely linked to reductive U(VI) immobilization at the OR-IFRC when ethanol was used as the electron donor, whereas the linkage was not as strong when acetate was used at Rifle. Iron(III) reduction and sulfate reduction occurred simultaneously and SRB were more abundant than FeRB during ethanol treatment, suggesting that SRB play a more important role in U(VI) immobilization (Wu et al., 2007). In agreement with this conclusion, bacteria from the genera *Desulfovibrio*, *Desulfosporosinus*, and *Desulfotomaculum*, all capable of ethanol utilization, were the predominant SRB species detected in OR-IFRC field experiments (Wu et al., 2007; Cardenas et al., 2008; Hwang et al., 2009). In environments co-contaminated with sulfate and nitrate, stimulation of *Desulfovibrio* may have a high bioremediation potential because the presence of sulfate represses nitrate reduction in this organism (Marietou et al., 2009).

Fermentative metal reduction should be considered further in U(VI) bioremediation strategies, especially when a fermentable substrate such as ethanol or glucose is used as electron donor for biostimulation. During biostimulation, carbon substrates will not be limiting and metal reduction may be catalyzed in parallel by respiratory and fermentative pathways (Edwards et al., 2007; Akob et al., 2008). Gram-positive fermentative organisms (including *Clostridia* and *Cellulomonas*) effectively reduce U(VI) in pure culture (Sani et al. 2002; Gao and Francis, 2008) and are present in the subsurface of DOE sites (North et al., 2004; N'Guessan et al., 2008; Hwang et al., 2009). The abundance of sequences of known fermentative metal-reducing bacteria (*Serratia, Clostridium*) doubled in response to biostimulation at the OR-IFRC (North et al., 2004). During field manipulations at both the Rifle and OR-IFRC sites, fermentative members of the *Firmicutes* (*Clostridium, Mollicutes*) were abundant and linked to sustained U(VI) immobilization under conditions of electron donor limitation (N'Guessan et al., 2008; Hwang et al., 2009). Field observations were supported by laboratory studies; in metal-reducing enrichment cultures and microcosms from U(VI)-contaminated subsurface sediments, fermentative organisms (*Paenibacillus, Brevibacillus, Anaerovibrio, Tolumonas, Clostridium*) were frequently detected (Petrie et al., 2003; Akob et al., 2008).

THE SUBSURFACE *GEOBACTER* CLADE

Of the described U(VI) reducers, the genus *Geobacter* stands out as the group most often detected in the U(VI)-contaminated subsurface (Table 1) (Lovley et al., 2008), and a recent study by Holmes et al. (2007) observed that the majority of sequences retrieved clustered in a phylogenetically coherent subsurface clade (*G. bemidjiensis, G. chapellei, G. daltonii,* and *G. uraniireducens*). In contrast, physiological studies in pure culture have focused on *Geobacter* species (*G. metallireducens, G. sulfurreducens*) for which the complete genome sequence and a genetic manipulation system have been available for a number of years. Much less physiological information is available from strains isolated from the subsurface, despite the fact that they are most often the predominant *Geobacter* species detected there.

A systems biology approach is being applied with the ultimate goal of developing in silico models that can predict the growth and metabolism of *Geobacter* species under a range of environmental conditions relevant to U(VI)-contaminated subsurface environments (Lovley et al., 2008). Genome sequences of a larger number of *Geobacter* species are now available, and genome-based models of several *Geobacter* species are able to predict physiological responses to environmental conditions. Furthermore, recent work shows that the subsurface *Geobacter* clade exhibits a remarkable genotypic and phenotypic plasticity and the 16S rRNA marker is not diagnostic for this plasticity. *G. uraniireducens* and *G. daltonii* were isolated from U(VI)-contaminated subsurface environments at the Rifle and OR-IFRC sites, respectively (Shelobolina et al., 2007; Prakash et al., 2010). Although these two *Geobacter* strains share 98.1% 16S rRNA gene sequence identity, their full genome sequences are highly divergent. In fact, few genes other than the rRNA genes show greater than 90% nucleotide identity, and the genomes differ in size by 0.8 Mb (Prakash et al., 2010). Limited physiological screening has begun to reveal substantial differences in electron acceptor and donor utilization within the subsurface clade of *Geobacter*. For example, *G. daltonii* and *G. toluenoxydans* conserve energy for growth with aromatic contaminants as the electron donor, while *G. uraniireducens* does not (Prakash et al., 2010; Kunapuli et al., 2009). This may be explained by the fact that both *G. daltonii* and *G. toluenoxydans* were isolated from subsurface sediments contaminated with aromatic hydrocarbons, whereas *G. uraniireducens* was isolated from groundwater that was not substantially impacted by organic contaminants. A number of additional features may provide a competitive advantage to *Geobacter* in the subsurface including the ability to utilize acetate, chemotaxis, and nitrogen fixation (Childers et al., 2002; Holmes et al., 2004). Perhaps most importantly, it was recently revealed that *Geobacter* maintains a maximal growth yield and the flexibility to switch electron acceptors at very low growth rates (Lin et al., 2009).

A POTENTIAL ROLE FOR MICROBIAL EUKARYOTES, THE FUNGI, IN THE U(VI)-CONTAMINATED SUBSURFACE?

Fungi are ubiquitous in soils and sediments where they often comprise a large portion of microbial biomass (Schmidt et al., 2008). These eukaryotic microorganisms mediate important reactions in the biogeochemical cycles of carbon, nutrients, and metals (Gadd, 2007). Even though the occurrence of fungi has been described for a few pristine and contaminated aquifers, to our knowledge, the identities and activities of these organisms remain virtually uncharacterized in subsurface environments (Bärlocher et al., 2006; Brad et al., 2008). Moreover, fungi contain many traits that may favor their persistence in the extreme conditions found in the U(VI)-contaminated subsurface including facultative anaerobic growth, resistance to acidic pH, resistance to toxic metals, and spore formation, and in some cases, radiation-enhanced growth (Dadachova et al., 2007; Hayatsu et al., 2008).

In surficial soils, biogeochemical evidence indicates that fungi can dominate the edaphic production of N_2O and N_2 from denitrification (Laughlin and Stevens, 2002). Yeasts as well as filamentous fungi are capable of denitrification. Many fungi use nitrate as an alternative electron acceptor and can denitrify under aerobic or microaerophilic conditions (Hayatsu et al., 2008). To date, the majority of examined fungi lack a N_2O reductase enzyme, and thus N_2O is the major denitrification product (e.g., Shoun et al., 1992).

Toxic metal species, including radionuclides, may be bound, accumulated, and precipitated by fungi (Gadd, 2007). Recent reports indicate that fungi have the ability to accumulate U in their biomass and even to transform soluble U into uranyl phosphate minerals (Fomina et al., 2008). However, the experiments were conducted under aerobic conditions only and the mechanisms of U precipitation or U speciation were not fully explored. Fungi are known to be metal resistant and occur in abundance in acidic, metal-rich environments, and thus could play an important role in U(VI) transformation in the subsurface of DOE sites. Future research should investigate the potential role of fungi in influencing U(VI) mobility directly through bioimmobilization or indirectly through the depletion of nitrate via denitrification in the subsurface.

ALTERNATIVE IMMOBILIZATION STRATEGIES: BIOMINERALIZATION OF U(VI) WITH ORGANOPHOSPHATE

As discussed above, field experiments at DOE sites have definitively demonstrated that, under specific geochemical conditions, U(VI) can be effectively removed from the groundwater aqueous phase and retained as a mineral phase (e.g., uraninite) in the subsurface. Although this process can be effective, there are a number of challenges that may limit its widespread application. These include the (i) inhibition of U(VI) reduction in the presence of nitrate, (ii) instability of the produced U(IV) mineral phase and the potential reoxidation of U(IV) under aerobic and denitrifying conditions, (iii) requirement for continual amendment of labile exogenous carbon to maintain reducing conditions, (iv) elaborate engineering requirements for bioreduction to proceed in highly contaminated subsurface environments, and (v) low biomass of activity of the "native" microbial communities in the low pH, highly contaminated areas that are prevalent at DOE sites (Istok et al., 2004; Wu et al., 2006b, 2006c, 2007). For example, in the highly engineered area of the OR-IFRC subsurface, pH has been elevated, nitrate intrusion limited, and high levels of aluminum removed. Such a highly engineered approach may not be feasible on a sitewide scale, as the extent of the plume ranges over 7 km, and in some cases to depths greater than 100 m (http://www.esd.ornl.gov/orifrc/), not to mention the heterogeneity of subsurface conditions found at the site. Furthermore, a bioremediation strategy dependent upon biostimulation of reductive

U(VI) immobilization by microorganisms is limited by the minor contribution of U(VI) to the total electron flow in the subsurface, and by the varied redox potential of U(VI) when complexed with common groundwater constituents. These factors make microbial U(VI) reduction dependent on site-specific geochemical conditions driving other metabolic processes, and make it difficult to differentially enrich U(VI) microorganisms. These challenges are the focus of current research, and they may be addressed provided that subsurface reducing conditions can be maintained long term. One area of active research advocates the use of slow-release substrate (SRS) to maintain reducing conditions in the subsurface (Borden, 2007). SRS is an emulsified vegetable oil mixture composed mainly of long-chain, unsaturated fatty acids that are hypothesized to be slowly degraded by a syntrophic consortium composed of fermentative and respiratory microorganisms. Even though the application of SRS appears to be effective in reducing levels of soluble U(VI), the subsurface will require periodic treatment to maintain reducing conditions, albeit much less frequently than when labile organic compounds such as ethanol are used as reductants.

What other options for effective removal of U(VI) from groundwater might be employed? According to past bioremediation research, treatment strategies should include the following characteristics: (i) they are redox *insensitive* because oxygen levels can fluctuate spatially and temporally, (ii) they produce a stable uranium mineral phase that cannot be readily remobilized, and (iii) they allow injection into the subsurface to reach contaminant zones throughout the site (i.e., a soluble delivery system). Along these lines, another active line of research has focused on U(VI) precipitation with phosphate. Phosphate-uranium minerals, such as autunite $[Ca(UO_2)_2(PO_4)_2]$, tend to have extremely positive thermodynamic stability constants, producing mineral phases that are stable under common groundwater conditions (e.g., Sowder et al., 2001; Raicevic et al., 2006; Wellman et al., 2007; Beazley, 2009).

However, free orthophosphate is generally absent from such environmental systems (e.g., Martinez et al., 2007), and must be introduced to the subsurface. Previous attempts to utilize phosphate remediation strategies have included the use of solid-phase phosphate in the form of hydroxyapatite (Fuller et al., 2002; Wellman et al., 2008a, 2008b). As described by Wellman et al. (2008a), such solid-phase applications are unsuitable for highly dispersed contaminant plumes, both at the OR-IFRC and other uranium-contaminated DOE sites. Furthermore, direct injections of soluble orthophosphate to the subsurface are inappropriate for technical reasons; namely, orthophosphate can rapidly precipitate in the presence of dissolved cations, resulting in the blockage of wells and pores within the subsurface and decreased hydraulic conductivity (Wellman et al., 2008a, 2008b; Shelobolina et al., 2009).

Soluble organophosphates have received attention as potential phosphate delivery vectors for the subsurface precipitation and stabilization of uranium (e.g., Seaman et al., 2003; Beazley et al., 2007; Martinez et al., 2007). The application of organophosphates to the subsurface may meet many of our requirements for a viable remedial technology. The advantages of this approach are due to the nonreductive precipitation of the uranyl [U(VI)] ion as a stable uranyl-phosphate-mineral phase, a precipitation process that is chemically, not biologically, catalyzed (although the release of phosphate is microbially mediated), and a semilabile soluble compound that can be dispersed via injection to the groundwater. Evidence derived from pure-culture work and sediment columns or microcosm studies has revealed the presence of nonspecific phosphatase enzyme activity in microorganisms and microbial communities derived from the contaminated subsurface (Beazley et al., 2007, 2009; Martinez et al., 2007). These enzymes appear to be expressed constitutively in microbes isolated from the subsurface, perhaps for the purpose of metal detoxification, and catalyze the release of orthophosphate from organophosphates such as glycerol-2-phosphate (G2P) (Beazley, 2009).

In addition, the activity of these enzymes is found under aerobic and anaerobic (nitrate-reducing) conditions, and is favorable under a range of pH conditions (5.5 to 7; Beazley, 2009). The addition of organophosphates to the subsurface introduces a combined carbon and phosphate source that stimulates microbial activity and biologically assisted uranium precipitation. Since many organisms appear capable of phosphatase production, particularly in metal-rich, phosphate-poor subsurface conditions, this process appears to be less sensitive to the microbial community structure in the subsurface than bioreduction of uranium and is not inhibited by the presence of nitrate.

Despite these advantages, the addition of organophosphates as a route to remediation of uranium-contaminated sites is problematic because of (i) protonation of orthophosphate at very low pH conditions present at some DOE sites, and (ii) the range of stability constants for various uranium-phosphate minerals formed under different groundwater conditions; the concentration of calcium is especially relevant (Shelobolina et al., 2009). The application of organophosphates for remediation may be best utilized in combination with multiple remediation strategies, or in areas where somewhat elevated pH conditions are found. Finally, the application of organophosphates such as G2P is unlikely to be feasible for sitewide remediation because of expense and the high C:P ratio in such molecules, and should be considered primarily as a model compound.

Inositol phosphates (IP) are a common group of organic compounds that are synthesized by plants and found frequently in the terrestrial environment (Turner et al., 2002). These compounds are found in a number of isomers, and the form *myo*-inositol hexakisphosphate is referred to as phytate (Turner et al., 2002). As a result of their strong anionic charge, they are capable of complexing polyvalent cations, and as they decompose, the release of orthophosphate can further result in uranium precipitation (Nash et al., 1998; Turner et al., 2002). Furthermore, phytate is more recalcitrant to degradation than other organophosphate molecules such as G2P, and is likely to migrate further in the subsurface (Turner et al., 2002). For inositol phosphates with six phosphate groups (IP6), two classes of enzymes catalyze desphosphorylation, including a microbial phytase and a plant-derived phytase (Turner et al., 2002). The rate of phosphatase activity appears to decrease with phosphate removal, and ultimately IP1 compounds are dephosphorylated with nonspecific phosphatase enzymes (e.g., Shan et al., 1993). Of particular relevance to the OR-IFRC, phytase enzymes appear to have low pH optima with low or no activity at circumneutral pH (Ullah and Gibson, 1987; McKelvie et al., 1995). However, under low pH conditions, the suppression of overall microbial activity may limit phytase activity (Ullah and Gibson, 1987), which does not appear to be constitutively expressed as with nonspecific phosphatases. Furthermore, some reports indicate that phytate amendments can, under some conditions, increase the solubility of contaminant metals (Seaman et al., 2003).

SUMMARY

Understanding uranium mobility in the sedimentary environment is not only essential for the protection of shallow subsurface aquifers, but also for predicting the fate of radionuclides from nuclear waste disposal in geological formations. Field studies indicate that microbially mediated, reductive immobilization is a promising strategy for the remediation of U(VI) contamination in subsurface environments. Although not yet tested in the field, the biomineralization of U(VI) upon amendment of the subsurface with organophosphate also shows promise as a U(VI) remediation strategy. Respiratory and fermentative members of the *Proteobacteria* and the *Firmicutes*, respectively, were shown to effectively catalyze U(VI) reduction in the laboratory and are predominant organisms in U(VI)-reducing environments. Members of the *Deltaproteobacteria* and the *Geobacter* group, in particular, have been strongly linked to the reductive immobilization of U, and the subsurface *Geobacter* clade shows a remarkable genotypic and phenotypic plasticity. However, the microbiology of the terrestrial subsurface is in its infancy and relatively little is known about subsurface microbial function. Interrogation of microbial populations that limit the fate and transport of U(VI) remains confined primarily to genetic targets for cultivated or described microorganisms. Microbial eukaryotes, the fungi, have been linked to U(VI) immobilization in the laboratory, but these organisms remain virtually unstudied in subsurface environments. To direct the function of subsurface microbial communities to achieve the aims of bioremediation and natural attenuation, genome-enabled studies are needed to directly link the phylogenetic structure with the metabolic activity of U(VI)-transforming microbial groups in situ. Expanded sequencing efforts will no doubt provide a clearer view of subsurface microbial community structure, but pure-culture studies are required for development of techniques to evaluate in situ function through quantification of gene expression patterns. The impact of complex physiochemical parameters, especially stressors such as acidity or metal toxicity, on the ecology of U(VI)-transforming populations in the subsurface must be further constrained. The systems approach that has been applied to the well-studied *Geobacter* group should be extended to other U(VI)-transforming organisms and processes such as biomineralization.

ACKNOWLEDGMENTS

The research for this article was supported by the Office of Science (BER), U.S. Department of Energy, grant DE-FG02-07ER64373, and by the Integrated Field-Scale Subsurface Research Challenge at Oak Ridge, operated by the Environmental Sciences Division, Oak Ridge National Laboratory (ORNL), under U.S. Department of Energy contract DEA-C0500OR22725. We particularly thank our colleagues at ORNL for their collaboration.

REFERENCES

Akob, D. M., H. J. Mills, T. M. Gihring, L. Kerkhof, J. W. Stucki, A. S. Anastacio, K. J. Chin, K. Kusel, A. V. Palumbo, D. B. Watson, and J. E. Kostka. 2008. Functional diversity and electron donor dependence of microbial populations capable of U(VI) reduction in radionuclide-contaminated subsurface sediments. *Appl. Environ. Microbiol.* **74:**3159–3170.

Akob, D. M., H. J. Mills, and J. E. Kostka. 2007. Metabolically active microbial communities in uranium-contaminated subsurface sediments. *FEMS Microbiol. Ecol.* **59:**95–107.

Anderson, R. T. 2006. DOE genomics: applications to in situ subsurface bioremediation. *Remediat. J.* **17:**23–38.

Anderson, R. T., H. A. Vrionis, I. Ortiz-Bernad, C. T. Resch, P. E. Long, R. Dayvault, K. Karp, S. Marutzky, D. R. Metzler, A. Peacock, D. C. White, M. Lowe, and D. R. Lovley. 2003. Stimulating the in situ activity of *Geobacter* species to remove uranium from the groundwater of a uranium-contaminated aquifer. *Appl. Environ. Microbiol.* **69:**5884–5891.

Bärlocher, F., L. G. Nikolcheva, K. P. Wilson, and D. D. Williams. 2006. Fungi in the Hyporheic Zone of a Springbrook. *Microb. Ecol.* **52:**708–715.

Baronofsky, J. J., W. J. A. Schreurs, and E. R. Kashket. 1984. Uncoupling by acetic-acid limits growth of and acetogenesis by *Clostridium-Thermoaceticum*. *Appl. Environ. Microbiol.* **48:**1134–1139.

Beazley, M. J. 2009. Non-reductive biomineralization of Uranium(VI) as a result of microbial phosphatase activity. Ph.D. thesis. Georgia Institute of Technology, Atlanta.

Beazley, M. J., R. J. Martinez, P. A. Sobecky, S. M. Webb, and M. Taillefert. 2007. Uranium biomineralization as a result of bacterial phosphatase activity: insights from bacterial isolates from a contaminated subsurface. *Environ. Sci. Technol.* **41:**5701–5707.

Beazley, M. J., R. J. Martinez, P. A. Sobecky, S. M. Webb, and M. Taillefert. 2009. Non-reductive biomineralization of uranium(VI) phosphate via microbial phosphatase activity in anaerobic conditions. *Geomicrobiol. J.* **26:**431–441.

Beller, H. R. 2005. Anaerobic, nitrate-dependent oxidation of U(IV) oxide minerals by the chemolithoautotrophic bacterium *Thiobacillus denitrificans*. *Appl. Environ. Microbiol.* **71:**2170–2174.

Blakeney, M. D., T. Moulaei, and T. J. DiChristina. 2000. Fe(III) reduction activity and cytochrome content of *Shewanella putrefaciens* grown on ten compounds as sole terminal electron acceptor. *Microbiol. Res.* **155:**87–94.

Borden, R. C. 2007. Concurrent bioremediation of perchlorate and 1,1,1-trichloroethane in an emulsified oil barrier. *J. Contam. Hydrol.* **94:**13–33.

Brad, T., M. Braster, B. M. van Breukelen, N. M. van Straalen, and W. F. M. Roling. 2008. Eukaryotic diversity in an anaerobic aquifer polluted with landfill leachate. *Appl. Environ. Microbiol.* **74**:3959–3968.

Brooks, S. C. 2001. *Waste Characteristics of the Former S-3 Ponds and Outline of Uranium Chemistry Relevant to NABIR Field Research Center Studies.* NABIR Field Research Center, Oak Ridge, TN.

Brooks, S. C., J. K. Fredrickson, S. L. Carroll, D. W. Kennedy, J. M. Zachara, A. E. Plymale, S. D. Kelly, K. M. Kemner, and S. Fendorf. 2003. Inhibition of bacterial U(VI) reduction by calcium. *Environ. Sci. Technol.* **37**:1850–1858.

Canfield, D. E., B. Thamdrup, and E. Kristensen. 2005. *Aquatic Geomicrobiology.* Elsevier Academic Press, San Diego, CA.

Cardenas, E., W. M. Wu, M. B. Leigh, J. Carley, S. Carroll, T. Gentry, J. Luo, D. Watson, B. Gu, M. Ginder-Vogel, P. K. Kitanidis, P. M. Jardine, J. Zhou, C. S. Criddle, T. L. Marsh, and J. A. Tiedje. 2008. Microbial communities in contaminated sediments, associated with bioremediation of uranium to submicromolar levels. *Appl. Environ. Microbiol.* **74**:3718–3729.

Catalano, J. G., and G. E. Brown. 2005. Uranyl adsorption onto montmorillonite: evaluation of binding sites and carbonate complexation. *Geochim. Cosmochim. Acta* **69**:2995–3005.

Chang, Y. J., P. E. Long, R. Geyer, A. D. Peacock, C. T. Resch, K. Sublette, S. Pfiffner, A. Smithgall, R. T. Anderson, H. A. Vrionis, J. R. Stephen, R. Dayvault, I. Ortiz-Bernad, D. R. Lovley, and D. C. White. 2005. Microbial incorporation of C-13-labeled acetate at the field scale: detection of microbes responsible for reduction of U(VI). *Environ. Sci. Technol.* **39**:9039–9048.

Chang, Y. J., A. D. Peacock, P. E. Long, J. R. Stephen, J. P. McKinley, S. J. Macnaughton, A. K. M. A. Hussain, A. M. Saxton, and D. C. White. 2001. Diversity and characterization of sulfate-reducing bacteria in groundwater at a uranium mill tailings site. *Appl. Environ. Microbiol.* **67**:3149–3160.

Chapelle, F. H. 2000. The significance of microbial processes in hydrogeology and geochemistry. *Hydrogeol. J.* **8**:41–46.

Childers, S. E., S. Ciufo, and D. R. Lovley. 2002. *Geobacter metallireducens* accesses insoluble Fe(III) oxide by chemotaxis. *Nature* **416**:767–769.

Coates, J. D., V. K. Bhupathiraju, L. A. Achenbach, M. J. McInerney, and D. R. Lovley. 2001. *Geobacter hydrogenophilus, Geobacter chapellei* and *Geobacter grbiciae,* three new, strictly anaerobic, dissimilatory Fe(III)-reducers. *Int. J. Syst. Evol. Microbiol.* **51**:581–588.

Cole, J. R., B. Chai, R. J. Farris, Q. Wang, A. S. Kulam-Syed-Mohideen, D. M. McGarrell, A. M. Bandela, E. Cardenas, G. M. Garrity, and J. M. Tiedje. 2007. The ribosomal database project (RDP-II): introducing myRDP space and quality controlled public data. *Nucleic Acids Res.* **35**:D169–D172.

Dadachova, E., R. A. Bryan, X. Huang, T. Moadel, A. D. Schweitzer, P. Aisen, J. D. Nosanchuk, and A. Casadevall. 2007. Ionizing radiation changes the electronic properties of melanin and enhances the growth of melanized fungi. *PLoS ONE* **2**:e457.

DeSantis, T. Z., P. Hugenholtz, K. Keller, E. L. Brodie, N. Larsen, Y. M. Piceno, R. Phan, and G. L. Andersen. 2006. NAST: a multiple sequence alignment server for comparative analysis of 16S rRNA genes. *Nucleic Acids Res.* **34**:W394–W399.

DiChristina, T. J., J. K. Fredrickson, and J. M. Zachara. 2005. Enzymology of electron transport: energy generation with geochemical consequences. *Rev. Mineral. Geochem.* **59**:27–52.

Duff, M. C., J. U. Coughlin, and D. B. Hunter. 2002. Uranium co-precipitation with iron oxide minerals. *Geochim. Cosmochim. Acta* **66**:3533–3547.

Edwards, L., K. Kusel, H. Drake, and J. E. Kostka. 2007. Electron flow in acidic subsurface sediments co-contaminated with nitrate and uranium. *Geochim. Cosmochim. Acta* **71**:643–654.

Elias, D. A., L. R. Krumholz, D. Wong, P. E. Long, and J. M. Suflita. 2003. Characterization of microbial activities and U reduction in a shallow aquifer contaminated by uranium mill tailings. *Microb. Ecol.* **46**:83–91.

Fields, M. W., T. F. Yan, S. K. Rhee, S. L. Carroll, P. M. Jardine, D. B. Watson, C. S. Criddle, and J. Z. Zhou. 2005. Impacts on microbial communities and cultivable isolates from groundwater contaminated with high levels of nitric acid-uranium waste. *FEMS Microbiol. Ecol.* **53**:417–428.

Finneran, K. T., M. E. Housewright, and D. R. Lovley. 2002. Multiple influences of nitrate on uranium solubility during bioremediation of uranium-contaminated subsurface sediments. *Environ. Microbiol.* **4**:510–516.

Fomina, M., S. Charnock, S. Hillier, R. Alvarez, F. Livens, and G. M. Gadd. 2008. Role of fungi in the biogeochemical fate of depleted uranium. *Curr. Biol.* **18**:R375–R377.

Francis, A. J., C. J. Dodge, F. L. Lu, G. P. Halada, and C. R. Clayton. 1994. XPS and XANES studies of uranium reduction by *Clostridium* sp. *Environ. Sci. Technol.* **28**:636–639.

Fredrickson, J. K., H. M. Kostandarithes, S. W. Li, A. E. Plymale, and M. J. Daly. 2000. Reduction of Fe(III), Cr(VI), U(VI), and Tc(VII) by *Deinococcus radiodurans* R1. *Appl. Environ. Microbiol.* **66**:2006–2011.

Fredrickson, J. K., and J. M. Zachara. 2008. Electron transfer at the microbe-mineral interface: a grand challenge in biogeochemistry. *Geobiology* **6:**245–253.

Fuller, C. C., J. R. Bargar, J. A. Davis, and M. J. Piana. 2002. Mechanisms of uranium interactions with hydroxyapatite: implications for groundwater remediation. *Environ. Sci. Technol.* **36:**158–165.

Gadd, G. M. 2007. Geomycology: biogeochemical transformations of rocks, minerals, metals and radionuclides by fungi, bioweathering and bioremediation. *Mycol. Res.* **111:**3–49.

Gao, W. M., and A. J. Francis. 2008. Reduction of uranium(VI) to uranium(IV) by clostridia. *Appl. Environ. Microbiol.* **74:**4580–4584.

Giovannoni, S. J., R. A. Foster, M. S. Rappe, and S. Epstein. 2007. New cultivation strategies bring more microbial plankton species into the laboratory. *Oceanography* **20:**62–69.

Griebler, C., and T. Lueders. 2009. Microbial biodiversity in groundwater ecosystems. *Freshw. Biol.* **54:**649–677.

Hayatsu, M., K. Tago, and M. Saito. 2008. Various players in the nitrogen cycle: diversity and functions of the microorganisms involved in nitrification and denitrification. *Soil Sci. Plant Nutr.* **54:**33–45.

Holmes, D. E., D. R. Bond, and D. R. Lovley. 2004. Electron transfer by *Desulfobulbus propionicus* to Fe(III) and graphite electrodes. *Appl. Environ. Microbiol.* **70:**1234–1237.

Holmes, D. E., R. A. O'Neil, H. A. Vrionis, L. A. N'Guessan, I. Ortiz-Bernad, M. J. Larrahondo, L. A. Adams, J. A. Ward, J. S. Nicoll, K. P. Nevin, M. A. Chavan, J. P. Johnson, P. E. Long, and D. R. Lovley. 2007. Subsurface clade of *Geobacteraceae* that predominates in a diversity of Fe(III)-reducing subsurface environments. *ISME J.* **1:**663–677.

Hua, B., H. F. Xu, J. Terry, and B. L. Deng. 2006. Kinetics of uranium(VI) reduction by hydrogen sulfide in anoxic aqueous systems. *Sci. Technol.* **40:**4666–4671.

Hwang, C. C., W. M. Wu, T. J. Gentry, J. Carley, G. A. Corbin, S. L. Carroll, D. B. Watson, P. M. Jardine, J. Z. Zhou, C. S. Criddle, and M. W. Fields. 2009. Bacterial community succession during in situ uranium bioremediation: spatial similarities along controlled flow paths. *ISME J.* **3:**47–64.

Ilton, E. S., N. P. Qafoku, C. X. Liu, D. A. Moore, and J. M. Zachara. 2008. Advective removal of intraparticle uranium from contaminated Vadose zone sediments, Hanford, US. *Environ. Sci. Technol.* **42:**1565–1571.

Istok, J. D., J. M. Senko, L. R. Krumholz, D. Watson, M. A. Bogle, A. Peacock, Y. J. Chang, and D. C. White. 2004. In situ bioreduction of technetium and uranium in a nitrate-contaminated aquifer. *Environ. Sci. Technol.* **38:**468–475.

Jakobsen, R., and D. Postma. 1999. Redox zoning, rates of sulfate reduction and interactions with Fe-reduction and methanogenesis in a shallow sandy aquifer, Romo, Denmark. *Geochim. Cosmochim. Acta* **63:**137–151.

Janssen, P. H. 2008. New cultivation strategies for terrestrial microorganisms, p. 173–192. *In* K. Zengler (ed.), *Accessing Uncultivated Microorganisms: from the Environment to Organisms and Genomes and Back*. ASM Press, Washington, DC.

Jeon, B. H., S. D. Kelly, K. M. Kemner, M. O. Barnett, W. D. Burgos, B. A. Dempsey, and E. E. Roden. 2004. Microbial reduction of U(VI) at the solid-water interface. *Sci. Technol.* **38:**5649–5655.

Kashefi, K., and D. R. Lovley. 2000. Reduction of Fe(III), Mn(IV), and toxic metals at 100°C by *Pyrobaculum islandicum*. *Appl. Environ. Microbiol.* **66:**1050–1056.

Kelly, S. D., K. M. Kemner, J. Carley, C. Criddle, P. M. Jardine, T. L. Marsh, D. Phillips, D. Watson, and W. M. Wu. 2008. Speciation of uranium in sediments before and after in situ biostimulation. *Environ. Sci. Technol.* **42:**1558–1564.

Khijniak, T. V., A. I. Slobodkin, V. Coker, J. C. Renshaw, F. R. Livens, E. A. Bonch-Osmolovskaya, N. K. Birkeland, N. N. Medvedeva-Lyalikova, and J. R. Lloyd. 2005. Reduction of uranium(VI) phosphate during growth of the thermophilic bacterium *Thermoterrabacterium ferrireducens*. *Appl. Environ. Microbiol.* **71:**6423–6426.

Kieft, T. L., J. K. Fredrickson, T. C. Onstott, Y. A. Gorby, H. M. Kostandarithes, T. J. Bailey, D. W. Kennedy, S. W. Li, A. E. Plymale, C. M. Spadoni, and M. S. Gray. 1999. Dissimilatory reduction of Fe(III) and other electron acceptors by a *Thermus* isolate. *Appl. Environ. Microbiol.* **65:**1214–1221.

Kumar, S., M. Nei, J. Dudley, and K. Tamura. 2008. MEGA: a biologist-centric software for evolutionary analysis of DNA and protein sequences. *Brief. Bioinformat.* **9:**299–306.

Kunapuli, U., M. K. Jahn, T. Lueders, R. Geyer, H. J. Heipieper, and R. U. Meckenstock. 2009. Anaerobic degradation of monoaromatic hydrocarbons by two novel iron-reducing bacteria: description of *Desulfitobacterium aromaticivorans* sp. nov., and *Geobacter toluenoxydans* sp. nov. *Int. J. Syst. Evol. Microbiol.* doi:10.1099/ijs.0.003525-0.

Laughlin, R. J., and R. J. Stevens. 2002. Evidence for fungal dominance of denitrification and codenitrification in a grassland soil. *Soil Sci. Soc. Am. J.* **66:**1540–1548.

Liger, E., L. Charlet, and P. Van Cappellen. 1999. Surface catalysis of uranium(VI) reduction by iron(II). *Geochim. Cosmochim. Acta* **63:**2939–2955.

Lin, B., H. V. Westerhoff, and W. F. M. Roling. 2009. How *Geobacteraceae* may dominate subsurface biodegradation: physiology of *Geobacter metallireducens* in slow-growth habitat-simulating retentostats. *Environ. Microbiol.* **11:**2425–2433.

Liu, C. X., B. H. Jeon, J. M. Zachara, Z. M. Wang, A. Dohnalkova, and J. K. Fredrickson. 2006. Kinetics of microbial reduction of solid phase U(VI). *Environ. Sci. Technol.* **40:**6290–6296.

Liu, C. X., J. M. Zachara, O. Qafoku, J. P. McKinley, S. M. Heald, and Z. M. Wang. 2004. Dissolution of uranyl microprecipitates in subsurface sediments at Hanford site, USA. *Geochim. Cosmochim. Acta* **68:**4519–4537.

Lloyd, J. R., and L. E. Macaskie. 2000. Bioremediation of radioactive metals, p. 277–327. *In* D. R. Lovley (ed.), *Environmental Microbe-Metal Interactions*. ASM Press, Washington, DC.

Loffler, F. E., and E. A. Edwards. 2006. Harnessing microbial activities for environmental cleanup. *Curr. Opin. Biotechnol.* **17:**274–284.

Lovley, D. R., D. E. Holmes, and K. P. Nevin. 2004. Dissimilatory Fe(III) and Mn(IV) reduction. *Adv. Microb. Physiol.* **49:**219–286.

Lovley, D. R., R. Mahadevan, and K. P. Nevin. 2008. Systems biology approach to bioremediation with extracellular electron transfer, p. 71–96. *In* E. Diaz (ed.), *Microbial Biodegradation: Genomics and Molecular Biology*. Horizon Scientific Press, Norwich, United Kingdom.

Lovley, D. R., and E. J. P. Phillips. 1992. Bioremediation of uranium contamination with enzymatic uranium reduction. *Environ. Sci. Technol.* **26:**2228–2234.

Lovley, D. R., E. J. P. Phillips, Y. A. Gorby, and E. R. Landa. 1991. Microbial reduction of uranium. *Nature* **350:**413–416.

Lovley, D. R., E. E. Roden, E. J. P. Philips, and J. C. Woodward. 1993. Enzymatic iron and uranium reduction by sulfate-reducing bacteria. *Mar. Geol.* **113:**41–53.

Luo, W. S., S. D. Kelly, K. M. Kemner, D. Watson, J. Z. Zhou, P. M. Jardine, and B. H. Gu. 2009. Sequestering uranium and technetium through co-precipitation with aluminum in a contaminated acidic environment. *Environ. Sci. Technol.* **43:**7516–7522.

Madden, A. S., A. C. Smith, D. L. Balkwill, L. A. Fagan, and T. J. Phelps. 2007. Microbial uranium immobilization independent of nitrate reduction. *Environ. Microbiol.* **9:**2321–2330.

Marietou, A., L. Griffiths, and J. Cole. 2009. Preferential reduction of the thermodynamically less favorable electron acceptor, sulfate, by a nitrate-reducing strain of the sulfate-reducing bacterium *Desulfovibrio desulfuricans* 27774. *J. Bacteriol.* **191:**882–889.

Markich, S. J. 2002. Uranium speciation and bioavailability in aquatic systems: an overview. *Sci. World J.* **2:**707–729.

Martinez, R. J., M. J. Beazley, M. Taillefert, A. K. Arakaki, J. Skolnick, and P. A. Sobecky. 2007. Aerobic uranium (VI) bioprecipitation by metal-resistant bacteria isolated from radionuclide- and metal-contaminated subsurface soils. *Environ. Microbiol.* **9:**3122–3133.

Mckelvie, I. D., B. T. Hart, T. J. Cardwell, and R. W. Cattrall. 1995. Use of immobilized 3-phytase and flow-injection for the determination of phosphorus species in natural-waters. *Anal. Chim. Acta* **316:**277–289.

Mohagheghi, A., D. M. Updegraff, and M. B. Goldhaber. 1985. The role of sulfate-reducing bacteria in the deposition of sedimentary uranium ores. *Geomicrobiol. J.* **4:**153–173.

Moon, J., Y. Roh, T. J. Phelps, D. H. Philips, D. B. Watson, Y.-J. Kim, and S. C. Brooks. 2005. Physicochemical and mineralogical characterization of soil-saprolite cores from a field research site, Tennessee. *J. Environ. Qual.* **35:**1731–1741.

Murphy, W. M., and E. L. Shock. 1999. Environmental aqueous geochemistry of actinides, p. 221–253. *In* P. C. Burns and R. Finch (ed.), *Uranium: Mineralogy, Geochemistry and the Environment*, vol. 38. Mineralogical Society of America, Washington, DC.

Muyzer, G., and A. J. M. Stams. 2008. The ecology and biotechnology of sulphate-reducing bacteria. *Nat. Rev. Microbiol.* **6:**441–454.

Nash, K. L., M. P. Jensen, and M. A. Schmidt. 1998. Actinide immobilization in the subsurface environment by in-situ treatment with a hydrolytically unstable organophosphorus complexant: uranyl uptake by calcium phytate. *J. Alloys Compd.* **271:**257–261.

National Research Council. 1993. *In Situ Bioremediation: When Does It Work?* National Academy Press, Washington, DC.

Natural and Accelerated Bioremediation Research (NABIR). 2003. *Bioremediation of Metals and Radionuclides: What It Is and How It Works*. Lawrence Berkeley National Laboratory, Berkeley, CA.

Nealson, K. H., A. Belz, and B. McKee. 2002. Breathing metals as a way of life: geobiology in action. *Antonie Van Leeuwenhoek* **81:**215–222.

N'Guessan, A. L., H. A. Vrionis, C. T. Resch, P. E. Long, and D. R. Lovley. 2008. Sustained removal of uranium from contaminated groundwater following stimulation of dissimilatory metal reduction. *Environ. Sci. Technol.* **42:**2999–3004.

North, N. N., S. L. Dollhopf, L. Petrie, J. D. Istok, D. L. Balkwill, and J. E. Kostka. 2004.

Change in bacterial community structure during in situ biostimulation of subsurface sediment cocontaminated with uranium and nitrate. *Appl. Environ. Microbiol.* **70:**4911–4920.

Payne, R. B., D. A. Gentry, B. J. Rapp-Giles, L. Casalot, and J. D. Wall. 2002. Uranium reduction by *Desulfovibrio desulfuricans* strain G20 and a cytochrome c3 mutant. *Appl. Environ. Microbiol.* **68:**3129–3132.

Peacock, A. D., Y. J. Chang, J. D. Istok, L. Krumholz, R. Geyer, B. Kinsall, D. Watson, K. L. Sublette, and D. C. White. 2004. Utilization of microbial biofilms as monitors of bioremediation. *Microb. Ecol.* **47:**284–292.

Petrie, L., N. N. North, S. L. Dollhopf, D. L. Balkwill, and J. E. Kostka. 2003. Enumeration and characterization of iron(III)-reducing microbial communities from acidic subsurface sediments contaminated with uranium(VI). *Appl. Environ. Microbiol.* **69:**7467–7479.

Pietzsch, K., and W. Babel. 2003. A sulfate-reducing bacterium that can detoxify U(VI) and obtain energy via nitrate reduction. *J. Basic Microbiol.* **43:**348–361.

Prakash, O., T. M. Gihring, D. D. Dalton, K.-J. Chin, S. J. Green, D. M. Akob, G. Wanger, and J. E. Kostka. 2010. *Geobacter daltonii* sp. nov., an iron(III)- and uranium(VI)-reducing bacterium isolated from the shallow subsurface exposed to mixed heavy metal and hydrocarbon contamination. *Int. J. Syst. Evol. Microbiol.* **60:**546–553. doi:10.1099/ijs.0.010843-0.

Raicevic, S., J. V. Wright, V. Veljkovic, and J. L. Conca. 2006. Theoretical stability assessment of uranyl phosphates and apatites: selection of amendments for in situ remediation of uranium. *Sci. Total Environ.* **355:**13–24.

Read, D., T. A. Lawless, R. J. Sims, and K. R. Butter. 1993. Uranium migration through intact sandstone cores. *J. Contam. Hydrol.* **13:**277–289.

Riley, R. G., and J. M. Zachara. 1992. *Chemical Contaminants on DOE Lands and Selection of Contaminant Mixtures for Subsurface Research*, vol. DOE/ER-0547T. U.S. Department of Energy, Washington, DC.

Roh, Y., S. V. Liu, G. S. Li, H. S. Huang, T. J. Phelps, and J. Z. Zhou. 2002. Isolation and characterization of metal-reducing *Thermoanaerobacter* strains from deep subsurface environments of the Piceance Basin, Colorado. *Appl. Environ. Microbiol.* **68:**6013–6020.

Sanford, R. A., Q. Wu, Y. Sung, S. H. Thomas, B. K. Amos, E. K. Prince, and F. E. Löffler. 2007. Hexavalent uranium supports growth of *Anaeromyxobacter dehalogenans* and *Geobacter* spp. with lower than predicted biomass yields. *Environ. Microbiol.* **9:**2885–2893.

Sani, R. K., B. M. Peyton, J. E. Amonette, and G. G. Geesey. 2004. Reduction of uranium(VI) under sulfate-reducing conditions in the presence of Fe(III)-(hydr)oxides. *Geochim. Cosmochim. Acta* **68:**2639–2648.

Sani, R. K., B. M. Peyton, W. A. Smith, W. A. Apel, and J. N. Petersen. 2002. Dissimilatory reduction of Cr(VI), Fe(III), and U(VI) by *Cellulomonas* isolates. *Appl. Microbiol. Biotechnol.* **60:**192–199.

Schmidt, S. K., K. L. Wilson, A. F. Meyer, C. W. Schadt, T. M. Porter, and J. M. Moncalvo. 2008. The missing fungi: new insights from culture-independent molecular studies of soil, p. 55–66. *In* K. Zengler (ed.), *Accessing Uncultivated Microorganisms: from the Environment to Organisms and Genomes and Back.* ASM Press, Washington, DC.

Scott, T. B., G. C. Allen, P. J. Heard, and M. G. Randell. 2005. Reduction of U(VI) to U(IV) on the surface of magnetite. *Geochim. Cosmochim. Acta* **69:**5639–5646.

Seaman, J. C., J. M. Hutchison, B. P. Jackson, and V. M. Vulava. 2003. In situ treatment of metals in contaminated soils with phytate. *J. Environ. Qual.* **32:**153–161.

Senko, J. M., J. D. Istok, J. M. Suflita, and L. R. Krumholz. 2002. In-situ evidence for uranium immobilization and remobilization. *Environ. Sci. Technol.* **36:**1491–1496.

Senko, J. M., Y. Mohamed, T. A. Dewers, and L. R. Krumholz. 2005. Role for Fe(III) minerals in nitrate-dependent microbial U(IV) oxidation. *Environ. Sci. Technol.* **39:**2529–2536.

Shan, Y., I. D. Mckelvie, and B. T. Hart. 1993. Characterization of immobilized *Escherichia coli* alkaline-phosphatase reactors in flow-injection analysis. *Anal. Chem.* **65:**3053–3060.

Shelobolina, E. S., H. Konishi, H. Xu, and E. E. Roden. 2009. U(VI) Sequestration in hydroxyapatite produced by microbial glycerol 3-phosphate metabolism. *Appl. Environ. Microbiol.* **75:**5773–5778.

Shelobolina, E. S., K. P. Nevin, J. D. Blakeney-Hayward, C. V. Johnsen, T. W. Plaia, P. Krader, T. Woodard, D. E. Holmes, C. G. VanPraagh, and D. R. Lovley. 2007. *Geobacter pickeringii* sp nov., *Geobacter argillaceus* sp nov and *Pelosinus fermentans* gen. nov., sp nov., isolated from subsurface kaolin lenses. *Int. J. Syst. Evol. Microbiol.* **57:**126–135.

Shelobolina, E. S., S. A. Sullivan, K. R. O'Neill, K. P. Nevin, and D. R. Lovley. 2004. Isolation, characterization, and U(VI)-reducing potential of a facultatively anaerobic, acid-resistant bacterium from low-pH, nitrate- and U(VI)-contaminated subsurface sediment and description of

Salmonella subterranea sp nov. *Appl. Environ. Microbiol.* **70:**2959–2965.

Shelobolina, E. S., C. G. Vanpraagh, and D. R. Lovley. 2003. Use of ferric and ferrous iron containing minerals for respiration by *Desulfitobacterium frappieri*. *Geomicrobiol. J.* **20:**143–156.

Shelobolina, E. S., H. A. Vrionis, R. H. Findlay, and D. R. Lovley. 2008. Geobacter uraniireducens sp nov., isolated from subsurface sediment undergoing uranium bioremediation. *Int. J. Syst. Evol. Microbiol.* **58:**1075–1078.

Shoun, H., D.-H. Kim, H. Uchiyama, and J. Sugiyama. 1992. Denitrification by fungi. *FEMS Microbiol. Lett.* **94:**281.

Singer, D. M., J. M. Zachara, and G. E. Brown. 2009. Uranium speciation as a function of depth in contaminated hanford sediments—a micro-XRF, micro-XRD, and micro- and bulk-XAFS study. *Environ. Sci. Technol.* **43:**630–636.

Sowder, A. G., S. B. Clark, and R. A. Fjeld. 2001. The impact of mineralogy in the U(VI)-Ca-PO4 system on the environmental availability of uranium. *J. Radioanal. Nucl. Chem.* **248:**517–524.

Spain, A. M., A. D. Peacock, J. D. Istok, M. S. Elshahed, F. Z. Najar, B. A. Roe, D. C. White, and L. R. Krumholz. 2007. Identification and isolation of a *Castellaniella* species important during biostimulation of an acidic nitrate- and uranium-contaminated aquifer. *Appl. Environ. Microbiol.* **73:**4892–4904.

Suzuki, Y., S. D. Kelly, K. M. Kemner, and J. F. Banfield. 2004. Enzymatic U(VI) reduction by *Desulfosporosinus* species. *Radiochim. Acta* **92:**11–16.

Suzuki, Y., and T. Suko. 2006. Geomicrobiological factors that control uranium mobility in the environment: update on recent advances in the bioremediation of uranium-contaminated sites. *J. Mineral. Petrol. Sci.* **101:**299–307.

Tebo, B. M., and A. Y. Obraztsova. 1998. Sulfate-reducing bacterium grows with Cr(VI), U(VI), Mn(IV), and Fe(III) as electron acceptors. *FEMS Microbiol. Lett.* **162:**193–198.

Thomas, S. H., E. Padilla-Crespo, P. M. Jardine, R. A. Sanford, and F. E. Löffler. 2009. Diversity and distribution of *Anaeromyxobacter* strains in a uranium-contaminated subsurface environment with a nonuniform groundwater flow. *Appl. Environ. Microbiol.* **75:**3679–3687.

Truex, M. J., B. M. Peyton, N. B. Valentine, and Y. A. Gorby. 1997. Kinetics of U(VI) reduction by a dissimilatory Fe(III)-reducing bacterium under non-growth conditions. *Biotechnol. Bioeng.* **55:**490–496.

Turner, B. L., M. J. Paphazy, P. M. Haygarth, and I. D. McKelvie. 2002. Inositol phosphates in the environment. *Philos. Trans. R. Soc. Lond. Ser. B* **357:**449–469.

Ullah, A. H. J., and D. M. Gibson. 1987. Extracellular phytase (EC 3.1.3.8) from *Aspergillus ficuum* Nrrl 3135—purification and characterization. *Prep. Biochem.* **17:**63–91.

Vrionis, H. A., R. T. Anderson, I. Ortiz-Bernad, K. R. O'Neill, C. T. Resch, A. D. Peacock, R. Dayvault, D. C. White, P. E. Long, and D. R. Lovley. 2005. Microbiological and geochemical heterogeneity in an in situ uranium bioremediation field site. *Appl. Environ. Microbiol.* **71:**6308–6318.

Wall, J. D., and L. R. Krumholz. 2006. Uranium reduction. *Annu. Rev. Microbiol.* **60:**149–166.

Wan, J. M., T. K. Tokunaga, E. Brodie, Z. M. Wang, Z. P. Zheng, D. Herman, T. C. Hazen, M. K. Firestone, and S. R. Sutton. 2005. Reoxidation of bioreduced uranium under reducing conditions. *Environ. Sci. Technol.* **39:**6162–6169.

Wang, Z. M., J. M. Zachara, P. L. Gassman, C. X. Liu, O. Qafoku, W. Yantasee, and J. G. Catalan. 2005. Fluorescence spectroscopy of U(VI)-silicates and U(VI)-contaminated Hanford sediment. *Geochim. Cosmochim. Acta* **69:**1391–1403.

Wellman, D. M., K. M. Gunderson, J. P. Icenhower, S. W. Forrester, and S. W. Forrester. 2007. Dissolution kinetics of synthetic and natural meta-autunite minerals, X-3-n((n)+) [(UO$_2$)(PO$_4$)](2) · xH(2)O, under acidic conditions. *Geochem. Geophys. Geosyst.* doi:10.1029/2007GC001695.

Wellman, D. M., E. M. Pierce, D. H. Bacon, M. Oostrom, K. M. Gunderson, S. M. Webb, C. C. Bovaird, E. A. Cordova, E. T. Clayton, K. E. Parker, R. M. Ermi, S. R. Baum, V. R. Vermeul, and J. S. Fruchter. 2008a. 300 Area treatability test: laboratory development of polyphosphate remediation technology for in situ treatment of uranium contamination in the vadose zone and capillary fringe, vol. PNNL-17818, Contract DE-AC05-76RL01830. U.S. Department of Energy, Pacific Northwest National Laboratory, Richland, WA.

Wellman, D. M., E. M. Pierce, V. R. Vermeul, S. V. Mattigod, E. L. Richards, M. D. Williams, J. S. Fruchter, and J. P. Icenhower. 2008b. In situ uranium stabilization through polyphosphate remediation: development and demonstration at the Hanford Site 300 Area, Washington State, p. 25–104. *In* T. V. Golush (ed.), *Waste Management Research Trends*. Nova Science Publishers, Inc., New York, NY.

Wilkins, M. J., F. R. Livens, D. J. Vaughan, and J. R. Lloyd. 2006. The impact of Fe(III)-reducing bacteria on uranium mobility. *Biogeochemistry* **78:**125–150.

Wu, Q., R. A. Sanford, and F. E. Löffler. 2006a. Uranium(VI) reduction by *Anaeromyxobacter dehalogenans* strain 2CP-C. *Appl. Environ. Microbiol.* **72:**3608–3614.

Wu, W. M., J. Carley, M. Fienen, T. Mehlhorn, K. Lowe, J. Nyman, J. Luo, M. E. Gentile, R. Rajan, D. Wagner, R. F. Hickey, B. H. Gu, D. Watson, O. A. Cirpka, P. K. Kitanidis, P. M. Jardine, and C. S. Criddle. 2006b. Pilot-scale in situ bioremediation of uranium in a highly contaminated aquifer. 1. Conditioning of a treatment zone. *Environ. Sci. Technol.* **40:**3978–3985.

Wu, W. M., J. Carley, T. Gentry, M. A. Ginder-Vogel, M. Fienen, T. Mehlhorn, H. Yan, S. Caroll, M. N. Pace, J. Nyman, J. Luo, M. E. Gentile, M. W. Fields, R. F. Hickey, B. H. Gu, D. Watson, O. A. Cirpka, J. Z. Zhou, S. Fendorf, P. K. Kitanidis, P. M. Jardine, and C. S. Criddle. 2006c Pilot-scale in situ bioremediation of uranium in a highly contaminated aquifer. 2. Reduction of U(VI) and geochemical control of U(VI) bioavailability. *Environ. Sci. Technol.* **40:**3986–3995.

Wu, W. M., J. Carley, J. Luo, M. A. Ginder-Vogel, E. Cardenas, M. B. Leigh, C. C. Hwang, S. D. Kelly, C. M. Ruan, L. Y. Wu, J. Van Nostrand, T. Gentry, K. Lowe, T. Mehlhorn, S. Carroll, W. S. Luo, M. W. Fields, B. H. Gu, D. Watson, K. M. Kemner, T. Marsh, J. Tiedje, J. Z. Zhou, S. Fendorf, P. K. Kitanidis, P. M. Jardine, and C. S. Criddle. 2007. In situ bioreduction of uranium (VI) to submicromolar levels and reoxidation by dissolved oxygen. *Environ. Sci. Technol.* **41:**5716–5723.

Yan, T. F., M. W. Fields, L. Y. Wu, Y. G. Zu, J. M. Tiedje, and J. Z. Zhou. 2003. Molecular diversity and characterization of nitrite reductase gene fragments (*nirK* and *nirS*) from nitrate- and uranium-contaminated groundwater. *Environ. Microbiol.* **5:**13–24.

Zengler, K. 2008. *Accessing Uncultivated Microorganisms: from the Environment to Organisms and Genomes and Back.* ASM Press, Washington, DC.

Zengler, K., G. Toledo, M. Rappe, J. Elkins, E. J. Mathur, J. M. Short, and M. Keller. 2002. Cultivating the uncultured. *Proc. Natl. Acad. Sci. USA* **99:**15681–15686.

DIRECT AND INDIRECT PROCESSES LEADING TO URANIUM(IV) OXIDATION

Rizlan Bernier-Latmani and Bradley M. Tebo

8

BACKGROUND

Extensive contamination of former uranium mining, milling, and processing sites in North America and Europe has prompted a search for remediation solutions. One such remediation approach entails the in situ immobilization of uranium in the subsurface by stimulating the native microbiota. Microbial activity has been shown to catalyze the reduction of the soluble and mobile contaminant U(VI) to the sparingly soluble, nanoparticulate mineral uraninite (UO_2) (Fig. 1).

Metal-reducing bacteria such as *Shewanella oneidensis* MR-1 or *Geobacter metallireducens*, sulfate-reducing bacteria such as *Desulfovibrio vulgaris* or *Desulfotomaculum reducens*, and bacteria such as *Clostridium sphenoides* and *Anaeromyxobacter dehalogenans* belonging to other metabolic groups have been shown to carry out enzymatic U(VI) reduction (Lovley et al., 1991; Lovley, 1993; Francis et al., 1994; Tebo and Obraztsova, 1998; Wu et al., 2006a). In addition, indirect or abiotic U(VI) reduction processes include reduction by sorbed Fe(II) (Liger et al., 1999; Boyanov et al., 2007) and Fe(II) minerals such as green rust (O'Loughlin et al., 2003), magnetite (Regenspurg et al., 2009; O'Loughlin et al., 2010), and ferrous sulfide (Wersin et al., 1994).

Biogenic UO_2 has been extensively characterized and compared with its abiotic counterpart, stoichiometric uraninite (Burgos et al., 2008; Schofield et al., 2008; Ulrich et al., 2008). It was found to be structurally similar to bulk UO_2 with a relatively well-ordered interior and an outer shell of more disordered crystal lattice. The most remarkable feature of biogenic uraninite is its nanoparticulate size (2 to 5 nm diameter), which confers to it a high surface area relative to bulk UO_2 (Bargar et al., 2008). Surprisingly, the uraninite minerals produced by *S. oneidensis*, *D. vulgaris*, and *Geobacter sulfurreducens* are consistent with stoichiometric UO_2 (Schofield et al., 2008; Sharp et al., 2009) in contrast to the generally cation-substituted or O-atom-enriched forms of uraninite observed in nature given by the stoichiometry of UO_{2+x} (where x may be as high as 0.25) (Allen and Tempest, 1986; Janeczek and Ewing, 1992; Finch and Murakami, 1999).

The long-term stability of reduced U in the subsurface is a crucial prerequisite for the feasibility of in situ reductive uranium

---

Rizlan Bernier-Latmani, Environmental Microbiology Laboratory, Ecole Polytechnique Fédérale de Lausanne, CH 1015 Lausanne, Switzerland. *Bradley M. Tebo*, Division of Environmental and Biomolecular Systems, Oregon Health & Science University, Beaverton, OR 97006.

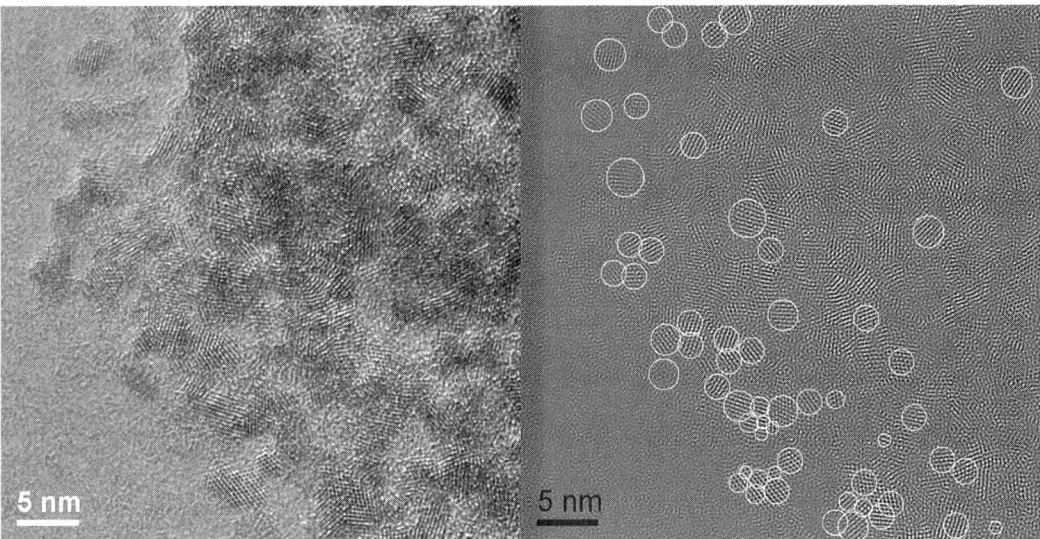

FIGURE 1 High-resolution transmission electron microscopy (HRTEM) (left) and Fourier-filtered HRTEM (right) image of biogenic nanoparticles of UO_2 obtained from *S. oneidensis* MR-1 under nongrowth conditions.

immobilization. Practically speaking, bioremediation involves the temporally limited supply of electron donor (Anderson et al., 2003; Wu et al., 2006b) to the subsurface which leads to the utilization of various electron acceptors (oxygen, nitrate) before the reduction of uranium [which is often concomitant with the reduction of Fe(III)]. After the electron donor amendments are halted, the potential for a reaction between environmental oxidants advected into the reduced zone and U(IV) increases.

Several environmental oxidants are thermodynamically favorable for oxidizing U(IV) including oxygen, nitrogen oxides, and solid-phase Fe(III) and Mn(IV) oxides (Fig. 2). Many of these oxidants can be formed biologically, and these various redox-active compounds can also react with each other. Thus, the potential oxidation of U(IV) in the subsurface is a complex and intricate series of direct and indirect coupled biological redox reactions. This review aims at presenting a comprehensive view of the collective understanding of the processes potentially responsible for U(IV) oxidation in the subsurface.

OXYGEN

Abiotic U(IV) Oxidation

Oxygen is an excellent oxidizing agent for UO_2 (Fig. 3). Several studies have reported the oxidation of biogenic U(IV) under fully aerobic conditions (21% O_2) and in the presence of 30 mM bicarbonate in a batch reactor (Senko et al., 2007; Burgos et al., 2008). In those studies, the oxidation of biogenic UO_2 by O_2 was observed to take place rapidly (about 200 μM in 6 h) when uraninite originated from *S. oneidensis* MR-1 (Burgos et al., 2008) and equally rapidly (about 100 μM in 2 h) when it originated from *Shewanella putrefaciens* CN32 (Senko et al., 2007).

Comparing the dissolution rate of biogenic UO_2 in the absence and in the presence (at 1 or 21%) of O_2 with 10^{-3} M dissolved inorganic carbon in a flowthrough dissolution reactor revealed a surface area-normalized dissolution rate (in $mol·m^{-2}·s^{-1}$) of 3.13×10^{-11} in the absence of O_2 in comparison with 9.83×10^{-11} and 1.50×10^{-9} for 1% and 21% O_2, respectively (Ulrich et al., 2009).

Most studies have focused on the most widely reported product of microbial U(VI)

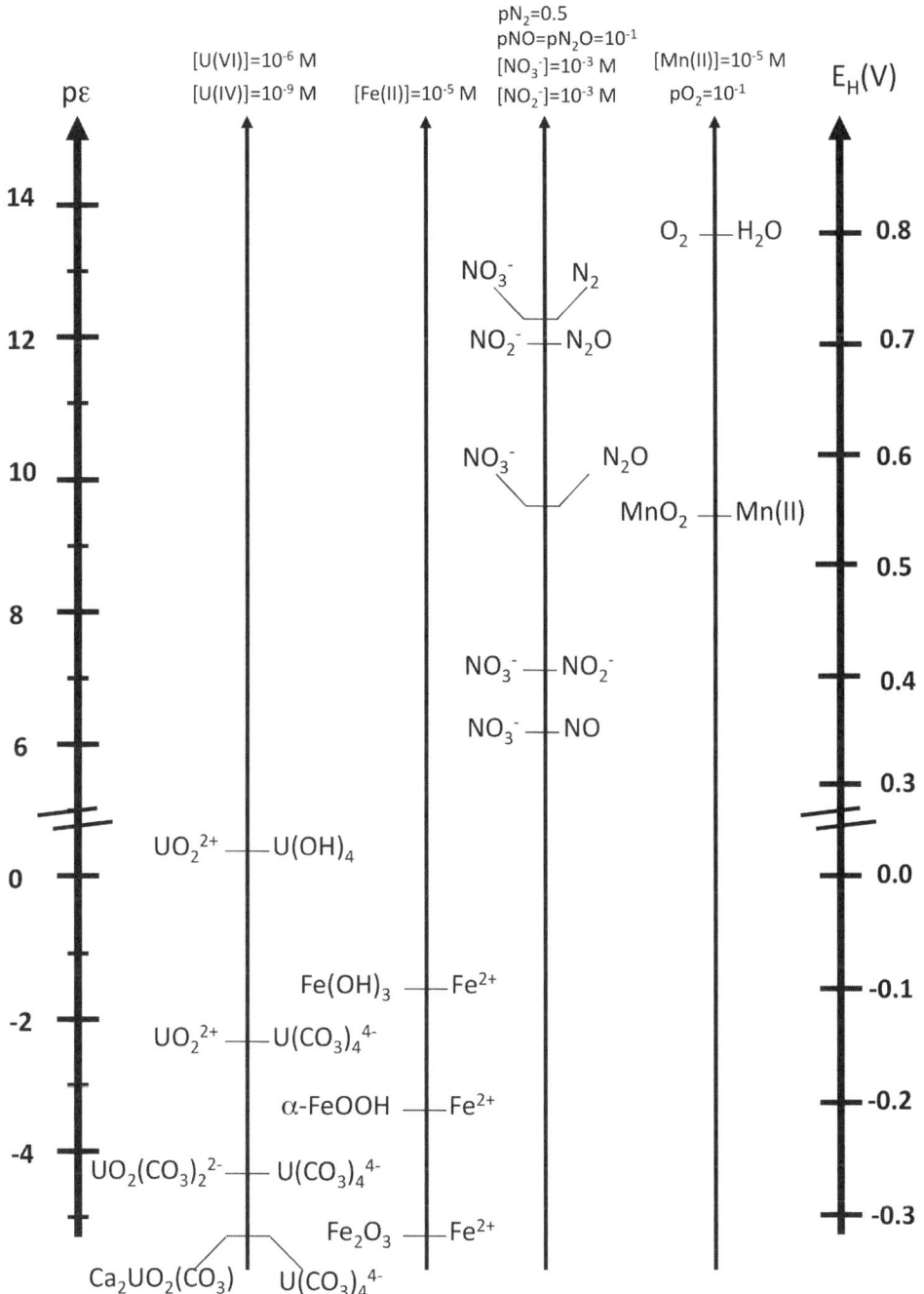

FIGURE 2 Redox potentials of species discussed in the text. Thermodynamic data were obtained from Ginder-Vogel et al. (2006) and Morel and Hering (1993). All calculations were carried out at pH 7.

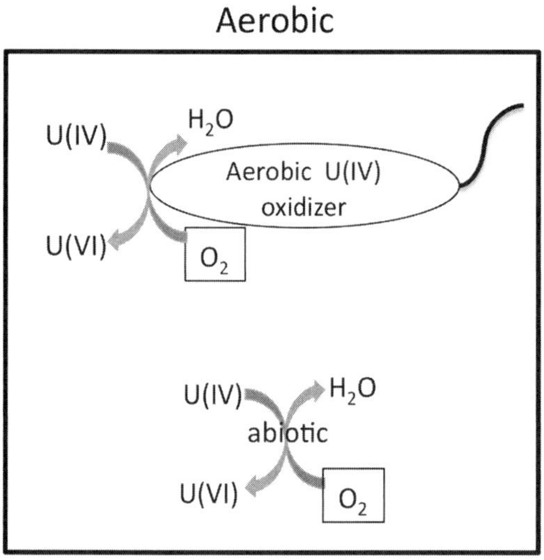

FIGURE 3 Schematic of the direct and indirect biological pathways involving oxygen for U(IV) oxidation. In this and subsequent figures, the boxed text corresponds to the electron acceptor in the processes considered.

reduction, namely the mineral uraninite. However, if humic acids or fulvic acids are present during the biological reduction of U(VI), the product of the reduction was reported to be a soluble complexed U(IV) species (Gu et al., 2005). This product was reoxidized by O_2 readily and at a significantly faster rate than uraninite. All complexed U(IV) was reoxidized in less than 1 h, while it took more than 6 h to oxidize U(IV) precipitates under the same conditions (Gu et al., 2005).

While the above studies show the ability of oxygen to readily oxidize U(IV) in laboratory settings, several investigations of subsurface environments or column studies reported that U(IV) oxidation by O_2 was slow. The first such study evaluated the reoxidation of U(IV) formed biologically in a flowthrough column containing sandstone (Abdelouas et al., 1999). Fe(III) in the sandstone and sulfate from the groundwater used for the experiment were also reduced in the column during the experiment leading to the formation of mackinawite ($FeS_{0.9}$). In an attempt to reoxidize U(IV), groundwater containing 7 mg/liter dissolved oxygen was pumped through the reduced column. The uranium concentration in the effluent did not exceed the solubility limit of UO_2, suggesting that uranium was only released from UO_2 by equilibrium dissolution rather than by oxidation processes. In contrast, the sulfate concentration in the effluent increased, pointing to mackinawite oxidation. The authors concluded that mackinawite was acting as an effective redox buffer that protected column uraninite from oxidation (Abdelouas et al., 1999).

However, in batch conditions, a higher solution-to-solid ratio with the same materials yielded oxidation of U(IV): for instance, for 7.5 g/liter of sandstone in groundwater, ~45% of the total U(IV) (which was present at 3.1 µg/g sand) was oxidized in about 10 days. In contrast, only ~5% of the total U(IV) was oxidized in the same amount of time with a 62.5 g/liter solids loading. For comparison, in the column experiments (which are representative of field conditions), the solution-to-solid ratio was approximately 6,600 g/liter sandstone (Abdelouas et al., 1999). Thus, the solid-to-solution ratio plays an important role in the rate of oxidation of U(IV) in subsurface environments.

More recent investigations have considered the oxidation of U(IV) in sediment columns under various conditions. In particular, the comparison of two studies allows an evaluation of the effect of sulfate on the reoxidation of U(IV) by oxygen (Moon et al., 2007, 2009). In the absence of sulfate, Fe(III) and U(VI) reduction took place simultaneously during the reduction phase in which an influent containing bicarbonate, U(VI), acetate, NH_4^+, phosphate, KCl, and trace elements flowed into the column containing Rifle Area Background Sediment (RABS) from a site in Old Rifle, Colorado, and seeded with *G. metallireducens*. In the presence of sulfate (6 mM), microorganisms catalyzed sulfate reduction and iron sulfide minerals precipitated in the column (Moon et al., 2009). During the reoxidation phase, the same influent was maintained but acetate, NH_4^+, and trace elements were not included and O_2 was supplied at a concentration of 0.27 mM (20% in the gas phase) while a control received no oxygen. In the sulfate-free column, the U(VI) effluent concentration rose above that of the control after ~2 days and remained elevated in comparison with the control for ~30 days. About 88% of U immobilized in the column during the reduction phase was recovered with the oxygen treatment. In contrast, in the presence of sulfate, no uranium remobilization was observed for the first 26 days of the experiment. A total of 0.8% of the uranium in the column was remobilized by 0.27 mM O_2 over the 54 days of the experiment (Moon et al., 2009).

The oxidation of U(IV) by oxygen was slow relative to that with nitrate (see next section) because O_2 reacts with many reduced species in the column sediments (including organic matter and ferrous iron) and, consequently, the oxygen front advances slowly through the column. This phenomenon was readily observable because the breakthrough of oxygen occurred after 30 days (40 days with sulfate), whereas that of nitrate coincided with the breakthrough of the conservative tracer bromide (Moon et al., 2007). These findings suggested that oxygen was consumed by reaction with several potential reductants including microbes and therefore is less effective at the oxidation of any single compound (Moon et al., 2009).

While oxygen is an excellent oxidizing agent, it is not always very effective at oxidizing U(IV) in systems that mimic the subsurface owing to its rapid consumption by reaction with the numerous competing abiotic and biological reductants. Thus, in the subsurface and column systems described above, the observed rate of oxidation of U(IV) was much slower than would be predicted from sediment-free batch or flowthrough U(IV) oxidation experiments (Senko et al., 2007; Burgos et al., 2008; Ulrich et al., 2009).

In some cases, however, rapid oxidation of U(IV) occurs even in porous media (Zhong et al., 2005; Wu et al., 2007; Komlos et al., 2008). In situ U(VI) reduction was carried out successfully by the addition of ethanol [chosen because it supported faster U(VI) reduction than acetate or lactate] at the Oak Ridge site in Tennessee. After ethanol addition was halted, the dissolved oxygen level increased in the system leading to a concomitant increase in the solution concentration of U(VI) (from the Environmental Protection Agency maximum contaminant level of ~0.13 μM to a maximum of 1.87 μM and stabilized at ~ 1 μM). This increase in U(VI) corresponded to the oxidation of solid-associated U(IV) in approximately 20 days (Wu et al., 2007).

A separate study in a column found similar results: U that had been reductively immobilized in a column packed with RABS and seeded with *G. metallireducens* exited rapidly (61% of the U in 21 days) after the influent was amended with 20% O_2. While rapid U(IV) oxidation took place, no O_2 breakthrough was observed for the duration of the experiment (Komlos et al., 2008). The authors attributed O_2 consumption in the first 21 days to cell decay, Fe(II) oxidation, and oxidation of U(IV). At 24 days, nitrate was detected in the effluent, and the authors suggest that the oxidation of the ammonium present in the influent (at 0.14 mM) is initiated at that point and acts as a sink for O_2 (Komlos et al., 2008).

A dichotomy is observed in the effect of O_2 on U(IV) stability in column or field studies because in some cases little oxidation is observed (Abdelouas et al., 1999; Moon et al., 2009), whereas in other cases extensive oxidation is observed (Wu et al., 2007; Moon et al., 2007; Komlos et al., 2008). These differences are likely attributable to the presence or absence of N and/or Fe species that act to augment the impact of O_2. In the presence of the appropriate species, amendment of O_2 unleashes a cascade of biogeochemical processes that promote U(IV) oxidation and that are explored more in-depth in following sections.

Biological U(IV) Oxidation

All of the studies described above focused on the abiotic oxidation of U(IV) by molecular oxygen. A single study reported the aerobic and enzymatic oxidation of U(IV) (Fig. 3).

Acidithiobacillus ferrooxidans is an acidophilic bacterium able to carry out the oxidation of aqueous U(IV) aerobically coupled to carbon fixation at pH 1.5 (DiSpirito and Tuovinen, 1982). Under those conditions, the chemical oxidation of U(IV) is negligible because it corresponds to less than 0.6% of the biological oxidation. Rusticyanin, the enzyme known to be responsible for Fe(II) oxidation in this bacterium, was implicated in the oxidation of the uranous ion. The maximum rate of oxidation of U(IV) for this organism was reported to be 0.77 μmol U/min·mg of protein^{-1} (DiSpirito and Tuovinen, 1982).

N OXIDES

Direct Biological U(IV) Oxidation

In contrast to aerobic microbially driven U(IV) oxidation, the direct anaerobic oxidation of U(IV) has received significant attention. In pure cultures of nitrate-grown *G. metallireducens*, solid-phase U(IV) was oxidized to U(VI) and nitrate reduced without the detectable production of nitrite at circumneutral pH (Fig. 4). The oxidation rate was faster when Fe(II) was added to the system (see "Fe oxides" below). These results suggest that *G. metallireducens* catalyzes the direct oxidation of U(IV) coupled to nitrate reduction (Finneran et al., 2002).

Thiobacillus denitrificans is a denitrifying bacterium capable of coupling the oxidation of Fe(II) and S compounds to nitrate reduction at circumneutral pH. *T. denitrificans* was reported to catalyze the oxidation of solid-phase U(IV) and to reduce nitrate concomitantly (Beller, 2005) (Fig. 4). Interestingly, only partial oxidation of U(IV) was observed (22% oxidized). Furthermore, U(IV) did not serve as the electron donor for the bacterium as the U(IV) oxidized corresponded to <2% of the amount required to reduce the amount of nitrate consumed (Beller, 2005). The author suggested that H_2 from the glove box atmosphere served as the electron donor and that U(IV) oxidation may require ongoing nitrate reduction coupled to a separate electron donor. Interestingly, a knockout of the H_2-oxidizing hydrogenase in *T. denitrificans* (Letain et al., 2007) was not defective in

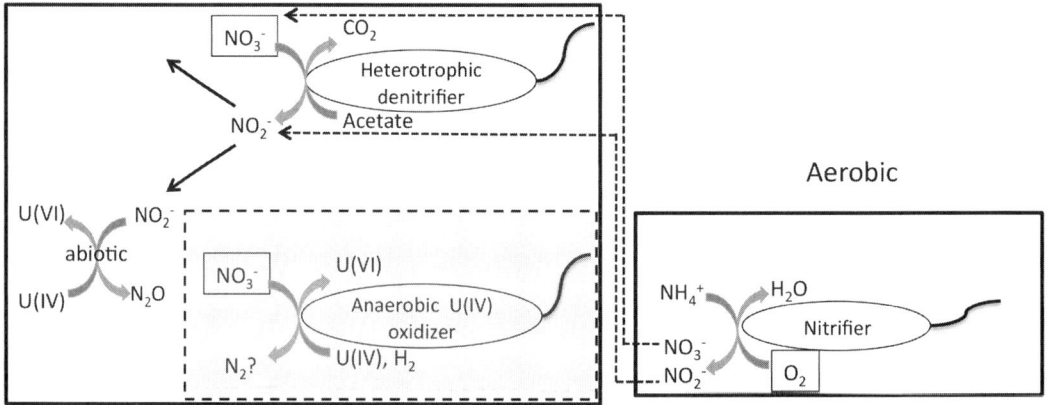

FIGURE 4 Schematic of the direct and indirect biological pathways involving N oxides for U(IV) oxidation. The product of nitrite reduction by U(IV) is taken to be NO_2^- (Senko et al., 2005a) by analogy to Fe(II) (Cooper et al., 2003). Only NO_2^- was considered as the product of heterotrophic denitrification for the sake of simplicity. In reality, N_2O and NO are also formed and also oxidize U(IV) abiotically (especially N_2O). For the anaerobic U(IV) oxidizer, the bacterium is capable of complete denitrification, but this was not measured (H. Beller, personal communication). In this and subsequent figures, dashed and shaded boxes are intended for purely aesthetic reasons to aid in the visual separation of the various processes in the diagram.

nitrate-dependent U(IV) oxidation (H. Beller, personal communication). Thus, the exact role of H_2 in U(IV) oxidation, if any, remains unknown. Nitrite was not detected in the suspension even in the absence of uraninite, and thus the abiotic oxidation of U(IV) by nitrite in this system was ruled out (Beller, 2005).

A third study from the same group investigated the proteins involved in U(IV) oxidation (Beller et al., 2009). The authors identified six c-type cytochromes expressed during denitrification with Fe(II) or thiosulfate as electron donors of which four were outer membrane proteins. Transposon mutants were constructed for the four proteins and tested for U(IV) oxidation activity. Two of the mutants were defective in U(IV) oxidation because they oxidized only ~50% of the U(IV) oxidized by the wild type. The two genes missing in the above mutants encoded diheme c-type cytochromes of unknown function. The proteins encoded by those genes correspond to membrane-associated, periplasm-facing cytochrome c_4 and c_5, respectively. The mechanism by which an insoluble extracellular electron donor could be reduced by membrane-bound periplasmic proteins remains unknown (Beller et al., 2009).

Indirect Biological and Abiotic U(IV) Oxidation

Several biogenic UO_2 samples prepared under various conditions were subjected to a systematic evaluation of their oxidation by various oxidants including nitrite (5 mM) and a combination of *T. denitrificans*, H_2, and nitrate (5 mM). The release of U(VI) was more rapid in the case of the biologically driven, nitrate-dependent U(IV) oxidation with *T. denitrificans*. Nitrite also catalyzed the chemical oxidation of U(IV) but at a slower rate (Senko et al., 2007). The results suggested that enzymatically catalyzed U(IV) oxidation is faster than the abiotic oxidation of U(IV) by nitrite. U(IV) was not completely oxidized in either case despite the great excess of either nitrate or nitrite in the system, which is consistent with the results of Beller (2005).

Experiments involving the addition of nitrate (7 mM) to sediments from a uranium mill tailings site in Shiprock, New Mexico, after complete reduction of U(VI) and Fe(III) showed near-complete reoxidation of U(IV) and the accumulation of nitrite (Finneran et al., 2002). The authors suggested that several processes may be taking place in the sediments: (i) direct enzymatic oxidation of U(IV) coupled to nitrate reduction as was described for *G. metallireducens* (Finneran et al., 2002) or for *T. denitrificans* (Beller, 2005); (ii) abiotic nitrite oxidation of U(IV). This possibility was not considered likely because nitrite-dependent U(IV) oxidation in heat-killed sediments is slower than U(IV) oxidation after nitrate addition to live sediments. This observation suggests that U(IV) oxidation involves a biological component; and (iii) the indirect oxidation of U(IV) via Fe(III) freshly formed from nitrate reduction coupled to Fe(II) oxidation (Finneran et al., 2002). The latter process will be explored in the next section.

Several other studies considered the oxidation of U(IV) in field sites and implicated N compounds. A push-pull test was designed to investigate the effect of nitrate on the oxidation of U(IV) and consisted of injecting groundwater amended with U(VI) or U(VI) and nitrate (1 or 5 mM) into a landfill leachate-impacted aquifer in which U(IV) had been previously immobilized. In the wells that received a nitrate amendment, substantial remobilization of U was observed during active nitrate reduction with about three times more U(VI) released at the higher nitrate concentration (Senko et al., 2002). After nitrate was consumed, U(VI) was promptly removed from solution. The authors hypothesized that denitrification intermediates were responsible for the mobilization of uranium in the sediments. To investigate the abiotic oxidation of U(IV) with denitrification intermediates, a heat-killed sediment amended with uraninite was incubated in the presence of nitrite, nitrous oxide (N_2O), or nitric oxide (NO) (Fig. 4). The most rapid rate of oxidation was obtained with nitrous oxide followed closely by nitrite, and then significantly slower was nitric oxide (Senko et al., 2002). However, the authors were not able to comment on the mechanism of oxidation of U(IV) in the presence of nitrite, N_2O, and NO, but they acknowledged the possibility that Fe(II) could be oxidized to Fe(III) by nitrite or nitrous oxide which in turn oxidizes U(IV). The authors considered the possible complicating role of Fe(II) in a subsequent publication (see "Fe oxides" below).

A second push-pull experiment carried out at the Oak Ridge field site (Tennessee) in the presence of added electron donor and high (~140 mM) nitrate showed the transient oxidation of U(IV) to U(VI), whereas low nitrate (1 mM) conditions yielded continuous U(VI) reduction with no evidence of reoxidation (Istok et al., 2004). Furthermore, the addition of acetylene—a known N_2O reductase inhibitor—in the amended water led to the accumulation of N_2O pointing to denitrification as the primary process of nitrate reduction (Istok et al., 2004). The exact mechanism by which denitrification leads to U(IV) oxidation was not explored directly in this work, but the authors suggested that oxidized denitrification intermediates play a role in the process either directly or indirectly via bacteria such as *Dechlorosomas* spp. that oxidize Fe(II) to Fe(III) by using nitrate (Istok et al., 2004).

The reoxidation of biogenic U(IV) by nitrate was investigated in parallel flowthrough studies with RABS and an influent that included either high (6 mM) or low (9 μM) sulfate during the reduction phase (Moon et al., 2007, 2009). In both cases, nitrate was present in the influent during the oxidation phase but not during the initial reduction phase. The introduction of nitrate through the column had a dramatic effect on the U(VI) effluent concentration albeit to a lesser extent in the high sulfate case. The U(VI) concentration in the effluent in both cases increased to a value much higher than that in the influent (a maximum of ~170 μM with low sulfate and 60 μM with high sulfate compared with an influent concentration of 20 μM). Furthermore, in contrast to what was described for O_2 (see "Oxygen" above), nitrate breakthrough occurred at about the same time as bromide in both cases suggesting that nitrate reacted with fewer compounds in the reduced column than O_2 and that nitrate was available throughout the column to react with reducing agents (Moon et al., 2007, 2009). Nonetheless, only a small fraction of total electrons transferred to nitrate came from U(IV) (about 6%) pointing to the presence of significant other sources of reducing power in the column. The major difference observed between the high and low sulfate conditions related to the amount of U(IV) oxidized. Significantly less U(IV) was remobilized in the case of high sulfate (60% of U recovered) compared with that in low sulfate (97%), and this discrepancy was due to the accumulation of iron sulfide phases (Moon et al., 2007, 2009). Furthermore, the mechanism of oxidation of U(IV) by nitrate was investigated. Nitrate was reduced to denitrification byproducts such as nitrite and nitrous oxide. In fact, the peak of U(VI) in the column effluent was consistent with nitrite serving as an oxidant for U(IV). The authors suggested that the oxidation of U(IV) in the nitrate-amended column is attributable to an indirect biological process involving nitrite obtained by microbially driven denitrification (Moon et al., 2009).

A comparison of the rate and extent of U(IV) oxidation by nitrite or nitrate that is reduced biologically in laboratory studies (Beller, 2005; Senko et al., 2007) to those in the field or involving field-derived sediments (e.g., Senko et al., 2002) yields an interesting observation: field-derived U(IV) oxidation is more rapid and proceeds to a greater extent than laboratory-based oxidation. This discrepancy is due to the importance of Fe in the cycling of both N compounds and U, the topic of the next section.

Fe OXIDES

Abiotic U(IV) Oxidation

When U(VI) was added to a suspension of *G. metallireducens* in the presence of synthetic poorly crystalline Fe(III) oxide, the rate of reduction of Fe(III) increased with increasing U(VI) concentration suggesting that U(VI) acted as an electron shuttle and that U(IV) reduced Fe(III) to produce Fe(II) while being itself oxidized in the process (Fig. 5) (Nevin and Lovley, 2000). This mechanism was confirmed by the observed abiotic reduction of poorly crystalline synthetic Fe(III) oxide by U(IV) [and concomitant oxidation of U(IV) to U(VI)] in bicarbonate buffer. However, U(IV) added to Fe(III)-containing sediments from Bemidji, Minnesota, under aquifer conditions (no added medium) did not reduce Fe(III). In contrast, Fe(III) reduction was observed when the same sediment was resuspended in bicarbonate buffer. The authors concluded that, while Fe(III) was able to oxidize U(IV) abiotically, the reaction was highly dependent on the nature of the Fe(III) phase and the geochemical conditions present (Nevin and Lovley, 2000).

A later study delving more deeply into the effect of Fe(III) speciation also found that abiotic oxidation of U(IV) by Fe(III) was dependent on Fe(III) speciation as well as the presence of bicarbonate (Ginder-Vogel et al., 2006). Experiments evaluating the oxidation of biogenic UO_2 by ferrihydrite, goethite, and hematite at pH 7 in 3 mM HCO_3^- showed that ferrihydrite reacted with UO_2 to produce U(VI) and Fe(II). The extent of the reaction was considerably smaller in the case of goethite, and no Fe(II) was detected with hematite. Thus, the first conclusion is that the nature of the Fe(III) oxide was critical in determining whether U(IV) oxidation can occur. In addition, the presence of Fe(II) formed through coupled U(IV) oxidation and Fe(III) reduction limited UO_2 oxidation because (i) it caused a decrease in the thermodynamic favorability of the reaction when it accumulates as a product and (ii) it led to the formation of goethite or magnetite which, as described above, are less favorable oxidants than ferrihydrite. Finally, the effect of bicarbonate on U(IV) oxidation by Fe(III) was investigated, and the results clearly showed that higher bicarbonate concentrations enhance the oxidation of U(IV). This is due to the shift in solution speciation at the higher bicarbonate concentration and to the increase in the concentration gradient that favored reaction products. Thus, geochemical conditions combining a low concentration of bicarbonate and a high concentration of Fe(II) led to the least U(IV) oxidation (Ginder-Vogel et al., 2006).

A third study considered the mechanistic details of uraninite oxidation by ferrihydrite. Three mechanisms were proposed and one retained: (i) direct solid-solid interaction was dismissed because the rate of UO_2 oxidation did not scale with the available surface area of ferrihydrite; (ii) dissolution of ferrihydrite followed by the interaction of soluble Fe(III) with the UO_2 surface was also dismissed for the same reason; (iii) finally, the dissolution and release of U(IV) from UO_2 followed by the interaction of soluble U(IV) with ferrihydrite was retained as the probable mechanism. This multistep process was expected to be rate limited by the dissolution step and, thus, the rate of oxidation of UO_2 was expected to increase under conditions where solubility was greater (Ginder-Vogel et al., 2010).

A column study of U(VI) reduction in contaminated Oak Ridge (Tennessee) Field Research Center sediment was carried out with lactate as an electron donor for over 400 days (Tokunaga et al., 2008). It revealed U(VI) reduction interrupted by transient U(IV) oxidation. Concomitantly, Fe redox cycling took place with the oxidation of U(IV) to U(VI) corresponding to the reduction of Fe(III) to Fe(II). However, only a small fraction of the overall Fe underwent such redox cycling. This small fraction (~21%) corresponded to the citrate-dithionite-extractable Fe, and this fraction was presumed to represent reactive Fe. This study complements the studies described above and confirms that the reactive fraction of sediment Fe can serve as a terminal electron acceptor for U(IV) oxidation (Tokunaga et al., 2008).

S Redox Cycling

As was observed by Ginder-Vogel et al. (2006) for biogenic UO_2, the addition of freshly

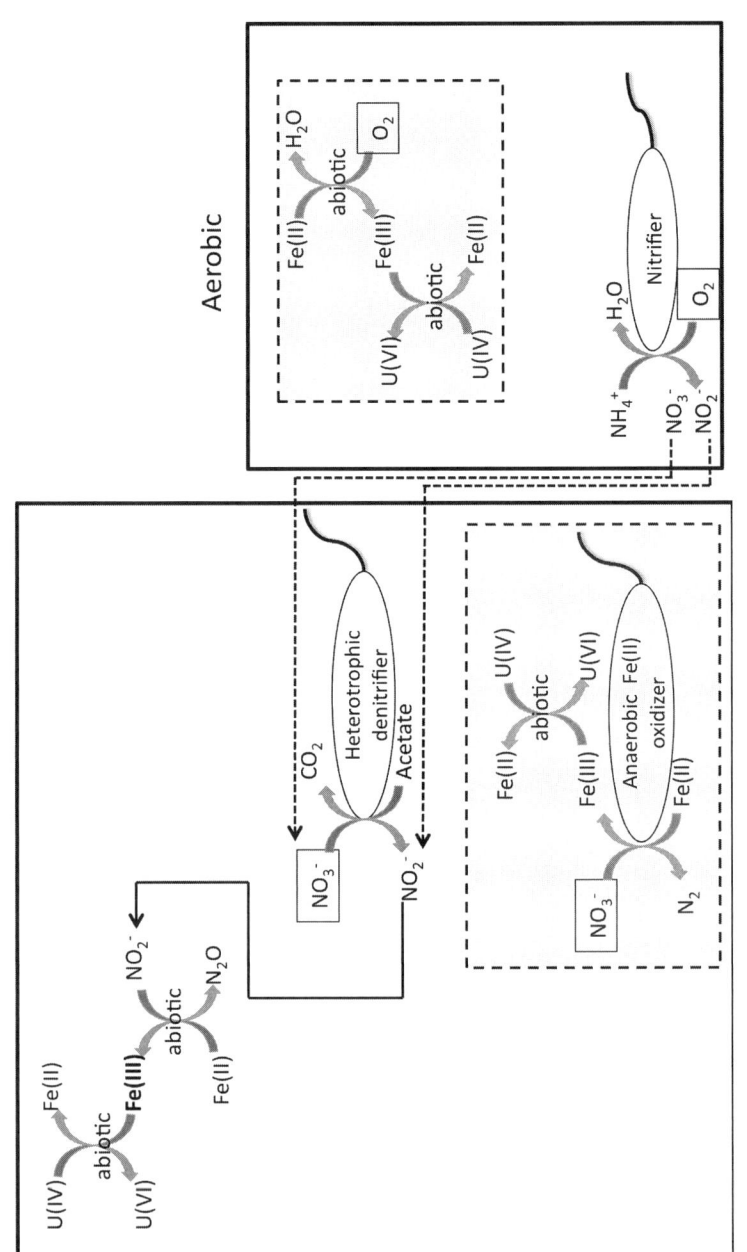

FIGURE 5 Schematic of the direct and indirect biological pathways involving iron oxides for U(IV) oxidation. The product of coupled Fe(II) oxidation and nitrate reduction is most commonly N_2 (Straub et al., 2001). Fe(III) labeled in bold corresponds to the most reactive form of Fe(III) produced by the processes considered.

precipitated Fe(III) led to the rapid oxidation of synthetic UO_2, whereas the addition of hematite at the same concentration registered very little U(IV) oxidation. However, when the microbial reduction of U(VI) in the presence of Fe(III) (hydr)oxides under sulfidogenic conditions (in the presence of *Desulfovibrio desulfuricans*) was evaluated in piperazine-*N, N'*-bis(2-ethanesulfonic acid) (PIPES) buffer, the nature of the Fe(III) phase provided affected the oxidation of U(IV) in the opposite manner than what was described in previous studies: in the presence of ferrihydrite, U(VI) reduction proceeded and the U(VI) concentration reached a low, constant level. In contrast, with goethite and hematite, the U(VI) concentration decreased but subsequently increased. Hematite yielded the greatest soluble U(VI) concentration, suggesting that the reoxidation of reduced U was greatest under those conditions (Sani et al., 2005). This is in complete contrast to the observation of Ginder-Vogel et al. (2006). The amount of sulfide generated by sulfate reduction that remained in solution may provide an explanation for this observation: the greatest amount of sulfide was measured in the goethite and hematite conditions and a decrease in sulfide concentration coincided with an increase in U(VI) in solution. The decrease in solution sulfide could be due to the sequestration of Fe(II) from solution. Unfortunately, Fe(II) was not measured so this hypothesis cannot be verified. However, it is conceivable that, due to the difference in the secondary iron sulfide produced for each Fe(III) oxide considered, more Fe(II) may be present in the ferrihydrite condition, leading to inhibited U(IV) oxidation. A similar experiment in bicarbonate confirmed the general trend except that reoxidation of U(IV) was observed with all three Fe(III) minerals considered. Thus, the type and the amount of the Fe(III) phase and the solution speciation significantly affect the rate of U(IV) oxidation. Additionally, a live *D. desulfuricans* culture enhanced the rate of oxidation of biogenic (or biologically produced) U(IV) after the addition of hematite, whereas a heat-killed control did not. This finding suggests that sulfate-reducing bacterial activity enhances U(IV) oxidation by an unknown mechanism (Sani et al., 2005).

U(IV) Oxidation in the Presence of O_2 and Fe

The addition of Fe(II) to sediment spiked with biogenic UO_2 and exposed to air led to a slower release of U(VI) at circumneutral pH than in the absence of added Fe(II). In the same sediment previously treated to remove Fe oxides, the same general trend held. While Fe(II) appeared to be effective at decreasing U(IV) oxidation, it was merely decreasing U mobilization (Zhong et al., 2005). In this case, the continued immobilization of uranium did not occur through the maintenance of a U(IV) phase. Rather, it took place through the formation of a fresh hydrous ferric oxide (HFO) precipitate resulting from the oxidation of Fe(II) by O_2 followed by the sorption of U(VI) onto this secondary phase (Zhong et al., 2005). While Fe(II) appeared to limit uranium mobilization, it did not seem to directly impact the oxidation of U(IV).

Cycling with N Oxides

The presence of Fe(II) in *G. metallireducens* cell suspensions amended with nitrate and U(IV) promoted the rate and extent of U(IV) oxidation to U(VI) (Finneran et al., 2002). Nitrate-dependent Fe(II) oxidation is known to be catalyzed by *G. metallireducens* and results in the formation of freshly precipitated Fe(III) oxides and no accumulation of nitrite. Thus, a proposed mechanism by which U(IV) oxidation probably occurs is the abiotic oxidation of U(IV) by Fe(III). Because of continuous nitrate-dependent Fe(II) oxidation, Fe(II) is continuously removed from solution and, as a result, does not inhibit U(IV) oxidation by Fe(III). Thus, the rate and extent of U(IV) oxidation is greater in the presence of Fe(III) and nitrate than in the presence of Fe(III) alone. In addition, in sediment incubations amended with nitrate after complete Fe(III) and U(VI) reduction, nitrite accumulated in solution. In this case, U(IV) oxidation probably took place through an additional mechanism, namely, abiotic oxidation by nitrite (Finneran et al., 2002).

The biological, nitrate-dependent oxidation of U(IV) can be catalyzed by three types of bacteria: (i) nitrate-dependent, Fe(II)-oxidizing bacteria that produce Fe(III) which oxidizes U(IV); (ii) organotrophic dissimilatory nitrate-reducing bacteria that produce nitrite which abiotically oxidizes Fe(II) to Fe(III) which, in turn, oxidizes U(IV); or (iii) nitrate-dependent U(IV)-oxidizing bacteria. Comparing the rate of oxidation of the electron donor for an enrichment for each type of metabolism led to the conclusion that direct enzymatic U(IV) oxidation by nitrate-reducing bacteria was not significant (Senko et al., 2005b).

Senko et al. (2005b) also investigated the effect of nitrate concentration on nitrate-dependent U(IV) oxidation by incubating sediments amended with biogenic U(IV) with varying nitrate concentrations. At low (0.1 to 4 mM) nitrate, no nitrite accumulation took place and U(IV) oxidation proceeded via Fe(III) produced by enzymatic Fe(II) oxidation. However, when higher nitrate concentrations were present, a second phase occurred in which nitrite-driven, abiotic Fe(II) oxidation was observed, leading to rapid U(IV) oxidation. Thus, nitrate concentrations determined whether Fe(III) oxidation occurred via a purely enzymatic process or via a combination of biotic and abiotic processes. Most importantly, the Fe(III) produced by the abiotic reaction of Fe(II) with nitrite was more reactive [i.e., oxidized U(IV) faster] than Fe(III) produced via nitrate-dependent enzymatic Fe(II) oxidation (Senko et al., 2005b).

Furthermore, soluble Fe(II) was found to be an effective inhibitor of nitrate-dependent U(IV) oxidation. Two hypotheses were put forth to explain this finding: (i) the sorption of Fe(II) onto Fe(III) minerals precluded contact between U(IV) and the Fe(III) surface or (ii) Fe(II) led to the formation of less reactive Fe(III) minerals that were ineffective at oxidizing U(IV). The authors suggested that crystalline Fe(III) minerals, less reactive toward U(IV), were formed in the presence of soluble Fe(II) (Senko et al., 2005b), which is consistent with the conclusions of Ginder-Vogel et al. (2006).

In a study exploring the abiotic coupling of N, Fe, and U redox processes, U(IV) was incompletely oxidized by poorly crystalline Fe(III), suggesting that the accumulation of Fe(II) at the surface of the Fe(III) oxide prevents further reaction (Senko et al., 2005a). In fact, an experiment in which Fe(II) was added to Fe(III) oxyhydroxide in the presence of U(IV) at various Fe(II) concentrations showed that U(IV) oxidation was indeed inhibited at the higher Fe(II) levels. However, the addition of nitrite resulted in the rapid oxidation of Fe(II) and concomitant oxidation of U(IV). In general, the combination of Fe(II) and nitrite addition resulted in the fastest and greatest oxidation of U(IV) because nitrite oxidized Fe(II) to Fe(III) which in turn oxidized U(IV) to U(VI). The Fe(II) produced by the latter reaction was also oxidized by excess nitrite. This result clearly showed that chemical oxidation of Fe(II) by nitrite and of U(IV) by Fe(III) are rapid and efficient processes (Fig. 5). Naturally, in live sediments, the source of nitrite is usually biological (Senko et al., 2005a).

An environmental isolate from the Oak Ridge site (Tennessee) that shared 97% 16S small-subunit rRNA sequence identity with *Klebsiella oxytoca* was found to oxidize Fe(II) or acetate and reduce nitrate to nitrite. This isolate (strain FW33AN) was tested for U(IV) oxidation under both organotrophic conditions [acetate oxidation and nitrate reduction but in the presence of Fe(II)] as well as lithotrophic conditions [Fe(II) oxidation and nitrate reduction] (Fig. 5) (Senko et al., 2005a). Under organotrophic conditions, Fe(II) oxidation took place abiotically with the reduction of nitrite as an electron acceptor, whereas under lithotrophic conditions, Fe(III) was produced enzymatically. Rapid Fe(II) oxidation took place in both cases. The rate and extent of U(IV) oxidation were greater under organotrophic conditions than under lithotrophic conditions. The authors tested the hypothesis that the Fe(III) minerals obtained from the organotrophic system were more reactive than those from the lithotrophic system. They found that HFO—the more reactive form of Fe(III)—was produced under organotrophic conditions, whereas goethite was obtained by enzymatic Fe(II) oxidation by strain FW33AN. Thus, the speciation of the Fe(III) mineral produced for U(IV) oxidation is a crucial parameter determining the rate of U(IV) oxidation (Senko et al., 2005a).

Fe cycling can play an important role in upholding or endangering U(IV) stability in the subsurface. Highly reactive, freshly precipitated Fe(III) (hydr)oxides have the potential to rapidly oxidize U(IV). The oxidation is typically incomplete unless there is a mechanism for the continuous cycling of Fe and the lack of accumulation of Fe(II). If such a mechanism is absent, more highly ordered Fe(III) phases such as goethite and hematite—with limited capacity of U(IV) oxidation—will form as a result of the interaction of Fe(II) with ferrihydrite. Continuous Fe cycling is most often obtained through coupling of the Fe and N redox cycles. Nitrite produced through the reduction of nitrate or the oxidation of ammonia oxidizes Fe(II) formed by the oxidation of U(IV) and, in this manner, prevents Fe(II) accumulation.

Mn OXIDES

Abiotic U(IV) Oxidation

Mn(III,IV) oxides, which occur in a great variety of mineral forms, are some of the most powerful naturally occurring environmental oxidants. The oxidation/reduction potential of Mn(IV) oxides under standard conditions are comparable to N oxides and greater than those of Fe(III) oxides and U(VI), although the actual redox potentials depend on environmental conditions and the mineral form. As such, Mn oxides are potent oxidants of U(IV) (Fig. 6) (Fredrickson et al., 2002; Liu et al., 2002; Chinni et al., 2008) and other reduced metals such as Fe(II), Cr(III), and Co(II) (Schroeder and Lee, 1975; Bartlett and James, 1979; Fendorf et al., 1999; Postma and Appelo, 2000; Murray and Tebo, 2007). Fredrickson et al. (2002) have demonstrated that synthetic Mn oxides (pyrolusite [β-MnO_2], birnessite [δ-MnO_2], and bixbyite [Mn_2O_3]) readily oxidize biogenic UO_2 under anaerobic conditions with bixbyite, a Mn(III) mineral, being the most rapid. A biogenic Mn(IV) oxide was an even better oxidant (Fredrickson et al., 2002), likely because of the fine particle size and vacancies in the mineral lattice (Villalobos et al., 2003; Bargar et al., 2005).

In the environment, Mn(III, IV) oxides and Fe(III) oxides frequently coexist and this is indeed the case at sites where U contamination is a problem, such as at the U.S. Department

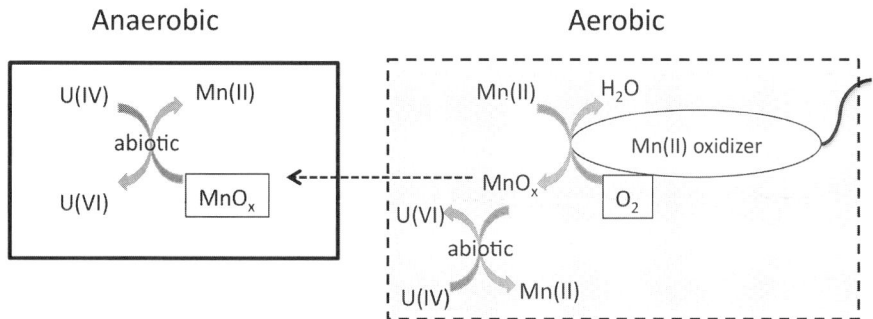

FIGURE 6 Schematic of the direct and indirect biological pathways involving Mn oxides for U(IV) oxidation.

of Energy Hanford Site in Washington and the Oak Ridge site in Tennessee (Fredrickson et al., 2002). Because of their redox reactions with Fe(II) in addition to UO_2 and their ability to sequester U(VI) (Webb et al., 2006), the presence of Mn(III,IV) oxides in the environment add another level of complexity to the processes involved in UO_2 oxidation and U cycling in general (Fig. 6).

U(IV) Oxidation in the Presence of O_2 and Mn

Although Mn oxides occur in geological deposits, in nature the most reactive Mn oxides are believed to be formed microbially (by bacteria and fungi) in a process that requires O_2 (Tebo et al., 2004). Microbially catalyzed Mn(II) oxidation will occur just about wherever Mn(II) and O_2 coexist (given the appropriate pH conditions), and in some environments Mn(II) oxidation can proceed rapidly at submicromolar O_2 and Mn(II) concentrations (Clement et al., 2009). Since Mn(III,IV) oxides readily oxidize a variety of reduced chemicals, Mn cycling can proceed with only catalytic amounts of Mn being present as long as some O_2 is available.

Chinni et al. (2008) have investigated the ability of the Mn(II)-oxidizing spores of the model organism *Bacillus* sp. strain SG-1 to oxidize UO_2 and the effect of different concentrations of U and Mn chemical species on the rates of UO_2 oxidation. UO_2 oxidation by *Bacillus* sp. SG-1 is a result of an indirect coupled process: the spores oxidize Mn(II) to MnO_2 which in turn oxidizes the UO_2 (Fig. 6). The Mn(II)-oxidizing spores did not oxidize UO_2 in the absence of Mn(II) or O_2. The SG-1 biogenic Mn oxide, a nanoparticulate X-ray amorphous material similar to synthetic δ-MnO_2 (Bargar et al., 2005), oxidized UO_2 about twice as fast as synthetic δ-MnO_2, probably because of its higher surface area (Chinni et al., 2008).

Strain SG-1 spores incubated under low O_2 conditions (5% air in the headspace) in the presence of Mn(II) oxidized UO_2 more slowly than was observed in identical experiments except with preformed biogenic Mn oxides (i.e., allowing the spores to form Mn oxides before adding the UO_2). This was because the preformed Mn oxides allowed for faster UO_2 oxidation than when starting with only Mn(II) and no Mn oxide. In any case, the Mn(II) produced by the reaction of UO_2 and Mn oxides could be reoxidized by the spores and this cycling of Mn, even with very low Mn concentrations (<5 μM), could drive more extensive UO_2 oxidation than was evident from measuring Mn(II) and Mn oxides. Under these low O_2 conditions with spores UO_2 oxidation was 1 to 2 orders of magnitude faster than UO_2 oxidation by O_2 alone (i.e., in the absence of spores) (Chinni et al., 2008).

The kinetics of UO_2 oxidation followed saturation-type kinetics with respect to the concentrations of Mn(II) or UO_2 in the active SG-1 spore-catalyzed system. In abiotic experiments the rate of UO_2 oxidation was controlled by both the UO_2 and Mn(IV) oxide concentrations. In all cases, the rate of UO_2 oxidation was shown to equal the rate of Mn(IV) oxide reduction demonstrating the stoichiometric coupling between UO_2 oxidation and Mn(IV) oxide reduction (Chinni et al., 2008). Overall these results illustrate the need to consider the importance of Mn-oxidizing bacteria, their metabolism, and resulting reaction rates for the formation of the powerful naturally occurring oxidants, biogenic Mn oxides, when predicting the potential for UO_2 oxidation.

CONCLUSION

The oxidation of U(IV) can proceed via a number of different processes, biological, indirectly biological, and chemical (Fig. 3 through 6). It emerges from this literature review that the most rapid and effective oxidants for U(IV) are amorphous, freshly formed Fe(III) or Mn(IV) oxides; however, there are still major gaps in our understanding of the role of Fe(III) and Mn(IV) oxides in the oxidation of UO_2 in the environment.

Although Fe(III) is formed both abiotically by nitrite oxidation of Fe(II) and enzymatically by nitrate-dependent Fe(II) oxidation, the product of the former is more reactive because it is HFO whereas the product of the latter is goethite. From the standpoint of in situ U(VI) bioremediation, the coupling of the N and Fe cycles is the most unfavorable combination for the stability of U(IV) in the subsurface. However, soluble Fe(II), the product of U(IV) oxidation by Fe(III), acts as an inhibitor of the oxidation reaction by catalyzing the formation of crystalline Fe(III) phases. Thus, incomplete U(IV) oxidation is often observed. Conditions under which soluble Fe(II) is rapidly scavenged (e.g., oxidation by excess nitrite or precipitation with sulfides) lead to the fastest and most extensive U(IV) oxidation.

To our knowledge, no work has been done examining the role of Mn(II) oxidation on uranium cycling in the environment. In subsurface environments where Mn and O_2 are present, such as in redox transition zones, Mn(II) oxidation is likely to proceed. If Mn(II) oxidation occurs in an area harboring reduced U(IV), such as in subsurface sediments following biostimulation of U(IV) reduction or through seasonal variations in groundwater O_2, Mn cycling may be an important driver of UO_2 oxidation. Accounting for this catalytic UO_2 oxidation coupled to Mn cycling is needed to evaluate the long-term stability of U(IV) at remediated sites.

A major aspect of the mechanism(s) of abiotic U(IV) oxidation by Fe(III) and Mn(III,IV) oxides remains that is poorly understood: both reactants are solid phases and it is unclear how the relative localization of the two minerals influences the oxidation of U(IV). If Fe(III) or Mn(III,IV) oxides are freshly formed, they may precipitate in the vicinity of U(IV). If the oxides are preexisting (such as in initially oxic sediments) and U(IV) is formed in the vicinity of the reactive fraction, it may be oxidized to U(VI) before being reduced again by microbial activity. This would lead to redox cycling of U in the subsurface until the reactive Fe(III) and Mn(III,IV) oxides phases are depleted.

ACKNOWLEDGMENTS

Funding for research on uranium in our laboratories is provided by the U.S. Department of Energy, Office of Basic Energy Sciences grants DE-FG02-06ER64227, DE-FG02-06ER64228, and DE-SC000532.

REFERENCES

Abdelouas, A., W. Lutze, and H. E. Nuttall. 1999. Oxidative dissolution of uraninite precipitated on Navajo Sandstone. *J. Contam. Hydrol.* **36:**353–375.

Allen, G. C., and P. A. Tempest. 1986. Ordered defects in the oxides of uranium. *Proc. R. Soc. Lond. Ser. A* **406:**325–344.

Anderson, R. T., H. A. Vrionis, I. Ortiz-Bernad, C. T. Resch, P. E. Long, R. Dayvault, K. Karp, S. Marutzky, D. R. Metzler, A. Peacock, D. C. White, M. Lowe, and D. R. Lovley. 2003. Stimulating the in situ activity of *Geobacter* species to remove uranium from the groundwater of a uranium-contaminated aquifer. *Appl. Environ. Microbiol.* **69:**5884–5891.

Bargar, J. R., R. Bernier-Latmani, D. E. Giammar, and B. M. Tebo. 2008. Biogenic uraninite nanoparticles and their importance for uranium remediation. *Elements* **4:**407–412.

Bargar, J. R., B. M. Tebo, U. Bergmann, S. M. Webb, P. Glatzel, V. Q. Chiu, and M. Villalobos. 2005. Biotic and abiotic products of Mn(II) oxidation by spores of the marine *Bacillus* sp. strain SG-1. *Am. Mineral.* **90:**143–154.

Bartlett, R., and B. James. 1979. Behavior of chromium in soils. 3. Oxidation. *J. Environ. Qual.* **8:**31–35.

Beller, H. R. 2005. Anaerobic, nitrate-dependent oxidation of U(IV) oxide minerals by the chemolithoautotrophic bacterium *Thiobacillus denitrificans*. *Appl. Environ. Microbiol.* **71:**2170–2174.

Beller, H. R., T. C. Legler, F. Bourguet, T. E. Letain, S. R. Kane, and M. A. Coleman. 2009. Identification of c-type cytochromes involved in anaerobic, bacterial U(IV) oxidation. *Biodegradation* **20:**45–53.

Boyanov, M. I., E. J. O'Loughlin, E. E. Roden, J. B. Fein, and K. M. Kemner. 2007. Adsorption of Fe(II) and U(VI) to carboxyl-functionalized microspheres: the influence of speciation on uranyl reduction studied by titration and XAFS. *Geochim. Cosmochim. Acta* **71:**1898–1912.

Burgos, W. D., J. T. McDonough, J. M. Senko, G. X. Zhang, A. C. Dohnalkova, S. D. Kelly,

Y. Gorby, and K. M. Kemner. 2008. Characterization of uraninite nanoparticles produced by *Shewanella oneidensis* MR-1. *Geochim. Cosmochim. Acta* **72**:4901–4915.

Chinni, S., C. R. Anderson, K. U. Ulrich, D. E. Giammar, and B. M. Tebo. 2008. Indirect UO_2 oxidation by Mn(II)-oxidizing spores of *Bacillus* sp. strain SG-1 and the effect of U and Mn concentrations. *Environ. Sci. Technol.* **42**:8709–8714.

Clement, B. G., G. W. Luther, and B. M. Tebo. 2009. Rapid, oxygen-dependent microbial Mn(II) oxidation kinetics at sub-micromolar oxygen concentrations in the Black Sea suboxic zone. *Geochim. Cosmochim. Acta* **73**:1878–1889.

Cooper, D. C., F. W. Picardal, A. Schimmelmann, and A. J. Coby. 2003. Chemical and biological interactions during nitrate and goethite reduction by *Shewanella putrefaciens* 200. *Appl. Environ. Microbiol.* **69**:3517–3525.

DiSpirito, A. A., and O. H. Tuovinen. 1982. Uranous ion oxidation and carbon dioxide fixation by *Thiobacillus ferrooxidans*. *Arch. Microbiol.* **133**:28–32.

Fendorf, S., P. M. Jardine, R. R. Patterson, D. L. Taylor, and S. C. Brooks. 1999. Pyrolusite surface transformations measured in real-time during the reactive transport of Co(II)EDTA$^{(2-)}$. *Geochim. Cosmochim. Acta* **63**:3049–3057.

Finch, R. J., and T. Murakami. 1999. Systematics and paragenesis of uranium minerals. *Rev. Mineral. Geochem.* **38**:91–179.

Finneran, K. T., M. E. Housewright, and D. R. Lovley. 2002. Multiple influences of nitrate on uranium solubility during bioremediation of uranium-contaminated subsurface sediments. *Environ. Microbiol.* **4**:510–516.

Francis, A. J., C. J. Dodge, F. Lu, G. P. Halada, and C. R. Clayton. 1994. XPS and XANES studies of uranium reduction by *Clostridium* sp. *Environ. Sci. Technol.* **28**:636–639.

Fredrickson, J. K., J. M. Zachara, D. W. Kennedy, C. Liu, M. C. Duff, D. B. Hunter, and A. Dohnalkona. 2002. Influence of Mn oxides on the reduction of U(VI) by the metal-reducing bacterium *Shewanella putrefaciens*. *Geochim. Cosmochim. Acta* **66**:3247–3262.

Ginder-Vogel, M., C. S. Criddle, and S. Fendorf. 2006. Thermodynamic constraints on the oxidation of biogenic UO_2 by Fe(III) (hydr) oxides. *Environ. Sci. Technol.* **40**:3544–3550.

Ginder-Vogel, M., B. Stewart, and S. Fendorf. 2010. Kinetic and mechanistic constraints on the oxidation of biogenic uraninite by ferrihydrite. *Environ. Sci. Technol.* **44**:163–169.

Gu, B. H., H. Yan, P. Zhou, D. B. Watson, M. Park, and J. Istok. 2005. Natural humics impact uranium bioreduction and oxidation. *Environ. Sci. Technol.* **39**:5268–5275.

Istok, J. D., J. M. Senko, L. R. Krumholz, D. Watson, M. A. Bogle, A. Peacok, Y.-J. Chang, and D. C. White. 2004. *In situ* bioreduction of technetium and uranium in a nitrate-contaminated aquifer. *Environ. Sci. Technol.* **38**:468–475.

Janeczek, J., and R. C. Ewing. 1992. Structural formula of uraninite. *J. Nucl. Mater.* **190**:128–132.

Komlos, J., A. Peacock, R. K. Kukkadapu, and P. R. Jaffe. 2008. Long-term dynamics of uranium reduction/reoxidation under low sulfate conditions. *Geochim. Cosmochim. Acta* **72**:3603–3615.

Letain, T. E., S. R. Kane, T. C. Legler, E. P. Salazar, P. G. Agron, and H. R. Beller. 2007. Development of a genetic system for the chemolithoautotrophic bacterium *Thiobacillus denitrificans*. *Appl. Environ. Microbiol.* **73**:3265–3271.

Liger, E., L. Charlet, and P. Van Cappellen. 1999. Surface catalysis of uranium(VI) reduction by iron(II). *Geochim. Cosmochim. Acta* **63**:2939–2955.

Liu, C. X., J. M. Zachara, J. K. Fredrickson, D. W. Kennedy, and A. Dohnalkova. 2002. Modeling the inhibition of the bacterial reduction of U(VI) by β-MnO$_{2(s)}$. *Environ. Sci. Technol.* **36**:1452–1459.

Lovley, D. R. 1993. Dissimilatory metal reduction. *Annu. Rev. Microbiol.* **47**:263–290.

Lovley, D. R., E. J. P. Phillips, Y. A. Gorby, and E. R. Landa. 1991. Microbial reduction of uranium. *Nature* **350**:413–416.

Moon, H. S., J. Komlos, and P. R. Jaffe. 2007. Uranium reoxidation in previously bioreduced sediment by dissolved oxygen and nitrate. *Environ. Sci. Technol.* **41**:4587–4592.

Moon, H. S., J. Komlos, and P. R. Jaffe. 2009. Biogenic U(IV) oxidation by dissolved oxygen and nitrate in sediment after prolonged U(VI)/Fe(III)/SO_4^{2-} reduction. *J. Contam. Hydrol.* **105**:18–27.

Morel, F. M. M., and J. G. Hering. 1993. *Principles and Applications of Aquatic Chemistry*. John Wiley & Sons, New York, NY.

Murray, K. J., and B. M. Tebo. 2007. Cr(III) is indirectly oxidized by the Mn(II)-oxidizing bacterium *Bacillus* sp strain SG-1. *Environ. Sci. Technol.* **41**:528–533.

Nevin, K. P., and D. R. Lovley. 2000. Potential for nonenzymatic reduction of Fe(III) via electron shuttling in subsurface sediments. *Environ. Sci. Technol.* **34**:2472–2478.

O'Loughlin, E. J., S. D. Kelly, R. E. Cook, R. Csencsits, and K. M. Kemner. 2003. Reduction of uranium(VI) by mixed iron(II)/iron(III) hydroxide (green rust): formation of UO_2 nanoparticles. *Environ. Sci. Technol.* **37**:721–727.

O'Loughlin, E. J., S. D. Kelly, and K. M. Kemner. 2010. XAFS investigation of the interactions of

U^{VI} with secondary mineralization products from the bioreduction of (Fe^{III}) oxides. *Environ. Sci. Technol.* **44**:1656–1661.

Postma, D., and C. A. J. Appelo. 2000. Reduction of Mn-oxides by ferrous iron in a flow system: column experiment and reactive transport modeling. *Geochim. Cosmochim. Acta* **64**:1237–1247.

Regenspurg, S., D. Schild, T. Schafer, F. Huber, and M. E. Malmstrom. 2009. Removal of uranium(VI) from the aqueous phase by iron(II) minerals in presence of bicarbonate. *Appl. Geochem.* **24**:1617–1625.

Sani, R. K., B. M. Peyton, A. Dohnalkova, and J. E. Amonette. 2005. Reoxidation of reduced uranium with iron(III) (hydr)oxides under sulfate reducing conditions. *Environ. Sci. Technol.* **39**:2059–2066.

Schofield, E. J., H. Veeramani, J. O. Sharp, E. Suvorova, R. Bernier-Latmani, A. Mehta, J. Stahlman, S. M. Webb, D. L. Clark, S. D. Conradson, E. S. Ilton, and J. R. Bargar. 2008. Structure of biogenic uraninite produced by *Shewanella oneidensis* strain MR-1. *Environ. Sci. Technol.* **42**:7898–7904.

Schroeder, D. C., and G. F. Lee. 1975. Potential transformations of chromium in natural waters. *Water Air Soil Pollut.* **4**:355–365.

Senko, J. M., J. D. Istok, J. M. Suflita, and L. R. Krumholtz. 2002. *In-situ* evidence for uranium immobilization and remobilization. *Environ. Sci. Technol.* **36**:1491–1496.

Senko, J. M., S. D. Kelly, A. C. Dohnalkova, J. T. Mcdonough, K. M. Kemner, and W. D. Burgos. 2007. The effect of U(VI) bioreduction kinetics on subsequent reoxidation of biogenic U(IV). *Geochim. Cosmochim. Acta* **71**:4644–4654.

Senko, J. M., Y. Mohamed, T. A. Dewers, and L. R. Krumholz. 2005a. Role for Fe(III) minerals in nitrate-dependent microbial U(IV) oxidation. *Environ. Sci. Technol.* **39**:2529–2536.

Senko, J. M., J. M. Suflita and L. R. Krumholz. 2005b. Geochemical controls on microbial nitrate-dependent U(IV) oxidation. *Geomicrobiol. J.* **22**:371–378.

Sharp, J. O., E. J. Schofield, H. Veeramani, E. I. Suvorova, D. W. Kennedy, M. J. Marshall, A. Mehta, J. R. Bargar, and R. Bernier-Latmani. 2009. Structural similarities between biogenic uraninites produced by phylogenetically and metabolically diverse bacteria. *Environ. Sci. Technol.* **43**:8295–8301.

Straub, K. L., M. Benz, and B. Schink. 2001. Iron metabolism in anoxic environments at near neutral pH. *FEMS Microbiol. Ecol.* **34**:181–186.

Tebo, B. M., J. R. Bargar, B. G. Clement, G. J. Dick, K. J. Murray, D. Parker, R. Verity, and S. M. Webb. 2004. Biogenic manganese oxides: properties and mechanisms of formation. *Annu. Rev. Earth Planet. Sci.* **32**:287–328.

Tebo, B. M., and A. Y. Obraztsova. 1998. Sulfate-reducing bacterium grows with Cr(VI), U(VI), Mn(IV), and Fe(III) as electron acceptors. *FEMS Microbiol. Lett.* **162**:193–198.

Tokunaga, T. K., J. M. Wan, Y. M. Kim, S. R. Sutton, M. Newville, A. Lanzirotti, and W. Rao. 2008. Real-time X-ray absorption spectroscopy of uranium, iron, and manganese in contaminated sediments during bioreduction. *Environ. Sci. Technol.* **42**:2839–2844.

Ulrich, K.-U., E. S. Ilton, H. Veeramani, J. O. Sharp, R. Bernier-Latmani, E. J. Schofield, J. R. Bargar, and D. E. Giammar. 2009. Comparative dissolution kinetics of biogenic and chemogenic uraninite under oxidizing conditions in the presence of carbonate. *Geochim. Cosmochim. Acta* **73**:6065–6083.

Ulrich, K. U., A. Singh, E. J. Schofield, J. R. Bargar, H. Veeramani, J. O. Sharp, R. Bernier-Latmani, and D. E. Giammar. 2008. Dissolution of biogenic and synthetic UO$_2$ under varied reducing conditions. *Environ. Sci. Technol.* **42**:5600–5606.

Villalobos, M., B. Toner, J. Bargar, and G. Sposito. 2003. Characterization of the manganese oxide produced by *Pseudomonas putida* strain MnB1. *Geochim. Cosmochim. Acta* **67**:2649–2662.

Webb, S. M., C. C. Fuller, B. M. Tebo, and J. R. Bargar. 2006. Determination of uranyl incorporation into biogenic manganese oxides using X-ray absorption spectroscopy and scattering. *Environ. Sci. Technol.* **40**:771–777.

Wersin, P., M. F. Hochella, P. Persson, G. D. Redden, J. O. Leckie, and D. W. Harris. 1994. Interaction between aqueous uranium(VI) and sulfide minerals: spectroscopic evidence for sorption and reduction. *Geochim. Cosmochim. Acta* **58**:2829–2843.

Wu, Q., R. A. Sanford, and F. E. Loffler. 2006a. Uranium(VI) reduction by *Anaeromyxobacter dehalogenans* strain 2CP-C. *Appl. Environ. Microbiol.* **72**:3608–3614.

Wu, W. M., J. Carley, T. Gentry, M. A. Ginder-Vogel, M. Fienen, T. Mehlhorn, H. Yan, S. Caroll, M. N. Pace, J. Nyman, J. Luo, M. E. Gentile, M. W. Fields, R. F. Hickey, B. H. Gu, D. Watson, O. A. Cirpka, J. Z. Zhou, S. Fendorf, P. K. Kitanidis, P. M. Jardine, and C. S. Criddle. 2006b. Pilot-scale in situ bioremedation of uranium in a highly contaminated aquifer. 2. Reduction of U(VI) and geochemical control of U(VI) bioavailability. *Environ. Sci. Technol.* **40**:3986–3995.

Wu, W. M., J. Carley, J. Luo, M. A. Ginder-Vogel, E. Cardenas, M. B. Leigh, C. C.

Hwang, S. D. Kelly, C. M. Ruan, L. Y. Wu, J. Van Nostrand, T. Gentry, K. Lowe, T. Mehlhorn, S. Carroll, W. S. Luo, M. W. Fields, B. H. Gu, D. Watson, K. M. Kemner, T. Marsh, J. Tiedje, J. Z. Zhou, S. Fendorf, P. K. Kitanidis, P. M. Jardine, and C. S. Criddle.** 2007. In situ bioreduction of uranium(VI) to submicromolar levels and reoxidation by dissolved oxygen. *Environ. Sci. Technol.* **41:**5716–5723.

Zhong, L. R., C. X. Liu, J. M. Zachara, D. W. Kennedy, J. E. Szecsody, and B. Wood. 2005. Oxidative remobilization of biogenic uranium(IV) precipitates: effects of iron(II) and pH. *J. Environ. Qual.* **34:**1763–1771.

ANAEROBIC RESPIRATORY IRON(II) OXIDATION

J. Cameron Thrash, Sarir Ahmadi, and John D. Coates

9

INTRODUCTION

Microorganisms significantly contribute to the iron biogeochemical cycle on Earth, because they are able to both oxidize and reduce this important element. Respiratory Fe(II) oxidation has been well studied in oxic environments at acidic pH, but it is only in the past decade that this metabolism was also discovered in anoxic, circumneutral environments. Research has shown that anaerobic, mesophilic Fe(II) oxidation can be coupled to electron acceptors such as nitrate and perchlorate, and that a variety of Fe(II) forms, including solid-phase Fe(II)-containing minerals, can be utilized by microorganisms as electron donors. Pure culture studies have identified a diversity of organisms capable of this metabolism, although to date only two have been successful in using it for growth in the absence of additional electron donors. Studies have revealed the biogenesis of a variety of mineral-phase end products as a result of bio-oxidation, and these biominerals are formed both within the periplasm and on the surface of Fe(II)-oxidizing cells. Biochemical and genomic studies provide evidence for the involvement of *c*-type cytochromes in nitrate-dependent Fe(II) oxidation; however, the exact mechanism of anaerobic Fe(II) oxidation is unknown. Chlorate and perchlorate [(per)chlorate]-dependent Fe(II) oxidation has been demonstrated, yet so far no organisms have been identified that are capable of coupling this metabolism to growth, potentially because of the differences between the nitrate- and perchlorate-reduction electron transport pathways. Through comparison of the various organisms capable of anaerobic, mesophilic Fe(II) oxidation, the variety of Fe(II) forms utilized, the biominerals formed, and electron acceptor-specific metabolic differences, researchers are beginning to understand the means by which these organisms make use of this vast electron donor reserve.

Iron is the fourth most abundant element in Earth's crust and exists primarily in the divalent ferrous [Fe(II)] and trivalent ferric [Fe(III)] oxidation states (Cornell and Schwertmann, 2003). At circumneutral pH (~7), these species form solid-phase iron minerals and soluble forms at acidic pH values (Stumm and Morgan, 1996). Fe(II) is stable at pH values below 4 as an aqueous species even in oxic environments.

J. Cameron Thrash and Sarir Ahmadi, Department of Plant and Microbial Biology, University of California, Berkeley, CA 94720. *John D. Coates,* Department of Plant and Microbial Biology, University of California, and Earth Sciences Division, Ernest Orlando Lawrence Berkeley National Laboratory, Berkeley, CA 94720.

It is now widely accepted that microbial respiration is responsible for the majority of iron redox transitions between the ferrous and ferric states (Weber et al., 2006a), making the study of organisms capable of these metabolisms vital to understanding the biogeochemical cycling of this important element.

Microorganisms can make use of both Fe(II) and Fe(III) as electron sources and sinks, respectively, for respiration. As reviewed comprehensively elsewhere in this volume, Fe(III) can be used as a terminal electron acceptor by a diversity of Fe(III)-reducing microorganisms that catalyze the redox transition of Fe(III) to Fe(II). Completing the microbial iron redox cycle, Fe(II)-oxidizing microorganisms utilize the reducing equivalents stored in Fe(II) species for a variety of metabolisms, and thereby transition Fe(II) to Fe(III) (Fig. 1). Aerobic Fe(II) oxidation in acidic environments has been well studied because of the role this metabolism plays in catalyzing acid mine drainage (Ghiorse, 1984; Harrison, 1984; Tyson et al., 2004), and microaerophilic Fe(II) oxidation has been shown to occur in a variety of environments (Edwards et al., 2003; Emerson and Moyer, 1997). Several groups have also identified and studied Fe(II) oxidation by anoxygenic phototrophs (Hegler et al., 2008; Poulain and Newman, 2009; Widdel et al., 1993).

Importantly, microorganisms can also oxidize Fe(II) in anoxic environments in the absence of electromagnetic radiation (Weber et al., 2006a). A variety of studies in the past 13 years have described anaerobic Fe(II) oxidation by mixed communities (Finneran et al., 2002; Nielsen and Nielsen, 1998; Ratering and Schnell, 2001; Straub and Buchholz-Cleven, 1998; Straub et al., 1996; Sun et al., 2009; Weber et al., 2006c) and pure cultures (Bruce et al., 1999; Chaudhuri et al., 2001; Edwards et al., 2003; Hafenbradl et al., 1996; Kumaraswamy et al., 2006; Lack et al., 2002; Miot et al., 2009a, 2009b; Muehe et al., 2009; Schadler et al., 2009; Shelobolina et al., 2003; Straub et al., 1998; Weber et al., 2006b, 2009). A diversity of organisms (Fig. 2) are now known to oxidize both soluble and solid-phase Fe(II) forms coupled to the reduction of nitrate (Chaudhuri et al., 2001; Edwards et al., 2003; Finneran et al., 2002; Hafenbradl et al., 1996; Lack et al., 2002; Miot et al., 2009a; Muehe et al., 2009; Schadler et al., 2009; Shelobolina et al., 2003; Weber et al., 2001), and (per)chlorate (Bruce et al., 1999; Lack et al., 2002), implicating them in environmental iron redox cycling where these terminal electron acceptors coexist with Fe(II). This chapter explores what is known about anaerobic, mesophilic Fe(II) oxidation in environmental samples and pure cultures, and includes an investigation of the mechanism of Fe(II) oxidation by dissimilatory (per)chlorate-reducing bacteria (DPRB), and a discussion on the oxidation products formed by these biomineralization processes, and emerging applications for the metabolism.

NITRATE-DEPENDENT Fe(II) OXIDATION

Environmental Samples

In situ geochemical analysis of flooded rice paddy soils counterintuitively indicated the environmental coexistence of both Fe(II) and nitrate, implying the natural occurrence of geochemical conditions conducive to anaerobic nitrate-dependent microbial Fe(II) oxidation (Ratering and Schnell, 2001). Subsequent studies with freshwater lake sediments confirmed this earlier result and demonstrated detectable concentrations of Fe(II) and nitrate even in highly reduced sediments in which both active microbial sulfate reduction and methanogenesis were occurring (Weber et al., 2006b). These samples were subsequently used to inoculate enrichment cultures from which the chemolithoautotrophic nitrate-dependent Fe(II) oxidizer *Pseudogulbenkiania* sp. strain 2002 was isolated (Weber et al., 2006b).

The first evidence for anaerobic, microbial respiratory Fe(II) oxidation was provided by Straub et al. in 1996. Enrichment cultures inoculated with sediment from ditches and brackish water in Germany were capable of coupling the oxidation of Fe(II) provided as $FeSO_4$ to

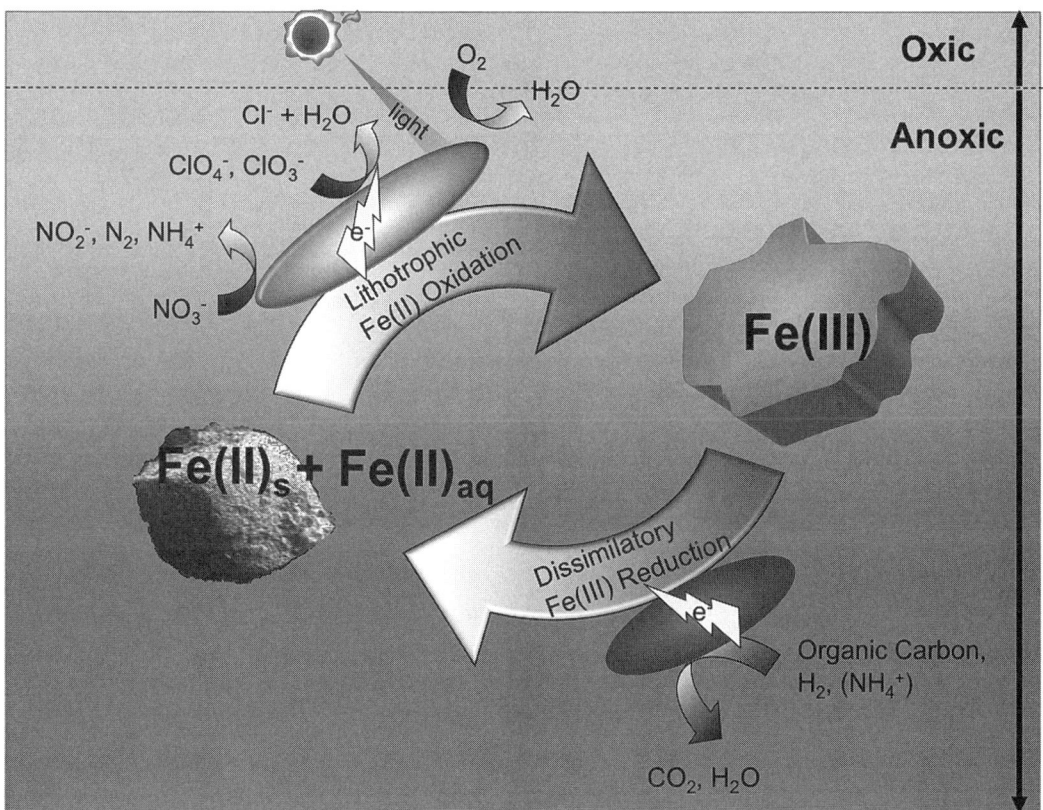

FIGURE 1 The microbially mediated iron redox cycle. Over the past two decades it has been unambiguously demonstrated that microorganisms play a central role in the environmental geochemical redox cycle of iron. Microbial Fe(III) reduction is mediated primarily through the activity of dissimilatory metal-reducing bacteria under anaerobic conditions, while Fe(II) oxidation can occur through the activity of both photolithotrophic Fe(II) oxidizers using Fe(II) as an electron source for CO_2 reduction, or chemolithotrophic respiratory Fe(II)-oxidizing bacteria (aerobic and anaerobic) using Fe(II) as an energy and electron source for carbon assimilation with a variety of alternative electron acceptors. Both Fe(III)-reducing and Fe(II)-oxidizing microorganisms have been shown to use either soluble or solid-phase iron sources, thus extending the biogeochemical cycle of iron to beyond the soluble form.

the reduction of nitrate. Several isolates were obtained from these enrichment cultures and, as a result of discovering these organisms, the authors tested and demonstrated that several other known strains were also capable of the metabolism. Brackish and freshwater enrichment cultures were able to oxidize the added Fe(II) and were dependent on or independent of added carbon sources, respectively; however, endogenous carbon was not reported, so it is unclear whether or not these cultures were capable of autotrophic Fe(II) oxidation. Regardless, this study demonstrated for the first time that nitrate-dependent Fe(II) oxidation could occur in anoxic, nonphototrophic environments, and that Fe(II) could serve as a suitable electron donor for denitrification.

Benz et al. (1998) reported both mixotrophic and autotrophic nitrate-dependent Fe(II) oxidation based on the successful transfer of these cultures for over a year; however, clear evidence of autotrophy was not presented.

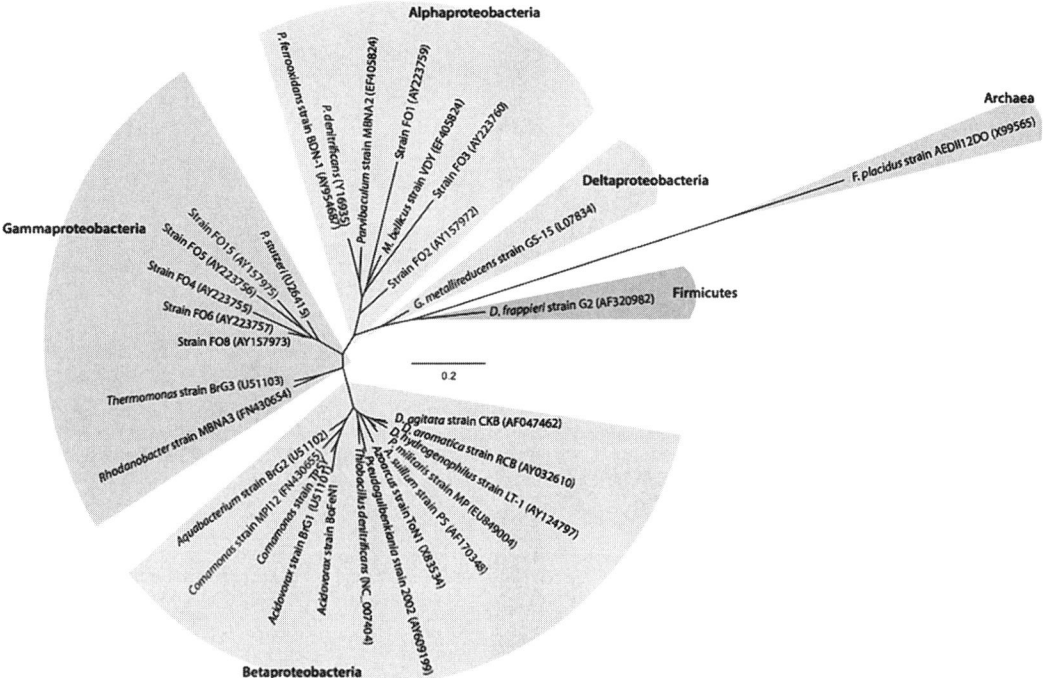

FIGURE 2 Phylogenetic diversity of anaerobic Fe(II)-oxidizing microorganisms. Available quality 16S rRNA gene sequences were aligned with MUSCLE (Edgar, 2004) and phylogeny was computed with MrBayes 3.2 (Ronquist and Huelsenbeck, 2003). Scale bar indicates 0.2 changes per position.

Both freshwater and marine media enrichments were capable of the metabolism in the presence of an amended labile carbon source, while one marine enrichment could complete it with no added carbon source. The metabolism was also observed in activated sludge (Nielsen and Nielsen, 1998). In that study the authors confirmed that the process was biologically catalyzed by showing that pasteurization, the addition of azide, and a decrease in incubation temperature all caused cessation of metabolism (Nielsen and Nielsen, 1998). Importantly, this group observed ammonium production coupled to Fe(II) oxidation, which was subsequently confirmed in studies performed with mixed and pure cultures, indicating an alternate pathway of nitrate reduction during Fe(II) oxidation compared with previously observed complete denitrification metabolisms (Weber et al., 2006a).

Given the data from various studies confirming the presence of both nitrate-dependent Fe(II)-oxidizing bacteria and dissimilatory Fe(III)-reducing bacteria in environmental samples, it seemed plausible that these organisms might be able to coexist and utilize the products of each other's iron metabolism. This was subsequently confirmed in studies in which biologically produced ferrihydrite by nitrate-dependent Fe(II)-oxidizing bacteria was used as the terminal electron acceptor to isolate novel dissimilatory Fe(III)-reducing bacteria (Straub et al., 1998). Weber et al. (2006c) showed that cycling of iron between the +2 and +3 valence states could occur in environmental samples with both nitrate-dependent Fe(II)-oxidizing bacteria and dissimilatory Fe(III)-reducing bacteria in the presence of nitrate and acetate. Thus, in environments where nitrate, iron, and organic matter are present, nitrate-dependent Fe(II)-oxidizing bacteria can replenish the terminal electron acceptors required by dissimilatory Fe(III)-reducing bacteria, and the reduction products generated by the dissimilatory Fe(III)-reducing bacteria can likewise furnish electron donors for the Fe(II)-oxidizing bacteria. Kinetic modeling of this microbially mediated iron redox cycling showed that, when the processes are coupled, the rate of iron metabolism increased relative to either Fe(II) oxidation or Fe(III) reduction in isolation (Blothe and Roden, 2008).

Pure Cultures

Many of the studies on nitrate-dependent Fe(II) oxidation in environmental samples have yielded an ever expanding list of pure cultures (Weber et al., 2006a). In addition, since the metabolism was first described, several previously known organisms isolated based on alternative metabolic capacities have also been shown capable of nitrate-dependent Fe(II) oxidation (Weber et al., 2006a) (Fig. 2). The first isolates, strain HidR2 and *Acidovorax* strain BrG1, were both reported capable of Fe(II) oxidation coupled to nitrate reduction under mixotrophic conditions with acetate as a suitable carbon source (Straub and Buchholz-Cleven, 1998; Straub et al., 1996). Subsequent studies revealed that acetate served an additional function to cell carbon assimilation because Fe(II) metabolism by strain HidR2 was dependent on its presence (Benz et al., 1998).

Although it was also reported that *Aquabacterium* strain BrG2, *Azoarcus* strain ToN1, *Thiobacillus denitrificans*, and *Pseudomonas stutzeri* were capable of chemolithoautotrophic Fe(II) oxidation with nitrate, no growth data were reported, so it is unknown whether or not these strains were capable of growth without an additional organic carbon source (Straub et al., 1996). Subsequent studies on metal oxidation by *T. denitrificans* demonstrated that, like strain HidR2, it also had an absolute requirement for an additional electron donor, this time in the form of H_2, indicating a biochemical and geochemical complexity of metal oxidation beyond that expected (Beller, 2005).

Other studies confirmed the capacity of organisms isolated on alternate substrates to be capable of nitrate-dependent Fe(II) oxidation. Several organisms isolated as dissimilatory (per)chlorate-reducing bacteria could carry out this metabolism, including *Dechloromonas agitata* strain CKB (Bruce et al., 1999), *Azospira suillum* strain PS (Chaudhuri et al., 2001; Lack et al., 2002), *Magnetospirillum bellicus* strain VDY (Thrash et al., 2010a), *Dechlorobacter hydrogenophilus* strain LT-1 (Thrash et al., 2010b), and *Propionivibrio militaris* strain MP (Thrash et al., 2010b). Growth information was only reported for strain PS, and, similarly to strain HidR2, this was dependent on an added carbon source and independent of Fe(II) oxidation (Lack et al., 2002). In addition to these organisms, several phylogenetically diverse Fe(III)-reducing bacteria, including the *Deltaproteobacteria* member *Geobacter metallireducens* GS-15 and the *Firmicutes* member *Desulfitobacterium frappieri* strain G2, were also capable of Fe(II) oxidation coupled to nitrate reduction (Finneran et al., 2002; Shelobolina et al., 2003). *D. frappieri* strain G2 grew by nitrate-dependent Fe(II) oxidation with acetate as a carbon source (Shelobolina et al., 2003). Although, in contrast, *G. metallireducens* was incapable of growth by this metabolism, it completely reduced nitrate to ammonium (Nielsen and Nielsen, 1998; Weber et al., 2006c) which represented the first pure-culture example of this metabolism and provided an explanation for the previously observed ammonium increases in mixed Fe(II)-oxidizing culture studies (Nielsen and Nielsen, 1998; Weber et al., 2006c).

Other strains isolated with the intention to identify new Fe(II)-oxidizing bacteria have increased the known diversity of these organisms. Edwards et al. (2003) isolated a number of organisms capable of nitrate-dependent Fe(II) oxidation in the *Alphaproteobacteria* and *Gammaproteobacteria*. Kappler et al. (2005) and Byrne-Bailey et al. (2009) independently described two new *Acidovorax* species strains BoFeN1

(Kappler et al., 2005) and TPSY (Byrne-Bailey et al., 2009) capable of nitrate-dependent Fe(II) oxidation. Weber et al. (Weber et al., 2006b, 2009) described a new betaproteobacterium, *Pseudogulbenkiania* sp. strain 2002, also capable of the metabolism, while enrichments initiated with [Fe(II)EDTA] and nitrate led to the isolation of *Paracoccus ferrooxidans* strain BDN-1 (Kumaraswamy et al., 2006). Finally, a recent characterization of the original enrichment culture used by Straub et al. (1996) led to the isolation of three more nitrate-dependent Fe(II) oxidizers, *Parvibaculum* strain MBNA2, *Rhodanobacter* strain MBNA3, and *Comamonas* strain MPI12 (Blothe and Roden, 2009).

While many of these organisms have been characterized as being capable of nitrate-dependent Fe(II) oxidation under autotrophic conditions, only two strains have been shown to grow by this metabolism. The hyperthermophile *Ferroglobus placidus* was described in 1996 by Hafenbradl et al. (1996), and growth data were reported under autotrophic conditions. Autotrophic nitrate-dependent Fe(II) oxidation was also described for the mesophilic *Pseudogulbenkiania* strain 2002 by Weber et al. in 2006 (Weber et al., 2006b, 2009). This study demonstrated both growth and uptake of $H^{14}CO_3^-$, showing conclusively for the first time that an organism could conserve enough energy by this metabolism to fix carbon for biomass production.

Diversity in Environmental Samples

Isolation and study of pure cultures have provided the needed starting points for investigating the presence of these organisms in environmental samples and to begin examining the microbial ecology of nitrate-dependent Fe(II) oxidation. Using oligonucleotide probes specific for the strains BrG1, BrG2, and BrG3 from previous studies, Straub and Buchholz-Cleven (1998) were able to determine the presence of these organisms in various enrichment cultures inoculated from a variety of sources. None of the organisms were present in all enrichments, but all were present in some. However, when these probes were used for in situ hybridization or to probe denaturing gradient gel electrophoresis bands from natural samples, all three groups of organisms were below the detection limit. Later denaturing gradient gel electrophoresis analysis of these enrichment cultures found evidence of *Geobacteraceae* and gram-positive organisms (Straub et al., 2004), and an even more comprehensive study done on the same enrichment culture identified sequences most similar to the betaproteobacterium *Sideroxydans lithotrophicus* in 62 to 72% of multiple clone libraries taken over the course of two years (Blothe and Roden, 2009).

Clone libraries were used to analyze the transition of the microbial community composition doing three successive changes of metabolism in the same enrichment culture (Weber et al., 2006c). As the culture changed from nitrate-reducing to iron(III)-reducing conditions with acetate as the electron donor, sequences similar to *Dechloromonas* and *Geobacter* spp. were significantly enriched. As the culture then switched to nitrate-dependent Fe(II) oxidation, further changes in the community were observed, notably with a decrease in the relative abundance of sequences similar to *Dechloromonas* and some deltaproteobacteria, and the appearance of sequences similar to *Azospira*, *Aquaspirillum*, and unclassified betaproteobacteria (Weber et al., 2006c).

Solid-Phase Fe(II) Oxidation Coupled to Nitrate

Although highly soluble in anoxic aqueous solution at neutral pH, the majority of naturally occurring Fe(II) is solid phase including surface-bound Fe(II), crystalline Fe(II) minerals (siderite, magnetite, pyrite, arsenopyrite, and chromite), and structural Fe(II) in nesosilicate (almandine and staurolite) and phyllosilicate (nontronite). As such, these Fe(II) mineral phases represent a significant energy source for organisms that might be able to access the electron-donating capacity within them. Microbial electron transfer to solid-phase Fe(III) forms is well documented for dissimilatory

Fe(III)-reducing bacteria, and it is now known that some nitrate-dependent Fe(II)-oxidizing bacteria are also capable of extracting electrons from Fe(II) in solid-phase minerals (Chaudhuri et al., 2001), although the mechanism for this ability is unknown (Weber et al., 2006a). This metabolism has even been implicated in the biogenesis of banded iron formations during the Precambrian era of Earth's early history (Chaudhuri et al., 2001). Mixed cultures have been demonstrated to oxidize the Fe(II) content of the crystalline Fe(II) minerals magnetite, siderite, goethite, and pyrite (Chaudhuri et al., 2001; Jorgensen et al., 2009; Weber et al., 2001). Pure-culture studies have revealed nitrate-dependent oxidation of the Fe(II) content of FeS by *F. placidus* and *A. sullum* (Chaudhuri et al., 2001; Hafenbradl et al., 1996); almandine, arsenopyrite, chromite, siderite, magnetite, and staurolite by *A. sullum* (Chaudhuri et al., 2001); nontronite by *D. frappieri* (Shelobolina et al., 2003); and vivianite by *Acidovorax* strain BoFeN1 (Miot et al., 2009a).

Biomineralization

The role of Fe(II)-oxidizing bacteria in the production of banded iron formations was hypothesized as a result of the observed iron oxidation products (Chaudhuri et al., 2001; Hafenbradl et al., 1996). Indeed, the biomineralization of iron as a result of nitrate-dependent Fe(II) oxidation has been the subject of a number of studies that have illuminated a variety of oxidation products and some curious effects of this process on the cells involved. Chaudhuri, Lack, and Coates used X-ray diffraction (XRD) to observe the biomineralization of a variety of Fe(III) oxides during Fe(II) oxidation by *A. sullum*, including hematite, green rust, maghemite, and magnetite (Chaudhuri et al., 2001). This same organism created amorphous ferrihydrite during nitrate-dependent Fe(II) oxidation under nongrowth, washed-cell conditions (Lack et al., 2002). This difference in end products was suggested to be the result of a faster rate of oxidation of resting cell suspensions compared with that observed in growth cultures (Chaudhuri et al., 2001). Weber et al. (2006c) observed the formation of goethite by an enrichment culture using both XRD and Mössbauer spectroscopy, while Hohmann et al. (2009) also observed goethite formation with Mössbauer.

Since Fe(II) oxidation at neutral pH necessarily leads to the formation of solid phase Fe(III) precipitates, microbial interaction with Fe(II) oxidation end products has been a subject of interest (Weber et al., 2006a). Several studies document the encrustation of strain BoFeN1 during Fe(II) oxidation, a phenomenon that seems counterproductive for the organism, assuming that this encrustation would inhibit diffusion of needed nutrients to the cell surface and may interfere with replication (Kappler et al., 2005; Miot et al., 2009a, 2009b; Muehe et al., 2009; Schadler et al., 2009). This encrustation was observed during experiments with both soluble and solid-phase Fe(II) forms, and occurred both on the surface and within the periplasm. This is consistent with observations that *c*-type cytochromes may be involved in nitrate-dependent Fe(II) oxidation in other organisms (Chaudhuri et al., 2001; Weber et al., 2009). Elegant redox-mapping with scanning transmission X-ray microscopy showed the evolution of Fe(III) forms on the surfaces of bacterial cells, and these precipitates were confirmed as amorphous Fe(III) oxides by a variety of techniques including XRD, X-ray absorption spectroscopy, and extended X-ray absorption fine structure (Miot et al., 2009a, 2009b). In contrast, anoxic phototrophic Fe(II) oxidation did not lead to the same kind of encrustation (Schadler et al., 2009). Why strain BoFeN1 cells become encrusted has yet to be resolved, but future work on other strains may reveal if this phenomenon is a common trait of nitrate-dependent Fe(II) oxidation or is specific to certain strains or species.

Applications: Metal Adsorption to Biomineralization Products

The mobility of trace metals and radionuclides released into aquatic and terrestrial environments by mining, industrial processes, and municipal waste disposal practices are areas of significant scientific, public health, and regulatory attention. The U.S. Environmental Protection Agency includes cadmium, chromium, copper, lead, mercury, nickel, silver, and zinc on its priority pollutant list for waste effluents. Toxic heavy metals such as uranium and arsenic are known to be transformable by microorganisms (Craft et al., 2004; Lloyd, 2003; Oremland et al., 2002; Smedley and Kinniburgh, 2002). In the case of uranium, the oxidized hexavalent U(VI) form is soluble at neutral pH and, therefore, precipitation/immobilization via microbial reduction to form solid-phase tetravalent U(IV) has been proposed as a remediation technology. Many studies have investigated the bioremediative potential of stimulating reducing bacteria to use soluble metal contaminants such as U(VI) as electron acceptors under anaerobic conditions and thus precipitate them out of solution (Bopp and Ehrlich, 1988; Gorby et al., 1998; Gorby and Lovley, 1992; Ishibashi et al., 1990; Lovley et al., 1991). However, many unknowns and potential limitations exist for this technique including:

1. What will be the fate of the reduced immobilized metal once the bioremediative process is complete, electron donor addition is halted, and the environment reverts back to an oxic state?
2. What is the potential for biooxidation and subsequent resolubilization of the reduced immobilized metal?
3. Reductive remediation may not be suitable for low-level, long-term contamination because there may be insufficient metal contaminants available to support a metal-reducing microbial community.
4. Many metals will be bound and solubilized by natural and anthropogenic organic matter present in most environments regardless of their valence state (Buffle, 1988; Coates et al., 2001; Moulin and Moulin, 1995; Thurman, 1985).

The judicious application of microbial nitrate-dependent Fe(II) oxidation has been suggested as a means to overcome many of these issues (Coates and Chakraborty, 2003). Primary geochemical controls that regulate the general trace metal concentrations in oxic natural waters include adsorption and coprecipitation by hydrous oxides of iron and manganese. Given that Fe is the fourth most abundant element in the Earth's crust, these reactions can be the dominant environmental control on the fate and transport of contaminating metals in most aqueous environments. Hydrous oxides of iron occur as discrete grains and as coatings on aquifer materials. They have been shown to be the major host minerals for many trace elements in soils, and for ^{60}Co, and isotopes of plutonium and americium in soils and sediments of a disposal area at Oak Ridge National Laboratory (Drever, 1997). Adsorption of heavy metals and radionuclides onto iron oxides has long been recognized as an important reaction for the immobilization of these compounds (Ainsworth et al., 1994; Ames et al., 1983; Dzombak and Morel, 1990; Hohl and Stumm, 1976; Means et al., 1978; Salomons and Forstner, 1984; Saunders et al., 1997). As such, the selective anaerobic bio-oxidation of added Fe(II) in situ (Lack et al., 2002; Lack, 2002) could be used as an effective means of "capping off" and completing the attenuation of heavy metals and radionuclides in a reducing environment. In such a scenario, once the contaminants have been initially immobilized through a bioreductive mechanism stimulated by the addition of a suitable electron donor such as acetate, the acetate additions would be halted and replaced with a single addition of Fe(II) and nitrate. The added Fe(II) would be bio-oxidized by the indigenous nitrate-dependent Fe(II)-oxidizing population and would precipitate over the immobilized reduced metals. At this point any further additions would be unnecessary and the system would be allowed to naturally revert back to an oxic state (Coates and Chakraborty, 2003; Lack et al., 2002).

Oxidized iron minerals represent a primary sink for heavy metals and metalloids in sedimentary environments and they regulate soluble metal concentrations, including those of uranium, chromium, and cobalt, in natural waters through adsorption and coprecipitation (Ainsworth et al., 1994; Hansel et al., 2002; Lack, 2002). Abiotic studies have shown that metals such as cobalt, chromium, cadmium, lead, uranium, and radium are rapidly adsorbed by these iron minerals (Ainsworth et al., 1994; Ames et al., 1983; Charlet and Manceau, 1992; Lack, 2002; Salomons and Forstner, 1984) and some of those metals with lower ionic radii (e.g., Co^{2+}, Cd^{2+}) are incorporated into the Fe(III)-oxide structure as it crystallizes with age. These metals become tightly bound into the Fe(III)-oxide crystal (Ainsworth et al., 1994) and are thus immobilized permanently. Lack et al. (2002) demonstrated that metals such as cobalt, cadmium, and uranium were rapidly adsorbed by biogenic iron hydrous oxides produced through the activity of nitrate-dependent Fe(II)-oxidizing bacteria. In these studies the microorganisms produced mixed-valence Fe(II)-Fe(III) minerals that rapidly crystallized with age. In the case of U(VI), it was preferentially bound (>80%) to the most crystalline phase and formed stable insoluble bidentate and tridentate inner sphere complexes (Lack, 2002). As such, these normally soluble metals become tightly bound into the most stable iron mineral form and are permanently immobilized.

Two recent studies have also demonstrated the ability of both pure and mixed cultures to successfully adsorb arsenic to iron(III) oxides generated during this process (Hohmann et al., 2009; Sun et al., 2009). Sun et al. (2009) showed that a mixed culture could oxidize both Fe(II) and As(III) with nitrate as the electron acceptor and coprecipitate these metals as Fe(III) and A(V). Furthermore, As(V) was better adsorbed than As(III), and thus the microbial oxidation of these metals simultaneously resulted in better arsenic removal than if arsenic were not microbially oxidizable. Hohmann et al. (2009) demonstrated similar arsenic sorption in both the As(III) and As(V) states [confirming more rapid adsorption of As(V)] to Fe(III) oxides created during nitrate-dependent Fe(II) oxidation by *Acidovorax* strain BoFeN1, and a mixed culture KS, as well as phototrophic Fe(II) oxidation by *Rhodobacter ferrooxidans* strain SW2. This study also showed that these cultures had resistance to high concentrations of As(III) and As(V). All three cultures could grow in up to 500 μM As(V), strain BoFeN1 could grow in the same amount of As(III), and strain SW2 and the KS culture were resistant to at least 100 μM As(III). Together, these studies show promise for this remediation technology which will undoubtedly be improved through research into the naturally occurring organisms that may be able to complete this process where metal contamination is present.

(Per)chlorate-Dependent Fe(II) Oxidation

Perchlorate has been widely used as a solid oxidant and, because of unregulated disposal before 1997, has become widespread throughout groundwater in the United States (Urbansky, 1998). Furthermore, recent studies have indicated significant ongoing natural neogenesis of perchlorate in the atmosphere with resultant global wet and dry deposition (Rajagopalan et al., 2006, 2009; Rao et al., 2007; Scanlon et al., 2008). The natural importance of this geochemical process was recently highlighted by the data from the Phoenix lander which provided evidence that perchlorate represents as much as 1% by mass of Martian soils (Hecht et al., 2009; Mumma et al., 2009). Perchlorate-reducing bacteria have been studied in considerable detail in response to the need for bioremediation of this important contaminant (Coates and Achenbach, 2004, 2006). Both perchlorate and chlorate can be utilized by microorganisms as terminal electron acceptors and reduced completely to chloride. The redox values for (per)chlorate reduction to chlorite are extremely favorable (Coates and Achenbach, 2004), and thus DPRB should readily couple the oxidation of Fe(II) to the reduction of these compounds. Most DPRB can alternatively reduce nitrate, and all strains tested to date have been capable of nitrate-dependent Fe(II) oxidation. However, with (per)chlorate as the terminal electron acceptor, Fe(II) has been inhibitory to growth of these same organisms at low concentrations (<1 mM FeCl$_2$) even if the cultures were amended with acetate as an additional electron donor and carbon source (Bruce et al., 1999; Lack et al., 2002). These data indicate that although some DPRB may be able to use Fe(II) as an electron donor, this physiology is electron acceptor-dependent, suggesting that the iron may interfere with electron transport to (per)chlorate.

A more in-depth investigation into this phenomenon using the DPRB *M. bellicus* strain VDY supported this hypothesis. Characterization of this recent isolate revealed it was capable of chemolithoautotrophic metabolism with H$_2$ or FeCl$_2$ oxidation coupled to (per)chlorate reduction (Thrash et al., 2010b). Resting cell suspensions incubated with 5 mM FeCl$_2$ and ClO$_4^-$, respectively, oxidized 1.17 mM Fe(II) with concomitant reduction of 1.58 mM ClO$_4^-$ over an 8-h period, while no Fe(II) oxidation occurred in heat-killed controls, indicating that this process was dependent on enzymatic activity. Interestingly, Fe(II) oxidation never went to completion although the cells were still replete with perchlorate as an electron acceptor. The stoichiometry for Fe(II) oxidation coupled to perchlorate reduction was 1.35 ± 0.52, which is significantly different from the predicted ratio of 0.125 according to the following:

$$8Fe(II) + ClO_4^- + 8H^+ \rightarrow 8Fe(III) + Cl^- + 4H_2O$$

Even if partial reduction of perchlorate was assumed to form chlorate or chlorite, the predicted ratio would be 0.5 or 0.25, respectively, according to these equations:

$$2Fe(II) + ClO_4^- + 2H^+ \rightarrow 2Fe(III) + ClO_3^- + H_2O$$

$$4Fe(II) + ClO_4^- + 4H^+ \rightarrow 4Fe(III) + ClO_2^- + 2H_2O$$

Thus, regardless of the assumed end product of perchlorate reduction, a considerable amount of perchlorate was inexplicably reduced suggesting the presence of an additional intrinsic electron donor (e.g., β-hydroxybutyrate or glycogen). While Fe(II) oxidation coupled to perchlorate reduction is prevalent in resting cell suspensions, iron either in the Fe(II) or Fe(III) form is inhibitory to growth in fresh cultures amended with acetate (Fig. 3). Inhibition is not observed when cells are grown with nitrate as the sole electron acceptor in the presence of iron suggesting that the inhibitory effect is specific to the perchlorate reductive pathway. Furthermore, and in contrast to Fe(II) oxidation with perchlorate, resting cell suspensions of strain VDY oxidized 101% of the theoretical expected Fe(II) relative to nitrate reduced according to the following equation:

$$10Fe(II) + 2NO_3^- + 12H^+ \rightarrow 10Fe(III) + N_2 + 6H_2O$$

Iron inhibition of perchlorate reduction was previously observed for other perchlorate reducing bacteria including *A. sullum* (Lack et al., 2002) indicating that this is not a species-specific effect and is likely based on inhibition of conserved enzymes in the perchlorate reduction pathway. The significant physiological differences associated with the two alternative electron acceptors therefore indicate a different mechanism for Fe(II) oxidation in each.

The observed data have been used to develop a model describing the specific effect of iron on the perchlorate-reduction pathway (Fig. 4). The results indicate that iron may be a competitive inhibitor, since perchlorate reduction was possible in washed cell suspensions where cell densities, in the order of 10^{11} to 10^{12}

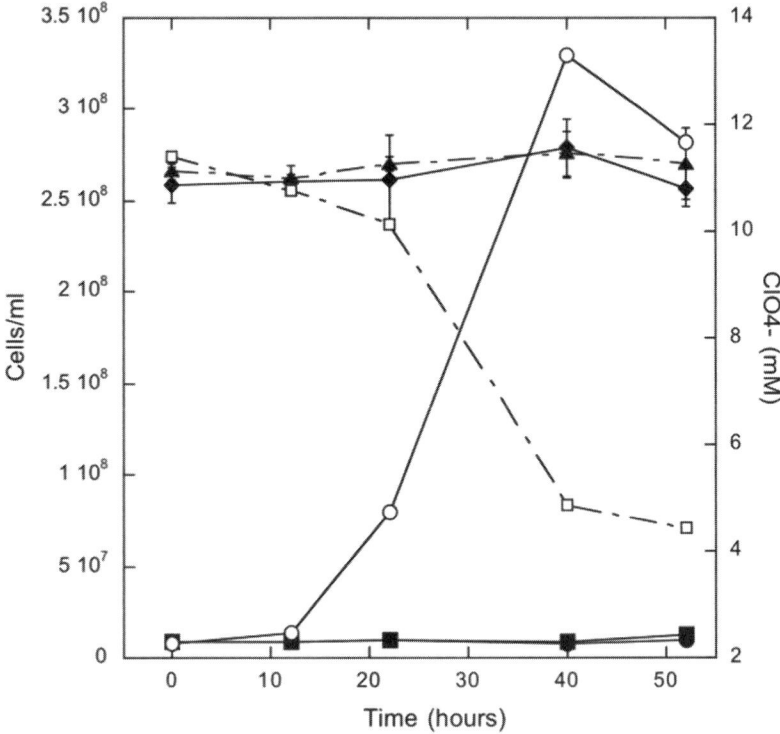

FIGURE 3 Effect of Fe(II) and Fe(III) on heterotrophic growth of strain VDY. ■, cells with added Fe(II); ●, cells with added Fe(III); ○, cells with no added iron; ▲, (ClO_4^-) for cells with added Fe(II); ◆, (ClO_4^-) for cells with added Fe(III); □, (ClO_4^-) for cells with no added iron.

cells·ml^{-1}, are significantly higher than growth cultures (~10^7 cells·ml^{-1}) in which perchlorate reduction was not observed. Removal of perchlorate suggests that the perchlorate reductase (Pcr) enzyme complex is functional; however, because no growth is observed, iron may be directly inhibiting the second enzyme in the reductive pathway, chlorite dismutase (Cld). This enzyme is critical to dissimilatory perchlorate reduction and is an outer membrane-bound protein (O'Connor and Coates, 2002). In further support of this, FeCl$_3$ is insoluble in the basal growth medium, making it unlikely to penetrate the outer membrane of the cells suggesting that its target of perchlorate-specific inhibition is the Cld. It has been shown that most energy conservation for growth during perchlorate respiration occurs not at the Pcr, but rather at the cytochrome oxidase step where the molecular O$_2$ produced through the activity of the Cld is reduced to H$_2$O (Fig. 4) (Sun, 2008). Electron transport chain inhibitor studies have confirmed the presence of complex 1, the bc_1 complex, and cytochrome oxidase in VDY (Sun, 2008), and difference-spectra absorption cytochrome scans (Coates and Achenbach, 2004) confirmed the presence of c-type cytochromes. Thus, if iron inhibited the Cld, oxygen biogenesis would not occur thereby preventing the establishment of a proton motive force from perchlorate respiration. Furthermore, inhibition of the Cld activity would result in a buildup of chlorite, which is toxic to the cells at concentrations above 0.8 mM and could abiotically react with the Fe(II) in solution producing Fe(III).

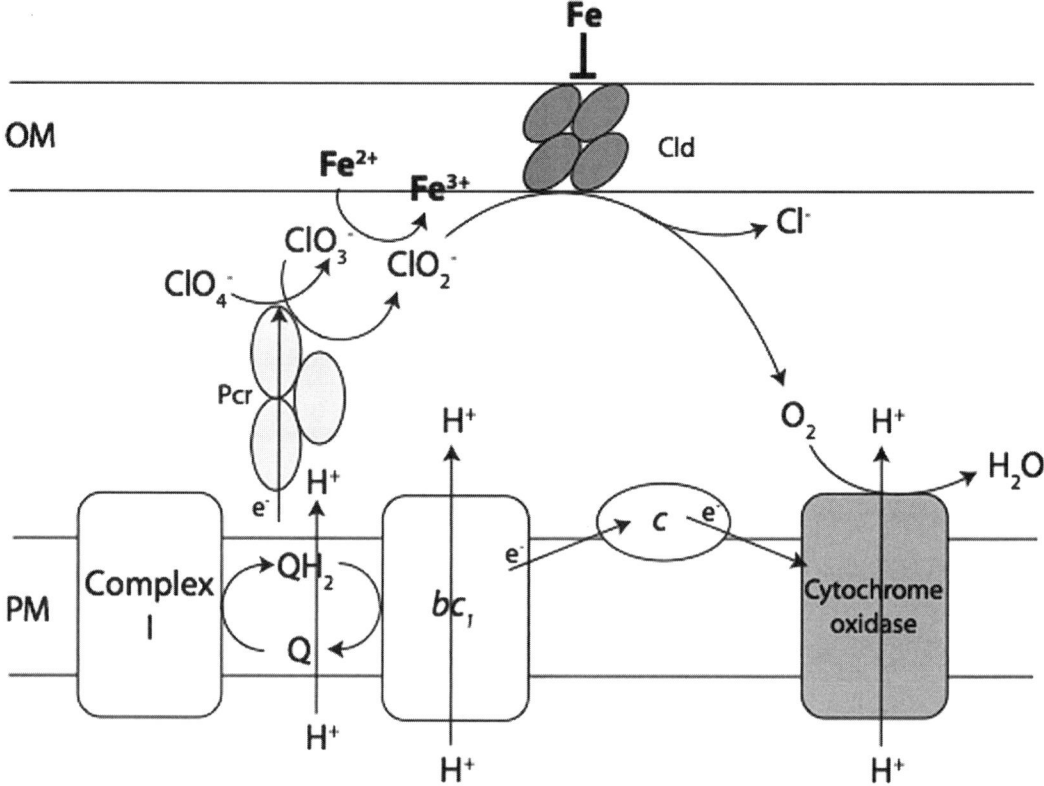

FIGURE 4 Model for possible iterations of iron with the perchlorate-reduction pathway of strain VDY. Both Fe(II) and Fe(III) are postulated to inhibit chlorite dismutase (Cld). This prevents the formation and subsequent reduction of oxygen, disabling the creation of a proton motive force. Cld inhibition would also cause a buildup of chlorite, which is toxic to the cell and can also abiotically react with iron(II). OM, outer membrane; PM, periplasmic membrane; Pcr, perchlorate reductase.

CONCLUDING REMARKS

It has become clear that a variety of organisms can catalyze the oxidation of various Fe(II) forms under anaerobic, mesophilic conditions. This process can be completed coupled to nitrate and perchlorate reduction, although, at present, nitrate is the only anaerobic electron acceptor that supports growth by this metabolism. The biochemistry and ecology of anaerobic Fe(II) oxidation is mostly unknown. Genome sequences of several known nitrate-dependent Fe(II)-oxidizing microorganisms (FOM) are now available or in the process of completion, and comparative genomics should yield potential targets for further genetic study of the metabolism. The variety of culture-independent techniques now in use for assessment of microbial ecology will inevitably lead to a better understanding of the prevalence, biogeography, and diversity of FOM in natural and contaminated settings. The combined efforts of these types of studies will help develop more accurate models regarding the role of FOM in the biogeochemical cycling of nitrogen, iron, and carbon.

REFERENCES

Ainsworth, C., J. Pilon, P. Gassman, and W. Van der Sluys. 1994. Cobalt, cadmium, and lead sorption to hydrous iron oxide: residence time effect. *Soil Sci. Soc. Am. J.* **58:**1615–1623.

Ames, L., J. McGarrah, B. Walker, and P. Salter. 1983. Uranium and radium sorption on amorphous ferric oxyhydroxide. *Chem. Geol.* **40:**135–148.

Beller, H. 2005. Anaerobic, nitrate-dependent oxidation of U(IV) oxide minerals by the chemolithoautotrophic bacterium *Thiobacillus denitrificans*. *Appl. Environ. Microbiol.* **71:**2170–2174.

Benz, M., A. Brune, and B. Schink. 1998. Anaerobic and aerobic oxidation of ferrous iron at neutral pH by chemoheterotrophic nitrate-reducing bacteria. *Arch. Microbiol.* **169:**159–165.

Blothe, M., and E. Roden. 2008. Microbial iron redox cycling in a circumneutral-pH groundwater seep. *Appl. Environ. Microbiol.* **75:**468–473.

Blothe, M., and E. Roden. 2009. Composition and activity of an autotrophic Fe(II)-oxidizing, nitrate-reducing enrichment culture. *Appl. Environ. Microbiol.* **75:**6937–6940.

Bopp, L. H., and H. L. Ehrlich. 1988. Chromate resistance and reduction in *Pseudomonas fluorescens* strain lb300. *Arch. Microbiol.* **150:**426–431.

Bruce, R. A., L. A. Achenbach, and J. D. Coates. 1999. Reduction of (per)chlorate by a novel organism isolated from a paper mill waste. *Environ. Microbiol.* **1:**319–331.

Buffle, J. 1988. *Complexation Reactions in Aquatic Systems.* Wiley, New York, NY.

Byrne-Bailey, K. G., K. A. Weber, A. H. Chair, S. Bose, T. Knox, T. L. Spanbauer, O. Chertkov, and J. D. Coates. 2009. Completed genome sequence of the anaerobic iron oxidizing bacterium *Acidovorax ebreus* strain TPSY. *J. Bacteriol.* **92:**1475–1476.

Charlet, L., and A. Manceau. 1992. X-ray absorption spectroscopic study of the sorption of Cr(III) at the oxide-water interface. *J. Colloid Interface Sci.* **148:**443–457.

Chaudhuri, S. K., J. G. Lack, and J. D. Coates. 2001. Biogenic magnetite formation through anaerobic biooxidation of Fe(II). *Appl. Environ. Microbiol.* **67:**2844–2848.

Coates, J. D., and L. A. Achenbach. 2004. Microbial perchlorate reduction: rocket-fuelled metabolism. *Nat. Rev. Microbiol.* **2:**569–580.

Coates, J. D., and L. A. Achenbach. 2006. The microbiology of perchlorate reduction and its bioremediative application, p. 279–291. *In* B. Gu and J. D. Coates (ed.), *Perchlorate, Environmental Occurrence, Interactions, and Treatment.* Springer, Berlin, Germany.

Coates, J. D., and R. Chakraborty. 2003. Anaerobic bioremediation—an emerging resource for environmental cleanup, p. 227–257. *In* I. Singleton, M. G. Milner, and I. M. Head (ed.), *Bioremediation: a Critical Review.* Horizon Press, Norfolk, United Kingdom.

Coates, J. D., R. Chakraborty, S. M. O'Connor, C. Schmidt, and J. Thieme. 2001. The geochemical effects of microbial humic substances reduction. *Acta Hydrochim. Hydrobiol.* **28:**420–427.

Cornell, R. M., and U. Schwertmann. 2003. *The Iron Oxides: Structure, Properties, Reactions, Occurrences and Uses.* Wiley-VCH, Weinheim, Germany.

Craft, E. S., A. W. Abu-Qare, M. M. Flaherty, M. C. Garofolo, H. L. Rincavage, and M. B. Abou-Donia. 2004. Depleted and natural uranium: chemistry and toxicological effects. *J. Toxicol. Environ. Health B Crit. Rev.* **7:**297–317.

Drever, J. L. 1997. *The Geochemistry of Natural Waters.* Prentice Hall, Upper Saddle River, NJ.

Dzombak, D. A., and F. M. M. Morel. 1990. *Surface Complexation Modeling: Hydrous Ferric Oxide.* John Wiley and Sons, New York, NY.

Edgar, R. 2004. Muscle: multiple sequence alignment with high accuracy and high throughput. *Nucleic Acids Res.* **32:**1792–1797.

Edwards, K. J., D. R. Rogers, C. O. Wirsen, and T. M. McCollom. 2003. Isolation and characterization of novel psychrophilic, neutrophilic, Fe-oxidizing, chemolithoautotrophic α- and γ-*Proteobacteria* from the deep sea. *Appl. Environ. Microbiol.* **69:**2906–2913.

Emerson, D., and C. L. Moyer. 1997. Isolation and characterization of novel iron-oxidizing bacteria that grow at circumneutral pH. *Appl. Environ. Microbiol.* **63:**4784–4792.

Finneran, K. T., M. E. Housewright, and D. R. Lovley. 2002. Multiple influences of nitrate on uranium solubility during bioremediation of uranium-contaminated subsurface sediments. *Environ. Microbiol.* **4:**510–516.

Ghiorse, W. C. 1984. Biology of iron-depositing and manganese-depositing bacteria. *Annu. Rev. Microbiol.* **38:**515–550.

Gorby, Y. A., F. Caccavo, Jr., and H. Bolton, Jr. 1998. Microbial reduction of cobalt^{iii}edta$^-$ in the presence and absence of manganese(IV) oxide. *Environ. Sci. Technol.* **32:** 244-250.

Gorby, Y. A., and D. R. Lovley. 1992. Enzymatic uranium precipitation. *Environ. Sci. Technol.* **26:**205–207.

Hafenbradl, D., M. Keller, R. Dirmeier, R. Rachel, P. Rossnagel, S. Burggraf, H. Huber, and K. O. Stetter. 1996. *Ferroglobus placidus* gen. Nov., sp. Nov., a novel hyperthermophilic archaeum that oxidizes Fe^{2+} at neutral pH under anoxic conditions. *Arch. Microbiol.* **166:**308–314.

Hansel, C. M., M. J. La Force, S. Fendorf, and S. Sutton. 2002. Spatial and temporal association of As and Fe species on aquatic plant roots. *Environ. Sci. Technol.* **36:**1988–1994.

Harrison, A. P. 1984. The acidophilic thiobacilli and other acidophilic bacteria that share their habitat. *Annu. Rev. Microbiol.* **38:**265–292.

Hecht, M. H., S. P. Kounaves, R. C. Quinn, S. J. West, S. M. M. Young, D. W. Ming, D. C. Catling, B. C. Clark, W. V. Boynton, J. Hoffman, L. P. DeFlores, K. Gospodinova, J. Kapit, and P. H. Smith. 2009. Detection of perchlorate and the soluble chemistry of Martian soil at the phoenix lander site. *Science* **325**:64–67.

Hegler, F., N. Posth, J. Jiang, and A. Kappler. 2008. Physiology of phototrophic iron(II)-oxidizing bacteria: implications for modern and ancient environments. *FEMS Microbiol. Ecol.* **66**:250–260.

Hohl, H., and W. Stumm. 1976. Interaction of Pb^{2+} with hydrous Al_3O_3. *J. Colloid Interface Sci.* **55**:281–288.

Hohmann, C., E. Winkler, G. Morin, and A. Kappler. 2009. Anaerobic fe(ii)-oxidizing bacteria show as resistance and immobilize as during Fe(III) mineral precipitation. *Environ. Sci. Technol.* **44**:94–101.

Ishibashi, Y., C. Cervantes, and S. Silver. 1990. Chromium reduction in *Pseudomonas putida*. *Appl. Environ. Microbiol.* **56**:2268–2270.

Jorgensen, J. C., O. S. Jacobsen, B. Elberling, and J. Aamand. 2009. Microbial oxidation of pyrite coupled to nitrate reduction in anoxic groundwater sediment. *Environ. Sci. Technol.* **43**:4851–4857.

Kappler, A., B. Schink, and D. Newman. 2005. Fe(III) mineral formation and cell encrustation by the nitrate-dependent Fe(II)-oxidizer strain BOFEN1. *Geobiology* **3**:235–245.

Kumaraswamy, R., K. Sjollema, G. Kuenen, M. Loosdrecht, and G. Muyzer. 2006. Nitrate-dependent $[Fe(II)EDTA]^{2-}$ oxidation by *Paracoccus ferrooxidans* sp. Nov., isolated from a denitrifying bioreactor. *Syst. Appl. Microbiol.* **29**:276–286.

Lack, J., S. Chaudhuri, R. Chakraborty, L. Achenbach, and J. Coates. 2002. Anaerobic biooxidation of Fe(II) by *Dechlorosoma suillum*. *Microb. Ecol.* **43**:424–431.

Lack, J. G. 2002. Immobilization of radionuclides and heavy metals through anaerobic biooxidation of Fe(II). *Appl. Environ. Microbiol.* **68**:2704–2710.

Lloyd, J. R. 2003. Microbial reduction of metals and radionuclides. *FEMS Microbiol. Rev.* **27**:411–425.

Lovley, D. R., E. J. P. Phillips, Y. A. Gorby, and E. R. Landa. 1991. Microbial reduction of uranium. *Nature* **350**:413–416.

Means, J. L., D. A. Crerar, and M. P. Borcsik. 1978. Adsorption of Co and selected actinides by Mn and Fe oxides in soils and sediments. *Geochim. Cosmochim. Acta* **42**:1763–1773.

Miot, J., K. Benzerara, G. Morin, S. Bernard, O. Beyssac, E. Larquet, A. Kappler, and F. Guyot. 2009a. Transformation of vivianite by anaerobic nitrate-reducing iron-oxidizing bacteria. *Geobiology* **7**:373–384.

Miot, J., K. Benzerara, G. Morin, A. Kappler, S. Bernard, M. Obst, C. Ferard, F. Skouripanet, J. Guigner, and N. Posth. 2009b. Iron biomineralization by anaerobic neutrophilic iron-oxidizing bacteria. *Geochim. Cosmochim. Acta* **73**:696–711.

Moulin, V., and C. Moulin. 1995. Fate of actinides in the presence of humic substances under conditions relevant to nuclear waste disposal. *Appl. Geochem.* **10**:573–580.

Muehe, E., S. Gerhardt, B. Schink, and A. Kappler. 2009. Ecophysiology and the energetic benefit of mixotrophic Fe(II) oxidation by various strains of nitrate-reducing bacteria. *FEMS Microbiol. Ecol.* **70**:3–11.

Mumma, M. J., G. L. Villanueva, R. E. Novak, T. Hewagama, B. P. Bonev, M. A. DiSanti, A. M. Mandell, and M. D. Smith. 2009. Strong release of methane on Mars in northern summer 2003. *Science* **332**:1041–1045.

Nielsen, J. L., and P. H. Nielsen. 1998. Microbial nitrate-dependent oxidation of ferrous iron in activated sludge. *Environ. Sci. Technol.* **32**:3556–3561.

O'Connor, S. M., and J. D. Coates. 2002. A universal immuno-probe for (per)chlorate-reducing bacteria. *Appl. Environ. Microbiol.* **68**:3108–3113.

Oremland, R. S., S. E. Hoeft, J. A. Santini, N. Bano, R. A. Hollibaugh, and J. T. Hollibaugh. 2002. Anaerobic oxidation of arsenite in mono lake water and by the facultative, arsenite-oxidizing chemoautotroph, strain mlhe-1. *Appl. Environ. Microbiol.* **68**:4795–4802.

Poulain, A., and D. Newman. 2009. *Rhodobacter capsulatus* catalyzes light-dependent Fe(II) oxidation under anaerobic conditions as a potential detoxification mechanism. *Appl. Environ. Microbiol.* **75**:6639–6646.

Rajagopalan, S., T. Anderson, S. Cox, G. Harvey, Q. Cheng, and W. A. Jackson. 2009. Perchlorate in wet deposition across North America. *Environ. Sci. Technol.* **43**:616–622.

Rajagopalan, S., T. A. Anderson, L. Fahlquist, K. A. Rainwater, M. Ridley, and W. A. Jackson. 2006. Widespread presence of naturally occurring perchlorate in high plains of Texas and New Mexico. *Environ. Sci. Technol.* **40**:3156–3162.

Rao, B., T. A. Anderson, G. J. Orris, K. A. Rainwater, S. Rajagopalan, R. M. Sandvig, B. R. Scanlon, D. A. Stonestrom, M. A. Walvoord, and W. A. Jackson. 2007. Widespread natural perchlorate in unsaturated zones of the southwest United States. *Environ. Sci. Technol.* **41**:4522–4528.

Ratering, S., and S. Schnell. 2001. Nitrate-dependent iron(II) oxidation in paddy soil. *Environ. Microbiol.* **3**:100–109.

Ronquist, F., and J. P. Huelsenbeck. 2003. MrBayes 3: Bayesian phylogenetic inference under mixed models. *Bioinformatics* **19**:1572–1574.

Salomons, W., and U. Forstner. 1984. *Metals in the Hydrocycle*. Springer-Verlag, New York, NY.

Saunders, J. A., M. A. Pritchett, and R. B. Cook. 1997. Geochemistry of biogenic pyrite and ferromanganese coatings from a small watershed: a bacterial connection. *Geomicrobiol. J.* **14**:203–217.

Scanlon, B. R., R. C. Reedy, W. A. Jackson, and B. Rao. 2008. Mobilization of naturally occurring perchlorate related to land-use change in the southern high plains, Texas. *Environ. Sci. Technol.* **42**:8648–8653.

Schadler, S., C. Burkhardt, F. Hegler, K. Straub, J. Miot, K. Benzerara, and A. Kappler. 2009. Formation of cell-iron-mineral aggregates by phototrophic and nitrate-reducing anaerobic Fe(II)-oxidizing bacteria. *Geomicrobiol. J.* **26**:93–103.

Shelobolina, E., C. Vanpraagh, and D. Lovley. 2003. Use of ferric and ferrous iron containing minerals for respiration by *Desulfitobacterium frappieri*. *Geomicrobiol. J.* **20**:143–156.

Smedley, P. L., and D. G. Kinniburgh. 2002. A review of the source behavior and distribution of arsenic in natural waters. *Appl. Geochem.* **17**:517–568.

Straub, K., and B. E. Buchholz-Cleven. 1998. Enumeration and detection of anaerobic ferrous iron-oxidizing, nitrate-reducing bacteria from diverse European sediments. *Appl. Environ. Microbiol.* **64**:4846–4856.

Straub, K., M. Hanzlik, and B. E. Buchholz-Cleven. 1998. The use of biologically produced ferrihydrite for the isolation of novel iron-reducing bacteria. *Syst. Appl. Microbiol.* **21**:442–449.

Straub, K., W. A. Schonhuber, B. Buchholz-Cleven, and B. Schink. 2004. Diversity of ferrous iron-oxidizing, nitrate-reducing bacteria and their involvement in oxygen-independent iron cycling. *Geomicrobiol. J.* **21**:371–378.

Straub, K. L., M. Benz, B. Schink, and F. Widdel. 1996. Anaerobic, nitrate-dependent microbial oxidation of ferrous iron. *Appl. Environ. Microbiol.* **62**:1458–1460.

Stumm, W., and J. J. Morgan. 1996. *Aquatic Chemistry: Chemical Equilibria and Rates in Natural Waters*. John Wiley & Sons, New York, NY.

Sun, W., R. Sierra-Alvarez, L. Milner, R. Oremland, and J. A. Field. 2009. Arsenite and ferrous iron oxidation linked to chemolithotrophic denitrification for the immobilization of arsenic in anoxic environments. *Environ. Sci. Technol.* **43**:6585–6591.

Sun, Y. 2008. Physiology of microbial perchlorate reduction. Ph.D. thesis. Plant and Microbial Biology. University of California, Berkeley, CA.

Thrash, J. C., B. J. Baker, S. Ahmadi, T. Torok, and J. D. Coates. 2010a. *Magnetospirillum bellicus* sp. nov., a novel dissimilatory perchlorate-reducing alphaproteobacterium isolated from a bioelectrical reactor. *Appl. Environ. Microbiol.* **76**:4730–4737.

Thrash, J. C., J. Pollock, T. Torok, and J. D. Coates. 2010b. Description of the novel perchlorate-reducing bacteria *Dechlorobacter hydrogenophilus* gen. Nov., sp. Nov., and *Propionivibrio militaris*, sp. Nov. *Appl. Microbiol. Biotechnol.* **86**:335–343.

Thurman, E. M. 1985. *Organic Geochemistry of Natural Waters*. Springer, Boston, MA.

Tyson, G., J. Chapman, P. Hugenholtz, E. E. Allen, R. J. Ram, P. M. Richardson, V. V. Solovyev, E. M. Rubin, D. S. Rokhsar, and J. F. Banfield. 2004. Community structure and metabolism through reconstruction of microbial genomes from the environment. *Nature* **428**:37–43.

Urbansky, E. T. 1998. Perchlorate chemistry: implications for analysis and remediation. *Bioremed. J.* **2**:81–95.

Weber, K. A., L. A. Achenbach, and J. D. Coates. 2006a. Microorganisms pumping iron: anaerobic microbial iron oxidation and reduction. *Nat. Rev. Microbiol.* **4**:752–764.

Weber, K. A., D. B. Hedrick, A. D. Peacock, J. C. Thrash, D. C. White, L. A. Achenbach, and J. D. Coates. 2009. Physiological and taxonomic description of the novel autotrophic, metal oxidizing bacterium, *Pseudogulbenkiania* sp. strain 2002. *Appl. Microbiol. Biotechnol.* **83**:555–565.

Weber, K. A., F. W. Picardal, and E. E. Roden. 2001. Microbially catalyzed nitrate-dependent oxidation of biogenic solid-phase Fe(II) compounds. *Environ. Sci. Technol.* **35**:1644–1650.

Weber, K. A., J. Pollock, K. A. Cole, S. M. O'Connor, L. A. Achenbach, and J. D. Coates. 2006b. Anaerobic nitrate-dependent iron(II) bio-oxidation by a novel lithoautotrophic betaproteobacterium, strain 2002. *Appl. Environ. Microbiol.* **72**:686–694.

Weber, K. A., M. M. Urrutia, P. F. Churchill, R. K. Kukkadapu, and E. E. Roden. 2006c. Anaerobic redox cycling of iron by freshwater sediment microorganisms. *Environ. Microbiol.* **8**:100–113.

Widdel, F., S. Schnell, S. Heising, A. Ehrenreich, B. Assmus, and B. Schink. 1993. Ferrous iron oxidation by anoxygenic phototrophic bacteria. *Nature* **362**:834–836.

# ACCENTUATE THE POSITIVE: DISSIMILATORY IRON REDUCTION BY GRAM-POSITIVE BACTERIA

Kelly C. Wrighton, Anna E. Engelbrektson, Iain C. Clark, Ryan A. Melnyk, and John D. Coates

10

INTRODUCTION

Bacterial reduction of metals is required for cell metabolism with considerable environmental consequences. The reduction of iron for biosynthetic functions impacts pathogen physiology and disease virulence in phylogenetically diverse bacteria (Brown and Holden, 2002; Schröder et al., 2003), while iron reduction has important implications in a variety of geochemical cycles (Lovley et al., 2004; Nealson and Saffarini, 1994). Iron reduction is also a promising system for bioremediation of organic and metal contaminants (Lloyd, 2003; Lovley et al., 2004) and corrosion control (Dubiel et al., 2002). Mechanisms employed for metal reduction have recently been shown to play a significant role for harvesting electrical current from waste materials in microbial fuel cells (Lovley, 2006; Pham et al., 2008). Despite the characterization of gram-positive dissimilatory Fe(III)-reducing bacteria (DIRB) in pure-culture and molecular studies (Kunapuli et al., 2007; Lovley et al., 2004; Slobodkin, 2005; Wrighton et al., 2008), the specific contribution of gram-positive bacteria to metal reduction has yet to be summarized.

Relative to gram-negative bacteria, gram-positive bacteria are defined as bacteria that lack an outer membrane, have a thicker cell wall (15 to 80 nm), and may be encased in a glycoprotein S-layer. The recent application of cryo-electron microscopy to thermophilic and mesophilic gram-positive bacteria has confirmed this architecture as well as revealed a periplasmic space (8 to 17 nm wide) located between the cytoplasmic membrane and the peptidoglycan cell wall network and associated proteins (Byrne-Bailey et al., 2010; Matias and Beveridge, 2005, 2006; Zuber et al., 2006). Not only does the lack of external membrane and thicker cell wall morphologically identify gram-positive bacteria, but this difference in physical structure also has important implications for extracellular electron transfer at the mechanistic and molecular levels (see the discussion below).

Both morphological and phylogenetic characters are required to identify gram-positive bacteria from gram-negative bacteria. Traditionally, the increased thickness of the cell wall was typically stained to distinguish the two. However, there are some problems with this approach because some gram-positive bacteria as in the class *Mollicutes* or the mycoplasmas lack cell walls (as

---

Kelly C. Wrighton, Anna E. Engelbrektson, Iain C. Clark, and Ryan A. Melnyk, Department of Plant and Microbial Biology, University of California, Berkeley CA 94720. *John D. Coates,* Department of Plant and Microbial Biology, University of California, and Earth Sciences Division, Ernest Orlando Lawrence Berkeley National Laboratory, Berkeley, CA 94720.

well as the outer membrane found in gram-negative bacteria) and therefore do not respond to Gram staining. Others, such as *Megasphaera, Pectinatus, Selenomonas,* and *Zymophilus,* have a porous pseudo-outer membrane that causes them to stain gram negative. Given these exceptions, today, gram-positive bacteria morphology is best confirmed via transmission electron microscopy or cryo-electron microscopy.

In addition to morphological differences, gram-positive bacteria are also phylogenetically distinct. Based on 16S rRNA gene analysis, the gram-positive bacteria are confined to two phyla, the *Firmicutes* and *Actinobacteria*. *Actinobacteria* is a phylum categorized by a high G+C content and includes only the class *Actinobacteria*, which is subdivided into various subclasses, of which the orders *Bifidobacteriales* and *Actinomycetales* contain the largest numbers of species and include the families *Streptomycetaceae, Mycobacteriaceae,* and *Corynebacteriaceae*. *Firmicutes* is a metabolically diverse phylum defined by low G+C content and divided into several classes, including bacilli, clostridia, mollicutes, and erysipelotrichi.

Gram-positive metal reduction is required for cell metabolism for both biosynthetic function (assimilation) and energy generation (dissimilation). Assimilatory metal reduction studies have often focused on iron, because this is a vital macroelement for most bacteria (Slobodkin, 2005). Ferric reductases reduce chelated iron prior to or after transport into the cell (Schröder et al., 2003). Following reduction, the resulting Fe^{2+} is available for intracellular incorporation into enzymes, cofactors, and the formation of magnetosomes (Lovley, 1991). Although it is not the focus of this chapter, iron assimilation by gram-positive bacteria, primarily pathogenic organisms, has been extensively reviewed (Brown and Holden, 2002; Schröder et al., 2003).

In contrast to the reduction of ferric iron for biosynthetic purposes, this work investigates gram-positive Fe(III) reduction for dissimilatory purposes, or bacteria that obtain energy for growth by concomitant reduction of Fe(III) as a respiratory electron acceptor. Dissimilatory metal reduction is either obligate or facultative. In obligate dissimilatory metal reduction, the metal functions as a terminal electron acceptor in anaerobic respiration and is reduced to support energy generation and organism growth. Alternatively, in facultative dissimilation, the organism grows predominantly by fermentation of sugars and amino acids, with Fe(III) acting as an electron sink for a minor portion of the reducing equivalents. Thus, gram-positive dissimilatory iron reduction includes not only metal-respiring organisms (obligate), but also fermenters that discharge part of the reducing equivalents from fermentation to Fe(III), resulting in a slight gain of additional energy and more oxidized fermentation products (facultative).

In this chapter we include an overview of gram-positive DIRB, with a focus toward obligate dissimilatory bacteria (Table 1). To complement the phylogeny, the physiology of obligate gram-positive DIRB is characterized, including the physical properties of the habitat, mineral forms of iron reduced, and alternative metal and nonmetal electron acceptors utilized. Unlike most soluble electron acceptors used by bacteria at circumneutral pH, Fe(III) exists primarily as insoluble oxyhydroxide minerals, which must be reduced externally to the cell. While the mechanisms used by gram-positive bacteria to reduce insoluble iron are poorly understood, this chapter begins to summarize current results and putative models for extracellular electron transfer by gram-positive bacteria (Color Plate 11). To support hypotheses for gram-positive extracellular electron transfer at the molecular level, this chapter incorporates both physiological and genomic information on several gram-positive DIRB.

ISOLATED GRAM-POSITIVE DISSIMILATORY IRON-REDUCING BACTERIA

Gram-positive DIRB belong to both the high and low G+C gram-positive phyla and include more than 16 unique genera (Table 1). These bacteria are not phylogenetically constrained, because the genera typically include both iron-reducing and non-iron-reducing

TABLE 1 Gram-positive bacteria capable of dissimilatory reduction of Fe(III) during growth[a]

Organism name	Type strain	Accession no.	Reference	pH	Temp (°C)	Dissim.	Metal electron acceptors	Electron acceptor	Electron donor	Source
Phylum *Actinobacteria* (Mesophilic)										
Acidimicrobium ferrooxidans TH3	DSM 10331	EF621760	Bridge & Johnson, 1998	2	45	Obligatory	Fe(III) sulfate	O_2	Gly	Copper leach dump, USA
Cellulomonas sp. strain WS01	ND	AY617102	Sani et al., 2002	ND	25	Fermentative	Cr (Methe et al., 2003), Fe(III), and U(VI)	ND	Lac	Hanford sediment core, USA
Cellulomonas sp. strain WS18	ND	AY617108	Sani et al., 2002	ND	25	Fermentative	Cr (Methe et al., 2003), Fe(III), and U(VI)	ND	Lac	Hanford sediment core, USA
Cellulomonas sp. strain ES6	ND	AY617100	Sani et al., 2002	ND	25	Fermentative	Cr (Methe et al., 2003), Fe(III), and U(VI)	ND	Lac	Hanford sediment core, USA
Ferrithrix thermotolerans Y005	DSM 19514	AY140237	Johnson et al., 2009	1.8	43	Fermentative	Fe(III) sulfate	ND	Gly	Geothermal site, Yellowstone NP
Ferromicrobium acidophilus T23	ND	AF251436	Johnson et al., 2009	2	35	Fermentative	Fe(III) sulfate	ND	Gly	Mine, UK
Phylum *Firmicutes* (Thermophilic)										
Alicyclobacillus sp. strain YOO4	ND	AY140236	Johnson et al., 2003	ND	55	Fermentative	Fe(III) sulfate	O_2	Glu	Acidic site, Yellowstone NP, USA
Anaerobranca gottschalkii LBS3	DSM 13577	AF203703	Prowe and Antranikian, 2001	9.5	55	Fermentative	Fe(III)	$S°, S_2O_3^{2-}$, Fum	Glu	Hot spring, Kenya
Anaerobranca horikoshii JW/YL_138	DSM 9786	U21809	Engle et al., 1995	8.5	57	Fermentative	Fe(III)	$S°, S_2O_3^{2-}$, Fum	Pep	Hot brook, Yellowstone NP, USA
Anaerobranca californiensis PAOHA_1	DSM 14826	AY064218	Gorlenko et al., 2004	9.2	58	Fermentative	Fe(III)-cit, HFO, Se (Methe et al., 2003)	$S°$	Pep	Mono Lake island sediment
Bacillus infernus TH-22	DSM 10277	U20384	Boone et al., 1995	7.5	61	Obligatory	Mn (Methe et al., 2003) oxide, $Fe(Cl_3)$	NO_3^-, TMAO	Lac, Form	Deep subsurface, USA

(*Continued*)

TABLE 1 Gram-positive bacteria capable of dissimilatory reduction of Fe(III) during growth[a]—Continued

Organism name	Type strain	Accession no.	Reference	pH	Temp (°C)	Dissim.	Metal electron acceptors	Electron acceptor	Electron donor	Source
Carboxydothermus hydrogenoformans Z-2901	DSM 6008	AF244579	Slobodkin et al., 2006a	7	67	Obligatory	HFO		H_2	Hot spring, Yellowstone NP, USA
Carboxydothermus ferrireducens JW/AS-Y7	DSM 11255	U76363	Slobodkin et al., 2006a	6.1	65	Obligatory	Fe(III)-cit, Fe(III)–EDTA, HFO	$S_2O_3^-$, Fum, AQDS	H_2, Gly, Lac, Glycerate, Pyr, Ye,	Hot spring, Yellowstone NP, USA
Carboxydothermus siderophilus	DSM 21278	EF542810	Slepova et al., 2009	6.7	70	Obligatory	Mn (Methe et al., 2003) oxide, HFO	AQDS, sulfite, thiosulfate, S^0, CO	Ye, Cas, starch, Gly, Pyr, H_2	Volcanic hot spring, Russia
Thermincola ferriacetica Z-0001	ND	AY631277	Zavarzina et al., 2007	7	59	Obligatory	HFO, magnetite, Mn (Methe et al., 2003) oxide	AQDS, Anode	Ac, CO/H_2	Terrestrial hydrothermal spring, Russia
Thermincola potens strain JR	ND	GU815244	Wrighton et al., 2008	7	60	Obligatory	HFO, Mn (Methe et al., 2003) oxide	AQDS, anode	CO/H_2, Ac, H_2	MFC anode biofilm, USA
Thermoanaerobacter acetoethylicus SL26	DSM 2359	ND	Slobodkin et al., 1999	ND	60	Obligatory	Fe(III)	$S_2O_3^{2-}$	Pep, H_2	Oil field, Russia
Thermoanaerobacter acetoethylicus SL28	DSM 2359	ND	Slobodkin et al., 1999	ND	65	Obligatory	Fe(III)	$S_2O_3^{2-}$	Pep, H_2	Oil field, Russia
Thermoanaerobacter brockii M739	DSM 1457	ND	Slobodkin et al., 1999	ND	65	Obligatory	Fe(III)	$S_2O_3^{2-}$	Pep, H_2	Oil field, Russia
Thermoanaerobacter sp. strain BSB-33	ND	EU368841	Bhowmick et al., 2009	6.5	60	Obligatory	HFO, Fe(III)-cit, Cr(VI), Mn (Methe et al., 2003) oxide	AQDS	Pep	Hot spring, India
Thermoanaerobacter wiegelii Rt8_B1	DSM 10319	X92513	Slobodkin et al., 1999	6.8	67	Obligatory	Fe(III)	ND	Pep	Cyanobacterial mat, Russia
Thermoanaerobacter siderophilus SR4	DSM12299	AF120479	Slobodkin et al., 1999	6.4	70	Obligatory	HFO, Mn (Methe et al., 2003) oxide	AQDS, thiosulfate, S^0	Pep, H_2	Hydrothermal vent, Russia
Thermoanaerobacter sulfurophilus L-64	DSM 11584	Y16940	Bonch-Osmolovskaya et al., 1997	7	55	Obligatory	HFO	$S_2O_3^{2-}$, S^0	Pep	Cyanobacterial mat, Russia

Organism	Strain	GenBank	Reference	pH	Temp	Metabolism	Electron acceptor	Alternative	Electron donor	Source
Thermoanaerobacter sp X513	ND	AF542520	Roh et al., 2002	7.4	60	Obligatory	HFO, Fe(III)-cit, cobalt(III), chromium (Methe et al., 2003), Mn (Methe et al., 2003), uranium (VI)	ND	Ac, Glu, Lac, Pyr, Suc, Xyl	Piceance basin Colorado, USA
Thermoanaerobacter sp X514	ND	AF542517	Roh et al., 2002	7.4	60	Obligatory	HFO, Fe(III)-cit, cobalt(III), chromium (Methe et al., 2003), Mn (Methe et al., 2003), uranium (VI)	ND	Ac, Glu, H$_2$, Lac, Pyr, Suc, Xyl	Piceance basin Colorado, USA
Thermoanaerobacter sp X561	ND	AF542518	Roh et al., 2002	7.4	60	Obligatory	HFO, Fe(III)-cit, cobalt(III), chromium (Methe et al., 2003), Mn (Methe et al., 2003), uranium (VI)	ND	Ac, Glu, Lac, Suc, Xyl	Piceance basin Colorado, USA
Thermovenabulum ferriorganovorum	ND	AY033493	Zavarzina et al., 2002	6.8	64	Obligatory	HFO, Mn (Methe et al., 2003) oxide	AQDS, NO$_3^-$, Fum, sulfite, thiosulfite, S^0	H$_2$	Hydrothermal vent, Russia
Thermolithobacter ferrireducens JW/JH-Fiji	DSM 13639	AF282252	Sokolova et al., 2007	7.3	73	Obligatory	HFO	AQDS, Fum, thiosulfite	H$_2$, Form	Sediment slurry, Fiji
Thermolithobacter ferrireducens JW/KA-2	DSM 13640	AF282253/ AF282254	Sokolova et al., 2007	7.3	73	Obligatory	HFO	AQDS, Fum, thiosulfite	H$_2$, Form	Sediment slurry, Yellowstone NP, USA
Thermosinus carboxydivorans NOR1	DSM 14886	AY519200	Sokolova et al., 2004	6.9	60	Fermentative	Fe(III)-cit, HFO	Thiosulfite	Suc, Lac, CO/H$_2$	Hot spring, Yellowstone NP, USA

(Continued)

TABLE 1 Gram-positive bacteria capable of dissimilatory reduction of Fe(III) during growth[a]—Continued

Organism name	Type strain	Accession no.	Reference	pH	Temp (°C)	Dissim.	Metal electron acceptors	Electron acceptor	Electron donor	Source
Tepidimicrobium ferriphilum	DSM 16624	AY656718	Slobodkin et al., 2006b	7.7	50	Fermentative	HFO, Fe(III)-cit, Fe(III)-NTA, Fe(III)-EDTA	AQDS, sulfite, thiosulfite, S^0, Fum	Pep, Val, Prop	Hot spring sediment, Russia
Phylum Firmicutes (Mesophilic)										
Alkaliphilus metalliredigens QMF	ND	AY137848	Ye et al., 2004	9.5	35	Obligatory	Fe(III)-EDTA, Fe(III)-cit, Co(III)-EDTA, Cr(VI)	AQDS	Ye, Tryp, Ac, Lac	Borax contaminated leachate ponds, USA
Alkaliphilus sp. strain ISO-W1	ND	DQ677018	Lin et al., 2007	7	35	Obligatory	Fe(III)-cit, HFO, Lepid, goethite, Fe(III)-phos	AQDS, NO_3^-, S^0	Lac, Ac	Sheldt estuarine sediment, Netherlands
Bacillus arsenicselenatis strain E1H	DSM 15340	AF064705	Switzer Blum et al., 1998	9.8	20	Obligatory	Fe-NTA, As(V), Se(VI)	NO_3^-, Fum	Lac, Mal	Mono Lake sediment, USA
Bacillus subterraneus	DSM 13966	AY672638	Kanso et al., 2002	7.3	38	Fermentative	HFO and Mn (Methe et al., 2003)	NO_3^-, Fum, NO_2^-	Ye, EtOH, Fru, Gly, Glu, Lac, Mat, starch, Xyl	Artesian basin. Australia
Bacillus sp. strain SFB	ND	AY669375	Pollock et al., 2007	9	30	Obligatory	Fe(III)-cit, Fe(III)-NTA, Fe(III)-pyro	NO_3^-, Fum, O_2	LB, Fruc, Glu, Suc, Lac, Cas, YE	Salt flat sediments soda lake
Clostridium sp. strain ISO-W6	ND	DQ677023	Lin et al., 2007	7	25	Fermentative	Fe(III)-cit, HFO, lepid, goethite, Fe(III)-phos	AQDS, NO_3^-, S^0	Ye	Sheldt estuarine sediment, Belgium
Clostridium sp. strain ISO-W7	ND	DQ677024	Lin et al., 2007	7	25	Fermentative	Fe(III)-cit, HFO, lepid, goethite, Fe(III)-phos	AQDS, NO_3^-, S^0	Ye	Sheldt estuarine sediment, Belgium
Clostridium butyricum sp. EG3	DSM 10702	AF226712	Park et al., 2001	7	37	Fermentative	Fe(III)-phos, HFO	Anode	Glu	MFC anode, Korea
Clostridium beijerinckii	DSM 791	X68179	Dobbin et al., 2001	7	30	Fermentative	Fe(III)-cit, Fe(III)-(maltol)3	ND	Glu	Freshwater sediment, UK

Species	DSM #	Accession	Reference	pH	Temp	Type	Electron acceptors	Electron donors	Source
Desulfitobacterium dehalogenans	DSM 9161	L28946	Villemur et al., 2006	7.5	35	Obligatory	Mn (Methe et al., 2003), Fe(III), Se(VI)	AQDS, humics, sufite, thiosulfite, Fum, NO_3^-	H_2, Form, Lac, Ye	Freshwater sediment, USA
Desulfitobacterium hafniense strain G2	DSM 1C664	AF320982	Shelobolina et al., 2003	7	30	Obligatory	HFO, U(VI), Fe(III)-NTA, Fe(III)-cit, smectite	TCE, PCE, AQDS, S_2O_3, SO_3^{2-} Fum, NO_3^-	Buty, BtOH, Cit, EtOH, Form H_2, Lac, Mal, Pyr	Subsurface smectite bedding, USA
Desulfitobacterium hafniense strain PCP-1	DSM 1C665	U40078	Niggemyer et al., 2001	7.5	38	Obligatory	Fe(III)-phos, HFO, As(V), Mn (Methe et al., 2003)	Fum, PCP, S^0, SO_3^{2-}, $S_2O_3^{2-}$, TCP	Lac	Methanogenic consortium, USA
Desulfitobacterium hafniense DCB-2	DSM 1C666	X94975	Niggemyer et al., 2001	7.5	37	Obligatory	As(V), Mn (Methe et al., 2003), Se(VI), HFO, Fe(II)P	Fum, NO_3^-, S^0, $S_2O_3^-$, PCP, TCP	Lac	Municipal sludge, USA
Desulfitobacterium hafniense DP7	DSM 1C667	AJ276701	Villemur et al., 2006	7.3	37	Obligatory	Mn (Methe et al., 2003), Fe(III), Se(VI), AQDS	SO_3^{2-}, thiosulfite, Fum, NO_3^-	H_2, Form, Lac, Pyr	Human fecal samples, USA
Desulfitobacterium hafniense GBFH	DSM 1C668	AJ207028	Niggemyer et al., 2001	7.5	37	Obligatory	As(V), Mn (Methe et al., 2003), Se(VI), HFO, Fe(III)-phos	Fum, S^0, SO_3^{2-}	Lac	Arsenic river sediments, USA
Desulfitobacterium hafniense TCE1	DSM 1C669	X95742	Villemur et al., 2006	7.2	35	Obligatory	Fe(III), As(V), Mn (Methe et al., 2003), Se(VI)	Sulfite, thiosulfite, fumarate, nitrate	H_2, Form, Lac, Pyr	Chloroethene-contaminated soil, USA

(Continued)

TABLE 1 Gram-positive bacteria capable of dissimilatory reduction of Fe(III) during growth[a]—Continued

Organism name	Type strain	Accession no.	Reference	pH	Temp (°C)	Dissim.	Metal electron acceptors	Electron acceptor	Electron donor	Source
Desulfitobacterium metallireducens 853-15A	DSM 15288	AF297871	Finneran et al., 2002	7	30	Obligatory	Fe(III)-cit, Fe(III)-NTA, Mn (Methe et al., 2003), Cr(VI)	AQDS, Humics, S°, $S_2O_3^{2-}$, selenite, TCE, PCE, 3-chloro-4-HPE	Lac, Form	Uranium mill tailings, USA Gasoline-contaminated groundwater, Australia
Desulfosporosinus meridiei S10	ND	AF076527	Robertson et al., 2000	6.1	28	Obligatory	Fe(III) Fe(III)-cit, Mn (Methe et al., 2003) oxide, Cr (Methe et al., 2003), U(VI)	SO_4^{2-}, SO_3^{2-}, $S_2O_3^{2-}$, S°, DMSO	Lac	Heavy metal contaminated sediment, USA
Desulfotomaculum reducens MI-1	ND	U95951	Tebo and Obraztsova, 1998	7.3	35	Obligatory		SO_4^{2-}	Buty, Lac, Val	
Sulfobacillus acidophilus ALV	DSM 10332	M79375	Bridge and Johnson, 1998	2	45	Obligatory	Fe(III) sulfate	O_2	Tetrathionate	Coal spoil heap, UK
Sulfobacillus acidophilus THWX	DSM 10333	ND	Bridge and Johnson, 1998	2	45	Obligatory	Fe(III) sulfate	O_2	Gly, tetrathionate	Coal spoil heap, UK
Sulfobacillus acidophilus YTF1	DSM 10334	AY007665		2	45	Obligatory	Fe(III) sulfate, HFO, geothite, jarosite	O_2	Gly	Hot spring, Yellowstone NP, USA
Sulfobacillus thermosulfidooxidans TH1	DSM 9293	ND	Bridge and Johnson, 1998	2	45	Obligatory	Fe(III) sulfate	O_2	Gly, tetrathionate	Thermal spring, Iceland

[a] The reference included is the citation for iron reduction, while the reported temperature and pH are the optimal growth characters for the bacterium. Dissim. stands for dissimilation and refers to whether an organism is capable of respiration with ferric iron (obligatory), not just fermentation. In cases where organisms are capable of both respiration and fermentation with ferric iron, it is just noted as respiration. The form of iron is included where provided in the reference; if not specified it is documented as Fe(III). Fe(III)-cit, ferric citrate; HFO, hydrous ferric oxide; Fe(III)-phos, ferric phosphate; Fe(III)-pyro, ferric pyrophosphate; Fe(III)-Cl₃, ferric chloride; AQDS, 2,6-anthraquinone disulfonic acid; DMSO, dimethyl sulfoxide; MFC, microbial fuel cell; ND, no data; NP, National Park; NTA, nitrilotriacetic acid; TMAO, trimethyamine oxide; TCP, trichlorophenol; PCP, pentachlorophenol; LB, Luria broth; Ye, yeast extract.

species (Table 1). The high G+C phylum *Actinobacteria* includes two types of iron reducers: iron-oxidizing bacteria that can also reduce iron, and those belonging to the fermentative genus *Cellulomonas*. The *Cellulomonas* spp. (strains WS01, WS18, and ES6) were isolated from subsurface sediment core contaminated with heavy metal and radionucleotides (Sani et al., 2002) and were demonstrated to reduce Fe(III)-nitrilotriacetic acid, Cr(V), and U(VI) in cell suspension with lactate as an electron donor. Later physiological studies using strains WSO1 and ES6 demonstrated chromium reduction under growth conditions with xylose as an electron donor. However, the lack of reported growth data coupled with the decrease in pH and accumulation of organic acids suggest that metal reduction was by fermentative reduction rather than obligate dissimilation (Viamajala et al., 2007). Unlike the mesophilic *Cellulomonas*, *Ferrimicrobium*, and *Ferrithrix* isolates, the acidophilic and moderately thermophilic *Acidimicrobium ferrooxidans* strain TH3 can couple glycerol oxidation to ferric reduction in an obligate dissimilatory fashion.

Akin to the *Actinobacteria*, the *Firmicutes* DIRB span a wide range of environmental conditions including broad temperature, pH, and salinity ranges. Of the mesophilic bacteria that grow under quotidian culture conditions, the genus *Desulfitobacterium* of the family *Peptococcaceae* contains the greatest number of isolated iron-reducing bacteria in pure culture with eight isolates spanning three species, *Desulfitobacterium dehalogenans*, *D. hafniense*, and *D. metallireducens*. All of these isolates can reduce chelated forms of iron, but *D. hafniense* strain G2 was the first *Desulfitobacterium* to grow by reduction of insoluble Fe(III) oxides (Table 1). Beyond the isolated bacteria, there is significant indirect support for *Desulfitobacterium* metal reduction from environmental molecular surveys, indicating that this metabolism may be more than a laboratory novelty. Petrie et al. (2003) found 16S rRNA gene sequences closely related to *Desulfitobacterium* spp. in Fe(III) enrichments taken from subsurface sediment contaminated with uranium and nitrate. Kostka et al. (2002) used 16S rRNA gene clone library analysis to demonstrate that the *Desulfitobacterium* were dominant members of an Fe(III)-reducing consortium from subsurface sediments. Reduction of Fe(III) was observed in a swine manure microcosm amended with $FeCl_3$, with reduction demonstrated to be biological and corresponding to the enrichment of *D. hafniense* strain PCP-1 and *D. metallireducens* (Castillo-Gonzalez and Bruns, 2005).

Bacillus infernus, a member of the *Firmicutes*, was one of the first gram-positive DIRB isolated and characterized (Boone et al., 1995). Since the isolation of this organism, research interests in the field of thermophilic gram-positive iron reduction have increased dramatically. Today, gram-positive bacteria capable of dissimilatory iron reduction have been isolated from most known types of thermal ecosystems including oil fields (*Thermoanaerobacter* spp.), terrestrial hydrothermal vents (*Carboxydothermus* spp.), geothermally heated subsurface waters and sediments (*Thermoanaerobacter* spp.), as well as microbial fuel cells operated at 55°C (*Thermincola* spp.). Within the thermophilic *Firmicutes*, the genus *Thermoanaerobacter* constitutes one of the most ubiquitously cultivated thermophilic taxa, with isolates enriched from a variety of thermophilic habitats (Table 1).

While the cell number of iron-reducing bacteria in thermophilic environments varies from 100 cells·ml^{-1} in high-temperature oil fields to 10^7 cells·ml^{-1} in terrestrial hot springs (Slobodkin et al., 1999; Slobodkin, 2005), the amounts of iron reduced or rates of this process in natural thermal ecosystems are unknown. Furthermore, until recently, there have been few data on the community composition and structure from natural thermophilic environments, so for many systems it is not known whether the organisms in pure culture are environmentally relevant. Interestingly, the *Firmicutes* constitute a large portion (48%) of known obligatory dissimilatory thermophilic iron reducers (Gavrilov et al., 2007), while gram-positive bacteria do not represent a dominant portion of isolated mesophilic iron reducers. Moreover, psychrophilic and hyperthermophilic gram-positive DIRB have yet to be isolated. The reasons for this gram-positive bias in thermophilic environments are unknown.

In addition to the DIRB adapted to elevated temperature, iron-reducing isolates from both the phyla *Actinobacteria* and *Firmicutes* have been isolated from extreme pH environments. Acidophilic iron reduction was demonstrated by aerobic iron-oxidizing bacteria when grown under anoxic conditions, including the thermophilic *Alicyclobacillus* sp. strain Y004 and mesophilic *Ferromicrobium acidophilus* and *Ferrithrix thermotolerans* (Table 1). Moreover, a broad phylogenetic diversity of alkaliphilic DIRB firmicutes have been described in pure culture. Under mesophilic conditions isolates have been isolated from natural haloalkaline environments like Mono Lake (*Anaerobranca* spp., *Bacillus arseniciselenatis*) and Soap Lake (*Bacillus* sp. strain SFB) as well as anthropogenic industrial borate leachate ponds (*Alkaliphilus* spp.). Identification of these organisms not only expanded the upper pH range for iron reduction to 11.0, but also implicated the contribution of microbially mediated Fe(III) reduction as a component of the iron redox cycling within alkaline environments (Pollock et al., 2007).

MICROBIAL FUEL CELL RESEARCH EXPANDS KNOWLEDGE OF GRAM-POSITIVE DIRB

Similar to iron reduction at neutral pH, current generation in a microbial fuel cell (MFC) depends on the capacity of bacteria to transfer electrons to a solid-phase external electron acceptor known as the anode. In the anoxic anodic chamber, respiratory bacteria oxidize electron donors to produce protons and electrons, with the electrons being transferred from the bacterial electron transport chain to the external anode electrode. The electrons produced in an MFC flow from the anode through an external electrical circuit to the cathode to generate electrical current. Given that many insoluble Fe(III) minerals and anodes share similar physical and redox properties, studies investigating current production in MFCs have also considerably advanced the phylogeny and physiology of respiratory gram-positive iron reducing bacteria that can utilize solid-phase electron acceptors.

Despite evidence from 16S rRNA gene-based molecular surveys which indicate that gram-positive bacteria are not only present but often dominant members (>50%) of electricity-generating communities (Aelterman et al., 2006; Rismani-Yazdi et al., 2007; Mathis et al., 2008; Pham et al., 2008; Rabaey et al., 2007; Wrighton et al., 2008), physiological and current generation research using gram-positive pure cultures is fairly limited. The first demonstration of electron transfer, or current, to an anode by a gram-positive bacterium was conducted by use of strain EG3 (Table 1), an organism closely related to *Clostridium butyricum*. With glucose as an electron donor, this organism was capable of electron transfer to soluble ferric iron (ferric pyrophosphate, ferric citrate), insoluble iron (HFO), and anodes. However, the amount of electrons in glucose recovered as current was less than 20%, indicating that this organism was using the electrode not for respiration but as a sink for excess reducing equivalents produced during fermentation (Park, 2001). Analogous to strain EG3, *Lactococcus lactis* is another gram-positive bacterium capable of coupling glucose fermentation to the reduction of hexacyanoferrate, ferric citrate, cupric chlorate, and anodes (Freguia et al., 2009).

D. hafniense strain DCB2, a gram-positive spore-forming bacterium, generated current in an MFC via respiration using exogenous electron shuttles. In this study, MFCs inoculated with formate as an electron donor and either humic acid or anthraquinone 2,6-disulfonic acid (AQDS) could produce current, while controls lacking the exogenous shuttle failed to produce current. Media replacement experiments and scanning electron microscopy images confirmed colonization of the anode surface and indicated that the organism was conserving energy from electron transfer to the anode surface. Interestingly, five other

D. hafniense strains tested could all reduce AQDS, but could not generate electricity continuously in an MFC.

The inability to produce, but the ability to use soluble electron shuttles is not confined to *D. hafniense* strain DCB2. The gram-positive bacterium *Brevibacillus* sp. PTH1 was isolated along with eight *Pseudomonas* strains from an MFC amended with acetate. Strain PTH1 was not able to produce current independently in an MFC using acetate as an electron donor. Given that previous research demonstrated that *Pseudomonas* spp. can produce redox-active compounds like phenazine-1-carboxamide (PCN), the possibility was explored that *Pseudomonas* sp. strain CMR12a was facilitating electron transport by *Brevibacillus* sp. PTH1. When *Brevibacillus* sp. strain PTH1 was inoculated into MFCs using media from cell-free supernatant from *Pseudomonas* sp. strain CMR12a, a sharp increase in current was observed, yet *Brevibacillus* failed to produce current when inoculated with supernatant from a mutant strain of CMR12a that could not produce PCN. Together these findings demonstrate that, despite their inability to independently transfer electrons to solid-phase electron acceptors like anodes, *Brevibacillus* sp. strain PTH1 can utilize redox-active compounds produced by other members of the community, even distantly related gram-negative organisms.

The operation of MFCs amended with acetate at elevated temperatures (55°C) demonstrated electricity production by anode communities dominated by respiratory gram-positive species and resulted in the isolation of three of the five most dominant populations identified in a 16S rRNA gene clone library (Wrighton et al., 2008). One of the isolates, *Geobacillus* strain S2E, a member of the *Firmicutes*, was the first reported member of this genus capable of respiration with extracellular electron acceptors like ferric oxides and anodes. However, current production by this bacterium was low (0.03 mA) and required the addition of AQDS as an electron shuttle. The enrichment of *Geobacillus* sequences in current-producing biofilms relative to no-current controls and the initial inoculum, combined with the fact that strain S2E failed to produce current independently, suggests that synergistic electron transfer activities are occurring within the biofilm. It is possible that *Geobacillus*, like *Brevibacillus* sp. strain PHT1, may exploit extracellular redox shuttles produced by electrochemically active biofilm members.

Like *Geobacillus* sp. strain S2E, *Thermincola potens* strain JR was also isolated from thermophilic MFC anode biofilms (Wrighton et al., 2008). Strain JR represented the first gram-positive bacterium to produce current (0.5 mA) without the addition of an electron shuttle as part of respiratory metabolism. Here, growth of strain JR was concomitant with 89% of the electrons in acetate recovered as current (Wrighton et al., 2008). Current production by strain JR was sustained at 90% of the original level over a 30°C difference in anode temperature, with a maximum between 60 and 70°C for a 1-h incubation and an optimum of 62°C for long-term operation. Scanning electron microscopy analysis of the anode surface after current production stabilized revealed heterogeneous colonization with half the fields containing areas with a monolayer of cells and other areas of the electrode sparsely covered.

Comparative sequence analysis using 16S rRNA gene sequence (1,228 nucleotides) indicated that *T. potens* strain JR belongs to the *Peptococcaceae* in the order *Clostridiales*. Strain JR is a member of the genus *Thermincola,* sharing 99% 16S rRNA gene sequence identity with the two previously characterized members, *Thermincola carboxydiphila* and *Thermincola ferriacetica,* both also thermophilic bacteria (Sokolova et al., 2005; Zavarzina et al., 2007). Akin to results from Wrighton et al. (2008), clone library sequences from current-producing anodes inoculated with marine sediment (Mathis et al., 2008) were dominated by sequences similar (93 to 99% identity) to strain *T. potens* JR. Moreover, it was also demonstrated that like *T. potens*, *T. ferriacetica* was capable of current production with use of acetate as an electron donor independent of an added mediator (Marshall and May, 2009). Beyond current production, sequences related (92%) to strain JR were identified using stable isotope probing and clone libraries from benzene-oxidizing and Fe(III)-reducing mesocosms at mesophilic temperatures (Kunapuli et al., 2007). Together these results demonstrate the selective enrichment of bacteria related to *Thermincola* in thermophilic environments with external electron acceptors, like anodes and iron hydroxide minerals, suggesting that *Thermincola* species may have a selective advantage in systems defined by respiration requiring external electron transfer.

MECHANISMS OF ELECTRON TRANSFER TO SOLID-PHASE ELECTRON ACCEPTORS

Unlike most soluble electron acceptors used by bacteria, Fe(III) exists primarily as insoluble oxyhydroxide minerals at circumneutral pH externally to the cell. As such, DIRB must transfer electrons from the electron transport chain on the cytoplasmic membrane to the electron acceptor on the cell surface. DIRB have evolved both mediated and direct strategies for such extracellular electron transfer. In a mediated electron transfer mechanism bacteria use either chelators or shuttles for respiration of external electron acceptors and thereby may alleviate the need for physical contact with the electron acceptor. Alternatively, in direct electron transfer bacteria require physical contact with the external electron acceptor, because electrons are transferred from the terminal respiratory enzyme via structural components consisting of multiheme c-type cytochromes (Shi et al., 2007; Weber et al., 2006) or conductive pili (Gorby et al., 2006; Reguera et al., 2005).

Researchers use a variety of experimental approaches to define extracellular electron transfer mechanisms in DIRB. One of the first reliable methods relied on physical exclusion of the bacterium from the electron acceptor using alginate or glass beads (Lies et al., 2005; Nevin and Lovley, 2000). The production of Fe(II) indicated that the bacterium was capable of transferring electrons to redox active shuttles that diffused through the beads to reduce the Fe(III) trapped inside. As an alternative method for detecting electron mediators, the ability of bacteria to excrete redox-active shuttles when grown on Fe(III) oxides was assessed by use of spent-medium assays (Bhowmick et al., 2009; Deneer et al., 1995; Junier et al., 2009; Nevin and Lovley, 2000, 2002). Here, the rate of Fe(III) oxide reduction was compared between fresh medium (negative control), medium spiked with a small amount of electron shuttle (positive control), and the supernatant from cells pregrown on iron oxide. When redox-active shuttles were secreted, the rates of Fe(III) reduction between spent medium and positive control were comparable.

In addition to Fe(III) oxide-based experiments, MFC anodes represent ideal research platforms for investigating bacterial mechanisms of extracellular electron transfer. Unlike studies using natural external electron acceptors such as Fe(III) or Mn(IV), MFC anodes do not participate in mineral dissolution reactions and electron transfer rates (current) can be quantified in real time. Researchers often assess the contribution of physical contact in extracellular electron transfer in MFCs by removing the spent medium surrounding a current-producing anode and replacing it with fresh amended medium while monitoring the time taken for current to return to original levels. In organisms that produce an electron shuttle, there is considerable lag in return of current production with delays from 3 to 10 days (Bond and Lovley, 2005; Marsili et al., 2008), while current returns to original levels in a manner of hours in organisms that use a direct mechanism of electron transfer because these organisms are bound onto the electrode in a tight biofilm (Bond and Lovley, 2003; Chaudhuri and Lovley, 2003). Cyclic voltammetry performed on anode biofilms and anode medium has also been used to identify redox-active components secreted into the medium and biofilm matrix in MFCs (Marshall and May, 2009; Marsili et al., 2008; Park et al., 1997; Rabaey et al., 2004; Xing et al., 2008).

Mediated Mechanisms

Mediated mechanisms of electron transfer include the production and/or use of chelators and electron shuttles. While the solubilization of Fe(III)-using ligands has been demonstrated to play an important role in iron assimilation by gram-positive bacteria (Brown and Holden, 2002), little is known about their role in gram-positive dissimilatory metal reduction. Circumstantial evidence for chelators was provided in current production studies with *Geobacillus* sp. strain S2E. This research revealed that, while the isolate could reduce insoluble Fe(III) oxides independently, it failed to utilize the anode as an electron acceptor unless the medium was amended with an exogenous electron shuttle (Wrighton et al., 2008). Reasons for the independent utilization of iron but not anodes were not explored, but could be explained by ligand production by strain S2E. It is plausible that strain S2E relies on endogenous ligands to solubilize solid-phase Fe(III) and subsequently utilize soluble Fe(III) as electron acceptor, but the bacterium lacks a mechanism for transferring electrons directly onto solid-phase electron acceptors that are immune to ligands such as a graphite anode surface.

Production of electron shuttles is not limited to gram-negative bacteria. *L. lactis* is a gram-positive bacterium that ferments glucose to lactose via glycolysis. *L. lactis* is also able to use an exogenous quinone 2-amino-3-dicarboxy-1,4-naphthoquinone to reduce hexacyanoferrate. Freguia et al. (2009) presented electrochemical evidence that *L. lactis* in the anodic chamber of an MFC can produce an endogenous quinone that shuttles electrons to the anode. Current production in the MFC increased sharply with the addition of glucose, and cyclic voltammetry revealed two redox-active compounds. One of the compounds had similar standard redox potential, absorbance spectrum, and retention time in a high-performance liquid chromatography-UV column to 2-amino-3-dicarboxy-1,4-naphthoquinone. This study provides evidence that gram-positive bacteria can produce electron shuttles to transfer electrons to an electrode, and also highlights the role that fermentative organisms can play in providing substrates (e.g., lactate, acetate) to other respiratory anodic bacteria. While there is no experimental evidence in gram-positive bacteria that redox shuttles are contained within anode or mineral biofilms, this finding has been observed in gram-negative bacteria and must be considered plausible (Lies et al., 2005; Marsili et al., 2008).

Interestingly, redox mediators have also recently been shown to promote reduction of insoluble metals by microbial spores. The sulfate reducer *Desulfotomaculum reducens* strain MI-1 is a gram-positive, spore-forming member of the *Firmicutes* that was originally isolated for its ability to reduce Cr(VI) in marine sediment. When strain MI-1 was tested for the ability to reduce U(VI) with pyruvate as the carbon and energy source, significant U(VI) reduction occurred after growth had ceased and protein concentration was decreasing. Microscopic observation during this later phase revealed the presence of spores, suggesting that spores were participating in U(VI) reduction. Sporulation was induced and it was demonstrated that MI-1 spores coupled uranyl(VI)-acetate or Fe(III)-citrate reduction to H_2 oxidation. In the case of uranium, this reduction was 100 times that of vegetative cells or pasteurized spores. Metal reduction required filtrate from spent fermentative medium, suggesting that an electron shuttle produced during vegetative growth facilitated U(VI) reduction by spores.

To understand the redox properties of the spent *D. reducens* strain MI-1 media, ultrafiltered spent medium was prereduced and the capacity of components in the spent medium to reduce U(VI) was tested. In sterile tubes amended with anoxic spent medium, U(VI) was reduced abiotically, but reduction was not sustained beyond 50 h. In the same experimental design, but with the addition of spores, U(VI) reduction lasted approximately 130 h and achieved three times higher levels of reduced uranium. Similarly, only partial reduction of Fe(III) was measured abiotically with the addition of prereduced spent medium filtrate, but full reduction was observed when spores were present. Transmission electron microscopy images showed dark precipitates surrounding the spores. Attempts to identify the spore-secreted shuttling compound were not successful; neither AQDS nor riboflavin in fresh media could rescue the uranium-reducing phenotype. However, based on its size (<3 kDa), it was speculated that the molecule is likely not a protein. This study characterized a novel mechanism of extracellular reduction where redox-active compounds secreted by vegetative cells of a gram-positive bacterium are subsequently utilized by their spores to reduce metals.

Direct Mechanism

One of the first studies of direct Fe(III) reduction and acquisition was conducted with *Listeria monocytogenes*, a gram-positive food-borne pathogen, where iron is required for growth during experimental infections. Dialysis membrane separation ruled out secreted mediators produced by *L. monocytogenes* and implicated cell surface ferric reductases in Fe(III) reduction by this pathogen (Coulanges et al., 1997; Deneer et al., 1995). In addition to iron assimilation by pathogens, contact-dependent mechanisms have been identified in two other genera of gram-positive dissimilatory metal reducing bacteria. In *Thermoanaerobacter* sp. strain BSB-33 the secretion of electron-shuttling compounds into the medium was evaluated by measuring the ability of spent culture filtrate and fresh medium to reduce amorphous Fe(III) oxyhydroxide. The observed rates of iron reduction were similar between fresh medium and medium amended with up to 100% filtrate. However, it is important to note that filtering as a method to remove cells from spent-medium experiments has recently been shown to result in loss of electron-shuttling compounds from the medium, because some mediators can be adsorbed to the filter (Lies et al., 2005; Marsili et al., 2008).

Thermincola spp. are the model gram-positive bacteria for direct mechanistic extracellular electron transfer research. Media-swap experiments and cyclic voltammetry demonstrated that *T. ferriacetica* employed a direct mechanism of electron transfer to MFC anodes when acetate was supplied as an electron donor (Marshall and May, 2009). Likewise, results from media-swap experiments, cyclic voltammetry, spent-medium experiments, and microscopic investigations confirmed that *T. potens* strain JR used a direct mechanism of electron transfer to Fe(III) oxides and anodes (K. C. Wrighton and J. D. Coates, unpublished data). The shared mechanism of electron transfer to anodes between *Thermincola* species is important, because differences in iron-reduction strategies within a single genus have been observed with both *Pyrobaculum* and *Shewanella* genera (Beliaev et al., 2001; Feinberg et al., 2008; Turick et al., 2003). Ultimately, these results prove that the physical structure of the gram-positive cell envelope does not preclude direct extracellular respiration.

BIOCHEMISTRY OF DIRECT EXTRACELLULAR ELECTRON TRANSFER

Gram-positive bacteria are capable of extracellular respiration by use of chelators and shuttles and through direct contact. However, little is known about the underlying molecular biology and biochemistry for direct electron transfer, because the molecular basis of Fe(III) reduction has only been studied in gram-negative DIRB to date, primarily *Shewanella oneidensis* and *Geobacter sulfurreducens*. It is reported that *S. oneidensis* can use either an electron shuttle, produce nanowires, or use *c*-type cytochrome chains connecting the inner membrane to the cell surface when transferring electrons to external electron acceptors, while *G. sulfurreducens* relies solely on direct electron transfer mechanisms to solid-phase acceptors with nanowires and chains of *c*-type cytochromes implicated in transmembrane electron transfer (Bretschger et al., 2007; Marsili et al., 2008; Reguera et al., 2005).

Given that gram-positive bacteria lack an outer membrane, have a thicker cell wall (20 to 80 nm), and may be encased in a glycoprotein S-layer, until recently, it was thought physically impossible for these organisms to participate in direct reduction of external electron transfer (Pham et al., 2008). However, recent physiological, electrochemical, and microscopic evidence demonstrates direct electron transfer to be compatible with a gram-positive envelope. In this section possible molecular explanations for how electrons are passed in a contact-dependent fashion from the inner membrane to the external electron acceptor on the cell surface via the cell wall are outlined. With use of gram-negative bacteria DIRB as a guide and accounting for structural differences in gram-positive envelopes, direct electron transfer to external electron acceptors by gram-positive DIRB can be rationalized using fully contained redox shuttle, nanowire, cytochrome chain, or conductive cell wall models (Color Plate 11).

There is little evidence for the first two models of direct electron transfer, yet both remain viable explanations for the evidence presented to date. It is possible that, similarly to gram-negative DIRB, some gram-positive bacteria may phenotypically appear to perform direct electron transfer in experimental assays, yet in reality these organisms are relying on redox mediators that are fully contained within the biofilm matrix (Color Plate 11a). Given the sensitivity of biofilm-based cyclic voltammetric methods in detecting redox-active components in gram-negative bacteria (Marsili et al., 2008), this option is plausible but unlikely. Similarly, there is no experimental evidence for conductive pili in gram-positive bacteria, yet this is a reasonable hypothesis given the circumstantial evidence for anode and Fe(III) reduction using these structures in gram-negative bacteria and the presence of homologous pili genes in gram-positive bacteria (Color Plate 11b).

Alternatively, a third model for direct electron transfer by conductive cell walls in gram-positive bacteria was proposed (Ehrlich, 2008). In this model, electrons from the inner membrane are transferred to a reductase located in the periplasmic space, which shuttles electrons to nonpeptide components of the cell wall, e.g., teichoic and teichuronic acids and associated metals. The electrons are conducted by these components across the cell wall to the cell surface (Color Plate 11c). Conceptually, this model fails to consider that gram-positive bacteria have extensive methods for covalently and noncovalently attaching proteins to the cell wall (Desvaux et al., 2006), and much of the circumstantial evidence for this model insufficiently distinguishes attached proteins from cell wall constituents. Furthermore, an initial structural examination of teichoic and teichuronic acids does not indicate features that could confer redox activity onto these components. However, ongoing studies evaluating the intrinsic redox capacity of native and proteinase-treated cell walls in gram-positive DIRB are warranted to confirm that cell wall structures independent of proteins are redox active.

In the fourth model, redox-active proteins, likely c-type cytochromes, facilitate electron transfer across the gram-positive cell envelope enabling direct extracellular electron transfer (Color Plate 11d). Several gram-negative DIRB have been shown to mediate direct extracellular electron transfer by multiheme c-type cytochromes (MHCs). These electrochemically active proteins are found as soluble electron carriers in the periplasm, as well as membrane-associated proteins in both the plasma and outer membranes. Model organisms such as *S. oneidensis* and *G. sulfurreducens* utilize several of these proteins to "complete the circuit" between the respiratory electron chain located in the plasma membrane and the insoluble electron acceptor located outside of the cell. The proteomes of these organisms reflect the importance of this class of proteins, because individual strains of *Shewanella* and *Geobacter* contain at least 20 genes predicted to encode MHCs, with some isolates containing upward of 70 genes. Only a few individual MHCs have been identified as being essential for extracellular electron transfer (Shi et al., 2007); thus, the purpose of such a large genetic reservoir of MHCs remains unexplained. However, a recent comparative genomics study of six *Geobacter* species suggests that this phenomenon is due to multiple recent duplication events, reflecting strong selective pressure for multiple genes encoding MHCs (Butler et al., 2010).

Evidence for *c*-type Cytochromes in Direct Extracellular Electron Transfer

It was proposed that gram-positive DIRB do not have an MHC-dependent mechanism of electron transfer because of their lack of both an outer membrane and a large periplasmic space (Ehrlich, 2008). However, recent physiological and genomic data suggest that MHCs may be an important component in gram-positive Fe(III) reduction.

Before the advent of genome sequences from gram-positive DIRB, spectral analysis confirmed the presence of *c*-type cytochromes in gram-positive bacteria and assessed the physiological capacity of these proteins for reducing insoluble Fe(III). Within the *Firmicutes*, *c*-type cytochrome(s) were identified in two ferrihydrite-reducing gram-positive bacteria, *Carboxydothermus ferrireducens* (Gavrilov et al., 2007) and *T. potens* strain JR (Byrne-Bailey et al., 2010). For both of these bacteria, reduced minus ferrihydrite-oxidized spectra of whole cells showed absorption peaks at 420, 526, and 552 nm, corresponding to the gamma, beta, and alpha bands of *c*-type cytochromes. The presence of *c*-type cytochrome(s) that could be oxidized by HFO implicates their functional role in the reduction of insoluble electron acceptors by this gram-positive bacterium. However, not all gram-positive DIRB contain detectable amounts of *c*-type cytochromes, because cytochrome spectra were reportedly not detected in whole cell extracts of *Desulfitobacterium metallireducens*, a mesophilic member of the *Firmicutes* capable of metal reduction (Finneran et al., 2002).

Recent genome sequencing of gram-positive DIRB has provided additional genetic evidence for *c*-type cytochromes (Table 1). In a recent cataloguing of MHCs across multiple gram-positive phyla, the Fe(III)-reducing *Firmicutes Carboxydothermus hydrogenoformans* and *D. hafniense* were the only gram-positive DIRB to contain *c*-type cytochromes, with 13 and 16, respectively (Sharma et al., 2010). In addition, the genome of *Thermincola potens* contains 36 MHCs, many of which are predicted to be localized to the inner membrane or cell wall (K. C. Wrighton and J. D. Coates, unpublished data). *Dethiobacter alkaliphilus* is an alkaliphilic gram-positive thiosulfate and polysulfide reducer that has 37 MHCs, and although it has not been shown to be a metal reducer, it may reduce polysulfide externally (Sorokin et al., 2008). Thus, the organism could also rely on a molecular wire of *c*-type cytochromes to transport electrons across the cell envelope (Table 2).

Of the four organisms that possess MHCs, only complete analysis has been performed in *T. potens* strain JR (Byrne-Bailey et al., 2010). Of the 36 MHCs, 74% contain multiple doubled heme domains, with one open reading frame (ThermJRDraft 1055) containing 56 heme-binding motifs. Moreover, 97% contain N-terminal signal peptides (SignalP 3.0) and an additional 37% contain a single N-terminal transmembrane domain (TMHMM 2.0), predicting that many cytochromes are embedded in or external to the gram-positive cell membrane. Initial analysis of these multiheme *c*-type cytochromes yielded gene models homologous to outer membrane *c*-type cytochromes identified as being involved in extracellular electron transfer in both *Geobacter* and *Shewanella* species. However, even more surprising was the large synteny (i.e., conserved nature of gene order on chromosomes of different species) between *Geobacter* and *Thermincola* genomes (Methe et al., 2003; K. C. Wrighton and J. D. Coates, unpublished data), as *c*-type cytochromes in the genomes occur in a nonrandom organization, with many of these MHCs localized in gene clusters encoding more than two multiheme *c*-type cytochrome proteins, cytochrome assembly proteins, proteins containing a NHL repeat domain, and proteins with at least one TPR repeat domain per cluster. The

TABLE 2 The genome gram-positive DIRB that have genome sequences and abundance of multiheme c-cytochromes[a]

Organism name	No. of MHCs	Obligatory dissimilation[b]	Solid-phase electron acceptors[b]	Mechanism
Alkaliphilus metalliredigens QMF	1	No	No	ND
Carboxydothermus hydrogenoformans Z-2901	16	Yes	HFO	ND
Clostridium beijerinckii	1	No	No	ND
Desulfitobacterium hafniense DCB-2	13	Yes	HFO	Exogenous shuttle[c]
Desulfotomaculum reducens MI-1	2	No	ND	Spore mediated
Dethiobacter alkaliphilus	36	Yes[d]	Polysulfide	ND
Thermincola potens JR	35	Yes	HFO and anodes	Direct
Thermoanaerobacter sp. strain X513	0	Yes	HFO	ND
Thermoanaerobacter sp. strain X514	0	Yes	HFO	ND
Thermoanaerobacter sp. strain X561	0	Yes	HFO	ND

[a]DIRB, dissimilatory Fe(III)-reducing bacteria; MHC, multiheme c-type cytochromes; AQDS, 2,6-anthraquinone disulfonic acid; MFC, microbial fuel cell; HFO, hydrous ferric oxide; ND, no data, tests not performed.
[b]No means organism was tested and found not capable of metabolism.
[c]Utilization of AQDS and humic substances as an electron shuttle but not capable of endogenous current production in MFC.
[d]This bacterium has not been demonstrated to reduce ferric iron compounds.

shared genomic organization between these two evolutionarily distinct organisms may denote a conserved functional role or selective pressure, and may warrant further comparative genomics between sequenced DMRB, resulting in hypotheses for the evolution of c-type cytochromes in metal-reducing bacteria.

If some gram-positive bacteria are indeed employing MHCs as a conduit for extracellular electron transfer, the process must be quite different than in the *Proteobacteria* because of the nature of the gram-positive cell wall architecture. Because of a much thicker peptidoglycan layer, it is unlikely that MHCs would freely diffuse between the cell membrane and the cell surface. To circumvent this, the cell wall could be "hard-wired" with cytochromes, because gram-positive bacteria have several methods for attaching proteins to the peptidoglycan layer both covalently and noncovalently (Desvaux et al., 2006). Beyond the cell wall, the S-layer is a protein matrix that can be functionalized with proteins that interact with the environment, such as the cellulosome in cellulolytic bacteria (Bayer et al., 2004). This layer could be studded with MHCs and proteins that mediate attachment and electron transfer to insoluble electron acceptors. In support of this model, several putative MHCs in both *T. potens* and *C. hydrogenoformans* were found to have the domain architectures characteristic of S-layer- and cell wall-associated proteins (R. A. Melnyk and J. D. Coates, unpublished data).

CONCLUDING REMARKS

Dissimilatory iron-reducing bacteria are located in both gram-positive radiations of the *Actinobacteria* and *Firmicutes*. Like gram-negative DIRB, these organisms can also utilize a wide range of alternative electron donors. While some hyperthermophilic DIRB have been demonstrated to show an obligate utilization of Fe(III) as an electron acceptor (Slobodkin, 2005), gram-positive DIRB are biochemically unconstrained with organisms capable of growth across a wide range of environmental conditions and capable of utilization of multiple terminal electron acceptors in addition to Fe(III) reduction. Commonly included in the physiological repertoire of gram-positive bacteria is the utilization of other metals (chromium, uranium, selenium, arsenic, and manganese), elemental sulfur, CO, and MFC anodes as terminal electron acceptors (Table 1).

Similarly to gram-negative counterparts, gram-positive DIRB are capable of extracellular respiration, with experimental evidence for the production of electron-shuttling compounds as well as the direct electron transfer mechanisms employed. While the molecular basis for this mechanism is presently not understood, physiological and genomic evidence suggests that c-type cytochromes may be integral to the physiology of some gram-positive DIRB. Future research is directed to understanding how c-type cytochromes are integrated into the physiology and ecology of gram-positive DIRB. To understand this process at a molecular level, continued mechanistic studies using phylogenetically distinct gram-positive DIRB are required, because it is quite possible that organisms have multiple methods for transferring electrons across the cell envelope. In addition, further characterization of obligatory DIRB is also required to expand the known phylogenetic, ecological, and physiological understanding of these organisms and their relationship to environmental processes. Together this information will improve understanding and modeling of gram-positive metabolism with applications to pathogen physiology, bioremediation, carbon cycling, and energy generation.

REFERENCES

Aelterman, P., K. Rabaey, H. T. Pham, N. Boon, and W. Verstraete. 2006. Continuous electricity generation at high voltages and currents using stacked microbial fuel cells. *Environ. Sci. Technol.* **40:**3388–3394.

Bayer, E. A., J. P. Belaich, Y. Shoham, and R. Lamed. 2004. The cellulosomes: multienzyme machines for degradation of plant cell wall polysaccharides. *Annu. Rev. Microbiol.* **58:**521–554.

Beliaev, A. S., D. A. Saffarini, J. L. McLaughlin, and D. Hunnicut. 2001. MTRC, an outer membrane decahaem c cytochrome required for metal reduction in *Shewanella putrefaciens* mr-1. *Mol. Microbiol.* **39:**722–730.

Bhowmick, D. C., B. Bal, N. S. Chatterjee, A. N. Ghosh, and S. Pal. 2009. A low-GC gram-positive thermoanaerobacter-like bacterium isolated from an Indian hot spring contains Cr(VI) reduction activity both in the membrane and cytoplasm. *J. Appl. Microbiol.* **106:**2006–2016.

Bonch-Osmolovskaya, E. A., M. L. Miroshnichenko, N. A. Chernykh, N. A. Kostrikina, E. V. Pikuta, and F. A. Rainey. 1997. Reduction of elemental sulfur by moderately thermophilic organotrophic bacteria and the description of Thermoanaerobacter sulfurophilus sp. nov. *Mikrobiologiya* **66:**483–489.

Bond, D. R., and D. R. Lovley. 2003. Electricity production by *Geobacter sulfurreducens* attached to electrodes. *Appl. Environ. Microbiol.* **69:**1548–1555.

Bond, D. R., and D. R. Lovley. 2005. Evidence for involvement of an electron shuttle in electricity generation by *Geothrix fermentans*. *Appl. Environ. Microbiol.* **71:**2186–2189.

Boone, D. R., Y. Liu, Z. J. Zhao, D. L. Balkwill, G. T. Drake, and T. O. Stevens. 1995. *Bacillus infernus* sp. Nov., an Fe(III)- and Mn(IV)-reducing anaerobe from the deep terrestrial subsurface. *Int. J. Syst. Bacteriol.* **45:**441–448.

Bretschger, O., A. Obraztsova, C. A. Sturm, I. S. Chang, Y. A. Gorby, S. B. Reed, D. E. Culley, C. L. Reardon, S. Barua, M. F. Romine, J. Zhou, A. S. Beliaev, R. Bouhenni, D. Saffarini, F. Mansfeld, B.-H. Kim, J. K. Fredrickson, and K. H. Nealson. 2007. Current production and metal oxide reduction by *Shewanella oneidensis* mr-1 wild type and mutants. **73:**7003–7012.

Bridge, T., and D. Johnson. 1998. Reduction of soluble iron and reductive dissolution of ferric iron-containing minerals by moderately thermophilic iron-oxidizing bacteria. *Appl. Environ. Microbiol.* **64:**2181–2186.

Brown, J. S., and D. W. Holden. 2002. Iron acquisition by gram-positive bacterial pathogens. *Microbes Infect.* **4:**1149–1156.

Butler, J. E., N. D. Young, and D. R. Lovley. 2010. Evolution of electron transfer out of the cell: comparative genomics of six *Geobacter* genomes. *BMC Genomics* **11:**40.

Byrne-Bailey, K. G., K. C. Wrighton, R. A. Melnyk, P. Agbo, T. C. Hazen, and J. D. Coates. 2010. Draft genome sequence of the electricity producing *Thermincola potens* strain JR. *J. Bacteriol.* **192:**4078–4079.

Castillo-Gonzalez, H. A., and M. A. Bruns. 2005. Dissimilatory iron reduction and odor indicator abatement by biofilm communities in swine manure microcosms. *Appl. Environ. Microbiol.* **71:**4972–4978.

Chaudhuri, S. K., and D. R. Lovley. 2003. Electricity generation by direct oxidation of glucose in mediatorless microbial fuel cells. *Nat. Biotechnol.* **21:**1229–1232.

Coulanges, V., P. Andre, O. Ziegler, L. Buchheit, and D. J. Vidon. 1997. Utilization of iron-catecholamine complexes involving ferric reductase activity in *Listeria monocytogenes*. *Infect. Immun.* **65:**2778–2785.

Deneer, H. G., V. Healey, and I. Boychuk. 1995. Reduction of exogenous ferric iron by a surface-associated ferric reductase of *Listeria* spp. *Microbiology* **141**(Pt. 8)**:**1985–1992.

Desvaux, M., E. Dumas, I. Chafsey, and M. Hebraud. 2006. Protein cell surface display in gram-positive bacteria: from single protein to macromolecular protein structure. *FEMS Microbiol. Lett.* **256:**1–15.

Dobbin, P. S., J. P. Carter, C. G. S. San Juan, M. von Hobe, A. K. Powell, and D. J. Richardson. 1999. Dissimilatory Fe(III) reduction by Clostridium beijerinckii isolated from freshwater sediment using Fe(III) maltol enrichment. *FEMS Microbiol. Lett.* **176:**131–138.

Dubiel, M., C. H. Hsu, C. C. Chien, F. Mansfeld, and D. K. Newman. 2002. Microbial iron respiration can protect steel from corrosion. *Appl. Environ. Microbiol.* **68:**1440–1445.

Ehrlich, H. L. 2008. Are gram-positive bacteria capable of electron transfer across their cell wall without an externally available electron shuttle? *Geobiology* **6:**220–224.

Engle, M., Y. Li, C Woese, and J. Wiegel. 1995. Isolation and characterization of a novel alkalitolerant thermophile, Anaerobranca horikoshii gen. nov., sp. nov. *Int. J. Syst. Bacteriol.* **45:**454–461.

Feinberg, L. F., R. Srikanth, R. W. Vachet, and J. F. Holden. 2008. Constraints on anaerobic respiration in the hyperthermophilic archaea Pyrobaculum islandicum and Pyrobaculum aerophilum. *Appl. Environ. Microbiol.* **74:**396–402.

Finneran, K., H. Forbush, C. Gaw VanPraagh, and D. Lovley. 2002. *Desulfitobacterium metallireducens* sp. Nov., an anaerobic bacterium that couples growth to the reduction of metals and humic acids as well as chlorinated compounds. *Int. J. Syst. Evol. Microbiol.* **52:**1929.

Freguia, S., M. Masuda, S. Tsujimura, and K. Kano. 2009. *Lactococcus lactis* catalyses electricity generation at microbial fuel cell anodes via excretion of a soluble quinone. *Bioelectrochemistry* **76:**14–18.

Gavrilov, S., A. Slobodkin, F. Robb, and S. de Vries. 2007. Characterization of membrane-bound Fe(III)-EDTA reductase activities of the thermophilic gram-positive dissimilatory iron-reducing bacterium *Thermoterrabacterium ferrireducens*. *Microbiology* **76:**139–146.

Gorby, Y. A., S. Yanina, J. S. McLean, K. M. Rosso, D. Moyles, A. Dohnalkova, T. J. Beveridge, I. S. Chang, B. H. Kim, K. S. Kim, D. E. Culley, S. B. Reed, M. F. Romine, D. A. Saffarini, E. A. Hill, L. Shi, D. A. Elias, D. W. Kennedy, G. Pinchuk, K. Watanabe, S. i. Ishii, B. Logan, K. H. Nealson, and J. K. Fredrickson. 2006. Electrically conductive bacterial nanowires produced by *Shewanella oneidensis* strain mr-1 and other microorganisms. *Proc. Natl. Acad. Sci. USA* **103:**11358–11363.

Gorlenko, V., A. Tsapin, Z. Namsaraev, T. Teal, T. Tourova, D. Engler, R. Mielke, and K. Nealson. 2004. *Anaerobranca californiensis* sp nov., an anaerobic, alkalithermophilic, fermentative bacterium isolated from a hot spring on Mono Lake. *Int. J. Syst. Evol. Microbiol.* **54:**739–743.

Johnson, D. B., P. Bacelar-Nicolau, N. Okibe, A. Thomas, and K. B. Hallberg. 2009. Ferrimicrobium acidiphilum gen. nov., sp. nov. and Ferrithrix thermotolerans gen. nov., sp. nov.: heterotrophic, iron-oxidizing, extremely acidophilic actinobacteria. *Int. J. Syst. Evol. Microbiol.* **59:**1082–1089.

Johnson, D. B., N. Okibe, and F. F. Roberto. 2003. Novel thermoacidophilic bacteria isolated from geothermal sites in Yellowstone National Park: physiological and phylogenetic characteristics. *Arch. Microbiol.* **180:**60–68.

Junier, P., M. Frutschi, N. S. Wigginton, E. J. Schofield, J. R. Bargar, and R. Bernier-Latmani. 2009. Metal reduction by spores of *Desulfotomaculum reducens*. *Environ. Microbiol.* **11:**3007–3017.

Kanso, S., A. Greene, and B. Patel. 2002. *Bacillus subterraneus* sp. nov., an iron-and manganese-reducing bacterium from a deep subsurface Australian thermal aquifer. *Int. J. Syst. Evol. Microbiol.* **52:**869.

Kostka, J. E., D. D. Dalton, H. Skelton, S. Dollhopf, and J. W. Stucki. 2002. Growth of iron(III)-reducing bacteria on clay minerals as the sole electron acceptor and comparison of growth yields on a variety of oxidized iron forms. *Appl. Environ. Microbiol.* **68:**6256–6262.

Kunapuli, U., T. Lueders, and R. U. Meckenstock. 2007. The use of stable isotope probing to identify key iron-reducing microorganisms involved in anaerobic benzene degradation. *ISME J.* **1:**643–653.

Lies, D. P., M. E. Hernandez, A. Kappler, R. E. Mielke, J. A. Gralnick, and D. K. Newman. 2005. *Shewanella oneidensis* mr-1 uses overlapping pathways for iron reduction at a distance and by direct contact under conditions relevant for biofilms. *Appl. Environ. Microbiol.* **71:**4414–4426.

Lin, B., C. Hycinthe, S. Bonneville, M. Braster, P. Van Cappellen, and W. F. M Rölling. 2007. Phylogenetic and physiological diversity of dissimilatory ferric iron reducers in sediments of the polluted Scheldt estuary, Northwest Europe. *Environ. Microbiol.* **9:**1956–1968.

Lloyd, J. R. 2003. Microbial reduction of metals and radionuclides. *FEMS Microbiol. Rev.* **27:**411–425.

Lovley, D. R. 1991. Dissimilatory Fe(III) and Mn(IV) reduction. *Microbiol. Rev.* **55:**259–287.

Lovley, D. R. 2006. Bug juice: harvesting electricity with microorganisms. *Nat. Rev. Microbiol.* **4:**497–508.

Lovley, D. R., D. E. Holmes, and K. Nevin. 2004. Dissimilatory Fe(III) and Mn(IV) reduction. *Adv. Microb. Physiol.* **49:**219–286.

Marshall, C., and H. May. 2009. Electrochemical evidence of direct electrode reduction by a thermophilic gram-positive bacterium, *Thermincola ferriacetica*. *Energy Environ. Sci.* **2:**699–705.

Marsili, E., D. B. Baron, I. D. Shikhare, D. Coursolle, J. A. Gralnick, and D. R. Bond. 2008. *Shewanella* secretes flavins that mediate extracellular electron transfer. *Proc. Natl. Acad. Sci. USA* **105:**3968–3973.

Mathis, B. J., C. W. Marshall, C. E. Milliken, R. S. Makkar, S. E. Creager, and H. D. May. 2008. Electricity generation by thermophilic microorganisms from marine sediment. *Appl. Microbiol. Biotechnol.* **78:**147–155.

Matias, V. R. F., and T. J. Beveridge. 2005. Cryo-electron microscopy reveals native polymeric cell wall structure in *Bacillus subtilis* 168 and the existence of a periplasmic space. *Mol. Microbiol.* **56:**240–251.

Matias, V. R. F., and T. J. Beveridge. 2006. Native cell wall organization shown by cryo-electron microscopy confirms the existence of a periplasmic space in *Staphylococcus aureus*. *J. Bacteriol.* **188:**1011–1021.

Methe, B., K. Nelson, J. Eisen, I. Paulsen, W. Nelson, J. Heidelberg, D. Wu, M. Wu, N. Ward, M. Beanan, R. Dodson, R. Madupu, L. Brinkac, S. Daugherty, R. DeBoy, A. Durkin, M. Gwinn, J. Kolonay, S. Sullivan, D. Haft, J. Selengut, T. Davidsen, N. Zafar, O. White, B. Tran, C. Romero, H. Forberger, J. Weidman, H. Khouri, T. Feldblyum, T. Utterback, S. Van Aken, D. Lovley, and C. Fraser. 2003. Genome of *Geobacter sulfurreducens*: metal reduction in subsurface environments. *Science* **302:**1967–1969.

Nealson, K. H., and D. Saffarini. 1994. Iron and manganese in anaerobic respiration: environmental significance, physiology, and regulation. *Annu. Rev. Microbiol.* **48:**311–343.

Nevin, K. P., and D. R. Lovley. 2000. Lack of production of electron-shuttling compounds or solubilization of Fe(III) during reduction of insoluble Fe(III) oxide by *Geobacter metallireducens*. *Appl. Environ. Microbiol.* **66:**2248–2251.

Nevin, K. P., and D. R. Lovley. 2002. Mechanisms for accessing insoluble Fe(III) oxide during dissimilatory Fe(III) reduction by *Geothrix fermentans*. *Appl. Environ. Microbiol.* **68:**2294–2299.

Niggemyer, A., S. Spring, E. Stackebrandt, and R. F. Rosenzweig. 2001. Isolation and characterization of a novel As(V)-reducing bacterium: implications for arsenic mobilization and the genus Desulfitobacterium. *Appl. Environ. Microbiol.* **67:**5568–5580.

Park, D., B. Kim, B. Moore, H. Hill, M. Song, and H. Rhee. 1997. Electrode reaction of *Desulfovibrio desulfuricans* modified with organic conductive compounds. *Biotechnol. Tech.* **11:**145–148.

Park, H. S. 2001. A novel electrochemically active and Fe(III)-reducing bacterium phylogenetically related to *Clostridium butyricum* isolated from a microbial fuel cell. *Anaerobe* **7:**297–306.

Petrie, L., N. N. North, S. L. Dollhopf, D. L. Blackwill, and J. E. Kostka. 2003. Enumeration and characterization of iron(III)-reducing microbial communities from acidic subsurface sediments contaminated with uranium(VI). *Appl. Environ. Microbiol.* **69:**7467–7479.

Pham, T. H., N. Boon, P. Aelterman, P. Clauwaert, L. De Schamphelaire, L. Vanhaecke, K. De Maeyer, M. Höfte, W. Verstraete, and K. Rabaey. 2008. Metabolites produced by *Pseudomonas* sp. enable a gram-positive bacterium to achieve extracellular electron transfer. *Appl. Microbiol. Biotechnol.* **77:**1119–1129.

Pollock, J., K. Weber, J. Lack, L. Achenbach, M. Mormile, and J. Coates. 2007. Alkaline iron(III) reduction by a novel alkaliphilic, halotolerant, Bacillus sp. isolated from salt flat sediments of Soap Lake. *Appl. Microbiol. Biotechnol.* **77:**927–934.

Prowe, S. G., and G. Antranikian. 2001. *Anaerobranca gottschalkii* sp. nov., a novel thermoalkaliphilic bacterium that grows anaerobically at high pH and temperature. *Int. J. Syst. Evol. Microbiol.* **51:**457–465.

Rabaey, K., N. Boon, S. D. Siciliano, M. Verhaege, and W. Verstraete. 2004. Biofuel cells select for microbial consortia that self-mediate electron transfer. *Appl. Environ. Microbiol.* **70:**5373–5382.

Rabaey, K., J. Rodríguez, L. L. Blackall, J. Keller, P. Gross, D. Batstone, W. Verstraete, and K. H. Nealson. 2007. Microbial ecology meets electrochemistry: electricity-driven and driving communities. *ISME J.* **1:**9–18.

Reguera, G., K. D. McCarthy, T. Mehta, J. S. Nicoll, M. T. Tuominen, and D. R. Lovley. 2005. Extracellular electron transfer via microbial nanowires. *Nature* **435:**1098–1101.

Rismani-Yazdi, H., A. Christy, B. Dehority, M. Morrison, Z. Yu, and O. Tuovinen. 2007. Electricity generation from cellulose by rumen microorganisms in microbial fuel cells. *Biotechnol. Bioeng.* **97:**1398–1407.

Robertson, W. J., P. D. Franzmann, and B. J. Mee. 2000. Spore-forming, Desulfosporosinus-like sulphate-reducing bacteria from a shallow aquifer contaminated with gasolene. *J. Appl. Microbiol.* **88:**248–259.

Roh, Y., S. V. Liu, G. Li, H. Huang, T. J. Phelps, and J. Zhou. 2002. Isolation and characterization of metal-reducing thermoanaerobacter strains from deep subsurface environments of the Piceance Basin, Colorado. *Appl. Environ. Microbiol.* **68:**6013–6020.

Sani, R., B. Peyton, W. Smith, W. Apel, and J. Petersen. 2002. Dissimilatory reduction of Cr(VI), Fe(III), and U(VI) by *Cellulomonas* isolates. *Appl. Microbiol. Biotechnol.* **60:**192–199.

Schröder, I., E. Johnson, and S. de Vries. 2003. Microbial ferric iron reductases. *FEMS Microbiol. Rev.* **27:**427–447.

Sharma, S., G. Cavallaro, and A. Rosato. 2010. A systematic investigation of multiheme c-type cytochromes in prokaryotes. *J. Biol. Inorg. Chem.* **15:**559–571.

Shelobolina, E. S., C. G. Vanpraagh, and D. R. Lovley. 2003. Use of ferric and ferrous iron containing minerals for respiration by Desulfitobacterium frappieri. *Geomicrobiol. J.* **20:**143–156.

Shi, L., T. C. Squier, J. M. Zachara, and J. K. Fredrickson. 2007. Respiration of metal (hydr)oxides by shewanella and geobacter: a key role for multihaem c-type cytochromes. *Mol. Microbiol.* **65:**12–20.

Slepova, T. V., T. G. Sokolova, T. V. Kolganova, T. P. Tourova, and E. A. Bonch-Osmolovskaya. 2009. Carboxydothermus siderophilus sp. nov., a thermophilic, hydrogenogenic, carboxydotrophic, dissimilatory Fe(III)-reducing bacterium from a Kamchatka hot spring. *Int. J. Syst. Evol. Microbiol.* **59:**213–217.

Slobodkin, A., C. Jeanthon, S. L'Haridon, T. Nazina, M. Miroshnichenko, and E. Bonch-Osmolovskaya. 1999. Dissimilatory reduction of Fe(III) by thermophilic bacteria and archaea in deep subsurface petroleum reservoirs of western Siberia. *Curr. Microbiol.* **39:**99–102.

Slobodkin, A. I. 2005. Thermophilic microbial metal reduction. *Mikrobiologiia* **74:**581–595.

Slobodkin, A. I., T. G. Sokolova, A. M. Lysenko, and J. Wiegel. 2006a. Reclassification of Thermoterrabacterium ferrireducens as Carboxydothermus ferrireducens comb. nov., and emended description of the genus Carboxydothermus. *Int. J. Syst. Evol. Microbiol.* **56:**2349–2351.

Slobodkin, A. I., T. P. Tourova, N. A. Kostrikina, A. M. Lysenko, K. E. German, E. A. Bonch-Osmolovskaya, and N.-K. Birkeland. 2006b. Tepidimicrobium ferriphilum gen. nov., sp. nov., a novel moderately thermophilic, Fe(III)-reducing bacterium of the order Clostridiales. *Int. J. Syst. Evol. Microbiol.* **56:**369–372.

Sokolova, T. G., N. A. Kostrikina, N. A. Chernyh, T. V. Kolganova, T. P. Tourova, and E. A. Bonch-Osmolovskaya. 2005. Thermincola carboxydiphila gen. Nov., sp. Nov., a novel anaerobic, carboxydotrophic, hydrogenogenic bacterium from a hot spring of the Lake Baikal area. *Int. J. Syst. Evol. Microbiol.* **55:**2069–2073.

Sorokin, D. Y., T. P. Tourova, M. Mussmann, and G. Muyzer. 2008. Dethiobacter alkaliphilus gen. nov. sp. nov., and Desulfurivibrio alkaliphilus gen. nov. sp. nov.: two novel representatives of reductive sulfur cycle from soda lakes. *Extremophiles* **12:**431–439.

Switzer Blum, J., A. Burns Bindi, J. Buzzelli, J. F. Stolz, and R. S. Oremland. 1998. Bacillus arsenicoselenatis, sp. nov., and Bacillus selenitireducens, sp. nov.: two haloalkaliphiles from Mono Lake, California that respire oxyanions of selenium and arsenic. *Arch. Microbiol.* **171:**19–30.

Tebo, B. M., and A. Y. Obraztsova. 1998. Sulfate-reducing bacterium grows with Cr(VI), U(VI), Mn(IV), and Fe(III) as electron acceptors. *FEMS Microbiol. Lett.* **162:**193–198.

Turick, C. E., F. Caccavo, and L. S. Tisa. 2003. Electron transfer from *Shewanella algae* BRY to hydrous ferric oxide is mediated by cell-associated melanin. *FEMS Microbiol. Lett.* **220:**99–104.

Viamajala, S., W. Smith, R. Sani, W. Apel, J. Petersen, A. Neal, F. Roberto, D. Newby, and B. Peyton. 2007. Isolation and characterization of Cr(VI) reducing *Cellulomonas* spp. from subsurface soils: implications for long-term chromate reduction. *Bioresource Technol.* **98:**612–622.

Villemur, R., M. Lanthier, R. Beaudet, and F. Lépine. 2006. The Desulfitobacterium genus. *FEMS Microbiol. Rev.* **30:**706–733.

Weber, K. A., L. A. Achenbach, and J. D. Coates. 2006. Microorganisms pumping iron: anaerobic microbial iron oxidation and reduction. *Nat. Rev. Microbiol.* **4:**752–764.

Wrighton, K. C., P. Agbo, F. Warnecke, K. A. Weber, E. L. Brodie, T. Z. Desantis, P. Hugenholtz, G. L. Andersen, and J. D. Coates. 2008. A novel ecological role of the firmicutes identified in thermophilic microbial fuel cells. *ISME J.* **2:**1146–1156.

Xing, D., Y. Zuo, S. Cheng, J. M. Regan, and B. E. Logan. 2008. Electricity generation by *Rhodopseudomonas palustris* DX-1. *Environ. Sci. Technol.* **42:**4146–4151.

Ye, Q., Y. Roh, S. L. Carroll, B. Blair, J. Zhou, C. L. Zhang, and M. W. Fields. 2004. Alkaline anaerobic respiration: isolation and characterization of a novel alkaliphilic and metal-reducing bacterium. *Appl. Environ. Microbiol.* **70:**5595–5602.

Zavarzina, D. G., T. G. Sokolova, T. P. Tourova, N. A. Chernyh, N. A. Kostrikina, and

E. A. Bonch-Osmolovskaya. 2007. *Thermincola ferriacetica* sp. Nov., a new anaerobic, thermophilic, facultatively chemolithoautotrophic bacterium capable of dissimilatory Fe(III) reduction. *Extremophiles* **11**:1–7.

Zavarzina, D. G., T. P. Tourova, B. B. Kuznetsov, E. A. Bonch-Osomolovskaya, and A. I. Slobodkin. 2002. Thermovenabulum ferriorganovorum gen. nov., sp. nov., a novel thermophilic, anaerobic, endospore-forming bacterium. *Int. J. Syst. Evol. Microbiol.* **52**:1737–1743.

Zuber, B., M. Haenni, T. Ribeiro, K. Minnig, F. Lopes, P. Moreillon, and J. Dubochet. 2006. Granular layer in the periplasmic space of gram-positive bacteria and fine structures of *Enterococcus gallinarum* and *Streptococcus gordonii* septa revealed by cryo-electron microscopy of vitreous sections. *J. Bacteriol.* **188**:6652–6660.

REGULATION OF ARSENIC METABOLIC PATHWAYS IN PROKARYOTES

Chad W. Saltikov

11

INTRODUCTION

Despite the toxicity of arsenic to most organisms, certain prokaryotes have the ability to grow on arsenic oxyanions by using them as alternative electron acceptors or electron donors. The common forms of arsenic used for these metabolic pathways are arsenate and arsenite. Because of its fairly positive midpoint potential of +130 mV, arsenate is a good electron acceptor and can therefore be respired and reduced to arsenite. Conversely, arsenite can be oxidized as an energy source and coupled to other types of respiration like oxygen and nitrate reduction. In some prokaryotes, arsenite oxidation does not appear to function in energy conservation reactions, but instead it is thought to occur for detoxification purposes. Considering that arsenate is less toxic than arsenite, this seems like a reasonable conclusion.

Arsenic-based metabolism has been investigated in several contexts: geochemical cycling of arsenic in aquatic environments (reviewed in Mukhopadhyay et al. [2002], Oremland and Stolz [2005], and Reyes et al. [2008]); exobiology and life in extreme environments (reviewed in Oremland et al. [2009]); and arsenic resistance and transformations important in biomedical research (reviewed in Xu et al. [1998], and Silver et al. [2002]). Geochemical investigations have become a major area of research because arsenic reduction has been implicated as an important mechanism that enhances arsenic mobility in groundwater environments (Reyes et al., 2008; Tufano et al., 2008). The end result of arsenic-based metabolism is that the organism couples oxidation/reduction of arsenite/arsenate to growth and/or arsenic resistance. The use of arsenic oxyanions as electron acceptors and donors presents a paradoxical question: why grow on compounds that are poisonous? Understanding how microorganisms metabolize arsenic may shed some light on the answer.

EARLY MICROBIOLOGY STUDIES

It has long been known that bacteria can resist high concentrations of arsenate and arsenite. This observation dates back to microbiology studies done in 1918 on arsenic-containing cattle-dipping tanks (Green, 1918). Several strains were isolated that exhibited tolerance to As(III) that were also found to oxidize As(III) to As(V). In this same environment, other bacteria were isolated that could reduce As(V) to As(III). In addition, As(III)-tolerant bacteria

---

Chad W. Saltikov, Department of Microbiology and Environmental Toxicology, University of California, Santa Cruz, Santa Cruz, CA 95064.

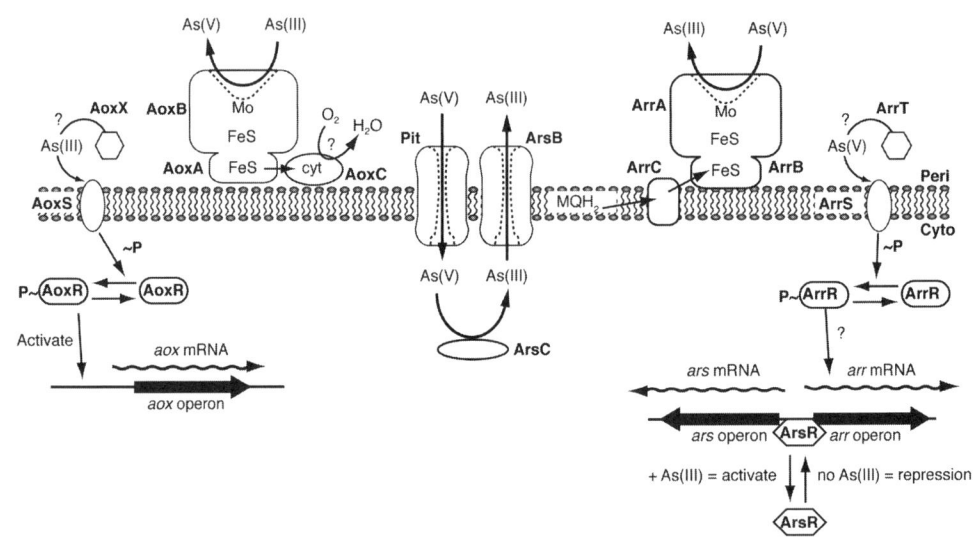

were isolated that neither reduced nor oxidized As compounds. Turner later reported on the isolation of 15 strains that could oxidize As(III) to As(V) (Turner, 1954). Similarly, these isolates originated from cattle-dip solutions and were identified as *Pseudomonadaceae* or *Achromobacter* (also named *Alcaligenes*). Other studies revealed that in one of their isolates a soluble enzyme affected the oxidation of As(III) to As(V) (Turner, 1949). These findings were the first biochemical reports of an arsenite oxidase, which would later be superseded by the purification of an arsenite oxidase enzyme in an As(III)-oxidizing strain of *Alcaligenes faecalis* (Anderson et al., 1992).

Both gram-negative and gram-positive bacteria have been found to contain plasmids that confer resistance to arsenicals (Hedges and Baumberg, 1973; Silver et al., 1981). *Escherichia coli* plasmids R773 and R46 (Mobley et al., 1984), *Staphylococcus* plasmids pI258 and pSX267 (Silver et al., 1981), and *Pseudomonas aeruginosa* plasmid pUM310 (Cervantes and Chavez, 1992) are among the many plasmids shown to confer resistance to As(V), As(III), and antimony [Sb(III)]. The gene locus conferring resistance was later found, indicating the existence of specific genetic systems governing resistance to arsenicals. Chromosomal and plasmid arsenic-specific resistance genes have also been found in mainly clinically significant bacteria (Diorio et al., 1995; Cai et al., 1998).

From 1970 to 1980 several avenues of research in phosphate metabolism in *E. coli* and studies with metal resistance plasmids would lead to the identification of the *ars* operon for arsenic resistance. Early phosphate transport studies in *E. coli* showed that the resistance plasmids also altered arsenate transport (Bennett and Malamy, 1970; Willsky and Malamy, 1974, 1980). In these studies arsenate was often used as a competitive inhibitor of phosphate uptake. Although arsenate was shown to be transported into the cell via phosphate transporters, arsenate uptake was decreased in *E. coli* strains expressing only the high-affinity inorganic phosphate transporter, *pst* (Willsky and Malamy, 1980). Moreover, *E. coli* strains harboring arsenic resistance plasmids accumulated significantly less arsenate compared with strains without an *ars*-containing plasmid (Chen et al., 1985). It was also thought that there were two separate regions on the *ars* operon that conferred separate resistances to arsenate and arsenite. It was speculated that ArsA and ArsB formed a complex that could pump out both arsenite and arsenate, and that ArsC modified the pump for arsenate efflux (Kaur and Rosen, 1992). Finally, it was shown that ArsC functioned as an arsenate reductase (Ji and Silver, 1992b). The model that emerged

FIGURE 1 Arsenic metabolic pathways. (A) Biochemical models for arsenite oxidation coupled to oxygen respiration; detoxifying reduction of arsenate by ArsC; the uptake of arsenate by the Pit, phosphate inorganic transporter; arsenite efflux by ArsB; and anaerobic respiration of arsenate by ArrAB. The regulatory pathways are also included for arsenite oxidation (AoxXSR), detoxifying arsenate reduction (ArsR), and arsenate respiratory reduction (ArsR and ArrTSR). Instead of ArrC, *Shewanella* sp. ANA-3 uses the membrane-bound tetraheme cytochrome CymA for transporting electrons from menaquinol to ArrAB. In other bacteria, there is genomic evidence that a transmembrane protein, ArrC, carries out this function. (B) Genomic regions of several bacteria that have *arr* gene clusters. The *arrSR* genes code for a two-component histidine kinase sensor and response regulators; *arrT* codes for a putative periplasmic phosphonate binding protein. The methyl-accepting chemotaxis-like protein gene of *Wolinella* codes for a putative chemotaxis protein. (C) Genomic regions of several bacteria that have *aox* gene clusters. Similar to *arrSR*, the *aoxSR* genes code for a two-component histidine kinase sensor and response regulators. Both ArrT and AoxX are predicted to bind either arsenite or arsenate in the periplasm and activate ArrS and AoxS phosphorylation, respectively. Activation of ArrS and AoxS would phosphorylate the cognate response regulators ArrR and AoxR to bind promoter regions of *arr* and *aox* operons, resulting in activation of transcription. The *Shewanella* species that have *arr* genes do not appear to have the *arrC* and *arrTSR* genes. The ArsR2 repressor mediates regulation of *arr* in *Shewanella* sp. ANA-3. *arr* transcription is activated in the presence of arsenite.

was that high-level arsenite resistance was due to an arsenite efflux pump (ArsB) aided by an ATPase subunit (ArsA) and that an intracellular arsenate reductase (ArsC) conferred resistance to arsenate (Fig. 1). Variations of this theme were also identified, such as ArsD, a protein that delivers arsenite to an ATPase ArsA (Lin et al., 2007), which helps accelerate arsenite efflux through ArsB. Together, the ArsDAB (pump complex) and ArsC (detoxifying arsenate reductase) comprise a highly efficient system for keeping arsenate and arsenite out of the cell. With the abundance of prokaryotic genome sequences, *ars* genes can be identified in almost every genome.

Some prokaryotes also contain arsenic transport genes that are similar to those identified in eukaryotes (Cai et al., 2009b). The arsenite-oxidizing strain *Pseudomonas* sp. TS44, which was isolated from highly arsenic contaminated soil, contains *ars*- and *acr3*-like genes in close proximity to an arsenite oxidation gene cluster (Cai et al., 2009b). The *acr3* gene was first identified in *Saccharomyces cerevisiae* and functions as an arsenite efflux pump, similarly to ArsB. Moreover, *acr3* is part of the *acr* (arsenic compound resistance) operon that also encodes an arsenic regulator (ACR1 or Yap8) and an arsenate reductase (ACR2) (Bobrowicz et al., 1997; Menezes et al., 2004). The arrangement of these arsenic metabolic genes in *Pseudomonas* sp. strain TS44 suggests that the organism is highly adapted to dealing with arsenic toxicity.

ARSENATE RESPIRATORY REDUCTION AND ARSENITE OXIDATION

Several reviews have been published on the subject of microbial arsenate respiration and oxidation (Mukhopadhyay et al., 2002; Silver et al., 2002; Oremland and Stolz, 2003; Stolz et al., 2006). A brief summary of the genes and enzymes for arsenate respiration and arsenite oxidation is presented here.

Arsenate respiratory reduction, also referred to as dissimilatory arsenate reduction, was first described in 1994 (Ahmann et al., 1994). Since then a variety of arsenate-respiring strains have been isolated and characterized; some have had their genomes sequenced. Biochemical (Krafft and Macy, 1998; Afkar et al., 2003; Malasarn et al., 2008) and genetic studies (Saltikov and Newman, 2003) have led to the identification of at least two key genes, *arrA* and *arrB*, that encode for the arsenate respiratory reductase. The *arrA* and *arrB* genes encode a large molybdenum-containing oxidoreductase and an Fe-S protein, respectively (Fig. 1A). Both ArrA and ArrB form a periplasmic reductase for arsenate. The ArrA protein sequences from a variety of bacteria are similar enough to form a distinct clade within the dimethyl sulfoxide (DMSO) reductase family of molybdenum-containing enzymes (Fig. 2).

The arsenite oxidizers span various branches of the proteobacteria (discussed in Oremland et al. [2009] and Stolz et al. [2010]). Most recently, anoxygenic photosynthetic arsenite oxidation has been described (Kulp et al., 2008). Unlike arsenate respiratory reduction, there has been some debate about the physiological role of arsenite oxidation: do microbes oxidize arsenite for energy-conserving purposes or for detoxification of arsenite? There is substantial evidence for both physiological outcomes of arsenite oxidation (Santini et al., 2000; Gihring and Banfield, 2001; Oremland et al., 2002; Donahoe-Christiansen et al., 2004; Cai et al., 2009b). There are also several naming conventions used to describe arsenite oxidase genes: *aro*, *aso*, and *aox*. Here, the *aox* convention is used to describe all of the arsenite oxidases. Like ArrA, a large molybdenum-containing oxidoreductase, AoxB, catalyzes arsenite oxidation. The smaller FeS subunit, AoxA, is also required for arsenite oxidation.

Both ArrA and AoxB enzymes have similar features such as their apparent size (~90 kDa), both have molybdopterin cofactors, and they interact with FeS subunits. Moreover, ArrA and AoxB are exported out of the cell and are either periplasmic or associated with the outer leaflet of the cytoplasmic membrane (Anderson et al., 1992; Macy et al., 1996; Malasarn et al., 2008). Interestingly, ArrA was shown to have bifunctional activity in vitro for arsenite oxidation (Richey et al., 2009). Last, the amino sequences of ArrA and AoxB are different enough such that they form two major clades on the DMSO reductase family of molybdenum oxidoreductases (Fig. 2). There appear to be some exceptions. For example, no *aoxB*-like genes can be identified in the genome sequence of the arsenite-oxidizing nitrate reducer *Alkalilimnicola ehrlichii* strain MLHE-1, a

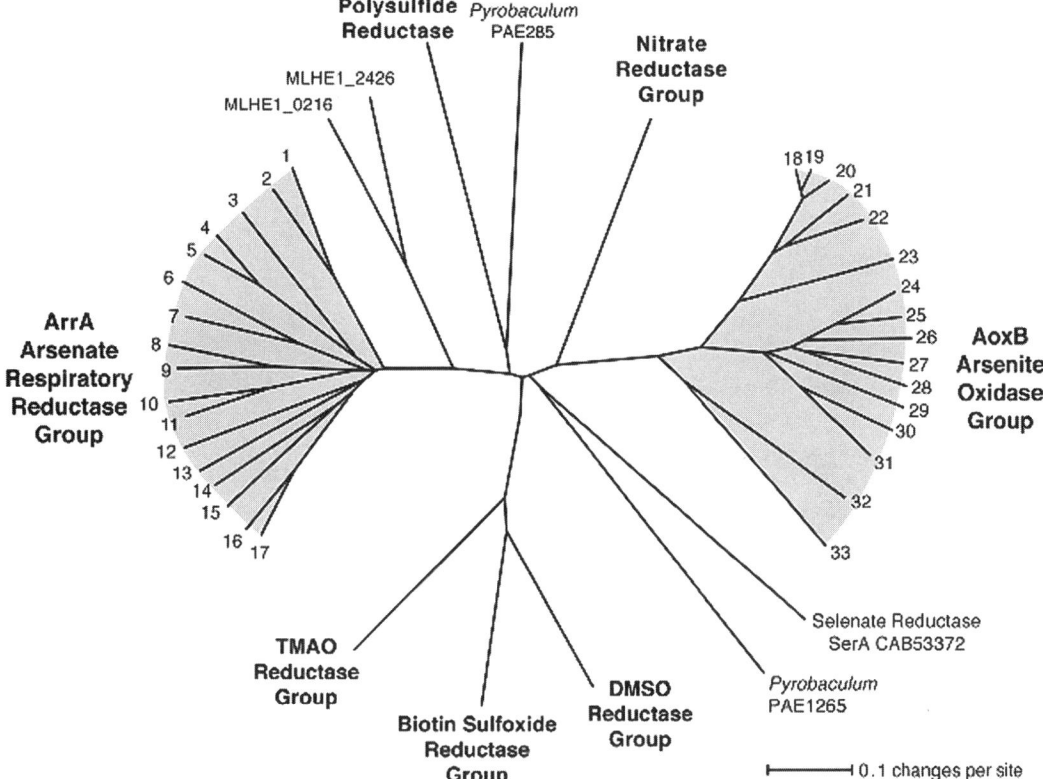

FIGURE 2 Phylogenetic analysis of arsenate respiratory reductases (ArrA), AoxB-type arsenite oxidases, and the ArxA-type arsenite oxidase of MLHE-1. The unrooted tree was constructed using a neighbor-joining method; gaps were ignored in the final phylogeny. The numbering refers to representative amino acid sequences ArrA and AoxB as described below. Arsenate respiratory reductase group (ArrA): 1, *Shewanella piezotolerans* (WP3, YP_002311519); 2, *Shewanella* sp. ANA-3 (★AAQ01672); 3, *Chrysiogenes* AAU11839 ★; 4, *Geobacter lovleyi* ZP_01593421; 5, *Geobacter uraniireducens* Rf4 ZP_01140714; 6, *Bacillus selenitireducens* str. MLS10 AAQ19491★; 7, *Bacillus arseniciselenatis* str. E1H AAU11841★; 8, *Sulfurospirillum barnesii* str. SES-3 AAU11840★; 9, *Wolinella succinogenes* NP_906980★; 10, *Desulfosporosinus* ABB02056★; 11, *Desulfitobacterium* ZP_01372404 ★; 12, MLMS-1 ZP_01288668 ★; 13, *Desulfonatronospira thiodismutans* ASO3-1 ZP_03737819; 14, *Natranaerobius thermophilus* YP_001916826; 15, *Halarsenatibacter silvermanii* SLAS-1 ACF74513 ★; 16, *Alkaliphilus metalliredigens* ZP_00800578; 17, *Alkaliphilus oremlandii* OhILAs ZP_01360543 ★. AoxB arsenite oxidase group: 18, NT26 AAR05656 ★★; 19, *Agrobacterium tumefaciens* ABB51928 ★★; 20, *Ochrobactrum tritici* ACK38267; 21, *Xanthobacter autotrophicus* Py2 ZP_01198801; 22, *Nitrobacter hamburgensis* YP_571843; 23, *Roseovarius* sp. 217 ZP_01034989; 24, *Alcaligenes* AAQ19838 ★★; 25, *Ralstonia* sp. 22 ACX69823; 26, *Herminiimonas arsenicoxydans* AAN05581 ★★; 27, *Burkholderia multivorans* ZP 0157266830; 28, *Rhodoferax ferrireducens* YP_524325; 29, *Thiomonas* sp. 3As CAM58792; 30, *Pseudomonas* sp. TS44 ACB05943; 31, *Halomonas* sp. HAL1 ACF77048; 32, *Chloroflexus aurantiacus* ZP_00356; 33, *Thermus thermophilus* YP_145366 ★★. The asterisks ★ and ★★ indicate that the organism is known to respire arsenate or to oxidize arsenite, respectively. TMAO, trimethylamine *N*-oxide.

haloalkaliphilic bacterium isolated from Mono Lake, CA (Oremland et al., 2002; Hoeft et al., 2007). Instead, there are two gene clusters that exhibit greater similarities to the arsenate respiratory reductases, *arrA* (Fig. 2). Richey et al. (2009) provided biochemical evidence that the gene *mlg_0216* codes for a new type of arsenite oxidase that was more like a reverse arsenate reductase. Genetic work by Saltikov and Zargar showed that *mlg_0216* is required for anaerobic arsenite oxidation coupled to nitrate reduction in MLHE-1 (Zargar et al., 2010). Gene disruption mutants were generated in *mlg_0216* and *mlg_2426*. Only the *mlg_0216* mutant strain did not grow chemoautotrophically with arsenite, whereas mlg_2426 strain grew similarly to the wild-type MLHE-1. It is unclear whether *mlg_2426* has any functional role in arsenic metabolism.

REGULATION OF ARSENIC-BASED METABOLIC PATHWAYS

The chemical differences between arsenate and arsenite manifest different biological responses (discussed in Cervantes et al. [1994] and Thomas et al. [2001]). Sensing arsenate poses a particular biochemical challenge because arsenate is chemically very similar to phosphate. The one major difference is that cells have specific reductases that can readily convert arsenate to arsenite. Because arsenite is considerably different than phosphate (and arsenate), it appears to be the form of arsenic that is preferentially sensed by arsenic-dependent regulators and arsenic efflux systems. By analogy to the phosphate transporters encoded by *pit* and *pst* genes, the existence of an arsenate efflux pump could be problematic because of the potential loss of phosphate from the cell. The only arsenate-specific proteins appear to be reductases such as ArsC and ArrA. Both *ars* and *arr* operons are induced by arsenite; although there is some evidence that *arr* is also induced by arsenate (Saltikov et al., 2005).

ars Regulation

The regulation of *ars* operon has been characterized in a variety of prokaryotes including the halophilic archaeon *Halobacterium* sp. strain NRC-1 (Wang et al., 2004). However, the best characterized *ars* operon is from the *E. coli ars* operon of the R773 plasmid (Chen et al., 1986). This *ars* operon is expressed as a 4,400-nucleotide polycistronic mRNA with a half-life of less than 4 min (reviewed in Kaur and Rosen [1992]). However, ArsA and ArsC are produced in greater abundance than ArsB, which is attributed to posttranscriptional regulation. The *ars* mRNA is initially synthesized with *arsRDABC* but then cleaved within the *arsB* gene into two mRNA molecules of 2,700 (*arsA*) and 500 nucleotides (*arsC*). The translation-initiation region of *arsB* gene is predicted to form a hairpin loop at the third codon. Moreover, the second codon encodes a rare leucine codon. Both of these molecular features could pause translation, resulting in an exposed region of mRNA with fewer ribosomes and, therefore, greater susceptibility to endonuclease activity. In addition, the two sub-mRNAs are more stable (10-min half-life), which would allow more time for translation of the *arsRDA* and *arsC* genes and differential accumulation of these gene products over ArsB. The physiological effect of differential expression of ArsB is a limiting factor on the level of arsenic resistance.

The first arsenic-sensing regulator, ArsR, was characterized in the *ars* operon of *Staphylococcus aureus* plasmid pI258 (Ji and Silver, 1992a) and *E. coli* R773 plasmid (Wu and Rosen, 1993). ArsR is an arsenite-dependent repressor that in the absence of arsenite binds to a specific operator region within the promoter of the *ars* operon (discussed in Busenlehner et al. [2003]). DNA binding of ArsR represses transcription of the *ars* operon. When ArsR binds arsenite, DNA binding is inhibited, leading to derepression of the *ars* operon and increased transcription of the *ars* genes. DNA binding activities of various ArsRs are mostly affected by the 3+ valence arsenic and antimony oxyanions, namely arsenite and antimonite. DNAse I and hydroxy radical footprinting studies revealed that the *E. coli* R773 ArsR protects the following sequence: 5'-TAA<u>TCATA</u>TGCGTT<u>TTT-GG</u>TTATGTGT-3'- with contact points to the DNA indicated in the underlined regions. With use of the chromosomal ArsR footprint and ArsR contact points (5'-ATTCGTTA-AG<u>TCATA</u>TATGTT<u>TTTG</u>ACTT-3'-), an *E. coli* ArsR consensus binding sequence (TCATNNNNNNNTTTG) was identified (Xu et al., 1996). DNAse I and hydroxy radical footprinting studies with the *Staphylococcus* ArsR revealed two protected regions within the promoter and more DNA contact points (5'-<u>AATC</u>TATATAGATGT<u>TA-AT</u>cTATT<u>AAC</u>TG-3'-; the c separates the two protected regions; underlined, ArsR-DNA contact points). This sequence does not occur in the *E. coli* ArsR consensus.

ArsR is part of the ArsR/SmtB family of winged-helix metalloregulatory proteins whose members are known to bind a variety of toxic metals (Busenlehner et al., 2003). These regulators share common features such as relatively small size, ~12 to 20 kDa, winged-helix protein domain, metal-binding motifs, and dimer formation. Several Cys residues usually mediate metal binding, which can either be located in the DNA binding domain or in the C-terminal dimer interface. The location of the Cys residues and specific motifs are often characteristic for the class of metals sensed by the particular regulator. For the *E. coli* ArsR, arsenite binds to three sulfur thiolates at conserved Cys residues in the motif Cys^{32}-Val-Cys^{34}-Asp-Leu-Cys^{37}. This motif occurs in the DNA binding domain. The binding of arsenite at this motif is thought to induce a structural changing in the protein that renders it unable to bind DNA. Although no structures exist for an ArsR, the Cd(II)/Pb(II)/Zn(II)-responsive regulator, CadC, has been used to model the ArsR from *Corynebacterium* (Cg) (Ordonez et al., 2008). Interestingly, the Cys residues of the CgArsR are located in different parts of the protein relative to *E. coli* ArsR. Mutational studies of CgArsR suggest that the regulator may bind arsenite through interthiolate sharing between a dimer pair.

Regulation of Arsenate Respiration

Little is known about how arsenate respiration genes are regulated in most of the known arsenate-respiring bacteria. Earlier biochemical work with the arsenate-respiring anaerobe *Chrysiogenes arsenatis* demonstrated that arsenate respiratory reductase activity (ARR) was regulated and required arsenate for induction (Krafft and Macy, 1998). Moreover, nitrate did not repress arsenate reductase activity. Selenium-grown cells of haloalkaliphiles like *Bacillus* strains E1H and MLS10 (Switzer Blum et al., 1998) could also reduce arsenate in cell suspensions. In strain MLS10 it also appears that ARR is regulated in response to arsenate. ARR is greatest in arsenate-grown cells and is also present in cells grown on other alternative electron acceptors, although to a lower activity level (Afkar et al., 2003).

In the facultative anaerobe *Shewanella* sp. strain ANA-3, *arrA* gene expression is inhibited by oxygen and nitrate and, surprisingly, induced by arsenite. The lowest amount of arsenite required to induce detectable *arrA* gene expression (compared with no-arsenic conditions) occurred at around 100 nM (~7.5 μg/liter). This concentration is within the maximum contaminant level for arsenic (10 μg/liter). For comparison, a detectable increase in expression of the detoxifying arsenate reductase, *arsC*, occurred at around 100 μM (7.5 mg/liter), which is nearly 1,000 times greater arsenite. It appears that the dynamics of *arrA* gene expression, at least in *Shewanella*, is well suited for responding to environmentally relevant arsenic concentrations.

The aerobic/anaerobic regulation of *arrA* of *Shewanella* follows similar patterns observed in *E. coli* for pathways of other alternative electron acceptors (e.g., fumarate, nitrate, and DMSO) (Tseng et al., 1996; Unden and Bongaerts, 1997). In addition to the well-known aerobic-anaerobic global regulators like Fnr and ArcA, components of the carbon catabolite repression system (cAMP, cAMP receptor protein, and adenylate cyclases) were also shown to be essential for the expression of *arr* genes and anaerobic growth (Murphy et al., 2009).

The induction of *arr* gene expression by arsenite was an intriguing observation and led to the hypothesis that an ArsR might serve as the regulator. The identification of a specific *arsR* gene in *Shewanella* was difficult because its genome has numerous genes annotated as *arsR*-like. Nevertheless, an ArsR-type regulator was eventually identified and shown to function as a repressor for the *arr* operon of *Shewanella* sp. strain ANA-3. Moreover, it appeared that this ArsR also coregulated the *ars* operon, and thus arsenic resistance (Murphy and Saltikov, 2009).

Arsenate-Dependent Regulation of ARR

Arsenate-dependent induction of arsenate respiration pathway is difficult to demonstrate conclusively because arsenite is produced once cells are provided arsenate. In most arsenate reducers, it is difficult to rule out the possibilities that arsenite could also be an inducer. In *Shewanella* sp. strain ANA-3, an arsenate reduction mutant was generated (strain lacking *arsC* and *arrA*) to address this question. By use of this arsenic reduction mutant, it was shown that arsenate could in fact induce *arr* gene expression; however, it was not as effective as arsenite. In *Shewanella,* the genetic basis for arsenate sensing is not clear because there are no regulators within the simple two-gene *arr* operon. The neighboring *ars* operon lacks an *arsR*-type arsenite regulator. However, there is an additional cluster of *ars* genes further upstream of *arr* that code for two different ArsR-like regulators (Fig. 1B). One of these was shown to play a significant role in regulation of arsenite-dependent *arr* gene expression (Murphy et al., 2009). Identification of the arsenate-dependent regulator in *Shewanella* remains elusive.

Observing the flanking regions of *arr* gene clusters in several of bacterial genomes reveals additional gene clusters that may code for regulators for the *arr* operon. Similarly to *Shewanella* sp. strain ANA-3, *ars* operons or *ars*-like genes are often located near some of the *arr* operons (Fig. 1B). Other *arr* gene clusters contain putative genes that encode for two-component regulatory systems (labeled as *arrTSR*). In some cases there are genes that code for both Ars proteins and two-component regulators. The latter regulators encode either for a sensory histidine kinase (ArrS), a phosphorylated response regulator (ArrR), a periplasmic ABC-type phosphonate binding periplasmic proteins (ArrT), and a methyl-accepting chemotaxis-like protein. Although speculative, ArrT may function to bind arsenate or arsenite in the periplasm. The activated ArrT could either (i) activate ArrS to phosphorylate ArrR leading to *arr* transcription; (ii) detoxify the periplasm of arsenate; or (iii) function as a periplasmic metallochaperone that delivers arsenate to ArrA. Further biochemical and genetic studies are needed to explore these hypotheses. Although it should be straightforward to determine whether the *arr* genes in other organisms are induced by arsenite, showing arsenate-dependent regulation of *arr* operons will be more challenging.

Regulation of Arsenite Oxidation

The *aox*-based arsenite oxidation pathway is also regulated in response to arsenite. However, the details are not well understood compared with the *ars* operon. Kashyap et al. (2006) were the first to identify the genetic basis for the regulation of arsenite oxidation using a soil isolate, *Agrobacterium tumefaciens* strain 5A. A transposon mutant was isolated that had a disruption in *aoxR*, predicted to encode a response regulator that is part of a two-component signal transduction system, *aoxSR*. These two genes were shown to be located next to the structural genes *aoxA* and *aoxB* that encode for the small and large subunits for the arsenite oxidase, respectively (Fig. 1C). Mutations of *aoxR* rendered the strain unable to oxidize arsenite. Mapping the *aox* mRNA by reverse transcriptase polymerase chain reaction (PCR) showed that the regulatory genes are most likely on the same mRNA as *aoxAB*. Gene expression studies showed that *aox* was induced with arsenite and interesting by unidentified quorum-sensing molecules, such as N-acylhomoserine lactones. The supernatants of non-arsenite-exposed late-log-phase-grown cultures could induce *aox* gene expression. It was hypothesized that AoxR activation might be affected by several sensory pathways such as one specific for arsenite and another involving autoinducers.

In contrast to ArsR, AoxR is predicated to positively regulate the *aox* operon in response to elevated arsenite. AoxR has several features that are typical of a response regulator (Stock et al., 2000). This includes conserved Asp residues that form an "acid pocket," a third Asp that could be phosphorylated, and a Lys that aids in changing conformation of the protein. However, AoxR most likely does not directly bind arsenite, but is part of a signal transduction cascade that originates from the periplasm (Fig 1A). Cai et al. (2009b) presented additional evidence for an arsenite-specific two-component regulatory system. In *Achromobacter* SY8, a soil isolate that can tolerate 13 mM arsenite, the *aox* operon is also likely regulated by an AoxSR system. However, a third gene was identified, *aoxX*. Like the putative ArrT of some *arr* operons, *aoxX* is predicted to encode a soluble periplasmic oxyanion binding protein. One hypothesis for its function is to bind arsenite in the periplasm and facilitate activation of AoxS. This would in turn modulate AoxR phosphorylation status, DNA binding activities, and *aox* transcription.

Whether mediated by AoxX or directly by arsenite interactions, AoxS is predicted to provide a signaling connection between the periplasmic space and the cytoplasm (Fig. 1A). AoxS has a putative signaling domain in its N terminus. A short exposed loop of amino acids (~14) presumably would interact with arsenite in the periplasm or provide a site for interacting with arsenite-bound AoxX. AoxS also has several domains typical of histidine kinases such as sites for phosphorylation and dimerization (reviewed in Stock et al. [2000] and Ashby [2004]). The identification of *aoxXSR* in isolates and in genome sequences expands the scope of proteins that can interact with arsenic.

PROTEOME AND TRANSCRIPTOME RESPONSES TO ARSENIC

In contrast to numerous single-pathway investigations, very few proteomic and transcriptomic approaches have been used for investigating global or systems level changes to the biology of arsenic-metabolizing prokaryotes. Only a few of these kinds of studies have been done with respect to arsenic-metabolizing bacteria, namely in *Herminiimonas arsenicoxydans*, a model arsenite-oxidizing bacterium isolated from an industrial wastewater treatment facility (Muller et al., 2006). This bacterium has an *aox* gene cluster similar to *Agrobacterium* (Fig. 1C) and appears to oxidize arsenite for detoxification purposes. In addition to *aox* genes, *H. arsenicoxydans* has four putative *ars* operons with varying combinations of *ars* and *acr* genes. Several proteomic and transcriptomic studies have been done with *H. arsenicoxydans* (Carapito et al., 2006; Muller et al., 2007). Arsenic exposure was shown to induce DNA recombination and repair pathways, motility and biofilm formation, and resistance (*ars*) pathways. Interestingly, arsenite oxidase was not detected in the proteome of arsenite-exposed cells. This was attributed to possible issues with detecting membrane proteins by two-dimensional polyacrylamide gel electrophoresis. In contrast, it was previously shown that an *aox* promoter transcription fusion to β-galactosidase gene was inducible by arsenite. Other interesting arsenite-induced proteomic changes were observed with respect to the high-affinity phosphate transport system. It appeared that arsenic exposure induced the expression of several *pst* genes, which signals a phosphate starvation-like response (Willsky and Malamy, 1980). Moreover, the *pst* genes are located in close proximity to one of the arsenite-induced *ars* clusters. These observations confirm the link between phosphate and arsenate metabolism as discussed earlier in the chapter.

The *Pyrobaculum* genus represents an interesting group of prokaryotes that are neutrophilic hyperthermophiles that can respire a variety of substrates including oxygen, arsenate, selenate, iron(III), thiosulfate, polysulfide, and nitrate. Although there are four genomes of *Pyrobaculum* species, nearly two-thirds of the genes in each genome remain hypothetical or cannot be reliably annotated. This is particularly troublesome for linking gene loci with a particular metabolic pathway or associating a particular hypothetical terminal oxidoreductase to a specific terminal electron acceptor. Moreover, the arsenate reductase in *Pyrobaculum aerophilum* has eluded detection with use of degenerate *arrA* primers (Malasarn et al., 2004). Cozen et al (2009) addressed this issue by employing transcriptomics as a tool to link gene expression patterns to particular metabolic pathways. In comparison with other electron acceptors, a gene cluster was identified that was strongly upregulated in arsenate-grown cells. This gene cluster (*pae1263-1265*) encodes for a predicted 135-kDa molybdenum oxidoreductase, a 23-kDa FeS, and 41-kDa membrane subunits. Of the nine predicted molybdenum-containing oxidoreductase homologs, PAE1265 appears to be more similar to tetrathionate reductases than to arsenate respiratory reductases. Interestingly, the closest ArrA homolog (PAE2859) was not induced with arsenate. A phylogenetic analysis of PAE1265 and PAE2859 showed that neither putative oxidoreductases cluster within the arsenate respiratory reductase group (Fig. 2). Further biochemical and/or genetic studies will be required to validate PAE1265 as an arsenate respiratory reductase.

ENVIRONMENTAL EXPRESSION OF ARSENIC METABOLIC GENES

A variety of microbial ecology studies have been carried out to investigate the abundance, distribution, and expression of *ars*, *arr*, and *aox* genes in a variety of environments. Before readily available genome sequences, the first use of PCR approaches to characterize arsenic metabolic genes in environmental isolates was done using specific primers designed against the *E. coli arsABC* genes (Saltikov and Olson, 2002). It was concluded that the *ars* genes in arsenite-resistant isolates from hot springs were too divergent from the *E. coli* model and that other *ars* reference genes were needed to develop better molecular detection tools. Since then, there have been a variety of studies on the diversity of *ars* and *acr3* genes in arsenic-resistant isolates and, to a lesser extent, in environmental samples (Macur et al., 2004; Achour et al., 2007; Chang et al., 2007, 2008; Cai et al., 2009a; Escalante et al., 2009; Kaur et al., 2009; Achour-Rokbani et al., 2010; Jareonmit et al., 2010). Most of these studies have not examined the metatranscriptome for the presence of these arsenic resistance genes. However, from gene regulation studies done with single organisms, it is reasonable to expect that *ars* and *acr3* genes would be induced in most organisms exposed to arsenic and thus detectable in environmental samples.

Environmental processes that are known to influence arsenic geochemistry have become a major focus of numerous studies. Studies by Malasarn et al. (2004) reported on the development of degenerate primers targeting the arsenate respiratory reductase, *arrA*. This work showed that the *arrA* gene could be used to detect diverse *arrA* sequences in both community DNA and cDNA from freshwater sediments with high levels of iron(III) and arsenic. Although the sequences were most similar to *arrA*, the environmental *arrA* sequences could not be associated with a known *arrA* from cultured microorganisms. Lear et al. (2007) applied stable isotope probing to determine the activities of arsenate respiratory reducers in Cambodian aquifer sediments. This approach revealed that the active *arrA* sequences were most similar to those in the *Sulfurospirillum* genus, which has cultured members that are known to respire arsenate, iron(III), and other alternative electron acceptors (Laverman et al., 1995).

As with many microbial ecology studies with functional genes, the primers are only as good as there are reference sequences for that particular function. Identifying new reference sequences increases the chances of detecting more diverse members of the particular functional gene. Although the Malasarn et al. (2004) PCR primers could detect *arrA* in a variety of bacterial strains, these primers were ineffective when used to detect *arrA* in hypersaline alkaline sediments of Mono Lake and Searles Lake, CA (Kulp et al., 2006). To overcome this problem, new ArrA alignments were generated and the degenerate primers were biased to have codons specific for the halophilic archaeon *Halobacterium*. This new archaeal-biased *arrA* primer set worked very well on detecting *arrA* in sediments collected from hypersaline alkaline environments. In fact, the partial *arrA* sequences did not match to any known cultured strains. The challenge now is to identify the genera associated with the archaeal-like *arrA* sequences. The application of stable isotope probing as demonstrated by Lear et al. (2007) would be useful to take on this challenge.

Similarly to *arrA*, more work has been done to characterize the microbial diversity of arsenite oxidase genes, however, mostly in geothermal environments (Inskeep et al., 2007; Clingenpeel et al., 2009; Hamamura et al., 2009) and in a few industrial sites (Quemeneur et al., 2008). There are a variety of arsenite-oxidizing bacteria in culture and many bacterial genome sequences that have *aox*-like genes (Fig. 2) (discussed in Oremland et al. [2009]). This has been useful for designing degenerate primers for detecting *aoxB* genes in various environmental samples and within newly cultured organism (Rhine et al., 2007). The first functional gene surveys for *aoxB* were done with a variety of arsenic-contaminated (natural and anthropogenic) environmental samples (Inskeep et al., 2007). These primers were used to retrieve *aoxB*-like sequences that represented a wide range of bacteria spanning α-, β-, and γ-proteobacteria. When applied to geothermal systems, these first-generation primers detect *aoxB* genes from dominant arsenite oxidizers such as *Hydrogenophaga* and *Thermus* species. Moreover, in situ *aoxB* expression was positively correlated with arsenite concentrations in certain geothermal waters.

Yellowstone geothermal systems have provided an interesting testing ground for overlaying in situ measurements of arsenite oxidation rates (Langner et al., 2001), microbial diversity, and *aoxB* functional gene analyses (Hamamura et al., 2009). Most of the studied hot springs are dominated by *Aquificales*. *Hydrogenobaculum* is one such *Aquificales* that occurs in many of the Yellowstone hot springs. Questions were raised whether this genus is the dominant arsenite oxidizer in these particular hot springs. Clingenpeel et al. (2009) addressed this question by tracking the expression of *Hydrogenobaculum aoxB* gene expression in one of the acidic geothermal springs of Yellowstone. For the first time, they were able to correlate in situ *aoxB* gene expression with the physicochemical gradients of arsenite.

Because of the diverse physicochemical nature of the geothermal habitats in Yellowstone, *aoxB* detection was not always successful despite the dominance of *Aquificales* (Inskeep et al., 2007; Hamamura et al., 2009). A second-generation *aoxB* primer set was constructed using the distant *aoxB* homolog of another Yellowstone *Aquificales*, *Sulfurihydrogenibium yellowstonense* (Hamamura et al., 2009). This new primer set was successfully used to detect *aoxB* in samples where none were previously detected. Moreover, these new *Aquificales*-like *aoxB* sequences were also present in cDNA reactions prepared from environmental RNA extracts from hot spring samples. Compared with the near-neutral low-sulfide-level sites dominated by *Thermocrinus aoxB* mRNA, the *aoxB* transcriptome shifted toward *Sulfurihydrogenibium* at sites with moderate levels of sulfide. At higher sulfide concentrations, the detection of *aoxB* with degenerate primers was unsuccessful. However, *Sulfurihydrogenibium*-specific *aoxB* could be detected in hot spring outflow channels using specific *aoxB* primers targeting that organism. These studies have shown that a variety of factors will influence the distribution of taxa and therefore the apparent effectiveness of a particular primer set. It is clear that having a detailed knowledge of the microbial diversity has been very useful in refining molecular detection methods for *aoxB*.

FUTURE PERSPECTIVES

Relative to the vast body of knowledge for *ars* arsenate-detoxifying reduction pathways, there is very little known about the regulation of arsenite oxidation and arsenate respiratory reduction pathways. Although background work has been done to establish the genetic basis for arsenic respiration and oxidation, this work has led to a variety of new questions regarding how biological systems recognize different forms of arsenic and distinguish arsenate from phosphate. Solving the latter dilemma will be insightful for understanding how cells can separate the toxic effects of arsenic versus the benefits of its metabolism (e.g., conserving energy through reduction or oxidation). Bacterial genome sequence analyses have also been extremely insightful for helping to identify new potential mechanisms for arsenic sensing and regulation. The identification of two-component histidine kinase sensor-response regulators in the *arr* and *aox* clusters open up possibilities of identifying novel arsenic binding proteins. Future research should focus on a mechanistic understanding of how these sensors sense arsenate and/or arsenite and the DNA binding activities of the response regulators.

Anoxygenic photoautotrophic arsenite oxidation is another exciting area of research recently identified (Kulp et al., 2008). The existence of a photosynthetic pathway for arsenite oxidation has important implications for the evolution of arsenic reduction and oxidation metabolic pathways. Preliminary evidence suggests that the oxidase for anoxygenic photosynthesis of arsenite oxidation is a DMSO reductase family enzyme that is more similar to the ArrA, arsenate respiratory reductase, than AoxB arsenite oxidases (Kulp et al., 2008). Future work should focus on uncovering the genetic and biochemical basis for this interesting metabolism.

Microbial ecology studies have resulted in an extensive database of *arr* and *aox* sequences from a variety of environments. The molecular tools are now available to carry out quantitative *aox* and *arr* gene expression studies in environmental samples. These approaches could be used, for example, to determine the occurrence of "hot spots" for biologically mediated arsenate reduction. This could help identify zones in a groundwater environment that are more prone to arsenic mobilization.

REFERENCES

Achour, A. R., P. Bauda, and P. Billard. 2007. Diversity of arsenite transporter genes from arsenic-resistant soil bacteria. *Res. Microbiol.* **158:**128–137.

Achour-Rokbani, A., A. Cordi, P. Poupin, P. Bauda, and P. Billard. 2010. Characterization of the *ars* gene cluster from extremely arsenic-resistant *Microbacterium* sp. strain A33. *Appl. Environ. Microbiol.* **76:**948–955.

Afkar, E., J. Lisak, C. Saltikov, P. Basu, R. S. Oremland, and J. F. Stolz. 2003. The respiratory arsenate reductase from *Bacillus selenitireducens* strain MLS10. *FEMS Microbiol. Lett.* **226:**107–112.

Ahmann, D., A. L. Roberts, L. R. Krumholz, and F. M. Morel. 1994. Microbe grows by reducing arsenic. *Nature* **371:**750.

Anderson, G. L., J. Williams, and R. Hille. 1992. The purification and characterization of arsenite oxidase from *Alcaligenes faecalis*, a molybdenum-containing hydroxlyase. *J. Biol. Chem.* **267:**23674–23682.

Ashby, M. K. 2004. Survey of the number of two-component response regulator genes in the complete and annotated genome sequences of prokaryotes. *FEMS Microbiol. Lett.* **231:**277–281.

Bennett, R. L., and M. H. Malamy. 1970. Arsenate resistant mutants of *Escherichia coli* and phosphate transport. *Biochem. Biophys. Res. Commun.* **40:**496.

Bobrowicz, P., R. Wysocki, G. Owsianik, A. Goffeau, and S. Ulaszewski. 1997. Isolation of three contiguous genes, ACR1, ACR2 and ACR3, involved in resistance to arsenic compounds in the yeast *Saccharomyces cerevisiae*. *Yeast* **13:**819–828.

Busenlehner, L. S., M. A. Pennella, and D. P. Giedroc. 2003. The SmtB/ArsR family of metalloregulatory transcriptional repressors: structural insights into prokaryotic metal resistance. *FEMS Microbiol. Rev.* **27:**131–143.

Cai, J., K. Salmon, and M. S. Dubow. 1998. A chromosomal *ars* operon homologue of *Pseudomonas aeruginosa* confers increased resistance to arsenic and antimony in *Escherichia coli*. *Microbiology* **144:**2705–2713.

Cai, L., G. H. Liu, C. Rensing, and G. J. Wang. 2009a. Genes involved in arsenic transformation and resistance associated with different levels of arsenic-contaminated soils. *BMC Microbiol.* **9:**4.

Cai, L., C. Rensing, X. Y. Li, and G. J. Wang. 2009b. Novel gene clusters involved in arsenite oxidation and resistance in two arsenite oxidizers: *Achromobacter* sp. SY8 and *Pseudomonas* sp. TS44. *Appl. Microbiol. Biotechnol.* **83:**715–725.

Carapito, C., D. Muller, E. Tarlin, S. Koechler, A. Danchin, A. Van Dorsselaer, E. Leize-Wagner, P. N. Bertin, and M. C. Lett. 2006. Identification of genes and proteins involved in the pleiotropic response to arsenic stress in *Caenibacter arsenoxydans*, a metalloresistant beta-proteobacterium with an unsequenced genome. *Biochimie* **88:**595–606.

Cervantes, C., and J. Chavez. 1992. Plasmid-determined resistance to arsenic and antimony in *Pseudomonas aeruginosa. Antonie Van Leeuwenhoek* **61:**333–337.

Cervantes, C., G. Ji, J. L. Ramairez, and S. Silver. 1994. Resistance to arsenic compounds in microorganisms. *FEMS Microbiol. Rev.* **15:**355–367.

Chang, J. S., J. H. Lee, and K. W. Kim. 2007. DNA sequence homology analysis of ars genes in arsenic-resistant bacteria. *Biotechnol. Bioprocess Eng.* **12:**380–389.

Chang, J. S., Y. H. Kim, and K. W. Kim. 2008. The *ars* genotype characterization of arsenic-resistant bacteria from arsenic-contaminated gold-silver mines in the Republic of Korea. *Appl. Microbiol. Biotechnol.* **80:**155–165.

Chen, C. M., H. L. T. Mobley, and B. P. Rosen. 1985. Separate resistances to arsenate and arsenite (antimonate) encoded by the arsenical resistance operon of R-Factor R773. *J. Bacteriol.* **161:**758–763.

Chen, C. M., T. K. Misra, S. Silver, and B. P. Rosen. 1986. Nucleotide sequence of the structural genes for an anion pump: the plasmid-encoded arsenical resistance operon. *J. Biol. Chem.* **261:**15030–15038.

Clingenpeel, S. R., S. D'Imperio, H. Oduro, G. K. Druschel, and T. R. McDermott. 2009. Cloning and in situ expression studies of the *Hydrogenobaculum* arsenite oxidase genes. *Appl. Environ. Microbiol.* **75:**3362–3365.

Cozen, A. E., M. T. Weirauch, K. S. Pollard, D. L. Bernick, J. M. Stuart, and T. M. Lowe. 2009. Transcriptional map of respiratory versatility in the hyperthermophilic Crenarchaeon *Pyrobaculum aerophilum. J. Bacteriol.* **191:**782–794.

Diorio, C., J. Cai, J. Marmor, R. Shinder, and M. S. DuBow. 1995. An *Escherichia coli* chromosomal ars operon homolog is functional in arsenic detoxification and is conserved in gram-negative bacteria. *J. Bacteriol.* **177:**2050–2056.

Donahoe-Christiansen, J., S. D'Imperio, C. R. Jackson, W. P. Inskeep, and T. R. McDermott. 2004. Arsenite-oxidizing *Hydrogenobaculum* strain isolated from an acid-sulfate-chloride geothermal spring in Yellowstone National Park. *Appl. Environ. Microbiol.* **70:**1865–1868.

Escalante, G., V. L. Campos, C. Valenzuela, J. Yanez, C. Zaror, and M. A. Mondaca. 2009. Arsenic-resistant bacteria isolated from an arsenic-contaminated river in the Atacama Desert (Chile). *Bull. Environ. Contam. Toxicol.* **83:**657–661.

Gihring, T. M., and J. F. Banfield. 2001. Arsenite oxidation and arsenate respiration by a new *Thermus* isolate. *FEMS Microbiol. Lett.* **204:**335–340.

Green, H. H. 1918. Description of a bacterium which oxidizes arsenite to arsenate and of one which reduces arsenate to arsenite isoated from a cattle-dipping tank. *S. Afr. J. Sci.* **14:**465–467.

Hamamura, N., R. E. Macur, S. Korf, G. Ackerman, W. P. Taylor, M. Kozubal, A. L. Reysenbach, and W. P. Inskeep. 2009. Linking microbial oxidation of arsenic with detection and phylogenetic analysis of arsenite oxidase genes in diverse geothermal environments. *Environ. Microbiol.* **11:**421–431.

Hedges, R. W., and S. Baumberg. 1973. Resistance to arsenic compounds conferred by a plasmid transmissible between strains of *Escherichia coli. J. Bacteriol.* **115:**459–460.

Hoeft, S. E., J. S. Blum, J. F. Stolz, F. R. Tabita, B. Witte, G. M. King, J. M. Santini, and R. S. Oremland. 2007. *Alkalilimnicola ehrlichii* sp. nov., a novel, arsenite-oxidizing haloalkaliphilic gammaproteobacterium capable of chemoautotrophic or heterotrophic growth with nitrate or oxygen as the electron acceptor. *Int. J. Syst. Evol. Microbiol.* **57:**504–512.

Inskeep, W. P., R. E. Macur, N. Hamamura, T. P. Warelow, S. A. Ward, and J. M. Santini. 2007. Detection, diversity and expression of aerobic bacterial arsenite oxidase genes. *Environ. Microbiol.* **9:**934–943.

Jareonmit, P., K. Sajjaphan, and M. J. Sadowsky. 2010. Structure and diversity of arsenic-resistant bacteria in an old tin mine area of Thailand. *J. Microbiol. Biotechnol.* **20:**169–178.

Ji, G. Y., and S. Silver. 1992a. Regulation and expression of the arsenic resistance operon from *Staphylococcus aureus* plasmid pI258. *J. Bacteriol.* **174:**3684–3694.

Ji, G. Y., and S. Silver. 1992b. Reduction of arsenate to arsenite by the ArsC protein of the arsenic resistance operon of *Staphylococcus aureus* plasmid pI258. *Proc. Natl. Acad. Sci. USA* **89:**9474–9478.

Kashyap, D. R., L. M. Botero, W. L. Franck, D. J. Hassett, and T. R. McDermott. 2006. Complex regulation of arsenite oxidation in *Agrobacterium tumefaciens. J. Bacteriol.* **188:**1081–1088.

Kaur, P., and B. P. Rosen. 1992. Plasmid-encoded resistance to arsenic and antimony. *Plasmid* **27:**29–40.

Kaur, S., M. R. Kamli, and A. Ali. 2009. Diversity of arsenate reductase genes (*arsC* genes) from arsenic-resistant environmental isolates of *E. coli*. *Curr. Microbiol.* **59:**288–294.

Krafft, T., and J. M. Macy. 1998. Purification and characterization of the respiratory arsenate reductase of *Chrysiogenes arsenatis*. *Eur. J. Biochem.* **255:**647–653.

Kulp, T. R., S. E., Hoeft, M. Asao, M. T. Madigan, J. T. Hollibaugh, J. C. Fisher, J. F. Stolz, C. W. Culbertson, L. G. Miller, and R. S. Oremland. 2008. Arsenic(III) fuels anoxygenic photosynthesis in hot spring biofilms from Mono Lake, California. *Science* **321:**967–970.

Kulp, T. R., S. E. Hoeft, L. G. Miller, C. Saltikov, J. N. Murphy, S. Han, B. Lanoil, and R. S. Oremland. 2006. Dissimilatory arsenate and sulfate reduction in sediments of two hypersaline, arsenic-rich soda lakes: Mono and Searles Lakes, California. *Appl. Environ. Microbiol.* **72:**6514–6526.

Langner, H. W., C. R. Jackson, T. R. McDermott, and W. P. Inskeep. 2001. Rapid oxidation of arsenite in a hot spring ecosystem, Yellowstone National Park. *Environ. Sci. Technol.* **35:**3302–3309.

Laverman, A. M., J. S., Blum, J. K. Schaefer, E. J. P. Phillips, D. R. Lovely, and R. S. Oremland. 1995. Growth of strain SES-3 with arsenate and other diverse electron acceptors. *Appl. Environ. Microbiol.* **61:**3556–3561.

Lear, G., B. Song, A. G. Gault, D. A. Polya, and J. R. Lloyd. 2007. Molecular analysis of arsenate-reducing bacteria within Cambodian sediments following amendment with acetate. *Appl. Environ. Microbiol.* **73:**1041–1048.

Lin, Y. F., J. Yang, and B. P. Rosen. 2007. ArsD: an As(III) metallochaperone for the ArsAB As(III)-translocating ATPase. *J. Bioenerg. Biomembr.* **39:**453–458.

Macur, R. E., C. R. Jackson, L. M. Botero, T. R. McDermott, and W. P. Inskeep. 2004. Bacterial populations associated with the oxidation and reduction of arsenic in an unsaturated soil. *Environ. Sci. Technol.* **38:**104–111.

Macy, J. M., K. Nunan, K. D. Hagen, D. R. Dixon, P. J. Harbour, M. Cahill, and L. I. Sly. 1996. *Chrysiogenes arsenatis* gen. nov., sp. nov., a new arsenate-respiring bacterium isolated from gold mine wastewater. *Int. J. Syst. Bacteriol.* **46:**1153–1157.

Malasarn, D., J. R. Keeffe, and D. K. Newman. 2008. Characterization of the arsenate respiratory reductase from *Shewanella* sp. strain ANA-3. *J. Bacteriol.* **190:**135–142.

Malasarn, D., C. W. Saltikov, K. M. Campbell, J. M. Santini, J. G. Hering, and D. K. Newman. 2004. *arrA* is a reliable marker for As(V) respiration. *Science* **306:**455.

Menezes, R. A., C. Amaral, A. S. Delaunay, M. Toledano, and C. Rodrigues-Pousada. 2004. Yap8p activation in *Saccharomyces cerevisiae* under arsenic conditions. *FEBS Lett.* **566:**141–146.

Mobley, H. L., S. Silver, F. D. Porter, and B. P. Rosen. 1984. Homology among arsenate resistance determinants of R factors in *Escherichia coli*. *Antimicrob. Agents Chemother.* **25:**157–161.

Mukhopadhyay, R., B. P. Rosen, L. Phung, and S. Silver. 2002. Microbial arsenic: from geocycles to genes and enzymes. *FEMS Microbiol. Rev.* **26:**311–311.

Muller, D., C. Medigue, S. Koechler, V. Barbe, M. Barakat, E. Talla, V. Bonnefoy, E. Krin, F. Arsène-Ploetze, C. Carapito, M. Chandler, B. Cournoyer, S. Cruveiller, C. Dossat, S. Duval, M. Heymann, E. Leize, A. Lieutaud, D. Lièvremont, Y. Makita, S. Mangenot, W. Nitschke, P. Ortet, N. Perdrial, B. Schoepp, P. Siguier, D. D. Simeonova, Z. Rouy, B. Segurens, E. Turlin, D. Vallenet, A. Van Dorsselaer, S. Weiss, J. Weissenbach, M. C. Lett, A. Danchin, and P. N. Bertin. 2007. A tale of two oxidation states: bacterial colonization of arsenic-rich environments. *PLoS Genet.* **3:**e53.

Muller, D., D. D. Simeonova, P. Riegel, S. Mangenot, S. Koechler, D. Lievremont, P. N. Bertin, and M. C. Lett. 2006. *Herminiimonas arsenicoxydans* sp nov., a metalloresistant bacterium. *Int. J. Syst. Evol. Microbiol.* **56:**1765–1769.

Murphy, J. N., K. J. Durbin, and C. W. Saltikov. 2009. Functional roles of *arcA*, *etrA*, cyclic AMP (cAMP)-cAMP receptor protein, and *cya* in the arsenate respiration pathway in *Shewanella* sp. strain ANA-3. *J. Bacteriol.* **191:**1035–1043.

Murphy, J. N., and C. W. Saltikov. 2009. The ArsR repressor mediates arsenite-dependent regulation of arsenate respiration and detoxification operons of *Shewanella* sp strain ANA-3. *J. Bacteriol.* **191:**6722–6731.

Ordonez, E., S. Thiyagarajan, J. D. Cook, T. L. Stemmler, J. A. Gil, L. M. Mateos, and B. P. Rosen. 2008. Evolution of metal(loid) binding sites in transcriptional regulators. *J. Biol. Chem.* **283:**25706–25714.

Oremland, R. S., S. E. Hoeft, J. M. Santini, N. Bano, R. A. Hollibaugh, and J. T. Hollibaugh. 2002. Anaerobic oxidation of arsenite in Mono Lake water and by a facultative, arsenite-oxidizing chemoautotroph, strain MLHE-1. *Appl. Environ. Microbiol.* **68:**4795–4802.

Oremland, R. S., C. W. Saltikov, F. Wolfe-Simon, and J. F. Stolz. 2009. Arsenic in the evolution of Earth and extraterrestrial ecosystems. *Geomicrobiol. J.* **26:**522–536.

Oremland, R. S., and J. F. Stolz. 2003. The ecology of arsenic. *Science* **300:**939–944.

Oremland, R. S., and J. F. Stolz. 2005. Arsenic, microbes and contaminated aquifers. *Trends Microbiol.* **13:**45–49.

Quemeneur, M., A. Heinrich-Salmeron, D. Muller, D. Livremont, M. Jauzein, P. N. Bertin, F. Garrido, and C. Joulian. 2008. Diversity surveys and evolutionary relationships of aoxB genes in aerobic arsenite-oxidizing bacteria. *Appl. Environ. Microbiol.* **74:**4567–4573.

Reyes, C., J. R. Lloyd, and C. W. Saltikov. 2008. Geomicrobiology of iron and arsenic in anoxic sediments, p. 123–146. *In* S. Ahuja (ed.), *Arsenic Contamination of Groundwater: Mechanisms, Analysis, and Remediation.* John Wiley and Sons, Inc., Hoboken, NJ.

Rhine, E. D., S. M. Ni Chadhain, G. J. Zylstra, and L. Y. Young. 2007. The arsenite oxidase genes (aroAB) in novel chemoautotrophic arsenite oxidizers. *Biochem. Biophys. Res. Commun.* **354:**662–667.

Richey, C., P. Chovanec, S. E. Hoeft, R. S. Oremland, P. Basu, and J. F. Stolz. 2009. Respiratory arsenate reductase as a bidirectional enzyme. *Biochem. Biophys. Res. Commun.* **382:**298–302.

Saltikov, C. W., and D. K. Newman. 2003. Genetic identification of a respiratory arsenate reductase. *Proc. Natl. Acad. Sci. USA* **100:**10983–10988.

Saltikov, C. W., and B. H. Olson. 2002. Homology of *Escherichia coli* R773 arsA, arsB, and arsC genes in arsenic-resistant bacteria isolated from raw sewage and arsenic-enriched creek waters. *Appl. Environ. Microbiol.* **68:**280–288.

Saltikov, C. W., R. A. Wildman, Jr., and D. K. Newman. 2005. Expression dynamics of arsenic respiration and detoxification in *Shewanella* sp. strain ANA-3. *J. Bacteriol.* **187:**7390–7396.

Santini, J. M., L. I. Sly, R. D. Schnagl, and J. M. Macy. 2000. A new chemolithoautotrophic arsenite-oxidizing bacterium isolated from a gold mine: phylogenetic, physiological, and preliminary biochemical studies. *Appl. Environ. Microbiol.* **66:**92–97.

Silver, S., K. Budd, K. M. Leahy, W. V. Shaw, D. Hammond, R. P. Novick, G. R. Willsky, M. H. Malamy, and H. Rosenberg. 1981. Inducible plasmid-determined resistance to arsenate, arsenite, and antimony (III) in *Escherichia coli* and *Staphylococcus aureus*. *J. Bacteriol.* **146:**983–996.

Silver, S., L. T. Phung, and B. P. Rosen. 2002. Arsenic metabolism: resistance, reduction, and oxidation, p. 254. *In* W.T. Frankenberger (ed.), *Environmental Chemistry of Arsenic.* Marcel Dekker, Inc., New York, NY.

Stock, A. M., V. L. Robinson, and P. N. Goudreau. 2000. Two-component signal transduction. *Annu. Rev. Biochem.* **69:**183–215.

Stolz, J. F., P. Basu, and R. S. Oremland. 2010. Microbial arsenic metabolism: new twists on an old poison. *Microbe* **5:**53–59.

Stolz, J. F., P. Basu, J. M. Santini, and R. S. Oremland. 2006. Arsenic and selenium in microbial metabolism. *Annu. Rev. Microbiol.* **60:**107–130.

Switzer Blum, J., A. Burns Bindi, J. Buzzelli, J. F. Stolz, and R. S. Oremland. 1998. *Bacillus arsenicoselenatis*, sp. nov., and *Bacillus selenitireducens*, sp. nov.: two haloalkaliphiles from Mono Lake, California that respire oxyanions of selenium and arsenic. *Arch. Microbiol.* **171:**19–30.

Thomas, D. J., M. Styblo, and S. Lin. 2001. The cellular metabolism and systemic toxicity of arsenic. *Toxicol. Appl. Pharmacol.* **176:**127–144.

Tseng, C. P., J. Albrecht, and R. P. Gunsalus. 1996. Effect of microaerophilic cell growth conditions on expression of the aerobic (cyoABCDE and cydAB) and anaerobic (narGHJI, frdABCD, and dmsABC) respiratory pathway genes in *Escherichia coli*. *J. Bacteriol.* **178:**1094–1098.

Tufano, K. J., C. Reyes, C. W. Saltikov, and S. Fendorf. 2008. Reductive processes controlling arsenic retention: revealing the relative importance of iron and arsenic reduction. *Environ. Sci. Technol.* **42:**8283–8289.

Turner, A. W. 1949. Bacterial oxidation of arsenite. *Nature* **164:**76–77.

Turner, A. W. 1954. Bacterial oxidation of arsenite. 1. Description of bacteria isolated from arsenical cattle-dipping fluids. *Aust. J. Biol. Sci.* **7:**452–478.

Unden, G., and J. Bongaerts. 1997. Alternative respiratory pathways of *Escherichia coli*: energetics and transcriptional regulation in response to electron acceptors. *Biochim. Biophys. Acta* **1320:**217–234.

Wang, G., S. P. Kennedy, S. Fasiludeen, C. Rensing, and S. DasSarma. 2004. Arsenic resistance in *Halobacterium* sp. strain NRC-1 examined by using an improved gene knockout system. *J. Bacteriol.* **186:**3187–3194.

Willsky, G., and M. Malamy. 1980. Effect of arsenate on inorganic phosphate transport in *Escherichia coli*. *J. Bacteriol.* **144:**366–374.

Willsky, G. R., and M. H. Malamy. 1974. Loss of PhoS periplasmic protein leads to a change in specificity of a constitutive inorganic-phosphate transport-system in *Escherichia coli*. *Biochem. Biophys. Res. Commun.* **60:**226–233.

Wu, J., and B. P. Rosen. 1993. Metalloregulated expression of the *ars* operon. *J. Biol. Chem.* **268:**52–58.

Xu, C., W. Shi, and B. P. Rosen. 1996. The chromosomal *arsR* gene of *Escherichia coli* encodes a trans-acting metalloregulatory protein. *J. Biol. Chem.* **271:**2427–2432.

Xu, C., T. Zhou, M. Kuroda, and B. P. Rosen. 1998. Metalloid resistance mechanisms in prokaryotes. *J. Biochem. (Tokyo)* **123:**16–23.

Zargar, K., S. Hoeft, R. S. Oremland, and C. W. Saltikov. 2010. Identification of a novel arsenite oxidase gene, *arxA*, in the haloalkaliphilic, arsenite-oxidizing bacterium *Alkalilimnicola ehrlichii* strain MLHE-1. *J. Bacteriol.* **192:**3755–3762.

NEW TECHNOLOGIES

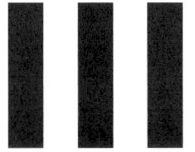

TRANSCRIPTOME ANALYSIS OF METAL-REDUCING BACTERIA

Dwayne A. Elias and Matthew W. Fields

12

INTRODUCTION

The major transition metals for biological cells include V, Mn, Ni, Mo, Fe, Cu, W, Co, and Zn (Wackett et al., 2004). Metal-reducing bacteria must interact with metals at concentrations that are conducive for utilization as energy sources in addition to the concentrations that most biological cells require for micronutrients. Namely, metal-reducing bacteria must tolerate high levels of metals that are known to be toxic to other cells, and in some cases the concentrations that are utilized would induce oxidative damage and disease in humans and other eukaryotic organisms (Jin et al., 2008). Iron is highly abundant in the soils and sediments of Earth, and microbial iron reduction was most likely an early form of anaerobic respiration on the planet (Vargas et al., 1998). The cellular systems used by microorganisms to reduce iron can also interact with other heavy metals; therefore, the microbial reduction of iron and other heavy metals greatly impacts biogeochemical cycles and bioremediation (Nealson et al., 2002; Shi et al., 2007). It is important to keep in mind that microorganisms that can directly and/or indirectly reduce metals are typically able to tolerate elevated levels, but that growth with the metal at such concentrations is still elucidating a stress response. This review will compare available data for more commonly studied bacteria that can transfer electrons to different metals in the context of carbon flow, electron flow, oxidative stress, and physiological states of the cell. A goal of the comparison is to identify similarities and differences in molecular mechanisms used by different metal-reducing bacteria to interact with metals and thus optimize their metabolism accordingly in order to tolerate "stressful" substrates. It should be noted that a systems biology view is taken with respect to cellular responses that could impact overall physiology and function (i.e., metal reduction).

Several heavy metals and metal interactions are essential for all domains of life, and many proteins function with essential metal cofactors. In addition, metal ions are crucial for nucleic acid function and maintenance, as well as for overall cell function and physiology. Iron is the most abundant transition metal ion in biological cells, and a large portion of the iron proteome (full complement of proteins that require Fe) was likely inherited from a last

---

Dwayne A. Elias, Biosciences Division, Oak Ridge National Laboratory, Oak Ridge, TN 37831-6036. *Matthew W. Fields,* Department of Microbiology, Center for Biofilm Engineering, Montana State University, Bozeman, MT 59717.

common ancestor in the course of evolution. A recent study demonstrated that the iron proteome constituted a higher fraction of the proteome in archaea (7.1 ± 2.1%) compared with bacteria (3.9 ± 1.6%) and eukaryota (1.1 ± 0.4%) (Andreini et al., 2007). Not surprisingly, large portions of iron proteins (proteins that contain Fe) were predicted to be involved in electron transfer and reduction-oxidation reactions. The study did not focus on metal-reducing bacteria, but among the metal-reducing bacteria included in the study, the average of iron proteins encoded in the genomes was 6.0 ± 1.2%. This comparison suggests that metal-respiring bacteria (e.g., *Shewanella*, *Geobacter*, *Desulfovibrio*) contain an enrichment of iron-containing proteins that are involved in metal interactions beyond the typical needs of other bacterial cells.

Most system-level studies in the literature involve metal-respiring bacteria that can reduce Fe, Cr, Co, Sr, and U; however, not all metal-reducing bacteria can utilize the respective metal as a terminal electron acceptor with concomitant growth. Therefore, a distinction must be made between metabolic and cometabolic processes, although it is more difficult to discern the possible generation of energy for maintenance. In addition, electron transfer to particular heavy metals may be the result of a detoxification mechanism instead of energy generation for growth, much in the same way that oxygen can be reduced by some "anaerobes" without growth. Heavy metal resistance mechanisms are diverse and microorganisms can display a combination of direct and indirect means that include biosorption, decreased influx, increased efflux, precipitation, or reduction via electron transfer. Many of the transcriptomic studies reviewed herein have identified genes involved in these mechanisms.

GENOMICS

Since the first published report of a sequenced genome in 1995, there has been an obvious explosion in available genomic data and exploitation via postgenomic technologies. As of May 2010, there were 1,273 completed bacterial genome sequences, 1,161 draft bacterial genome sequences, 1,094 in progress projects, and 511 that have been targeted for near-future work (www.genomesonline.org). The phylogenetic groups of *Proteobacteria*, *Firmicutes*, and *Actinobacteria* account for a vast majority of the sequenced genomes through the beginning of 2009, and only 38% of the projects could be classified as biotechnology and/or environmentally related. With respect to well-studied metal-reducing bacteria, the genus *Shewanella* has 22 reported genome projects, *Geobacter* has 10 genome projects, *Desulfovibrio* has 25 genome projects, *Thermoanaerobacter* has 12 genome projects, *Desulfitobacterium* and *Desulfotomaculum* each have 9 genome projects, the *Anaeromyxobacter* genus has 4 genome projects, and the *Desulfuromonas* genus has 2 genome projects. As high-throughput methodology continues to advance our understanding of biological systems, one must keep in mind the representative nature of a single genome, even for a bacterial species, and appreciate the undiscovered diversity that still occupies countless niches in natural and engineered systems. Therefore, this review cannot be exhaustive for all organisms within a single genus nor all metals for which direct or indirect interactions are known. This chapter will focus on transcriptomic studies in the more well-studied systems of metal-reducing bacteria represented by *Shewanella*, *Geobacter*, and *Desulfovibrio* that have both genome sequence data and postgenomic tools (e.g., gene microarrays) available.

Interestingly, two of our three model organisms are classified as δ-*Proteobacteria*. Based upon six δ-*Proteobacteria* genomes in 2006, Karlin et al. estimated the unique characteristics of the δ-*Proteobacteria* with predictions of highly expressed proteins using bioinformatics and codon/amino acid frequencies. The unique attributes included major ribosomal protein gene clusters located near the chromosome replication terminus instead of the origin, multiple copies of rubrerythrin, numerous σ^{54}-activators, multiple TonB receptors, and Fe^{2+} siderophore receptors.

TRANSCRIPTOMICS

Microarrays have been a powerful tool for monitoring dynamic gene expression under various conditions and have been widely used for genome-wide transcriptional analyses in microorganisms (e.g., DeRisi et al., 1997; He et al., 2005; Clark et al., 2006; Elias et al., 2009) and environmental studies (Zhou and Thompson, 2002; Taroncher-Oldedburg et al., 2003; Rhee et al., 2004). Typically, microarrays have been constructed with either PCR-generated amplicons (200 to 2,000 bp) or oligonucleotide probes (20 to 70 nt) with each providing trade-offs for sensitivity and specificity (He et al., 2005). Commercial companies can now synthesize desired microarrays relatively inexpensively for organisms with sequenced genomes, but proper experimental design is crucial for interpretation of results in a physiological context. For example, transcriptomic studies with different substrates or stresses need to consider the potential role of growth-rate-dependent responses as well as responses to counterions or changes in geochemistry. Another consideration is the method of cultivation itself, because uncontrolled batch growth (i.e., flasks and tubes) can introduce variation in the cell-to-cell growth state as well as between biological replicates compared with steady-state chemostatic cultivation (Elias et al., 2008). However, batch modes can allow the elucidation of temporal gene expression in the context of dynamic growth parameters (Clark et al., 2006). In addition, normalization and standardization across experiments and collaborations (e.g., culturing medium composition, temperature, pH) are crucial for valid interpretation of the data between laboratories.

Methodologies for transcriptomic analysis are being impacted by the same technologies that have vastly advanced mass nucleic acid sequence acquisition, namely the use of Roche/454 (http://www.454.com), Illumina (http://www.illumina.com), and SOLiD (http://marketing.appliedbiosystems.com) platforms for transcriptome analysis. More recently, two different technologies that are reported to not require amplification are Helicos Heliscope (www.helicosbio.com) and Pacific Biosciences SMRT (www.pacificbiosciences.com), although both are not yet commercially available. The Roche/454 technology has been used for transcriptome sequencing (e.g., Torres et al., 2007), and more recently the shorter-read platforms have been used (referred to in Wilhelm and Landry, 2009). New software (RNA-seq) to analyze transcriptomic sequence data has been developed and tested (Wilhelm and Landry, 2009), and a recent study developed an in silico simulator (ESTcalc) to estimate which platforms to use for the project of interest (Kerr-Wall et al., 2009). The simulations for *Arabidopsis* suggested a combination of FLX and Solexa sequencing for optimal transcriptome coverage with respect to cost (Kerr-Wall et al., 2009). Undoubtedly, these and emerging technologies will continue to enhance the ability of researchers to probe cellular systems.

With respect to metal-reducing bacteria, the high-throughput nature of microarrays has allowed for the direct measurement of most genes in an organism in response to different growth conditions, various stresses, and mutant strains with almost all work being done in *Shewanella*, *Geobacter*, and *Desulfovibrio*. The field of transcriptomics has become increasingly robust in the past 10 years, and more detailed analyses along temporal and spatial scales have provided great insight into the cellular systems that are associated with a particular set of conditions (e.g., regulons and regulatory networks). Most of the studies discussed in this chapter have used microarrays validated via quantitative PCR.

As more studies use combinations of systems biology methods (e.g., transcriptomics and proteomics), a mostly unanswered question remains as to the correlation between mRNA and protein levels and the overall effects on activity as a result of complex regulatory and metabolic networks. Thus far, studies have shown that mRNA-protein correlations have not been good and that analytical variations for protein abundance might account for between 34 and 44% of total mRNA-protein variation (Washburn et al., 2003; Nie et al., 2006). Recently, Nie et al. (2006) examined other contributors that could affect translational efficiency in *Desulfovibrio vulgaris* such as initiation, Shine-Delgarno sequences, codon usage (bias), amino acid usage, and stop codon identity. The study reported that mRNA-protein correlation was affected most by codon usage and amino acid composition based on the ability to explain 5 to 15% and 5 to 12% of the variability, respectively. While the study provided an estimate of translation-related sequence features to protein expression, it is still not known to what degree different mRNA and protein molecules differ with respect to in vivo half-lives, although Nie et al. (2006) predicted that protein stability only accounted for 5% of the variation.

The following sections will focus on available data for the most intensively studied metal-reducing bacteria, *Shewanella*, *Geobacter*, and *Desulfovibrio*, with data from other organisms added when available. In contrast to an organism-centric discussion, we have attempted to categorize the known cellular responses to electron acceptors and donors of metal-reducing bacteria. In addition, the cellular responses to oxidative stress will be considered in light of the ability of some metals to form hydroxyl radicals, metallo-oxo, and metalloperoxo species (Kasprzak, 2002). We will also consider the physiological state of biofilm cells, with the assumption that most of the available data have been compiled with planktonic cells. The biofilm state will be important to consider in the context of populations adhered to subsurface particles and electrodes in fuel cells in which electrons are being transferred from biological cells to a solid, abiotic surface.

Over the past few decades, many bioremediation projects have studied the biogeochemical interactions that control heavy metal and radionuclide contamination. Within this area of research, subsurface bacteria have received much attention because of the ability to reduce and immobilize heavy metals and radionuclides in subsurface sediments (Lovley et al., 1993a; Barton et al., 1996). While a number of field studies have successfully identified key geochemical or biogeochemical aspects that either enhance or inhibit in situ bioremediation (Anderson et al., 2003; Chang et al., 2001; Wu et al., 2006, 2007), many other studies have focused on understanding metal reduction at the population, organismal, protein, or gene level.

The in situ reduction and immobilization of heavy metals and radionuclides is presently the most promising form of site cleanup as well as the most cost-effective. A number of the microorganisms that utilize metals as terminal electron acceptors have been isolated and identified, and include members of three major anaerobic groups: the denitrifying, sulfate-reducing, and Fe(III)-reducing bacteria. However, nitrate reduction tends to predominate when nitrate is available (Smith, 2002), followed by Fe(III), and then sulfate reduction (Chapelle and Lovley, 1992) because of thermodynamic constraints. It is unknown how U reduction might integrate into such generalizing patterns of microbial activity since the potential for uranium bioremediation is a function of multiple factors (Elias et al., 2003). These include the presence of suitable microorganisms, the availability of electron donors and acceptors, as well as sediment components that may complex with uranium (McKinley et al., 1995). Bacteria grouped into the classical guilds of dissimilatory Fe(III) and sulfate reducers have been the focus of many studies. Given these environmental parameters, researchers have attempted to use system biology (understanding a biological cell as a system) approaches to understand organismal networks at the cellular, population, and community levels, and the following sections summarize transcriptomic studies in these metal-reducing bacteria.

ELECTRON TRANSPORT SYSTEM ORGANIZATION

One important aspect to consider when investigating the metal-reduction mechanisms within a given species is the organization of vectoral electron transport. Herein lies a difference between bacteria that are capable of coupling metal reduction with growth (e.g., *Shewanella* and *Geobacter*) versus those that can only reduce the metals (e.g., *Desulfovibrio*). In both *Shewanella* and *Geobacter*, electron transfer is believed to begin with oxidation of the carbon/electron donor in the cytoplasm and the electrons are transported via membrane-bound/membrane-associated proteins across the inner membrane and into the periplasm. From here, they are shuttled through the periplasm to the outer membrane where *b*-type and *c*-type cytochromes located within, and outside, the outer membrane can transfer the electrons to the external terminal electron acceptor (Shi et al., 2007). For *Desulfovibrio*, while electron donor oxidation also occurs in the cytoplasm, the electrons are believed to take one of a few routes depending upon the organic carbon/electron donor. The major difference exists in that the electrons are not routed to the outer membrane, but rather are either transported along the inner membrane, or through it to the periplasm, but then cross back into the cytoplasm. This is believed to be via one or more multienzyme, transmembrane complexes. The electrons are then used to reduce an oxidized sulfur species (i.e., sulfate, sulfite, thiosulfate), and the resulting H_2S effluxes either passively or actively from the cell. To date, reduction-oxidation-reactive proteins have not been observed outside the cell proper in the sulfate-reducing bacteria, and may be a result of selection for the utilization of a soluble electron acceptor (i.e., sulfate). Excellent figures for depiction of these electron transfer systems can be found for *Shewanella* and *Geobacter* (within Shi et al., 2007; Wall and Krumholz, 2006) and for *Desulfovibrio* (Hieldelberg et al., 2004).

The specific architecture of electron transfer for each of these organisms varies quite extensively. For *Shewanella* during metal reduction, electrons within a quinol pool located in the inner membrane are transferred to CymA, a periplasmic *c*-type cytochrome. From here, one operon, in particular, becomes essential, the *mtr* operon, which includes *mtrABC* and *omcA*. Some reports in the literature also refer to *mtrC* as *omcB*. It is believed that MtrA is also periplasmic and can further transfer the electrons to the trans-outer membrane MtrB and then finally to MtrC and OmcA, both of which are located on the outer face of the outer membrane when a complex is formed. *Geobacter* also has *c*-type cytochromes that are present on the outer face of the outer membrane, but while the overall schema of vectoral electron transport is similar to that of *Shewanella*, the roles of particular proteins appear to vary. In general, during metal reduction for *Geobacter*, electrons are transferred from the quinol pool to the inner membrane protein MacA and then to a series of electron transfer proteins known as PpcA-E in order to transfer electrons to the outer membrane. In the outer membrane a variety of *c*-type cytochromes are used to direct electrons to terminal electron acceptors. While the role(s) of some of these have been distinguished, much work remains to be done. The gene *omcE* appears to be used for U(VI) but not Fe(III) reduction, while loss of *omcC* did not affect metal reduction (Shelobolina et al., 2007). OmcB is apparently only used for Fe(III) reduction. Other proteins are essential for metal reduction overall, such as the outer membrane porin, OmpJ (Afkar et al., 2005). While this protein, and likely others, serve analogous functions in *Geobacter* to GspD in *Shewanella*, these are very different proteins both structurally as well as with regard to nucleotide and amino acid sequences.

For *Desulfovibrio*, the path of electron transfer is not nearly so straightforward and depends a great deal on the source of electrons (see the discussions on electron donors and H_2 cycling below). A number of periplasmic c_3-type cytochromes are utilized for electron transfers from the quinol pool, and the cytochrome used is dependent upon the electron donor. In addition, the possible formation of H_2 from protons and electrons with a hydrogenase or via a carbon monoxide dehydrogenase has been previously discussed (Voordouw, 2002). From the quinol pool, several *c*-type cytochromes could be used to shuttle electrons back into the cytoplasm via one of several trans-inner membrane multisubunit complexes derived from polycistronic operons such as *rnf*, *hmc*, *tmc*, or *qmo*. These complexes accept the electrons from the periplasm and channel them back into the cytoplasm for donation to the sulfate-reduction mechanism of adenylyl sulfate reductase (APS) and the dissimilatory sulfite reductase (DSR) complexes. For metal reduction, it appears that the electrons are donated primarily by the periplasmic tetraheme cytochrome c_3 with reduction occurring in the periplasm rather than outside the cell (Lovley et al., 1993a, 1993b). Electrons are preferentially transferred to the metals because of the higher oxidation-reduction potentials of U(VI) and Fe(III) compared with sulfate, and this is likely why energy cannot be conserved in these cells to support growth via metal-reduction (Elias et al., 2004; Wall and Krumholz, 2006).

Although the basic organization of the electron transport system in the representative microorganisms is reasonably well understood, it is also important to understand the support system behind these essential frameworks for electron transfer. A number of chaperone and secretion systems are essential to maintain the efficient and balanced transport of electrons as well as optimized carbon and nitrogen acquisition. While it is outside the scope of this chapter to delve into the detail of these support systems, essential work has revealed the importance of secretion systems to the ability to respire metals and so understanding them within this context is essential. The two main secretion systems used by the gram-negative bacteria are the twin-arginine translocation system (Tat) and the general secretory (Sec) system.

Apocytochromes c and the heme cofactors are each translocated across the inner membrane to the periplasm by means of the Sec pathway after which they are assembled and folded into mature c-type cytochromes (Thöny-Meyer, 1997). In contrast, reduction-oxidation proteins that participate in periplasmic electron transfer reactions and bind cofactors such as Ni, Fe, Se, flavin adenine dinucleotide, Cu, and molybdopterin are assembled and folded in the cytoplasm and are believed to be secreted by the specialized Tat pathway that is able to accommodate these prefolded proteins (Berks et al., 2003). Substrates of the Tat pathway have a characteristic leader peptide that is distinct from those recognized by the Sec pathway. Tat leaders are often longer and generally less hydrophobic than Sec leaders and often have a positively charged Sec avoidance signal near the carboxy termini (Berks et al., 2000). Tat leaders are most easily recognized by the presence of a characteristic motif, RRXFXK, in which the consecutive arginine residues are nearly always conserved and the other motif residues are conserved greater than 50% of the time (Berks, 1996). A program called TAT-FIND 1.1 (Rose et al., 2002) was designed to detect these leader peptides, using the predicted haloarchaeal Tat signal peptides to establish the pattern-matching rules. These rules were later modified to enable detection of Tat signal peptides in *Pseudomonas aeruginosa* (Dilks et al., 2003). Once within the periplasm, type II or type IV systems are used to move proteins to the outside of the cell. The type IV system is primarily used for pilin and flagella, while the type II, or General Type II Secretion System (Gsp) system is used for other proteins. In *Shewanella oneidensis* MR-1, the well-studied MtrABC and OmcAB that are essential for metal-reduction (Myers and Myers, 1997, 2002; Beliaev et al., 2001, 2005; Pitts et al., 2003; Shi et al., 2006) appear to be Sec dependent (Shi et al., 2008).

Several reports have determined that in these metal-reducing bacteria, the deletion of certain suspected transporters resulted in decreased metal reduction, while others had no effect. In *S. oneidensis* MR-1, deletion of *gspDG* as well as *pilD* ceased certain forms of Fe^{3+} and Mn^{4+} reduction (Bretschger et al., 2007; Shi et al., 2008), as did the removal of *mtrAB* or *omcA*. Similar results were observed in *Shewanella putrefaciens* strain 200 when *gspE* was deleted (DiChristina et al., 2002). These results suggest that the Gsp system is primarily responsible for the correct placement and orientation of the outer membrane proteins in *Shewanella*. With respect to *Geobacter*, while there appears to be a functioning Gsp system in this bacterium (Methe et al., 2003), many of the amino acid sequences diverge from those in *Shewanella* and may suggest some level of specialization. However, to date, little work has been done either to characterize this part of the system or to determine which of the numerous reduction-oxidation-reactive, outer membrane proteins dependent on the Gsp system. Once again, *Desulfovibrio* diverge in this respect from *Shewanella* and *Geobacter* in that it does not possess any of the classical Gsp genes, with the exception of a degenerate copy of the *gspE* gene (Heidelberg et al., 2004). Since there are no known outer membrane redox-reactive proteins, it makes sense that the Gsp system is likely not needed in *Desulfovibrio* spp.; however, a functioning Tat and Sec system in all other respects is present. Interestingly, *D. vulgaris* Hildenborough appears to possess the genes for type III secretion systems on a native plasmid, which are usually observed in pathogens and symbionts for the transport of bacterial proteins into eukaryotic cells to influence host responses (Heidelberg et al., 2004).

IRON RESPONSES

It is important to keep in mind that under anaerobic conditions (i.e., lower redox potentials) Fe(II) typically predominates over Fe(III) and that many metal-reducing organisms can transport ferrous iron via an ATP-dependent transport system (FeoAB). Interestingly, many δ-*Proteobacteria* do contain paralogs for *feoAB* genes but do not have easily identifiable siderophore synthesis systems, although siderophore transport genes are predicted from gene models of sequenced genomes. This might suggest that either novel, as yet unknown, siderophore systems exist or the organisms utilize siderophores synthesized and excreted by other members of the community. The fact that metal-reducing bacteria have a large number of iron-containing proteins predicted from the genome sequences (i.e., cytochromes and Fe-S proteins), and the inherent ability to utilize ferric iron as a terminal electron acceptor during anaerobic respiration, makes transition metal ion homeostasis a key target to provide insight into the structure and function of anaerobic bacterial communities. Albeit there are exceptions; for example, *Alkaliphilus metalliredigens* is a strict anaerobe that can reduce Fe, Co, Cr, and U (Ye et al., 2004), yet the *A. metalliredigens* genome sequence contains very few putative *c*-type cytochromes (http://img.jgi.doe.gov/cgi-bin/pub/main.cgi).

Several years ago, the genomes of seven metal-reducing δ-*Proteobacteria* were compared in terms of global regulons and regulators for transition metal homeostasis (Rodionov et al., 2004). In many cases, the transcriptional regulators are homologous to known factors in *Escherichia coli*; however, the DNA signals are entirely different, and there are differences in the details of how the response is coordinated within the respective organism. In *Desulfovibrio* and *Geobacter* species, one to three Fur (ferric uptake regulator) orthologs are present and form a distinct cluster in a phylogenetic tree of Fur, Zur, and Per proteins (Rodionov et al., 2004). The strongest detected signal in the genomes was a 17-bp palindrome (WTGAAAATNATTTTCAW) that was observed upstream of multiple *feoAB* operons and *fur* genes. It was also observed that within the δ-*Proteobacteria*, *Geobacter* and *Desulfuromonas* had smaller Fur regulons compared with those of *Desulfovibrio*.

A recent study by Yang et al. (2009) characterized iron depletion and repletion in *S. oneidensis* MR-1 and predicted gene expression networks. Typically, Fur and the regulatory RyhB are common controllers for iron homeostasis in bacteria (including metal-reducing bacteria); however, metal-reducing bacteria have an increased need for iron because of the higher than normal content of heme-containing proteins. The analyses identified not only iron acquisition systems, but also genes related to aerobic and anaerobic metabolism. Interestingly, some aerobic genes in *S. oneidensis* MR-1 were upexpressed under iron depletion conditions, and included genes annotated as a methylcitrate synthase, methylisocitrate lyase, methyl citrate dehydratase, isocitrate dehydrogenase, and malate dehydrogenase (Yang et al., 2009). These enzymes are part of the methylcitrate pathway that is also present in *E. coli*. It is possible that an alternative part of the citrate cycle is used under an iron-depleted state when proteins with more iron content are downexpressed, although the pathway is used in other organisms to deal with 4-carbon propyl intermediates. A subset of genes involved with anaerobic metabolism was not upexpressed upon iron depletion, and the authors proposed that a subset of anaerobic metabolism genes could function as iron storage proteins in the event of iron depletion (Yang et al., 2009). The use of iron-cofactor proteins for iron homeostasis has been shown in *E. coli* (Masse et al., 2005) and may occur in *S. oneidensis* MR-1. A similar function may be present in the metal-reducing δ-*Proteobacteria* and *Firmicutes* that do have Fur homologs, but the storage mechanism has not been specifically studied. For *D. vulgaris*, an iron-depleted state did not significantly change expression of genes involved in the citrate cycle, gluconeogenesis, or pyruvate metabolism for wild-type cells or a Fur homolog mutant (Bender et al., 2007; J. Zhou, unpublished results).

In the Yang et al. (2009) study, putative transcriptional factors were also classified into functional gene modules that responded to iron levels, and some also varied expression levels under iron-reducing conditions. For example, the transcriptional regulator SO0490 was induced under iron-reducing conditions compared with oxygen-reducing conditions, but it was also induced under uranium-reducing conditions (Bencheikh-Latmani et al., 2005). A mutant in SO1415, a TetR family putative regulator, was deficient in iron reduction but also trimethylamine oxide- and thiosulfate-reduction (Yang et al., 2009). Another global regulator that has been shown to be involved in metal-reduction is cAMP receptor protein (CRP). Once considered to be solely a regulator of catabolite repression, CRP and cAMP have been shown to be involved in iron and manganese reduction in *S. oneidensis* MR-1, and an adenylate cyclase (SO1329) was recently shown to regulate anaerobic respiration in *S. oneidensis* MR-1 (Saffarini et al., 2003; Charania et al., 2009). These regulators may play an important role in coordinating iron homeostasis under metal-reducing conditions for *Shewanella* species and possibly other metal-reducers. *D. vulgaris* has three homologs and *Geobacter sulfurreducens* has two homologs annotated as CRPs. *S. oneidensis* has an annotated EtrA that is classified in the CRP and Fnr protein families, but is specifically a regulator of electron transport proteins. Mutant analysis indicated that EtrA was involved in regulation of dimethyl sulfoxide reductase, fumarate reductase, formate dehydrogenase, a cytochrome oxidase system, most of the tricarboxylic acid (TCA) cycle enzymes, and the transcription of other regulatory genes (Beliaev et al., 2002a; Wan et al., 2004).

With respect to electron transfer proteins in *S. oneidensis*, one study compared cells grown aerobically versus cultures transitioned to growth with nitrate, fumarate, or Fe(III) as the sole electron acceptor (Beliaev et al., 2002b). Curiously, a number of cytochrome *c*- and *d*-containing oxidases were downexpressed under all anaerobic conditions while quinones, reductases (including those for dimethyl sulfoxide and fumarate), dehydrogenases, and hydrogenases, including the *nap* operon, were highly upexpressed. The well-studied *mtrABC* and *omcAB* were also upexpressed, but the trend showed that the greatest degree of expression was observed with fumarate rather than with Fe(III), and induction with nitrate was the least obvious. These results suggest that the Mtr and Omc proteins are regulated along with the other members of the electron transport family and not specifically for Fe(III) reduction. Indeed, this would make sense in that these proteins are required for electron transfer in general, and one would expect that the trend of differential expression between the anaerobic electron acceptors may be different for different forms of Fe(III). In a more recent study, Beliaev et al. (2005) used an expanded number of soluble electron acceptors, including O_2, Co(III)-EDTA, MnO_2, ferric citrate, S_2O_3, NO_3, trimethylamine N-oxide, and dimethyl sulfoxide. This time, however, they also included solid forms of oxidized metals including hydrous ferric oxide and colloidal Mn(IV). Once again, nitrate was observed to invoke the greatest number of differentially regulated genes as well as the highest degree of change in regulation of those genes and spanned virtually all role categories. Remarkably, the *mtrABC* and *omcAB* genes displayed a 2- to 8-fold decrease in expression when exposed to the solid-phase metals. This is most surprising given the body of evidence that identifies these proteins as being necessary for metal reduction. However, the study was designed for a 3.5-h exposure to each electron acceptor so that cell division was not a factor. While this leaves a substantial amount of time to capture any changes at the mRNA level, the full expression of genes needed for utilization of solid-phase minerals may require longer periods of time (G. Geesey, personal communication). The similarly decreased expressional response for clusters of flagellar and amino acid biosynthesis genes along with other periplasmic electron transfer genes (*nuo*, *nqr*, *cyo*, *coo*) suggests that the cells downexpressed essential carbon and electron fluxes during the growth transition.

A similar study conducted with *G. sulfurreducens* showed a clear distinction for the genes used for anaerobic respiration but not metal reduction, and few genes could be classified as metal specific (Chin et al., 2004). The cultures were limited for acetate (the sole carbon and electron donor), fumarate, or ferric citrate. When fumarate was limiting, the fumarate reductase (*frdA*) transcripts were approximately 3-fold higher than with acetate limitation, and this result suggested that the cells attempted to scavenge for any available fumarate. The protein OmcB has been shown to be essential for Fe(III) reduction in *G. sulfurreducens*, while another related protein OmcC is not essential even though *omcB* and *omcC* are 79% identical at the nucleotide level and within the same operon (Leang et al., 2003; Methe et al., 2003). While there was little to no difference in transcript abundance for either *omcB* or *omcC* with respect to acetate versus fumarate limitation, *omcB* transcripts were 60-fold higher with Fe(III) limitation versus acetate limitation. In contrast, *omcC* transcripts were reduced by >60-fold under Fe(III) limitation, clearly showing that OmcB, but not OmcC, is intimately involved in Fe(III) reduction (Chin et al., 2004).

With use of quantitative PCR, it was shown that *Geobacter* species downexpressed *feo* transcripts [Fe(II) transporter] in response to increasing Fe(II) levels during in situ biostimulation (O'Neil et al., 2008); however, the effects on iron-containing proteins were not monitored. In the case of *D. vulgaris*, Fe(II) levels decline during growth because of the reduced conditions and precipitation of iron species, and the *feo* system was significantly upexpressed at later growth phases to deal with self-imposed iron limitation during sulfate reduction (Clark et al., 2006). Pathogens experience a similar situation when limited for Fe in the host and can use the *feo* system for iron acquisition (Robey and Cianciotto, 2002). The *feo* system appears to play a similar role in Fe(II) acquisition for environmental microorganisms under iron-limited conditions although the *feo* system most likely plays a role in assimilation as opposed to dissimilatory iron reduction. When growth with ferric citrate was compared with fumarate in *G. sulfurreducens*, a proteomic approach was used to identify the iron upexpressed proteins that included OmpA, superoxide dismutase, catalase, and a rubredoxin oxidoreductase (Khare et al., 2006). The results suggested that differences existed in the oxidative environment possibly because of the presence of elevated iron levels.

The capacity for Fe(III) reduction was substantially diminished in a RelA homolog mutant in *G. sulfurreducens*, and this growth defect may be the result of downexpression of genes involved in electron transport (e.g., rubrerythrin, cytochrome *c*, cytochrome c_{551}, and iron reductase), although other cytochrome *c* genes were upexpressed (DiDonato et al., 2006). Some carbon metabolism genes were also downexpressed in the mutant (e.g., α-ketoglutarate dehydrogenase and a ribose-5-P isomerase) suggesting that ppGpp plays a role in coordinating carbon and electron flux. The α-ketoglutarate dehydrogenase supplies the cell with succinyl-coenzyme A (CoA) as a backbone for protoporphyrin heme biosynthesis, and the ribose-5-P isomerase is important for nucleic acid biosynthesis. The results suggested that oxidative stress and nutrient deprivation could be linked to iron respiration in *Geobacteraceae* via ppGpp as a way to coordinate electron transport systems with exogenous energy sources yet minimize potential oxidative stress. Interestingly, the mutant displayed faster growth rates than wild-type cells and might suggest that ppGpp plays a role in regulating balanced growth and not solely stress responses per se. Using the *relA* mutant, it was determined that ppGpp can be a signal for Fur- and RpoS-dependent gene expression (Krushkal et al., 2007).

When *S. oneidensis* MR-1 cells grown on different electron acceptors were compared via transcriptomics, a majority of Clusters of Orthologous Genes displayed similar expression levels. However, there were some differences for genes involved in amino acid biosynthesis, protein synthesis, and energy metabolism (Beliaev et al., 2005). Interestingly, genes involved with the biosynthesis of glutamate and pyruvate family amino acids were downexpressed 2- to 20-fold under metal-reducing conditions (i.e., Fe, Co, Mn). Perhaps this is a consequence of less available energy (or ATP) per transferred electron during metal reduction, and thus the cells alter carbon flux and biomass yields. The two sigma factor genes, *rpoS* and *rpoN*, displayed downexpression in the presence of metals and thiosulfate, while *rpoH* and *rpoE* were upexpressed (Beliaev et al., 2005a). In addition, two putative extracytoplasmic function-type sigma factors (SO1986, SO3096) were significantly upexpressed under metal-reducing conditions, and SO1986 was upexpressed in response to strontium exposure (Beliaev et al., 2005a; Brown et al., 2006). Recently, two separate studies have shown the involvement of extracytoplasmic function-type factors in metal and stress response in *Bradyrhizobium* and *Caulobacter* (Lourenco and Gomes, 2009; Gourion et al., 2009). For *Shewanella*, functions important for iron homeostasis were predicted to be a part of the Fur modulon and included the *tonB* system, *irgA*, hemin and siderophore transport, and ferritin (Wan et al., 2004).

When *Geobacter uraniireducens* was grown with fumarate and acetate and compared with cells grown with iron oxides in the presence of sediments, the sediment-grown cells displayed 1,084 and 882 up- and downexpressed genes, respectively, and a majority of the significant changers coded for proteins involved in energy metabolism (e.g., *c*-type cytochromes, hydrogenases, Fe-S proteins), hypothetical, and conserved hypothetical proteins (Holmes et al., 2009). However, a distinction between a sediment effect or iron oxide effect is difficult to discern. Putative outer membrane *c*-type cytochromes were upexpressed in cells grown with iron-containing sediments, and some of the same genes were upexpressed in *Geobacter* cells grown with iron oxides (Mehta et al., 2005), electrodes (Holmes et al., 2006), and U(VI) (Shelobolina et al., 2007). Two outer membrane proteins in *S. oneidensis* MR-1, OmcA and MtrB, have been shown to be involved with iron oxide reduction and the OmcA is involved in attachment to iron oxide surfaces (Xiong et al., 2006). Interestingly, OmcA has weak homology to conserved hypothetical proteins in the metal-reducing bacteria *Geobacter lovleyi*, *Geobacter metallireducens*, *Anaeromyxobacter dehalogenans*, and *Rhodoferax ferrireducens*.

Both nitrite reductase genes in *G. uraniireducens* were upexpressed in cells grown in the presence of sediments, and others have speculated that nitrogen limitation can cause the upexpression of nitrite reductase for the incorporation of nitrogen into cellular material (Hoffman et al., 1998). In addition, the sediment-grown cells upexpressed a number of heavy metal-related transporters annotated as general efflux and ABC transport systems and the sediments were known to contain relatively high amounts of Ar, Ba, Mo, Ra, Se, and U (Holmes et al., 2009). A select group of upexpressed genes from sediment-grown cells were also observed to be upexpressed during in situ stimulation for U(VI) reduction, and the test case with *G. uraniireducens* demonstrated that transcriptomic methods could provide environmentally relevant data dependent upon the extent of community diversity.

Based upon putative Fur-binding signatures in DNA sequences across *Geobacter*, *Desulfovibrio*, and *Desulfuromonas*, genes of unknown function were predicted to be iron regulated and included presumptive porins (Rodionov et al., 2004). *D. vulgaris* Hildenborough and *Desulfovibrio desulfuricans* G20 appeared to have an expanded Fur regulon that included different types of electron carriers, and this arrangement might allow the cell to use iron-independent electron carriers during iron-depleted conditions. A subsequent transcriptomic study in *D. vulgaris* showed that Fur does act as a traditional repressor for *feoAB* and other genes that were predicted to contain a Fur binding site, including putative siderophore transport systems (Bender et al., 2007). Surprisingly, a Fur mutant was not affected by iron availability, although the mutant did have increased sensitivity to nitrite and osmotic stress, and these results indicated that Fur was not essential for dissimilatory sulfate reduction in *D. vulgaris* under iron-replete or -depleted conditions. However, the results suggested that iron homeostasis might be linked to the presence of nitrite and/or osmotic stress. These data suggest that iron homeostasis is coordinated with multiple stimuli that can impact iron solubility and availability.

Furthermore, a recent survey of the hypothetical and conserved genes in *D. vulgaris* showed that, along with the extensive network of known Fur genes, many hypothetical protein genes showed differential expression under stress conditions in a Fur mutant compared with wild type (Elias et al., 2009). These results suggest that there may be an even more extensive influence of the Fur regulon in sulfate-reducing bacteria. Similar to *D. vulgaris*, a Fur mutant in *S. oneidensis* MR-1 did not display significant growth defects under anaerobic, iron-depleted conditions (Thompson et al., 2002). The osmotic stresses of either NaCl or KCl caused the upexpression of the Fur regulon in *D. vulgaris* (Mukhopadhyay et al., 2006), and the Fur mutant had increased sensitivity to salt stress, although the exact mechanism that links osmotic stress is not known.

Both *G. sulfurreducens* and *G. metallireducens* and *D. vulgaris* and *D. desulfuricans* contain a putative nickel-responsive regulator, NikR, that is predicted to control expression of nickel transporters (Rodionov et al., 2004) (nickel is crucial for hydrogenases and carbon monoxide dehydrogenases). Both *D. vulgaris* and *D. desulfuricans*, but only *G. sulfurreducens* are predicted to contain a Zur homolog that might control the expression of zinc transporters. As mentioned above, the δ-*Proteobacteria* seem to be lacking the OxyR and SoxR regulators of *E. coli* and instead use paralogs of the PerR regulator that have been extensively studied in *Bacillus subtilis*. The main oxygen detoxification systems for aerobes and anaerobes alike involve superoxide dismutase, catalase, and peroxidases. In addition, *Desulfovibrio* species have desulfoferrodoxin:oxidoreductase that can help protect against oxidative stress. *D. vulgaris* and *D. desulfuricans* were also predicted to contain a CooA regulon that controls carbon monoxide dehydrogenase expression as well as a novel fumarate/nitrate reduction regulatory protein-like regulator that might help control the expression of genes involved with sulfate assimilation (Rodionov et al., 2004). When considering physiological responses to Ni and Zn, *Geobacter* and *Desulfovibrio* spp. use similar regulators, but oxygen response regulators appear to be different. The presence of trace metals and oxygen are important effectors for iron homeostasis and metal reduction.

In summary for iron responses, the *feo* system is important for Fe(II) acquisition as in other bacteria, while siderophore systems are less known in *Geobacter* and *Desulfovibrio* compared with *Shewanella*. The discussed bacteria have Fur, Pur, and Zur homologs, but the exact regulatory networks may be used differently compared with organisms such as *E. coli* and *B. subtilis*. The regulators, Crp, Fnr, and Etr, are also involved in iron homeostasis in *Shewanella*. *Shewanella* may also use non-Fe proteins during iron-depleted conditions as well as Fe-containing proteins as a Fe storage mechanism. Many data are available on the role of the *mtr* and *omc* systems in *Shewanella* for iron reduction, while some *omc* genes are involved in *Geobacter* but few genes have been designated as metal specific. The sulfate-reducing bacteria, such as *Desulfovibrio*, commonly face iron-depleted environments as a consequence of sulfide production and use multiple regulators to coordinate cellular response, although little is known about sulfide responses. A recent transcriptomic study monitored gene expression in response to high and low sulfide levels during *D. vulgaris* growth, and, as expected, genes involved in iron accumulation, stress response, and protein turnover were upexpressed at higher sulfide levels (Caffrey and Voordouw, 2010). Other factors involved in iron responses appear to include different sigma factors, ppGpp, and carbon flux.

CHROMATE RESPONSES

Chromium is the third most common pollutant at hazardous waste sites and the second most common inorganic contaminant after Pb. Cr(VI) is more soluble than Cr(III) and is considered to be 1,000-fold more toxic than Cr(III) in humans (ATSDR 2004). Cr(VI) can easily pass through cell membranes where it can be reduced to reactive species [Cr(III) and Cr(V)] and can generate free radicals that can damage nucleic acids and proteins. As with other heavy metals, the toxicity of chromate is thought to be the generation of reactive oxygen species during reduction to Cr(V) and then onto Cr(III) via one-electron transfers (Cervantes et al., 2001). In eukaryotes, chromate toxicity is mediated via oxidative damage to DNA and/or proteins (Stearns et al., 1995; Vasant et al., 2001; Sumner et al., 2005). Therefore, many of the cellular responses to heavy metals, including chromate, consist of upexpressed antioxidant proteins and enzymes. Expression of *copA* and *cusA* in *S. oneidensis* MR-1 increased during aerobic and anaerobic growth in the presence of copper, and *cusA* expression was detected in the presence of cadmium (Toes et al., 2008).

When *S. oneidensis* MR-1 cells were grown in the presence of U(VI) or Cr(VI) and compared with nitrate-grown cells, a large fraction of the upexpressed genes (approximately 30%) encoded hypothetical proteins (Bencheikh-Latmani et al., 2005). A difference between U(VI) and Cr(VI) responses was the upexpression of efflux pump genes by Cr(VI) and not U(VI). In contrast, when MR-1 cells grown in rich media were exposed to 0.3 mM Cr(VI) under aerobic conditions, genes involved in DNA metabolism, cell division, cell wall maintenance, membrane stress, and general stress response were upexpressed and genes involved in chemotaxis, motility, and transport were downexpressed during a 24-h exposure to chromate (Chourey et al., 2006). In contrast to an acute exposure (1 mM for 90 min), a predominant response during the chronic, low exposure was observed in the upexpression of prophage related genes. The acute exposure in rich medium caused an upexpression in iron transport (*tonB*-related), hemin transport, sulfate transport, and cellular detoxification (Brown et al., 2006).

Caulobacter crescentus also showed upexpression of a *tonB*-like gene in response to chromate, dichromate, and uranium when grown in a minimal medium (Hu et al., 2005). Interestingly, an azoreductase (SO3585) was upexpressed in response to chromate, and a mutant was recently shown to have increased sensitivity to Cr(VI), Cd(II), Cu(II), and Zn(II), and the initial rate of Cr(VI) reduction was impaired (Mugerfeld et al., 2009).

Because of the potential for oxidative damage to DNA and protein during heavy metal exposure, it is not surprising that *S. oneidensis* MR-1 upexpressed an annotated catalase/peroxidase, a catalase, and a presumptive antioxidant protein although the predicted iron-type superoxide dismutase did not display a change in expression (Brown et al., 2006). In the non-metal-reducing bacterium *C. crescentus*, the major responses to Cr, Cd, Se, and U exposure were the upexpression of a Mn-dependent superoxide dismutase (Hu et al., 2005). With respect to Cr and Cd, genes that coded for glutathione transferases, thioredoxin, and glutaredoxins were strongly upexpressed.

A group of predicted ABC sulfate transporters were upexpressed in *S. oneidensis* MR-1 and could be a cellular response of a decline in sulfate uptake via chromate inhibition of the transporters. To the contrary, in *C. crescentus*, sulfate transporters were downexpressed to perhaps reduce the nonspecific influx of chromate (Hu et al., 2005). Iron homeostasis was also affected by the acute chromate exposure, and siderophore, ferritin, and hemin transporter genes were upexpressed. In addition, a presumptive response regulator was shown to help control expression of genes involved with siderophore-mediated iron acquisition and iron storage and may provide a linkage between iron homeostasis and chromate-induced stress (Chourey et al., 2008). However, for both sulfate and iron, it is unknown whether chromate causes a nutrient limitation with respect to sulfate and iron and/or whether sulfate and iron are needed for reductases and thiol-containing compounds, respectively, to deal with the chromate stress. It is also possible that elevated Cr(VI)/Cr(III) levels displace intended metal ions from cognate regulators and cause derepression or repression of respective genes (e.g., Fur, Fe, and Mn) (Guedon et al., 2003).

Almost all studies with *S. oneidensis* MR-1 chromate responses and transcriptomics have used aerobically grown cells, and previous work has shown that the chromate response of MR-1 depends upon the physiological state of the cells, including aerobic versus anoxic conditions. Under anoxic conditions, MR-1 cells are much more susceptible to chromate exposure, both in minimum inhibitory concentrations and degree of growth arrest (Viamajala et al., 2004). Therefore, a comparison of transcriptomic responses in MR-1 is warranted between aerobic and anoxic conditions combined with chromate exposure.

On the contrary, the other "model" metal-reducing bacteria (i.e., *Geobacter* and *Desulfovibrio*) are considered to be "strict" anaerobes, although both have been shown to have varying levels of oxygen tolerance. To our knowledge, the combination of heavy metal stress and low-level oxygen exposure has not been characterized in obligately anaerobic, metal-reducing bacteria. In addition, the use of complex medium could affect overall cellular responses to heavy metals given the opportunity for complexants and increased biomass yields in the presence of easily used substrates that do not necessarily represent natural conditions.

Previous work has shown that *D. vulgaris* requires hydrogen sulfide, hydrogenases, and cytochrome c_3 for the reduction of Cr(VI) (Chardin et al., 2002), and that Cr(III) can be detected on the cell surface as well as between the cytoplasmic and outer membranes (Goulhen et al., 2005). Microcalorimetry was used to observe energy production without growth in the presence of Cr(VI) (Chardin et al., 2002), but acetate and sulfate levels were not reported. When *D. vulgaris* was exposed to a sublethal level of Cr(VI) in mid-exponential growth in defined medium, the major differentially expressed genes at 60 min posttreatment were a nitroreductase, a thioredoxin reductase, and a *clp* protease adaptor (Klonowska et al., 2006). The nitroreductase does not appear to have significant homology to the azoreductase that is upexpressed in *S. oneidensis*. In addition, four predicted metal or drug transporter genes were upexpressed, and included presumptive *merP*, *acrA*, and *chrA*. The *chrA* is located on pDV1, a 200-kb plasmid, and a strain that was cured of the plasmid was more susceptible to Cr(VI) exposure (Klonowska et al., 2006).

In summary, for chromate responses, *S. oneidensis* upexpressed metal efflux proteins, iron transport, and sulfate transport. In addition, an azoreductase was upexpressed, and an azoreductase mutant was more sensitive to chromate. In *D. vulgaris*, a nitroreductase that does not share significant sequence similarity to the azoreductase was upexpressed in response to chromate. *D. vulgaris* also had elevated gene expression for *chrA*, a known chromate efflux mechanism reported in *P. aeruginosa* and *Ralstonia metallireducens*.

URANIUM RESPONSES

Uranium contamination, as a result of processes associated with uranium extraction for the production of nuclear weapons, remains a significant environmental problem. In addition, the use of depleted uranium and other heavy metals in nonnuclear weapons will continue to create environmental hazards around the world. Depleted uranium is weakly radioactive and can also damage the kidneys because of heavy metal toxicity (Craft et al., 2004). The oxidized form of uranium [i.e., U(VI)] is soluble, and thus mobile in groundwater, and particular microorganisms can use metals and metalloids such as U(VI) as an electron acceptor (Lloyd, 2003). In terms of biostimulation, lower than predicted growth yields with U(VI) as the electron acceptor were determined for *G. sulfurreducens*, *G. lovleyi*, and *A. dehalogenans* (Sanford et al., 2007). These results coincide with some of the general stress responses detected for uranium-exposed cells. The formation of U(IV) via reduction forms a less soluble precipitate that is much less likely to contaminate surrounding water supplies. Interestingly, a recent study reported the use of diheme *c*-type cytochromes in *Thiobacillus denitrificans* and *G. metallireducens* for nitrate-dependent, anaerobic U(IV) oxidation (Beller et al., 2009). Uranium bioreduction is certainly plausible but must be considered within the context of environmentally relevant conditions (i.e., presence of mixed contaminants and geochemical conditions).

Indicators of membrane stress were more apparent after U(VI) exposure in *S. oneidensis* MR-1, and the membrane stress and damage may be related to interactions with solid-phase metals once metal ions have been reduced. In addition, cytoplasmic stress seemed to be more prevalent during U(VI) reduction compared with Cr(VI) and may be the result of the lack of a specific detoxification system for U(VI) (Bencheikh-Latmani et al., 2005), although general membrane stress responses could also be a general response to slower growth and/or nutrient deprivation associated with stationary phases of growth. It is curious, however, that one apparent trend in this study was that for both U(VI) and Cr(VI) exposure the most downexpressed genes appeared to be porins, transporters/symporters, and carbon/fatty acid metabolism-associated genes, while the most upexpressed genes were involved in electron transport. The latter included both *b*-type and *c*-type cytochromes and the well-studied *mtrABC* and *omcAB*.

Using proteomics on in situ groundwater poststimulation with acetate, populations of *Geobacter* spp. were identified. The metaproteome was predominated by peptides indicative of enzymes for the conversion of acetate to acetyl-CoA and pyruvate, tricarboxylic acid cycle proteins, and ATP synthase subunits (Wilkens et al., 2009). In addition, an abundance of pyruvate ferredoxin oxidoreductase was detected, and is likely used to synthesize acetyl-CoA for amino acid synthesis (Wilkens et al., 2009). Both *ompB* and *ompC* have been shown to be required for Fe(III) oxide reduction, and an increase in *ompC* transcripts was shown in situ during biostimulation for U(VI) bioreduction (Holmes et al., 2008). The addition of acetate to organic- and nitrogen-poor subsurface sediments stimulated the growth of *Geobacteraceae* and Fe(III) reduction, and this coincided with increased transcript levels of *nifD* (Methe et al., 2005; Holmes et al., 2004).

In comparison, recent work has shown that U(VI) inhibited sulfate depletion in *D. vulgaris*, and that both Fe(III) and U(VI) inhibited lactate-mediated sulfate reduction (Elias et al., 2004). A similar inhibition of sulfate reduction was observed with chromate (Klonowska et al., 2008). However, when H_2 was the electron donor, sulfate reduction was only slowed, and these results suggested that diverse molecular mechanisms may be involved in electron transport. The possibilities include that electron transfer mechanisms can be electron donor dependent as previously suggested (Elias et al., 2004), or the precipitation of these reduced metals within the cell causes a loss of function in a primary electron transport system component as observed for the type II tetraheme cytochrome c_3 during U(VI) reduction in *D. desulfuricans* strain G20 (Payne et al., 2004). In the latter case, the cell may then increase transcription and translation to compensate for the loss of total activity or either activate or upregulate "backup" systems. These strategies lead to a plausible conclusion that, at least in the sulfate-reducing bacteria that cannot conserve energy for growth from the reduction of metals, the reduction of U(VI) and expulsion of the resulting U(IV) is more of a detoxification event than the "use" of an alternate electron acceptor. This may explain the observation of uraninite being extracellular in *D. vulgaris* (Fig. 1). The exact location of U(VI) reduction has still not been fully resolved. Unfortunately, there is a lack of transcriptomic data with respect to the metabolism/reduction of U(VI) in the sulfate reducers and so, to this point in time, we are left to draw conclusions based primarily upon the observations comparing the rate and extent of U(VI) reduction using deletion mutants versus wild-type cells and electron donor/acceptor combinations.

Membrane stress was more prevalent in *S. oneidensis* cells exposed to U(VI) compared with other metals, and the cells upexpressed electron transport proteins, including

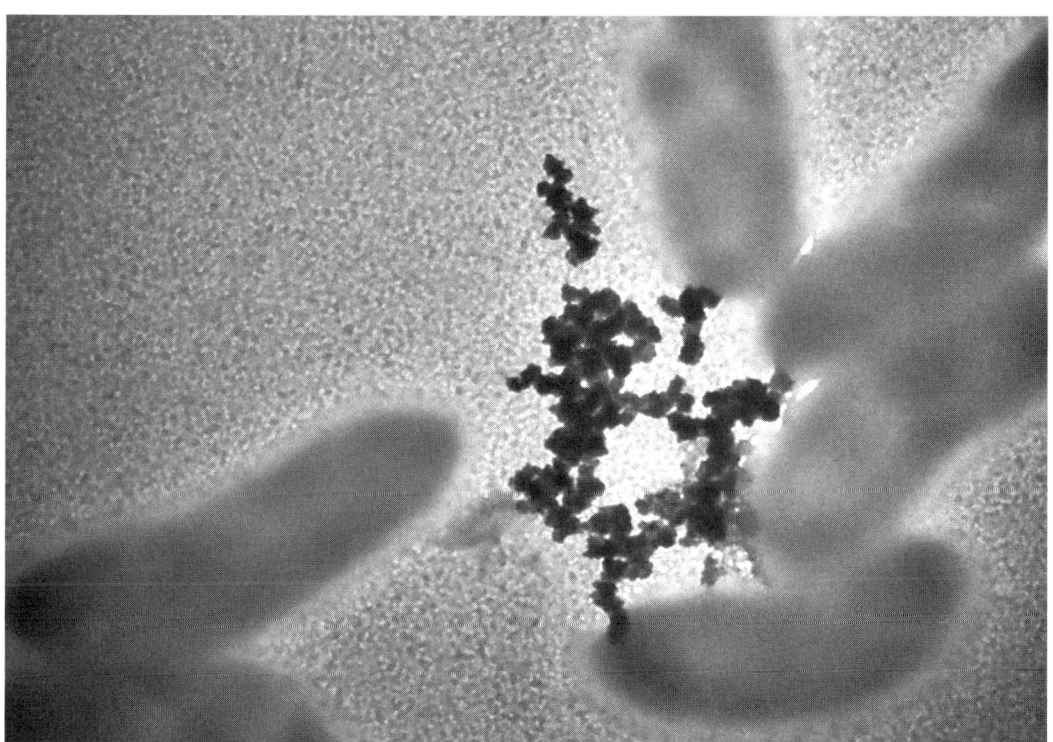

FIGURE 1 Transmission electron micrograph of *D. vulgaris* Hildenborough after 24 h of incubation with 200 mM U(VI). The cells are unstained and show the extracellular nature of U(IV), whereas none was present inside the cells (D. Elias, unpublished data).

mtr and *omc* genes. An increase in *Geobacter ompC* transcripts was observed in situ during biostimulation for U(VI) reduction. While transcriptomic studies for U(VI) exposure in *Desulfovibrio* have not been reported, physiological data have shown that, like Fe(III) and Cr(VI), U(VI) decouples lactate oxidation and sulfate reduction.

STRONTIUM RESPONSES

When *S. oneidensis* MR-1 cells were grown aerobically in the presence of 180 mM strontium chloride and sampled at different times postexposure, siderophore and iron transport–related genes were highly upexpressed over the entire time course (Brown et al., 2006). The siderophore system was related to aerobactin, and the iron-related systems were putative iron, heme, and hemin transport systems. To a lesser extent, transcripts that coded efflux pumps, osmotic adaptation genes, sulfate transporters, and assimilatory sulfur metabolism genes were also upexpressed. A mutant in the siderophore system (siderophore synthetase) was hypersensitive to strontium chloride exposure (Brown et al., 2006). Wild-type cells formed a precipitate that the mutant cells did not, and spectroscopy and mass spectrometry indicated that the precipitates contained strontium (Brown et al., 2006). The *mtrAB* and *omcAB* genes in MR-1 have been shown to be important for iron oxide utilization, and were downexpressed in the strontium exposure study, but downexpression may be due to cell growth in aerobic, undefined medium. It is unknown whether the induction of the iron and heme acquisition systems by strontium is through direct or indirect mechanisms: (i) iron limitation due to strontium precipitates; (ii) derepression of Fur regulon due to interactions with strontium; or (iii) a defense mechanism to sequester toxic heavy metals. Interestingly, when Fe(III) was added back to cells being exposed to strontium, tolerance was not improved, and this result suggested that strontium toxicity was not solely due to iron limitations. In addition, the authors postulated that the upexpression of sulfur acquisition systems could be a result of an increased need for cysteine to synthesize the upexpressed siderophore (Brown et al., 2006).

BIOFILM RESPONSES

It has been commonly accepted that bacterial counts can range from 10^4 to 10^9 cells/g of sediment, depending on the environmental conditions, and that the subsurface microbial biomass is greater than all other microbial environments (Whitman et al., 1998). Specifically for the saturated subsurface, bacterial numbers can typically range from 10^5 to 10^7 cells/g. The distribution of attached versus suspended bacteria in the subsurface is an issue of considerable controversy (Rittman, 1993) and must be considered when dynamic flow can affect mass transport. Oligotrophic groundwater and sediments had different populations and activities at a Savannah River site, with more organisms attached to matrix surfaces (Hazen et al., 1991). At a more eutrophic site, increased levels of bacteria were attached to surfaces in a sewage-contaminated aquifer (Harvey et al., 1984). Studies have used the term "biofilm" to describe bacterial growth in the subsurface (Taylor and Jaffe, 1990; Cunningham et al., 1991), while others have not used the term "film" because of the possible presence of bacterial aggregates. Recent work has shown that surfaces are colonized when placed downwell at contaminated sites at the Oak Ridge-FRC (Reardon et al., 2004; Boyd et al., 2007) and Hanford (Fields et al., unpublished), and these results indicated that in situ microorganisms have the ability to adhere to surfaces. It has also been shown that model organisms for the genera of *Shewanella*, *Geobacter,* and *Desulfovibrio* can adhere to surfaces and grow as biofilms (Wilkens et al., 2007; Clark et al., 2007; McLean et al., 2008).

Regardless of the terms used, subsurface sediments contain bacterial populations, and these populations can be associated with the solid phase. In fact, multiple studies have shown that reduction-oxidation transformations of uranium and other metals are controlled by kinetic factors that are strongly influenced by microbial activity (Suzuki et al., 2005). It is the microbes that are associated with the subsurface matrix that pose challenges in manipulating and/or predicting the mobility and fate of heavy metals in complex and heterogeneous environmental settings (Suzuki and Suko, 2006). Whether microbial growth associated with the subsurface sediments is desired (bioreduction) or deleterious (aquifer clogging), the surface-associated populations represent a portion of the system that we know the least. It is assumed that microorganisms are present, and activity measures can indicate whether biostimulation can occur, but little is known about the nature of activity in relation to the surfaces (where most of the heavy metals and/or radionuclides are located), the effects on hydrology, and the different physiological states of surface-adhered populations.

It is well accepted that nutrient levels are elevated at the liquid-surface interface and can be dynamic, and it is known that the majority of the heavy metals and radionuclides of interest are associated with the solid phase. It has been previously shown that surface-attached populations can differ functionally and physiologically from planktonic populations, and additional studies showed that surface-attached populations displayed increased activities when compared with planktonic cells (Marshall, 1992; Van Loosdrecht et al., 1990; Kolbel-Boelke et al., 1988; Lehman and O'Connell, 2002). In addition, preliminary data with *D. vulgaris* indicated that biofilms had increased resistance to Cr(VI) exposure (Fields et al., unpublished). Even if surface-adhered cells were only a portion of the total subsurface activity, the surface-adhered cells alter the porous media and how it reacts with aqueous geochemistry and heavy metals.

When *D. vulgaris* DSM 664 was grown as a biofilm on a steel surface compared with planktonic cells from the same reactor (40 mM lactate and 20 mM sulfate), many of the upexpressed genes included flagellins, chemotaxis proteins, outer membrane proteins, and exopolysaccharide biosynthesis (Zhang et al., 2007). The biofilm cells downexpressed genes involved in protein synthesis, energy metabolism, and sulfate reduction, and the authors suggested a similarity to a slower growth mode as reported for other organisms (Beloin and Ghigo, 2005). The biofilm grown on steel was reported to be 1 to 2 mm, although the method of determination was not provided and uncovered areas were visible in the scanning electron microscopy images (Zhang et al., 2007). The reported downexpressed genes included genes for the ATP synthase and *ech* dehydrogenase (Zhang et al., 2007). However, a recent study of *D. vulgaris* Hildenborough biofilms grown under sulfate-reducing conditions on glass slides showed that transcripts for ATP synthase genes were downexpressed, but that proteins were upexpressed while transcript and protein levels were upexpressed for the *ech* dehydrogenase genes (Fields et al., unpublished). Some genes predicted to be involved in exopolymer production displayed upexpression, and the biofilm on steel was reported to have a "slimy" appearance, although carbohydrate levels were not measured (Zhang et al., 2007). A separate study characterized *D. vulgaris* biofilms on glass slides under sulfate-reducing conditions, and significant amounts of carbohydrates (hexose, pentose, or uronic acids) were not detected; *D. vulgaris* biofilms on an electrode did not upexpress as many carbohydrate-related genes (Clark et al., 2007; Caffrey et al., 2008). However, it is feasible that *Desulfovibrio* biofilms alter extracellular matrices dependent upon the surface, carbon/energy sources, and other environmental parameters, as other sulfate-reducing bacterium (SRB) biofilms have been reported to contain significant carbohydrate (Beech et al., 1996).

Recently, gene expression was compared between *D. vulgaris* Hildenborough grown on cathodic hydrogen versus that of cells grown with directly supplied hydrogen gas. The electrode cells upexpressed two particular hydrogenases, *hyn*-1 and *hyd*, a [NiFe] and [Fe] hydrogenase, respectively, as well as genes for the *hmc* (high-molecular-weight cytochrome) complex (Caffrey et al., 2008). The Hmc complex was needed for hydrogen oxidation (Dolla et al., 2000) but was not strongly upexpressed in gaseous-hydrogen-grown cells compared with lactate-grown cells (Caffrey et al., 2007). The cells grown directly with gaseous hydrogen upexpressed the *hys* gene for the [NiFeSe] hydrogenase, and both expression profiles were significantly different from that of cells grown with sulfate and lactate (Caffrey et al., 2008). In comparison, *D. vulgaris* did not upexpress the same hydrogenases or the *hmc* genes when grown as biofilms with sulfate and lactate (Fields et al., unpublished). As expected, the electrode-grown cells upexpressed genes assumed to be associated with biofilm formation, including putative signal transduction, carbohydrate transfer, and different flagellar genes, and the authors hypothesized that the biofilm was in a surface-associated, motile state as opposed to a nonmotile, mature biofilm state (Caffrey et al., 2008). With respect to flagella, *D. vulgaris* has six presumptive flagellins, and distinct flagellar filaments may be produced in response to different conditions, may contain a mixture of different flagellins, and may not be used exclusively for motility per se.

Similar to heterogenous biofilms in aerobic bacteria, *S. oneidensis* biofilms stratify with respect to metabolism and size (Teal et al., 2006), and these results underlie the important effects of mass transport. For mature *S. oneidensis* biofilms, regions of cells in an inactive state (depicted via fluorescent reporter proteins) typically upexpressed genes associated with anaerobic growth (Teal et al., 2006). These results showed that slowly metabolizing biofilm cells are not inactive. When iron hydroxide was encapsulated in nanoporous beads, it could be shown that *S. oneidensis* biofilms could reduce the iron without direct contact, and the reduction could be mediated by menaquinone that was associated with the biofilm matrix (Lies et al., 2005). When insertion mutants were screened for the ability to form and maintain biofilms on glass in *S. oneidensis*, 41% of the affected biofilm mutants had a nonmotile phenotype (Thormann et al., 2004). Contrary to results in *Escherichia* and *Pseudomonas*, disruption of particular flagellar genes (i.e., *flhB*, *fliK*, and *pomA*) did not affect initial attachment, but rather the development of mature biofilm, whereas type IV pilin biosynthesis genes appeared to be more specifically involved in adhesion to abiotic surfaces (Thormann et al., 2004).

Transposon mutagenesis was also used to identify and differentiate genes important for biofilm formation and extracellular electron transfer in *G. sulfurreducens* (Rollefson et al., 2009). A mutant in GSU2505 was defective in electrode reduction but demonstrated wild-type biofilm formation. GSU2505 has putative NHL domains (peptide sequence homologies in NCl−1, HT2A, Lin−41 proteins), a transmembrane region, a signal peptide, and therefore, may be a transmembrane serine/threonine protein kinase. A putative ATP-dependent transporter (GSU1501) could not colonize the electrode but could reduce Fe(III) citrate. The data indicated that most insertions were in transporters, signaling proteins, and hypothetical proteins (see below) (Rollefson et al., 2009). When *G. sulfurreducens* was grown on graphite cloth anodes, OmcZ was predicted to act as an electron transfer mediator within the biofilm matrix, while OmcB mediated electron transfer between the biofilm and the abiotic surface (Richter et al., 2009). Type IV pili were important for both types of electron transfer, probably via formation and maintenance of anode biofilms.

Biofilms can be heterogeneous and contain a mixture of cells in different physiological states, and the effects of these physiological state(s) are becoming better appreciated in many aspects of microbial structure and function. The importance of the biofilm state is certainly important for many of the topics discussed in this chapter dealing with metal-reducing bacteria in light of subsurface growth, metal mineral interactions, and electrodes. Data thus far suggest that the studied metal-reducing bacteria express different genes in biofilms related to extracellular appendages, electron transfer, and stress responses.

OXYGEN/NITRATE RESPONSES

Sulfate reduction and iron reduction are typically anaerobic functions that can be observed in habitats with periodic oxic conditions. SRB have developed several defense strategies to survive exposure to oxygen that include aerotaxis and dedicated enzymatic systems for the reduction of oxygen (Dolla et al., 2006, and references therein). In addition, there have been reports for both *Desulfovibrio* and *Geobacter* species that could tolerate and utilize low to high levels of molecular oxygen (Cypionka, 2000; Lobo et al., 2007; Johnson et al., 1997; Lin et al., 2004; Holman et al., 2009). However, the use of complex media (e.g., yeast extract) at any level can alter cellular responses to perturbations (i.e., oxygen) in laboratory experiments, and preliminary data indicate that *D. vulgaris* viability during oxygen exposure is greatly enhanced in the presence of yeast extract (Fields et al., unpublished). The situation is obviously different for *Shewanella* species because of the facultative nature of this genus, and different species of the *Anaeromyxobacter* can tolerate or grow with varying levels of oxygen.

The exposure of *D. vulgaris* to 0.1% oxygen in the headspace caused the upexpression of only 12 genes (Mukhopadhyay et al., 2007), while exposure to air for 1 h upexpressed 130 genes without a decline in cell viability (Zhang et al., 2006a). Exposure to air for 4 h caused the upexpression of almost 400 genes and the downexpression of 450 genes with a 90% decline in cell viability (Mukhopadhyay et al., 2007). When mid-exponential *D. vulgaris* cells were exposed to approximately 1 mM dissolved oxygen, there was an approximate 30% decline in cell viability and the periplasmic [NiFeSe] hydrogenase (*hys*A), formate dehydrogenases, and *hmc* membrane complex were upexpressed with respect to electron transfer mechanisms (Pereira et al., 2008b). The *hys*A hydrogenase has been hypothesized to be able to function as an oxygen-reducing system in the periplasm (Baumgarten et al., 2001) and has been shown to be less sensitive to oxygen compared with the other hydrogenases (Valente et al., 2005). A proteomic analysis of oxygen exposure in *D. vulgaris* (mid-exponential culture sparged with pure oxygen) displayed the upexpression of thiol-specific peroxidases, that included a peroxiredoxin that is associated with bacterioferritin and putative glutaredoxin, and reverse transcriptase PCR showed the corresponding upexpression of transcripts (Fournier et al., 2006).

The *D. vulgaris* data also suggested that Fur-regulated genes were downexpressed perhaps as a means to restrict the availability of free iron and the possible generation of reactive oxygen species via the Fenton reaction. Along those lines, it is possible that the upexpression of *nifU* and *nifS* is due to the need for repair and biosynthesis of iron-sulfur clusters damaged by reactive oxygen species (Pereira et al., 2008b). Interestingly, the PerR regulon did not show a significant change in expression for the Pereira et al. (2008b) study but was upexpressed in nitrite and heat shock studies in *D. vulgaris* (He et al., 2006; Chhabra et al., 2006). A number of energy metabolism genes were downexpressed, including genes for sulfate reduction, and this result suggested a limitation for Fe and/or competition for electrons between oxygen and sulfate (Mukhopadhyay et al., 2007; Pereira et al., 2008). A similar result was observed with nitrite-exposed cells in *D. vulgaris* (He et al., 2006) in which a major cellular response was the turnover of damaged metalloproteins as evidenced by the upexpression of chaperones and proteases.

Regarding the increased nitrite sensitivity in the Fur mutant, a previous study had shown that nitrite exposure in *D. vulgaris* wild-type cells upexpressed the Fur regulon (He et al., 2006). Interestingly, nitrite sensitivity was exacerbated in the Fur mutant, and a similar result was observed in *E. coli* (Mukhopadhyay et al., 2004). This could be a direct result of nitrosylation of Fur-Fe complexes that inactivate repressor and/or activation activities; however, direct evidence has not been presented. In addition, the *D. vulgaris* nitrite transcriptome indicated that nitrite exposure caused a shift of reducing equivalents to nitrite reduction to ammonia (He et al., 2006), and Fur may be an important regulator to help in coordinating iron homeostasis during nitrite stress. The possible connection between oxidative stress and iron homeostasis is mainly mediated via the ability of iron to generate reactive oxygen species. Although *Desulfovibrio* and *Geobacter* are considered to be strict anaerobes, members of these genera are known to survive oxygen exposure and contain oxidative stress response genes. The δ-*Proteobacteria* also have a PerR homolog, and PerR is known to respond to oxidative stress but also iron in *B. subtilis* (Chen et al., 1995).

Two oxidative stress proteins from *Geobacter* spp., CydA and SodA, were monitored during in situ biostimulation for uranium bioreduction. The gene *cydA* is a presumptive ubiquinol oxidase, and *sodA* is a superoxide dismutase. Despite the predicted low levels of dissolved oxygen, both genes were upexpressed (Mouser et al., 2009). An exposure to 5% oxygen for 8 h did cause upexpression of oxidative stress proteins, including *cydA* and *sodA*, in *G. uraniireducens* as well as the downexpression in genes involved in anaerobic respiration and motility (Mouser et al., 2009). However, upexpression of *cydA* and *sodA* was also observed in *G. sulfurreducens* grown anaerobically with Fe(III) compared with fumarate (Mouser et al., 2009). The expression of genes/proteins related to oxidative stress in metal-reducing bacteria may be related to the physiological state of the cells in response to the combination of geochemical parameters and not necessarily oxygen exposure per se, as can be observed with the data for both *Geobacter* and *Desulfovibrio* species. More work is needed to fully understand the physiological responses to combinations of geochemical conditions experienced by organisms that interact with metals.

Unlike *Desulfovibrio* species, *Geobacter* has a *rpoS* gene, and data indicated the RpoS acted as both a positive and negative regulator in *G. sulfurreducens* (Nunez et al., 2006). In a fashion similar to other gram-negative bacteria, the RpoS in *G. sulfurreducens* activated genes involved in oxidative stress and nutrient deprivation. The expression of the cytochrome *c* oxidase system, necessary for use of oxygen as an electron acceptor, was RpoS dependent, as was that of at least 13 *c*-type cytochromes, including *macA* (Nunez et al., 2006). A mutant in *rpoS* was more sensitive to oxygen exposure and could not grow when low levels of oxygen were sequentially added to the culture (Nunez et al., 2006). It is interesting to note that the stress- and stationary-phase-associated RpoS is involved in the oxygen response in *G. sulfurreducens*.

For the facultative anaerobe *S. oneidensis*, the versatility of electron acceptors covers a wide spectrum of the redox potential, and even for central carbon metabolism, recent studies have shown some surprises. It was thought that *S. oneidensis* used the serine-isocitrate pathway to incorporate carbon in the form of formaldehyde from pyruvate. However, metabolite labeling showed that this complete pathway is most likely not used, and that *S. oneidensis* uses a complete TCA cycle under certain anaerobic growth conditions along with a malic enzyme and phosphoenolpyruvate carboxylase (Tang et al., 2007). When transcriptomes with different electron acceptors were compared, many of the metals had similar expression profiles compared to fumarate or oxygen. Interestingly, only one *c*-type cytochrome, SO3300, was upexpressed with all tested metals, and distinct transcriptomes for nitrate, thiosulfate, and trimethylamine oxide were more obvious (Beliaev et al., 2005). The oxygen-grown cells had a larger component of upexpressed genes involved in energy transduction along with thiosulfate compared with the other acceptors (Beliaev et al., 2005a). With respect to oxygen utilization, the following genes were upexpressed compared with the other electron acceptors: a *c*-type cytochrome (SO0264), a *b*-type cytochrome (SO2726), ferritin, and genes involved in type II general secretion (Beliaev et al., 2005a).

In the presence of 3.5 mg/liter dissolved oxygen and 0.7 mM $CaCl_2$, *S. oneidensis* cells formed aggregates, and the cells disaggregated when oxygen levels were below detection (McLean et al., 2008). Transcriptome analysis indicated that increased oxygen levels upexpressed genes involved in cell-cell and cell-surface interactions, and the aggregated cells also upexpressed genes associated with anaerobic growth (*mtrDEF*) (McLean et al., 2008). Mutant analysis showed that *pilD* was required for efficient aggregate formation as well as protection from oxidative agents.

For nitrate, *S. oneidensis* upexpressed cytochromes involved with nitrate and nitrite reduction, but also had higher expression of a Na-pumping NADH oxidoreductase and TCA enzymes (Beliaev et al., 2005). In conjunction with the notion that *S. oneidensis* may use a complete TCA cycle under some nonaerobic conditions (trimethylamine oxide), nitrate may also provide enough soluble electron acceptors to allow this type of electron flow. Proteomic data for nitrate exposure in *D. vulgaris* showed the upexpression of oxidative stress proteins (SodB, Rbr2, and Rbo) as well as nigerythrin and thiol peroxidase. Not surprisingly, NrfA (nitrite reductase) was upexpressed (A. M. Redding and A. Mukhopadhyay, unpublished results). Nitrite exposure in *D. vulgaris* caused the upexpression of transcripts for *nrfA*, cytochrome c_3, formate dehydrogenase, fumarate reductase, and glutamine synthetase (He et al., 2006). Nitrite exposure caused the downexpression of subunits for *dsr* (sulfate-reducing complex) and subunits of the ATP synthase. Interestingly, when multiple stress responses were analyzed via hierarchical clustering, nitrate and nitrite gene expression were different (Color Plate 12). Based upon gene expression, nitrate clustered more with general stationary-phase growth, while the nitrite response clustered more with general exponential-phase growth, and the chromate response was distinct compared with this subset of gene expression data. Nitrite was a stronger inducer of rubrerythrin genes and iron acquisition via the *feo* system compared with nitrate. However, nitrate caused an upexpression for a putative siderophore transporter and pyruvate-formate lyase. The authors proposed that electron transfer to sulfate declined in order to transfer electrons onto nitrite for ammonia formation that was utilized by the glutamine synthetase (He et al., 2006). It would appear that, for *D. vulgaris*, nitrite causes a redirection of electron flow away from sulfate to the formation of ammonia, while nitrate slows down electron flow overall and that pyruvate-formate lyase is used to accommodate altered carbon flow.

Oxygen and nitrates can impact the function of metal-reducing bacteria by competition for electrons as well as the effects on physiological status of the cells. For *D. vulgaris*, the upexpression of hydrogenases and *nif* genes and the downexpression of Fur-related genes were denoted with oxygen exposure. Transcriptomes also indicated that *D. vulgaris* cells were starved for Fe, and a competition occurred between sulfate and oxygen. In *Geobacter*, *cydA* and *sodA* responded to oxygen exposure, but a similar response may be a result of oxidative damage associated with metal respiration. RpoS in *Geobacter* appears to help control oxygen responses as well as the expression of particular *c*-type cytochromes. For *Shewanella*, distinct transcriptomes for nitrate, thiosulfate, and trimethylamine oxide were more obvious than for Fe, Mn, or Co. For *D. vulgaris*, nitrate and nitrite exposure shifted the flow of electrons away from sulfate (He et al., 2006, 2010).

ELECTRON DONOR RESPONSES

Many studies have been performed over the past several decades to gain a better and more comprehensive understanding of the metabolism and physiology of a particular microorganism. This is mostly done by building on previous studies that look at very discrete points within a given biochemical pathway or set of pathways, and then attempting to elucidate the functional pathway and any flux of carbon and electrons between intersecting pathways. This is usually accomplished by comparing different physiological states, or mutants versus wild-type cells. However, one of the most telling approaches for accomplishing these goals is the choice or choices of electron donors to be used in the study. With respect to the organisms discussed in this chapter, virtually all can utilize H_2 for growth, while *Shewanella* and *Desulfovibrio* incompletely oxidize the organic donors lactate, pyruvate, and formate, and *Geobacter* species can completely oxidize acetate.

With respect to sulfate-reducing bacteria, the theory of H_2 cycling was originally proposed by Odom and Peck (1981) based upon both physiological studies along with the isolation and characterization of hydrogenases and cytochrome c_3, and was later reviewed in the context of all known electron transfer reactions at that time (Odom and Peck, 1984). The rationale was at least partially based upon the observation that when *D. vulgaris* used lactate as the sole carbon and electron source, (i) a transient production of H_2 was observed, which relatively quickly decreased to below detectable concentrations, and (ii) both H_2-utilizing methanogens and syntrophs could grow in coculture with *D. vulgaris* via interspecies H_2 transfer. These findings suggested that *D. vulgaris*, in particular, and perhaps sulfate-reducing bacteria, in general, reconstituted molecular H_2 in the cytoplasm from the electrons of lactate oxidation and the available cytoplasmic protons, essentially using a hydrogenase in reverse. The H_2 was then transported to the periplasm where it was treated the same as exogenous H_2, oxidized by the periplasmic hydrogenase(s) and the electrons used to reduce sulfate. This phenomenon was further explored by assessing the "hydrogen burst" observed in the early stages of lactate-sulfate-mediated growth in *D. vulgaris* by using a deletion mutant in the [Fe] hydrogenase (Dolla et al., 2000; Pohorelic et al., 2002). It was found that the absence of this enzyme resulted in substantial H_2 accumulation relative to wild-type cells and that molecular H_2 is indeed produced from lactate oxidation, while also suggesting that the [Fe] hydrogenase is not responsible for the reconstitution of H_2. However, it should be noted that different forms of hydrogen cycling are still debated for the SRB, and the importance of molecular hydrogen in electron movement may be time and donor dependent.

More recently, it was discovered that there are relatively discrete roles for particular hydrogenases within the sulfate-reducing bacteria. This was done using deletion mutants in each of the [Fe] hydrogenase (*hyd*), [NiFeSe] hydrogenase (*hys*), and two [NiFe] hydrogenases (*hyn1*and *hyn2*) in *D. vulgaris* Hildenborough in combination with 5% or 50% H_2, or lactate (Caffrey et al., 2007). The authors observed that Δ*hyd* cells were negatively impacted for growth with 50% H_2 or lactate, while the Δ*hys* cells showed less growth with 5% H_2. These results suggested that the [NiFeSe] hydrogenase was preferentially utilized for low H_2 concentrations, while the [Fe] hydrogenase was used for higher intracellular H_2 concentrations such as when 50% H_2 or lactate was supplied. Furthermore, whole genome microarrays revealed that >600 genes were differentially expressed in either H_2 concentration versus lactate, and 263 genes had altered expression between the higher and lower H_2 concentrations. Essentially, the expression pattern for the *hys* genes followed the order 5% hydrogen > 50% hydrogen > lactate, but the *hyd* genes showed the opposite trend. The authors suggested upexpression of several of the *rnf*, *hmc*, and *ech* genes, and according to the original H_2-cycling theory and the proposition of the *ech* hydrogenase being the reverse hydrogenase, the *ech* genes would be expected to be upexpressed along with the [Fe] hydrogenase.

There have also been a number of other studies using sulfate reducers that compare different energy donors and acceptors (hydrogen/sulfate, pyruvate/sulfate, and lactate/thiosulfate) compared with a control of lactate and sulfate. It should be noted that a modified Postgate C medium was used that contained yeast extract (1 g/liter) and could have impacted physiological responses; however, all treatments and the control contained the same amount of yeast extract, although this should be considered when compared with other experimental results that did not have complex nutrients in the medium. Cells grown with hydrogen as the electron donor displayed the largest changes in gene expression, and the periplasmic (*hysAB*) and *ech* hydrogenases were significantly upexpressed (Pereira et al., 2008b). There was also evidence for formate cycling when hydrogen was the electron donor, but the CO hydrogenase was upexpressed when cells were grown with lactate or pyruvate (Pereira et al., 2008b). The *hmc* complex genes were downexpressed in hydrogen-grown cells compared with lactate-grown cells, but the cytochrome *c* gene (*tmcA*, DVU0263) for the Tmc complex (homologous to Hmc) was upexpressed (Pereira et al., 2008b).

Zhang et al. (2006b) also used a medium with yeast extract and data suggested that the *hynBA* hydrogenase was the primary hydrogenase linked to lactate growth, while the *hynBA* and *hyd* hydrogenases performed hydrogen oxidation during formate growth. When *D. vulgaris* cells were grown with different levels of exogenous hydrogen, expression analysis indicated that the [Fe] hydrogenase was a low-affinity, high-activity enzyme and the [NiFeSe] hydrogenase was a high-affinity, low-activity enzyme that the cell will use at lower intracellular hydrogen concentrations (Caffrey et al., 2008). Different alcohol dehydrogenases have been shown to be upexpressed during lactate or formate growth, and Haveman et al. (2003) proposed a hydrogen-cycling mechanism via reductants from cytoplasmic ethanol oxidation and a heterodisulfide oxidoreductase in *D. vulgaris*. When compared with lactate-grown cells, formate-grown cells upexpressed formate dehydrogenases (DVU0588/DVU2481), alcohol dehydrogenase (DVU2405), four *ech* genes, and cytochrome c_{553} (Zhang et al., 2006b). The cytochrome c_{553} is needed for the electron transfer mechanisms of a formate dehydrogenase in *D. vulgaris* (Kreuzer et al., 1998). Deletion mutants of Δ*hyn*, Δ*hmc*, and Δ*hyd* were constructed in a strain of *D. vulgaris* that lacks the 200-kb native plasmid (pDV1) and tested using this strain as the "wild type" for comparing soluble and insoluble Fe(III) reduction (Park et al., 2008). The data suggested that ferric nitrilotriacetic acid reduction is a periplasmic process that requires a periplasmic hydrogenase, and that the reduction of complexed iron was not linked to growth. Further attempts to reduce solid-phase ferric iron failed with this strain and the mutants, and these results suggested that structures such as bacterial nanowires were absent in *D. vulgaris*. However, similar experiments with wild-type *D. vulgaris* that does contain the native plasmid contradict this observation, in that wild-type *D. vulgaris* can reduce ferrihydrate (D. Elias, unpublished results). Whether other structures are present is currently unknown.

In comparison, sulfate/lactate-grown biofilm cells upexpressed *ech* hydrogenase, formate dehydrogenase (DVU2482), and *rnf* oxidoreductase genes in comparison with planktonic cells (Fields et al., unpublished). However, biofilm cells also upexpressed *hypA* hydrogenase, and this gene was downexpressed in hydrogen-grown cells (Fields et al., unpublished; Pereira et al., 2008b). Pyruvate- and sulfate-grown cells displayed few significant gene expression changes compared with lactate- and sulfate-grown cells, but *ech* hydrogenase genes were also upexpressed in the pyruvate cells, similarly to hydrogen-grown cells, and it is difficult to ascertain the direction of electrons through the *ech* complex (formation versus utilization of molecular hydrogen). Interestingly, growth with thiosulfate and lactate caused the downexpression of several metabolism genes similar to a nitrite exposure and might be a consequence of higher accumulation of sulfite- compared with sulfate-based growth (Pereira et al., 2008b; He et al., 2006). Pereira et al. (2008b) hypothesized, based on an accumulation of data, that formate cycling and the Tmc complex are more important for hydrogen growth, while carbon monoxide cycling and the Qmo and Dsr complexes are more important for lactate growth in planktonic cells.

Four putative acetate transporters (AplA, AplB, AplC, and AplD) were identified in *G. sulfurreducens* and expression was monitored via quantitative PCR in acetate- and fumarate-limited cultures. The results indicated that *aplA,B,C* were upexpressed under acetate-limited conditions, while *aplD* expression did not change significantly under the tested conditions (Risso et al., 2008). Likewise, *D. vulgaris* has six presumptive lactate permease genes (DVU2110, DVU2285, DVU2451, DVU2683, DVU3284, and DVU3026), and expression was followed throughout the transition to electron-donor depletion. As the cells depleted lactate levels, two permeases had unchanged expression, three were downexpressed, and one permease was upexpressed (Clark et al., 2006). In the same study, three sulfate permeases (DVU0053, DVU1999, and DVU0279) displayed different expression profiles as lactate (and sulfate) was depleted relative to one another (Clark et al., 2006). These results suggest that, as expected, the studied metal-reducing bacteria have multiple transporters that respond to varying concentrations of exogenous carbon and energy sources.

A recent study in *G. sulfurreducens* used multiple microarray experiments that consisted of oxidative stress, biofilm, planktonic, and various electron acceptors to identify genetic regulatory networks for central metabolism, sulfate metabolism, hydrogen metabolism, and acetate uptake within the context of in silico metabolic models (Mahadevan et al., 2008). K-means clustering analysis of the expression data identified 13 clusters that contained 256 genes in different functional categories, and a high fraction of genes were related to energy metabolism, transport, and nucleotide synthesis. For example, clustering analysis identified potential coregulation of the *hyb* operon with *omcS* and *omcT* operons, along with putative transcriptional factor binding sequences. The *hyp* genes code for hydrogenases that are essential for hydrogen-dependent iron reduction, and the Omc proteins are essential for the reduction of iron oxides as well as electrodes (Mahadevan et al., 2008). A potential transcriptional factor, GSU1072, was predicted to control gene expression of a gene cluster that included *hya* (hydrogenase), two acetate permeases, and a pyruvate ferredoxin:oxidoreductase (Mahadevan et al., 2008).

Different from the typical conditions of abiotic electron donors and acceptors, the syntrophic growth of microorganisms that cross-feed each other represent different sources of electron donors and acceptors and have implications for many anaerobic environments (Galushko and Schink, 2000; Stams and Plugge, 2009). Several recent studies have used transcriptomics to study the syntrophic growth of *D. vulgaris* with *Methanococcus maripaludis* when *D. vulgaris* oxidized lactate and produced hydrogen gas in the absence of sulfate and *M. maripaludis* consumed the hydrogen gas. The consumption of hydrogen gas by the methanogen was confirmed via metabolic modeling and showed that formate was not a significant contribution as an interspecies electron shuttle (Stolyar et al., 2007). Comparative transcriptomics showed that *D. vulgaris* altered the electron transfer apparatus during syntrophic growth compared with lactate plus sulfate growth (Walker et al., 2009). Namely, the *coo* hydrogenase, two periplasmic hydrogenases (*hyd* and *hyn*), and *hmc* were highly upexpressed, and mutants in these genes were deficient for lactate oxidation during syntrophic growth (Walker et al., 2009). In addition, some genes involved in lactate transport and oxidation were upexpressed. These data coincide with other studies that showed upexpression of *hmc*, *hyn*, and *hyd* when cells were grown with cathodic hydrogen (Caffrey et al., 2008) as well as the downexpression of *hmc* in hydrogen-grown cells (Pereira et al., 2008b).

Just as metal-reducing bacteria alter electron flow in response to electron acceptors, the presence and/or utilization of different electron donors can change gene expression and physiology. Hydrogen is a crucial intermediate for both metal- and sulfate-reducing bacteria. In particular, electron donors can affect electron flow in *Desulfovibrio*.

HYPOTHETICAL AND CONSERVED HYPOTHETICAL GENE RESPONSES

While transcriptomic, proteomic, and other measures of quantifying gene regulation at either the mRNA or protein level have yielded unprecedented insight into understanding bacterial metabolism, the scientific community is still struggling with the issue of hypothetical and conserved genes. These genes are essentially unknown and are classified as such upon genome sequence determination. Hypotheticals are defined as genes with no significant nucleotide sequence similarity (i.e., homology) to any characterized or uncharacterized predicted genes. Conserved hypotheticals are genes that have significant similarity to a predicted protein in another species or strain without direct evidence of gene expression (Elias et al., 2009).

Even as additional genomes are sequenced, approximately one-third of all of the genes within a given genome are typically predicted to encode hypothetical or conserved hypothetical genes (Kolker et al., 2004). In fact, such genes in *D. vulgaris* Hildenborough, *G. sulfurreducens* PCA, *G. metallireducens* GS-15, *G. uraniireducens* Rf4, and *S. oneidensis* MR-1 collectively comprise 34.5%, 28.8%, 21.9%, 29.1%, and 41.0% of the respective genomes. In the context of transcriptional regulation and its relation to metal reduction, a common theme has been the significant proportion of genes with altered expression is hypothetical or conserved hypothetical proteins. A recent survey of these genes in *D. vulgaris* transcriptomic data from a variety of stress conditions showed that all but 17 of 1,234 such genes were expressed, and virtually all of them displayed differential regulation in one or more stress conditions, including oxygen exposure, iron starvation, chromate exposure, and biofilm growth (Elias et al., 2009). Furthermore, deletion of one gene (DVUA0095) that was only upregulated with chromate exposure resulted in delayed growth in the presence of chromium and a reduced rate of Cr(VI) reduction. Other studies such as a Fur deletion in *D. vulgaris* revealed that many of the most differentially regulated genes were hypotheticals (Bender et al., 2007), with many more hypotheticals appearing to be influenced by Fur regulation to some degree as described above (Elias et al., 2009).

In the case of *Shewanella*, transcriptomic studies have revealed that many hypothetical and conserved hypothetical genes were differentially regulated during metal reduction. In fact, using a 3-fold cutoff for such regulation, approximately one-third of the upexpressed genes during U(VI) and Cr(VI) reduction, respectively, were hypothetical genes (Bencheikh-Latmani et al., 2005). In fact, of the 32 most highly upexpressed genes with both metals, 12 were cytochromes and 10 were hypotheticals. Similar results were observed with respect to the differential regulation of hypothetical genes during Cr(VI) exposure in *S. oneidensis* MR-1 (Chourey et al., 2006).

In a more comprehensive manner, three separate studies have attempted to classify uncharacterized genes. The first focused on the hypothetical genes and used differential transcription as well as comprehensive proteomics before and after the perturbations of UV irradiation and suboxic growth in *Shewanella* (Kolker et al., 2004). The results indicated that fewer than half of the differentially expressed genes could be assigned a specific biochemical function. More recent studies showed that virtually all of the hypothetical and conserved hypothetical genes in *S. oneidensis* MR-1 were expressed according to mass spectrometry-based proteomics (Elias et al., 2005, 2006). These results are similar to the confirmed expression of all but 17 predicted genes in *D. vulgaris* (Elias et al., 2009).

SUMMARY

Based upon the many different transcriptomic studies of stress and growth responses in the three model metal-reducing bacteria, *S. oneidensis*, *G. sulfurreducens*, and *D. vulgaris*, some similarities and differences can be discerned. Obviously, expression of genes involved with energy and electron flow is altered with respect to the nature of both electron acceptors and donors, particularly reduction-oxidation potentials and solubility. As noted in previous reviews, a major difference between these model bacteria is the location of electron transfer to the terminal electron acceptor, namely, outer membrane for *Shewanella* and *Geobacter* species, as opposed to the cytoplasm and periplasm for *Desulfovibrio* species. Given the unique metabolism and nature of metal interactions unique to organisms that depend on iron, not only for a high complement of cytochromes and iron-sulfur proteins, but also as an electron acceptor for energy generation, the coordination of metal homeostasis appears to be a major theme for the studies done to date. One must keep in mind a major inherent difference between iron reducers, such as *Shewanella* and *Geobacter,* and sulfate reducers (SRB), such as *Desulfovibrio,* is that iron reducers utilize iron as a terminal electron acceptor compared with SRB that still need the iron for large numbers of cytochromes, yet experience iron starvation as a consequence of sulfide precipitation.

Shewanella, Geobacter, and *Desulfovibrio* contain genes with homology to typical metal-responsive regulators (e.g., *mer, cad, cop, ars, znt, fur, pur, zur, fnr,* and *crp*). However, more work is needed to elucidate the respective regulatory networks for these regulators in anaerobic microorganisms and how the regulatory networks are coordinated. For example, Fur is a global regulator that typically represses gene expression in the presence of Fe(II) in many different bacteria. Fur has been shown to have a role in oxidative stress in *E. coli,* and recent work with the model bacteria discussed in this chapter has shown the importance of iron homeostasis. In addition, the speciation of available iron is more complex for anaerobic microorganisms that directly and/or indirectly reduce different iron compounds. Heavy metal exposure and utilization can cause oxidative damage and iron starvation, and could possibly alter the activity of metal-sensitive regulators. The data suggest that, although the regulators are common in these bacteria, the stimuli and coordinated networks are different, but crucial under many of the different conditions studied to date.

Another major theme appeared to be related to the localization of electron transfer in terms of transporters. Both influx and efflux transport systems responded to different environmental stimuli and partly depended on the organism and localization of metal interactions. In addition, the importance of hypothetical and conserved hypothetical proteins was evident from multiple studies that observed the up- and downexpression of these putative genes. The unknown nature of their involvement underscores the exquisite and intricate control of bacterial metabolism to optimize activity within the context of environmental parameters. As systems biology methodology advances, the elucidation of genetic regulatory networks that control metabolism will provide great insight into the system-wide mechanisms that allow biological cells to synchronize physiology with the cellular environment.

ACKNOWLEDGMENTS

We thank collaborators who have shared unpublished data as well as Prof. G. Geesey for comments and discussion. M.W.F. thanks the members of his laboratory for comments and discussions. M.W.F. and D.A.E. are members of ENIGMA (http://enigma.lbl.gov) supported by the U.S. Department of Energy, Office of Biological and Environmental Research, Genomic Sciences program, through contract DE-AC02-05CH11231 between Lawrence Berkeley National Laboratory and the U.S. Department of Energy and contract DE-ACO5-00OR22725 between Oak Ridge National Laboratory (managed by UT Battelle, LLC) and the U.S. Department of Energy.

REFERENCES

Afkar, E., G. Reguera, M. Schiffer, and D. R. Lovley. 2005. A novel *Geobacteraceae*-specific outer membrane protein J (OmpJ) is essential for electron transport to Fe(III) and Mn(IV) oxides in *Geobacter sulfurreducens. BMC Microbiol.* **5:**41.

Agency for Toxic Substances and Disease Registry (ATSDR). 2004. *Toxicological Profile for Chromium.* Agency for Toxic Substances and Disease Registry, Atlanta, GA.

Anderson, R. T., H. A. Vrionis, I. Ortiz-Bernad, C. T. Resch, P. E. Long, R. Dayvault, K. Karp, S. Marutzky, D. R. Metzler, A. Peacock, D. C. White, M. Lowe, and D. R. Lovley. 2003. Stimulating the in situ activity of *Geobacter* species to remove uranium from the groundwater of a uranium-contaminated aquifer. *Appl. Environ. Microbiol.* **69:**5884–5891.

Andreini, C., L. Banci, I. Bertini, S. Elmi, and A. Rosato. 2007. Non-heme iron through the three domains of life. *Proteins* **67:**317–324.

Barton, L. L., K. Choudhury, B. M. Thomsom, K. Steenhoudt, and A. R. Groffman. 1996. Bacterial reduction of soluble uranium: the first step of *in situ* immobilization of uranium. *Radioact. Waste Manage. Environ. Rest.* **20:**141–151.

Baumgarten, A., I. Redenius, J. Kranczoch, and H. Cypionka. 2001. Periplasmic oxygen reduction by *Desulfovibrio* species. *Arch. Microbiol.* **176:**306–309.

Beech, I. B., C. W. S. Cheung, D. B. Johnson, and J. R. Smith. 1996. Comparative studies of bacterial biofilms on steel surfaces using atomic force microscopy and environmental scanning electron microscopy. *Biofouling* **10:**65.

Beliaev, A. S., D. M. Klingeman, J. A. Klappenbach, L. Wu, M. F. Romine, J. M. Tiedje, K. H. Nealson, J. K. Fredrickson, and J. Zhou. 2005. Global transcriptome analysis of *Shewanella oneidensis* MR-1 exposed to different terminal electron acceptors. *J. Bacteriol.* **187:**7138–7145.

Beliaev, A. S., D. A. Saffarini, J. L. McLaughlin, and D. Hunnicutt. 2001. MtrC, an outer membrane decahaem c cytochrome required for metal reduction in *Shewanella putrefaciens* MR-1. *Mol. Microbiol.* **39:**722–730.

Beliaev, A. S., D. K. Thompson, M. W. Fields, L. Wu, D. P. Lies, K. H. Nealson, and J. Zhou. 2002a. Microarray transcription profiling of a *Shewanella oneidensis* etrA mutant. *J. Bacteriol.* **184:**4612–4616.

Beliaev, A. S., D. K. Thompson, T. Khare, H. Lim, C. C. Brandt, G. Li, A. E. Murray, J. F. Heidelberg, C. S. Giometti, J. Yates III, K. H. Nealson, J. M. Tiedje, and J. Zhoui. 2002b. Gene and protein expression profiles of *Shewanella oneidensis* during anaerobic growth with different electron acceptors. *OMICS* **6:**39–60.

Beller, H. R., T. C. Legler, F. Bourguet, T. E. Letain, S. R. Kane, and M. A. Coleman. 2009. Identification of c-type cytochromes involved in anaerobic, bacterial U(IV) oxidation. *Biodegradation* **20:**45–53.

Beloin, C., and J. M. Ghigo. 2005. Finding geneexpression patterns in bacterial biofilms. *Trends Microbiol.* **13:**16–19.

Bencheikh-Latmani, R., S. M. Williams, L. Haucke, C. S. Criddle, L. Wu, J. Zhou, and B. M. Tebo. 2005. Global transcriptional profiling of *Shewanella oneidensis* MR-1 during Cr(VI) and U(VI) reduction. *Appl. Environ. Microbiol.* **71:**7453–7460.

Bender, K. S., H.-C. Yen, C. L. Hemme, Z. Yang, Z. He, Q. He, J. Zhou, K. H. Huang, E. J. Alm, T. C. Hazen, A. P. Arkin, and J. D. Wall. 2007. Analysis of a ferric uptake regulator (Fur) mutant of *Desulfovibrio vulgaris* Hildenborough. *Appl. Environ. Microbiol.* **73:**5389–5400.

Berks, B. C. 1996. A common export pathway for proteins binding complex redox cofactors? *Mol. Microbiol.* **22:**393–404.

Berks, B. C., T. Palmer, and F. Sargent. 2003. The Tat protein translocation pathway and its role in microbial physiology. *Adv. Microb. Physiol.* **47:**187–254.

Berks, B. C., F. Sargent, and T. Palmer. 2000. The TAT protein export pathway. *Mol. Microbiol.* **35:**260–274.

Boyd, E. S., D. E. Cummings, and G. G. Geesey. 2007. Mineralogy influences structure and diversity of bacterial communities associated with geological substrata in a pristine aquifer. *Microb. Ecol.* **54:**170–182.

Bretschger, O., A. Obraztsova, C. A. Sturm, I. S. Chang, Y. A. Gorby, S. B. Reed, D. E. Culley, C. L. Reardon, S. Barua, M. F. Romine, J. Zhou, A. S. Beliaev, R. Bouhenni, D. Saffarini, F. Mansfeld, B. H. Kim, J. K. Fredrickson, and K. H. Nealson. 2007. Current production and metal oxide reduction by *Shewanella oneidensis* MR-1 wild type and mutants. *Appl. Environ. Microbiol.* **73:**7003–7012.

Brown, S. D., M. Martin, S. Deshpande, S. Seal, K. Huang, E. Alm, Y. Yang, L. Wu, T. Yan, X. Liu, A. P. Arkin, K. Chourey, J. Zhou, and D. K. Thompson. 2006. Cellular response of *Shewanella oneidensis* to strontium stress. *Appl. Environ. Microbiol.* **72:**890–900.

Brown, S. D., M. R. Thompson, N. C. VerBerkmoes, K. Chourey, M. Shah, J. Zhou, R. L. Hettich, and D. K. Thompson. 2006. Molecular dynamics of the *Shewanella oneidensis* response to chromate stress. *Mol. Cell. Proteomics* **5:**1054–1071.

Caffrey, S. M., H.-S. Park, J. Been, P. Gordon, C. W. Sensen, and G. Voordouw. 2008. Gene expression by the sulfate-reducing bacterium *Desulfovibrio vulgaris* Hildenborough grown on an iron electrode under cathodic protection conditions. *Appl. Environ. Microbiol.* **74:**2404–2413.

Caffrey, S. M., and G. Voordouw. 2010. Effect of sulfide on growth physiology and gene expression of *Desulfovibrio vulgaris* Hildenborough. *Antonie van Leeuwenhoek* **97:**11–20.

Caffrey, S. M., H.-S. Park, J. K. Voordouw, Z. He, J. Zhou, and G. Voordouw. 2007. Function of periplasmic hydrogenases in the sulfate-reducing bacterium *Desulfovibrio vulgaris* Hildenborough. *J. Bacteriol.* **189:**6159–6167.

Cervantes, C., J. Campos-Garca, S. Devars, F. Gutierrez-Corona, H. Loza-Tavera, J. C. Torres-Guzman, and R. Moreno-Sanchez. 2001. Interactions of chromium with microorganisms and plants. *FEMS Microbiol. Rev.* **25:**335–347.

Chang, Y., A. D. Peacock, P. E. Long, J. R. Shephen, J. P. McKinley, S. J. MacNaughton,

A. K. M. Anwar-Hussain, A. M. Saxton, and D. C. White. 2001. Diversity and characterization of sulfate-reducing bacteria in groundwater at a uranium mill tailings site. *Appl. Environ. Microbiol.* **67:**3149–3160.

Chapelle, F. H., and D. R. Lovley. 1992. Competitive exclusion of sulfate-reduction by iron(III)-reducing bacteria: a mechanism for producing discrete zones of high-iron ground water. *Ground Water* **30:**29–36.

Charania, M. A., K. L. Brockman, Y. Zhang, A. Banerjee, G. E. Pinchuk, J. K. Fredrickson, A. S. Beliaev, and D. A. Saffarini. 2009. Involvement of a membrane-bound class III adenylate cyclase in regulation of anaerobic respiration in *Shewanella oneidensis* MR-1. *J. Bacteriol.* **191:**4298–4306.

Chardin, B., A Dolla, F. Chaspoul, M. L. Fardeau, P. Gallice, and M. Bruschi. 2002. Bioremediation of chromate: thermodynamic analysis of the effects of Cr(VI) on sulfate-reducing bacteria. *Appl. Microbiol. Biotechnol.* **60:**352–360.

Chen, L., L. Keramati, and J. D. Helmann. 1995. Coordinate regulation of *Bacillus subtilis* peroxide stress genes by hydrogen peroxide and metal ions. *Proc. Natl. Acad. Sci. USA* **92:**8190–8194.

Chhabra, S. R., Q. He, K. H. Huang, S. P. Gaucher, E. J. Alm, Z. He, M. Z. Hadi, T. C. Hazen, J. D. Wall, J. Zhou, A. P. Arkin, and A. K. Singh. 2006. Global analysis of heat shock response in *Desulfovibrio vulgaris* Hildenborough. *J. Bacteriol.* **188:**1817–1828.

Chin, K. J., A. Esteve-Nunez, C. Leang, and D. R. Lovley. 2004. Direct correlation between rates of anaerobic respiration and levels of mRNA for key respiratory genes in *Geobacter sulfurreducens*. *Appl. Environ. Microbiol.* **70:**5183–5189.

Chourey, K., M. R. Thompson, J. Morrell-Falvey, N. C. VerBerkmoes, S. D. Brown, M. Shah, J. Zhou, M. Doktycz, R. L. Hettich, and D. K. Thompson. 2006. Global molecular and morphological effects of 24-hour chromium(VI) exposure on *Shewanella oneidensis* MR-1. *Appl. Environ. Microbiol.* **72:**6331–6344.

Chourey, K., W. Wei, X.-F.Wan, and D. K. Thompson. 2008. Transcriptome analysis reveals response regulator SO2426-mediated gene expression in *Shewanella oneidensis* MR-1 under chromate challenge. *BMC Genomics* **9:**395.

Clark, M. E., R. E. Edelmann, M. L. Duley, J. D. Wall, and M. W. Fields. 2007. Biofilm formation in *Desulfovibrio vulgaris* Hildenborough is dependent upon protein filaments. *Environ. Microbiol.* **9:**2844–2854.

Clark, M. E., Q. He, Z. He, E. J. Alm, K. H. Huang, T. C. Hazen, A. P. Arkin, J. D. Wall, J. Zhou, and M. W. Fields. 2006. Temporal transcriptomic analyses of *Desulfovibrio vulgaris* Hildenborough during electron donor depletion. *Appl. Environ. Microbiol.* **72:**5578–5588.

Craft, E. S., A. W. Abu-Qare, M. M. Flaherty, M. C. Garofolo, H. L. Rincavage, and M. B. Abou-Donia. 2004. Depleted and natural uranium: chemistry and toxicological effects. *J. Toxicol. Environ. Health Part B: Crit. Rev.* **7:**297–317.

Cunningham, A., W. G. Characklis, F. Abedeen, and D. Crawford. 1991. Influence of biofilm accumulation on porous media hydrodynamics. *Environ. Sci. Technol.* **25:**1305–1311.

Cypionka, J. 2000. Oxygen respiration by *Desulfovibrio* species. *Annu. Rev. Microbiol.* **54:**827–848.

DeRisi, J. L., V. R. Iyer, and P. O. Brown. 1997. Exploring the metabolic and genetic control of gene expression on a genomic scale. *Science* **278:**680–686.

DiChristina, T. J., C. M. Moore, and C. A. Haller. 2002. Dissimilatory Fe(III) and Mn(IV) reduction by *Shewanella putrefaciens* requires *ferE*, a homolog of the *pulE* (*gspE*) type II protein secretion gene. *J. Bacteriol.* **184:**142–151.

DiDonato, L. N., S. A. Sullivan, B. A. Methé, K. P. Nevin, R. England, and D. R. Lovley. 2006. Role of Rel$_{Gsu}$ in stress response and Fe(III) reduction in *Geobacter sulfurreducens*. *J. Bacteriol.* **188:**8469–8478.

Dilks, K., R. W. Rose, E. Hartmann, and M. Pohlschroder. 2003. Prokaryotic utilization of the twin-arginine translocation pathway: a genomic survey. *J. Bacteriol.* **185:**1478–1483.

Dolla, A., B. K. Pohorelic, J. K. Voordouw, and G. Voordouw. 2000. Deletion of the hmc operon of *Desulfovibrio vulgaris* subsp. vulgaris Hildenborough hampers hydrogen metabolism and low-redox-potential niche establishment. *Arch. Microbiol.* **174:**143–151.

Dolla, A., M. Fourniera, and Z. Dermoun. 2006. Oxygen defense in sulfate-reducing bacteria. *J. Biotechnol.* **126:**87–100.

Elias, D. A., E. C. Drury, A. M. Redding, A. Mukhopadyay, H.-C. B. Yen, M. W. Fields, T. C. Hazen, A. P. Arkin, J. D. Keasling, and J. D. Wall. 2009. Expression profiling of hypothetical genes in *Desulfovibrio vulgaris* leads to improved functional annotation. *Nucleic Acids Res.* **37:**2926–2939.

Elias, D. A., M. E. Monroe, M. J. Marshall, M. F. Romine, A. S. Belieau, J. K. Fredrickson, G. A. Anderson, R. D. Smith, and M. S. Lipton. 2005. Global detection and characterization of hypothetical proteins in *Shewanella oneidensis* MR-1 using LC-MS-based proteomics. *Proteomics* **5:**3120–3130.

Elias, D. A., M. E. Monroe, R. D. Smith, J. K. Fredrickson, and M. S. Lipton. 2006. Confirmation of the expression of a large set of conserved hypothetical proteins in *Shewanella oneidensis* MR-1. *J. Microbiol. Methods* **66**:223–233.

Elias, D. A., L. R. Krumholz, D. Wong, P. E. Long, and J. M. Suflita. 2003. Characterization of microbial activities and U reduction in a shallow aquifer contaminated by uranium mill tailings. *Microb. Ecol.* **46**:83–91.

Elias, D. A., J. M. Suflita, M. J. McInerney, and L. R. Krumholz. 2004. Periplasmic cytochrome c3 of *Desulfovibrio vulgaris* is directly involved in H_2-mediated metal but not sulfate reduction. *Appl. Environ. Microbiol.* **70**:413–420.

Elias, D. A., S. L. Tollakson, D. W. Kennedy, H. M. Mottaz, C. S. Giometti, J. S. McLean, E. A. Hill, G. E. Pinchuk, M. S. Lipton, J. K. Fredrickson, and Y. A. Gorby. 2008. The influence of cultivation methods on *Shewanella oneidensis* physiology and proteome expression. *Arch. Microbiol.* **189**:313–324.

Fournier, M., C. Aubert, Z. Dermoun, M.-C. Durand, D. Moinier, and A. Dolla. 2006. Response of the anaerobe *Desulfovibrio vulgaris* Hildenborough to oxidative conditions: proteome and transcript analysis. *Biochimie* **88**:85–94.

Galushko, A. S., and B. Schink. 2000. Oxidation of acetate through reactions of the citric acid cycle by *Geobacter sulfurreducens* in pure culture and in syntrophic coculture. *Arch. Microbiol.* **174**:314–321.

Goulhen, F., A. Gloter, F. Guyot, and M. Bruschi. 2005. Cr(VI) detoxification by *Desulfovibrio vulgaris* strain Hildenborough: microbe–metal interactions studies. *Appl. Microbiol. Biotechnol.* **71**:892–897.

Gourion, B., S. Sulser, J. Frunzke, A. Francez-Charlot, P. Stiefel, G. Pessi, J. A. Vorholt, and H. M. Fischer. 2009. The PhyR-sigma(EcfG) signalling cascade is involved in stress response and symbiotic efficiency in *Bradyrhizobium japonicum*. *Mol. Microbiol.* **73**:291–305.

Guedon, E., C. M. Moore, Q. Que, T. Wang, R. W. Ye, and J. D. Helmann. 2003. The global transcriptional response of *Bacillus subtilis* to manganese involves the MntR, Fur, TnrA and σB regulons. *Mol. Microbiol.* **49**:1477–1491.

Harvey, R. W., R. L. Smith, and L. George. 1984. Effect of organic contamination upon microbial distributions and heterotrophic uptake in a Cape Cod, Mass, aquifer. *Appl. Environ. Microbiol.* **48**:1197–1202.

Haveman, S. A., V. Brunelle, J. K. Voordouw, G. Voordouw, J. F. Heidelberg, and R. Rabus. 2003. Gene expression analysis of energy metabolism mutants of *Desulfovibrio vulgaris* Hildenborough indicates an important role for alcohol dehydrogenase. *J. Bacteriol.* **185**:4345–4353.

Hazen, T. C., L. Jimenez, G. L. d. Victoria, and C. B. Fliermans. 1991. Comparison of bacteria from deep subsurface sediment and adjacent groundwater. *Microbial Ecol.* **22**:293–304.

He, Q., Z. He, D. C. Joyner, M. Joachimiak, M. N. Price, Z. K. Yang, H.-C. Yen, C. L. Hemme, R. Chakraborty, W. Chen, M. W. Fields, D. A. Stahl, J. D. Keasling, M. Keller, A. P. Arkin, T. C. Hazen, J. D. Wall, and J. Zhou. Impact of elevated nitrate on sulfate-reducing bacteria: implications of inhibitory mechanisms in addition to osmotic stress. *J. Bacteriol.*, in review.

He, Q., et al. 2010. Impact of elevated nitrate on sulfate-reducing bacteria: a comparative study of *Desulfovibrio vulgaris*. *ISME J.* **4**:1386–1397.

He, Q., K. H. Huang, Z. He, E. J. Alm, M. W. Fields, T. C. Hazen, A. P. Arkin, J. D. Wall, and J. Zhou. 2006. Energetic consequences of nitrite stress in *Desulfovibrio vulgaris* Hildenborough, inferred from global transcriptional analysis. *Appl. Environ. Microbiol.* **72**:4370–4381.

He, Z., L. Wu, M. W. Fields, and J. Zhou. 2005. Comparison of microarrays with different probe sizes for monitoring gene expression. *Appl. Environ. Microbiol.* **71**:5154–5162.

Heidelberg, J. F., et al. 2004. The genome sequence of the anaerobic sulfate-reducing bacterium *Desulfovibrio vulgaris* Hildenborough. *Nat. Biotechnol.* **22**:554–559.

Hoffmann, T., N. Frankenberg, M. Marino, and D. Jahn. 1998. Ammonification in *Bacillus subtilis* utilizing dissimilatory nitrite reductase is dependent on resDE. *J. Bacteriol.* **180**:186–189.

Holman, H. Y. N., E. Wozei, L. Comolli, Z. Lin, S. Boglin, K. H. Downing, M. W. Fields, and T. C. Hazen. 2009. Real-time monitoring of hydrogen-bond dynamics during oxygen-stress adaptive response in strict anaerobes. *Proc. Natl. Acad. Sci. USA* **106**:12599–12604.

Holmes, D. E., S. K. Chaudhuri, K. P. Nevin, T. Mehta, B. A. Methe, A. Liu, J. E. Ward, T. L. Woodard, J. Webster, and D. R. Lovley. 2006. Microarray and genetic analysis of electron transfer to electrodes in *Geobacter sulfurreducens*. *Environ. Microbiol.* **8**:1805–1815.

Holmes, D. E., T. Mester, R. A. O'Neil, L. A. Perpetua, M. J. Larrahondo, R. Glaven, M. L. Sharma, J. E. Ward, K. P. Nevin, and D. R. Lovley. 2008. Genes for two multicopper proteins required for Fe(III) oxide reduction in *Geobacter sulfurreducens* have different expression patterns both in the subsurface and on energy-harvesting electrodes. *Microbiology* **154**:1422–1435.

Holmes, D. E., K. P. Nevin, and D. R. Lovley. 2004. *In situ* expression of *nif*D in *Geobacteraceae* in subsurface sediments. *Appl. Environ. Microbiol.* **70:**7251–7259.

Holmes, D. E., R. A. O'Neil, M. A. Chavan, L. A. N'Guessan, H. A. Vrionis, L. A. Perpetua, M. J. Larrahondo, R. DiDonato, A. Liu, and D. R. Lovley. 2009. Transcriptome of *Geobacter uraniireducens* growing in uranium-contaminated subsurface sediments. *ISME J.* **3:**216–230.

Hu, P., E. L. Brodie, Y. Suzuki, H. H. McAdams, and G. L. Andersen. 2005. Whole-genome transcriptional analysis of heavy metal stresses in *Caulobacter crescentus*. *J. Bacteriol.* **187:**8437–8449.

Jin, Y. H., P. E. Dunlap, S. J. McBride, H. Al-Refai, P. R. Bushel, and J. H. Freedman. 2008. Global transcriptome and deletome profiles of yeast exposed to transition metals. *PLoS Genet.* **4:**e1000053

Johnson, M. S., I. B. Zhulin, M.-E. R. Gapuzan, and B. L. Taylor. 1997. Oxygen-dependent growth of the obligate anaerobe *Desulfovibrio vulgaris* Hildenborough. *J. Bacteriol.* **179:**5598–5601.

Karlin, S., L. Brocchieri, J. Mrazek, and D. Kaiser. 2006. Distinguishing features of δ-*Proteobacterial* genomes. *Proc. Natl. Acad. Sci. USA* **103:**11352–11357.

Kasprzak, K. S. 2002. Oxidative DNA and protein damage in metal-induced toxicity and carcinogenesis. *Free Radic. Biol. Med.* **32:**958–967.

Kerr-Wall, P., J. Leebens-Mack, A. S. Chanderbali, A. Barakat, E. Wolcott, H. Liang, L. Landherr, L. P. Tomsho, Y. Hu, J. E. Carlson, H. Ma, S. C. Schuster, D. E. Soltis, P. S. Soltis, N. Altman, and C. W. dePamphilis. 2009. Comparison of next generation sequencing technologies for transcriptome characterization. *BMC Genomics* **10:**347.

Khare, T., A. Esteve-Núñez, K. P. Nevin, W. Zhu, J. R. Yates, D. Lovley and C. S. Giometti. 2006. Differential protein expression in the metal-reducing bacterium *Geobacter sulfurreducens* strain PCA grown with fumarate or ferric citrate. *Proteomics* **6:**632–640.

Klonowska, A., M. E. Clark, S. B. Thieman, B. J. Giles, J. D. Wall, and M.W. Fields. 2008. Hexavalent chromium reduction in *Desulfovibrio vulgaris* Hildenborough causes transitory inhibition of sulfate reduction and cell growth. *Appl. Microbiol. Biotechnol.* **78:**1007–1016.

Klonowska, A., Z. He, Q. He, M. E. Clark, S. B. Thieman, T. C. Hazen, E. J. Alm, H.-Y. Holman, A. P. Arkin, J. D. Wall, J. Zhou, and M. W. Fields. 2006. Global transcriptomic analysis of chromium(VI) exposure of *Desulfovibrio vulgaris* Hildenborough under sulfate-reducing conditions. Abstract K-052. Gen. Meet. Am. Soc. Microbiol. American Society for Microbiology, Washington, DC.

Kolbel-Boelke, J., E. M. Anders, and A. Nehrkorn. 1988. Microbial communities in the saturated groundwater environment. Diversity of bacterial communities in a pleistocene sand aquifer and their in vitro activities. *Microb. Ecol.* **16:**31–48.

Kolker, E., K. S. Makarova, S. Shabalina, A. F. Picone, S. Purvine, T. Holzman, T. Cherny, D. Armbruster, R. S. Munson, G. Kolesov, D. Frishman, and M. Y. Galperin. 2004. Identification and functional analysis of hypothetical genes expressed in *Haemophilus influenzae*. *Nucleic Acids Res.* **32:**2353–2361.

Kreuzer, C. S., M. Blackledge, A. Dolla, D. Marion, and F. Guerlesquin. 1998. Tyrosine 64 of cytochrome c553 is required for electron exchange with formate dehydrogenase in *Desulfovibrio vulgaris* Hildenborough. *Biochemistry* **37:**8331–8340.

Krushkal, J., B. Yan, L. N. DiDonato, M. Puljic, K. P. Nevin, T. L. Woodard, R. M. Adkins, B. A. Methé, and D. R. Lovley. 2007. Genome-wide expression profiling in *Geobacter sulfurreducens*: identification of Fur and RpoS transcription regulatory sites in a relGsu mutant. *Funct. Integr. Genomics* **7:**229–255.

Leang, C., M. V. Coppi, and D. R. Lovley. 2003. OmcB, a c-type polyheme cytochrome, involved in Fe(III) reduction in *Geobacter sulfurreducens*. *J. Bacteriol.* **185:**2096–2103.

Lehman, R. M., and S. P. O'Connell. 2002. Comparison of extracellular enzyme activities and community composition of attached and free-living bacteria in porous medium columns. *Appl. Environ. Microbiol.* **68:**1569–1575.

Lies, D. P., M. E. Hernandez, A. Kappler, R. E. Mielke, J. A. Gralnick, and D. K. Newman. 2005. *Shewanella oneidensis* MR-1 uses overlapping pathways for iron reduction at a distance and by direct contact under conditions relevant for biofilms. *Appl. Environ. Microbiol.* **71:**4414–4426.

Lin, W. C., M. V. Coppi, and D. R. Lovley. 2004. *Geobacter sulfurreducens* can grow with oxygen as a terminal electron acceptor. *Appl. Environ. Microbiol.* **70:**2525–2528.

Lloyd, J. R. 2003. Microbial reduction of metals and radionuclides. *FEMS Microbiol. Rev.* **27:**411–425.

Lobo, S. A. L., A. M. P. Melo, J. N. Carita, M. Teixeira, and L. M. Saraiva. 2007. The anaerobe *Desulfovibrio desulfuricans* ATCC 27774 grows at nearly atmospheric oxygen levels. *FEBS Lett.* **581:**433–436.

Lourenco, R. F., and S. L. Gomes. 2009. The transcriptional response to cadmium, organic hydroperoxide, singlet oxygen and UV-A mediated by the sigma(E)-ChrR system in *Caulobacter crescentus*. *Mol. Microbiol.* **72:**1159–1170.

Lovley, D. R., E. E. Roden, E. J. P. Phillips, and J. C. Woodward. 1993a. Enzymatic iron and uranium reduction by sulfate-reducing bacteria. *Marine Geol.* **113**:41–53.

Lovley, D. R., P. K. Widman, J. C. Woodward, and E. J. P. Phillips. 1993b. Reduction of uranium by cytochrome c_3 of *Desulfovibrio vulgaris*. *Appl. Environ. Microbiol.* **59**:3572–3576.

Mahadevan, R., B. Yan, B. Postier, K. P. Nevin, T. L. Woodard, R. O'Neil, M. V. Coppi, B. A. Methé, and J. Krushkal. 2008. Characterizing regulation of metabolism in *Geobacter sulfurreducens* through genome-wide expression data and sequence analysis. *OMICS* **12**:33–59.

Marshall, K. C. 1992. Planktonic versus sessile life of prokaryotes, p. 262–275. *In* A. Balows, H. G. Truper, M. Dworkin, W. Harder, and K.-H. Schleifer (ed.), *The Prokaryotes: a Handbook on the Biology of Bacteria: Ecophysiology, Isolation, Identification, Applications*. Springer-Verlag, Berlin, Germany.

Masse, E., C. K. Vanderpool, and S. Gottesman. 2005. Effect of RyhB small RNA on global iron use in *Escherichia coli*. *J. Bacteriol.* **187**:6962–6971.

McLean, J. S., P. D. Majors, C. L. Reardon, C. L. Bilskis, S. B. Reed, M. F. Romine, and J. K. Fredrickson. 2008. Investigations of structure and metabolism within *Shewanella oneidensis* MR-1 biofilms. *J. Microbiol. Methods* **74**:47–56.

McKinley, J. P., J. M. Zachara, S. Smith, and G. D. Turner. 1995. The influence of uranyl hydrolysis and multiple site-binding reactions on adsorption of U(VI) to montmorillonite. *Clays Clay Miner.* **43**:586–598.

Mehta, T., M. V. Coppi, S. E. Childers, and D. R. Lovley. 2005. Outer membrane c-type cytochromes required for Fe(III) and Mn(IV) oxide reduction in *Geobacter sulfurreducens*. *Appl. Environ. Microbiol.* **71**:8634–8641.

Methe, B. A., K. E. Nelson, J. A. Eisen, I. T. Paulsen, W. Nelson, J. F. Heidelberg, D. Wu, M. Wu, N. Ward, M. J. Beanan, R. J. Dodson, R. Madupu, L. M. Brinkac, S. C. Daugherty, R. T. DeBoy, A. S. Durkin, M. Gwinn, J. F. Kolonay, S. A. Sullivan, D. H. Haft, J. Selengut, T. M. Davidsen, N. Zafar, O. White, B. Tran, C. Romero, H. A. Forberger, J. Weidman, H. Khouri, T. V. Feldblyum, T. R. Utterback, S. E. Van Aken, D. R. Lovley, and C. M. Fraser. 2003. Genome of *Geobacter sulfurreducens*: metal reduction in subsurface environments. *Science* **302**:1967–1969.

Methe, B. A., J. Webster, K. P. Nevin, J. Butler, and D. R. Lovley. 2005. DNA microarray analysis of nitrogen fixation and Fe(III) reduction in *Geobacter sulfurreducens*. *Appl. Environ. Microbiol.* **71**:2530–2538.

Mouser, P. J., D. E. Holmes, L. A. Perpetua, R. DiDonato, B. Postier, A. Liu, and D. R. Lovley. 2009. Quantifying expression of *Geobacter* spp. oxidative stress genes in pure culture and during in situ uranium bioremediation. *ISME J.* **3**:454–465.

Mugerfeld, I., B. A. Law, G. S. Wickham, and D. K. Thompson. 2009. A putative azoreductase gene is involved in the *Shewanella oneidensis* response to heavy metal stress. *Appl. Microbiol. Biotechnol.* **82**:1131–1141.

Mukhopadhyay, A., Z. He, E. J. Alm, A. P. Arkin, E. E. Baidoo, S. C. Borglin, W. Chen, T. C. Hazen, Q. He, H. Y. Holman, K. Huang, R. Huang, D. C. Joyner, N. Katz, M. Keller, P. Oeller, A. Redding, J. Sun, J. D. Wall, J. Wei, Z. Yang, H.-C. Yen, J. Zhou, and J. D. Keasling. 2006. Salt stress in *Desulfovibrio vulgaris* Hildenborough: an integrated genomics approach. *J. Bacteriol.* **188**:4068–4078.

Mukhopadhyay, A., A. M. Redding, M. P. Joachimiak, A. P. Arkin, S. E. Borglin, P. S. Dehal, R. Chakraborty, J. T. Geller, T. C. Hazen, Q. He, D. C. Joyner, V. J. J. Martin, J. D. Wall, Z. K. Yang, J. Zhou, and J. D. Keasling. 2007. Cell-wide responses to low-oxygen exposure in *Desulfovibrio vulgaris* Hildenborough. *J. Bacteriol.* **189**:5996–6010.

Mukhopadhyay, P., M. Zheng, L. A. Bedzyk, R. A. LaRossa, and G. Storz. 2004. Prominent roles of the NorR and Fur regulators in the *Escherichia coli* transcriptional response to reactive nitrogen species. *Proc. Natl. Acad. Sci. USA* **101**:745–750.

Myers, C. R., and J. M. Myers. 1997. Cloning and sequence of *cym*A, a gene encoding a tetraheme cytochrome c required for reduction of iron(III), fumarate, and nitrate by *Shewanella putrefaciens* MR-1. *J. Bacteriol.* **179**:1143–1152.

Myers, C. R., and J. M. Myers. 2002. MtrB is required for proper incorporation of the cytochromes OmcA and OmcB into the outer membrane of *Shewanella putrefaciens* MR-1. *Appl. Environ. Microbiol.* **68**:5585–5594.

Nealson, K. H., A. Belz, and B. McKee. 2002. Breathing metals as a way of life: geobiology in action. *Antonie Van Leeuwenhoek* **81**:215–222.

Nie, L., G. Wu, and W. Zhang. 2006. Correlation between mRNA and protein abundance in *Desulfovibrio vulgaris*: a multiple regression to identify sources of variations. *Biochem. Biophys. Res. Commun.* **339**:603–610.

Nunez, C., A. Esteve-Nunez, C. Giometti, S. Tollaksen, T. Khare, W. Lin, D. R. Lovley, and B. A. Methe. 2006. DNA microarray and proteomic analyses of the RpoS regulon in *Geobacter sulfurreducens*. *J. Bacteriol.* **188**:2792–2800.

Odom, J. M., and H. D. Peck. 1981. Hydrogen cycling as a general mechanism for energy coupling in the sulfate-reducing bacteria, *Desulfovibrio* sp. *FEMS Microbiol. Lett.* **12:**47–50.

Odom, J. D., and H. D. Peck. 1984. Hydrogenase, electron transfer proteins, and energy coupling in the sulfate-reducing bacteria *Desulfovibrio*. *Annu. Rev. Microbiol.* **38:**551–592.

O'Neil, R. A., D. E. Holmes, M. V. Coppi, L. A. Adams, M. J. Larrahondo, J. E. Ward, K. P. Nevin, T. L. Woodard, H. A. Vrionis, A. L. N'Guessan, and D. R. Lovley. 2008. Gene transcript analysis of assimilatory iron limitation in *Geobacteraceae* during groundwater bioremediation. *Environ. Microbiol.* **10:**1218–1230.

Park, H. S., S. Lin, and G. Voordouw. 2008. Ferric iron reduction by *Desulfovibrio vulgaris* Hildenborough wild type and energy metabolism mutants. *Antonie Van Leeuwenhoek* **93:**79–85.

Payne, R. B., L. Casalot, T. Rivere, J. H. Terry, L. Larsen, B. J. Giles, and J. D. Wall. 2004. Interaction between uranium and the cytochrome c_3 of *Desulfovibrio desulfuricans* strain G20. *Arch. Microbiol.* **181:**398–406.

Pereira, P. M., Q. He, F. M. A. Valente, A. V. Xavier, J. Zhou, I. A. C. Pereira, and R. O. Louro. 2008a. Energy metabolism in *Desulfovibrio vulgaris* Hildenborough: insights from transcriptome analysis. *Antonie Van Leeuwenhoek* **93:**347–362.

Pereira, P. M., Q. He, A. V. Xavier, J. Zhou, I. A. C. Pereira, and R. O. Louro. 2008b. Transcriptional response of *Desulfovibrio vulgaris* Hildenborough to oxidative stress mimicking environmental conditions. *Arch. Microbiol.* **189:**451–461.

Pitts, K. E., P. S. Dobbin, F. Reyes-Ramirez, A. J. Thomson, D. J. Richardson, and H. E. Seward. 2003. Characterization of the *Shewanella oneidensis* MR-1 decaheme cytochrome MtrA: expression in *Escherichia coli* confers the ability to reduce soluble Fe(III) chelates. *J. Biol. Chem.* **278:**27758–27765.

Pohorelic, B. K. J., J. K. Voordoouw, E. Lojou, A. Dolla, J. Harder, and G. Voordouw. 2002. Effects of deletion of genes encoding Fe-only hydrogenase of *Desulfovibrio vulgaris* Hildenborough on hydrogen and lactate metabolism. *J. Bacteriol.* **184:**679–686.

Reardon, C. L., D. E. Cummings, L. M. Petzke, B. L. Kinsall, D. B. Watson, B. M. Peyton, and G. G. Geesey. 2004. Composition and diversity of microbial communities recovered from surrogate minerals incubated in an acidic uranium-contaminated aquifer. *Appl. Environ. Microbiol.* **70:**6037–6046.

Rhee, S. K., X. Liu, L. Wu, S. C. Chong, X. Wan, and J. Zhou. 2004. Detection of biodegradation and biotransformation genes in microbial communities using 50-mer oligonucleotide microarrays. *Appl. Environ. Microbiol.* **70:**4303–4317.

Richter, H., K. P. Nevin, H. F. Jia, D. A. Lowy, D. R. Lovley, and L. M. Tender. 2009. Cyclic voltammetry of biofilms of wild type and mutant *Geobacter sulfurreducens* on fuel cell anodes indicates possible roles of OmcB, OmcZ, type IV pili, and protons in extracellular electron transfer. *Energy Environ. Sci.* **2:**506–516.

Risso, C., B. A. Methe, H. Elifantz, D. E. Holmes, and D. R. Lovley. 2008. Highly conserved genes in *Geobacter* species with expression patterns indicative of acetate limitation. *Microbiology* **154:**2589–2599.

Rittman, B. E. 1993. The significance of biofilms in porous media. *Water Res.* **29:**2195–2202.

Robey, M., and N. P. Cianciotto. 2002. *Legionella pneumophila feoAB* promotes ferrous iron uptake and intracellular infection. *Infect. Immun.* **70:**5659–5669.

Rodionov, D. A., I. Dubchak, A. P. Arkin, E. Alm, and M. S. Gelfand. 2004. Reconstruction of regulatory and metabolic pathways in metal-reducing δ-proteobacteria. *Genome Biol.* **5:**R90.

Rollefson, J. B., C. E. Levar, and D. R. Bond. 2009. Identification of genes involved in biofilm formation and respiration via mini-Himar transposon mutagenesis of *Geobacter sulfurreducens*. *J. Bacteriol.* **191:**4207–4217.

Rose, R. W., T. Bruser, J. C. Kissinger, and M. Pohlschroder. 2002. Adaptation of protein secretion to extremely high-salt conditions by extensive use of the twin-arginine translocation pathway. *Mol. Microbiol.* **45:**943–950.

Saeed, A. I., V. Sharov, J. White, J. Li, W. Liang, N. Bhagabati, J. Braisted, M. Klapa, T. Currier, M. Thiagarajan, A. Sturn, M. Snuffin, A. Rezantsev, D. Popov, A. Ryltsov, E. Kostukovich, I. Borisovsky, Z. Liu, A. Vinsavich, V. Trush, and J. Quackenbush. 2003. TM4: a free, open-source system for microarray data management and analysis. *Biotechniques* **34:**374–378.

Saffarini, D. A., R. Schultz, and A. Beliaev. 2003. Involvement of cyclic AMP (cAMP) and cAMP receptor protein in anaerobic respiration of *Shewanella oneidensis*. *J. Bacteriol.* **185:**3668–3671.

Sanford, R. A., Q. Wu, Y. Sung, S. H. Thomas, B. K. Amos, E. K. Prince, and F. E. Loffler. 2007. Hexavalent uranium supports growth of *Anaeromyxobacter dehalogenans* and *Geobacter* spp. with lower than predicted biomass yields. *Environ. Microbiol.* **9:**2885–2893.

Shelobolina, E. S., M. V. Coppi, A. A. Korenevsky, L. N. DiDonato, S. A. Sullivan, H. Konishi, H. Xu, C. Leang, J. E. Butler, B. C. Kim, and D. R. Lovley. 2007. Importance of *c*-type cytochromes for U(VI) reduction by *Geobacter sulfurreducens*. *BMC Microbiol.* **7:**16.

Shi, L., B. Chen, Z. Wang, D. A. Elias, M. U. Mayer, Y. A. Gorby, S. Ni, B. H. Lower, D. W. Kennedy, D. S. Wunschel, H. M. Mottaz, M. J. Marshall, E. A. Hill, A. S. Beliaev, J. M. Zachara, J. K. Fredrickson, and T. C. Squier. 2006. Isolation of high-affinity functional protein complex between OmcA and MtrC: two outer membrane decaheme c-type cytochromes of *Shewanella oneidensis* MR-1. *J. Bacteriol.* **188:**4705–4714.

Shi, L., S. Deng, M. J. Marshall, Z. Wang, D. W. Kennedy, A. Dohnalkova, H. M. Mottaz, E. A. Hill, Y. A. Gorby, A. S. Beliaev, D. J. Richardson, J. M. Zachara, and J. K. Fredrickson. 2008. Direct involvement of type II secretion system in extracellular translocation of *Shewanella oneidensis* outer membrane cytochromes MtrC and OmcA. *J Bacteriol.* **190:**5512–5516.

Shi, L., T. C. Squier, J. M. Zachara, and J. K. Fredrickson. 2007. Respiration of metal (hydr)oxides by *Shewanella* and *Geobacter*: a key role for multihaem c-type cytochromes. *Mol. Microbiol.* **65:**12–20.

Smith, R. L. 2002. Determining the terminal electron-accepting reaction in the saturated subsurface, p. 743–752. *In* C. J. Hurst (ed.), *Manual of Environmental Microbiology*, 2nd ed. American Society for Microbiology, Washington, DC.

Stams, A. J. M., and C. M. Plugge. 2009. Electron transfer in syntrophic communities of anaerobic bacteria and archaea. *Nat. Rev. Microbiol.* **7:**568–577.

Stearns, D. M., L. J. Kennedy, K. D. Courtney, P. H. Giangrande, L. S. Phieffer, and K. E. Wetterhahn. 1995. Reduction of chromium(VI) by ascorbate leads to chromium–DNA binding and DNA strand breaks in vitro. *Biochemistry* **34:**910–919.

Stolyar, S., S. Van Dien, K. L. Hillesland, N. Pinel, T. J. Lie, J. A. Leigh, and D. A. Stahl. 2007. Metabolic modeling of a mutualistic microbial community. *Mol. Syst. Biol.* **3:**92.

Sumner, E. R., A. Shanmuganathan, T. C. Sideri, S. A. Willetts, J. E. Houghton, and S. V. Avery. 2005. Oxidative protein damage causes chromium toxicity in yeast. *Microbiology* **151:**1939–1948.

Suzuki, Y., S. D. Kelly, K. M. Kemner, and J. F. Banfield. 2005. Direct microbial reduction and subsequent preservation of uranium in natural near-surface sediment. *Appl. Environ. Microbiol.* **71:**1790–1797.

Suzuki, Y., and T. Suko. 2006. Geomicrobiological factors that control uranium mobility in the environment: update on recent advances in the bioremediation of uranium-contaminated sites. *J. Mineral. Petrol. Sci.* **101:**299–307.

Tang, Y. J., A. L. Meadows, J. Kirby, and J. D. Keasling. 2007. Anaerobic central metabolic pathways in *Shewanella oneidensis* MR-1 reinterpreted in the light of isotopic metabolite labeling. *J. Bacteriol.* **189:**894–901.

Taroncher-Oldedburg, G., E. M. Griner, C. A. Francis, and B. B. Ward. 2003. Oligonucleotide microarray for the study of functional gene diversity in the nitrogen cycle in the environment. *Appl. Environ. Microbiol.* **69:**1159–1171.

Taylor, S. W., and P. R. Jaffe. 1990. Biofilm growth and the related changes in the physical properties of a porous medium. 1. Experimental investigation. *Water Res.* **26:**2153–2159.

Teal, T. K., D. P. Lies, B. J. Wold, and D. K. Newman. 2006. Spatiometabolic stratification of *Shewanella oneidensis* biofilms. *Appl. Environ. Microbiol.* **72:**7324–7330.

Thompson, D. K., A. S. Beliaev, C. S. Giometti, S. L. Tollaksen, T. Khare, D. P. Lies, K. H. Nealson, H. Lim, J. Yates III, C. C. Brandt, and J. M. Tiedje. 2002. Transcriptional and proteomic analysis of a ferric uptake regulator (fur) mutant of *Shewanella oneidensis*: possible involvement of fur in energy metabolism, transcriptional regulation, and oxidative stress. *Appl. Environ. Microbiol.* **68:**881–892.

Thöny-Meyer, L. 1997. Biogenesis of respiratory cytochromes in bacteria. *Microbiol. Mol. Biol. Rev.* **61:**337–376.

Thormann, K. M., R. M. Saville, S. Shukla, D. A. Pelletier, and A. M. Spormann. 2004. Initial phases of biofilm formation in *Shewanella oneidensis* MR-1. *J. Bacteriol.* **186:**8096–8104.

Toes, A.-C. M., M. H. Daleke, J. G. Kuenen, and G. Muyzer. 2008. Expression of copA and cusA in *Shewanella* during copper stress. *Microbiology* **154:**2709–2718.

Torres, T. T., M. Metta, B. Ottenwalder, and C. Schlotterer. 2007. Gene expression profiling by massively parallel sequencing. *Genome Res.* **18:**172–77.

Valente, F. M., A. S. F. Oliveira, N. Gnadt, I. Pacheco, A. V. Coelho, A. V. Xavier, M. Teixeira, C. M. Soares, and I. A. C. Pereira. 2005. Hydrogenases in *Desulfovibrio vulgaris* Hildenborough: structural and physiologic characterisation of the membrane-bound [NiFeSe] hydrogenase. *J. Biol. Inorg. Chem.* **10:**667–682.

Van Loosdrecht, M. C. M., J. Lyklema, W. Norde, and A. J. B. Zehnder. 1990. Influence of interfaces on microbial activity. *Microbiol. Rev.* **54:**75–87.

Vargas, M., K. Kashefi, E. L. Blunt-Harris, and D. R. Lovley. 1998. Microbiological evidence for Fe(III) reduction on early Earth. *Nature* **395:**65–67.

Vasant, C., K. Balamurugan, R. Rajaram, and T. Ramasami. 2001. Apoptosis of lymphocytes in the presence of Cr(V) complexes: role in Cr(VI)-

induced toxicity. *Biochem. Biophys. Res. Commun.* **285:**1354–1360.

Viamajala, S., B. M. Peyton, R. K. Sani, W. A. Apel, and J. N. Petersen. 2004. Toxic effects of chromium(VI) on anaerobic and aerobic growth of *Shewanella oneidensis* MR-1. *Biotechnol. Prog.* **20:**87–95.

Voordouw, G. 2002. Carbon monoxide cycling by *Desulfovibrio vulgaris* Hildenborough. *J. Bacteriol.* **184:**5903–5911.

Wackett, L. P., A. G. Dodge, and L. B. M. Ellis. 2004. Microbial genomics and the periodic table. *Appl. Environ. Microbiol.* **70:**647–655.

Walker, C. B., Z. He, Z. K. Yang, J. A. Ringbauer, Q. He, J. Zhou, G. Voordouw, J. D. Wall, A. P. Arkin, T. C. Hazen, S. Stolyar, and D. A. Stahl. 2009. The electron transfer system of syntrophically grown *Desulfovibrio vulgaris*. *J. Bacteriol.* **191:**5793–5801.

Wall, J. D., and L. R. Krumholz. 2006. Uranium reduction. *Annu. Rev. Microbiol.* **60:**149–166.

Wan, X.-F., N. C. VerBerkmoes, L. A. McCue, D. Stanek, H. Connelly, L. J. Hauser, L. Wu, X. Liu, T. Yan, A. Leaphart, R. L. Hettich, J. Zhou, and D. K. Thompson. 2004. Transcriptomic and proteomic characterization of the Fur modulon in the metal-reducing bacterium *Shewanella oneidensis*. *J. Bacteriol.* **186:**8385–8400.

Washburn, M. P., A. Koller, G. Oshiro, R. R. Ulaszek, D. Plouffe, C. Deciu, E. Winzeler, and J. R. Yates III. 2003. Protein pathway and complex clustering of correlated mRNA and protein expression analyses in *Saccharomyces cerevisiae*. *Proc. Natl. Acad. Sci. USA* **100:**3107–3112.

Whitman, W. B., D. C. Coleman, and W. J. Wiebe. 1998. Prokaryotes: the unseen majority. *Proc. Natl. Acad. Sci. USA.* **95:**6578–6583.

Wilhelm, B. T., and J.-R. Landry. 2009. RNA-Seq—quantitative measurement of expression through massively parallel RNA-sequencing. *Methods* **48:**249–257.

Wilkins, M. J., N. C. VerBerkmoes, K. H. William, S. J. Callister, P. J. Mouser, H. Elifantz, A. L. N'Guessan, B. C. Thomas, C. D. Nicora, M. B. Shah, P. Abraham, M. S. Lipton, D. R. Lovley, R. L. Hettich, P. E. Long, and J. F. Banfield. 2009. Proteogenomic monitoring of *Geobacter* physiology during stimulated uranium bioremediation. *Appl. Environ. Microbiol.* **75:**6591–6599.

Wilkins, M. J., P. L. Wincott, D. J. Vaughan, F. R. Livens, and J. R. Lloyd. 2007. Growth of *Geobacter sulfurreducens* on poorly crystalline Fe(III) oxyhydroxide coatings. *Geomicrobiol. J.* **24:**199–204.

Wu, W., J. Carley, T. J. Gentry, M. A. Ginder-Vogel, M. Fienen, T. Mehlhorn, S. L. Carroll, M. N. Pace, J. Nyman, J. Luo, M. Gentile, M. W. Fields, R. F. Hickey, B. Gu, D. B. Watson, O. Cirpka, J. Zhou, S. Fendorf, P. Kitanidis, P. M. Jardine, and C. S. Criddle. 2006. Pilot-Scale *in situ* bioremedation of uranium in a highly contaminated aquifer. 2. Reduction of U(VI) and geochemical control of U(VI) bioavailability. *Environ. Sci. Technol.* **40:**3986–3995.

Wu, W., J. Carley, J. Luo, M. A. Ginder-Vogel, E. Cardenas, M. Leigh, C. Hwang, S. D. Kelly, C. Ruan, L. Wu, T. J. Gentry, K. Lowe, T. Mehlhorn, S. L. Carroll, M. W. Fields, B. Gu, D. B. Watson, K. M. Kemner, T. Marsh, J. Tiedje, J. Zhou, S. Fendorf, P. Kitanidis, P. M. Jardine, and C. S. Criddle. 2007. Bioreduction of uranium (VI) *in situ* and stability of immobilized uranium: impact of dissolved oxygen. *Environ. Sci. Technol.* **41:**5716–5723.

Xiong, Y., L. Shi, B. Chen, M. U. Mayer, B. H. Lower, Y. Londer, S. Bose, M. F. Hochella, J. K. Fredrickson, and T. C. Squier. 2006. High-affinity binding and direct electron transfer to solid metals by the *Shewanella oneidensis* MR-1 outer membrane c-type cytochrome OmcA. *J. Am. Chem. Soc.* **128:**13978–13979.

Yang, Y., D. P. Harris, F. Luo, W. Xiong, M. Joachimiak, L. Wu, P. Dehal, J. Jacobsen, Z. Yang, A. V. Palumbo, A. P. Arkin, and J. Zhou. 2009. Snapshot of iron response in *Shewanella oneidensis* by gene network reconstruction. *BMC Genomics* **10:**131.

Ye, Q., Y. Roh, B. B. Blair, C. Zhang, J. Zhou, and M. W. Fields. 2004. Isolation and characterization of a novel, alkaliphilic, metal-reducing bacterium, and possible implications for alkaline chemotrophy. *Appl. Environ. Microbiol.* **70:**5595–5602.

Zhang, W., D. E. Culley, M. Hogan, L. Vitiritti, and F. J. Brockman. 2006a. Oxidative stress and heat-shock responses in *Desulfovibrio vulgaris* by genome-wide transcriptomic analysis. *Antonie Van Leeuwenhoek* **90:**41–55.

Zhang, W., D. E. Culley, L. Nie, and J. C. M. Scholten. 2007. Comparative transcriptome analysis of *Desulfovibrio vulgaris* grown in planktonic culture and mature biofilm on a steel surface. *Appl. Microbiol. Biotechnol.* **76:**447–457.

Zhang, W., D. E. Culley, J. C. M. Scholten, M. Hogan, L. Vitiritti, and F. J. Brockman. 2006b. Global transcriptomic analysis of *Desulfovibrio vulgaris* on different electron donors. *Antonie Van Leeuwenhoek* **89:**221–237.

Zhou, J., and D. K. Thompson. 2002. Challenge in applying microarrays to environmental studies. *Curr. Opin. Biotechnol.* **13:**204–207.

APPLICATION OF PROTEOMICS IN BIOREMEDIATION

Peter Chovanec, Partha Basu, and John F. Stolz

13

INTRODUCTION

Industrial activities, energy production, and national security needs over the past century have contributed to an improved quality of life, but often at the cost of environmental degradation. The production of steel, plastics, novel compounds, solvents, insulators, and lubricants has resulted in organic and inorganic pollutants that are recalcitrant to decomposition. The need for energy has left in its wake orphaned oil and gas wells, abandoned coal mines, acid mine drainage, fly ash, and nuclear waste. The Cold War with its proliferation of nuclear arms has left a legacy of radioactive and heavy metal contamination. The U.S. Environmental Protection Agency currently lists 1,281 CERCLA (e.g., Superfund) sites where known or threatened releases of hazardous compounds exist (http://www.epa.gov/superfund/sites/npl). Since the establishment of the National Priority List, 1,098 sites have been remediated by various methods. The discovery of microorganisms capable of degrading, transforming, or accumulating a wide range of hazardous compounds (e.g., petroleum, polyaromatic hydrocarbons, polychlorinated biphenyls, radionuclides, and heavy metals) continues to generate interest in microbial biodegradation and remediation. While the concept has been around for several decades, natural attenuation, the stimulation of the in situ microbial population to bioremediate a specific contaminant, is in its relative infancy. This approach avoids the use of genetically modified or nonindigenous species, but relies on a fundamental understanding of the geochemistry and microbiology of the site. Identification of the indigenous microbial population is not only essential for site assessment, but also to provide a baseline for monitoring changes and activity of the population during and after the stimulation. The use of culture-independent methods, such as 16S rRNA techniques (e.g., denaturing gradient gel electrophoresis, terminal restriction fragment length polymorphism, automated rRNA intergenic spacer analysis), metagenomics, and gene arrays (both phylogenetic and functional), have been valuable tools in this regard. The advent of high-throughput sequencing has resulted in the exponential growth of available genomic data. The Joint Genome Institute's Integrated Microbial Genomes page now lists over 1,200 complete genomes, with over four times that many in production. Many of these organisms have

---

Peter Chovanec and John F. Stolz, Department of Biological Sciences, Duquesne University, Pittsburgh, PA 15282. *Partha Basu*, Department of Chemistry and Biochemistry, Duquesne University, Pittsburgh, PA 15282.

been chosen for genome sequencing based on their unique physiology (e.g., heavy metal and organic metabolism) and relevance to Department of Energy contaminated sites. This bolus of data has resulted in a better understanding of microbial-mediated bioremediation and the ability to monitor microbial metabolism in contaminated environments.

The molecular data, however, only tell part of the story. Information about the protein and lipid composition as well as metabolites requires additional tools. An important development in this regard has been the recent advances in mass spectrometry, with increased resolution, sensitivity, and throughput. This has led to the development of the "-omic" sciences such as proteomics (proteins), lipidomics (fats), metabolomics (metabolites), and metallomics (metals). The proliferation of genomics data, which can be translated into protein sequence, has facilitated the identification of expressed gene products (i.e., proteins) in both prokaryotic and eukaryotic systems, cultures, organelles, cells, and tissues under different conditions, thus providing valuable insight into the molecular machinery and the functional network of the cell. Proteome analysis can be used to identify and compare the suite of expressed proteins from a particular sample in a particular state (e.g., healthy versus dis-

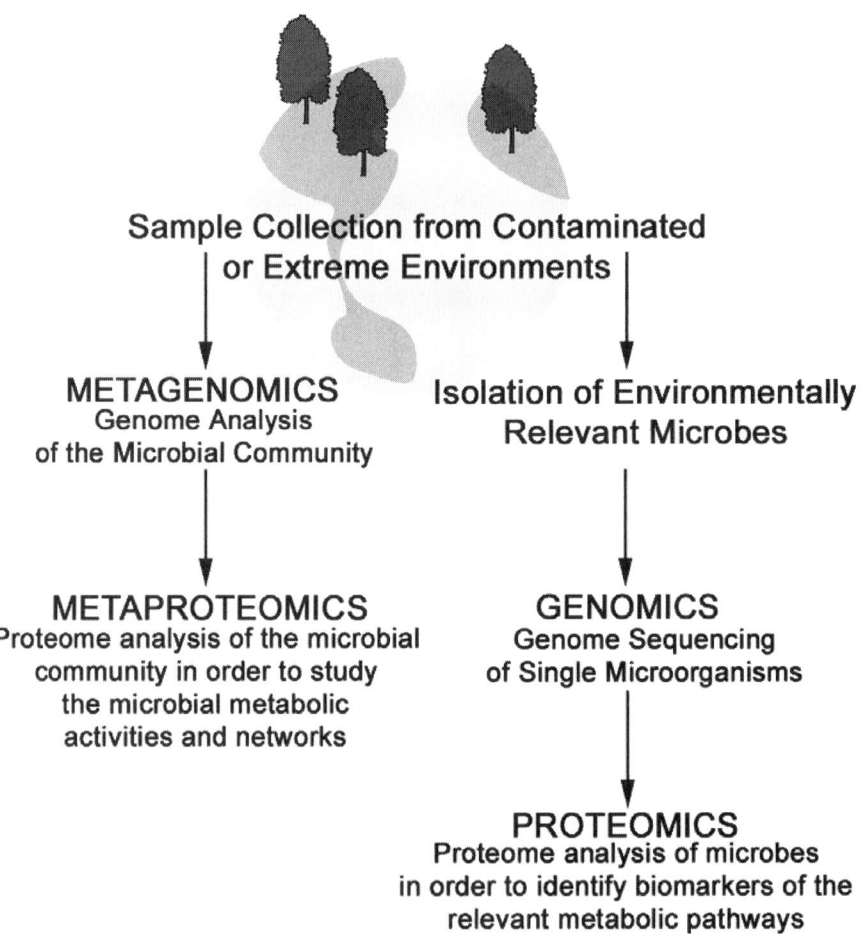

FIGURE 1 Application of "-omic" approaches to both pure cultures and microbial communities for assessing the bioremediation of contaminated environments.

ease state), in order to elucidate a metabolic pathway or analyze protein posttranslational modifications. Proteomics is being applied extensively across different scientific disciplines, including medicine, pharmacology, biology, and environmental studies. For example, analyzing proteome changes in response to toxic contaminants can potentially lead to the discovery of biomarkers for exposure to hazardous substances (López-Barea and Gómez-Ariza, 2006; Nesatyy and Suter, 2007, 2008). Thus, proteomics used in environmental studies has the potential to provide relevant protein identifications to enhance our understanding of microbial-mediated bioremediation.

The purpose of this chapter is to present proteomics in the context of bioremediation by presenting how this approach has helped elucidate mechanisms of survival in contaminated and extreme environments, the transformation of toxic compounds, and the metabolic pathways and enzymes involved in these processes. We provide a brief review of the techniques and approaches for protein separation and identification that are employed and include case studies of microbial community proteomics and the transformation of metals and metalloids in pure cultures.

ENVIRONMENTAL PROTEOMICS

The proteome may be defined as the protein composition of an organism expressed under a specific growth condition, and proteomics as the study of the proteome. Proteomics not only provides a catalog of the proteins, but also a means to discover posttranscriptional and posttranslational modification that may not be ascertained from the genomic data. Proteomics is especially useful for examining organisms that possess a wide variety of metabolic and energetic pathways, because certain proteins may be expressed only under specific growth conditions. The application of this approach to an environmental sample, "environmental proteomics," can provide information on the proteome of the dominant microbial species or the metaproteome of the microbial community under specific environmental conditions (Fig. 1). In addition it may be used to detect changes that occur as a result of the response to stress and adaptation to environmental changes. Since the introduction of the term "proteome" in the early nineties (Wasinger et al., 1995) the field has rapidly proliferated and a number of reviews providing overviews of proteomics applied research in the environmental field have already been published (Dowling and Sheehan, 2006; Singh and Nagaraj, 2006; Kim et al., 2007; Kumar et al., 2007; Maron et al., 2007; Monsinjon and Knigge, 2007; Schweder et al., 2008; Wood, 2008; Zhao and Poh, 2008; Keller and Hettich, 2009; Lacerda et al., 2007; Lacerda and Reardon, 2009; Wilmes and Bond, 2009).

GEL-BASED AND GEL-FREE PROTEOMICS

Early proteomics studies applied to microbial species relevant to bioremediation used two-dimensional (2D)-gel electrophoresis for the separation of proteins (e.g., Kim et al., 2003, 2004); 2D-gel electrophoresis involves two steps of separation (Fig. 2). The first step is isoelectric focusing where the proteins are separated in a pH gradient according to their isoelectric point, the pH at which protein has no net charge. The second step separates proteins by their molecular mass using sodium dodecyl sulfate polyacrylamide gel electrophoresis (Fig. 3). In this way, proteins are resolved into single spots allowing for the separation of proteins of similar mass but different isoelectric points. Gel analysis involves spot detection by staining followed by differentiating the spot patterns from different but related samples by quantifying and normalizing the spot volumes, and finally matching the spot patterns on the gel set to an image selected as the master. Gel-to-gel variation, which can impair this alignment, has been overcome through the use of precast gels and a technique known as differential in gel electrophoresis (DIGE) analysis. In DIGE, two different samples can be run on the same gel as each is labeled with a different fluorescent dye (e.g., Cy3, Cy5), and mixed before isoelectric focusing (Color Plate 13). An internal standard is created by staining equal aliquots of both proteins samples with another fluorochrome (usually Cy2). The protein spots are excised and subjected to trypsin digestion. The masses of the trypsin-produced peptide fragments are then measured by matrix-assisted laser desorption/ionization-time of flight mass spectrometry (MALDI-TOF/MS). The digested sample is coprecipitated with a UV-light-absorbing matrix (e.g., α-cyano-4-hydroxycinnamic acid) and irradiated by a nanosecond laser pulse. Unwanted fragmentation of the sample is prevented by the matrix absorption of the laser energy. The ionized peptide fragments are accelerated in an electric field and enter the flight tube where different fragments are separated according to their mass to charge (m/z) ratio reaching the detector at different times. Short measuring time (few minutes) and little sample consumption (less than 1 pmol) are the advantages of protein identification by this technique. The identification of proteins is based on a search of measured masses determined from peptide mass fingerprints and their comparison against masses calculated from theoretical digests from the database (e.g., NCBI, SwissProt) using the on-line search engine MASCOT (www.matrixscience.com). Either internal (autolytic tryptic peptides) or external (calibration mixture) calibration, monoisotopic mode of search with selected peaks containing nonisotopic elements and removal of false masses that are not related to the protein (e.g., keratin contamination) enhance the probability of protein identification.

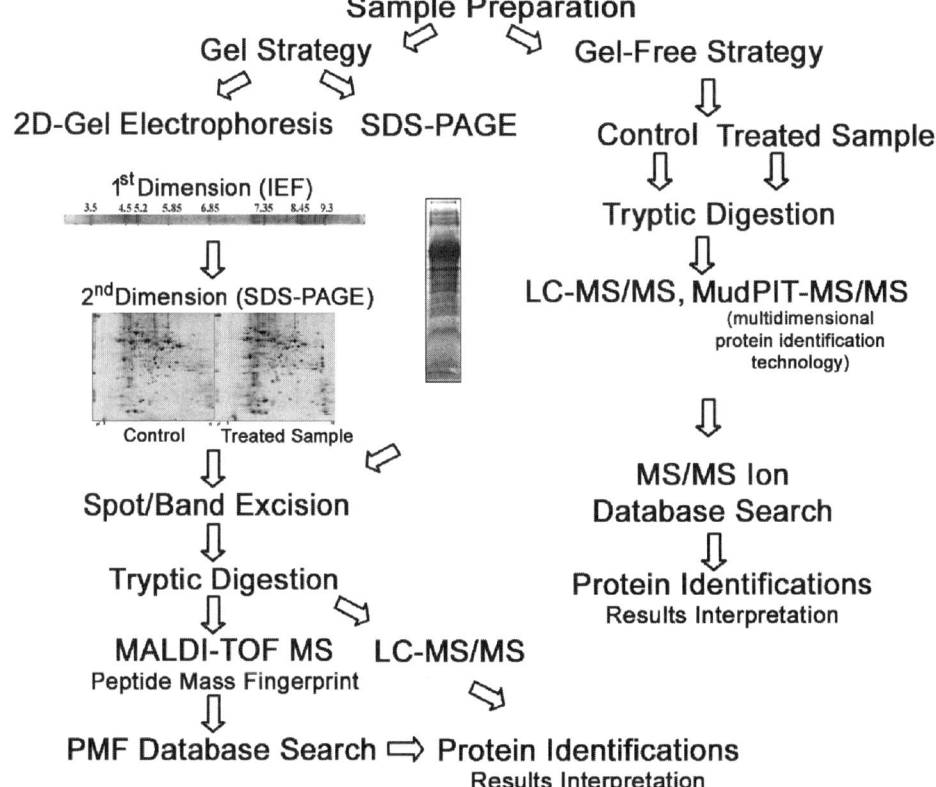

FIGURE 2 Workflow of MS-driven proteomics in bioremediation studies. Gel (1D and 2D separation) versus gel-free separation of proteins followed by tryptic digestion and MS analyses (MALDI-TOF/MS, LC-MS/MS, multidimensional protein identification technology [MudPIT] coupled with tandem MS).

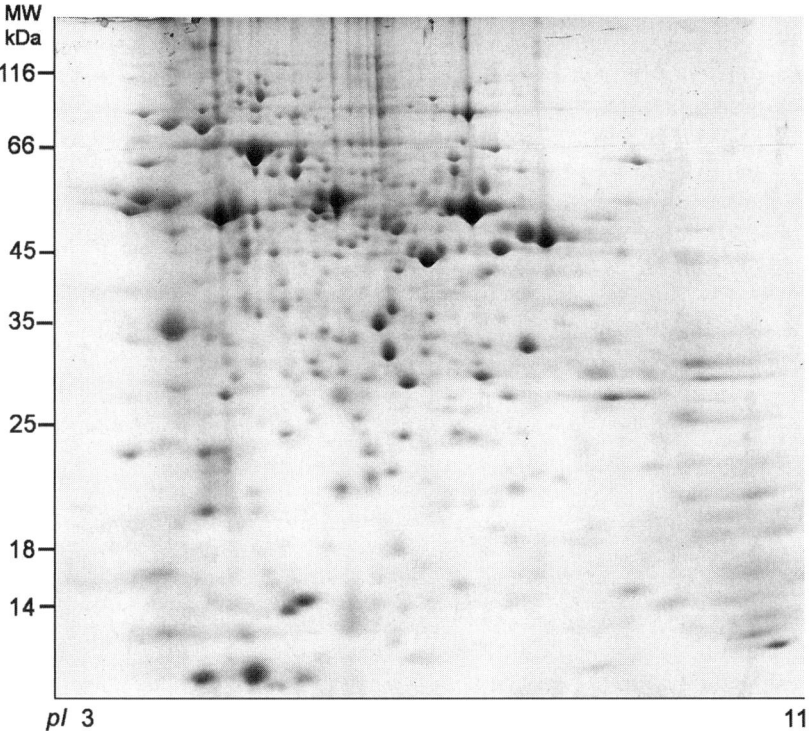

FIGURE 3 Two-dimensional gel electrophoresis (18 cm, pH 3 to 11) of proteins from nitrate grown cultures of *Desulfovibrio desulfuricans* (250 μg of cell lysate) and visualized with Coomassie brilliant blue. The proteins are first separated by charge through isoelectric focusing (first dimension), then by molecular weight using sodium dodecyl sulfate polyacrylamide gel electrophoresis (second dimension). Several hundreds of individual proteins can be resolved and quantified, including posttranslationally modified isoforms.

We have used DIGE to examine the proteome of the arsenite-oxidizing bacterium *Alkalilimnicola ehrlichii*. We compared cells grown under heterotrophic conditions, aerobically on acetate with cells grown under chemolithoautotrophic conditions, anaerobically with NO_3^-, AsO_3^-, and CO_2 (Richey et al., 2009). The aerobic cell proteins were labeled with Cy3 (excitation at 553 nm) and the anaerobic cell proteins were labeled with Cy5 (excitation at 645 nm). The gel was scanned under two different excitation wavelengths providing images of the gel for Cy5 (green) and Cy3 (red). When the Cy5 and Cy3 images are overlaid, protein spots with equal protein abundance expressed under both conditions appear yellow (Color Plate 13). The abundant proteins were identified by MALDI-TOF/MS. Proteins expressed only under anaerobic chemolithotrophic conditions included proteins involved in nitrate metabolism (e.g., nitrous oxide reductase, nitrate transport protein), Calvin-Benson carbon fixation (e.g., fructose-1,6-bisphosphate aldolase, phosphoribulokinase), and arsenic resistance. Conversely, superoxide dismutase was only expressed under aerobic conditions. Proteins expressed under both conditions included those involved in central metabolism (e.g., tricarboxylic acid cycle).

Visualization of bacterial proteomes by 2D-gel electrophoresis separation followed by MALDI-TOF/MS, however, does not provide the characterization of all components. Thus, more recent proteomics in bioremediation relevant studies have relied on gel-free technologies (Fig. 2). The use of liquid chromatography (LC) and the development of other MS techniques have enabled scientists performing proteomic analyses to increase the total number of proteins as well as less abundant proteins more confidently than ever. Integration of the separation power of LC (one or multidimensional) and tandem MS represents a "shotgun" proteomics approach allowing for the routine identification of several thousands of proteins. Tandem mass spectrometry (MS/MS) provides superior mass accuracy, resolution, dynamic range, sensitivity and speed, generating exceptional data quality. Identification of proteins from sequenced organisms is achieved by comparing fragmentation data of peptides to predicted peptides from the genome sequence information. In the case of unknown proteins or ones from a microorganism for which genomic data are not available, the MS/MS data are analyzed by de novo computational methods using partial genome sequences or genomes from related species. The high-throughput nature of LC-MS/MS allows advanced data analysis of replicate samples. For example, we have compared the proteome of *Desulfovibrio desulfuricans* grown on nitrate (control) to cells grown on nitrate and chromate (treatment 1), and cells grown on nitrate and exposed to chromate for 2 hours (treatment 2). The data can be analyzed by hierarchical clustering and presented as a heat map (Color Plate 14).

ISOLATION, CHARACTERIZATION, AND PROTEOMIC STUDIES ON PURE CULTURES

The isolation and characterization of bacteria from contaminated environments is often the starting point for proteomic analyses. Pure cultures allow for in-depth physiological investigations into the mechanisms of resistance and contaminant biotransformation. Mass spectrometry, such as MALDI-TOF, can be used to assist in the characterization of new isolates (van Baar, 2000; Fenselau and Demirev, 2001; Lay, 2001; Dworzanski and Snyder, 2005). When genome information is not available, mass spectrometry-based proteomics provides a means to identify proteins (Shevchenko et al., 2001). These studies, however, are enhanced by the availability of genomic data coupled with proteomics, because they can provide valuable information about the enzymes involved in the degradation and subsequently the establishment of the pathway. The Integrated Microbial Genomes web site of the Joint Genome Institute (http://imgweb.jgi-psf.org) currently lists 2,290 *Bacteria*, 94 *Archaea*, and 76 *Eukarya* genomes either finished or in draft, providing an invaluable resource of web-based information for genome content and analytical capability (Markowitz et al., 2010).

The completed genome sequences of environmentally significant microorganisms obtained in pure culture have revealed their metabolic versatility and provided clues to pathways involved in the bacterial degradation of hydrocarbons (Nelson et al., 2002; Rabus et al., 2005; Schneiker et al., 2006), chlorinated organic compounds (Seshadri et al., 2005; Chain et al., 2006; McLeod et al., 2006; Pérez-Pantoja et al., 2008), and toxic metals (Heidelberg et al., 2002, 2004; Methé et al., 2003). *Shewanella* species are an excellent example of this metabolic versatility in contaminant metabolism that includes uranium (Wall and Krumholz, 2006), technetium (Marshall et al., 2008), mercury (Wiatrowski et al., 2006), cobalt (Gorby et al., 1998), arsenic (Saltikov et al., 2005), chromium (Bencheikh-Latmani et al., 2005), nitramines (Zhao et al., 2004), and halogenated organics (Picardal et al., 1995). Use of an integrated approach combining genome modeling with transcriptomic and proteomic analyses can facilitate improved annotation through the verification of "hypothetical" proteins as was done for *Shewanella oneidensis* strain MR-1 (Kolker et al., 2005). Further studies of strain MR-1 include 2D-gel electrophoresis coupled with MS to identify upregulated proteins in biofilms (De Vriendt et al., 2005), LC-MS-based proteomics for detection of hypothetical proteins expressed under different conditions (Elias et al., 2005), and comparative temporal proteomics to investigate molecular response and regulation from exposure to chromate (Chourey et al., 2009).

Gel-based visualization of protein profiles (e.g., 1D and 2D gel electrophoresis) of single microbial cultures can provide phylogenetic information (Dopson et al., 2004) as well as the differential expression of proteins under different conditions yielding insights into how bacteria metabolize different aromatic substrates, nuclides, and metals. Two-dimensional gel electrophoresis coupled with mass spectrometry has been used in many investigations for the protein characterizations relevant to bioremediation processes (Reardon and Kim, 2002; Liang et al., 2006; Tam Le et al., 2006; Tomás-Gallardo et al., 2006; Martínez et al., 2007; Mazzoli et al., 2007; Cao et al., 2008; Lee et al., 2008; Jennings et al., 2009; Kabiri et al., 2009). Although this popular approach for protein separation has been supplanted by newer separation techniques (e.g., nanoscale high-performance LC, multidimensional electrophoresis), 2D-gel electrophoresis as a "classic" technique of choice is still very applicable to proteomic research in bioremediation. Indeed, it is still a valuable technique for the visual detection of proteins and the quantification of their expression levels, and in many situations complementary to gel-free separation systems.

Proteomics, among other "-omic" technologies, have helped establish a basis for understanding cellular processes involved in the degradation of pollutants, such as polycyclic aromatic hydrocarbons (Kim et al., 2004, 2009; Kweon et al., 2007), aromatics (Reardon and Kim, 2002; Kim et al., 2003, 2006; Kühner et al., 2005; Zhao et al., 2005; Tomás-Gallardo et al., 2006; Santos et al., 2007; Wöhlbrand et al., 2007; Chovanec et al., 2010), chlorate (Bansal et al., 2009), and metals (Sharma et al., 2006; Ding et al., 2008), or how bacteria tolerate toxic compounds (Lupi et al., 1995; Krayl et al., 2003; Santos et al., 2004; Segura et al., 2005). The study of hexadecane metabolism of *Geobacillus thermodenitrificans* NG80-2, a thermophilic bacillus isolated from a deep oil reservoir in China, for example, revealed the function of the long-chain alkanes monooxygenase gene *ladA* (Feng et al., 2007). This protein is involved in the degradation of long-chain alkanes, a major component of oil, and thus strain NG80-2 could see application in the bioremediation of oil-contaminated environments.

MICROBIAL COMMUNITIES AND THEIR BENEFIT IN BIOREMEDIATION STUDIES

Taking into account that more than 99% of the microbes present in the environment cannot be readily cultured, great emphasis has been placed on molecular approaches and high-throughput sequencing technologies (Streit and Schmitz, 2004). Recent developments in high-throughput LC-MS technology also allow whole community protein analyses. A metaproteomic approach was used to investigate microbial phosphorus removal in a wastewater despite the lack of the metagenomic sequences (Wilmes and Bond, 2004). Two-dimensional gel electrophoresis combined with MS and de novo sequencing were used to separate and identify over a hundred proteins from a microbial community exposed to cadmium in a continuous-flow wastewater treatment reactor (Lacerda et al., 2007). ATPases, oxidoreductases, and transport proteins were among differentially expressed proteins as a result of cadmium shock. Shotgun proteomics has been successfully applied in the analysis of two natural biofilms present in an acid mine drainage (Ram et al., 2005). More than 2,000 proteins from the five most abundant species were identified with an impressive 48% of the predicted proteins from the metagenome of a strain belonging to *Leptospirillum* group II. A metaproteomics approach was also used in the monitoring of *Geobacter*-dominated microbial community members during the dissimilatory reduction of uranium (Wilkins et al., 2009). Several other metaproteomics studies are a growing testament to the utility of this approach (Kan et al., 2005; Schulze et al., 2005; Benndorf et al., 2007, 2009; Lo et al., 2007; Park and Helm, 2008; Wilmes et al., 2008a, 2008b; Goltsman et al., 2009; Sowell et al., 2009).

FUTURE DIRECTIONS

In the past few years, the "-omic" technologies have shown great potential for providing valuable information relevant to bioremediation. There are, nevertheless, several issues that pose considerable obstacles to this field. One is the lack of genomic information for uncultured microorganisms. The metagenome data generated from environmental microbial communities samples (65 environments to date) will help, but efforts to obtain environmentally important strains in pure culture need to be continued. Better annotation of the genomes of cultured organisms is also needed. Proteomics can be a power tool in this endeavor when coupled with physiological, biochemical, and enzymological studies. Increased computational power and new data analysis programs will also be critical for sifting through the reams of information. With more rapid development of analytical tools and bioinformatics comes the promise of identifying novel biomarkers relevant to bioremediation. In addition to further improvements in mass spectrometry, development of more efficient and cost-effective methods for protein extraction from microbial communities and improvement in bioinformatics tools will be needed for more accurate identification of proteins from shotgun proteomics.

ACKNOWLEDGMENTS

Our proteomics work was supported in part by a grant from the DOE Environmental Remediation Science Program. The Mass Spectrometry facility at Duquesne University was supported in part by an MRI grant from the National Science Foundation.

REFERENCES

Bansal, R., L. A. Deobald, R. L. Crawford, and A. J. Paszczynski. 2009. Proteomic detection of proteins involved in perchlorate and chlorate metabolism. *Biodegradation* **20:**603–620.

Bencheikh-Latmani, R., S. M. Williams, L. Haucke, C. S. Criddle, L. Wu, J. Zhou, and B. M. Tebo. 2005. Global transcriptional profiling of *Shewanella oneidensis* MR-1 during Cr(VI) and U(VI) reduction. *Appl. Environ. Microbiol.* **71:**7453–7460.

Benndorf, D., G. U. Balcke, H. Harms, and M. von Bergen. 2007. Functional metaproteome analysis of protein extracts from contaminated soil and groundwater. *ISME J.* **1:**224–234.

Benndorf, D., C. Vogt, N. Jehmlich, Y. Schmidt, H. Thomas, G. Woffendin, A. Shevchenko, H. H. Richnow, and M. von Bergen. 2009. Improving protein extraction and separation methods for investigating the metaproteome of anaerobic benzene communities within sediments. *Biodegradation* **20:**737–750.

Cao, B., A. Geng, and K. C. Loh. 2008. Induction of ortho- and meta-cleavage pathways in *Pseudomonas* in biodegradation of high benzoate concentration: MS identification of catabolic enzymes. *Appl. Microbiol. Biotechnol.* **81:**99–107.

Caraux, G., and S. Pinloche. 2005. PermutMatrix: a graphical environment to arrange gene expression profiles in optimal linear order. *Bioinformatics* **21:**1280–1281.

Chain, P. S., V. J. Denef, K. T. Konstantinidis, L. M. Vergez, L. Agulló, V. L. Reyes, L. Hauser, M. Córdova, L. Gómez, M. González, M. Land, V. Lao, F. Larimer, J. J. LiPuma, E. Mahenthiralingam, S. A. Malfatti, C. J. Marx, J. J. Parnell, A. Ramette, P. Richardson, M. Seeger, D. Smith, T. Spilker, W. J. Sul, T. V. Tsoi, L. E. Ulrich, I. B. Zhulin, and J. M. Tiedje. 2006. *Burkholderia xenovorans* LB400 harbors a multi-replicon, 9.73-Mbp genome shaped for versatility. *Proc. Natl. Acad. Sci. USA* **103:**15280–15287.

Chourey, K., M. R. Thompson, M. Shah, B. Zhang, N. C. Verberkmoes, D. K. Thompson, and R. L. Hettich. 2009. Comparative temporal proteomics of a response regulator (SO2426)-deficient strain and wild-type *Shewanella oneidensis* MR-1 during chromate transformation. *J. Proteome Res.* **8:**59–71.

Chovanec, P., P. Basu, and J. F. Stolz. 2010. A proteome investigation of roxarsone degradation by *Alkaliphilus oremlandii* strain OhILAs. *Metallomics* **2:**133–139.

De Vriendt, K., S. Theunissen, W. Carpentier, L. De Smet, B. Devreese, and J. Van Beeumen. 2005. Proteomics of *Shewanella oneidensis* MR-1 biofilm reveals differentially expressed proteins, including AggA and RibB. *Proteomics* **5:**1308–1316.

Ding, Y. H., K. K. Hixson, M. A. Aklujkar, M. S. Lipton, R. D. Smith, D. R. Lovley, and T. Mester. 2008. Proteome of *Geobacter sulfurreducens* grown with Fe(III) oxide or Fe(III) citrate as electron acceptor. *Biochim. Biophys. Acta* **1784:**1935–1941.

Dopson, M., C. Baker-Austin, and P. L. Bond. 2004. First use of two-dimensional polyacrylamide gel electrophoresis to determine phylogenetic relationships. *J. Microbiol. Methods* **58**:297–302.

Dowling, V. A., and D. Sheehan. 2006. Proteomics as a route to identification of toxicity targets in environmental toxicology. *Proteomics* **6**:5597–5604.

Dworzanski, J. P., and A. P. Snyder. 2005. Classification and identification of bacteria using mass spectrometry-based proteomics. *Expert Rev. Proteomics* **2**:863–878.

Elias, D. A., M. E. Monroe, M. J. Marshall, M. F. Romine, A. S. Belieav, J. K. Fredrickson, G. A. Anderson, R. D. Smith, and M. S. Lipton. 2005. Global detection and characterization of hypothetical proteins in *Shewanella oneidensis* MR-1 using LC-MS based proteomics. *Proteomics* **5**:3120–3130.

Feng, L., W. Wang, J. Cheng, Y. Ren, G. Zhao, C. Gao, Y. Tang, X. Liu, W. Han, X. Peng, R. Liu, and L. Wang. 2007. Genome and proteome of long-chain alkane degrading *Geobacillus thermodenitrificans* NG80-2 isolated from a deep-subsurface oil reservoir. *Proc. Natl. Acad. Sci. USA* **104**:5602–5607.

Fenselau, C., and P. A. Demirev. 2001. Characterization of intact microorganisms by MALDI mass spectrometry. *Mass Spectrom. Rev.* **20**:157–171.

Goltsman, D. S., V. J. Denef, S. W. Singer, N. C. VerBerkmoes, M. Lefsrud, R. S. Mueller, G. J. Dick, C. L. Sun, K. E. Wheeler, A. Zemla, B. J. Baker, L. Hauser, M. Land, M. B. Shah, M. P. Thelen, R. L. Hettich, and J. F. Banfield. 2009. Community genomic and proteomic analyses of chemoautotrophic iron-oxidizing "*Leptospirillum rubarum*" (Group II) and "*Leptospirillum ferrodiazotrophum*" (Group III) bacteria in acid mine drainage biofilms. *Appl. Environ. Microbiol.* **75**:4599–4615.

Gorby, Y. A., F. Caccavo, Jr., and H. Bolton, Jr. 1998. Microbial reduction of cobalt III EDTA⁻ in the presence and absence of manganese (IV) oxide. *Environ. Sci. Technol.* **32**:244–250.

Heidelberg, J. F., I. T. Paulsen, K. E. Nelson, E. J. Gaidos, W. C. Nelson, T. D. Read, J. A. Eisen, R. Seshadri, N. Ward, B. Methe, R. A. Clayton, T. Meyer, A. Tsapin, J. Scott, M. Beanan, L. Brinkac, S. Daugherty, R. T. DeBoy, R. J. Dodson, A. S. Durkin, D. H. Haft, J. F. Kolonay, R. Madupu, J. D. Peterson, L. A. Umayam, O. White, A. M. Wolf, J. Vamathevan, J. Weidman, M. Impraim, K. Lee, K. Berry, C. Lee, J. Mueller, H. Khouri, J. Gill, T. R. Utterback, L. A. McDonald, T. V. Feldblyum, H. O. Smith, J. C. Venter, K. H. Nealson, and C. M. Fraser. 2002. Genome sequence of the dissimilatory metal ion-reducing bacterium *Shewanella oneidensis*. *Nat. Biotechnol.* **20**:1118–1123.

Heidelberg, J. F., R. Seshadri, S. A. Haveman, C. L. Hemme, I. T. Paulsen, J. F. Kolonay, J. A. Eisen, N. Ward, B. Methe, L. M. Brinkac, S. C. Daugherty, R. T. Deboy, R. J. Dodson, A. S. Durkin, R. Madupu, W. C. Nelson, S. A. Sullivan, D. Fouts, D. H. Haft, J. Selengut, J. D. Peterson, T. M. Davidsen, N. Zafar, L. Zhou, D. Radune, G. Dimitrov, M. Hance, K. Tran, H. Khouri, J. Gill, T. R. Utterback, T. V. Feldblyum, J. D. Wall, G. Voordouw, and C. M. Fraser. 2004. The genome sequence of the anaerobic, sulfate-reducing bacterium *Desulfovibrio vulgaris* Hildenborough. *Nat. Biotechnol.* **22**:554–559.

Jennings, L. K., M. M. Chartrand, G. Lacrampe-Couloume, B. S. Lollar, J. C. Spain, and J. M. Gossett. 2009. Proteomic and transcriptomic analyses reveal genes upregulated by *cis*-dichloroethene in *Polaromonas* sp. strain JS666. *Appl. Environ. Microbiol.* **75**:3733–3744.

Kabiri, M., M. A. Amoozegar, M. Tabebordbar, K. Gilany, and G. H. Salekdeh. 2009. Effects of selenite and tellurite on growth, physiology, and proteome of a moderately halophilic bacterium. *J. Proteome Res.* **8**:3098–3108.

Kan, J., T. E. Hanson, J. M. Ginter, K. Wang, and F. Chen. 2005. Metaproteomic analysis of Chesapeake Bay microbial communities. *Saline Syst.* **1**:7.

Keller, M., and R. Hettich. 2009. Environmental proteomics: a paradigm shift in characterizing microbial activities at the molecular level. *Microbiol. Mol. Biol. Rev.* **73**:62–70.

Kim, S. I., J. S. Choi, and H. Y. Kahng. 2007. A proteomics strategy for the analysis of bacterial biodegradation pathways. *OMICS* **11**:280–294.

Kim, S. I., S. Y. Song, K. W. Kim, E. M. Ho, and K. H. Oh. 2003. Proteomic analysis of the benzoate degradation pathway in *Acinetobacter* sp. KS-1. *Res. Microbiol.* **154**:697–703.

Kim, S. J., R. C. Jones, C. J. Cha, O. Kweon, R. D. Edmondson, and C. E. Cerniglia. 2004. Identification of proteins induced by polycyclic aromatic hydrocarbon in *Mycobacterium vanbaalenii* PYR-1 using two-dimensional polyacrylamide gel electrophoresis and *de novo* sequencing methods. *Proteomics* **4**:3899–3908.

Kim, S. J., O. Kweon, and C. E. Cerniglia. 2009. Proteomic applications to elucidate bacterial aromatic hydrocarbon metabolic pathways. *Curr. Opin. Microbiol.* **12**:301–309.

Kim, Y. H., K. Cho, S. H. Yun, J. Y. Kim, K. H. Kwon, J. S. Yoo, and S. I. Kim. 2006.

Analysis of aromatic catabolic pathways in *Pseudomonas putida* KT 2440 using a combined proteomic approach: 2-DE/MS and cleavable isotope-coded affinity tag analysis. *Proteomics* **6**:1301–1318.

Kolker, E., A. F. Picone, M. Y. Galperin, M. F. Romine, R. Higdon, K. S. Makarova, N. Kolker, G. A. Anderson, X. Qiu, K. J. Auberry, G. Babnigg, A. S. Beliaev, P. Edlefsen, D. A. Elias, Y. A. Gorby, T. Holzman, J. A. Klappenbach, K. T. Konstantinidis, M. L. Land, M. S. Lipton, L. A. McCue, M. Monroe, L. Pasa-Tolic, G. Pinchuk, S. Purvine, M. H. Serres, S. Tsapin, B. A. Zakrajsek, W. Zhu, J. Zhou, F. W. Larimer, C. E. Lawrence, M. Riley, F. R. Collart, J. R. Yates III, R. D. Smith, C. S. Giometti, K. H. Nealson, J. K. Fredrickson, and J. M. Tiedje. 2005. Global profiling of *Shewanella oneidensis* MR-1: expression of hypothetical genes and improved functional annotations. *Proc. Natl. Acad. Sci. USA* **102**:2099–2104.

Krayl, M., D. Benndorf, N. Loffhagen, and W. Babel. 2003. Use of proteomics and physiological characteristics to elucidate ecotoxic effects of methyl tert-butyl ether in *Pseudomonas putida* KT2440. *Proteomics* **3**:1544–1552.

Kühner, S., L. Wöhlbrand, I. Fritz, W. Wruck, C. Hultschig, P. Hufnagel, M. Kube, R. Reinhardt, and R. Rabus. 2005. Substrate-dependent regulation of anaerobic degradation pathways for toluene and ethylbenzene in a denitrifying bacterium, strain EbN1. *J. Bacteriol.* **187**:1493–1503.

Kumar, R., S. Singh, and O. V. Singh. 2007. Bioremediation of radionuclides: emerging technologies. *OMICS* **11**:295–304.

Kweon, O., S. J. Kim, R. C. Jones, J. P. Freeman, M. D. Adjei, R. D. Edmondson, and C. E. Cerniglia. 2007. A polyomic approach to elucidate the fluoranthene-degradative pathway in *Mycobacterium vanbaalenii* PYR-1. *J. Bacteriol.* **189**:4635–4647.

Lacerda, C. M., L. H. Choe, and K. F. Reardon. 2007. Metaproteomic analysis of a bacterial community response to cadmium exposure. *J. Proteome Res.* **6**:1145–1152.

Lacerda, C. M., and K. F. Reardon. 2009. Environmental proteomics: applications of proteome profiling in environmental microbiology and biotechnology. *Brief. Funct. Genomic Proteomic* **8**:75–87.

Lay, J. O., Jr. 2001. MALDI-TOF mass spectrometry of bacteria. *Mass Spectrom. Rev.* **20**:172–194.

Lee, B. U., S. C. Park, Y. S. Cho, and K. H. Oh. 2008. Exopolymer biosynthesis and proteomic changes of *Pseudomonas* sp. HK-6 under stress of TNT (2,4,6-trinitrotoluene). *Curr. Microbiol.* **57**:477–483.

Liang, Y., D. R. Gardner, C. D. Miller, D. Chen, A. J. Anderson, B. C. Weimer, and R. C. Sims. 2006. Study of biochemical pathways and enzymes involved in pyrene degradation by *Mycobacterium* sp. strain KMS. *Appl. Environ. Microbiol.* **72**:7821–7828.

Lo, I., V. J. Denef, N. C. Verberkmoes, M. B. Shah, D. Goltsman, G. DiBartolo, G. W. Tyson, E. E. Allen, R. J. Ram, J. C. Detter, P. Richardson, M. P. Thelen, R. L. Hettich, and J. F. Banfield. 2007. Strain-resolved community proteomics reveals recombining genomes of acidophilic bacteria. *Nature* **446**:537–541.

López-Barea, J., and J. L. Gómez-Ariza. 2006. Environmental proteomics and metallomics. *Proteomics* **6**(Suppl. 1):S51–S62.

Lupi, C. G., T. Colangelo, and C. A. Mason. 1995. Two-dimensional gel electrophoresis analysis of the response of *Pseudomonas putida* KT2442 to 2-chlorophenol. *Appl. Environ. Microbiol.* **61**:2863–2872.

Markowitz, V. M., I. M. Chen, K. Palaniappan, K. Chu, E. Szeto, Y. Grechkin, A. Ratner, I. Anderson, A. Lykidis, K. Mavromatis, N. N. Ivanova, and N. C. Kyrpides. 2010. The integrated microbial genomes system: an expanding comparative analysis resource. *Nucleic Acids Res.* **38**:D382–D390.

Maron, P. A., L. Ranjard, C. Mougel, and P. Lemanceau. 2007. Metaproteomics: a new approach for studying functional microbial ecology. *Microb. Ecol.* **53**:486–493.

Marshall, M. J., A. E. Plymale, D. W. Kennedy, L. Shi, Z. Wang, S. B. Reed, A. C. Dohnalkova, C. J. Simonson, C. Liu, D. A. Saffarini, M. F. Romine, J. M. Zachara, A. S. Beliaev, and J. K. Fredrickson. 2008. Hydrogenase- and outer membrane c-type cytochrome-facilitated reduction of technetium(VII) by *Shewanella oneidensis* MR-1. *Environ. Microbiol.* **10**:125–136.

Martínez, P., L. Agulló, M. Hernández, and M. Seeger. 2007. Chlorobenzoate inhibits growth and induces stress proteins in the PCB-degrading bacterium *Burkholderia xenovorans* LB400. *Arch. Microbiol.* **188**:289–297.

Mazzoli, R., E. Pessione, M. G. Giuffrida, P. Fattori, C. Barello, C. Giunta, and N. D. Lindley. 2007. Degradation of aromatic compounds by *Acinetobacter radioresistens* S13: growth characteristics on single substrates and mixtures. *Arch. Microbiol.* **188**:55–68.

McLeod, M. P., R. L. Warren, W. W. Hsiao, N. Araki, M. Myhre, C. Fernandes, D. Miyazawa, W. Wong, A. L. Lillquist, D. Wang, M. Dosanjh, H. Hara, A. Petrescu, R. D. Morin, G. Yang, J. M. Stott, J. E. Schein,

H. Shin, D. Smailus, A. S. Siddiqui, M. A. Marra, S. J. Jones, R. Holt, F. S. Brinkman, K. Miyauchi, M. Fukuda, J. E. Davies, W. W. Mohn, and L. D. Eltis. 2006. The complete genome of *Rhodococcus* sp. RHA1 provides insights into a catabolic powerhouse. *Proc. Natl. Acad. Sci. USA* **103:**15582–15587.

Methé, B. A., K. E. Nelson, J. A. Eisen, I. T. Paulsen, W. Nelson, J. F. Heidelberg, D. Wu, M. Wu, N. Ward, M. J. Beanan, R. J. Dodson, R. Madupu, L. M. Brinkac, S. C. Daugherty, R. T. DeBoy, A. S. Durkin, M. Gwinn, J. F. Kolonay, S. A. Sullivan, D. H. Haft, J. Selengut, T. M. Davidsen, N. Zafar, O. White, B. Tran, C. Romero, H. A. Forberger, J. Weidman, H. Khouri, T. V. Feldblyum, T. R. Utterback, S. E. Van Aken, D. R. Lovley, and C. M. Fraser. 2003. Genome of *Geobacter sulfurreducens*: metal reduction in subsurface environments. *Science* **302:**1967–1969.

Monsinjon, T., and T. Knigge. 2007. Proteomic applications in ecotoxicology. *Proteomics* **7:**2997–3009.

Nelson, K. E., C. Weinel, I. T. Paulsen, R. J. Dodson, H. Hilbert, V. A. Martins dos Santos, D. E. Fouts, S. R. Gill, M. Pop, M. Holmes, L. Brinkac, M. Beanan, R. T. DeBoy, S. Daugherty, J. Kolonay, R. Madupu, W. Nelson, O. White, J. Peterson, H. Khouri, I. Hance, P. Chris Lee, E. Holtzapple, D. Scanlan, K. Tran, A. Moazzez, T. Utterback, M. Rizzo, K. Lee, D. Kosack, D. Moestl, H. Wedler, J. Lauber, D. Stjepandic, J. Hoheisel, M. Straetz, S. Heim, C. Kiewitz, J. A. Eisen, K. N. Timmis, A. Düsterhöft, B. Tümmler, and C. M. Fraser. 2002. Complete genome sequence and comparative analysis of the metabolically versatile *Pseudomonas putida* KT2440. *Environ. Microbiol.* **4:**799–808.

Nesatyy, V. J., and M. J. Suter. 2007. Proteomics for the analysis of environmental stress responses in organisms. *Environ. Sci. Technol.* **41:**6891–6900.

Nesatyy, V. J., and M. J. Suter. 2008. Analysis of environmental stress response on the proteome level. *Mass Spectrom. Rev.* **27:**556–574.

Park, C., and R. F. Helm. 2008. Application of metaproteomic analysis for studying extracellular polymeric substances (EPS) in activated sludge flocs and their fate in sludge digestion. *Water Sci. Technol.* **57:**2009–2015.

Pérez-Pantoja, D., R. De la Iglesia, D. H. Pieper, and B. González. 2008. Metabolic reconstruction of aromatic compounds degradation from the genome of the amazing pollutant-degrading bacterium *Cupriavidus necator* JMP134. *FEMS Microbiol. Rev.* **32:**736–794.

Picardal, F., R. G. Arnold, and B. B. Huey. 1995. Effects of electron donor and acceptor conditions on reductive dehalogenation of tetrachloromethane by *Shewanella putrefaciens* 200. *Appl. Environ. Microbiol.* **61:**8–12.

Rabus, R., M. Kube, J. Heider, A. Beck, K. Heitmann, F. Widdel, and R. Reinhardt. 2005. The genome sequence of an anaerobic aromatic-degrading denitrifying bacterium, strain EbN1. *Arch. Microbiol.* **183:**27–36.

Ram, R. J., N. C. Verberkmoes, M. P. Thelen, G. W. Tyson, B. J. Baker, R. C. Blake II, M. Shah, R. L. Hettich, and J. F. Banfield. 2005. Community proteomics of a natural microbial biofilm. *Science* **308:**1915–1920.

Reardon, K. F., and K. H. Kim. 2002. Two-dimensional electrophoresis analysis of protein production during growth of *Pseudomonas putida* F1 on toluene, phenol, and their mixture. *Electrophoresis* **23:**2233–2241.

Richey, C., P. Chovanec, S. E. Hoeft, R. S. Oremland, P. Basu, and J. F. Stolz. 2009. Respiratory arsenate reductase as a bidirectional enzyme. *Biochem. Biophys. Res. Commun.* **382:**298–302.

Saltikov, C. W., R. A. Wildman, Jr., and D. K. Newman. 2005. Expression dynamics of arsenic respiration and detoxification in *Shewanella* sp. strain ANA-3. *J. Bacteriol.* **187:**7390–7396.

Santos, P. M., D. Benndorf, and I. Sá-Correia. 2004. Insights into *Pseudomonas putida* KT2440 response to phenol-induced stress by quantitative proteomics. *Proteomics* **4:**2640–2652.

Santos, P. M., V. Roma, D. Benndorf, M. von Bergen, H. Harms, and I. Sá-Correia. 2007. Mechanistic insights into the global response to phenol in the phenol-biodegrading strain *Pseudomonas* sp. M1 revealed by quantitative proteomics. *OMICS* **11:**233–251.

Schneiker, S., V. A. Martins dos Santos, D. Bartels, T. Bekel, M. Brecht, J. Buhrmester, T. N. Chernikova, R. Denaro, M. Ferrer, C. Gertler, A. Goesmann, O. V. Golyshina, F. Kaminski, A. N. Khachane, S. Lang, B. Linke, A. C. McHardy, F. Meyer, T. Nechitaylo, A. Pühler, D. Regenhardt, O. Rupp, J. S. Sabirova, W. Selbitschka, M. M. Yakimov, K. N. Timmis, F. J. Vorhölter, S. Weidner, O. Kaiser, and P. N. Golyshin. 2006. Genome sequence of the ubiquitous hydrocarbon-degrading marine bacterium *Alcanivorax borkumensis*. *Nat. Biotechnol.* **24:**997–1004.

Schulze, W. X., G. Gleixner, K. Kaiser, G. Guggenberger, M. Mann, and E. D. Schulze. 2005. A proteomic fingerprint of dissolved organic carbon and of soil particles. *Oecologia* **142:**335–343.

Schweder, T., S. Markert, and M. Hecker. 2008. Proteomics of marine bacteria. *Electrophoresis* **29**:2603–2616.

Segura, A., P. Godoy, P. van Dillewijn, A. Hurtado, N. Arroyo, S. Santacruz, and J. L. Ramos. 2005. Proteomic analysis reveals the participation of energy- and stress-related proteins in the response of *Pseudomonas putida* DOT-T1E to toluene. *J. Bacteriol.* **187**:5937–5945.

Seshadri, R., L. Adrian, D. E. Fouts, J. A. Eisen, A. M. Phillippy, B. A. Methe, N. L. Ward, W. C. Nelson, R. T. Deboy, H. M. Khouri, J. F. Kolonay, R. J. Dodson, S. C. Daugherty, L. M. Brinkac, S. A. Sullivan, R. Madupu, K. E. Nelson, K. H. Kang, M. Impraim, K. Tran, J. M. Robinson, H. A. Forberger, C. M. Fraser, S. H. Zinder, and J. F. Heidelberg. 2005. Genome sequence of the PCE-dechlorinating bacterium *Dehalococcoides ethenogenes*. *Science* **307**:105–108.

Sharma, S., C. S. Sundaram, P. M. Luthra, Y. Singh, R. Sirdeshmukh, and W. N. Gade. 2006. Role of proteins in resistance mechanism of *Pseudomonas fluorescens* against heavy metal induced stress with proteomics approach. *J. Biotechnol.* **126**:374–382.

Shevchenko, A., S. Sunyaev, A. Loboda, A. Shevchenko, P. Bork, W. Ens, and K. G. Standing. 2001. Charting the proteomes of organisms with unsequenced genomes by MALDI-quadrupole time-of-flight mass spectrometry and BLAST homology searching. *Anal. Chem.* **73**:1917–1926.

Singh, O. V., and N. S. Nagaraj. 2006. Transcriptomics, proteomics and interactomics: unique approaches to track the insights of bioremediation. *Brief. Funct. Genomic Proteomic* **4**:355–362.

Sowell, S. M., L. J. Wilhelm, A. D. Norbeck, M. S. Lipton, C. D. Nicora, D. F. Barofsky, C. A. Carlson, R. D. Smith, and S. J. Giovannoni. 2009. Transport functions dominate the SAR11 metaproteome at low-nutrient extremes in the Sargasso Sea. *ISME J.* **3**:93–105.

Streit, W. R., and R. A. Schmitz. 2004. Metagenomics—the key to the uncultured microbes. *Curr. Opin. Microbiol.* **7**:492–498.

Tam Le, T., C. Eymann, D. Albrecht, R. Sietmann, F. Schauer, M. Hecker, and H. Antelmann. 2006. Differential gene expression in response to phenol and catechol reveals different metabolic activities for the degradation of aromatic compounds in *Bacillus subtilis*. *Environ. Microbiol.* **8**:1408–1427.

Tomás-Gallardo, L., I. Canosa, E. Santero, E. Camafeita, E. Calvo, J. A. López, and B. Floriano. 2006. Proteomic and transcriptional characterization of aromatic degradation pathways in *Rhodoccocus* sp. strain TFB. *Proteomics* **6**(Suppl. 1): S119–S132.

van Baar, B. L. 2000. Characterisation of bacteria by matrix-assisted laser desorption/ionisation and electrospray mass spectrometry. *FEMS Microbiol. Rev.* **24**:193–219.

Wall, J. D., and L. R. Krumholz. 2006. Uranium reduction. *Annu. Rev. Microbiol.* **60**:149-166.

Wasinger, V. C., S. J. Cordwell, A. Cerpa-Poljak, J. X. Yan, A. A. Gooley, M. R. Wilkins, M. W. Duncan, R. Harris, K. L. Williams, and I. Humphery-Smith. 1995. Progress with gene-product mapping of the Mollicutes: *Mycoplasma genitalium*. *Electrophoresis* **16**:1090–1094.

Wiatrowski, H. A., P. M. Ward, and T. Barkay. 2006. Novel reduction of mercury (II) by mercury-sensitive dissimilatory metal reducing bacteria. *Environ. Sci. Technol.* **40**:6690–6696.

Wilkins, M. J., N. C. Verberkmoes, K. H. Williams, S. J. Callister, P. J. Mouser, H. Elifantz, A. L. N'Guessan, B. C. Thomas, C. D. Nicora, M. B. Shah, P. Abraham, M. S. Lipton, D. R. Lovley, R. L. Hettich, P. E. Long, and J. F. Banfield. 2009. Proteogenomic monitoring of *Geobacter* physiology during stimulated uranium bioremediation. *Appl. Environ. Microbiol.* **75**:6591–6599.

Wilmes, P., and P. L. Bond. 2004. The application of two-dimensional polyacrylamide gel electrophoresis and downstream analyses to a mixed community of prokaryotic microorganisms. *Environ. Microbiol.* **6**:911–920.

Wilmes, P., M. Wexler, and P. L. Bond. 2008a. Metaproteomics provides functional insight into activated sludge wastewater treatment. *PLoS One* **3**:e1778.

Wilmes, P., A. F. Andersson, M. G. Lefsrud, M. Wexler, M. Shah, B. Zhang, R. L. Hettich, P. L. Bond, N. C. VerBerkmoes, and J. F. Banfield. 2008b. Community proteogenomics highlights microbial strain-variant protein expression within activated sludge performing enhanced biological phosphorus removal. *ISME J.* **2**:853–864.

Wilmes, P., and P. L. Bond. 2009. Microbial community proteomics: elucidating the catalysts and metabolic mechanisms that drive the Earth's biogeochemical cycles. *Curr. Opin. Microbiol.* **12**:310–317.

Wöhlbrand, L., B. Kallerhoff, D. Lange, P. Hufnagel, J. Thiermann, R. Reinhardt, and R. Rabus. 2007. Functional proteomic view of metabolic regulation in "*Aromatoleum aromaticum*" strain EbN1. *Proteomics* **7**:2222–2239.

Wood, T. K. 2008. Molecular approaches in bioremediation. *Curr. Opin. Biotechnol.* **19**:572–578.

Zhao, B., and C. L. Poh. 2008. Insights into environmental bioremediation by microorganisms through functional genomics and proteomics. *Proteomics* **8:**874–881.

Zhao, B., C. C. Yeo, and C. L. Poh. 2005. Proteome investigation of the global regulatory role of sigma 54 in response to gentisate induction in *Pseudomonas alcaligenes* NCIMB 9867. *Proteomics* **5:**1868–1876.

Zhao, J. S., J. Spain, S. Thiboutot, G. Ampleman, C. Greer, and J. Hawari. 2004. Phylogeny of cyclic nitramine-degrading psychrophilic bacteria in marine sediment and their potential role in the natural attenuation of explosives. *FEMS Microbiol. Ecol.* **49:**349–357.

MONITORING MICROBIAL ACTIVITY WITH GEOCHIP

Joy D. Van Nostrand, Sanghoon Kang, Ye Deng, Yuting Liang, Zhili He, and Jizhong Zhou

14

INTRODUCTION

Microorganisms are the most diverse, phylogenetically and functionally, group of organisms on the planet, with an estimated 2,000 to 50,000 microbial species per gram of soil (Torsvik et al., 1990; Schloss and Handelsman, 2006; Hong et al., 2006; Roesch et al., 2007). In part because of this functional diversity, they are vitally important to ecosystem functioning and are involved in the biogeochemical cycling of carbon, nitrogen, sulfur, phosphorus, and metals, as well as degradation or stabilization of contaminants in the environments. However, studying microbial communities can be challenging since >99% of microorganisms are uncultured (Amann et al., 1995; Fuhrman and Campbell, 1998; Whitman et al., 1998). As such, culture-independent approaches are necessary to study the vast majority of microorganisms. Several methods have been used, including 16S rRNA gene-based cloning, denaturing gradient gel electrophoresis, terminal-restriction fragment length polymorphism, quantitative PCR, and in situ hybridization. However, most of these approaches require a PCR amplification step, which introduces well-known biases (Warnecke et al., 1997; Lueders and Friedrich, 2003; Suzuki and Giovannoni, 1996). While functional genes can be examined using these methods, only a limited amount of information regarding functional potential and/or activity of a microbial community can be obtained, since conserved PCR primers cannot be designed for many functional genes because of a lack of sequence homology or an insufficient number of sequences. In addition, these assays can be time consuming and expensive, especially if multiple genes are examined.

Microarray technology has overcome many of these limitations since they can be used to examine thousands of genes at one time. Microarrays were first developed to study gene expression in the plant *Arabidopsis thaliana* (Schena et al., 1995). Hundreds of such gene expression arrays have since been developed and are now commonly used to study individual microorganisms. The utility of microarrays was further expanded when Guschin and colleagues (1997) proposed and tested the use of microarrays to study microbial communities.

Several new types of microarrays have since been developed for the study of microbial communities, and new uses for these arrays have

---

Joy D. Van Nostrand, Ye Deng, Yuting Liang, Zhili He, and Jizhong Zhou, Institute for Environmental Genomics, Department of Botany and Microbiology, University of Oklahoma, Norman, OK 73019. *Sanghoon Kang*, School of Science and Computer Engineering, University of Houston—Clear Lake, Houston, TX 77058.

also been discovered (Zhou and Thompson, 2002; Zhou, 2003; Gentry et al., 2006). Phylogenetic oligonucleotide arrays (POAs) are designed to examine phylogenetic relatedness or community composition using 16S rRNA or other conserved genes (Small et al., 2001; Loy et al., 2002; Wilson et al., 2002; Brodie et al., 2006). The most comprehensive POA to date is the PhyloChip, which can detect 842 subfamilies (8,741 taxa) from 121 bacterial and archaeal orders at the family to subfamily levels (Brodie et al., 2006, 2007; DeSantis et al., 2007). Other POAs have been developed to study specific environments such as compost (Franke-Whittle et al., 2005) or specific groups such as the "*Rhodocyclales*" order (Loy et al., 2005). Community genome arrays (CGAs) are used to determine the relatedness of microbial species or strains or to identify community members. These arrays use whole genomic DNA as probes, eliminating the need for probe design or PCR amplification (Wu et al., 2004; Zhang et al., 2004). CGAs have been used to compare microbial communities from different environments (Wu et al., 2004), to examine communities from acid mine drainage and bioleaching communities (Chen et al., 2009), and to determine the relatedness of different *Escherichia coli* strains (Zhang et al., 2004) and 55 different *Azoarcus, Pseudomonas,* and *Shewanella* strains (Wu et al., 2008b). Metagenomic arrays use environmental clone library inserts as probes and can be used as a high-throughput screening method (Sebat et al., 2003; Mockler and Ecker, 2005; Gresham et al., 2008). Metagenomic arrays have been used to examine communities from marine environments (Rich et al., 2008). In general, whole-genome ORF arrays (WGAs) are used for analysis of gene expression for individual microorganisms and have probes for all ORFs in one or more genomes (Wilson et al., 1999). However, WGAs can also be used for comparative genomics (Murray et al., 2001) and several studies have used WGAs for this purpose. *Shewanella* strains were compared using a WGA comprising 192 ORFs from *Shewanella oneidensis* MR-1 (Murray et al., 2001), an *E. coli* K-12 WGA was used to examine similar genes in *Klebsiella pneumoniae* (Dong et al., 2001), *Pseudomonas syringae* strains were compared using a WGA of 353 virulence factors (Sarkar et al., 2006), and a *Pyrococcus furiosus* WGA was used to examine seven *Pyrococcus* isolates (White et al., 2008). A DNA fingerprinting array has also been developed (Kingsley et al., 2002). Functional gene arrays (FGAs) have probes for key genes involved in microbial functional processes of interest (Wu et al., 2001; Gentry et al., 2006; He et al., 2007). This type of array can be used to simultaneously examine multiple functional genes at one time (Wu et al., 2001, 2006; Zhou and Thompson, 2002; Gentry et al., 2006; He et al., 2007; Wagner et al., 2007; Zhou et al., 2008; Wang et al., 2009). The most comprehensive FGA available is the GeoChip (He et al., 2007, 2010a). Other types of arrays have also been developed. For example, a fingerprinting array was developed that contained short (nonamer) probes randomly generated from the *E. coli* K-12 genome and was designed for tracking pathogens. A "bar code" was generated from the hybridization results, which was then used to distinguish similar strains. This method showed a higher resolution power than traditional gel electrophoresis. The remainder of this chapter will focus on FGAs.

FUNCTIONAL GENE ARRAYS

Since the first development of FGAs, several novel FGAs have been reported. These arrays have generally focused on specific functional groups or locations. Bodrossy and colleagues (2003) designed an array containing oligonucleotide probes for 59 *pmoA/amoA* genes to study methanotrophs. This array was later expanded to cover 68 genes and was used to study methanotrophic communities from simulated landfills using DNA (Stralis-Pavese et al., 2004) and mRNA (Bodrossy et al., 2006).

Several nitrogen-cycling arrays have been developed. Two arrays covering *amoA*, *nifH*, *nirS*, and *nirK* genes were developed to examine denitrification, N fixation, ammonia oxidation, and nitrite reduction (Taroncher-Oldenburg et al., 2003). Steward and colleagues (2004) developed a nylon membrane macroarray to study nitrogenase gene diversity using PCR amplicons of *nifH* as probes. This array was expanded and was then used to study Chesapeake Bay diazotrophs (Jenkins et al., 2004). Zhang and colleagues (2007) developed a *nifH* array to study nitrogen fixation.

FGAs have also been developed for clinical applications. Kostić and colleagues (2005) developed an array to detect virulence factors specific to *Salmonella* (*invA* and *sopB*). Miller et al. (2008) designed a virulence and marker gene array to detect waterborne pathogens, which targeted 67 virulence and marker genes from 17 pathogens. Call and colleagues (2003) developed an antibiotic resistance array for high-throughput screening of clinical isolates. A prototype pathogen detection array was developed that contained probes for virulence, antibiotic resistance, and metabolic or structural genes specific to *E. coli*, *Staphylococcus aureus*, and *Pseudomonas aeruginosa* (Cleven et al., 2006). A larger pathogen detection array was designed to detect genes from *Acinetobacter* spp., *Candida albicans*, *Enterobacter* spp., *Enterococcus* spp., *E. coli*, *Klebsiella* spp., *Proteus* spp., *P. aeruginosa*, *Staphylococcus* spp., *Stenotrophomonas* spp., and *Streptococcus* spp. as well as antimicrobial resistance genes (Palka-Santini et al., 2009). These pathogen detection arrays are still in the early phase of development but show promise as high-throughput, rapid tools for clinical diagnosis.

Other FGAs have also been designed to focus on specific communities. For example, a *nodC* array was designed to characterize rhizobial isolates (Bontemps et al., 2005), and an acid mine drainage-specific array was designed to cover both 16S rRNA and functional genes relevant to acid mine drainage and bioleaching systems (Yin et al., 2007).

The first FGA, a prototype of the GeoChip, was created using PCR-amplicon probes and was designed to target four N-cycling genes (*nirS*, *nirK*, *amoA*, and *pmoA*) from approximately 30 bacterial strains and 50 laboratory clones (Wu et al., 2001). However, the use of PCR primers greatly limits the number of genes that can be included on the array since conserved primers can only be designed for a few functional genes. To overcome this obstacle, oligonucleotide probes, which are more specific (Zhou, 2003), can be easily customized for a more targeted design (Denef et al., 2003; Zhou, 2003; Gentry et al., 2006), and are relatively inexpensive, were used for later GeoChip versions (Rhee et al., 2004; Tiquia et al., 2004; He et al., 2007, 2010a). The GeoChip 1.0 was reported in two studies (Rhee et al., 2004; Tiquia et al., 2004). This array has 2,006 oligonucleotide probes (50-mers) and covered genes involved in nitrification, denitrification, nitrogen fixation, methane oxidation, sulfate reduction (Tiquia et al., 2004), organic contaminant degradation, and metal resistance (Rhee et al., 2004). This was the first FGA to have such a diverse number of functional groups and has been used in several studies that have shown the usefulness of a comprehensive FGA for environmental microbiology studies (Wu et al., 2006, 2008b; Zhang et al., 2007; Waldron et al., 2009).

GeoChip 2.0 was developed to fill a gap in FGA coverage by providing a truly comprehensive probe set for multiple functional gene categories and to provide an increased level of specificity for highly homologous gene variants (He et al., 2007). GeoChip 2.0 contains 24,243 (50-mer) oligonucleotide probes targeting ~10,000 functional genes from 150 gene families involved in the geochemical cycling of C, N, and P cycling, sulfate reduction, metal reduction and resistance, and organic contaminant degradation. Numerous studies have used this comprehensive array (Rodríguez-Martínez et al., 2006; Leigh et al., 2007; Yergeau et al., 2007; Gao et al., 2007; Zhou et al., 2008; Liang et al., 2009a, 2009b; Van Nostrand et al., 2009; Wang et al., 2009; Mason et al., 2009; Parnell et al., 2010; Taş et al., 2009; Kimes et al., 2009; Van Nostrand et al., in press).

The latest version of the GeoChip is version 3.0, which covers 56,990 sequences from 292 gene families (He et al., 2010a). An image of the GeoChip 3.0 is shown in Color Plate 15. This version greatly increased the number of genes covered (~doubled from GeoChip 2.0) and expanded gene category coverage to include antibiotic resistance, energy processing, and phylogenetic markers (i.e., *gyrB*). A number of control features have been added or expanded. This includes a set of 16S rRNA gene probes as positive controls, human, plant, or hyperthermophile gene probes as negative controls, and a common oligo reference standard (CORS) for data normalization and comparison. The CORS probe is an artificial sequence cospotted with each gene probe on the GeoChip (Liang et al., 2010). The complementary CORS target is labeled with Cy-3 and spiked into each sample before hybridization. The signal intensity of the CORS probe is then used to normalize the signal intensity of the sample and allows a comparison of hybridization results among different samples. A computational pipeline has also been developed for GeoChip probe design and data analysis. The GeoChip 3.0 has already been used for microbial ecology studies (He et al., 2010b).

GEOCHIP DESIGN

Probe Design

GeoChip was designed to cover as many genes and gene variants as possible. Once functions or processes of interest are selected for inclusion, key enzymes or proteins, vital to the process, are chosen. Then key words are used to search public sequence databases (e.g., GenBank) and the resulting sequences are downloaded. The key words chosen should be as broad as possible to include as many sequences as possible. It is important to remember that enzymes from different microorganisms may be annotated differently or have more general or specific annotations. Next, the sequences are confirmed using the HMMER program (http://hmmer.wustl.edu/) with seed sequences, which have had protein identity and function experimentally confirmed. Seed sequence selection is a critical step in probe design and should be chosen carefully. The sequences verified by HMMER are then used to design gene- or group-specific 50-mer oligonucleotide probes using the CommOligo software (Li et al., 2005) and experimentally determined criteria (He et al., 2005c). Probe design criteria have been experimentally determined and are based on sequence homology ($\leq$90% identity for gene-specific probes, and $\geq$96% for group-specific probes), continuous stretch length ($\leq$20 bases for gene-specific probes, and $\geq$35 for group-specific probes), and free energy (≥ -35 kJ mol^{-1} for gene-specific probes, and ≤ -60 kJ mol^{-1} for group-specific probes) (He et al., 2005c; Liebich et al., 2006). All designed probes are then validated against the GenBank database to ensure their specificity. All key words, downloaded sequences, seed sequences, HMMER-confirmed sequences, and designed probes are stored in their corresponding databases for later array updates.

The final set of probes is then commercially synthesized and used for array construction. Probes can be spotted onto either glass slides (Taroncher-Oldenburg et al., 2003; Tiquia et al., 2004; Rhee et al., 2004) or nylon membranes (Steward et al., 2004), although glass slides are usually used since they produce less background fluorescence (Schena et al., 1995, 1996) and allow higher probe density (Ehrenreich, 2006). GeoChips are currently fabricated using contact printing, although other printing formats can be used, such as bubble Jet printing (Okamoto et al., 2000), laser-induced forward transfer (Serra et al., 2004), or photolithography (Chen et al., 2009).

Target Preparation

While many nucleic acid extraction methods are available, a well-established freeze-grind method (Zhou et al., 1996; Hurt et al., 2001) is generally used to extract DNA for GeoChip analysis since it results in high-molecular-weight DNA fragments that are important in subsequent amplification steps. RNA can also be used with GeoChip to monitor microbial activity (Gao et al., 2007). However, using RNA for GeoChip analysis presents some significant challenges, including the low abundance of environmental mRNA and the short turnover rates of mRNA. Several methods have been reported for extraction of environmental RNA (Hurt et al., 2001; Burgmann et al., 2003; McGrath et al., 2008; Poretsky et al., 2009).

Since relatively large amounts of DNA (2 to 5 μg) or RNA (10 to 20 μg) are needed for GeoChip hybridization, nucleic acid amplification is often necessary. For DNA, a whole community genome amplification (WCGA) method has been developed using the Templiphi 500 amplification kit (phi 29 DNA polymerase, GE Healthcare, Piscataway, NJ) with a modified amplification buffer (Wu et al., 2006). This method uses a small amount of DNA (1 to 100 ng) and provides a sensitive (10 fg detection limit) and representative amplification (<0.5% of amplified genes showed >2-fold different from unamplified) (Wu et al., 2006). A whole community RNA amplification method has also been evaluated and shown to provide a representative amplification using 50 to 100 ng of starting material (Gao et al., 2007).

The amplified DNA or RNA is directly labeled with a fluorescent dye such as Cy3 or Cy5 using random priming with the Klenow fragment of DNA polymerase for DNA (Wu et al., 2006) or Superscript™ II/III RNase H-reverse transcriptase for RNA (He et al., 2005c). The labeled DNA/RNA is then purified and dried for hybridization.

Hybridization

Hybridization buffer (50% formamide, 3× SSC, 0.3% sodium dodecyl sulfate, 0.7 μg μl^{-1} Herring sperm DNA, 0.86 mM dithiothreitol) is added to the labeled nucleic acid in preparation for hybridization. GeoChips are generally hybridized at 42 to 50°C and 50% formamide (He et al., 2007; Mason et al., 2009; Liang et al., 2009a, 2009b; Van Nostrand et al., 2009; Waldron et al., 2009). The specificity of the GeoChip is controlled in part by the hybridization stringency, which can be adjusted by altering the hybridization temperature and formamide concentration (effective hybridization temperature increases by 0.6°C for every 1% of formamide).

Glass slide-based arrays can be hybridized manually or using semi- or fully automated hybridization stations. Manual hybridizations are performed using special coverslips (e.g., LifterSlips; Erie Scientific), which have raised edges so that solutions wick between the array and coverslip. A minimal volume of hybridization solution should be used to provide the best sensitivity (Shalon et al., 1996). This assembly is then placed into a hybridization chamber, which helps to maintain humidity levels so the array does not dry out during hybridization. The hybridization chambers are incubated in a water bath or hybridization oven for 10 to 20 h. Mixing during hybridization has been shown to increase sensitivity (Adey et al., 2002), and several hybridization stations have been developed that provide mixing with incubation at controlled temperatures (e.g., Mai Tai® from SciGene, SlideBooster from Advalytix, Hybridization Station from Roche Nimblegen). Posthybridization washing can be done manually through a series of buffer transfers with mixing or via an automated wash station (e.g., Maui Wash Station, BioMicro Systems). Completely automated hybridization stations are also available which perform all hybridization steps from prehybridization to posthybridization washes (e.g., Tecan HS4800Pro, TECAN US).

Image Analysis

The arrays are scanned and analyzed by quantifying the pixel density (intensity) of each spot using image analysis software, GenePix Pro (Molecular Devices, Sunnyvale, CA), GeneSpotter (MicroDiscovery, San Diego, CA), or ImaGene (BioDiscovery, El Segundo, CA). The raw image data are then uploaded to the GeoChip data analysis pipeline (http://ieg.ou.edu/) and evaluated. The quality of each hybridization is determined by both evenness of control spot hybridization signals and the background signal across the slide surface. Poor- and low-quality spots, evaluated by predetermined criteria, and outliers are removed. Outliers are determined based on Grubbs' test of outliers (Grubbs, 1969) among the signal intensities of replicate arrays. The signal intensities of all spots are then normalized. Positive spots can be determined using signal-to-noise ratio [SNR = (signal mean − background mean)/background standard deviation]. Most studies use a cutoff of ≥2 for a positive signal, although lower cutoffs have been used, such as 1.2 (Taş et al., 2009) or 1.5 (Van Nostrand et al., 2009). Signal-to-background ratio (SBR = signal mean/background mean) has also been used (Loy et al., 2002). Another calculation is the signal-to-both-standard-deviations ratio [SSDR = (signal mean − background mean)/(signal standard deviation − background standard deviation)], which was recently proposed, and using SSDR resulted in fewer false positives and false negatives than using SNR (He and Zhou, 2008).

Data Analysis

Like other high-throughput technologies, GeoChip analysis generates a large amount of data, making effective and efficient data analysis a challenge. GeoChip data have a multivariate structure and the number of variables is much larger than the number of observations (p >> n). Frequently used data analysis methods include relative abundance of gene groups based on gene number or total signal intensity, richness and diversity indices based on gene number, percentage of gene overlap between samples, and response ratios. Methods commonly used for descriptive and exploratory data analysis include unconstrained ordination, hierarchical cluster analysis, and neural network analysis (He et al., 2008). Unconstrained ordination methods such as principal component analysis (PCA) and correspondence analysis (CA) reduce the dimensionality of variables to maximize the visible variability of the data. Because of the differences in the assumed data structures, CA may be preferred over PCA for ecological studies. Nonmetric multidimensional scaling (NMDS) represents the relative interrelatedness of samples on a priori dimensions. Together with analysis of similarity, NMDS can serve as both descriptive and hypothesis-driven analyses. Hierarchical cluster analysis groups communities based on the similarity of their gene profiles and provides insight into the structure and function of a given microbial community. Response ratios compare gene levels or signal intensity between conditions (e.g., treatment versus control, contaminated versus uncontaminated) (Luo et al., 2006). Response ratios have been used to examine the community response to oil contamination (Liang et al., 2009a). Interrelationships among microbial communities and other abiotic and biotic factors can be pursued by constrained ordination, such as canonical correspondence analysis (CCA) (ter Braak, 1986), distance-based redundancy analysis (db-RDA) (Legendre and Anderson, 1999), variation partitioning analysis (VPA) (Økland and Eilertsen, 1994; Ramette and Tiedje, 2007), and the Mantel test. CCA is a commonly used analysis method in GeoChip-based studies and is used to better understand how environmental factors impact and drive community structure (Yergeau et al., 2007; Wu et al., 2008a; Zhou et al., 2008; Waldron et al., 2009; Van Nostrand et al., 2009). Canonical VPA, based on the results of CCA and partial CCA, is used to examine the relative influence of individual biological and environmental parameters on the microbial community structure. The Mantel test has been used to compare environmental factors with functional genes detected by GeoChip (He et al., 2007; Wu et al., 2008a; Van Nostrand et al., 2009; Waldron et al., 2009).

Neural network analysis is a relatively new technique for GeoChip data analysis that can be used to examine gene relationships and is based on the random matrix theory, a novel mathematical theory used to distinguish random and system-specific properties of complex systems (Mehta, 1990). In nature, biological community members interact through the exchange of energy, matter, and information, forming complex ecological networks (Montoya et al., 2006). These networks have been intensively studied in macroecology, but little is known regarding microbial ecological networks (Fuhrman and Steele, 2008; Raes and Bork, 2008). The massive amounts of data available through metagenomic technologies, such as high-throughput sequencing and microarrays (Handelsman et al., 2007; He et al., 2007), provide an unprecedented opportunity to examine network interactions within microbial communities (Raes and Bork, 2008). A random matrix theory-based approach has been used to construct microbial ecological networks using GeoChip data from various habitats, such as forest soils, grasslands, groundwater, lake sediment, and human intestine, and the results show that the constructed networks had consistent general features with many other cellular networks. Further network analyses on the network structures of microbial communities from a long-term field experiment indicated that elevated CO_2 dramatically altered the network interactions of the belowground microbial communities, and network connectivity significantly correlated with soil carbon and nitrogen dynamics and plant productivity (Zhou et al., submitted).

Important Issues for Microarray Application

NUCLEIC ACID QUALITY

One of the most important steps in GeoChip analysis is to obtain high-quality DNA. Depending on how "clean" a sample is, the DNA may need to be purified. This is especially important for microarray analysis since contaminants in the DNA may inhibit later amplification, labeling, and hybridization steps. The raw DNA extract can be purified using an agarose gel purification followed by phenol-chloroform-butanol extraction (Xie et al., 2007; Liang et al., 2009b) if relatively large amounts of DNA are available. This method works best with fresh DNA since DNA degrades over time. Otherwise, a column purification method can be used. While these purification methods work well and often provide high-quality DNA, there is some loss of DNA with each method that can be problematic for samples with very low abundance. In addition, even with purification, it is difficult to obtain the necessary level or purity, so better methods of purification are needed. The purified DNA quality is evaluated based on ratios of the absorbance values at 230, 260, and 280 nm (A_{230}, A_{260}, A_{280}). Ideal values are A_{260}:A_{280} >1.8 and A_{260}:A_{230} >1.7. The A_{260}:A_{230} ratio has been shown to be most critical for hybridization success (Ning et al., 2009). Purified RNA should have A_{260}:A_{280} >1.9 and A_{260}:A_{230} >1.7.

PROBE COVERAGE

One objective in designing the GeoChip 2.0 was to provide a truly comprehensive probe set (He et al., 2007). Most of the previous FGAs covered a single gene or gene group and covered only a few gene variants (generally <100). The GeoChip 2.0 provided a greatly expanded probe set, covering ~10,000 genes from >150 gene groups (He et al., 2007), providing the most comprehensive array reported. GeoChip 3.0 covers ~57,000 genes from ~300 gene groups (He et al., 2010a). However, a challenge in maintaining such comprehensive probe sets is that there is a continual increase in the number of functional gene groups and the number of sequences for each gene available since sequences are constantly being submitted to public databases. To remain relevant, GeoChip must be updated regularly. The GeoChip design pipeline (ieg.ou.edu) has an automatic update feature that uses predetermined key words and seed sequences to download new sequences and design appropriate probes. Even with the automated design, this is still a time-consuming process because of the large volume of sequences and probes that must be tested.

SPECIFICITY

An important quality of microarray probes used for environmental studies is specificity. Environmental communities can have extremely vast diversities. However, obtaining specific hybridization results can be difficult since so many environmental sequences are unknown and so many gene variants have high homologies. Several methods are available for increasing specificity. The first is designing specific probes. Experimentally determined design criteria are used to design gene- and group-specific GeoChip probes (He et al., 2005c; Liebich et al., 2006) as mentioned before. Designed probes are then screened against a public sequence database to confirm specificity. A very small number of false positives (0.002 to 0.004%) were observed for probes designed using these criteria (He et al., 2007).

As mentioned previously, hybridization conditions (temperature and formamide concentration) affect specificity by varying the stringency. For example, using PCR probes, Wu et al. (2001) found that at 45°C, sequences with 70 to 75% similarity to a probe hybridized; while at 65°C, similarities >87% were required for hybridization. Using 50-mer oligonucleotide probes designed with the criteria described above, hybridizations at 50°C and 50% formamide were able to discriminate sequences with <88 to 94% similarities (Rhee et al., 2004; Liebeich et al., 2006; Deng et al., 2008).

In addition to sequence abundance, the signal intensity of probes can also be affected by sequence divergence. Methods to determine whether a signal is true can increase specificity as well. Nonspecific hybridization is usually caused by highly homologous sequences, such as sequences from the same gene families or multispliced variants (Evertsz et al., 2001; Modrek and Lee, 2002). One option to handle nonspecific hybridization is the use of perfect match (PM) and mismatch (MM) probes for each gene (Affymetrix, 2001; Relogio et al., 2002; Chou et al., 2004). Any signal from the MM probe would be from nonspecific hybridization, which could then be subtracted from the corresponding PM probe. Theoretically, this strategy would remove the nonspecific signal and the remaining signal would then be directly proportional to the concentration of the real target (Lockhart et al., 1996; Lipshutz et al., 1999). PM/MM probe sets have been widely used in short oligonucleotide microarrays such as the Affymetrix GeneChip®, which usually has 11 pairs of PM/MM probes (25-mer) per gene (Affymetrix, 2001). The PhyloChip 16S rRNA array uses this strategy as well (Brodie et al., 2006; DeSantis et al., 2007). However, GeoChip uses long oligonucleotide probes (50-mer), so to test the use of PM/MM probes an oligonucleotide microarray was constructed to evaluate and optimize parameters for a 50-mer mismatch probe design (Deng et al., 2008). The experimental results showed that probes with evenly distributed mismatches were generally more distinguishable than those with randomly distributed mismatches. MM probes with 3, 4, and 5 mismatched nucleotides were differentiated for 50-mer oligonucleotide probes hybridized at 50, 45, and 42°C, respectively. A modified positional dependent nearest-neighbor (MPDNN) model was then established to better predict hybridization signals for long oligonucleotide microarrays and could provide general guidance for long oligonucleotide probe design.

SENSITIVITY

Sensitivity is another important quality for microarrays. This is especially true for environmental samples since these microbial communities can be very complex with many low-abundance strains. The current detection limit for FGAs is 5% or 25 ng of the microbial community (Wu et al., 2001; Bodrossy et al., 2003; Rhee et al., 2004; Tiquia et al., 2004). Several methods and strategies can be used to increase microarray sensitivity, although some of these also decrease specificity. These include increasing probe length (Denef et al., 2003; He et al., 2005b) and increasing probe concentration per spot (Cho and Tiedje, 2002; Relógio et al., 2002; Zhou and Thompson, 2002). Probe concentration is thought to be the reason that arrays printed onto membranes are more sensitive than those printed onto glass slides, since >1 μg/spot of probe is used for membranes, while <20 pg/spot is used for glass slides (Cho and Tiedje, 2002). However, both of these options do decrease specificity through either decreasing overall probe signal intensity (Denef et al. 2003) or allowing a greater chance of probe mismatch (Relógio et al., 2002).

Amplification of community DNA has been used to increase sensitivity by increasing the concentration of low-abundance sequences. For example, WCGA can representatively amplify 1 to 250 ng of community DNA (Wu et al., 2006), and could increase the detection limit from 25 ng to 10 pg (two bacterial cells), although this level of sensitivity did result in a much higher amplification bias than observed with 1 ng of DNA. PCR-based amplification has also been used for some types of FGAs. For example, multiplex PCR amplification, which uses primers for all genes contained on an array, increased the sensitivity of a pathogen array (Palka-Santini et al., 2009). PCR-based amplifications could work for FGAs that contain a relatively small number of genes but would not work for more comprehensive arrays like the GeoChip.

Improved labeling techniques can also improve sensitivity. Cyanine dye-doped nanoparticles have been shown to increase sensitivity 10-fold (Zhou and Zhou, 2004). A tyramide signal amplification labeling also has been shown to increase sensitivity (Denef et al., 2003). Other strategies include development of more sensitive signal detection systems (Cho and Tiedje, 2002; Zhou and Thompson, 2002), decreasing background fluorescence by the use of unmodified array slides (Kumar et al., 2000; Gudnason et al., 2008) or improved washing protocols, or decreasing ozone levels, which can cause degradation of Cy-dye signal (Branham et al., 2007).

QUANTITATIVE CAPABILITY

A great advantage of GeoChip analysis over other PCR-based methods, such as PhyloChip or 454 sequencing, is the ability to provide quantitative information. PCR amplification steps produce well-known biases that prohibit quantification (Warnecke et al., 1997; Lueders and Friedrich, 2003; Suzuki and Giovannoni, 1996). GeoChips, however, have been shown to provide quantitative information with or without WCGA for both DNA and RNA (Wu et al., 2001, 2006; Tiquia et al., 2004; Rhee et al., 2004; Gao et al., 2007). For example, hybridization of DNA over a concentration of 8 to 1,000 ng showed a linear relationship ($r = 0.98$ to 0.99) (Tiquia et al., 2004).

ACTIVITY

Until now, most GeoChips were hybridized with DNA that can only provide information regarding gene abundance. While these changes can be used to infer microbial activity, it does not provide direct proof of functional gene activity. In addition, some community members may be very active, but may not change in number. Therefore, other strategies are needed to monitor microbial activity using GeoChip. Since each probe is designed from coding sequences, RNA can be used for FGA analysis (Dennis et al., 2003; Bodrossy et al., 2006; Gao et al., 2007). However, as mentioned previously, the use of environmental mRNA presents some special challenges, including low abundance and instability. The use of whole community RNA amplification can increase the abundance of the mRNA (Gao et al., 2007). A few different methods are available for mRNA amplification, although these methods require multiple steps and can be time consuming, so improved methods are needed.

Another strategy for monitoring community activity with DNA is combining GeoChip hybridization with stable isotope probing (SIP) (Leigh et al., 2007). In these experiments, microcosms were set up using soil collected from the root zone of a tree growing in a PCB-contaminated site. The sediments were fed ^{13}C-labeled or unlabeled biphenyl, incubated at 25°C, and sampled at 14 days for GeoChip analysis. The ^{13}C-labeled DNA was separated from the extracted DNA by use of equilibrium density gradient centrifugation. Total community DNA, ^{13}C-labeled DNA, and DNA from nonincubated sediment (background DNA) were then amplified, labeled, and hybridized to GeoChip 2.0. After genes detected in the background sample were subtracted out, 30 genes involved in organic contaminant degradation were detected in the ^{13}C-labeled sample.

Application of GeoChip for Microbial Community Analysis

LAKE DePue, A METAL-CONTAMINATED SITE

Microbes are known for their metabolic versatility, which is particularly useful in bioremediation applications. However, while microorganisms can be useful in bioremediation (biodegrading organic contaminants or reducing and sequestering metal contaminants); they can be negatively impacted by the presence of these contaminants. A number of studies on soil microbes have shown decreases in biomass (Konopka et al., 1999; Hartmann et al., 2005), diversity (Hu et al., 2007; Wang et al., 2007), and activity (Ramsey et al., 2005; Lazzaro et al., 2008), as well as significant shifts in microbial community structure (Pennanen, 2001; Frey et al., 2006) under metal stress. While lake sediments are often the final repository of metal contaminants and play critical roles in aquatic system health, microbial communities associated with these metal-contaminated sediments have been rarely studied. It was our goal to study metal-contaminated lake sediment microbial communities, and identify the significant controlling factors on microbial communities.

Lake DePue (Illinois, USA), a U.S. EPA Superfund site (ILD062340641), has been contaminated with metals for about 80 years because of the adjacent zinc-smelting activities. Sediments were collected in triplicate from five areas of the lake, which had been previously identified as having varied metal contamination levels (Gough et al., 2008), and are located along a transect from near a creek inlet to the main body of the lake. The sediments were first treated with EDTA to remove abundant divalent metal ions, since they prevent DNA from precipitating (Gough and Stahl, 2011). Community DNA extraction and subsequent procedures including purification, random amplification, and labeling with Cy5-dUTP dye were described previously (Zhou et al., 2008). The labeled community DNA was hybridized with GeoChip 2.0 using a Tecan 4800Pro hybridization station.

Lake sediment characteristics and pore water and total metal concentrations were reported in a previous study (Gough et al., 2008). All variables, except for pore water metal concentration, showed a significant spatial structure. Analyses were performed using all functional genes detected and three functional subgroups, selected based on their relevance to metal contamination (sulfate-reducing, metal resistance and reduction, and C-cycling genes). Proximity of sampling sites to the contamination source was an apparent factor in shaping the microbial community structure. The two sites closest to the contamination source (sites 1 and 2) clustered together, as measured by NMDS and hierarchical clustering. Interestingly, these two sites also showed higher percentages of unique genes resulting in the highest diversity indices (Shannon's H', Simpson's $-\ln D$ and Fisher's α). However, proximity of sampling locations alone could not explain community variability because samples from site 4, located between sites 3 and 5, were more similar to samples from sites 1 and 2. Unlike the environmental variables, all functional genes and the three functional subgroups were not spatially structured except for a very significant patch at ~250 m. Multivariate correlation analysis selected pore water metal concentration as the most relevant environmental variable to all three microbial community subgroups.

Polyphasic approaches were used to select the most important 2 to 4 metals (e.g., zinc, arsenic, lead, etc.) in determining community

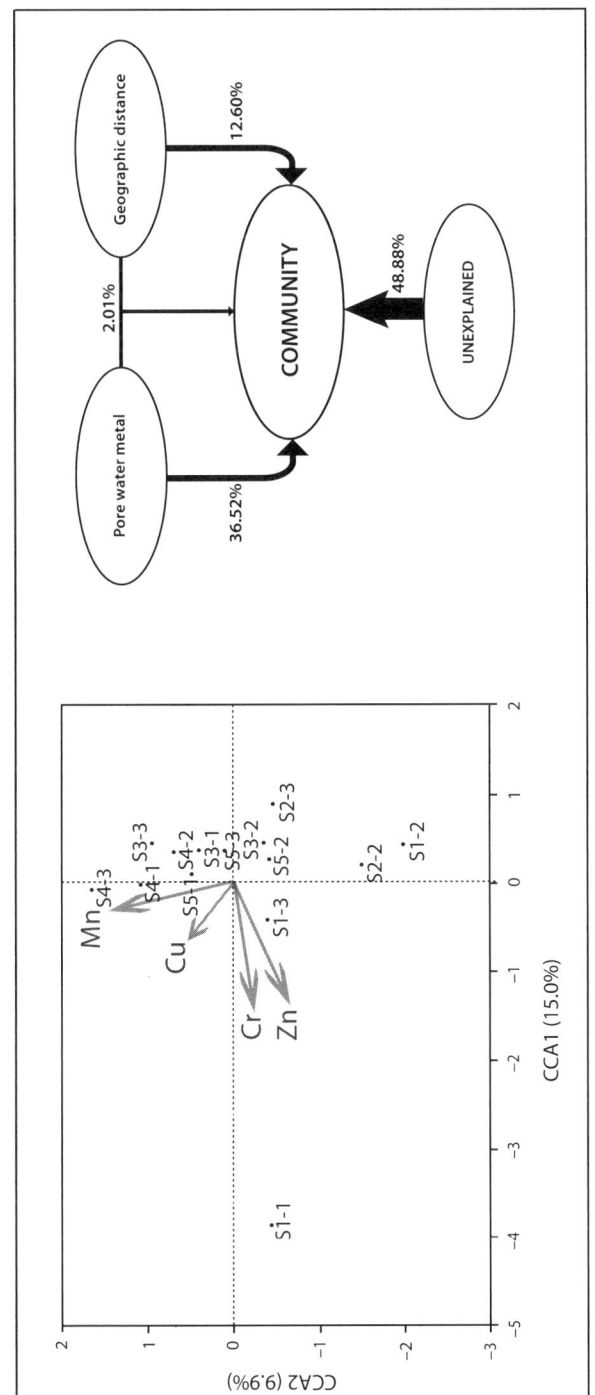

FIGURE 1 CCA and VPA results of metal resistance and reduction microbial communities. CCA model was most significant ($P = 0.017$).

structure based on all functional genes detected and for the three functional subgroups. When geographic distance was held constant, the partial Mantel test indicated that selected pore water metals showed a fairly significant correlation with all groups of microbial functional genes tested ($P = 0.108$), especially with sulfate reducing genes ($P = 0.036$). CCA models using the same sets of pore water metals were also very significant based on the Monte Carlo permutation ($P < 0.075$) (Fig. 1). VPA based on simple and partial CCA results indicated that selected pore water metals explained more variance than geographic distance, and that pore water metals alone were much more significant than geographic distance for all microbial communities (Fig. 1). Consistently smaller (<2.01%) shared portions of the variance between pore water metals and geographic distance indicate near-independence of the two components, which is also demonstrated by the small differences in the results between the simple and partial Mantel tests. The metal resistance and reduction genes showed the most significance with pore water metals like copper, zinc, chromium, and manganese based on results from both CCA and VPA. This is probably because C-cycling and sulfate-reducing genes are indirectly influenced by metal-contaminated environments.

In conclusion, microbial communities in the metal-contaminated lake sediments varied significantly from the source of contamination and downstream. The variation among microbial communities was primarily the result of pore water metal concentrations rather than geographic distance between samples.

OTHER METAL-CONTAMINATED SITES

Another study examined contaminated groundwater communities at the U.S. Department of Energy Oak Ridge Integrated Field Research Challenge site. Groundwater samples with varying levels of contamination and an uncontaminated background sample were compared (Wu et al., 2006). Overall, functional gene diversity and gene numbers decreased as contaminant levels increased and genes for denitrification, organic contaminant degradation, metal resistance, and U(VI) reduction (*dsr*) were detected as expected based on contaminants present at this site. These samples and an additional sample were analyzed in greater detail in a subsequent study (Waldron et al., 2009). Sulfate pH, U, and Tc showed the greatest influence on the community structure. VPA analysis was then used to determine the relative influence of each geochemical variable and results indicated that pH and the combination of U and Tc had the greatest influence, explaining ~21% of the variance or 29 to 40% if sulfate were included.

Microbial communities from a pilot-scale field-test system for bioremediation and immobilization of U(VI) have been extensively examined with GeoChip. During the initial phases of operation, U(VI) was actively reduced and functional genes associated with denitrification, sulfate reduction, and Fe(III) reduction showed a concomitant increase, indicating that these populations were important in U(VI) reduction (Van Nostrand et al., in press). In later operational phases, the stability of the bioreduced U(IV) was examined by stopping ethanol injections and allowing dissolved O_2 to enter the system. GeoChip results showed a shift in the community structure after ethanol injection was resumed (Van Nostrand et al., 2009). While gene numbers changed before and after ethanol addition, the percentage of genes in each functional group remained relatively stable, indicating a functionally diverse community that could be stimulated after adverse conditions. Chemical oxygen demand (COD, i.e., ethanol), temperature, sulfate, and U(VI) were found to be the most important drivers in this system, with COD having the greatest effect.

Microbial communities in a hypersaline lake containing high levels of chromium and sulfate were evaluated using GeoChip 2.0 and PhyloChip to examine horizontal gene transfer within these communities (Parnell et al., 2010). A high diversity of *dsr* and chromium resistance genes was noted in all samples, suggesting that horizontal gene transfer was occurring due to selection pressure of chromium and sulfate (Parnell et al., 2010).

OIL- AND PESTICIDE-CONTAMINATED SITES

The microbial community associated with a fluidized bed reactor designed for the bioremediation of diesel-contaminated groundwater was examined with the GeoChip (Rodríguez-Martínez et al., 2006). Genes involved in the degradation of diesel fuel and other organic contaminants (acetylene, aniline, benzoate, biphenyl, cyclohexanol, methyl *tert*-butyl ether, naphthalene, phthalate, protocatechuate, and toluene) were detected. A shift toward anaerobiosis over time was indicated by increased signal intensities for genes involved in anaerobic benzoate degradation. This conclusion was supported by other experimental evidence.

Liang et al. (2009b) examined how bioremediation treatments affected microbial communities using laboratory-scale bioremediation systems using sediment from contaminated oil fields and inoculated with oil-degrading enrichment cultures. The bioremediation system was incubated 242 days, treated with ozone, and incubated a further 125 days. Many genes involved in oil degradation (benzene, benzoate, catechol, polyaromatic hydrocarbon aromatics, protocatechuate, phthalate) were detected. Ozonation treatment resulted in an almost 50% reduction in the number of functional genes detected, although gene numbers did increase after a recovery period and the community retained the ability to degrade the oil.

Another study characterized microbial communities along an oil-contaminant gradient. As contaminant levels increased, the number and diversity of functional genes detected decreased (Liang et al., 2009a). However, genes involved in the degradation of biphenyl, catechol, and protocatechuate increased in samples with a moderate level of contamination. Oil concentration and soil available nitrogen were found to be strong drivers in determining community structure at these sites.

Liebich et al. (2009) examined a background and three atrazine-contaminated aquifers. More genes were detected in the background site than in the contaminated sites. Among the contaminated sites, the aquifer with the highest level of contamination had the highest number of genes detected, with a majority involved in contaminant degradation. Genes involved in atrazine degradation were detected in all contaminated wells, and the presence of these genes was validated by PCR. These results indicated that even small amounts of contaminants were enough to select for specific degrading populations.

Industrial pollutant- and pesticide-contaminated river sediments were characterized with the GeoChip and the results indicated that contaminant level was not a major driver in these systems (Taş et al., 2009). Rather, C/N ratio, depth, total Kjeldahl N, and location were the strongest drivers in determining the community structure. Results also indicated that most reductive dehalogenation genes detected from the river sediments were from *Dehalococcoides* spp., suggesting that this microorganism may play an important role in contaminant degradation in this system.

MARINE ENVIRONMENTS

Microbial communities from Puget Sound sediments were examined with a partial version of GeoChip 1.0 containing 763 50-mer oligonucleotide probes for *amoA*, *nirS*, *nirK*, *nifH*, *pmoA*, and *dsrAB* (Tiquia et al., 2006). PCA ordination plots based on the functional genes detected indicated that communities from 0 to 25 cm clustered together, while communities from 50 to 85 cm clustered separately, indicating that the community structure changed as the depth increased. Lower functional diversity was also observed for communities at 50 to 50.5 cm depth, where nutrient levels were lower than communities from 0 to 0.5 cm depths, which had higher nutrient levels.

Microbial communities from different sediment depths in the Gulf of Mexico were characterized with GeoChip and results indicated that the community structure changed as depth increased, becoming more unique with depth (Wu et al., 2008a). CCA indicated that depth, porosity, and ammonium, Mn(II), phosphate, and silicic acid concentrations were important drivers in determining the community structure in this environment.

GeoChip has also been used to examine several marine environments. Deep-sea basalt communities were characterized with GeoChip and carbon fixation, methane oxidation, methanogenesis, and nitrogen fixation genes were detected, indicating that these processes, which had not been previously associated with this environment may be important (Mason et al., 2009). Another study examined deep-sea hydrothermal vent communities and found that communities from the inner chimney were less diverse than those from the outer portion of the 5-day-old chimney or the mature chimney (Wang et al., 2009).

Microbial communities associated with the surface mucopolysaccharide layer and tissue from healthy and yellow band diseased coral, *Montastraea faveolata*, were examined, and results indicated that cellulose degradation and nitrification genes were increased in diseased samples suggesting that these functional activities may provide a competitive advantage to coral pathogens (Kimes et al., 2009).

TERRESTRIAL SITES

The effects of different land use strategies on microbial communities were evaluated with this array (Zhang et al., 2007). Samples from a primitive fir forest, two spruce plantations (established for ~20 and ~40 years, respectively), and cropland were analyzed. Functional diversity and gene numbers were positively correlated with soil organic carbon in the disturbed sites (spruce plantation, cropland), suggesting that maintaining organic carbon level in disturbed soils is critical for functional diversity of the microbial communities.

Antarctic sediment communities involved in N and C cycling were examined with the GeoChip 2.0 and results indicated that pH, the number of freeze-thaw cycles, and the C:N ratio were important environmental factors in shaping these communities (Yergeau et al., 2007). In addition, soil temperature was important for cellulose degradation and denitrification genes. Zhou et al. (2008) used GeoChip to examine gene-area relationships in forest soils and found that the microbial community had a relatively flat gene-area relationship with less turnover than observed for plants and animals.

OTHER SAMPLES

While GeoChip is generally used to examine the presence of functional genes in the environment, some studies have examined microbial activity. Amplified community mRNA from a denitrifying fluidized bed reactor was used with GeoChip, and genes for nitrate and nitrite reduction, organic contaminant degradation, *dsr*, and polyphosphate kinase were detected (Gao et al., 2007). Leigh et al. (2007) used stable isotope probing coupled to GeoChip to detect active microbial populations. GeoChip 2.0 has also been used to probe four, Ni-resistant gram-positive actinomycetes for metal resistance genes (Van Nostrand et al., 2007). Genes associated with resistance to Al, As, Cd, Cr, Cu, Hg, Ni, Te, and Zn were detected.

GEOCHIP 3.0 FOR GLOBAL CHANGE-RELATED STUDIES

GeoChip 3.0 is the newest expansion of GeoChip, and several global change-related studies are now under way using this version. One study is focused on the multifactor grassland experiment site, BioCON (Biodiversity, CO_2, and Nitrogen deposition) located at the Cedar Creek Ecosystem Science Reserve, MN, to determine the effects of elevated CO_2 on soil microbial communities. Results indicate that elevated CO_2 does significantly affect soil microbial communities, and that soil microbes formed distinct microbial community structures at ambient and elevated CO_2 conditions (He et al., 2010b).

SUMMARY

While many advances have been made in FGA technology in the past decade, there are still challenges that need to be addressed in the future. FGAs, specifically, the GeoChips, have been shown to be powerful tools in linking microbial function to ecosystem processes and are able to provide sensitive, specific, and potentially quantitative information. This technology is expected to transform the field of microbial ecology and the study of microbial community functional structure and dynamics.

ACKNOWLEDGMENTS

The effort for preparing this review was supported by the Virtual Institute for Microbial Stress and Survival (http://VIMSS.lbl.gov) supported by the U.S. Department of Energy, Office of Science, Office of Biological and Environmental Research, Genomics Program: GTL through contract DE-AC02-05CH11231 between Lawrence Berkeley National Laboratory and the U.S. Department of Energy, Environmental Remediation Science Program (ERSP), Office of Biological and Environmental Research, Office of Science, and Oklahoma Applied Research Support (OARS), Oklahoma Center for the Advancement of Science and Technology (OCAST), the State of Oklahoma through the Project AR062-034.

REFERENCES

Adey, N. B., M. Lei, M. T. Howard, J. B. Jenson, D. A. Mayo, D. L. Butel, S. C. Coffin, T. C. Moyer, D. E. Slade, M. K. Spute, A. M. Hancock, G. T. Eisenhohoffer, B. K. Dalley, and M. R. McNeely. 2002. Gains in sensitivity with a device that mixes microarray hybridization solution in a 25-μm-thick chamber. *Anal. Chem.* **74:**6413–6417.

Affymetrix. 2001. GeneChip arrays provide optimal sensitivity and specificity for microarray expression analysis. *User Guide.* Affymetrix Inc., Santa Clara, CA.

Amann, R. I., W. Ludwig, and K. H. Schleifer. 1995. Phylogenetic identification and in situ detection of individual microbial cells without cultivation. *Microbiol. Rev.* **59:**143–169.

Bodrossy, L., N. Stralis-Pavese, M. Konrad-Köszler, A. Weilharter, T. G. Reichenauer, D. Schöfer, and A. Sessitsch. 2006. mRNA-based parallel detection of active methanotroph populations by use of a diagnostic microarray. *Appl. Environ. Microbiol.* **72:**1672–1676.

Bodrossy, L., N. Stralis-Pavese, J. C. Murrell, S. Radajewski, A. Weilharter, and A. Sessitsch. 2003. Development and validation of a diagnostic microbial microarray for methanotrophs. *Environ. Microbiol.* **5:**566–582.

Bontemps, C., G. Goldier, C. Gris-Liebe, S. Carere, L. Talini, and C. Boivin-Masson. 2005. Microarray-based detection and typing of rhizobium nodulation gene *nodC*: the potential of DNA arrays to diagnose biological functions of interest. *Appl. Environ. Microbiol.* **71:**8042–8048.

Branham, W. S., C. D. Melvin, T. Han, V. G. Gesai, C. L. Moland, A. T. Scully, and J. C. Fuscoe. 2007. Elimination of laboratory ozone leads to a dramatic improvement in the reproducibility of microarray gene expression measurements. *BMC Biotechnol.* **7:**8.

Brodie, E. L., T. Z. DeSantis, D. C. Joyner, S. M. Baek, J. T. Larsen, G. L. Andersen, T. C. Hazen, P. M. Richardson, D. J. Herman, T. K. Tokunaga, J. M. Wan, and M. K. Firestone. 2006. Application of a high-density oligonucleotide microarray approach to study bacterial population dynamics during uranium reduction and reoxidation. *Appl. Environ. Microbiol.* **72:**6288–6298.

Brodie, E. L., T. Z. DeSantis, J. P. Moberg Parker, I. X. Zubietta, Y. M. Piceno, and G. L. Andersen. 2007. Urban aerosols harbor diverse and dynamic bacterial populations. *Proc. Natl. Acad. Sci. USA* **104:**299–304.

Burgmann, H., F. Widmer, W. V. Sigler, and J. Zeyer. 2003. mRNA extraction and reverse transcription-PCR protocol for detection of *nifH* gene expression by *Azotobacter vinelandii* in soil. *Appl. Environ. Microbiol.* **69:**1928–1935.

Call, D. R., M. K. Bakko, M. J. Krug, and M. C. Roberts. 2003. Identifying antimicrobial resistance genes with DNA microarrays. *Antimicrob. Agents Chemother.* **47:**3290–3295.

Chen, Q., Y. Huaqun, H. Luo, M. Xie, G. Qiu, and X. Liu. 2009. Micro-array based whole-genome hybridization for detection of microorganisms in acid mine drainage and bioleaching systems. *Hydrometallurgy* **95:**96–103.

Cho, J. C., and J. M. Tiedje. 2002. Quantitative detection of microbial genes by using DNA microarrays. *Appl. Environ. Microbiol.* **58:**1425–1430.

Chou, C. C., C. H. Chen, T. T. Lee, and K. Peck. 2004. Optimization of probe length and the number of probes per gene for optimal microarray analysis of gene expression. *Nucleic Acids Res.* **32:**e99.

Cleven, B. E. E., M. Palka-Santini, J. Gielen, S. Meembor, M. Krönke, and O. Krut. 2006. Identification and characterization of bacterial pathogens causing bloodstream infections by DNA microarray. *J. Clin. Microbiol.* **44:**2389–2397.

Denef, V. J., J. Park, J. L. M. Rodrigues, T. V. Tsoi, S. A. Hashsham, and J. M. Tiedje.

2003. Validation of a more sensitive method for using spotted oligonucleotide DNA microarrays for functional genomics studies on bacterial communities. *Environ. Microbiol.* **5:**933–943.

Deng, Y., Z. He, J. D. Van Nostrand, and J. Zhou. 2008. Design and analysis of mismatch probes for long oligonucleotide microarrays. *BMC Genomics* **9:**491–503.

Dennis, P., E. A. Edwards, S. N. Liss, and R. Fulthorpe. 2003. Monitoring gene expression in mixed microbial communities by using DNA microarrays. *Appl. Environ. Microbiol.* **69:**769-778.

DeSantis, T. Z., E. L. Brodie, J. P. Moberg, I. X. Zubieta, Y. M. Piceno, and G. L. Andersen. 2007. High-density universal 16S rRNA microarray analysis reveals broader diversity than typical clone library when sampling the environment. *Microb. Ecol.* **53:**371–383.

Dong, Y., J. D. Glasner, F. R. Blattner, and E. W. Triplett. 2001. Genomic interspecies microarray hybridization: rapid discovery of three thousand genes in the maize Endophyte, *Klebsiella pneumoniae* 342, by microarray hybridization with *Escherichia coli* K-12 open reading frames. *Appl. Environ. Microbiol.* **67:**1911–1921.

Ehrenreich, A. 2006. DNA microarray technology for the microbiologist: an overview. *Appl. Microbiol. Biotechnol.* **73:**255–273.

Evertsz, E. M., J. Au-Young, M. V. Ruvolo, A. C. Lim, and M. A. Reynolds. 2001. Hybridization cross-reactivity within homologous gene families on glass cDNA microarrays. *BioTechniques* **31:**1182, 1184, 1186 passim.

Franke-Whittle, I. H., S. H. Klammer, and H. Insam. 2005. Design and application of an oligonucleotide microarray for the investigation of compost microbial communities. *J. Microbiol. Methods* **62:**37–56.

Frey, B., M. Stemmer, F. Widmer, J. Luster, and C. Sperisen. 2006. Microbial activity and community structure of a soil after heavy metal contamination in a model forest ecosystem. *Soil Biol. Biochem.* **38:**1745–1756.

Fuhrman, J. A., and L. Campbell. 1998. Microbial microdiversity. *Nature* **393:**410–411.

Fuhrman, J. A., and J. A. Steele. 2008. Community structure of marine bacterioplankton: patterns, networks, and relationships to function. *Aquat. Microb. Ecol.* **53:**69–81.

Gao, H., Z. K. Yang, T. J. Gentry, L. Wu, C. W. Schadt, and J. Zhou. 2007. Microarray-based analysis of microbial community RNAs by whole-community RNA amplification. *Appl. Environ. Microbiol.* **73:**563–571.

Gentry, T. J., G. S. Wickham, C. W. Schadt, Z. He, and J. Zhou. 2006. Microarray application in microbial ecology research. *Microb. Ecol.* **52:**159–175.

Gough, H. L., A. L. Dahl, M. A. Nolan, J.-F. Gaillard, and D. A. Stahl. 2008. Metal impacts on microbial biomass in the anoxic sediments of a contaminated lake. *J. Geophys. Res.* **113:**G2017.

Gough, H. L., and D. A. Stahl. 2011. Microbial community structures in anoxic freshwater lake sediment along a metal contamination gradient. *ISME J.* **5:**543–558.

Gresham, D., M. J. Dunham, and D. Botstein. 2008. Comparing whole genomes using DNA microarrays. *Nat. Rev.* **9:**291–302.

Grubbs, F. 1969. Procedures for detecting outlying observations in samples. *Technometrics* **11:**469–479.

Gudnason, H., M. Dufva, D. D. Bang, and A. Wolff. 2008. An inexpensive and simple method for thermally stable immobilization of DNA on an unmodified glass surface: UV linking of poly(T)10-poly(C)10-tagged DNA probes. *BioTechniques* **45:**261–271.

Guschin, D. Y., B. K. Mobarry, D. Proudnikov, D. A. Stahl, B. E. Rittmann, and A. D. Mirzabekov. 1997. Oligonucleotide microchips as genosensors for determinative and environmental studies in microbiology. *Appl. Environ. Microbiol.* **63:**2397–2402.

Handelsman, J., J. M. Tiedje, L. Alvarez-Cohen, M. Ashburner, I. K. O. Cann, E. F. DeLong, W. F. Doolittle, C. M. Fraser-Liggett, A. Godzik, J. I. Gordon, M. Riley, M. B. Schmid, and A. H. Reid. 2007. *Committee on Metagenomics: Challenges and functional applications.* National Academy of Sciences, Washington, DC.

Hartmann, M., B. Frey, R. Kölliker, and F. Widmer. 2005. Semi-automated genetic analyses of soil microbial communities: comparison of T-RFLP and RISA based on descriptive and discriminative statistical approaches. *J. Microbiol. Methods* **61:**349–360.

He, J., Z. Xu, and J. Hughes. 2005a. Pre-lysis washing improves DNA extraction from a forest soil. *Soil Biol. Biochem.* **37:**2337–2341.

He, Z., Y. Deng, J. D. Van Nostrand, Q. Tu, M. Xu, C. L. Hemme, X. Li, L. Wu, T. J. Gentry, Y. Yin, J. Liebich, T. C. Hazen, and J. Zhou. 2010a. GeoChip 3.0 as a high-throughput tool for analyzing microbial community structure, composition and functional activity. *ISME J.* **4:**1167–1179.

He, Z., T. J. Gentry, C. W. Schadt, L. Wu, J. Liebich, S. C. Chong, Z. Huang, W. Wu, B. Gu, P. Jardine, C. Criddle, and J. Zhou. 2007. GeoChip: a comprehensive microarray for investigating biogeochemical, ecological and environmental processes. *ISME J.* **1:**67–77.

He, Z., J. D. Van Nostrand, L. Wu, and J. Zhou. 2008. Development and application of functional gene arrays for microbial community analysis. *Trans. Nonferrous Met. Soc. China* **18:**1319–1327.

He, Z., L. Wu, M. W. Fields, and J. Zhou. 2005b. Use of microarrays with different probe sizes for monitoring gene expression. *Appl. Environ. Microbiol.* **71:**5154–5162.

He, Z., L. Y. Wu, X. Y. Li, M. W. Fields, and J. Z. Zhou. 2005c. Empirical establishment of oligonucleotide probe design criteria. *Appl. Environ. Microbiol.* **71:**3753–3760.

He, Z., M. Xu, Y. Deng, S. Kang, L. Kellogg, L. Wu, J. D. Van Nostrand, S. E. Hobbie, P. B. Reich, and J. Zhou. 2010b. Metagenomic analysis reveals a marked divergence in the functional structure of belowground microbial communities at elevated CO_2. *Ecol. Lett.* **13:**564–575.

He, Z., and J. Zhou. 2008. Empirical evaluation of a new method for calculating signal to noise ratio (SNR) for microarray data analysis. *Appl. Environ. Microbiol.* **74:**2957–2966.

Hong, S.-H., J. Bunge, S.-O. Jeon, and S. S. Epstein. 2006. Predicting microbial species richness. *Proc. Natl. Acad. Sci. USA* **103:**117–122.

Hu, Q., H. Y. Qi, J. H. Zeng, and H. X. Zhang. 2007. Bacterial diversity in soils around a lead and zinc mine. *J. Environ. Sci.* **19:**74–79.

Hurt, R. A., X. Qiu, L. Wu, Y. Roh, A. V. Palumbo, J. M. Tiedje, and Z. Zhou. 2001. Simultaneous recovery of RNA and DNA from soils and sediments. *Appl. Environ. Microbiol.* **67:**4495–4503.

Jenkins, B. D., G. F. Steward, S. M. Short, B. B. Ward, and J. P. Zehr. 2004. Fingerprinting diazotroph communities in the Chesapeake Bay. *Appl. Environ. Microbiol.* **70:**1767–1776.

Kimes, N. E., J. D. Van Nostrand, E. Weil, J. Zhou, and P. J. Morris. 2009. The microbial functional structure of *Montastraea faveolata*, an important Caribbean reef-building coral, differs between healthy and yellow-band diseased (YBD) colonies. *Environ. Microbiol.* **12:**541–556.

Kingsley, M. T., T. M. Straub, D. R. Call, D. S. Daly, S. C. Wunschel, and D. P. Chandler. 2002. Fingerprinting closely related *Xanthomonas* pathovars with random nonamer oligonucleotide microarrays. *Appl. Environ. Microbiol.* **68:**6361–6370.

Konopka, A., T. Zakharova, M. Bischoff, L. Oliver, C. Nakatsu, and R. F. Turco. 1999. Microbial biomass and activity in lead-contaminated soil. *Appl. Environ. Microbiol.* **65:**2256–2259.

Kostic´, T., A. Weilharter, A. Sessitsch, and L. Bodrossy. 2005. High-sensitivity, polymerase chain reaction-free detection of microorganisms and their functional genes using 70-mer oligonucleotide diagnostic microarray. *Anal. Biochem.* **346:**333–335.

Kumar, A., O. Larsson, D. Parodi, and Z. Liang. 2000. Silanized nucleic acids: a general platform for DNA immobilization. *Nucleic Acids Res.* **28:**e71.

Lazzaro, A., F. Widmer, C. Sperisen, and B. Frey. 2008. Identification of dominant bacterial phylotypes in a cadmium-treated forest soil. *FEMS Microbiol. Ecol.* **63:**143–155.

Legendre, P., and M. J. Anderson. 1999. Distance-based redundancy analysis: testing multi-species responses in multi-factorial ecological experiments. *Ecol. Monogr.* **69:**1–24.

Leigh, M. B., V. H. Pellizari, O. Uhlík, R. Sutka, J. Rodrigues, N. E. Ostrom, J. Zhou, and J. M. Tiedje. 2007. Biphenyl-utilizing bacteria and their functional genes in a pine root zone contaminated with polychlorinated biphenyls (PCBs). *ISME J.* **1:**134–148.

Li, X., Z. He, and J. Zhou. 2005. Selection of optimal oligonucleotide probes for microarrays using multiple criteria, global alignment and parameter estimation. *Nucleic Acids Res.* **33:**6114–6123.

Liang, Y., Z. He, L. Wu, Y. Deng, G. Li, and J. Zhou. 2010. Development of a common oligo reference standard (CORS) for microarray data normalization and comparison across different microbial communities. *Appl. Environ. Microbiol.* **76:**1088–1094.

Liang, Y., G. Li, J. D. Van Nostrand, Z. He, L. Wu, Y. Deng, X. Zhang, and J. Zhou. 2009a. Microarray-based analysis of microbial functional diversity along an oil contamination gradient in oilfield. *FEMS Microbiol. Ecol.* **70:**324–333.

Liang, Y., J. Wang, J. D. Van Nostrand, J. Zhou, X. Zhang, and G. Li. 2009b. Microarray-based functional gene analysis of soil microbial communities in ozonation and biodegradation of crude oil. *Chemosphere* **75:**193–199.

Liebich, J., C. W. Schadt, S. C. Chong, Z. He, S. K. Rhee, and J. Zhou. 2006. Improvement of oligonucleotide probe design criteria for functional gene microarrays in environmental applications. *Appl. Environ. Microbiol.* **72:**1688–1691.

Lipshutz, R. J., S. P. Fodor, T. R. Gingeras, and D. J. Lockhart. 1999. High density synthetic oligonucleotide arrays. *Nat. Genet.* **21:**20–24.

Lockhart, D. J., H. Dong, M. C. Byrne, M. T. Follettie, M. V. Gallo, M. S. Chee, M. Mittmann, C. Wang, M. Kobayashi, H. Horton, and E. L. Brown. 1996. Expression monitoring by hybridization to high-density oligonucleotide arrays. *Nat. Biotechnol.* **14:**1675–1680.

Loy, A., A. Lehner, N. Lee, J. Adamcsyk, H. Meier, J. Ernst, K.-H. Schleifer, and M. Wagner. 2002. Oligonucleotide microarray for 16S rRNA gene-based detection of all recognized lineages of sulfate-reducing prokaryotes in the environment. *Appl. Environ. Microbiol.* **68:** 5064–5081.

Loy, A., C. Schulz, S. Lücker, A. Schöpfer-Wendels, K. Stoecker, C. Baranyi, A. Lehner, and M. Wagner. 2005. 16S rRNA gene-based oligonucleotide microarray for environmental monitoring of the Betaproteobacterial order "Rhodocyclales." *Appl. Environ. Microbiol.* **71:**1373–1386.

Lueders, T., and M. W. Friedrich. 2003. Evaluation of PCR amplification bias by terminal restriction fragment length polymorphism: analysis of small-subunit rRNA and *mcrA* genes by using defined template mixtures of methanogenic pure cultures and soil DNA extracts. *Appl. Environ. Microbiol.* **69:**320–326.

Luo, J., W. Wu, M. N. Fienen, P. M. Jardine, T. L. Mehlhorn, D. B. Watson, O. A. Cirpka, C. S. Criddle, and P. K. Kitanidis. 2006. A nested-cell approach for in situ remediation. *Ground Water* **44:**266–274.

Mason, O. U., C. A. DiMeo-Savoie, J. D. Van Nostrand, J. Zhou, M. R. Fisk, and S. J. Giovannoni. 2009. Prokaryotic diversity, distribution, and preliminary insights into their role in biogeochemical cycling in marine basalts. *ISME J.* **3:**231–242.

McGrath, K. C., S. R. Thomas-Hall, C. T. Cheng, L. Leo, A. Alexa, S. Schmidt, and P. M. Schenk. 2008. Isolation and analysis of mRNA from environmental microbial communities. *J. Microbiol. Methods* **75:**172–176.

Mehta, M. L. 1990. *Random Matrix Theory*. Springer, New York, NY.

Miller, S. M., D. M. Tourlousse, R. D. Stedtfeld, S. W. Baushke, A. B. Herzog, L. M. Wick, J. M. Rouillard, E. Gulari, J. M. Tiedje, and S. A. Hashsham. 2008. In situ-synthesized virulence and marker gene biochip for detection of bacterial pathogens in water. *Appl. Environ. Microbiol.* **74:**2200–2209.

Mockler, T. C., and J. R. Ecker. 2005. Applications of DNA tiling arrays for whole-genome analysis. *Genomics* **85:**1–15.

Modrek, B., and C. Lee. 2002. A genomic view of alternative splicing. *Nat. Genet.* **30:**13–19.

Montoya, J. M., S. L. Pimm, and R. V. Sole. 2006. Ecological networks and their fragility. *Nature* **442:**259–264.

Murray, A. E., D. Lies, G. Li, K. Nealson, J. Zhou, and J. M. Tiedje. 2001. DNA–DNA hybridization to microarrays reveals gene-specific differences between closely related microbial genomes. *Proc. Natl. Acad. Sci. USA* **98:**9853–9858.

Ning, J., J. Liebich, M. Kästner, J. Zhou, A. Schäffer, and P. Burauel. 2009. Different influences of DNA purity indices and quantity on PCR-based DGGE and functional gene microarray in soil microbial community study. *Appl. Microbiol. Biotechnol.* **82:**983–993.

Okamoto, T., T. Suzuki, and N. Yamamoto. 2000. Microarray fabrication with covalent attachment of DNA using Bubble Jet technology. *Nat. Biotechnol.* **18:**438–441.

Økland, R. H., and O. Eilertsen. 1994. Canonical correspondence analysis with variation partitioning: some comments and an application. *J. Veg. Sci.* **5:**117–126.

Palka-Santini, M., B. E. Cleven, L. Eichinger, M. Krönke, and O. Krut. 2009. Large scale multiplex PCR improves pathogen detection by DNA microarrays. *BMC Microbiol.* **9:**1.

Parnell, J. J., G. Rompato, L. C. Latta IV, M. E. Pfrender, J. D. Van Nostrand, Z. He, J. Zhou, G. Andersen, P. Champine, B. Ganesan, and B. C. Weimer. 2010. Functional biogeography and gene transfer among hypersaline microbial communities. *PLoS One* **5:**e12919.

Pennanen, T. 2001. Microbial communities in boreal coniferous forest humus exposed to heavy metals and change in soil pH—a summary of the use of phospholipid fatty acids, Biolog and ^{3}H-thymidine incorporation methods in field studies. *Geoderma* **100:**91–126.

Poretsky, R. S., S. Gifford, J. Rinta-Kanto, M. Vila-Costa, and M. A. Moran. 2009. Analyzing gene expression from marine microbial communities using environmental transcriptomics. *JoVE.* 24. http://www.jove.com/index/Details.stp?ID=1086, doi: 10.3791/1086.

Raes, J., and P. Bork. 2008. Molecular eco-systems biology: towards an understanding of community function. *Nat. Rev. Microbiol.* **6:**693–699.

Ramette, A., and J. M. Tiedje. 2007. Multiscale responses of microbial life in spatial distance and environmental heterogeneity in a patchy ecosystem. *Proc. Natl. Acad. Sci. USA* **104:**2761–2766.

Ramsey, P. W., M. C. Rillig, K. P. Feris, N. S. Gordon, J. N. Moore, W. E. Holben, and J. E. Gannon. 2005. Relationship between communities and processes; new insights from a field study of a contaminated ecosystem. *Ecol. Lett.* **8:**1201–1210.

Relógio, A., C. Schwager, A. Richter, W. Ansorge, and J. Valcárcel. 2002. Optimization of oligonucleotide-based DNA microarrays. *Nucleic Acids Res.* **30:**e51.

Rhee, S. K., X. Liu, L. Wu, S. C. Chong, X. Wan, and J. Zhou. 2004. Detection of genes involved in biodegradation and biotransformation in microbial communities by using 50-mer oligonucleotide microarrays. *Appl. Environ. Microbiol.* **70:** 4303–4317.

Rich, V. I., K. Konstantinidis, and E. F. DeLong. 2008. Design and testing of 'genome-proxy' microarrays to profile marine microbial communities. *Environ. Microbiol.* **10:**506–521.

Rodríguez-Martínez, E. M., E. X. Pérez, C. W. Schadt, J. Zhou, and A. A. Massol-Deyá. 2006. Microbial diversity and bioremediation of a hydrocarbon-contaminated aquifer (Vega Baja, Puerto Rico). *Int. J. Environ. Res. Public Health* **3:**292–300.

Roesch, L. F. W., R. R. Fulthorpe, A. Riva, G. Casella, A. K. M. Hadwin, A. D. Kent, S. H. Daroub, F. O. A. Camargo, W. G. Farmerie, and E. W. Triplett. 2007. Pyrosequencing enumerates and contrasts soil microbial diversity. *ISME J.* **1:**283–290.

Sarkar, S. F., J. S. Gordon, G. B. Martin, and D. S. Guttman. 2006. Comparative genomics of host-specific virulence in *Pseudomonas syringae*. *Genetics* **174:**1041–1056.

Schena, M., D. Shalon, R. W. Davis, and P. O. Brown. 1995. Quantitative monitoring of gene expression patterns with a complementary DNA microarray. *Science* **270:**467–470.

Schena, M., D. Shalon, R. Heller, A. Chai, and P. O. Brown. 1996. Parallel human genome analysis: microarray-based expression monitoring of 1,000 genes. *Proc. Natl. Acad. Sci. USA* **93:**10614–10619.

Schloss, P. D., and J. Handelsman. 2006. Toward a census of bacteria in soil. *PLOS Comput. Biol.* **2:**786–793.

Sebat, J. L., F. S. Colwell, and R. L. Crawford. 2003. Metagenomic profiling: microarray analysis of an environmental genomic library. *Appl. Environ. Microbiol.* **69:**4927–4934.

Serra, P., M. Colina, J. M. Fernández, L. Sevilla, and J. L. Morenza. 2004. Preparation of functional DNA microarrays through laser-induced forward transfer. *Appl. Physics Lett.* **85:**1639–1641.

Shalon, D., S. J. Smith, and P. O. Brown. 1996. A DNA microarray system for analyzing complex DNA samples using two-color fluorescent probe hybridization. *Genome Res.* **6:**639–645.

Small, J., D. R. Call, F. J. Brockman, T. M. Straub, and D. P. Chandler. 2001. Direct detection of 16S rRNA in soil extracts by using oligonucleotide microarrays. *Appl. Environ. Microbiol.* **67:**4708–4716.

Steward, G. F., B. D. Jenkins, B. B. Ward, and J. P. Zehr. 2004. Development and testing of a DNA macroarray to assess nitrogenase (*nifH*) gene diversity. *Appl. Environ. Microbiol.* **70:**1455–1465.

Stralis-Pavese, N., A. Sessitsch, A. Weilharter, T. Reichenauer, J. Riesing, J. Csontos, J. C. Murrell, and L. Bodrossy. 2004. Optimization of diagnostic microarray for application in analysing landfill methanotroph communities under different plant covers. *Environ. Microbiol.* **6:**347–363.

Suzuki, M. T., and S. J. Giovannoni. 1996. Bias caused by template annealing in the amplification of mixtures of 16S rRNA genes by PCR. *Appl. Environ. Microbiol.* **62:**625–630.

Taroncher-Oldenburg, G., E. M. Griner, C. A. Francis, and B. B. Ward. 2003. Oligonucleotide microarray for the study of functional gene diversity in the nitrogen cycle in the environment. *Appl. Environ. Microbiol.* **69:**1159–1171.

Tas, N., M. H. A. van Eekert, G. Schraa, J. Zhou, W. M. de Vos, and H. Smidt. 2009. Tracking functional guilds: *Dehalococcoides* spp. in European river basins contaminated with hexachlorobenzene. *Appl. Environ. Microbiol.* **75:**4696–4704.

ter Braak, C. J. F. 1986. Canonical correspondence analysis: a new eigenvector technique for multivariate direct gradient analysis. *Ecology* **67:**1167–1179.

Tiquia, S. M., S. Gurczynski, A. Zholi, and A. Devol. 2006. Diversity of biogeochemical cycling genes from Puget Sound sediments using DNA microarrays. *Environ. Technol.* **27:**1377–1389.

Tiquia, S. M., L. Wu, S. C. Chong, S. Passovets, D. Xu, Y. Xu, and J. Zhou. 2004. Evaluation of 50-mer oligonucleotide arrays for detecting microbial populations in environmental samples. *BioTechniques* **36:**1–8.

Torsvik, V., J. Goksoyr, and F. L. Daae. 1990. High diversity in DNA of soil bacteria. *Appl. Environ. Microbiol.* **56:**782–787.

Van Nostrand, J. D., T. V. Khijniak, T. J. Gentry, M. T. Novak, A. G. Sowder, J. Z. Zhou, P. M. Bertsch, and P. J. Morris. 2007. Isolation and characterization of four Gram-positive nickel-tolerant microorganisms from contaminated sediments. *Microb. Ecol.* **53:**670–682.

Van Nostrand, J. D., L. Wu, W. M. Wu, T. J. Gentry, Z. Huang, Y. Deng, J. Carley, S. Carroll, Z. He, B. Gu, J. Luo, C. S. Criddle, D. B. Watson, P. M. Jardine, T. L. Marsh, J. M. Tiedje, T. C. Hazen, and J. Zhou. Dynamics of microbial community composition and function during in situ bioremediation of a uranium-contaminated aquifer. *Appl. Environ Microbiol.*, in press.

Van Nostrand, J. D., W. M. Wu, L. Wu, Y. Deng, J. Carley, S. Carroll, Z. He, B. Gu, J. Luo, C. Criddle, D. B. Watson, P. M. Jardine, T. L. Marsh, J. M. Tiedje, T. C. Hazen, and J. Zhou. 2009. GeoChip-based analysis of functional microbial communities during the reoxidation of a bioreduced uranium-contaminated aquifer. *Environ. Microbiol.* **11:**2611–2626.

Wagner, M., H. Smidt, A. Loy, and Z. Zhou. 2007. Unravelling microbial communities with DNA-microarrays: challenges and future directions. *Microb. Ecol.* **53:**498–506.

Waldron, P. J., L. Wu, J. D. Van Nostrand, C. Schadt, D. Watson, P. Jardine, T. Palumbo, T. C. Hazen, and J. Zhou. 2009. Functional gene array-based analysis of microbial community structure in groundwaters with a gradient of contaminant levels. *Environ. Sci. Technol.* **43:**3529–3534.

Wang, F., H. Zhou, J. Meng, X. Peng, L. Jiang, P. Sun, C. Zhang, J. D. Van Nostrand, Y. Deng, Z. He, L. Wu, J. Zhou, and X. Xiao. 2009. GeoChip-based analysis of metabolic diversity of microbial communities at the Juan de Fuca Ridge hydrothermal vent. *Proc. Natl. Acad. Sci. USA* **106:**4840–4845.

Wang, Y., J. Shi, H. Wang, Q. Lin, X. Chen, and Y. Chen. 2007. The influence of soil heavy metals pollution on soil microbial biomass, enzyme activity, and community composition near a copper smelter. *Ecotox. Environ. Saf.* **67:**75–81.

Warnecke, P. M., C. Stirzaker, J. R. Melki, D. S. Millar, C. L. Paul, and S. J. Clark. 1997. Detection and measurement of PCR bias in quantitative methylation analysis of bisulphite-treated DNA. *Nucleic Acids Res.* **25:**4422–4426.

White, J. R., P. Escobar-Paramo, E. F. Mongodin, K. E. Nelson, and J. DiRuggiero. 2008. Extensive genome rearrangements and multiple horizontal gene transfers in a population of *Pyrococcus* isolates from Vulcano Island, Italy. *Appl. Environ. Microbiol.* **74:**6447–6451.

Whitman, W. B., D. C. Coleman, and W. J. Wiebe. 1998. Prokaryotes: the unseen majority. *Proc. Natl. Acad. Sci. USA* **95:**6578–6583.

Wilson, M., J. DeRisi, H. H. Kristensen, P. Imboden, S. Rane, P. O. Brown, and G. K. Schoolnik. 1999. Exploring drug-induced alterations in gene expression in Mycobacterium tuberculosis by microarray hybridization. *Proc. Natl. Acad. Sci. USA* **96:**12833–12838.

Wilson, W. J., C. L. Strout, T. Z. DeSantis, J. L. Stilwell, A. V. Carrano, and G. L. Andersen. 2002. Sequence-specific identification of 18 pathogenic microorganisms using microarray technology. *Mol. Cell. Probes* **16:**119–127.

Wu, L., L. Kellogg, A. H. Devol, J. M. Tiedje, and J. Zhou. 2008a. Microarray-based characterization of microbial community functional structure and heterogeneity in marine sediments from the Gulf of Mexico. *Appl. Environ. Microbiol.* **74:**4516–4529.

Wu, L., X Liu, M. W. Fields, D. K. Thompson, C. E. Bagwell, J. M. Tiedje, T. C. Hazen, and J. Zhou. 2008b. Microarray-based whole-genome hybridization as a tool for determining prokaryotic species relatedness. *ISME J.* **6:**642–655.

Wu, L., X. Liu, C. W. Schadt, and J. Zhou. 2006. Microarray-based analysis of submicrogram quantities of microbial community DNAs by using whole-community genome amplification. *Appl. Environ. Microbiol.* **72:**4931–4941.

Wu, L., D. K. Thompson, G. Li, R. A. Hurt, J. M. Tiedje, and J. Zhou. 2001. Development and evaluation of functional gene arrays for detection of selected genes in the environment. *Appl. Environ. Microbiol.* **67:**5780–5790.

Wu, L., D. K. Thompson, X. Liu, M. W. Fields, C. E. Bagwell, J. M. Tiedje, and J. Zhou. 2004. Development and evaluation of microarray-based whole genome hybridization for detection of microorganisms within the context of environmental applications. *Environ. Sci. Technol.* **38:**6775–6782.

Xie, J., L. Wu, X. Liu, G. Qiu, and J. Zhou. 2007. Improved procedure for DNA extraction and purification from soil. 108th General Meeting of the American Society for Microbiology, June 1 to 5, Boston, MA. American Society for Microbiology, Washington, DC.

Yergeau, E., S. Kang, Z. He, J. Zhou, and G. A. Kowalchuk. 2007. Functional microarray analysis of nitrogen and carbon cycling genes across an Antarctic latitude transect. *ISME J.* **1:**1–17.

Yin, H., L. Cao, G. Qiu, D. Wang, L. Kellogg, J. Zhou, Z. Dai, and X. Liu. 2007. Development and evaluation of 50-mer oligonucleotide arrays for detecting microbial populations in Acid Mine Drainages and bioleaching systems. *J. Microbiol. Methods* **70:**165–178.

Zhang, L., T. Hurek, and B. Reihold-Hurek. 2007. A *nifH*-based oligonucleotide microarray for functional diagnostics of nitrogen-fixing microorganisms. *Microb. Ecol.* **53:**456–470.

Zhang, L., U. Srinivasan, C. F. Marrs, D. Ghosh, J. R. Gilsdorf, and B. Foxman. 2004. Library on a slide for bacterial comparative genomics. *BMC Microbiol.* **4:**12–18.

Zhang, Y., X. Zhang, X. Liu, Y. Xiao, L. Qu, L. Wu, and J. Zhou. 2007. Microarray-based analysis of changes in diversity of microbial genes involved in organic carbon decomposition following land use/cover changes. *FEMS Lett.* **266:**144–151.

Zhou, J., M. A. Bruns, and J. M. Tiedje. 1996. DNA recovery from soils of diverse composition. *Appl. Environ. Microbiol.* **62:**316–322.

Zhou, J., S. Kang, C. W. Schadt, and C. T. Garten, Jr. 2008. Spatial scaling of functional gene diversity across various microbial taxa. *Proc. Natl. Acad. Sci. USA* **105:**7768–7773.

Zhou, J., and D. K. Thompson. 2002. Challenges in applying microarrays to environmental studies. *Curr. Opin. Biotechnol.* **13:**204–207.

Zhou, J. 2003. Microarrays for bacterial detection and microbial community analysis. *Curr. Opin. Microbiol.* **6:**288–294.

Zhou, X., and J. Zhou. 2004. Improving the signal sensitivity and photostability of DNA hybridizations on microarrays by using dye-doped core-shell silica nanoparticles. *Anal. Chem.* **76:**5302–5312.

METHODS FOR DETECTION OF ARSENATE-RESPIRING BACTERIA: ADVANCES, CAUTIONS, AND CAVEATS

*John F. Stolz, Mahmoud M. Berekaa, Edward Fisher,
Ganna Polshyna, Mirunalni Thangavelu, Rishu Dheer,
Antonio Garcia Moyano, Samy El Assar, and Partha Basu*

15

INTRODUCTION

Arsenic speciation, mobility, and toxicity can be greatly influenced by microbial activity. Whether for purposes of detoxification or energy generation and growth, a growing number of bacteria have been recognized to transform arsenic (Oremland and Stolz, 2003; Oremland et al., 2005; Rhine et al., 2005). Specific microbial metabolic processes include arsenate reduction, arsenite oxidation, methylation, and demethylation (Stolz et al., 2006). Recent studies have focused on the environmental impact of these organisms (see Chapter 5). Microbes that are typically associated with poultry and poultry litter have been shown to readily metabolize the organoarsenical roxarsone, releasing inorganic arsenic in the process (Stolz et al., 2007; Fisher et al., 2008). Thus, the routine practice of using organoarsenicals in feed as both a prophylactic treatment for coccidiosis and a growth stimulant has been called into question. Bacteria capable of dissimilatory arsenate reduction, in particular, have been linked with arsenic mobilization and contaminated drinking water (Harvey et al., 2002; Oremland and Stolz, 2005). It has also been hypothesized that arsenate-respiring bacteria residing in the gastrointestinal tract of animals (including humans) may pose an additional health risk especially if they generate arsenite, As(III) (Herbel et al., 2002). Thus, the ability to readily identify arsenic-metabolizing bacteria and their activities in a variety of environments (e.g., reservoirs, subsurface aquifers, farms) and animal systems (e.g., gastrointestinal tract) is of great interest.

Assessment of microbial arsenic transformation has typically been done using in situ incubations and microcosms, enrichment and pure culture, and molecular analyses (see Chapters 5 and 11). Early on, in situ incubations with As(V) amendments were used to establish dissimilatory arsenate reduction subsurface and surface sediments (Dowdle et al., 1996; Islam et al., 2004). More recently, the use of radiolabeled ^{73}As(V) has extended this approach to probe arsenic cycling in Mono Lake and Searles Lake, California (Oremland et al., 2000, 2002, 2005), and arsenate respiration by the gut microbiota of the Syrian hamster (Herbel et al.,

---

John F. Stolz, Edward Fisher, Mirunalni Thangavelu, and Rishu Dheer, Department of Biological Sciences, Duquesne University, Pittsburgh, PA 15282. *Mahmoud M. Berekaa,* Environmental Sciences Department, Faculty of Science, Alexandria University, Alexandria, Egypt. *Ganna Polshyna and Partha Basu,* Department of Chemistry and Biochemistry, Duquesne University, Pittsburgh, PA 15282. *Antonio Garcia Moyano,* Centro de Biologia Molecular, Universidad Autonoma de Madrid, Madrid, Spain. *Samy El Assar,* Botany Department, Faculty of Science, Alexandria University, Alexandria, Egypt.

2002). Enrichment cultures have been used extensively in a wide range of environments to elucidate arsenite oxidation (e.g., Gihring et al., 2001; Oremland et al., 2002; Rhine et al., 2006; Santini et al., 2002), arsenate reduction associated with resistance (Jackson et al., 2005), and respiratory arsenate reduction (e.g., Ahmann et al., 1994; Macy et al., 1996; Switzer Blum et al., 1998; Hoeft et al., 2004). These studies have resulted in the description of more than 30 species of arsenate-respiring prokaryotes and 30 strains of arsenite-oxidizing bacteria (as recently reviewed in Stolz et al., 2006, 2010). More recently, molecular approaches have been used to probe the activities of these organisms. Primers for PCR amplification include the arsenite-specific efflux pump of the resistance system (*arsB*, ACR3) (Achour et al., 2007) and arsenate reductase (*arsC*) involved in resistance (Sun et al., 2004), the catalytic subunit of arsenite oxidase (*aoxB*) (Inskeep et al., 2007; Rhine et al., 2007; Quemeneur et al., 2008), and the catalytic subunit of respiratory arsenate reductase (*arrA*) (Malasarn et al., 2004; Kulp et al., 2007; Lear et al., 2007). Here, we present the results of our efforts to develop and examine the efficacy of culture medium formulation, biochemical probes (using polyclonal antibodies that recognize ArrA), and molecular probes (PCR primers suitable for denaturing gradient gel electrophoresis and restriction fragment length polymorphism analyses) for the detection of arsenate-respiring bacteria. The results suggest that, while each approach can be used to assess the presence and activity of these organisms, it is prudent to employ multiple approaches concurrently in order to address the wide range of metabolic capabilities and arsenic tolerances; these organisms also have gene sequence diversity in the functional genes.

ENRICHMENT MEDIA AND THE CULTIVATION OF ARSENATE-RESPIRING BACTERIA

The great phylogenetic diversity and metabolic versatility of arsenate-respiring bacteria poses a challenge for their enrichment and culture. Environments with sufficient concentrations of arsenic can harbor a variety of species. For example, both heterotrophic *Firmicutes Bacillus selenitireducens* and *B. arseniciselenatis*, and chemolithoautotrophic *Deltaproteobacteria* strain MLMS-1 were isolated from sediments of Mono Lake, California (Switzer Blum et al., 1998; Hoeft et al., 2004). Whereas *B. selenitireducens* and *B. arseniciselenatis* utilize lactate as an electron donor and carbon source, strain MLMS-1 can grow chemoautotrophically with H_2S as the electron donor, CO_2 as the carbon source, and As(V) as the electron acceptor. In addition, different laboratories may have a preference for a particular medium. As such, a variety of media formulations have been used (Ahmann et al., 1994; Switzer Blum et al., 1998; Jones et al., 2000; Islam et al., 2004; Macur et al., 2004; Rhine et al., 2006). We investigated the effect of arsenate concentration and electron donor on the enrichment of arsenate-respiring bacteria from sediments of the Ohio River in Pittsburgh, PA. The base medium contained (in 1 liter of deionized water) 0.095 g of $MgCl$, 4.2 g of $NaHCO_3$, 0.225 g of K_2HPO_4, 0.225 g of KH_2PO_4, 1.0 g of NaCl, 0.178 g of NH_4Cl, 0.1 g of Na_2S, 1.0 g of yeast extract, 10 ml of vitamin mix, and 10 ml of mineral mix (Stolz et al., 1997). The medium was amended with different electron donors (e.g., 10 mM acetate, formate, pyruvate, or lactate) and different concentrations of sodium arsenate (1, 5, 10, or 20 mM). The medium was brought to a pH of 7.3 and dispensed into 125-ml Wheaton bottles, degassed with oxygen-free N_2 and CO_2 (80:20 ratio), sealed with butyl rubber septa and aluminum crimp tops, and autoclaved. For testing hydrogen as the electron donor, the medium was bubbled for 5 min with a flow of hydrogen before sealing, and acetate (10 mM) was provided as the carbon source. Thus, for any given electron donor, such as acetate, there were bottles that contained 0 (control), 1, 5, 10, or 20 mM sodium arsenate. Likewise, for any given concentration of sodium arsenate, there were bottles with a different electron donor. The bottles were inoculated (3 ml) with a sediment slurry (made up in base medium and prepared in an anaerobic glove box) and incubated at ambient temperatures (~20°C). Heat-killed controls were autoclaved after inoculation.

Because it had been previously determined that, under circumneutral pH, the generation of sulfide and As(III) results in the rapid formation of arsenic trisulfide (Newman et al., 1997), visible inspection of the cultures was done on a daily basis. Arsenic trisulfide formation occurred within 3 to 7 days. Enrichment cultures with no added As(V) or 1 mM As(V), regardless of electron donor, had a strong odor of hydrogen sulfide, and the bottles turned from tan to black in color (Color Plate 16). Enrichment cultures provided H_2 as the electron donor and acetate as the carbon source, generated negative pressure suggesting the consumption of H_2. In cultures with 5 mM As(V), the rapid (within 3 days) formation of a copious yellow precipitate indicated rigorous arsenic trisulfide production (Color Plate 16). Arsenic trisulfide was not formed in any of the heat-killed controls. Interestingly, little if any arsenic trisulfide was evident in enrichment cultures with the higher concentrations of As(V) (e.g., 10 mM and 20 mM) (Color Plate 16), and together with the noticeable absence of the hydrogen sulfide smell, this suggests that sulfate reduction was inhibited. This conclusion was supported when arsenic speciation was determined and revealed that most of the As(V) in these cultures had nevertheless been reduced to As(III) (Fig. 1).

As(V) and As(III) were determined using high-performance liquid chromatography (HPLC) by comparison of the retention times of standard solutions and quantified with calibration curves as described in Fisher et al. (2008). Loss of As(V) and increase in As(III) was observed in all enrichment cultures within 7 days (Fig. 1 and 2). A comparison of the different electron donors with 10 mM As(V) is shown in Fig. 1. As(V) or As(III) concentration in the uninoculated or heat-killed controls did not occur during the 7 days (Fig. 1); however, the abiotic reduction of As(V) as a result of heat sterilization was noted (in subsequent experiments, arsenic stocks were prepared separately and added after autoclaving). As(V) was below detection limits at day 7 in cultures where formate, lactate, or hydrogen was provided as the electron donor. Ninety percent of the As(V) was reduced to As(III) with pyruvate, while about 75% was reduced in the enrichments with acetate or live controls with no added donor (Fig. 1). It was assumed that the arsenate reduction measured in the enrichments with no added donor was due to endogenous organics in the sediment. A comparison of different concentrations of As(V) with hydrogen as the electron donor and acetate as the exogenous carbon source is shown in Fig. 2. As(V) reduction in the 1 mM cultures was of the same magnitude as seen in the uninoculated and heat-killed controls. For both the 5 mM and 10 mM As(V) cultures, As(V) was reduced to below detection limits (Fig. 2). We attributed the discrepancy in mass balance of the As(V) reduced to the As(III) produced to the formation of arsenic trisulfide. A similar pattern was seen in enrichments that were provided with acetate, formate, or lactate as the electron donor. For the 20 mM enrichment cultures, the amount of As(V) reduced with hydrogen as the electron donor and acetate as the carbon source was twice that of enrichment cultures with acetate alone (Fig. 2).

These results indicate that both the concentration of As(V) and electron donor can impact the enrichment of arsenate-respiring microorganisms. Low concentrations (e.g., 1 mM) may not be enough to support respiratory growth, whereas 10 mM and 20 mM may be inhibitory to some species of arsenate-respiring bacteria as well as other physiological groups (e.g., sulfate-reducing bacteria). The most suitable concentration for enrichment cultures and incubations appears to be 5 mM As(V). This assessment is further supported by pure culture work that has demonstrated a wide range of As(V) concentrations for optimal growth (Stolz et al., 2006). For example, *Sulfurospirillum* species grow optimally at 5 mM As(V), whereas 10 mM is inhibitory

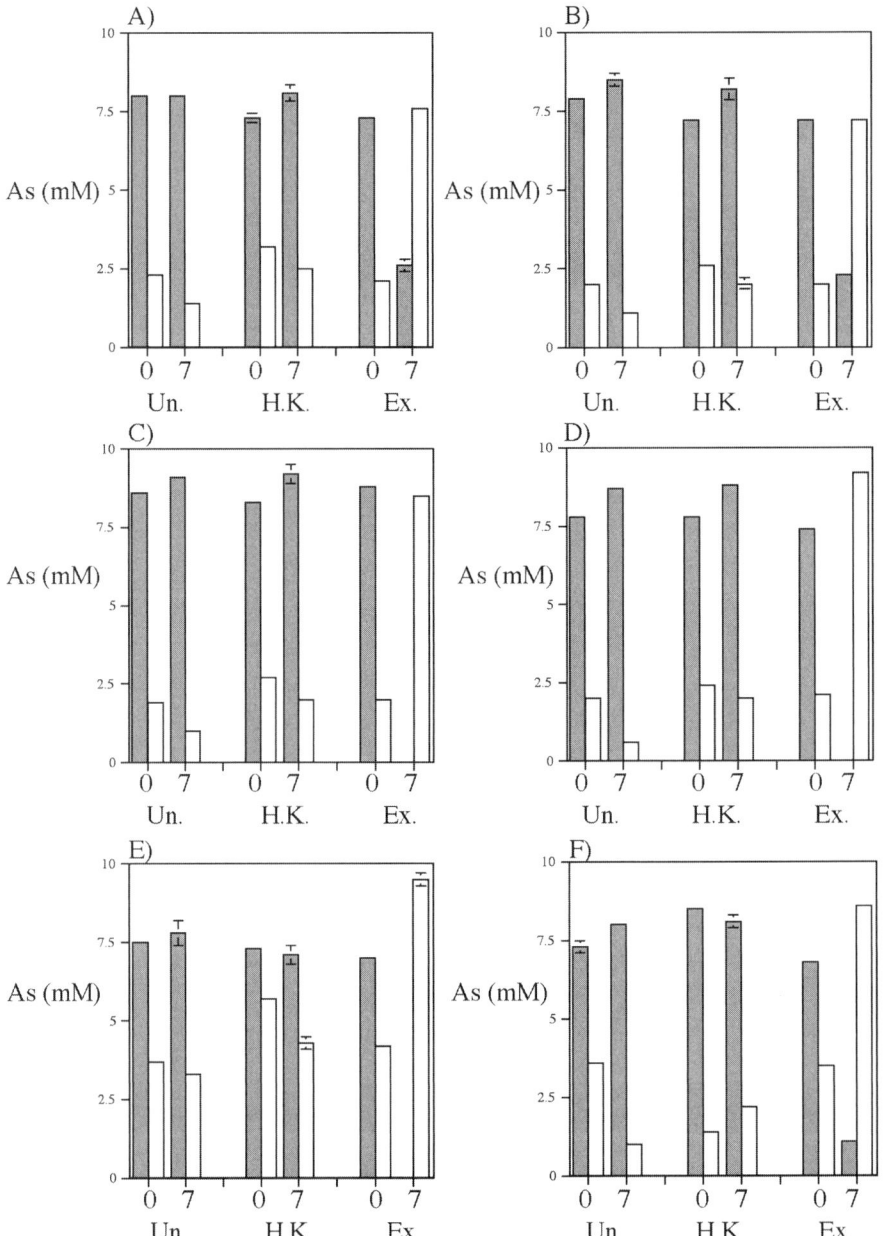

FIGURE 1 Arsenic speciation (as determined by HPLC) in the Ohio River sediment matrix containing 10 mM arsenate with different electron donors: (A) control with no added donor, (B) acetate, (C) hydrogen plus acetate, (D) formate, (E) lactate, and (F) pyruvate. Initial (day 0) and day 7 are shown for uninoculated control (Un.), heat-killed control (H.K.), and experimental (Ex.). Arsenate results are displayed in the gray bars, and arsenite results are displayed in the white bars.

(Laverman et al., 1995; Stolz et al., 1999). *Alkaliphilus oremlandii*, an arsenate-respiring firmicute, also grows optimally at 5 mM As(V) but can tolerate concentrations in excess of 40 mM (Fisher et al., 2008). Although active As(V) reduction (and presumably respi-

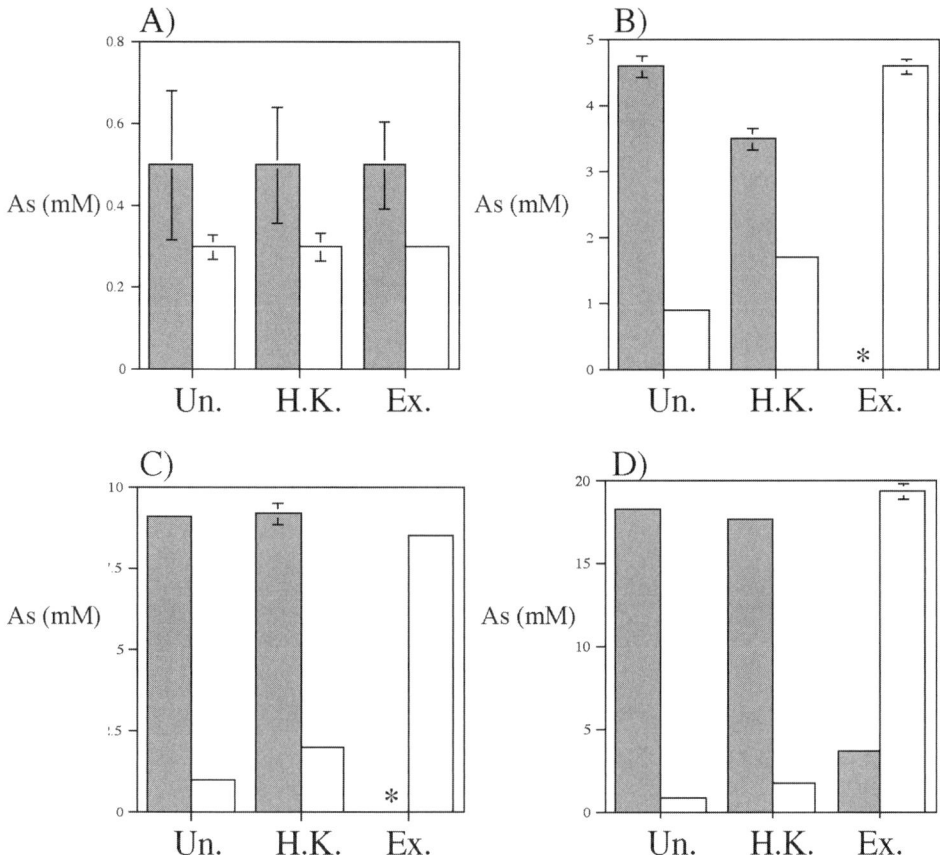

FIGURE 2 Arsenate speciation (as determined by HPLC) after 7 days in the Ohio River sediment matrix using hydrogen as the electron donor and acetate as the carbon source with different concentrations of As(V): 1 mM (A), 5 mM (B), 10 mM (C), and 20 mM sodium arsenate (D). Un., uninoculated control; H.K., heat-killed control; Ex., experimental. Results are displayed in the gray bars, and arsenite results are displayed in the white bars. An asterisk (*) indicates a concentration below the level of detection.

ration) occurred regardless of electron donor provided (e.g., acetate, formate, lactate, pyruvate), hydrogen appeared to support the most robust activity at all concentrations of As(V), at least for these river enrichments.

ARSENATE REDUCTASE ACTIVITY ASSAY

Arsenic resistance using the ars system and arsenate respiration both involve the reduction of As(V) to As(III). It is possible, however, to differentiate the two mechanisms enzymatically. ArsC, the small arsenate reductase in the resistance mechanism, requires either reduced glutathione or reduced thioredoxin, as the electron donor (Mukhopadhyay et al., 2002). Arr, on the other hand, may accept electrons from reduced viologen species such as methyl or benzyl viologen (Krafft and Macy, 1998; Afkar et al., 2003). Another interesting characteristic we have discovered is that Arr from many species may maintain enzymatic activity even in the presence of detergents (e.g., sodium dodecyl sulfate). Thus, the active enzyme may be separated using conventional sodium dodecyl sulfate polyacrylamide gel electrophoresis (SDS-PAGE) if the loading buffer does not contain reducing agents such as β-mercaptoethanol and the sample is not heated (Fig. 3). We have tested a number of species using the in-gel

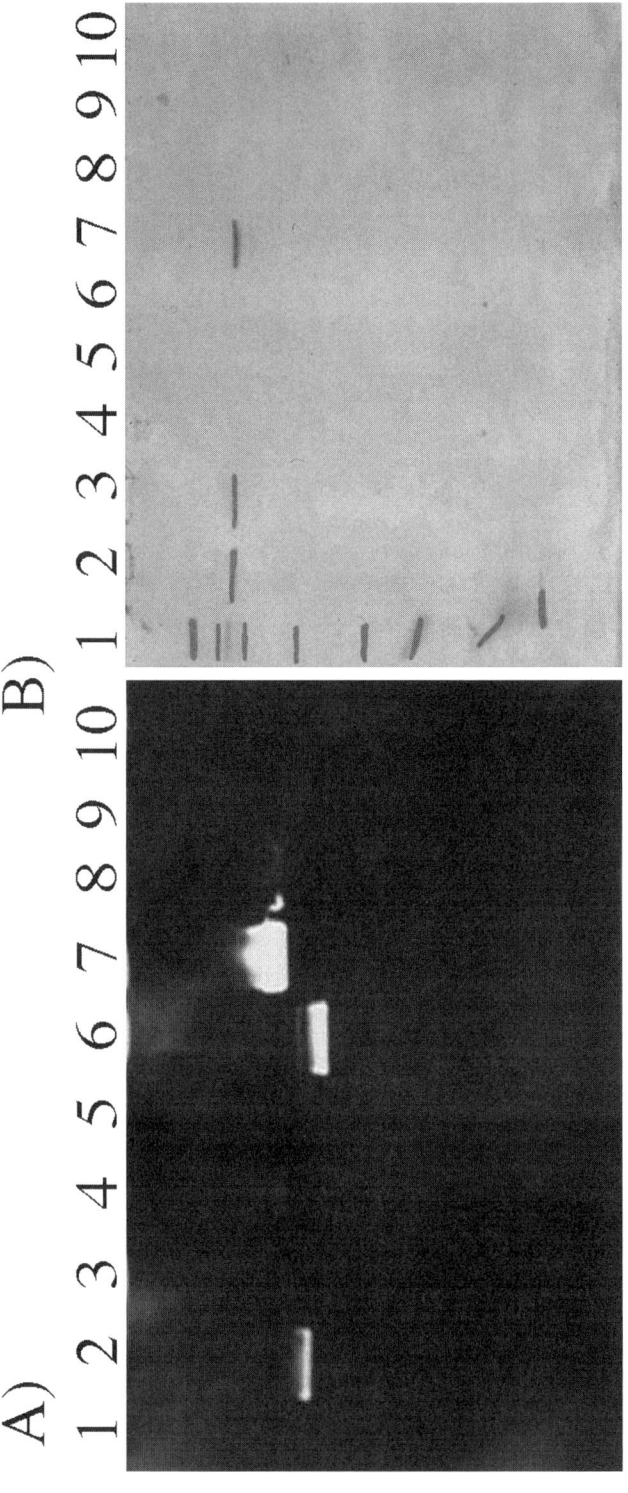

FIGURE 3 Arsenate reductase activity and cross-reactivity with anti-ArrA-1 affinity-purified antibodies. (A) Zymogram (nondenaturing PAGE) showing arsenate reductase activity in *B. selenitireducens* (lane 2), Mono Lake strain SLAS-1 (lane 6), and *A. oremlandii* (lane 7), but not *B. arseniciselenatis* (lane 3), Mono Lake strain MLMS1 (lane 4), *B. macyae* strain JMM-4 (lane 5), *S. barnesii* strain SES-3 (lane 8), *S. deleyianum* (lane 9), and *S. arsenophilum* strain MIT-13 (lane 10). Lane 1 displays molecular mass standards (206, 115, 98, 54, 37, 29, 20, 7 kDa). (B) Western blot analysis (SDS-PAGE gel) showing cross-reactivity with the anti-ArrA-1 antisera with *B. selenitireducens*, *B. arseniciselenatis*, and *A. oremlandii*, but not Mono Lake strain SLAS-1, Mono Lake strain MLMS1, *B. macyae* strain JMM4, *S. barnesii*, *S. deleyianum*, or *S. arsenophilum*. The lanes are the same as in panel A.

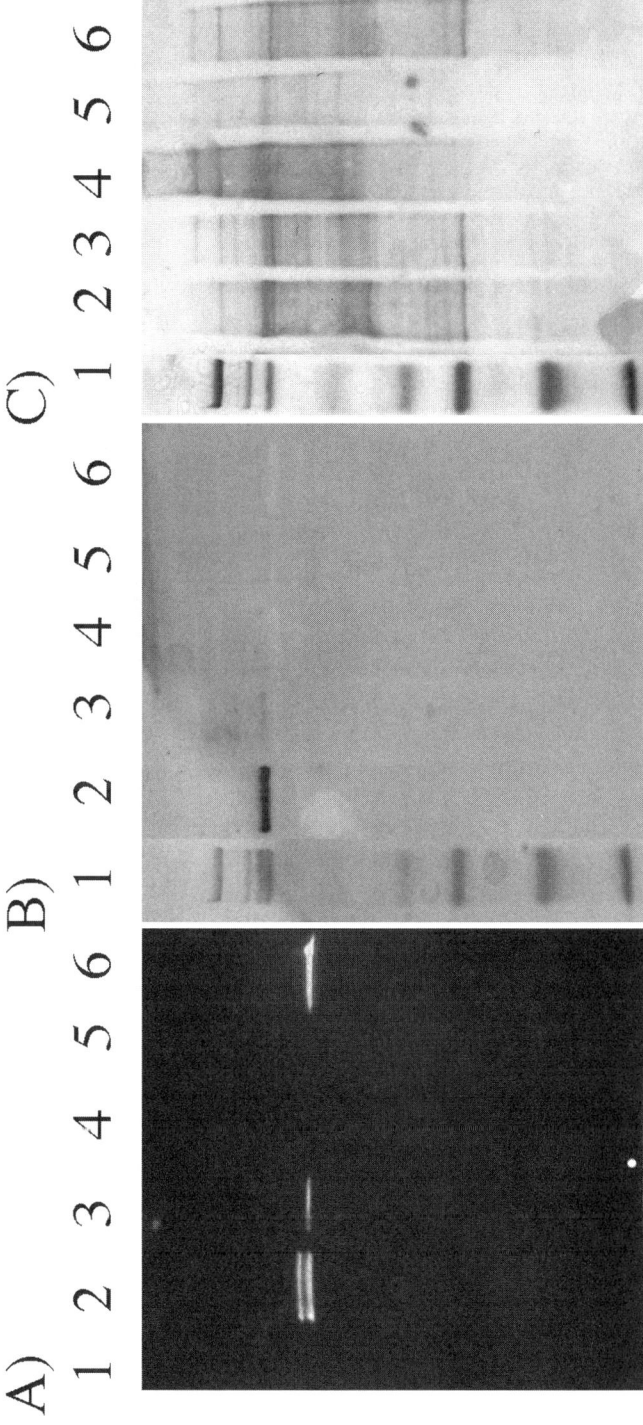

FIGURE 4 Detection of ArrA in cells of *B. selenitireducens* grown with different terminal electron acceptors. The results show that, although arsenate reductase activity can be detected in nitrate– and DMSO-grown cells, only arsenate-grown cells express ArrA. (A) Zymogram (nondenaturing PAGE) showing arsenate reductase activity in cells grown in arsenate (lane 2), nitrate (lane 3), and DMSO (lane 6) but not fumarate (lane 4) or selenite (lane 5). Lane 1 displays molecular mass standards (206, 115, 98, 54, 37, 29, 20, 7 kDa). (B) Western blot analysis (SDS-PAGE gel) showing presence of ArrA only in *B. selenitireducens* cells grown on arsenate. The lanes are the same as in panel A. (C) Duplicate gel of panel B stained with Coomassie blue (indicating that comparable amount of protein was loaded to each lane). Lanes are the same as in panels A and B.

assay and have found that *B. selenitireducens*, *A. oremlandii*, Searles Lake strain SLAS-1 maintain activity, while *Bacillus macyae*, Mono Lake strain MLMS-1, and the *Sulfurospirillum* species (*S. barnesii*, *S. arsenophilus*, *S. deleyianum*) do not (Fig. 3A). It is interesting to note that the active ArrAB complex exhibits greater mobility than the large subunit alone (migrating further down the gel, the relative distance of which may vary from organism to organism, Fig. 4A) and in some preparations, the activity shows up as a pair of bands (Fig. 4A, lane 2). We attribute the former to the globular nature of the intact protein complex (Richey et al., 2009) and the latter to its partial denaturation. In addition, related molybdoenzymes can also exhibit arsenate reductase activity when methyl or benzyl viologen are provided as the electron donor. We have indeed found this to be the case for *B. selenitireducens* as cell lysates of *B. selenitireducens* grown on nitrate or dimethyl sulfoxide give a positive results for methyl viologen oxidation coupled with As(V) reduction in the in-gel assay (Fig. 3A). Nitrate reductase and dimethyl sulfoxide (DMSO) reductase are also members of the large family of mononuclear molybdoenzymes (Stolz et al., 2006), are similar in structure and molecular mass, and exhibit similar mobility in the gel (e.g., they migrate the same distance in the gel, Fig. 4A). For this reason we began exploring the possibility of using biochemical probes to detect Arr.

BIOCHEMICAL PROBES

Biochemical probes were developed with the idea of being able to use them to identify Arr in PAGE and to fluorescently label arsenate-respiring bacteria in environmental samples and cultures. The biochemical probes were designed using the sequence alignments of ArrA from *B. selenitireducens* (Afkar et al., 2003), *Shewanella* strain ANA-3 (Saltikov and Newman, 2003), *Desulfitobacterium hafniense*, and *Wolinella succinogenes* (the latter two sequences obtained from the U.S. Department of Energy's Joint Genome Institute). Two stretches of sequence, each 15 amino acids in length (ArrA1: KCYGQGHWAYGHIAS, ArrA2: AIAHVILTEGVWYKP), that we believed to be highly conserved, were synthesized and used to raise polyclonal antibodies in New Zealand White rabbits (Anti-ArrA1, Anti-ArrA2). Anti-ArrA2 was found to cross-react with many different proteins in the cell lysate of *B. selenitireducens* and therefore was not affinity purified. Anti-ArrA1, however, gave a strong reaction with ArrA from *B. selenitireducens*. It was subsequently affinity purified and found to be specific to ArrA (Fig. 4B). The antibody was used to demonstrate the presence of ArrA only in cells grown on arsenate (Fig. 4B). Thus, this particular antibody will be quite useful in further studies of arsenic metabolism in *B. selenitireducens*. Anti-ArrA1 was also found to also cross-react with ArrA from *B. arseniciselenatis*, and *A. oremlandii*; however, it did not recognize ArrA from *B. macyae*, any of the *Sulfurospirillum* species tested (*S. barnesii*, *S. arsenophilus*, *S. deleyianum*) or the haloalkaliphilic strains MLMS-1 and SLAS-1 (Fig. 3B). These results can be explained by the fact that the inferred amino acid sequence of ArrA from *S. barnesii*, *S. deleyianum*, SLAS-1, and MLMS-1 show a striking lack of sequence identity. Thus, unfortunately, this particular amino acid sequence is not as conserved as originally believed.

MOLECULAR PROBES

The use of molecular approaches for the detection of arsenic-metabolizing bacteria is seeing greater application. Several groups have designed and tested primers targeting *arrA* for detecting the presence of arsenate respiring bacteria (Malasarn et al., 2004; Kulp et al., 2006, 2007; Lear et al., 2007). Each, however, has seen limited application. The primer set reported by Malasarn and colleagues (2004) amplifies a relatively small product (165 to 205 bp) and shows an efficacy for *Shewanella* species. Nevertheless, it established that *arrA* is a suitable target. Kulp and colleagues created a different primer set that amplifies a 500-bp product, but also modified the set to reflect the coding bias of the extreme halophiles in Mono Lake and Searles Lake (Kulp et al.,

TABLE 1 PCR primers for *arrA* amplification and DGGE

Primer	Sequence[a]
*arrA*UF1-GC	CGCCCGCCGCGCCCCGCGCCCGTCCGCCGCCCCCGCCCGTGTCAAGGHTGTACB-DCHTGG
*arrA*UR2	GTATCVATHGCTTCNTCCCA
*arrA*UR3	GCWGCCCAYTCVGGNGT
*arrA*UF5-GC	CGCCCGCCGCGCCCCGCGCCCGTCCGCCGCCCCCGCCCGADTTTGAGTTYTAY-AGYGAAAC
*arrA*UR6	CGGAAACANGGNGTGATACT
*arrA*UR8	GGGTTVAWTTTNGCBACATC

[a]N, A+C+G+T; B, T+C+G; D, A+T+G; H, A+T+C; V, A+C+G; W, A+T; Y, C+T.

2006, 2007). We designed a set of primers specifically for use in denaturing gradient gel electrophoresis (DGGE) (Table 1). Of these, *arrA*UF2 has a 14-base overlap with the AS2F primer designed by Song (Lear et al., 2007; Song et al., 2009), and *arrA*UF3 contains the first 17 bases of the ARR2 primer reported by Malasarn and colleagues (2004).

Primers have also been designed that target the catalytic subunit of arsenite oxidase (e.g., *aoxB*) (Rhine et al., 2006; Inskeep et al., 2007; Quemeneur et al., 2008) and the arsenite-specific efflux pump of the resistance system (e.g., *arsB*, ACR3) (Achour et al., 2007). The applications have included cloned libraries (Lear et al., 2007), denaturing gradient gel electrophoresis (DGGE; Kulp et al., 2006), reverse transcriptase PCR (Malasarn et al., 2004), and restriction fragment length polymorphism. We have expanded the range of primer sets for *arrA*. With the availability of more gene sequences, it has become apparent that *arrA* is conserved (Stolz et al., 2006, 2010), but that enough sequence variability exists to necessitate the use of multiple primer sets. This point is further underscored by the recent discovery of a homolog of Arr that actually functions as arsenite oxidase (Richey et al., 2009; Zargar et al., 2010). This homolog, designated Arx, was identified as such in *Alkalilimnicola ehrlichii* (Richey et al., 2009; Zargar et al., 2010), but has also been found in phototrophs that use As(III) as the electron donor in anoxygenic photoautotrophy (Kulp et al., 2008).

Degenerate primers were designed based on multiple sequence alignments of *arrA* sequences from *B. selenitireducens*, *B. arseniciselenatis*, *S. barnesii*, *Shewanella* sp. strain ANA-3, *D. hafniense*, and *W. succinogenes*. We initially designed six forward (*arrA*UF1-6) and seven reverse (*arrA*UR2-8) primers and tested them for their efficacy to amplify a portion of *arrA*. Two forward primers with a GC clamp (*arrA*UF1 and *arrA*UF5) and four reverse primers with the GC clamp (*arrA*UR2, *arrA*UR3, *arrA*UR6, and *arrA*UR8) were chosen (Table 1). Two primers, *arrA*UR4 and *arrA*UR7, failed initial screening and were not tested further. The location of the primers relative to the whole gene is shown in Fig. 5. They were specifically designed to include enough degeneracy to amplify most known *arrA* genes (Table 1). Based on actual sequence data, four primer sets were predicted to generate amplicons of differing size and %GC content from a variety of species. The primer set *arrA*UF1/*arrA*UR2 was predicted to amplify a ~245-base pair fragment near the 5′ end of the gene. The primer set *arrA*UF1/*arrA*UR3 was predicted to amplify a ~850-base pair fragment near the 5′ end of the gene. The primer set *arrA*UF5/*arrA*UR6 was predicted to amplify a ~138-base pair fragment near the 3′ end of the gene. The primer set *arrA*UF5/*arrA*UR8 was predicted to amplify a ~300-base pair fragment near the 3′ end of the gene. The reaction parameters for primer pair *arrA*UF1GC and *arrA*UR2 consisted of an initial denaturation step at 94°C

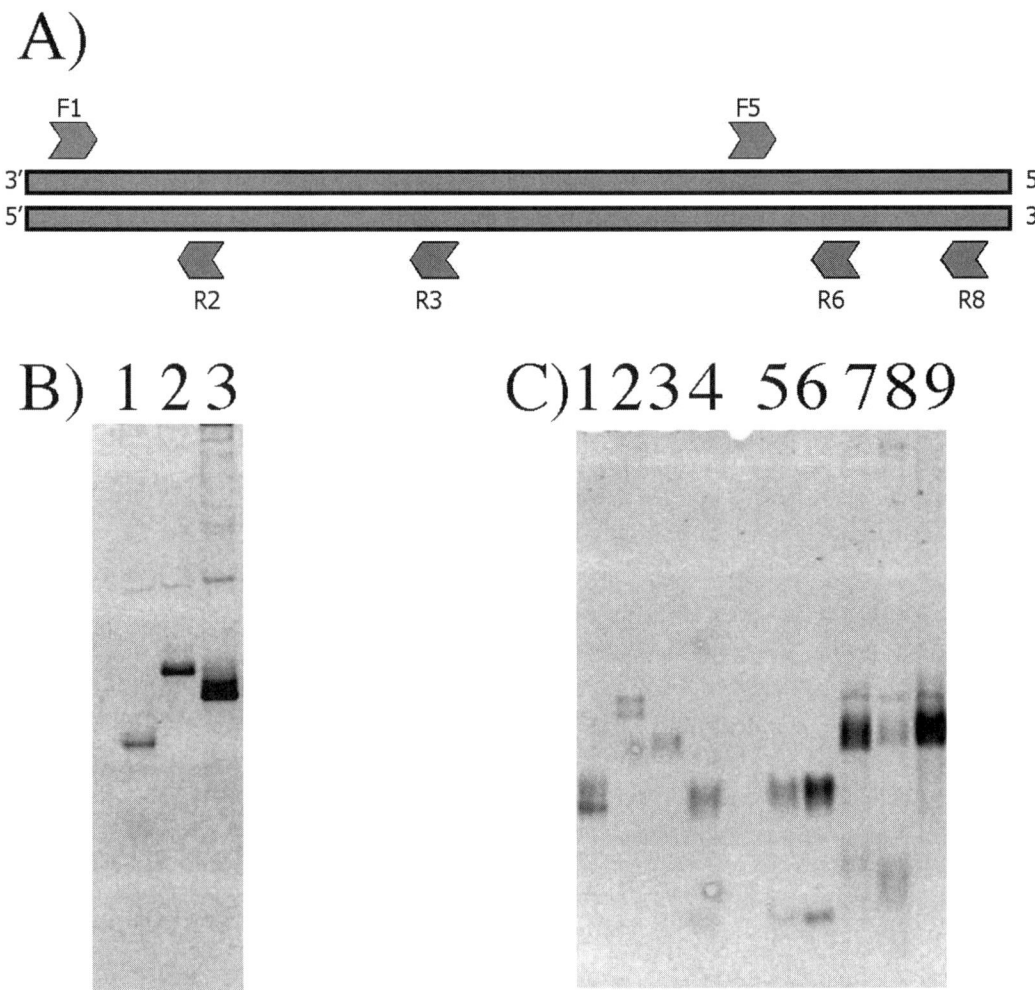

FIGURE 5 Primers and DGGE of *arrA*. (A) Gene map of *arrA* showing the relative locations of the *arrA* primers. (B) DGGE of PCR amplicons using *arrA*UF1GC/*arrA*UR3 on a 30 to 70% gradient. Lane 1, *B. selenitireducens*; lane 2, *A. oremlandii*; lane 3, *S. barnesii*. (C) DGGE of PCR amplicons using *arrA*UF1GC/*arrA*UR2 (lanes 1 to 3), *arrA*UF5GC/*arrA*UR6 (lanes 4 to 6), and *arrA*UF5GC/*arrA*UR8 (lanes 7 to 9) on a 30 to 70% gradient. Lanes 1, 4, and 7 represent *B. selenitireducens*; lanes 2, 5, and 8, *A. oremlandii*; lanes 3, 6, and 9, *S. barnesii*.

for 5 min, followed by two cycles of denaturation at 94°C for 45 s, and annealing at 56°C for 45 s, and extension at 72°C for 1 min. At each subsequent loop (for up to four loops) the annealing temperature was decreased by 2°C. This was followed by 20 cycles of denaturation at 94°C for 45 s, annealing at 48°C for 45 s, extension at 72°C for 1 min, final extension at 72°C for 5 min, and the final hold at 4°C. For primer pair *arrA*UF5GC/*arrA*UR6 and *arr*AUF5GC/ *arrA*UR8, the PCR cycling conditions were similar to those mentioned above except that the initial annealing temperature was 54°C. The annealing temperature was reduced by 2°C/cycle with the final annealing temperature being 42°C in the seventh cycle. Primer pair *arrA*UF1GC and *arrA*UR3 was used with touchdown PCR. The reaction parameters for touchdown PCR were as follows: initial denaturation at 94°C for 5 min, followed by cycles of denaturation at 94°C for 1 min, annealing starting at 60°C for 1 min and subsequently decreased at 0.5°C/cycle (ending at 50°C), extension at 72°C for 3 min. This was followed by 15 cycles of denaturation at 94°C for 1 min, annealing at 50°C for 1 min, extension at 72°C for 3 min (with final step for 7 min), and a final hold at 4°C .

The four primer sets amplified *arrA* from phylogenetically diverse arsenate-respiring bacteria, providing products of different mole %GC content to allow for separation by DGGE. Figure 5 shows the results for three different species (*S. barnesii*, *A. ormelandii*, *B. selenitireducens*) for the four primer sets using a 30 to 70% denaturing gel. The amplicons generated using the *arr*AUF1/*arr*AUR2 and *arr*AUF1/*arr*AUR3 primer sets were easily resolved and migrated relative to their predicted %GC content (Fig. 5B and C). The amplicons generated using the *arr*AUF5/*arr*AUR6 and *arr*AUF5/*arr*AUR8, however, showed little separation of the amplicons under the same denaturing conditions despite the predicted differences in %GC content (Fig. 5C). Thus, adjustments to the running conditions (e.g., denaturing conditions, percentage of acrylamide) are necessary to provide separation. Nevertheless, because the different primer sets give amplicons of different sizes, they may be applied to other techniques. For example, the *arr*AUF5/*arr*AUR6 primer set should be useful for reverse transcriptase PCR. The predicted restriction fragment maps of the *arr*AUF5/*arr*AUR8 amplicon digested with Taq1 suggest this primer set should be useful for terminal-restriction fragment length polymorphism analysis.

CONCLUSIONS

Different approaches have been used for identifying arsenate-respiring bacteria in the environment. Our results indicate that media composition can have an impact on the types of organisms that can be enriched for and cultured from arsenic-impacted environments. Ideally, a "matrix" consisting of different media formulations that include an assortment of electron donors (e.g., acetate, formate, lactate, pyruvate, or H_2) and different arsenic concentrations should be employed. Enzyme assays using artificial electron donors such as methyl or benzyl viologen can be useful in both spectrophotometric and gel enzyme assays; however, some arsenate reductases are more labile and may not always remain active. The ever increasing database of *arrA* gene sequences has revealed less conservation, thus limiting the utility of specific biochemical probes (e.g., anti-ArrA antibodies) and primer sets for molecular detection of arsenate-respiring bacteria. Again, using more than one functional gene target and several different primer sets is warranted. More importantly, this study underscores the need to use a combination of approaches that include geochemical analysis, incubations, and enrichment culture, in concert with biochemical (e.g., enzymological and immunological) and molecular approaches. Nevertheless, identifying arsenic-metabolizing bacteria and their role in the speciation, mobility, and ultimately toxicity of arsenic remains a priority for both microbial ecology (e.g., biogeochemical cycling) and environmental health.

ACKNOWLEDGMENTS

This work was supported in part by the USGS National Institutes of Water Resources (to J.F.S. and P.B.), NASA Exobiology (to J.F.S.), and NSF U.S.-Egypt Cooperative Research (to J.F.S., M.M.B., S.A.). A. Garcia Moyano was supported by a NASA Planetary Biology Internship.

We thank C. Saltikov and R. S. Oremland for helpful discussion.

REFERENCES

Achour, A. R., P. Bauda, and P. Billard. 2007. Diversity of arsenite transporter genes from arsenic-resistant soil bacteria. *Res. Microbiol.* **158:**128–137.

Afkar, E., J. Lisak, C. Saltikov, P. Basu, R. S. Oremland, and J. F. Stolz. 2003. The respiratory arsenate reductase from *Bacillus selenitireducens* strain MLS10. *FEMS Microbiol. Lett.* **226:**107–112.

Ahmann, D., L. R. Krumholz, H. F. Hemond, D. R. Lovley, and F. M. M. Morel. 1997. Microbial mobilization of arsenic from sediments of the Aberjona Watershed. *Environ. Sci. Technol.* **31:**2923–2930.

Ahmann, D., A. L. Roberts, L. R. Krumholz, and F. M. M. Morel. 1994. Microbe grows by reducing arsenic. *Nature* **371:**750.

Dowdle, P. R., A. M. Laverman, and R. S. Oremland. 1996. Bacterial reduction of arsenic(V) to arsenic(III), in anoxic sediments. *Appl. Environ. Microbiol.* **62:**1664–1669.

Fisher, E., A. M. Dawson, G. Polshnya, J. Lisak, B. Crable, E. Perera, M. Ranganathan, M. Thangavelu, P. Basu, and J. F. Stolz. 2008. Transformation of inorganic and organic arsenic by *Alkaliphilus oremlandii* sp. nov. strain OhILAs, p. 230–241. *In* J. Wiegel, R. Maier, and M. Adams (ed.), *Incredible Anaerobes: From Physiology to Genomics to Fuels*. Annals of the New York Academy of Sciences, New York, NY.

Gihring, T. M., G. K. Druschel, R. B. McCleskey, R. J. Hamers, and J. F. Banfield. 2001. Rapid arsenite oxidation by *Thermus aquaticus* and *Thermus thermophilus*: field and laboratory observations. *Environ. Sci. Technol.* **35:**3857–3862.

Harvey, C. F., C. H. Swartz, A. B. M. Badruzzaman, N. Keon-Blute, W. Yu, M. A. Ali, J. Jay, R. Beckie, V. Niedan, D. Brabander, P. M. Oates, K. N. Ashfaque, S. Islam, H. F. Hemond, and M. F. Ahmed. 2002. Arsenic mobility and groundwater extraction in Bangladesh. *Science* **298:**1602–1606.

Herbel, M. J., J. Switzer Blum, S. E. Hoeft, S. M. Cohen, L. L. Arnold, J. Lisak, J. F. Stolz, and R. S. Oremland. 2002. Dissimilatory arsenate reductase activity and arsenate-respiring bacteria in bovine rumen fluid, hamster feces, and the termite hindgut. *FEMS Microbiol. Ecol.* **41:**59–67.

Hoeft, S. E., T. R. Kulp, J. F. Stolz, T. Hollibaugh, and R. S. Oremland. 2004. Dissimilatory arsenate reduction with sulfide as the electron donor: experiments with Mono Lake water and isolation of strain MLMS-1, a chemoautotrophic arsenate respirer. *Appl. Environ. Microbiol.* **70:**2741–2747.

Inskeep, W. P., R. E. Macur, N. Hamamura, T. P. Warelow, S. A. Ward, and J. M. Santini. 2007. Detection, diversity and expression of aerobic bacterial arsenite oxidase genes. *Environ. Microbiol.* **9:**934–943.

Islam, F. S., A. G. Gault, C. Boothman, D. A. Polya, D. Chatterjee, and J. R. Lloyd. 2004. Role of metal-reducing bacteria in arsenic release from Bengal delta sediments. *Nature* **430:**68–71.

Jackson, C. R., K. G. Harrison, and S. L. Dugas. 2005. Enumeration and characterization of culturable arsenate resistant bacteria in a large estuary. *Syst. Appl. Microbiol.* **28:**727–734.

Jones, C. A., H. W. Langner, K. Anderson, T. R. McDermott, and W. P. Inskeep. 2000. Rates of microbially mediated arsenate reduction and solubilization. *Soil Sci. Soc. Am. J.* **64:**600–608.

Krafft, T., and J. M. Macy. 1998. Purification and characterization of the respiratory arsenate reductase of *Chrysiogenes arsenatis*. *Eur. J. Biochem.* **255:**647–653.

Kulp, T. R., S. Han, C. W. Saltikov, B. D. Lanoil, K. Zargar, and R. S. Oremland. 2007. Effects of imposed salinity gradients on dissimilatory arsenate reduction, sulfate reduction, and other microbial processes in sediments from two California soda lakes. *Appl. Environ. Microbiol.* **73:**5130–5137.

Kulp, T. R., S. E. Hoeft, M. Asao, M. T. Madigan, J. T. Hollibaugh, J. C. Fisher, J. F. Stolz, C. W. Culbertson, L. G. Miller, and R. S. Oremland. 2008. As(III) fuels anoxygenic photosynthesis in hot spring biofilms from Mono Lake, California. *Science* **321:**967–970.

Kulp, T. R., S. E. Hoeft, L. G. Miller, C. Saltikov, J. N. Murphy, S. Han, B. Lanoil, and R. S. Oremland. 2006. Dissimilatory arsenate and sulfate reduction in sediments of two hypersaline arsenic-rich soda lakes: Mono and Searles Lakes, California. *Appl. Environ. Microbiol.* **72:**6514–6526.

Laverman, A. M., J. S. Blum, J. K. Schaefer, E. J. P. Phillips, D. R. Lovley, and R. S. Oremland. 1995. Growth of strain SES-3 with arsenate and other diverse electron acceptors. *Appl. Environ. Microbiol.* **61:**3556–3561.

Lear, G., B. Song, A. G. Gault, D. A. Polya, and J. R. Lloyd. 2007. Molecular analysis of arsenate-reducing bacteria within Cambodian sediments following amendment with acetate. *Appl. Environ. Microbiol.* **73:**1041–1048.

Macur, R. E., C. R. Jackson, L. M. Botero, T. R. McDermott, and W. P. Inskeep. 2004. Bacterial populations associated with the oxidation and reduction of arsenic in an unsaturated soil. *Environ. Sci. Technol.* **38:**104–111.

Macy, J. M., K. Nunan, K. D. Hagen, D. R. Dixon, P. J. Harbour, M. Cahill, and L. Sly. 1996. *Chrysiogenes arsenatis*, gen. nov. sp. nov., a new arsenate-respiring bacterium isolated from gold mine wastewater. *Int. J. Syst. Bacteriol.* **46:**1153–1157.

Malasarn, D., C. W. Saltikov, K. M. Campbell, J. M. Santini, J. G. Hering, and D. K. Newman. 2004. *arrA* is a reliable marker for As(V) respiration. *Science* **306:**455.

Mukhopadhyay, R., B. P. Rosen, L. T. Phung, and S. Silver. 2002. Microbial arsenic: from geocycles to genes and enzymes. *FEMS Microbiol. Rev.* **26:**311–325.

Newman, D. K., T. J. Beveridge, and F. M. M. Morel. 1997 Precipitation of arsenic trisulfide by *Desulfotomaculum auripigmentum*. *Appl. Environ. Microbiol.* **63:**2022–2028.

Oremland, R. S., P. R. Dowdle, S. Hoeft, J. O. Sharp, J. K. Schaefer, L. G. Miller, J. Blum, R. L. Smith, N. S. Bloom, and D. Wallschlaeger. 2000. Bacterial dissimilatory reduction of arsenate and sulfate in meromictic Mono Lake, CA. *Geochim. Cosmochim. Acta* **64:**3073–3084.

Oremland, R. S., S. E. Hoeft, J. M. Santini, N. Bano, R. A. Hollibaugh, and J. T. Hollibaugh. 2002. Anaerobic oxidation of arsenite in Mono Lake Water and by a facultative arsenite-oxidizing chemoautotroph, strain MLHE-1. *Appl. Environ. Microbiol.* **68:**4795–4802.

Oremland, R. S., T. R. Kulp, J. Switzer Blum, S. E. Hoeft, S. Baesman, L. G. Miller, and J. F. Stolz. 2005. A microbial arsenic cycle in a salt-saturated, extreme environment: Searles Lake, California. *Science* **308:**1305–1308.

Oremland, R. S., and J. F. Stolz. 2003. Ecology of arsenic. *Science* **300:**939–944.

Oremland, R. S., and J. F. Stolz. 2005. Arsenic, microbes, and contaminated aquifers. *Trends Microbiol.* **13:**45–49.

Quemeneur, M., A. Heinrich-Salmeron, D. Muller, D. Lievremont, M. Jauzein, P. N. Bertin, F. Garrido, and C. Joulian. 2008. Diversity surveys and evolutionary relationships of *aoxB* genes in aerobic arsenite-oxidizing bacteria. *Appl. Environ. Microbiol.* **74:**4567–4573.

Rhine, D. E., E. Garcia-Dominguez, C. D. Phelps, and L. Y. Young. 2005. Environmental microbes can speciate and cycle arsenic. *Environ. Sci. Technol.* **39:**9569–9573.

Rhine, E. D., S. M. Ni Chadhain, G. J. Zylstra, and L. Y. Young. 2007. The arsenite oxidase genes (*aroAB*) in novel chemoautotrophic arsenite oxidizers. *Biochem. Biophys. Res. Commun.* **354:**662–667.

Rhine, E. D., C. D. Phelps, and L. Y. Young. 2006. Anaerobic arsenite oxidation by novel denitrifying isolates. *Environ. Microbiol.* **8:**899–908.

Richey, C., P. Chovanec, S. E. Hoeft, R. S. Oremland, P. Basu, and J. F. Stolz. 2009. Respiratory arsenate reductase as a bidirectional enzyme. *Biochem. Biophy. Res. Commun.* **382:**298–302.

Saltikov, C. W., and D. K. Newman. 2003. Genetic identification of a respiratory arsenate reductase. *Proc. Natl. Acad. Sci. USA* **100:**10983–10988.

Santini, J. M., L. I. Sly, A. Wen, D. Comrie, P. De Wulf-Durand, and J. M. Macy. 2002. New arsenite-oxidizing bacteria isolated from Australian gold mining environments: phylogenetic relationships. *Geomicrobiol. J.* **19:**67–76.

Song, B., E. Chyun, P. R. Jaffe, and B. B. Ward. 2009. Molecular methods to detect and monitor dissimilatory arsenate-respiring bacteria (DARB) in sediments. *FEMS Microbiol. Ecol.* **68:**108–117.

Stolz, J. F., P. Basu, and R. S. Oremland. 2010. The microbial transformation of arsenic: new twists on an old poison. *Microbe* **5:**53–59.

Stolz, J. F., P. Basu, J. M. Santini, and R. S. Oremland. 2006. Arsenic and selenium in microbial metabolism. *Annu. Rev. Microbiol.* **60:**107–130.

Stolz, J. F., D. J. Ellis, J. S. Blum, D. Ahmann, R. S. Oremland, and D. R. Lovley. 1999. *Sulfurospirillum barnesii* sp. nov. and *Sulfurospirillum arsenophilum* sp. nov., new members of the *Sulfurospirillum* clade of the epsilon Proteobacteria. *Int. J. Syst. Bacteriol.* **49:**1177–1180.

Stolz, J. F., T. Gugliuzza, J. S. Blum, R. Oremland, and F. M. Murillo. 1997. Differential cytochrome content and reductase activity in *Geospirillum barnesii* strain SeS3. *Arch. Microbiol.* **167:**1–5.

Stolz, J. F., E. Perera, B. Kilonzo, B. Kail, B. Crable, E. Fisher, M. Ranganathan, L. Wormer, and P. Basu. 2007. Biotransformation of 3-nitro-4-hydroxybenzene arsonic acid and release of inorganic arsenic by *Clostridium* species. *Environ. Sci. Technol.* **41:**818–823.

Switzer Blum, J., A. B. Bindi, J. Buzzelli, J. F. Stolz, and R. S. Oremland. 1998. *Bacillus arsenicoselenatis* sp. nov., and *Bacillus selenitireducens* sp. nov.: two haloalkaliphiles from Mono Lake, California, which respire oxyanions of selenium and arsenic. *Arch. Microbiol.* **171:**19–30.

Sun, Y., E. A. Polishchuk, U. Radoja, and W. R. Cullen. 2004. Identification and quantification of *arsC* genes in environmental samples by using real-time PCR. *J. Microbiol. Methods* **58:**335–349.

Zargar, K., S. Hoeft, R. Oremland, and C. W. Saltikov. 2010. Identification of a novel arsenite oxidase gene, *arxA*, in the haloalkaliphilic, arsenite-oxidizing bacterium *Alkalilimnicola ehrlichii* strain MLHE-1. *J. Bacteriol.* **192:**3755–3762.

NANOPARTICLES FORMED FROM MICROBIAL OXYANION REDUCTION OF TOXIC GROUP 15 AND GROUP 16 METALLOIDS

Carolyn I. Pearce, Shaun M. Baesman, Jodi Switzer Blum, Jonathan W. Fellowes, and Ronald S. Oremland

16

INTRODUCTION

Environmental Significance of Group 15 and 16 Toxic Metalloids

Selenium, tellurium, and arsenic are present naturally in aquatic and terrestrial environments and share many similar biogeochemical characteristics. These elements are released into the environment through the weathering and decomposition of minerals contained within a variety of lithologies, with slow release rates resulting in low environmental concentrations. Selenium, tellurium, and arsenic occur in several oxidation states as oxyanions (e.g., selenate [SeO_4^{2-}], selenite [SeO_3^{2-}], tellurate [TeO_4^{2-}], tellurite [TeO_3^{2-}], arsenate [$HAsO_4^{2-}$], and arsenite [$HAsO_3^{2-}$]) in their native elemental states [e.g., Se(0), Te(0)] or in their most reduced states as selenide (-II) and telluride (-II) or arsenide/arsines (-III). These elements can be methylated through microbial activity to form compounds such as dimethylselenide (Ehrlich, 2002; Masscheleyn, et al., 1990), dimethyltelluride (Basnayake et al., 2001; Fleming and Alexander, 1972), and methylarsonous acid (Dopp et al., 2004) as well as a variety of toxic methylated arsine gases (Yuan et al., 2008). These elements are also found as analogues of sulfurous proteins such as selenocysteine and selenomethionine (Bock et al., 1991; Jones et al., 1979; Stolz et al., 2006; Zannoni et al., 2008), tellurocysteine, telluromethionine (Zannoni et al., 2008), and the arsenic-containing amino acid, arsenomethionine (Dembitsky and Levitsky, 2004).

In aerobic environments, these three elements all occur as readily soluble oxyanions. In anoxic environments, Se and Te are predominantly present in their (IV) oxidation states or as their insoluble elemental forms [i.e., Se(0) and Te(0)]. Arsenic in anoxic environments is predominantly present as the (III) state, as $H_3AsO_3^0$ or $H_2AsO_3^-$ (Stolz et al., 2006), which can also form thioarsenate/thioarsenite complexes in lieu of the oxyanions when there is reactive sulfide present (Planer-Friedrich et al., 2006, 2009). Despite their low crustal abundances, the potential toxic and teratogenic effects of these elements are of major concern and, in aquatic environments, especially those under certain evapoconcentrative conditions, they can attain relatively high concentrations.

Carolyn I. Pearce, Pacific Northwest National Laboratory, Richland, WA 99352. *Shaun M. Baesman* and *Jodi Switzer Blum*, U.S. Geological Survey, Water Resources Division, Menlo Park, CA 94025. *Jonathan W. Fellowes*, School of Earth, Atmospheric & Environmental Sciences, University of Manchester, Manchester, M13 9PL, United Kingdom. *Ronald S. Oremland*, U.S. Geological Survey, Water Resources Division, Menlo Park, CA 94025.

In such locales, they can accumulate to micromolar (Oremland et al., 1989, 2000) or even millimolar levels (Oremland et al., 2005). During recent decades, anthropogenic activities such as mining, irrigated agriculture, petrochemical refining, and industrial manufacturing operations have exacerbated the problems associated with these elements in the environment (Lemly, 2004). This has resulted in several high-profile pollution incidents, including wildfowl deaths at the Kesterson reservoir (California) due to selenium contamination (Presser, 1994), and human arsenicosis as a result of arsenic contamination in drinking water wells in Bangladesh and West Bengal (e.g., Ahmed et al. [2006]). Consequently, there has been substantial interest in the cycling of these elements in the environment, particularly with respect to changes in speciation as a result of microbial activity.

A surprisingly wide range of environmentally and clinically isolated microbes are capable of altering the chemical state of Se, Te, and As by a variety of oxidation, reduction, or methylation reactions. These are achieved for purposes of respiration, detoxification, and the maintenance of redox poise and, in some cases, to serve as inorganic electron donors for chemo- and photoautotrophic growth (Table 1). Thus, microbes play an important role in the cycling of these elements between reservoirs in the environment (Oremland et al., 2004; Shrift, 1964). The full redox cycle of Se speciation observed in nature is influenced by microbial activity, with microbes controlling both the oxidation and reduction of Se (Dowdle and Oremland, 1998). To date, no environmental cycle has been developed for Te, although it has been reported for nearly a century that microbes can reduce tellurite to elemental tellurium (Klett, 1900) and can methylate tellurium oxyanions (Basnayake et al., 2001; Fleming and Alexander, 1972). The activities of As-metabolizing microbes can affect the speciation and mobility of As in the environment, through arsenate respiration and nitrate-linked anaerobic oxidation of arsenite.

The toxicity of the metalloids Se, Te, and As is due to the disruption of thiol intracellular biochemistry through the formation of stable, long-lived sulfur complexes (Zannoni et al., 2008) and, in the case of As, by substitution for phosphorus, thereby disrupting cellular metabolism. However, both Se and As are readily assimilated by microbes, and Se is an essential trace element present in naturally occurring proteins such as selenocysteine (Bock et al., 1991; Stolz et al., 2006; Zannoni et al., 2008). No biochemical use for Te has been as yet described and it is significantly more toxic than Se, disrupting metabolic processes through its strong oxidizing potential (Salminen et al., 2006). Nonetheless, microorganisms have evolved a variety of resistance mechanisms to the presence of Te-oxyanions (Chasteen et al., 2009).

Microbial resistance to these three toxic elements has been well-documented and relies upon a number of different strategies, depending upon the element in question and its chemical state. The basis of the resistance to As(V) is the reductive expulsion from the cytoplasm as As(III) (Bhattacharjee and Rosen, 2007), while for Se and Te the reduction of cytoplasmic oxyanions to their insoluble and nontoxic elemental states [Se(0), Te(0)] establishes resistance. In some microorganisms this results in the accumulation of external biominerals associated with the outer cell envelope. Due to a range of biotic and abiotic factors associated with the templating environment in which these biominerals are formed, the biomineral phases are often nanoscale in dimension. Selenium, tellurium, and arsenic have optoelectrical properties, that is, they have the potential to convert light energy into electricity (and vice versa); thus, these bionanominerals have possible applications in novel photonic devices (Stolz and Oremland, 1999). The production of such desirable nanoscale materials using a biosynthetic route, thereby eliminating toxic organic solvents and minimizing expensive high-temperature processing, is, as yet, a largely unexplored and unexploited area of considerable potential. Indeed, the possibility

TABLE 1 Some examples of microbial interaction with Se, Te, and As

Microbe	Environment	Starting material	End product	Reference(s)
Pterotrigonia brevicula	Mine water, PA, USA	Te^{IV}	Te^0, $(CH_3)_2Te$ gas	Smithers and Krouse, 1967
Acidithiobacillus ferrooxidans		Se^{-II}	Se^0	Torma and Habashi, 1972
Corynebacterium	Clinical isolate, human skin	Se^0, Se^{IV}, Se^{VI}, org-Se	Se^{-II} (volatile alkylselenides)	Doran and Alexander, 1977
Methanococcus vannieli	San Francisco Bay mud flat, CA, USA	Se^{IV}	Se^{-II} (organic-Se)	Jones et al., 1979
Desulfovibrio desulfuricans		$Se^{VI}, Se^{IV} Te^{IV}$	Aqueous HSe^-, Te^0	Lloyd et al., 2001; Zehr and Oremland, 1987
Pseudomonas stutzeri	Drainage water	Se^{VI}	Se^0	Lortie et al., 1992
Citrobacter, Wolinella succinogenes	Termite gut and bovine rumen fluid, respectively	Se^{VI}, Se^{IV}, As^V and $S_2O_3^-$	Se^0, As^{III}, AsS-like mineral	Herbel et al., 2002; Tomei et al., 1992
Thauera selenatis	San Joaquin Valley, CA, USA	Se^{VI}	Se^{IV}, Se^0 in presence of nitrate	Macy et al., 1993
Chrysiogenes arsenatis	Gold mine wastewater	As^V	As^{III}	Macy et al., 1996
Desulfotomaculum auripigmentum	Upper Mystic Lake, MA, USA	As^V	As^{III}	Newman et al., 1997
Enterobacter cloacae SLD1a-1	San Joaquin Valley, CA, USA	Se^{VI}, Se^{IV}	Se^0 nanospheres	Losi and Frankenberger, 1997
Rhodobacter sphaeroides	Pond water, Hyderabad, India	Se^{VI}, Se^{IV}	Se^{-II} (organic Se and alkylselenides), Se^0 (red amorphous and black vitreous)	Van Fleet-Stalder et al., 2000; Yamada et al., 1997
Bacillus selenitireducens	Mono Lake, CA, USA	Se^{IV}, Te^{IV}, As^V	Se^0 nanospheres, Te^0 nanorods/rosettes	Baesman et al., 2007; Blum et al., 1998
Bacillus arsenicoselenatis	Mono Lake, CA, USA	Se^{VI}, As^V	Se^{IV}	Blum et al., 1998

(Continued)

TABLE 1 Some examples of microbial interaction with Se, Te, and As—*Continued*

Microbe	Environment	Starting material	End product	Reference(s)
Thiobacillus ASN-1 *Leptothrix* MnB1	Marine sediment, Sapelo Island, GA, USA	Se^0	Se^{VI}	Dowdle and Oremland, 1998
Aeromonas hydrophila	"...Tin of milk with a fishy odor"	Se^{VI}	Se^{IV}	Knight and Blakemore, 1998
Geobacillus stearothermophilus V		Se^{IV}, Te^{IV}	Se^0, Te^0	Moscoso et al., 1998
Sulfurospirillum arsenophilum	Aberjona watershed, MA, USA	As^V	As^{III}	Stolz et al., 1999
Sulfurospirillum barnesii	Massie Slough, NV, USA	Se^{VI}, Te^{IV}	Se^0, Te^0 (nanospheres)	Baesman et al., 2007; Stolz et al., 1999
Rhodospirillum rubrum		Se^{IV}	Se^0	Kessi et al., 1999
Pseudomonas aeruginosa, Agrobacterium tumefaciens, Escherichia coli		Te^{IV}	Te^0 granules, $(CH_3)_2Te$ gas	Trutko et al., 2000
Pyrobaculum arsenaticum, Pyrobaculum aerophilum	Hot spring, Naples, Italy	Se^{VI}	Se^0	Huber et al., 2000
Pseudomonas fluorescens K27	Kesterson Reservoir, CA, USA	Te^{IV}, Te^{VI}	$Te^0, (CH_3)_2Te$ gas	Basnayake et al., 2001
Selenihalanaerobacter shriftii	Dead Sea sediments, Israel	Se^{VI}	Se^{IV}, Se^0	Blum et al., 2001
Pseudoalteromonas	Hydrothermal vents, Juan de Fuca Ridge, Pacific Ocean	Se^{IV}, Te^{IV}	Se^0 granules, Te^0 granules	Rathgeber et al., 2002
Stenotrophomonas maltophilia	Agricultural evaporation pond, Tulare Lake, CA, USA	Se^{VI}, Se^{IV}	Se^0 nanospheres and Se^{-II} (volatile alkylselenides)	Dungan et al., 2003
Shewanella oneidensis	Sediment from Oneida Lake, NY, USA	Se^{IV}, Te^{IV}	Se^0 (extracellular nanospheres), Te^0 (cytoplasmic needles)	Klonowska et al., 2005
Desulfomicrobium sp.	Sulfate-reducing biofilm	Se^{VI}	Se^0, Se^{-II} (FeS/Se)	Hockin and Gadd, 2006

Organism	Location	Substrate	Product	Reference
Desulfobulbacea	Mono Lake, CA, USA	As^V, S^{-II}	AsS-like mineral	Hollibaugh et al., 2006
Shewanella sp. strains ER-Se-17L, ER-Te-48, and ER-V-6	Hydrothermal vent fields, Eastern Pacific Ocean	Se^{IV}, Te^{IV}, Te^{VI}	Se^0, Te^0	Csotonyi et al., 2006
Bacillus selenatarsenatis	Effluent drain of glass manufacturing plant, Japan	Se^{VI}	Se^0	Yamamura et al., 2007)
Shewanella sp. strain HN-41	Tidal flats, Haenam Jeollanam-do, Republic of Korea	As^V and $S_2O_3^-$	AsS and As_4S	Lee et al., 2007
Clostridiaceae	Alvord Basin, OR, USA	As^V	β-AsS	Ledbetter et al., 2007
Dechloromonas sp.	Rifle, CO, USA	Se^{VI}	Se^0	Williams et al., 2007
Ectothiorhodospira	Mono Lake, CA, USA	As^{III}	As(V)	Budinoff and Hollibaugh, 2008; Kulp et al., 2008
Veillonella atypica	Clinical isolate, human tonsils	Se^{IV}, Te^{IV}	Se^0 (nanospheres), Te^0 (nanorods), aqueous HSe^-	Pearce et al., 2008; this work
Bacillus beveridgei MLTeJB	Mono Lake, CA, USA	Te^{IV}	Te^0 nanorods and nanogranules	Baesman et al., 2009
Bacillus sp. strain NS3	Punjab, India	Se^{IV}	Se^0 (nanospheres),	Prakash et al., 2009
Geobacter sulfurreducens	Surface sediments, Norman, OK, USA	Se^{IV}, Te^{VI}	Se^0 (nanospheres), Te^0 (nanorods)	Pearce et al., 2009; this work

of using industrial waste as the starting material for synthesis of microbially generated nanomaterials is of particular interest.

Microbial Interaction with Group 15 and 16 Toxic Metalloids

The reduction of selenium oxyanions as terminal electron acceptors is energetically favorable, with the free energy ($\Delta G_o'$) of the reduction of SeO_4^{2-} and $HSeO_3^-$ coupled with the oxidation of H_2, respectively, yielding -15.53 kcal $mol^{-1}e^{-1}$ (-65.0 kJ mol^{-1} e^{-1}) and -8.93 kcal $mol^{-1}e^{-1}$ (-37.4 kJ mol^{-1} e^{-1}) (Newman et al., 1998). Selenium oxyanion reduction occurs in a wide range of microbes, including representatives of the *Wolinella, Pseudomonas, Sulfurospirillum, Enterobacter, Thaurea, Bacillus, Shewanella,* and *Citrobacter* genera (see also Table 1). The various mechanisms of selenium oxyanion reduction by microbes have been reviewed by Zannoni et al. (2008). There are four modeled pathways through which the reduction of selenium oxyanions may occur: (i) a painter-type reaction in which the selenium oxyanions are reduced to elemental selenium by glutathione via several intermediates such as selenodiglutathione and glucothioselenol; (ii) oxyanion reduction by cytosolic and periplasmic oxidoreductases; (iii) abiotic reactions with metabolic products, e.g., during the reduction of sulfate to sulfide, the sulfide can react with selenite to produce elemental selenium and sulfur; and (iv) the production of elemental selenium by a reaction with the biologically derived chelating agent pyridine-2,6-bisthiocarboxylic acid (Cortese et al., 2002; Hockin and Gadd, 2003; Stolz and Oremland, 1999). Microbes can also reduce the solid elemental selenium phase to produce volatile alkylselenides, with dimethyl selenide and dimethyl diselenide as the most common forms in which Se is present as Se(-II). Aqueous selenide (HSe^-) can also be a product of Se(0) reduction, which reacts with metal cations to form a solid precipitate, e.g., FeSe. Gaseous H_2Se formed by cultured anaerobic bacteria will rapidly auto-oxidize back to Se(0) if exposed to O_2, typically caused by changing the incubation headspace from N_2 to air. The production of Se(-II) after Se(0) formation can be either biphasic, as with the Se(IV)-respiring gram-positive haloalkaliphile, *Bacillus selenitireducens* (Herbel et al., 2003) or continuous, as reported for *Geobacter sulfurreducens* and *Shewanella oneidensis* (Pearce et al., 2009).

The biochemistry of dissimilatory selenate reduction has been studied most extensively in the Se(VI) respirer, *Thauera selenantis* (Macy et al., 1993; Rech and Macy, 1992). Selenate reductase from this organism has been purified and characterized (Krafft et al., 2000; Maher et al., 2004; Schröder et al., 1997), and is placed in the class of Mo-containing reductase enzymes (McEwan et al., 2002). A number of microbes can carry out the reduction of Se oxyanions without a direct linkage to energy conservation. Among these *Enterobacter cloacae* has been best characterized (Leaver et al., 2008; Losi and Frankenberger, 1997) because it has proven amenable to genetic manipulation via knockout mutagenesis (Ma et al., 2007, 2009; Yee and Kobayashi, 2008). *E. cloacae* is a facultative anaerobe and forms external accumulations of Se nanospheres (Yee et al., 2007) that appear outwardly identical to those produced by fastidious Se-respiring anaerobes (Oremland et al., 2004). This makes *E. cloacae* a more practical model organism to work with. In contrast, surprisingly little is known as yet about the biochemistry of dissimilatory Se(IV) reduction.

Microbial respiration of tellurate has been claimed to be achieved by an as yet unidentified microbe isolated from hydrothermal vents located in the eastern Pacific Ocean (Csotonyi et al., 2006). Work by Baesman et al. (2007) showed that *Sulfurospirillum barnesii* and *B. selenitireducens* were capable of respiratory growth using tellurate and tellurite oxyanions, respectively, which was quantitatively coupled to their oxidation of lactate to acetate, together with CO_2. Although both these strains grew well on millimolar levels of selenium oxyanions (as well as arsenate), they both proved sensitive to tellurium oxyanions, with concentrations of ≥1 mM completely inhibiting growth. Hence Te-dependent growth had to be achieved by the sustained pulsing of cultures with sublethal quantities (~0.6 mM) of Te(IV) or Te(VI). Most recently, a new microbial species was isolated from Mono Lake by using ~10 mM Te(IV) routinely in the enrichment/isolation process. This organism, *Bacillus beveridgei,* is capable of growth using millimolar concentrations of Te(VI) or Te(IV) (Baesman et al., 2009). Microbial reduction of tellurate and tellurite results in the formation of elemental tellurium or the more reduced methylated form, but, in marked contrast to the case for selenium, not in the formation of aqueous telluride anions. An ability to form biotelluride anions for the production of telluride nanomaterials (e.g., CdTe) would be of great significance because of their use in the development of photonic devices. The fact that this can be achieved for selenium but not for tellurium suggests a possible divergence (or limitation) in how these oxyanions are metabolized biochemically during their respective dissimilatory reductions. Overall, the mechanisms of microbial tellurium oxyanion reduction are thought to be similar to those for selenium. However, two *Pseudomonas aeruginosa* strains and *Rhodobacter sphaeroides* (Moore and Kaplan, 1992) have been shown to use *c*-type cytochromes and cytochrome oxidases during reduction of tellurite, and tellurite reduction by *Escherichia coli* involves a quinol oxidase system. None of these three organisms is a "true" Te-respirer that can actually conserve energy for growth as do *B. selenitireducens, B. beveridgei,* or *S. barnesii*. The biochemical pathways for dissimilatory Te-oxyanion reduction and how they conform to or diverge from those for Se oxyanions is a research question that has yet to be addressed. Indeed, only a few novel microorganisms have been examined to date for their ability to conduct dissimilatory reductions of Te oxyanions. Hence, the possibility that some as yet undiscovered microbe exists with the ability to form telluride from more oxidized forms of Te is still a distinct possibility.

There are four basic processes of microbial arsenic transformation: methylation, demethylation, oxidation, and reduction (Stolz et al., 2006). Methylation involves the reduction of arsenate followed by the oxidative addition of a methyl group to form a range of different compounds with As in the V, III or –III state. Little is known about the mechanisms of demethylation, but if the pathway is the reverse of that described for methylation, it would involve reductive elimination and oxidation of the center, restricting demethylation to As(V) species (Stolz et al., 2006). Arsenate reductases and arsenite oxidases have been purified for a limited number of organisms and are members of the dimethyl sulfoxide reduction family of molybdenum enzymes. Both the respiratory arsenate reductase and arsenite oxidase are heterodimers with similar structures and molecular weights, but are thought to be unidirectional in their mode of action (Silver and Phung, 2005). However, a reverse functionality for the arsenate reductase of *Alkalilimnicola ehrlichii* was discovered (Hoeft et al., 2007; Richey et al., 2009) whereby it oxidized As(III). This mechanism was also found in a photosynthetic bacterium of the *Ectothiorhodospira* clade of *Gammaproteobacteria* as the means by which As(III) served as an electron donor for anoxygenic photosynthesis (Kulp et al., 2008). The implications that this discovery has for microbial evolution on Earth have been recently reviewed (Oremland et al., 2009).

Applications of Group 15 and Group 16 Metalloid Bionanoparticles

Microbial interaction with Se, Te, and As results in the production of nanoparticles (>100 nm) with unique physical, optical, and electrical properties that are not representative of their bulk equivalents as a result of (i) the ratio of surface area to volume and (ii) unique surface properties in particles <10 nm, and (iii) the onset of quantum confinement where the particle's band gap is size dependent. Of particular interest are the optoelectronic properties exhibited by nanominerals composed of

Group 15 and 16 elements, which result in the conversion of incident light energy into electricity and vice versa (Baesman et al., 2009). Microbially produced nano-sized materials comprising elemental Se, Te, and As, as well as those in combination with metallic/nonmetallic elements (e.g., CdSe, As_2S_3), have a range of technological, medical, and environmental applications. The process of microbial nanomineral formation itself can be used to efficiently remove Se, Te, and As oxyanions from drinking water, groundwater, or wastewater. This type of anaerobic process offers advantageous over aerobic processes in terms of low sludge production, smaller reactor volumes, and cost savings in aeration and nutrient supply (Lenz et al., 2008). Nanoparticles of elemental Se can be used in a photocatalytic process to reduce environmentally harmful organic dye contaminants, as shown by the reduction of methylene blue (Nath et al., 2004). Se(0) nanospheres have also been shown to be the cheapest and most efficient sorbents of mercury vapor with potential applications in the safe disposal of compact fluorescent lamps (Johnson et al., 2008).

Technological applications of Se(0) and Te(0) nanoparticles include their use in photocopiers, microelectronic circuits, and solar cells as a result of their photo-optical and semiconducting physical properties (Baesman et al., 2007; Oremland et al., 2004). The fluorescence absorption and emission of CdSe and ZnSe quantum dots are conveniently tunable by their size, resulting in a range of applications including optoelectronic devices, light sensors, and high-purity emission lasers (Dettmer, 1988; Pickett and O'Brien, 2001) and as powerful probes for the labeling of biological components (Giepmans et al., 2006). CdSe nanoparticles have also been shown to act as very efficient and highly selective catalysts for the reduction of aromatic azides to aromatic amines when activated by light (Warrier et al., 2004). Pearce et al. (2008) have shown that nanoscale, luminescent CdSe/ZnSe can be produced by *Veillonella atypica* using an environmentally friendly, aqueous-based synthesis route. Polycrystalline phases of the chalcogenide minerals realgar (AsS) and duranusite (As_4S), produced by the *Shewanella* sp. strain HN-41 in the form of nanotubes, behave as metals and semiconductors in terms of their electrical and photoconductive properties, respectively, and have potential applications in nano- and optoelectronic devices (Lee et al., 2007).

With regard to medical applications, much recent attention has been given to the antioxidant properties of Se(0) nanospheres, which have the capacity to limit the damage caused by free radicals and reactive oxygen species (Gao et al., 2002). People from many parts of the world do not consume enough selenium in their diets, and several recent studies have suggested a link between cancer and selenium deficiency. Nanoparticulate Se(0) is the least toxic of all Se supplements and is effective as a chemopreventive agent (Peng et al., 2007).

The process of microbial nanoparticle formation may also result in the production of novel materials with possible commercial applications that cannot be synthesized by traditional inorganic approaches. Thus, primary research investigations into how these nanoparticles are formed are well justified. Such efforts would include the scrutiny of the proteins, lipids, and polysaccharides that unite cellular processes with inorganic substrates, thereby allowing their initial adhesion to the cell envelope, and subsequent electron transfer and dissimilatory reduction of the metalloids in question. Oremland et al. (2004) showed that microbial Se(0) can have unique structural and spectral features that vary not only from chemically derived materials, but also between bacterial species. Electron microscopy revealed that the Se(0) nanoparticles were encapsulated by a biologically derived exopolymer. Redox-active proteins in the form of *c*-type cytochromes and ferredoxin were also found to be associated with postreduction mineral phases during the biotransformation of Se oxyanions by *G. sulfurreducens* and may therefore be involved in the formation of Se(0) nanospheres

(Pearce et al., 2009). Microbially produced Te(0) has distinct characteristics not replicated by chemical methods (Baesman et al., 2007). Fourier transform infrared spectroscopy of Te(0) nanorods produced by *B. selenitireducens* revealed the presence of functional amide groups on the Te(0), suggesting the attachment of cell wall proteins to the Te(0) (Baesman et al., 2007). As-S nanotubes, formed via the reduction of As(V) and $S_2O_3^-$ by *Shewanella* sp. strain HN-41 were shown to be associated with EPS-containing polysaccharides (Lee et al., 2007). It is possible that specific proteins serve as templates for assembly/precipitation of Se, Te, and As as nanoparticles and that differences in protein structure could produce the geometric and spectral variations between nanoparticles formed by diverse prokaryotes that interact with oxyanions of these elements. What follows is a brief survey that gives some specific examples of the types and variety of microbially produced Se, Te, and As nanominerals.

CHARACTERISTICS OF ELEMENTAL Se AND Te BIONANOPARTICLES

A range of *Bacteria* and *Archaea* can link the oxidation of organic substrates or H_2 to the reduction of selenium oxyanions under a variety of conditions (Table 1). Despite the variety of reduction conditions, the end products of these reactions are predominantly red, amorphous, or monoclinic allotropes of Se(0), in the form of small (~100- to 200-nm) spheres (Fig. 1). The spheres in Fig. 1A were formed under anaerobic conditions at 25°C by the reduction of Na_2SeO_4 (3 mM) using growing cells of the Mono Lake isolate *B. selenitireducens*, in carbonate-bicarbonate-buffered media (pH 9.8, salinity 56 g/liter), containing lactate as the electron donor, with vitamins and yeast extract added as growth supplements (Oremland et al., 2004). *Bacillus* sp. NS3 cells, isolated from Punjab, India, grown in tryptone soy broth at neutral pH and 30°C, reduced Na_2SeO_4 (5mM) to form Se(0) nanospheres under aerobic conditions are shown in Fig. 1B (Prakash et al., 2009). "Resting" cells of the environmental isolate *G. sulfurreducens* at 30°C, and the medical isolate *V. atypica* at 37°C, in 3-(*N*-morpholino)propanesulfonic acid buffer (20 mmol × l^{-1}, pH 7.5), reduced Na_2SeO_4 (5mM) under anaerobic conditions at pH 7.5 formed the Se(0) nanospheres in Fig. 1C and D, respectively (Pearce et al., 2009). Although the morphologies of the Se(0) nanospheres are similar, Oremland et al. (2004) have shown that the Se molecular chain structural orientation and the optical properties of these materials can be very different, depending on the organism, possibly because of the involvement of different reductase enzymes for the different microbes. The spectral properties of these biogenic Se(0) materials are also substantially different from their chemically formed counterparts (Oremland et al., 2004). There is evidence that morphologies of Se(0) other that the red amorphous nanospheres in Fig. 1A to D can be also be produced by microbial communities that exist in extreme environments, such as the mineral-rich, highly alkaline (pH = 9.8) and hypersaline (salinity = ~75 to 90 g/liter) ecosystem of Mono Lake, California. Sediment (4 ml) collected from the hot spring area of Paoha Island, was added to 16 ml sterile, basal salts medium (Switzer Blum et al., 2009) at pH 9.3 and dispensed into serum bottles (30 ml bottle volume; 20 ml slurry volume) under anaerobic conditions. Sodium selenite (2 mM) was added to all bottles as an electron acceptor, and molecular hydrogen was provided as the electron donor. Killed controls were autoclaved (250 kPa, 121°C, 60 min). The slurries were incubated in the dark at 43°C. The live sediment slurries turned black within 24 h (Fig. 1E, inset), without progressing through an intermediary red phase. No such color change was observed in the killed controls. Scanning electron micrographs (Fig. 1E) show that the Se(0) precipitates formed after incubation of Se(IV)-amended sediment slurries with hydrogen have a plate-like structure and are relatively large (200 to 300 nm) compared with the amorphous red Se(0) nanospheres in Fig. 1A to D. Energy dispersive X-ray spectroscopy (EDS) confirmed that the observed

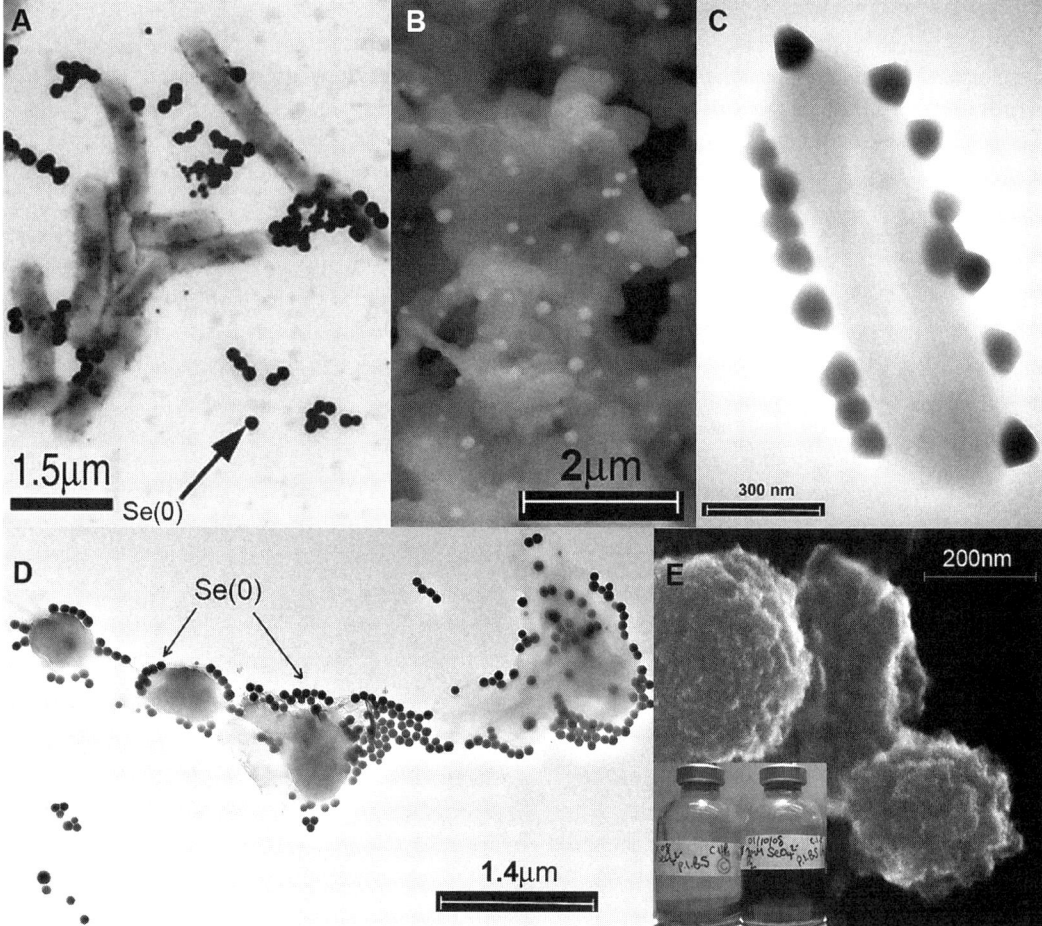

FIGURE 1 Electron micrographs of elemental Se(0) precipitates. (A) TEM image of nanospheres produced by *B. selenitireducens*. (Image taken by Sean Langley.) (B) SEM image of nanospheres produced by *Bacillus* sp. strain NS3. (C) TEM image of nanospheres produced by *G. sulfurreducens*. (D) TEM image of nanospheres produced by *V. atypica*. (E) SEM image of nanoplatelets produced by indigenous microbes present in Mono Lake sediments (inset shows black precipitate production).

color change was the result of the development of Se(0) precipitates.

The microbial diversity of Se-reducing prokaryotes, employing a correspondingly vast range of enzymatic reactions to form Se(0), offers the intriguing potential to tailor the properties of biologically based Se bionanominerals for specific applications in the field of nanotechnology. Relatively facile washing treatments employed to remove the biomass from microbially produced Se(0) nanospheres have also resulted in the production of novel Se(0) phases, including clusters of hexagonal nanorods, further increasing the "toolbox" of Se(0) nanomaterials that can be made available from inexpensive starting materials and low cost bio-manufacturing procedures (Prakash et al., 2009). Figure 2 shows how Se(0) nanospheres (Fig. 1B) produced by either growing cells of *Bacillus* sp. strain NS3 in tryptone soy broth or by "resting" cells of *Bacillus* sp. strain NS3 in 3-(*N*-morpholino)propanesulfonic acid buffer

(20 mmol × l⁻¹, pH 7.5), can be manipulated by postpreparative acetone-washing procedures to form structures with completely different sizes and shapes (Fig. 2A and B).

In contrast to Se-oxyanion reduction, the reduction of Te oxyanions results in the production of Te(0) with a range of basic morphologies including Te rosettes, Te granules, Te nanospheres, and Te nanorods, the latter being the most commonly formed precipitate. Figure 3 shows examples of some of these different morphologies. The medical isolate *V. atypica* cannot use Te oxyanions as terminal electron acceptors for anaerobic respiration. However, this organism is capable of forming ~100-nm Te(0) nanorods, which can be seen protruding from the spherical cells, as well as aggregating into extracellular clusters (Fig. 3A), when grown under anaerobic conditions in the presence of Te(IV) at 37°C in a rich medium containing yeast extract and lactate. The freshwater isolate *S. barnesii* forms both intracellular and extracellular irregularly shaped, ~20-nm-sized nanospheres of crystalline Te(0) that coalesce together to form larger, ~500- to 1,000-nm clusters attached to the cell surface when using pulsed additions of Te(VI) as a terminal electron acceptor for anaerobic respiration, in a defined medium containing lactate at 28°C (Fig. 3B). The Te(0) precipitates produced by the haloalkaliphilic Mono Lake isolates, *B. beveridgei* strain MLTeJB (Fig. 3C) and *B. selenitireducens*, during Te(IV)-dependent growth in a basal salts medium containing yeast extract and lactate at 28°C, are rod shaped but are much larger than those formed by *V. atypica* (Fig. 3A). It is possible that either the differences in Te(0) morphology, shown in Fig. 3, can be explained in terms of rate of reduction (reduction rate for *V. atypica* > *S. barnesii* > *B. selenitireducens*) or the Te(0) morphology may be controlled by different rate-limiting steps along the specific reduction pathways; Te(0) production by *S. barnesii* is a two-step process involving reduction of Te(VI) to Te(IV), followed by reduction of the latter to Te(0). Despite the differences in external morphology, analysis of various Te(0) precipitates using Raman spectroscopy revealed that they have similar internal structures, possibly as a result of the fundamental trigonal alignment features of the Te chains (Baesman et al., 2007). This is in contrast to biogenic Se nanospheres, which have a similar outward appearance but differ with respect to their spectral properties.

Control of biogenic Te(0) precipitate morphology is possible, not only through the application of different Te-oxyanion-reducing organisms, but also via alteration of the growth conditions for a particular organism. Figure 4 shows the Te(0) precipitates formed by the subsurface isolate *G. sulfurreducens*. In an attempt to determine whether *G. sulfurreducens* could couple growth to the reduction of Te(IV), cells were grown in an acetate-amended modified freshwater medium with 1 mM Te(IV) as the only available terminal electron acceptor. The number of cells did not increase over time, indicating that *G. sulfurreducens* was unable to use Te(IV) as a terminal electron acceptor for growth; however, reduction of Te(IV) by the organism was revealed by the formation of a fine, black precipitate (Fig. 4A, inset). The whole mount transmission electron microscopy (TEM) image of the Te(0) precipitates produced by *G. sulfurreducens* under these "growing" conditions (Fig. 4A) shows ~100-nm-sized nanospheres on the cell surface. Confirmation of the Te(0) composition of the nanophase produced was confirmed by EDS (data not shown). High-resolution TEM of these nanospheres (Fig. 4B) reveals that they are composed of much smaller (<10 nm) particles. In contrast, nongrowing or "resting" anaerobic cells of *G. sulfurreducens*, resuspended in tricine buffer and exposed to 1 mM Te (IV), in the presence of anthraquinone disulfonic acid as an electron shuttle with hydrogen as the electron donor at 30°C, produced a dense, black precipitate (Fig. 4C, inset). The scanning electron microscopy (SEM) and high-resolution TEM images of these precipitates in Fig. 4C clearly show an entirely different morphology consisting of ~100-nm-sized nanorods of Te(0). The composition and highly crystalline nature of the Te(0) nanorods

FIGURE 2 Electron micrographs of elemental Se precipitates produced by *Bacillus* sp. strain NS3. (A) SEM image of solvent-washed Se precipitates produced by "growing" cells. (B) SEM image of solvent-washed Se precipitates produced by "resting" cells and (C) a representative EDX spectrum.

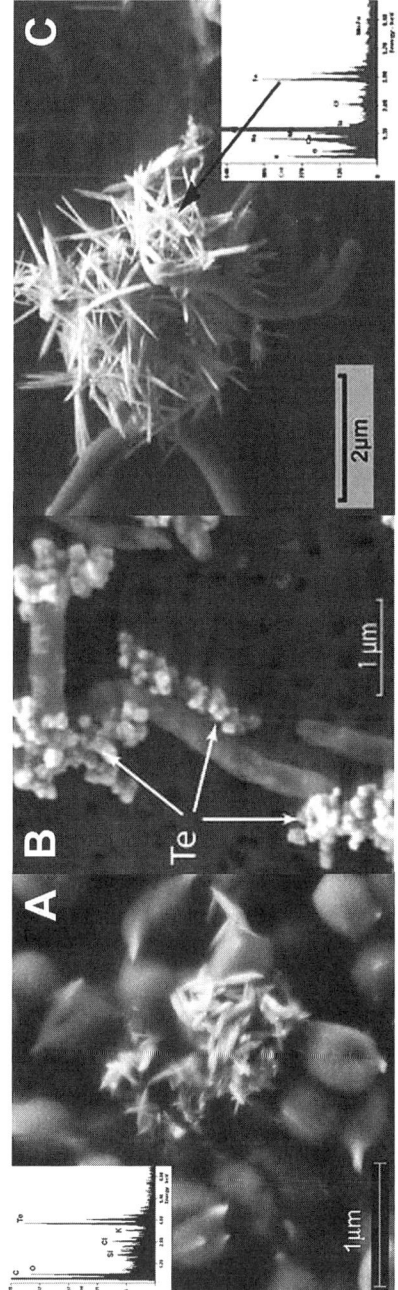

FIGURE 3 Electron micrographs of elemental Te precipitates. (A) SEM images of Te nanorods produced by *V. atypica*. (B) SEM images of Te nanogranules produced by *S. barnesii*. (C) SEM images of Te nanorods produced by *B. beveridgei* strain MLTeJB (inset shows EDX spectrum).

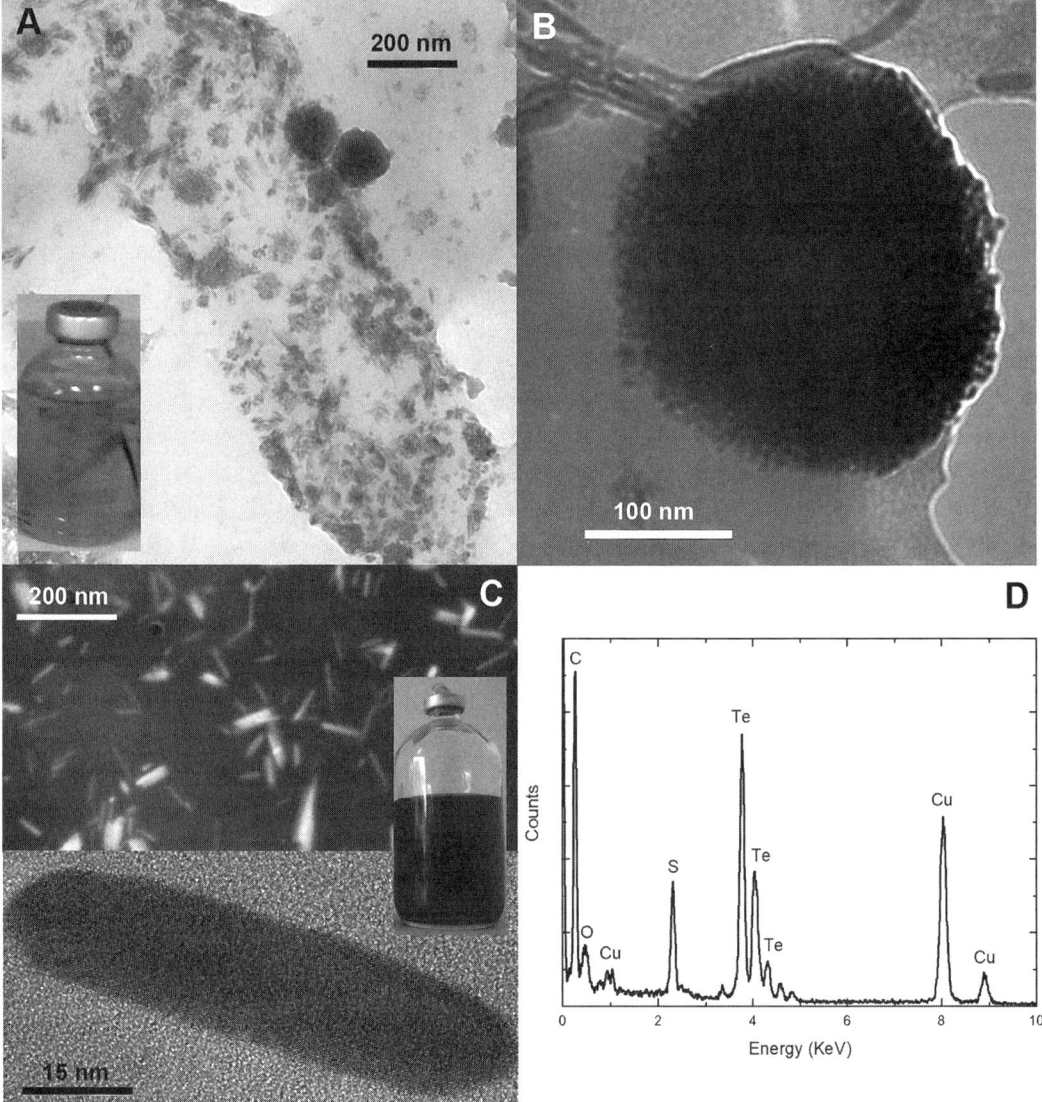

FIGURE 4 Electron micrographs of elemental Te precipitates produced by *G. sulfurreducens*. (A) TEM image showing whole cell "grown" in the presence of tellurate (inset shows limited precipitate production). (B) High-resolution TEM image of Te nanospheres. (C and D) SEM image and high-resolution TEM image of Te nanorods formed by "resting" cells (inset shows extensive precipitate production) with accompanying EDS spectrum.

were confirmed by EDS (Fig. 4D) and high-resolution TEM (Fig. 4C) respectively. The presence of sulfur in association with the Te(0) nanorods could result from thiol groups on the surface, indicating a role for the glutathione and thioredoxin redox system in Te(IV) reduction by *G. sulfurreducens* (Turner et al., 2001). Baesman et al. (2007) also noted differences between the Te(0) nanorods formed by cell suspensions of *B. selenitireducens* and those formed during growth experiments. This was attributed to the fact that cell wall proteins remained firmly attached to the Te(0) nanorods produced by the cell suspensions, but this strong attachment of the Te(0) nanorods to the cell surface was eliminated by the presence of cysteine in the medium for the growing cells.

As shown in Table 1, many organisms are capable of reducing both Se(IV) and Te(IV); however, only the *B. beveridgei* strain MLTeJB, along with certain isolates from marine hydrothermal vents (Rathgeber et al., 2002), are capable of reducing significant quantities of both of these oxyanions. During the study of the growth of strain MLTeJB under different electron donor and acceptor conditions, it was found that cells grew on up to 10 mM Te(IV), with excess lactate, and reduced it to Te(0) as the sole reduction product. This result contrasts with growth on Se(IV), where Se(−II) was observed as an end product, as well as Se(0). Subsequent addition of Se(IV) (5 mM) to the reduced suspension of cells and Te(0) resulted in reduction to Se(0) and Se(−II), along with the development of a yellow color in the media over time (Fig. 5C, inset). Figure 5A shows cells of strain MLTeJB, grown in the presence of Te(IV) and Se(IV), encrusted with mixed Se/Te precipitate. The yellow solution was separated from the biomass mass by filtration though a 0.22-μm filter and then centrifuged at 30,000 × g for 5 min to remove any remaining nanoparticles. Oxidation of the yellow solution resulted in the formation of a dark-red precipitate (Fig. 5C inset). SEM images of this precipitate showed that it had nanospherical morphology and was composed of both Se and Te (Fig. 5B and C). These results suggest that, despite the lack of Te(−II) formation during reduction of Te(IV), strain MLTeJB is able to produce a reduced soluble Te phase under certain circumstances, i.e., in the presence of a mixture of Te(0), Se(0), and Se(−II). The mechanism of production of this reduced Se/Te phase is not clear, but it is possible that the addition of Se(IV) to the Te(0) suspension stimulated the strain MLTeJB cells to produce a reducing environment in which the further reduction of Te(0) to Te(−II) was possible.

CHARACTERISTICS OF MSe (M = Cd, Zn) BIONANOPARTICLES

For certain organisms, Se(0) is not the end of the reduction pathway for Se oxyanions, and they can mediate the biomethylation and volatilization of Se(0) to form organo-Se

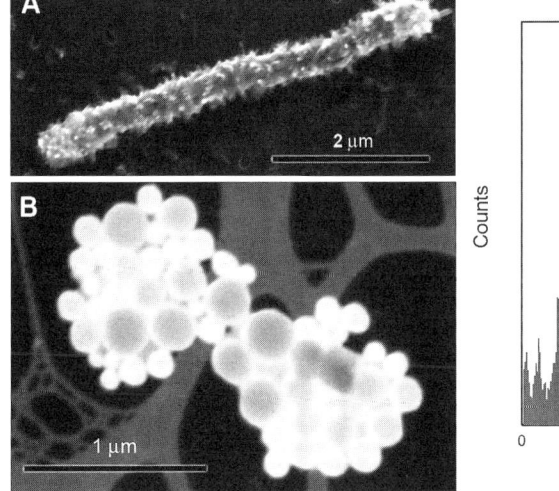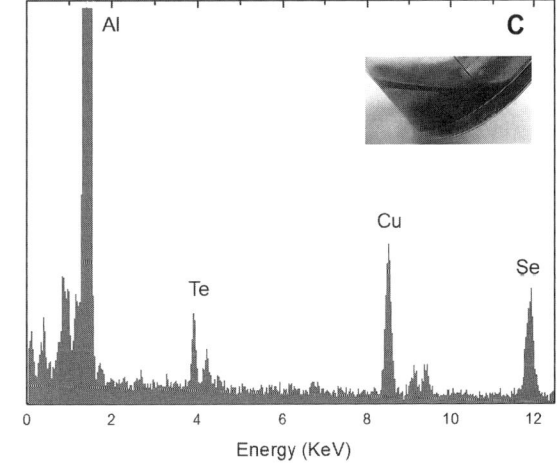

FIGURE 5 Electron micrographs of mixed Se/Te precipitate formed by *B. beveridgei* strain MLTeJB grown in the presence of tellurite and selenite (inset shows black precipitate; a color illustration would show a yellow solution). (A) SEM image showing whole cell encrusted with Se/Te precipitate. (B) TEM image showing mixed Se/Te nanosphere precipitates produced upon oxidation of (yellow) solution and (C) accompanying EDS spectrum. (The inset shows Se/Te precipitate, which would be red in a color illustration.)

compounds with the selenium present as Se(–II) (Dungan et al., 2003; Herbel et al., 2003). A limited number of organisms can also extend the reduction pathway beyond Se(0) to form an aqueous selenide end product. Over a period of several hours, *V. atypica* cells can entirely reduce 5 mM Se(IV) to the soluble Se(–II), present as dissociated HSe– surrounded by water molecules (Pearce et al., 2008). This biogenic Se(–II) solution is produced under ambient conditions, as opposed to the hazardous, expensive production of NaHSe from Al_2Se_3, and can be employed in an aqueous-based, wet chemical synthesis for the fabrication of CdSe/ZnSe quantum dots (Pearce et al., 2008). Precipitation of the biogenic selenide solution with a suitable metal cation, such as $ZnCl_2$, results in the formation of ~30-nm ZnSe particles. These particles are too large for quantum dot-type applications, because semiconductor quantum dots exhibit tunable optical properties as a result of quantum confinement in the regime below 10 nm (Alivisatos, 1996). However, the employment of simple thiol molecules, such as 2-mercaptoethanol or reduced glutathione, as agents to stabilize the metal cation during synthesis results in the formation of stable metal selenide colloids which, upon heating to 110°C for 2 h, produce quantum-sized (~5 nm), luminescent metal selenide nanoparticles. Color Plate 17A shows the fluorescence spectrum of size-fractionated glutathione-stabilized CdSe quantum dots, illustrating how quantum dots with different emission wavelengths can be formed via a postpreparative size-selective precipitation, involving a nonsolvent such as 2-propanol. Confirmation of the crystallinity and chemical purity of the CdSe quantum dots is provided by the high-resolution TEM image and EDS spectrum in Color Plate 17B.

CHARACTERISTICS OF As_xS_y BIONANOPARTICLES

Arsenic sulfides are important infrared transparent materials for a variety of applications such as sensors, waveguides, photonic crystals, and photolithography template materials (Johnson et al., 2005). The fabrication of As_xS_y in the form of nanowires paves the pathway for application in advanced integrated nanophotonic structures and devices. The recent discovery of microbial production of filamentous arsenic-sulfide (As-S) nanowires by *Shewanella* sp. strain HN-41 (Lee et al., 2007), grown in the presence of As(V) and S_2O_3 under anaerobic conditions, again offers an environmentally friendly alternative to present high-temperature fabrication routes. Figure 6A shows an electron micrograph of the yellow biogenic As-S precipitate (Fig. 6A, inset) produced by strain HN-41, grown in anaerobic basal medium at pH 7.5, supplemented with 20 mM lactate, 10 mM thiosulfate, and 5 mM arsenate at 30°C, revealing that they are long (up to ~30 μm), variable diameter (20 to 100 nm), extracellular As-S filamentous nanotubes. These biogenic As-S precipitates were shown to behave as metals and semiconductors in terms of their electrical properties and were also photoconductive (Lee et al., 2007).

Investigations into other organisms capable of growing in environments with high arsenic and sulfide concentrations have revealed further examples of biogenic As-S nanowire formation. Strain TSA-1, a *Citrobacter* species isolated from the hindgut of the subterranean wood-feeding termite *Reticulitermes flavipes* (Kollar), grows in a freshwater mesophilic basal salts medium (Oremland et al., 1994) with cysteine-sulfide present as a reducing agent, and was originally isolated based on its Se(IV)-respiring ability. The organism was later found to have the ability to grow via H_2-driven respiration of As(V) as part of a study to evaluate arsenic speciation within the gastrointestinal tract (Herbel et al., 2002), with As (III) determined to be the product of this respiration. High-performance liquid chromatography analysis revealed the presence of a yellow thioarsenite precipitate (Color Plate 18A, inset) within the culture medium resulting from the presence of both arsenite (5 mM) and cysteine sulfide (0.5 mM). Electron micrographs (Color Plate 18A to C) showed that the As-S tube-like precipitates are several

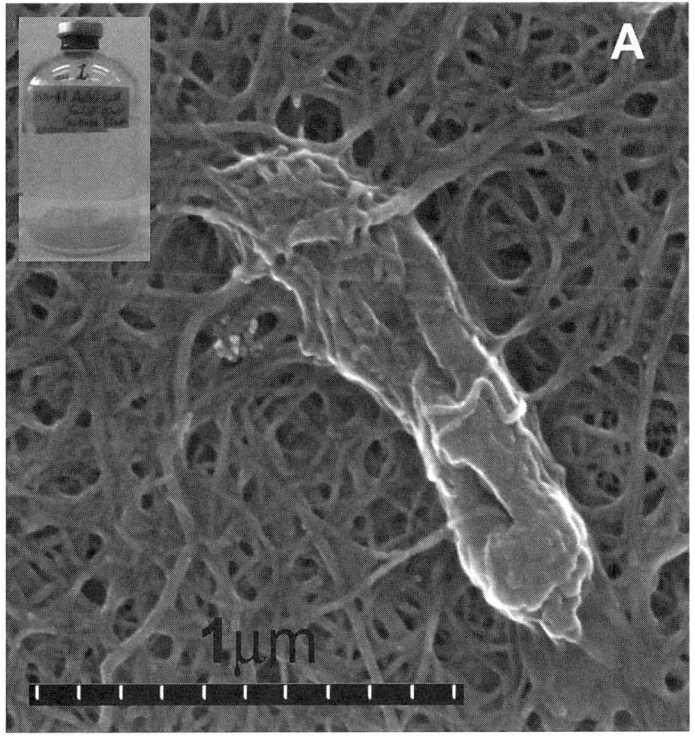

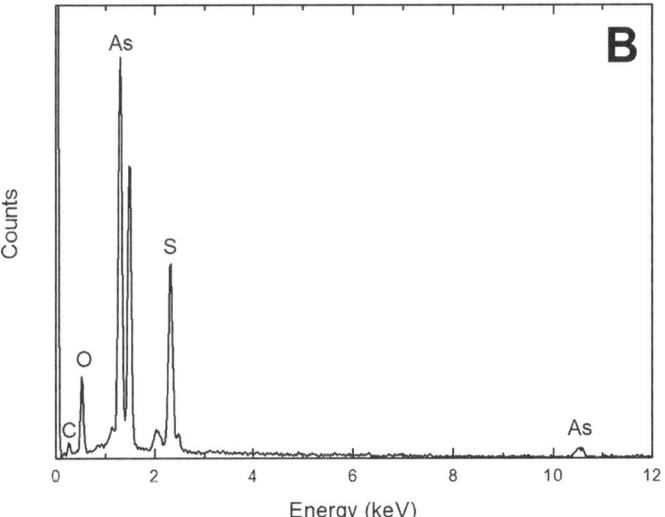

FIGURE 6 Electron micrographs of biogenic As-S precipitates. (A) SEM image of As-S nanotubes produced by *Shewanella* strain HN-41 (inset shows precipitate, which would be yellow in a color illustration) and (B) accompanying EDS spectrum (Lee et al., 2007).

micrometers long with a diameter of <100 nm and are composed of very small (~10 nm) crystallites. Scanning transmission electron microscope-EDS mapping images and spectrum (Color Plate 18C and D) confirmed the composition of the As-S nanotubes, with an As:S ratio suggesting that orpiment (As_2S_3) was the dominant phase. A mixed enrichment culture obtained from sediments of Searles Lake, California, using an extremely halophilic (salinity, 346 g/liter) and alkaliphilic (pH 9.5) culture medium (Searles

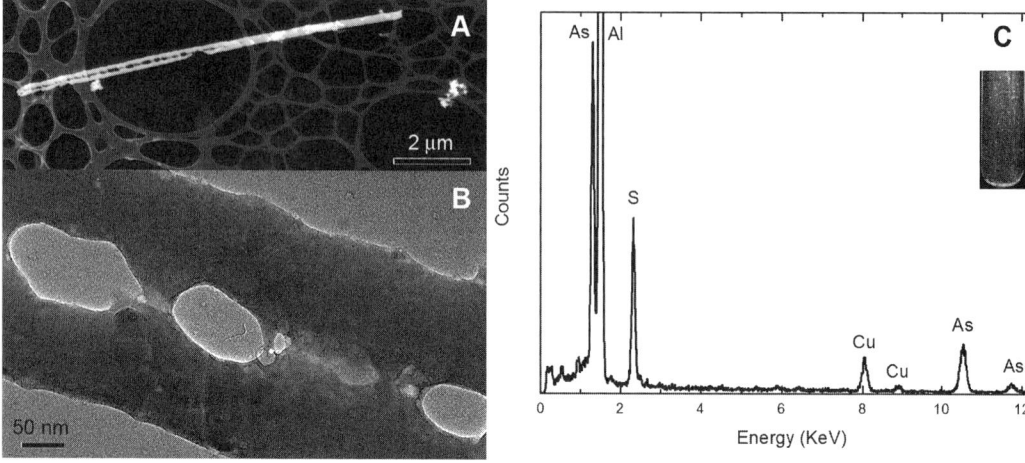

FIGURE 7 Electron micrographs of As-S precipitate formed by Searles Lake mixed antibiotic-fed enrichment culture (Ab-1) grown in the presence of arsenate and sulfate (inset shows precipitate, which would be yellow in a color illustration). (A) SEM image. (B) TEM image. (C) Accompanying EDS spectrum.

Lake Ab1) was also investigated in terms As(V) respiration and S(VI) reduction in extreme environments. This particular antibiotic-containing enrichment was, by design, a stratagem for the isolation of novel species from the domain *Archaea* with the ability to respire arsenate (Switzer Blum et al., 2009). An unexpected result of the antibiotic amendment was the proliferation in the culture of an S(VI)-reducing bacterium. The appearance of both an arsenate respirer and a sulfate reducer during the growth of this culture resulted in the formation of both arsenite (5 mM) and high levels of sulfide (5 to 10 mM). As(III) levels decreased at the end of the exponential phase of growth as thioarsenite species were formed. Figure 7C (inset) shows that the (yellow) precipitate formed under these conditions was substantially less dense than that observed with *Shewanella* sp. strain HN-41 (Fig. 6A, inset) and *Citrobacter* strain TSA-1 (Color Plate 18A, inset). This is to be expected when taking into consideration the pH of 9.5, at which thioarsenite species are expected to be soluble, with precipitation occurring under these circumstances as a result of the high arsenite and sulfide concentrations in the media. The less abundant As-S precipitates formed by Searles Lake Ab1 (Fig. 7A and B) were less filamentous and more structured than those formed by *Citrobacter* strain TSA-1 (Color Plate 18A), adopting a "nanoladder" morphology (Fig. 7A and 7B). The chemical composition of the As-S "nanoladders" was confirmed by EDS (Fig. 7C).

CONCLUSIONS

The ability of a large range of microbial species to produce nanoparticles relevant to one of the world's most important technological frontiers makes their continued study of both intrinsic and commercial interest. In this chapter, we have presented some examples of these nanoparticles formed by only a few microbial species that are cultivated in only a handful of laboratories worldwide. Thus, the investigations so far have just scratched the surface of the potential of the natural world to yield bionanomineral producers. Indeed, examples of nanoscaled precipitates of group XV and XVI elements produced by representatives of the domain *Archaea* are entirely lacking. While future research should involve screening surveys of the prokaryotes for this biomineralizing phenomenon, more detailed investigations are justified. These should include more thorough physiological and biochemical investigations with known species, including variations in growth conditions and postpreparative techniques for the harvested nanoparticles. The ability to genetically manipulate the organisms (e.g., via knockout mutagenesis), especially if conducted with microbes having undergone full genomic sequencing and annotation, also holds promise to tune the fabrication of an enormous range of different nanoparticles by these novel biological pathways. The initial results highlighted here begin to reveal this potential for using a novel biological approach to produce functional nanoparticles, in what is an environmentally friendly methodology. A wide range of Se and Te nanostructures have already been identified that can be matched to several potential uses. The precipitation from bioselenide of capped, luminescent, semiconducting CdSe with chemical, optical, and electronic properties governed by composition, size, and shape takes this a step further, and direct comparisons with inorganically produced, similar materials can be made. The studies have also highlighted that novel biogenic fabrication routes can produce functional bielemental materials such as As-S nanostructures, for instance, for use in the construction of the next generation of nanoscale optoelectronic materials. However, the potential of bionanoparticle fabrication is largely untapped. Nonetheless, once novel Se, Te, and As bionanoparticles are identified as having significant technical applications, applied research into their practical commercial production will without doubt ensue rapidly.

ACKNOWLEDGMENTS

The work described in this review was supported by funding from Research Councils UK, NERC, USGS, and NASA. Portions of this work were performed at the Molecular Foundry, Lawrence Berkeley National Laboratory, which is supported by the Office of Science, Office of Basic Energy Sciences, U.S. Department of Energy, under contract no. DE-AC02-05CH11231, with special thanks to R. Zuckermann (Biological Nanostructures Facility) and V. Altoe (Imaging and Manipulation of Nanostructures Facility). We thank N. Hylton for his assistance with photoluminescence measurements.

We are grateful to J.-H. Lee for providing the images in Figure 7 and T. Prakash for invaluable discussions on the production of Se-based nanomaterials. We also thank R. A. D. Pattrick and A. E. Plymale for their constructive comments on the manuscript.

REFERENCES

Ahmed, M. F., S. Ahuja, M. Alauddin, S. J. Hug, J. R. Lloyd, A. Pfaff, T. Pichler, C. Saltikov, M. Stute, and A. van Geen. 2006. Ensuring safe drinking water in Bangladesh. *Nature* **314:**1687.

Alivisatos, P. 1996. Semiconductor clusters, nanocrystals, and quantum dots. *Science* **271:**933–937.

Baesman, S. M., T. D. Bullen, J. Dewald, D. H. Zhang, S. Curran, F. S. Islam, T. J. Beveridge, and R. S. Oremland. 2007. Formation of tellurium nanocrystals during anaerobic growth of bacteria that use Te oxyanions as respiratory electron acceptors. *Appl. Environ. Microbiol.* **73:**2135–2143.

Baesman, S. M., J. F. Stolz, T. R. Kulp, and R. S. Oremland. 2009. Enrichment and isolation of *Bacillus beveridgei* sp. nov., a facultative anaerobic haloalkaliphile from Mono Lake, California that respires oxyanions of tellurium, selenium, and arsenic. *Extremophiles* **13:**695–705.

Basnayake, R. S. T., J. H. Bius, O. M. Akpolat, and T. G. Chasteen. 2001. Production of dimethyl telluride and elemental tellurium by bacteria amended with tellurite or tellurate. *Appl. Organomet. Chem.* **15:**499–510.

Bhattacharjee, H., and B. P. Rosen (ed.). 2007. *Arsenic Metabolism in Prokaryotic and Eukaryotic Microbes*, vol. 6. Springer-Verlag, Berlin, Germany.

Blum, J. S., A. B. Bindi, J. Buzzelli, J. F. Stolz, and R. S. Oremland. 1998. *Bacillus arsenicoselenatis*, sp nov, and *Bacillus selenitireducens*, sp nov: two haloalkaliphiles from Mono Lake, California that respire oxyanions of selenium and arsenic. *Arch. Microbiol.* **171:**19–30.

Blum, J. S., J. F. Stolz, A. Oren, and R. S. Oremland. 2001. *Selenihalanaerobacter shriftii* gen. nov., sp nov., a halophilic anaerobe from Dead Sea sediments that respires selenate. *Arch. Microbiol.* **175:**208–219.

Bock, A., K. Forchhammer, J. Heider, W. Leinfelder, G. Sawers, B. Veprek, and F. Zinoni. 1991. Selenocysteine—the 21st amino-acid. *Mol. Microbiol.* **5:**515–520.

Budinoff, C. R., and J. T. Hollibaugh. 2008. Arsenite-dependent photoautotrophy by an *Ectothiorhodospira*-dominated consortium. *ISME J.* **2:**340–343.

Chasteen, T. G., D. E. Fuentes, J. C. Tantaleán, and C. C. Vásquez. 2009. Tellurite: history, oxidative stress, and molecular mechanisms of resistance. *FEMS Microbiol. Rev.* **33:**820–832.

Cortese, M. S., A. Paszczynski, T. A. Lewis, J. L. Sebat, V. Borek, and R. L. Crawford. 2002. Metal chelating properties of pyridine-2,6-bis(thiocarboxylic acid) produced by Pseudomonas spp. and the biological activities of the formed complexes. *Biometals* **15:**103–120.

Csotonyi, J. T., E. Stackebrandt, and V. Yurkov. 2006. Anaerobic respiration on tellurate and other metalloids in bacteria from hydrothermal vent fields in the eastern Pacific Ocean. *Appl. Environ. Microbiol.* **72:**4950–4956.

Dembitsky, V. M., and D. O. Levitsky. 2004. Arsenolipids. *Prog. Lipid Res.* **43:**403–448.

Dettmer, R. 1988. The Quest for the quantum dot. *IEE Rev.* **34:**395–397.

Dopp, E., L. M. Hartmann, A. M. Florea, U. von Recklinghausen, R. Pieper, B. Shokouhi, A. W. Rettenmeier, A. V. Hirner, and G. Obe. 2004. Uptake of inorganic and organic derivatives of arsenic associated with induced cytotoxic and genotoxic effects in Chinese hamster ovary (CHO) cells. *Toxicol. Appl. Pharmacol.* **201:**156–165.

Doran, J. W., and M. Alexander. 1977. Microbial transformations of selenium. *Appl. Environ. Microbiol.* **33:**31–37.

Dowdle, P. R., and R. S. Oremland. 1998. Microbial oxidation of elemental selenium in soil slurries and bacterial cultures. *Environ. Sci. Technol.* **32:**3749–3755.

Dungan, R. S., S. R. Yates, and W. T. Frankenberger. 2003. Transformations of selenate and selenite by *Stenotrophomonas maltophilia* isolated from a seleniferous agricultural drainage pond sediment. *Environ. Microbiol.* **5:**287–295.

Ehrlich, H. L. 2002. *Geomicrobiology*, 4th ed. Marcel Dekker, Inc., New York, NY.

Fleming, R. W., and M. Alexander. 1972. Dimethylselenide and dimethyltelluride formation by a strain of *Penicillium*. *Appl. Microbiol.* **24:**424–429.

Gao, X., J. Zhang, and L. Zhang. 2002. Hollow sphere selenium nanoparticles: their in-vitro anti hydroxyl radical effect. *Adv. Mater.* **14:**290–293.

Giepmans, B. N. G., S. R. Adams, M. H. Ellisman, and R. Y. Tsien. 2006. The fluorescent toolbox for assessing protein location and function. *Science* **312:**217–224.

Herbel, M. J., J. S. Blum, R. S. Oremland, and S. E. Borglin. 2003. Reduction of elemental selenium to selenide: experiments with anoxic sediments and bacteria that respire Se-oxyanions. *Geomicrobiol. J.* **20:**587–602.

Herbel, M. J., J. Switzer Blum, S. E. Hoeft, S. M. Cohen, L. L. Arnold, J. Lisak, J. F. Stolz, and R. S. Oremland. 2002. Dissimilatory arsenate reductase activity and arsenate-respiring bacteria in bovine rumen fluid, hamster feces, and the termite hindgut. *FEMS Microbiol. Ecol.* **41:**59–67.

Hockin, S., and G. M. Gadd. 2006. Removal of selenate from sulfate-containing media by sulfate-reducing bacterial biofilms. *Environ. Microbiol.* **8:**816–826.

Hockin, S. L., and G. M. Gadd. 2003. Linked redox precipitation of sulfur and selenium under anaerobic conditions by sulfate-reducing bacterial biofilms. *Appl. Environ. Microbiol.* **69:**7063–7072.

Hoeft, S. E., J. Switzer Blum, J. F. Stolz, F. R. Tabita, B. Witte, G. M. King, J. M. Santini, and R. S. Oremland. 2007. *Alkalilimnicola ehrlichii*, sp. nov., a novel, arsenite-oxidizing haloalkaliphilic γ-Proteobacterium capable of chemoautotrophic or heterotrophic growth with nitrate or oxygen as the electron acceptor. *Int. J. Syst. Evol. Microbiol.* **57:**504–512.

Hollibaugh, J. T., C. Budinoff, R. A. Hollibaugh, B. Ransom, and N. Bano. 2006. Sulfide oxidation coupled to arsenate reduction by a diverse microbial community in a soda lake. *Appl. Environ. Microbiol.* **72:**2043–2049.

Huber, R., M. Sacher, A. Vollmann, H. Huber, and D. Rose. 2000. Respiration of arsenate and selenate by hyperthermophilic archaea. *Syst. Appl. Microbiol.* **23:**305–314.

Johnson, B. R., M. J. Schweiger, and S. K. Sundaram. 2005. Chalcogenide nanowires by evaporation-condensation. *J. Non-Crystalline Solids* **351:**1410–1416.

Johnson, N. C., S. Manchester, L. Sarin, Y. M. Gao, I. Kulaots, and R. H. Hurt. 2008. Mercury vapor release from broken compact fluorescent lamps and in situ capture by new nanomaterial sorbents. *Environ. Sci. Technol.* **42:**5772–5778.

Jones, J. B., G. L. Dilworth, and T. C. Stadtman. 1979. Occurrence of selenocysteine in the selenium-dependent formate dehydrogenase of *Methanococcus vannielii*. *Arch. Biochem. Biophys.* **195:**255–260.

Kessi, J., M. Ramuz, E. Wehrli, M. Spycher, and R. Bachofen. 1999. Reduction of selenite and detoxification of elemental selenium by the phototrophic bacterium *Rhodospirillum rubrum*. *Appl. Environ. Microbiol.* **65:**4734–4740.

Klett, A. 1900. Zur kenntniss der reducirenden eigenschaften der bakterian. *Z. Hyg. Infektionskr.* **33:**137.

Klonowska, A., T. Heulin, and A. Vermeglio. 2005. Selenite and tellurite reduction by *Shewanella oneidensis*. *Appl. Environ. Microbiol.* **71:**5607–5609.

Knight, V. V., and R. Blakemore. 1998. Reduction of diverse electron acceptors by *Aeromonas hydrophila*. *Arch. Microbiol.* **169:**239–248.

Krafft, T., A. Bowen, F. Theis, and J. M. Macy. 2000. Cloning and sequencing of the genes encoding the periplasmic-cyctochrome B-containing selenate reductase for *Thauera selenatis*. *DNA Seq.* **10**:365–377.

Kulp, T. R., S. E. Hoeft, M. Asao, M. T. Madigan, J. T. Hollibaugh, J. C. Fisher, J. F. Stolz, C. W. Culbertson, L. G. Miller, and R. S. Oremland. 2008. Arsenic(III) fuels anoxygenic photosynthesis in hot spring biofilms from Mono Lake, California. *Science* **321**:967–970.

Leaver, J. T., D. J. Richardson, and C. S. Butler. 2008. *Enterobacter cloacea* SLD1a-1 gains a selective advantage from selenate reduction when growing in nitrate-depleted anaerobic environment. *J. Ind. Microbiol. Biotechnol.* **35**:863–873.

Ledbetter, R. N., S. A. Connon, A. L. Neal, A. Dohnalkova, and T. S. Magnuson. 2007. Biogenic mineral production by a novel arsenic-metabolizing thermophilic bacterium from the Alvord Basin, Oregon. *Appl. Environ. Microbiol.* **73**:5928–5936.

Lee, J. H., M. G. Kim, B. Y. Yoo, N. V. Myung, J. S. Maeng, T. Lee, A. C. Dohnalkova, J. K. Fredrickson, M. J. Sadowsky, and H. G. Hur. 2007. Biogenic formation of photoactive arsenic-sulfide nanotubes by *Shewanella* sp. strain HN-41. *Proc. Natl. Acad. Sci. USA* **104**:20410–20415.

Lemly, A. D. 2004. Aquatic selenium pollution is a global environmental safety issue. *Ecotoxicol. Environ. Saf.* **59**:44–56.

Lenz, M., E. D. Van Hullebusch, G. Hommes, P. F. X. Corvini, and P. N. L. Lens. 2008. Selenate removal in methanogenic and sulfate-reducing upflow anaerobic sludge bed reactors. *Water Res.* **42**:2184–2194.

Lloyd, J. R., A. N. Mabbett, D. R. Williams, and L. E. Macaskie. 2001. Metal reduction by sulphate-reducing bacteria: physiological diversity and metal specificity. *Hydrometallurgy* **59**:327–337.

Lortie, L., W. D. Gould, S. Rajan, R. G. L. Mccready, and K. J. Cheng. 1992. Reduction of selenate and selenite to elemental selenium by a *Pseudomonas stutzeri* isolate. *Appl. Environ. Microbiol.* **58**:4042–4044.

Losi, M. E., and W. T. Frankenberger, Jr. 1997. Reduction of selenium oxyanions by *Enterobacter cloacae* SLD1a-1: isolation and growth of the bacterium and its expulsion of selenium particles. *Appl. Environ. Microbiol.* **63**:3079–3084.

Ma, J., D. Y. Kobayashi, and N. Yee. 2007. Chemical kinetic and molecular genetic study of selenium oxyanion reduction by *Enterobacter cloacae* SLD1a-1. *Environ. Sci. Technol.* **41**:7795–7801.

Ma, J., D. Y. Kobayashi, and N. Yee. 2009. Role of menaquinone biosynthesis genes in selenate reduction by *Enterobacter cloacae* SLD1a-1 and *Escherichia coli* K12. *Environ. Microbiol.* **11**:149–158.

Macy, J. M., K. Nunan, K. D. Hagen, D. R. Dixon, P. J. Harbour, M. Cahill, and L. I. Sly. 1996. *Chrysiogenes arsenatis* gen. nov., sp. nov., a new arsenate-respiring bacterium isolated from gold mine wastewater. *Int. J. Syst. Bacteriol.* **46**:1153–1157.

Macy, J. M., S. Rech, G. Auling, M. Dorsch, E. Stackebrandt, and L. I. Sly. 1993. *Thauera selenatis* gen. nov, sp. nov, a member of the Beta-subclass of Proteobacteria with a novel type of anaerobic respiration. *Int. J. Syst. Bacteriol.* **43**:135–142.

Maher, M. J., J. Santini, I. J. Pickering, R. Prince, J. M. Macy, and G. N. George. 2004. X-ray absorption spectroscopy of selenate reductase. *Inorg. Chem.* **43**:402–404.

Masscheleyn, P. H., R. D. Delaune, and W. H. Patrick. 1990. Transformations of selenium as affected by sediment oxidation reduction potential and pH. *Environ. Sci. Technol.* **24**:91–96.

McEwan, A. G., J. P. Ridge, C. A. McDevitt, and P. Hugenholtz. 2002. The DMSO reductase family of microbial molybdenum enzymes: molecular properties and role in the dissimilatory reduction of toxic elements. *Geomicrobiol. J.* **19**:3–22.

Moore, M. D., and S. Kaplan. 1992. Identification of intrinsic high-level resistance to rare-earth-oxides and oxyanions in members of the class Proteobacteria—characterization of tellurite, selenite, and rhodium sesquioxide reduction in *Rhodobacter sphaeroides*. *J. Bacteriol.* **174**:1505–1514.

Moscoso, H., C. Saavedra, C. Loyola, S. Pichuantes, and C. Vásquez. 1998. Biochemical characterization of tellurite-reducing activities of *Bacillus stearothermophilus* V. *Res. Microbiol.* **149**:389–397.

Nath, S., S. K. Ghosh, S. Panigahi, T. Thundat, and T. Pal. 2004. Synthesis of selenium nanoparticle and its photocatalytic application for decolorization of methylene blue under UV irradiation. *Langmuir* **20**:7880–7883.

Newman, D. K., D. Ahmann, and F. M. M. Morel. 1998. A brief review of microbial arsenate reduction. *Geomicrobiol. J.* **15**:255–268.

Newman, D. K., E. K. Kennedy, J. D. Coates, D. Ahmann, D. J. Ellis, D. R. Lovley, and F. M. Morel. 1997. Dissimilatory arsenate and sulfate reduction in *Desulfotomaculum auripigmentum* sp. nov. *Arch. Microbiol.* **168**:380–388.

Oremland, R. S., J. S. Blum, C. W. Culbertson, P. T. Visscher, L. G. Miller, P. Dowdle, and F. E. Strohmaier. 1994. Isolation, growth, and metabolism of an obligately anaerobic, selenate-respiring bacterium, strain SES-3. *Appl. Environ. Microbiol.* **60**:3011–3019.

Oremland, R. S., P. R. Dowdle, S. Hoeft, J. O. Sharp, J. K. Schaefer, L. G. Miller, J. Blum, R. L. Smith, N. S. Bloom, and D. Wallschlaeger. 2000. Bacterial dissimilatory reduction of arsenate and sulfate in meromictic Mono Lake, California. *Geochim. Cosmochim. Acta* **64**:3073–3084.

Oremland, R. S., M. J. Herbel, J. S. Blum, S. Langley, T. J. Beveridge, P. M. Ajayan, T. Sutto, A. V. Ellis, and S. Curran. 2004. Structural and spectral features of selenium nanospheres produced by Se-respiring bacteria. *Appl. Environ. Microbiol.* **70**:52–60.

Oremland, R. S., J. T. Hollibaugh, A. S. Maest, T. S. Presser, L. G. Miller, and C. W. Culbertson. 1989. Selenate reduction to elemental selenium by anaerobic bacteria in sediments and culture—biogeochemical significance of a novel, sulfate-independent respiration. *Appl. Environ. Microbiol.* **55**:2333–2343.

Oremland, R. S., T. R. Kulp, J. S. Blum, S. E. Hoeft, S. Baesman, L. G. Miller, and J. F. Stoltz. 2005. A microbial arsenic cycle in a salt-saturated, extreme environment. *Science* **308**:1305–1308.

Oremland, R. S., F. Wolfe-Simon, C. W. Saltikov, and J. F. Stolz. 2009. Arsenic in the evolution of earth and extraterrestrial ecosystems. *Geomicrobiol. J.* **26**:522–536.

Pearce, C. I., V. S. Coker, J. M. Charnock, R. A. D. Pattrick, J. F. W. Mosselmans, N. Law, T. J. Beveridge, and J. R. Lloyd. 2008. Microbial manufacture of chalcogenide-based nanoparticles via the reduction of selenite using *Veillonella atypica*: an in situ EXAFS study. *Nanotechnology* **19**:155–603.

Pearce, C. I., R. A. D. Pattrick, N. Law, J. C. Charnock, V. S. Coker, J. F. Fellowes, R. S. Oremland, and J. R. Lloyd. 2009. Investigating different mechanisms for biogenic selenite transformations: *Geobacter sulfurreducens*, *Shewanella oneidensis* and *Veillonella atypica*. *Environ. Technol.* **30**:1313–1326.

Peng, D., J. Zhang, Q. Liu, and E. W. Taylor. 2007. Size effect of elemental selenium nanoparticles (nano-Se) at supranutritional levels on selenium accumulation and glutathione S-transferase activity. *J. Inorg. Biochem.* **101**:1457–1463.

Pickett, N. L., and P. O'Brien. 2001. Syntheses of semiconductor nanoparticles using single-molecular precursors. *Chem. Rec.* **1**:467–479.

Planer-Friedrich, B., J. C. Fischer, J. T. Hollibaugh, E. Süß, and D. Wallschläger. 2009. Oxidative transformation of trithioarsenate along alkaline geothermal drainages: abiotic versus microbially mediated processes. *Geomicrobiol. J.* **26**:339–350.

Planer-Friedrich, B., C. Lehr, J. Matschullat, B. J. Merkel, D. K. Nordstrom, and M. W. Sandstrom. 2006. Speciation of volatile arsenic at geothermal features in Yellowstone National Park. *Geochim. Cosmochim. Acta* **70**:2480–2491.

Prakash, N. T., N. Sharma, R. Prakash, K. K. Raina, J. F. Fellowes, C. I. Pearce, J. R. Lloyd, and R. A. D. Pattrick. 2009. Aerobic microbial manufacture of nanoscale selenium: exploiting nature's bio-nanomineralization potential. *Biotechnol. Lett.* **31**:1857–1862. doi: 10.1007/s10529-009-0096-0.

Presser, T. S. 1994. The Kesterson Effect. *Environ. Manage.* **18**:437–454.

Rathgeber, C., N. Yurkova, E. Stackebrandt, J. T. Beatty, and V. Yurkov. 2002. Isolation of tellurite- and selenite-resistant bacteria from hydrothermal vents of the Juan de Fuca Ridge in the pacific ocean. *Appl. Environ. Microbiol.* **68**:4613–4622.

Rech, S. A., and J. M. Macy. 1992. The terminal reductases for Se(VI) and nitrate respiration in *Thauera selenatis* are two distinct enzymes. *J. Bacteriol.* **174**:7316–7320.

Richey, C., P. Chovanec, S. E. Hoeft, R. S. Oremland, P. Basu, and J. F. Stolz. 2009. Respiratory arsenate reductase as a bidirectional enzyme. *Biochem. Biophys. Res. Commun.* **382**:298–302.

Salminen, R., M. J. Batista, M. Bidovec, A. Demetriades, B. De Vivo, W. De Vos, M. Duris, A. Gilucis, V. Gregorauskiene, J. Halamic, P. Heitzmann, A. Lima, G. Jordan, G. Klaver, P. Klein, J. Lis, J. Locutura, K. Marsina, A. Mazreku, P. J. O'Connor, S. A. Olsson, R. T. Ottesen, V. Petersell, J. A. Plant, S. Reeder, I. Salpeteur, H. Sandstrom, U. Siewers, A. Steenfelt, and T. Tarvainen. 2006. Geochemical Atlas of Europe. *In* R. Salminen (ed.), Part 1: Background Information, Methodology and Maps. EuroGeoSurveys-FOREGS Geochemical Baseline Mapping Programme. http://www.gtk.fi/publ/foregsatlas/.

Schröder, I., S. R. Rech, T. Krafft, and J. M. Macy. 1997. Purification and characterization of the selenate reductase from *Thauera selenatis*. *J. Biol. Chem.* **272**:23765–23768.

Shrift, A. 1964. Selenium cycle in nature. *Nature* **201**:1304–1305.

Silver, S., and L. T. Phung. 2005. Genes and enzymes involved in bacterial oxidation and reduction of inorganic arsenic. *Appl. Environ. Microbiol.* **71**:599–608.

Smithers, R. M., and H. R. Krouse. 1967. Tellurium isotope fractionation study. *Can. J. Chem.* **46**:583–591.

Stolz, J. E., P. Basu, J. M. Santini, and R. S. Oremland. 2006. Arsenic and selenium in microbial metabolism. *Annu. Rev. Microbiol.* **60**:107–130.

Stolz, J. F., D. J. Ellis, J. S. Blum, D. Ahmann, D. R. Lovley, and R. S. Oremland. 1999. *Sulfurospirillum barnesii* sp. nov. and *Sulfurospirillum arsenophilum* sp. nov., new members of the *Sulfuro-*

spirillum clade of the epsilon Proteobacteria. *Int. J. Syst. Bacteriol.* **49:**1177–1180.

Stolz, J. F., and R. S. Oremland. 1999. Bacterial respiration of arsenic and selenium. *FEMS Microbiol. Rev.* **23:**615–627.

Switzer Blum, J., S. Han, B. Lanoil, C. Saltikov, B. Witte, F. R. Tabita, S. Langley, T. J. Beveridge, L. Jahnke, and R. S. Oremland. 2009. Ecophysiology of "*Halarsenatibacter silvermanii*" strain SLAS-1^T, gen. nov., sp. nov., a facultative chemoautotrophic arsenate respirer from salt-saturated Searles Lake, California. *Appl. Environ. Microbiol.* **75:**1950–1960.

Tomei, F. A., L. L. Barton, C. L. Lemanski, and T. G. Zocco. 1992. Reduction of selenate and selenite to elemental selenium by *Wolinella succinogenes*. *Can. J. Microbiol.* **38:**1328–1333.

Torma, A. E., and F. Habashi. 1972. Oxidation of copper (II) selenide by *Thiobacillus ferrooxidans*. *Can. J. Microbiol.* **18:**1780–1781.

Trutko, S. M., V. K. Akimenko, N. E. Suzina, L. A. Anisimova, M. G. Shlyapnikov, B. P. Baskunov, V. I. Duda, and A. M. Boronin. 2000. Involvement of the respiratory chain of gram-negative bacteria in the reduction of tellurite. *Arch. Microbiol.* **173:**178–186.

Turner, R. J., Y. Aharonowitz, J. H. Weiner, and D. E. Taylor. 2001. Glutathione is a target in tellurite toxicity and is protected by tellurite resistance determinants in *Escherichia coli*. *Can. J. Microbiol.* **47:**33–40.

Van Fleet-Stalder, V., T. G. Chasteen, I. J. Pickering, G. N. George, and R. C. Prince. 2000. Fate of selenate and selenite metabolized by *Rhodobacter sphaeroides*. *Appl. Environ. Microbiol.* **66:**4849–4853.

Warrier, M., M. K. Lo, H. Monbouquette, and M. A. Garcia-Garibay. 2004. Photocatalytic reduction of aromatic azides to amines using CdS and CdSe nanoparticles. *Photochem. Photobiol. Sci.* **3:**859–863.

Williams, K. H., A. L. N'Guessan, J. Druhan, M. J. Wilkins, D. Holmes, P. E. Long, and D. R. Lovley. 2007. Field-scale evidence for selenium bioremediation in a uranium-contaminated aquifer. 107th General Meeting of the American Society for Microbiology, Toronto, Canada.

Yamada, A., N. Miyagishima, and T. Matsunaga. 1997. Tellurite removal by marine photosynthetic bacteria. *J. Mar. Biotechnol.* **5:**46–49.

Yamamura, S., M. Yamashita, N. Fujimoto, M. Kuroda, M. Kashiwa, K. Sei, M. Fujita, and M. Ike. 2007. *Bacillus selenatarsenatis* sp. nov., a selenate- and arsenate-reducing bacterium isolated from the effluent drain of a glass-manufacturing plant. *Int. J. Syst. Evol. Microbiol.* **57:**1060–1064.

Yee, N., and D. Y. Kobayashi. 2008. Molecular genetics of selenate reduction by *Enterobacter cloacae* SLD-1. *Adv. Appl. Microbiol.* **64:**107–123.

Yee, N., J. Ma, A. Dalia, T. Boonfueng, and D. Y. Kobayashi. 2007. Se(VI) reduction and the precipitation of Se(0) by the facultative bacterium *Enterobacter cloacae* SLD1a-1 are regulated by FNR. *Appl. Environ. Microbiol.* **73:**1914–1920.

Yuan, C., X. Lu, J. Qin, B. P. Rosen, and X. C. Le. 2008. Volatile arsenic species released from *Escherichia coli* expressing the AsIII S-adenosylmethionine methyltransferase gene. *Environ. Sci. Technol.* **42:**3201–3206.

Zannoni, D., F. Borsetti, J. J. Harrison, and R. J. Turner. 2008. The bacterial response to the chalcogen metalloids Se and Te. *Adv. Microb. Physiol.* **53:**1–71.

Zehr, J. P., and R. S. Oremland. 1987. Reduction of selenate to selenide by sulfate-respiring bacteria: experiments with cell-suspensions and estuarine sediments. *Appl. Environ. Microbiol.* **53:**1365–1369.

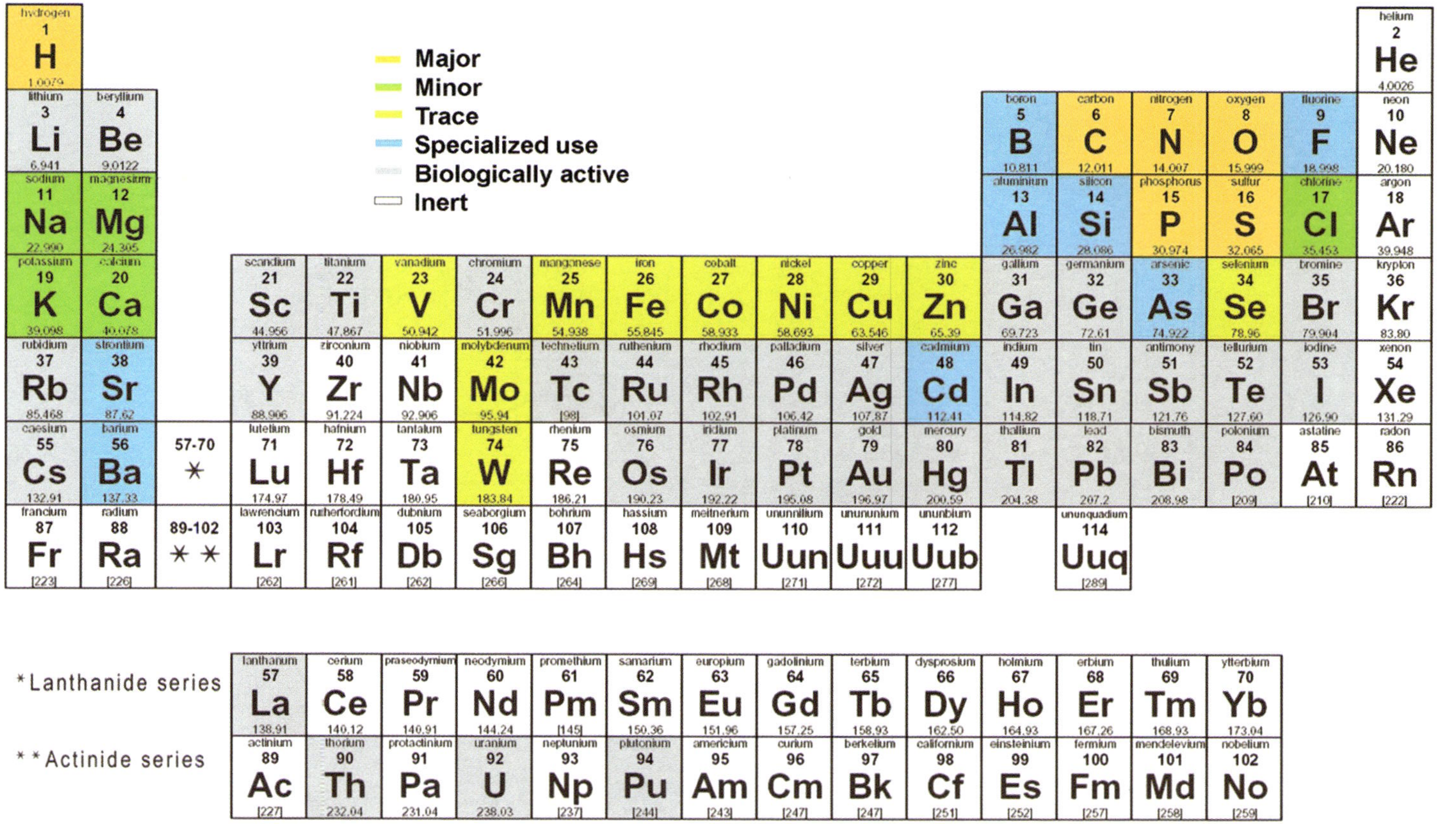

COLOR PLATE 1 (PREFACE) Periodic table of the elements highlighting the currently known major, minor, trace, and biologically active elements. Updated from Wackett et al. (2004).

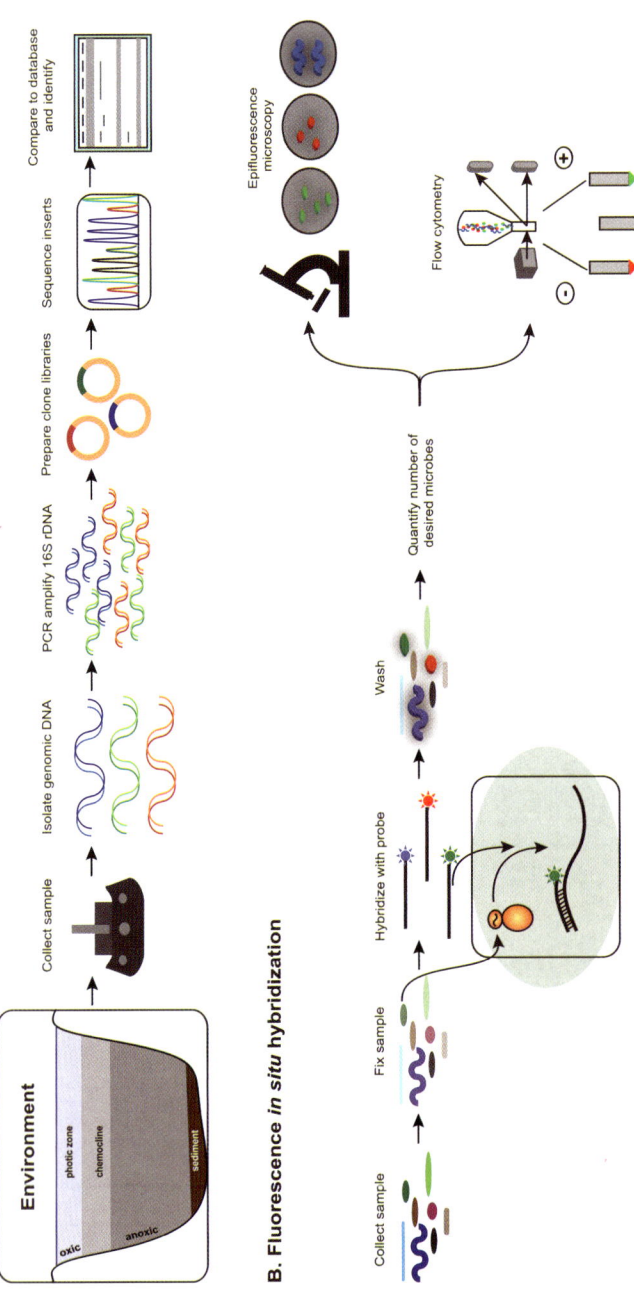

COLOR PLATE 2 (CHAPTER 2) Culture-independent approaches to understand community structure. (A) Strategy for construction of clone libraries. Samples are collected from a given environment such as Lake Matano. The total DNA is isolated from the sample with use of commercial kits. This DNA is used for PCR amplification using degenerate 16S rDNA primers. The PCR products are then cloned into appropriate cloning vectors that are also commercially available. The inserts of each plasmid are then sequenced using primers that are specific for regions of the plasmid. The sequence obtained is then compared with comprehensive databases that have 16S rDNA sequence data to determine the organism with the closest 16S rDNA sequence. (B) General strategy for performing FISH. A collected sample is fixed to preserve the natural structure and physiological state of the cells and to permeabilize the cells. The samples are then hybridized to a fluorescently labeled probe that targets a desired group of organisms. The excess probe is washed, and cells that are now fluorescently labeled can be visualized and quantified using epifluorescence microscopy or sorted and quantified using flow cytometry.

A. FISH - Microautoradiography (FISH-MAR)

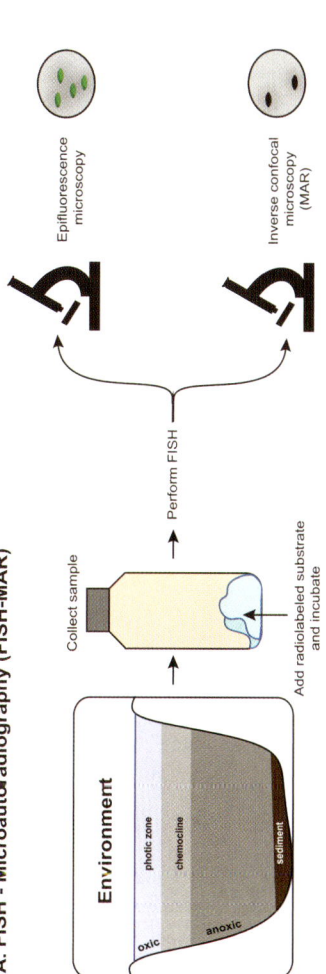

B. Isotope array

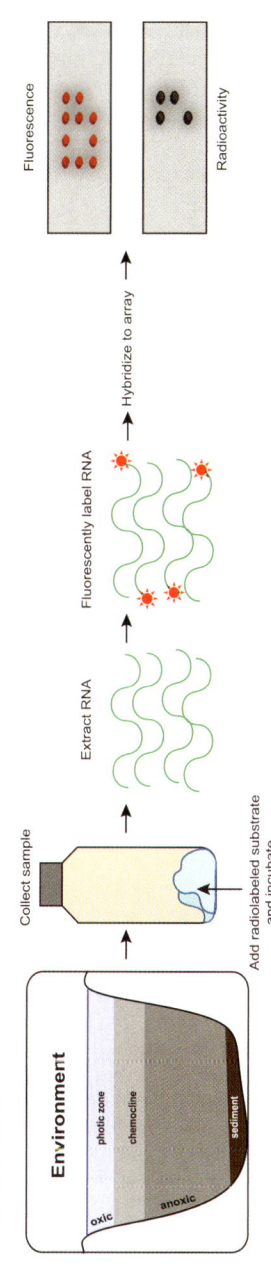

COLOR PLATE 3 (CHAPTER 2) FISH-MAR and isotope arrays as methods for understanding the function of specific organisms in a community. (A) FISH-MAR involves collecting a sample and incubating the sample with a desired radiolabeled substrate. The sample is then used to perform FISH as described in Color Plate 2. The same sample is then treated with a photographic emulsion and the cells are then visualized by inverse confocal microscopy. Comparison of the FISH image and the photographic image reveals organisms that incorporated the radiolabel into cell material using the substrate provided. (B) Isotope array is a modification of the DNA microarray approach that involves the incubation of the sample with a radiolabeled substrate. The RNA from the sample is then isolated and fluorescently labeled. This labeled RNA sample is then hybridized to a DNA microarray that has 16S rDNA oligos for a number of predetermined microbial species spotted onto a glass slide. The fluorescence indicates the organisms that are present in a given sample and comparison with the radiographic image confirms which of these organisms incorporated the label into their RNA.

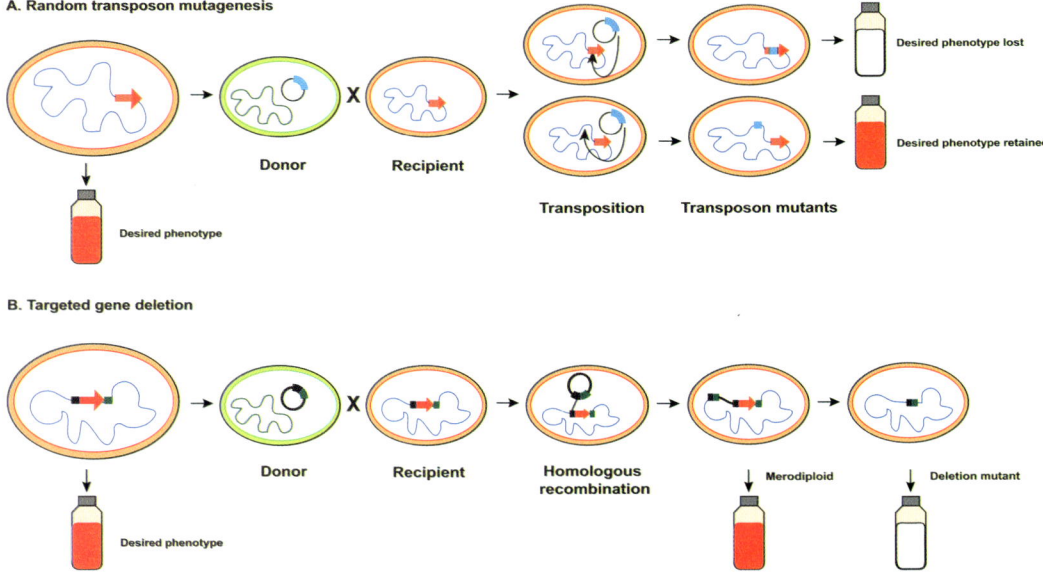

COLOR PLATE 4 (CHAPTER 2) "Loss-of-function" genetic strategies to determine what genes are responsible for a particular phenotype in an environmental isolate. (A) Random transposon mutagenesis. This approach involves the use of randomly inserting transposons to find a desired genetic locus. A plasmid carrying a transposon and a selectable marker usually for antibiotic resistance is introduced into an environmental isolate either via conjugation or other means. The plasmid carrying the transposon cannot replicate in the environmental isolate. However, once the plasmid is transferred to the isolate, a transposition event occurs randomly into the chromosome of the isolate, conferring a selectable phenotype. Subsequent selection and search for the loss of the desired phenotype results in identification of a genetic locus likely responsible for the phenotype. Later complementation experiments confirm that the genetic locus predicted to confer the phenotype is indeed due to the identified locus. (B) Targeted gene deletion via homologous recombination. Bioinformatic or other means allow the prediction of a genetic locus that likely confers a desired phenotype. The upstream and downstream region of the desired locus is cloned into a plasmid vector that carries both a selectable marker (usually resistance to an antibiotic) and a counterselectable marker (a gene whose presence causes the cell to die when a certain selection is applied). This vector is then transferred to the isolate and selection is applied. The inability of the carrier plasmid to replicate in the isolate forces the plasmid to integrate into the chromosome via homologous recombination when selected for antibiotic resistance. Subsequent segregation and counterselection lead to deletion of the desired gene.

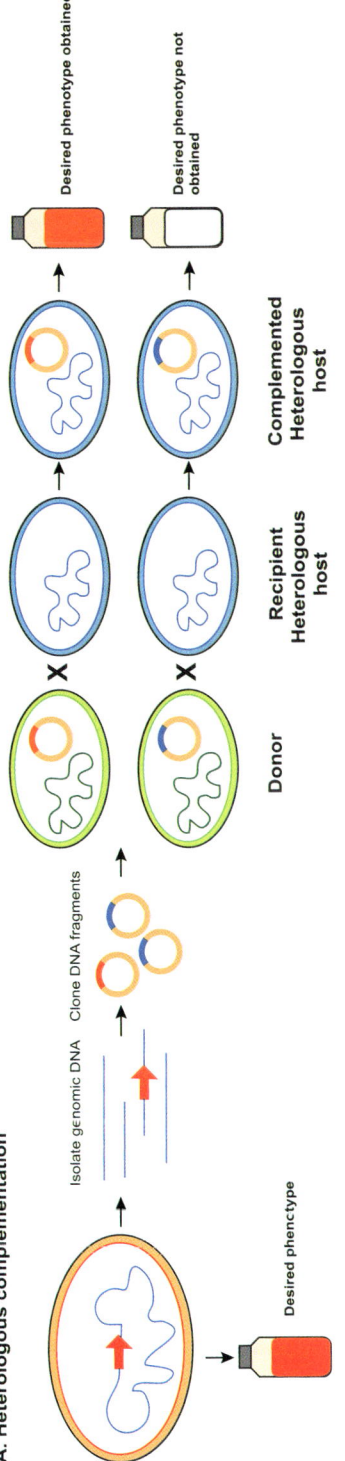

COLOR PLATE 5 (CHAPTER 2) Gain-of-function genetic strategy to determine the genes responsible for a particular phenotype. Heterologous complementation involves isolation of genomic DNA from an environmental isolate. This DNA is then cloned into a plasmid that can replicate in an organism that is closely related to the environmental isolate being studied that lacks the phenotype specific to the environmental isolate (a heterologous host). The plasmid is then transferred to the heterologous host using conjugation. If a genetic locus can confer the phenotype being sought, then we determine the sequence of the inserted DNA. The identity of the genetic locus can then be revealed by homology searches in publicly available sequence databases.

COLOR PLATE 6 (CHAPTER 3) Black smokers venting 296°C fluid on the Inferno metal sulfide deposit (top) and 25°C diffuse fluid venting in the Marker 113/62 hydrothermal field (bottom), both within the Axial Volcano caldera in the northeastern Pacific Ocean. Photos courtesy of NOAA Vents Program, NeMO Project.

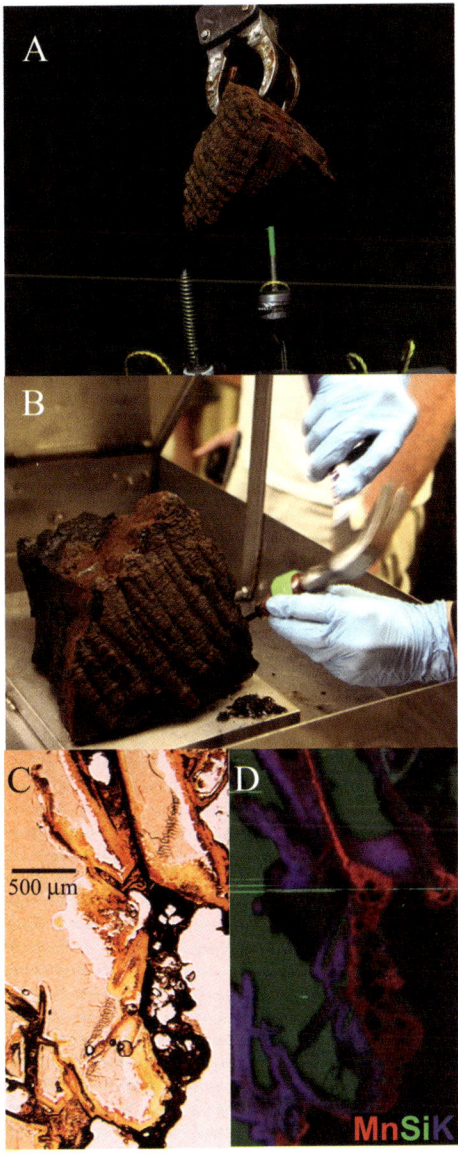

COLOR PLATE 7 (CHAPTER 4) (A) Seafloor pillow basalt from the Loihi Seamount immediately after collection by ROV *Jason*. (B) Seafloor pillow basalt being sampled later on board the ship. (C and D) Paired petrographic thin section and synchrotron-based μX-ray fluorescence images. These show the transition through basalt glass (left; green), the palagonite weathering rind (blue/purple, infilling cracks), and the precipitation rind that formed on the exterior of the rock during interaction with hydrothermal fluids (red).

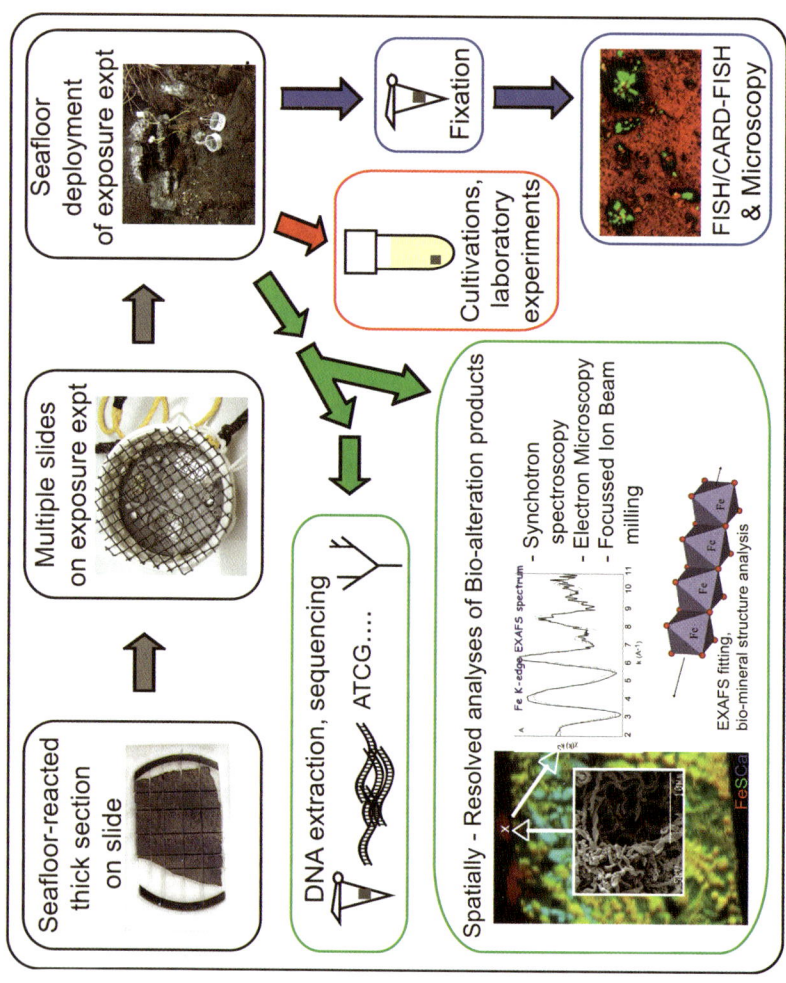

COLOR PLATE 8 (CHAPTER 4) Flow chart of typical seafloor exposure experiment, including preparation of the experiment (top row) and different downstream analyses carried out after recovery of experiment, weeks to years after initial deployment. Note that the initial thick sections may be derived from rocks collected from the seafloor (as in this case–"seafloor reacted" indicates rocks that were collected from the seafloor and later deployed in exposure experiments), or standard rocks purchased from suppliers. Different colored arrows and boxes indicate different analysis pipelines. EXAFS, extended X-ray absorption fine structure.

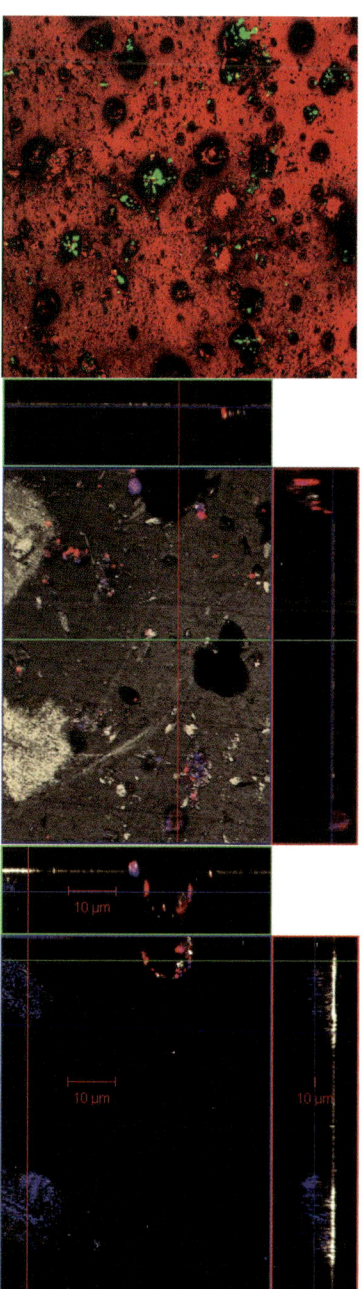

COLOR PLATE 9 (CHAPTER 4) Confocal imaging of catalyzed reported deposition–fluorescence in situ hybridization on polished basalt chip incubated at the seafloor on the EPR. The top panel is a composite of multiple layers (slices). The middle and bottom panels are two individual slices, with three-dimensional renderings of the slice surface on the side and above each panel. Stained cells appear in green in the top panel and in red in the middle and bottom panels. Yellow arrows in the middle panel indicate polishing scratches that are preferably colonized by microbes. Note that these features are not naturally occurring; rather, they are an artifact of the sample preparation. The side panels of the three-dimensional structure illustrate colonization of pits and pores.

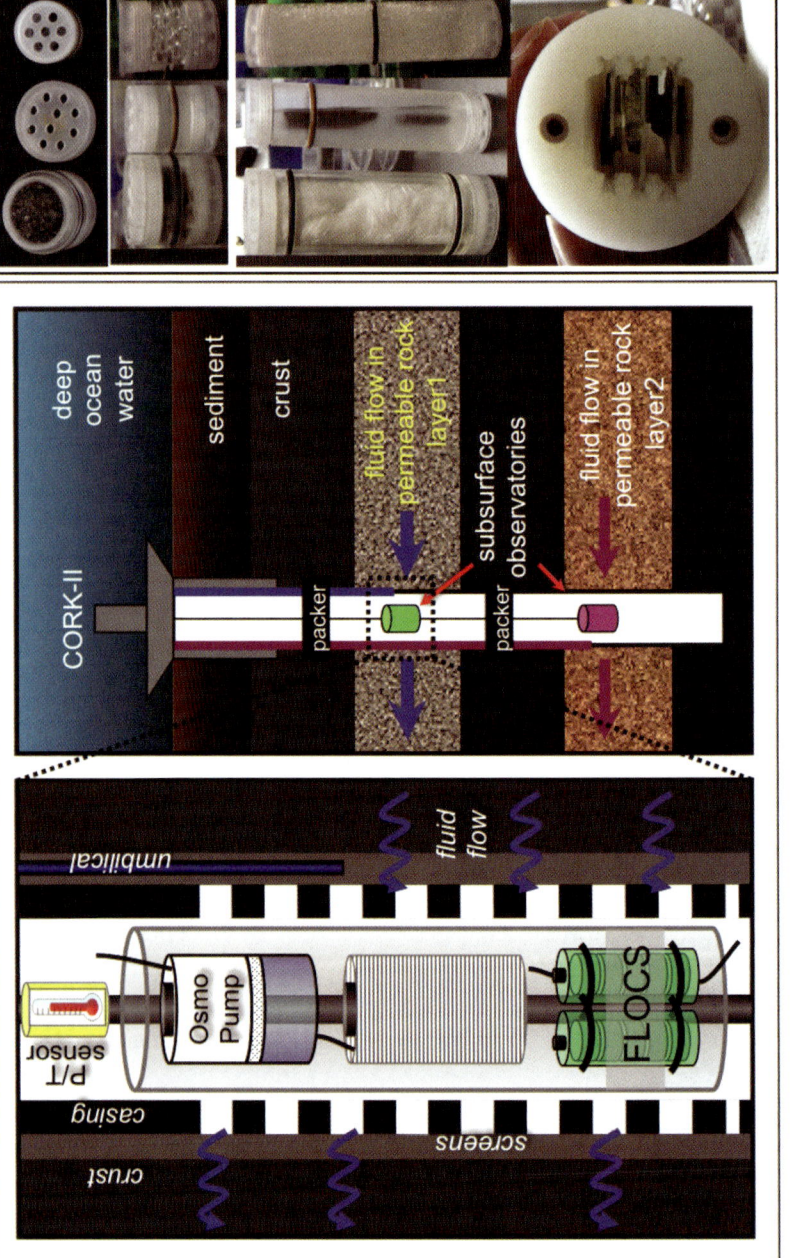

COLOR PLATE 10 (CHAPTER 4) CORK observatory colonization chambers (FLOCS) with a schematic diagram of their deployment within a sealed hole in the subseafloor. The left panel depicts FLOC chambers, which are designed to be used in conjunction with passive geochemical sampling systems and pumps (OSMO pump; Wheat et al. [2000]) and other sensors. Center panel depicts how they can be deployed in sealed hydrological and geological horizons (sealed by packers) at different depths for conducting discrete experiments under different basement conditions. Right panel shows the modular modern FLOC (Orcutt et al., 2010) chamber design, which can also be a variety of colonization materials, for example, rock chips, polished sections, glass wool, and so on. Figure adapted with permission from Orcutt et al. (2010).

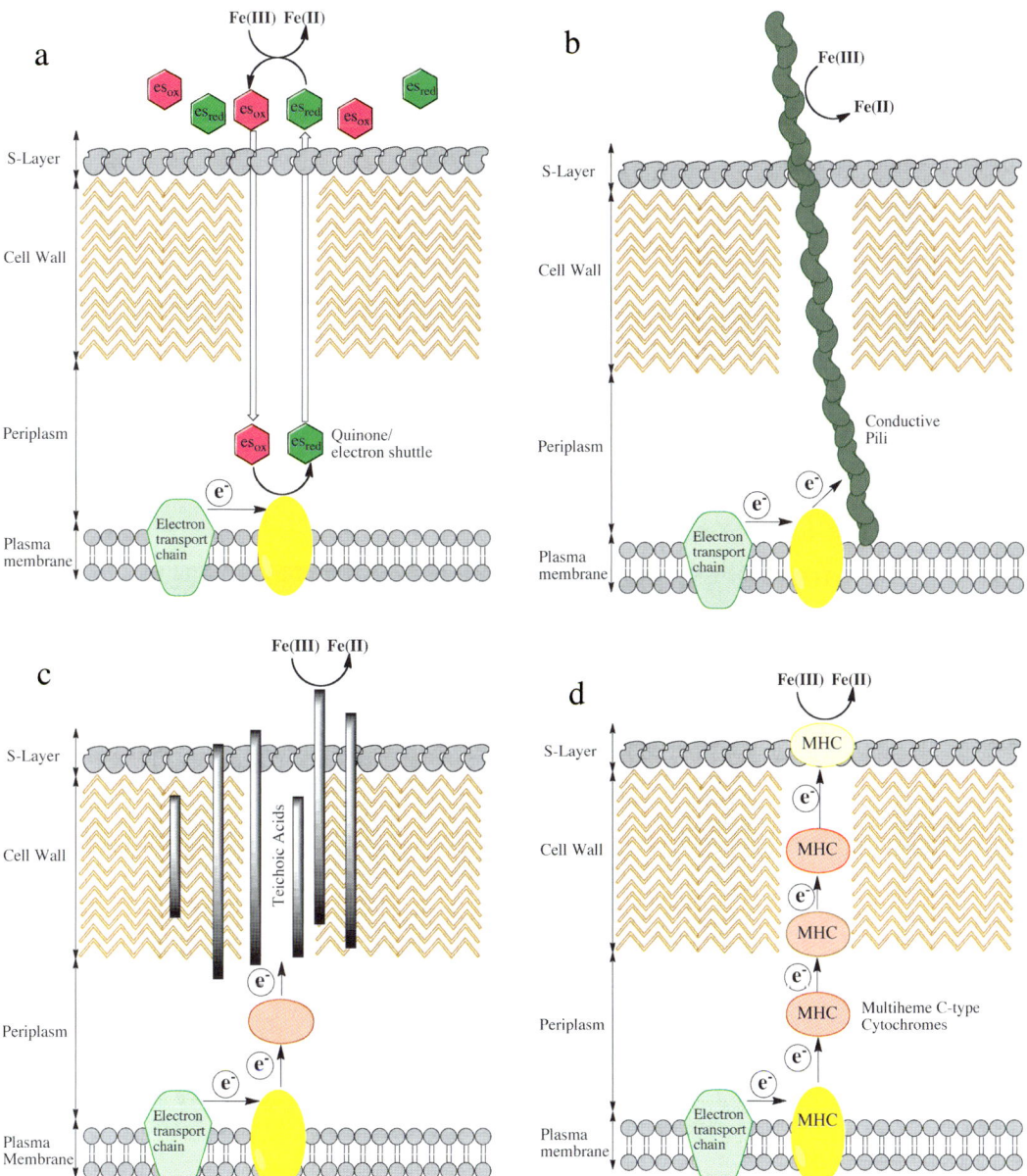

COLOR PLATE 11 (CHAPTER 10) Illustration of the four models for direct electron transfer by a gram-positive bacterium. Biofilm contained redox mediator (a), "nanowires" (b), conductive cell walls (c), or cytochrome chain linking the inner membrane to cell surface (d). Ferric iron is shown as a representative exogenous electron acceptor. Image courtesy of Dr. H. K. Carlson.

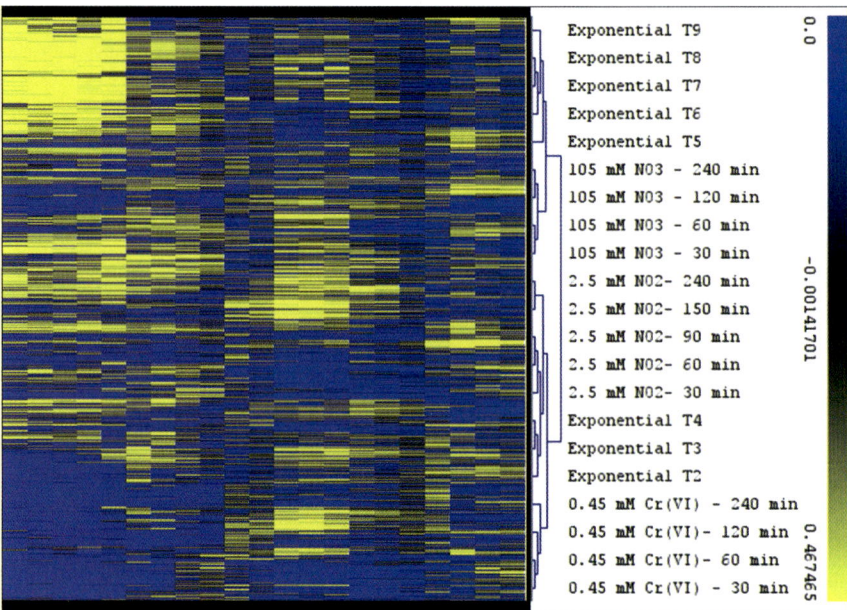

COLOR PLATE 12 (CHAPTER 12) Hierarchical cluster analysis of whole-genome, temporal gene expression for *D. vulgaris* cells exposed to nitrate (He et al., in review), nitrite (He et al., 2006), or Cr(VI) (Klonowska et al., 2006). The gene expression data from exponential- and stationary-growth phases (Clark et al., 2006) were also compared with the different stress conditions. The analysis was done with Multiple Experiment Viewer v4.4 (Saeed et al., 2003) with a Pearson correlation as the distance metric and a complete linkage clustering.

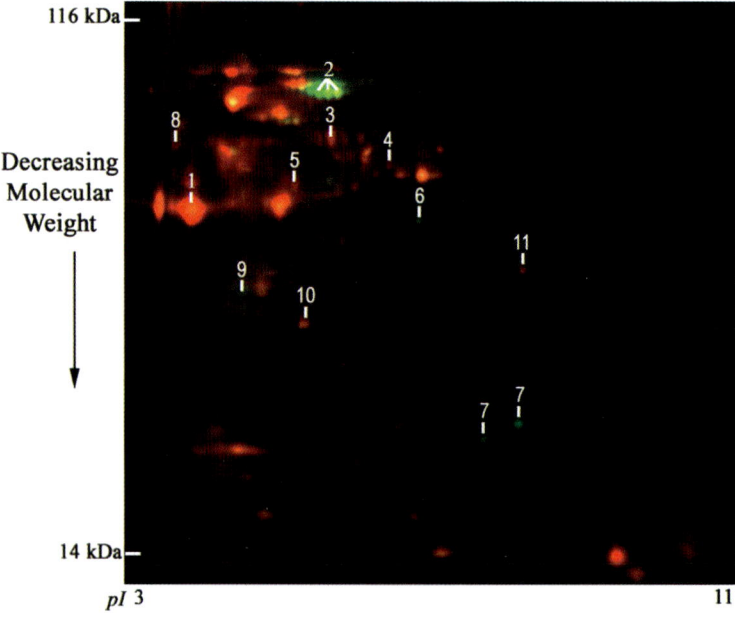

COLOR PLATE 13 (CHAPTER 13) Differential in gel expression analysis (DIGE) of proteins from *A. ehrlichii* grown aerobically with acetate (labeled with Cy3, red fluorescence) and anaerobically with NO_3^-, AsO_3^-, and CO_2 (labeled with Cy5, green fluorescence). Protein spots with equal protein abundance expressed under both conditions appear yellow. The numbered spots on the gel correspond to the list of numbered proteins identified by MALDI-TOF/MS. 1, porin; 2, nitrous-oxide reductase; 3, citrate synthase; 4, nitrate transporter; 5, putative nitrate transporter; 6, phosphoribulokinase; 7, fructose-1,6-bisphosphate aldolase; 8, branched-chain amino acid ABC transporter; 9, triosephosphate isomerase; 10, superoxide dismutase; 11, 2-oxoglutarate dehydrogenase.

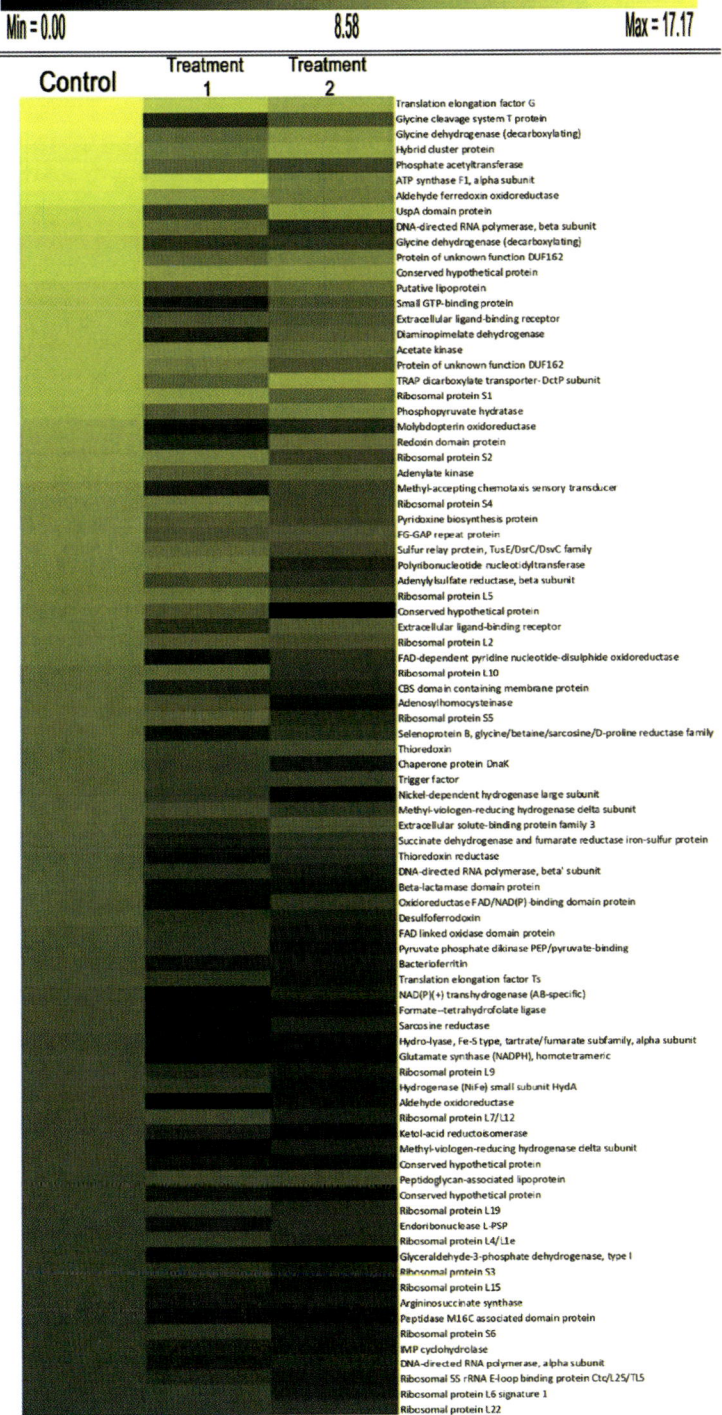

COLOR PLATE 14 (CHAPTER 13) Heat map visualization of the relative differences in protein abundances obtained by LC-MS/MS of *D. desulfuricans* cells exposed to chromium (average of $n = 3$). Control, growth on nitrate alone; Treatment 1, growth on nitrate with chromate (100 µM); Treatment 2, two-hour exposure to chromate (100 µM). Scale runs from yellow (high abundance) to black (low abundance). Hierarchical clustering was done using the PermutMatrix program (Caraux and Pinloche, 2005).

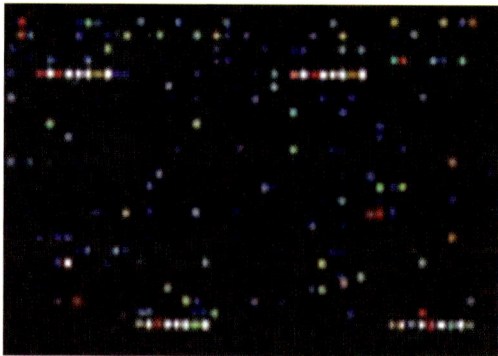

COLOR PLATE 15 (CHAPTER 14) GeoChip 3.0. Microbial community DNA was labeled with Cy5 fluorescent dye and hybridized to the array. Brighter colors represent higher signal intensities.

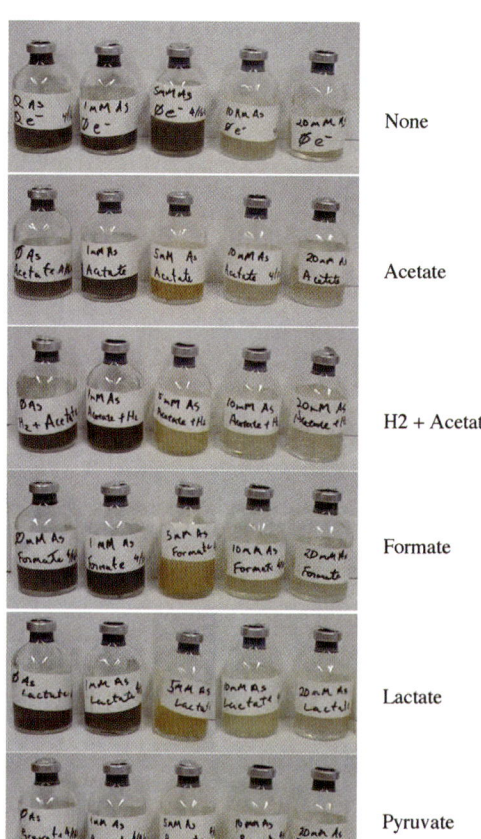

COLOR PLATE 16 (CHAPTER 15) Medium matrix of Ohio River sediment enrichment cultures after 7 days of incubation. Electron donor, from top row to bottom row: no donor, acetate, hydrogen plus acetate, formate, lactate, pyruvate. Concentrations of As(V), from left to right, are 0, 1, 5, 10, and 20 mM. Copious quantities of arsenic trisulfide were produced in the 5 mM cultures where electron donor was provided.

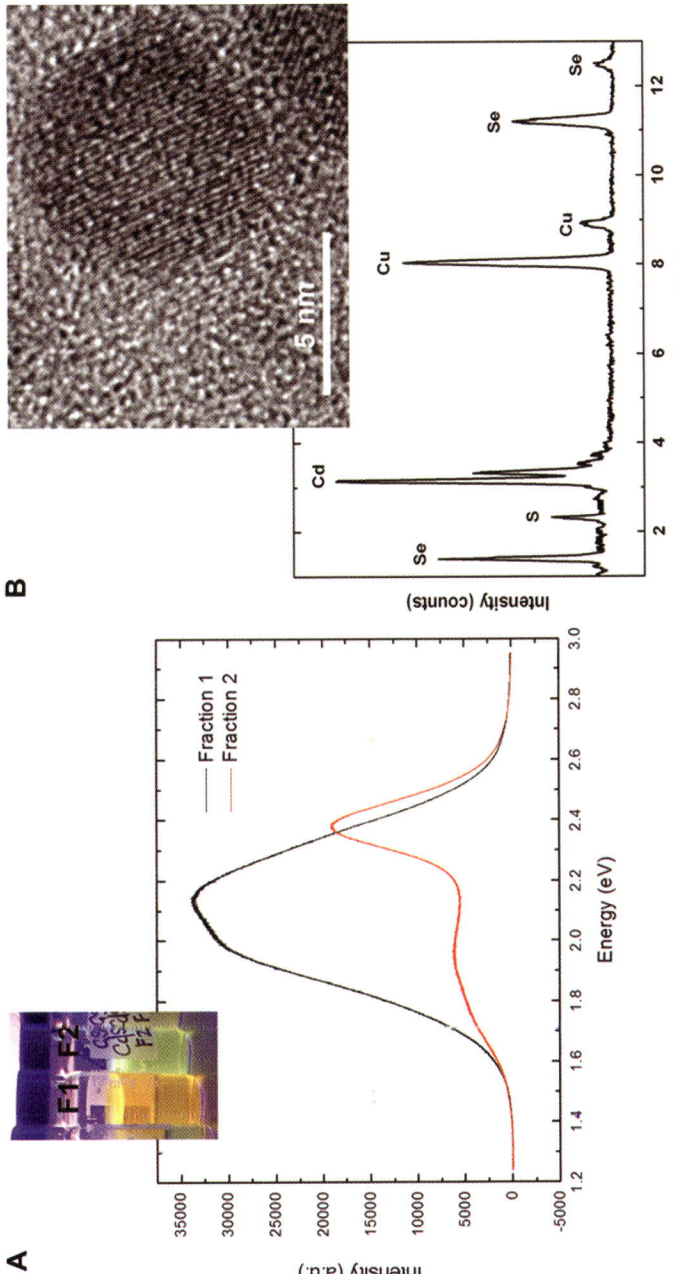

COLOR PLATE 17 (CHAPTER 16) (A) Fluorescence spectra of size-fractionated glutathione-stabilized CdSe produced by *Veillonella atypica* (inset shows fractions 1 and 2 under UV light). (B) High-resolution TEM of glutathione-stabilized CdSe and accompanying EDX spectrum of fraction 1.

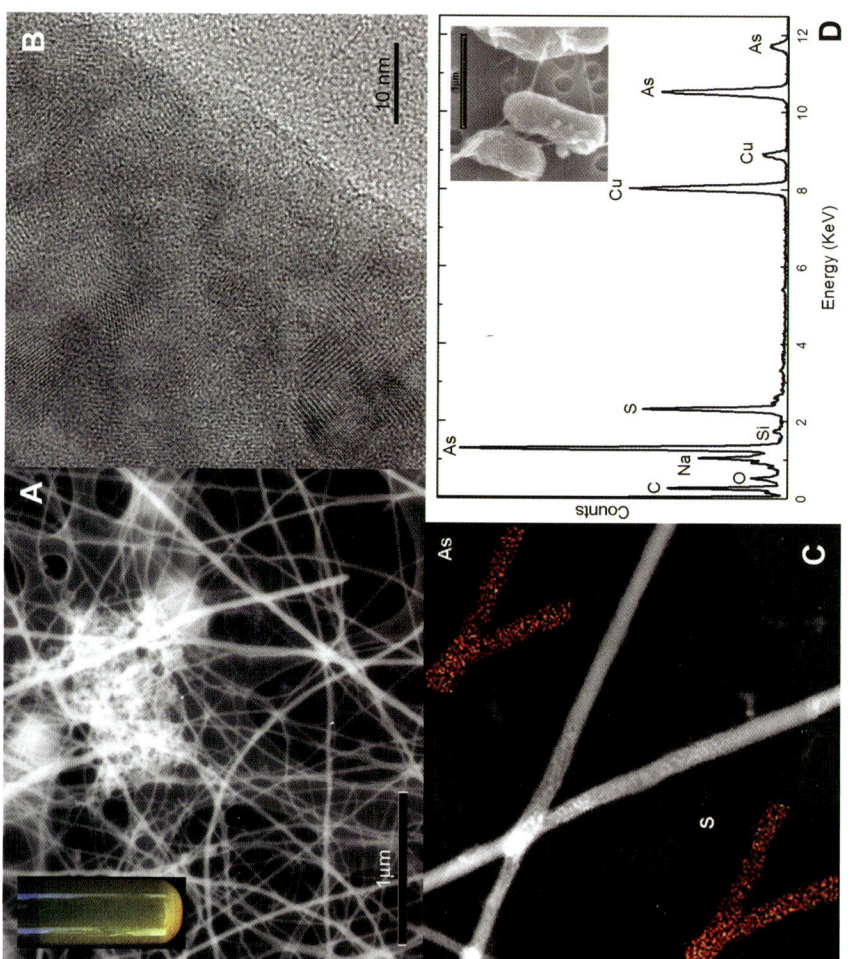

COLOR PLATE 18 (CHAPTER 16) Electron micrographs of As–S precipitate formed by *Citrobacter* strain TSA-1 grown in the presence of arsenate in media with cysteine sulfide as a reducing agent and trace levels of sulfate (inset shows yellow precipitate). (A) SEM image. (B) High-resolution TEM image. (C and D) Scanning transmission electron microscope–EDX mapping images and spectrum (inset shows SEM image of strain TSA-1).

MICROBIAL RESPIRATION OF ANODES AND CATHODES IN ELECTROCHEMICAL CELLS

Kelvin B. Gregory and Dawn E. Holmes

17

Bioelectrochemical cells, in all their forms, are devices that take advantage of the catalytic activity of microorganisms to enable or accelerate electrochemical reactions at electrodes. Central to the advantage of microbial catalysis at the electrode is the broad range of fuel and oxidant substrates that are made possible through microbial colonization of the electrodes and biocatalysis of otherwise slow abiotic electrochemical reactions. For example, oxidation of complex assemblages of organic matter such as biomass is not electrochemically reactive at the anode of a fuel cell without the aid of microbial catalysis. The microbial fuel cell for bioenergy production from renewable fuels is one of many potential applications of microbial catalysis in electrochemical cells. Numerous investigators have proposed analogous bioelectrochemical cells for engineering applications in bioremediation and biohydrogen production.

The advantages of microbial biocatalysis in electrochemical cells has been recognized for 100 years. However, the mechanisms of electron transfer to and from the bacteria in electrochemical cells have only recently been examined. Pure-culture studies have helped identify different mechanisms of electron transfer between electrodes and bacteria. Of late, genome-enabled inquiry of microbial physiology has revealed the molecular basis for some of these electron transfer mechanisms. Successful optimization of electrode-based biocatalysis for energy production and environmental remediation has largely transpired through modification of the electrochemical cell architecture and operational parameters. The sole design criterion of this approach is to minimize electrochemical and mass-transport-related losses, largely without regard for the molecular microbial basis for electron transfer. It is likely that future opportunities for further optimization of bioelectrochemical cells will emerge through an engineering approach that is guided by microbial physiology.

In this chapter, we summarize and discuss the interactions between microbes and electrodes for energy generation and environmental applications. The electrode of an electrochemical cell may serve as either an electron acceptor or an electron donor, depending on the needs of the application. We emphasize the microbial communities that develop on both anodes and cathodes of electrochemical cells, the known bacteria which conserve energy and grow on

---

Kelvin B. Gregory, Department of Civil and Environmental Engineering, Carnegie Mellon University, Pittsburgh, PA 15213. *Dawn E. Holmes*, Physical and Biological Sciences, School of Arts and Sciences, Western New England College, Springfield, MA 01119.

electrodes, and the current state of understanding for the molecular basis of electron transfer between bacteria and electrodes of electrochemical cells.

ELECTROCHEMICAL CELL OVERVIEW

Electrochemical cells are devices used for the study of chemical and electrical phenomena that are induced by the flow of current during the consumption or generation of electricity from chemical and biological reactions. Stripped to its most basic components, an electrochemical cell is simply two electrodes, separated by an electrolyte phase (Bard and Faulkner, 2001). If these two electrodes are connected by a voltmeter, the open-circuit cell potential may be measured in volts (V). This potential represents the sum of the electric potentials that arise between all of the different phases in the cell, whether electrical or chemical, and may be interpreted as a measure of the energy available to drive the transfer of charge through the external connection between the two electrodes. When charge is transferred between the two electrodes, the electrochemical cell may be classified as either *galvanic* or *electrolytic*. The difference may be generally summarized by whether chemical energy is being converted into electrical energy or vice versa.

Electrolytic Cell

The electrolytic cell occurs when an external electrical potential is applied between the two electrodes (usually via a power supply or battery) which is greater than the open-circuit potential of the cell (Bard and Faulkner, 2001). When this condition is imposed, a chemical reaction at the electrodes is driven against its ΔG by the external potential (Fig. 1 [top]). Thus, as current flows between the electrodes, driving the chemical reaction, electrical energy is consumed. Electrolytic processes are familiar and used commonly during chemical synthesis. Figure 1 (top) shows the electrolysis of water to produce H_2 at the cathode and oxygen at the anode. Electrolysis cells are also used during chemical synthesis of chlorine gas and electroplating. A car battery becomes, for the time being, an electrolytic cell during a recharge cycle to ready it for its more common duty cycle as a galvanic cell for electricity production.

Galvanic Cell

When chemical energy is converted to electrical energy via an electrochemical cell, the galvanic cell is formed. The chemical reactions at the electrodes occur spontaneously as the reaction proceeds downhill with its ΔG, releasing energy that passes in the form of electrical current between the two electrodes through an external conductor (Bard and Faulkner, 2001). The galvanic cell is exemplified in Fig. 1 (bottom) by the common hydrogen fuel cell. Electricity is generated by oxidation of H_2 at the anode and reduction of O_2 at the cathode.

Current Direction and Polarity of Electrochemical Cells

The anode and cathode distinction in electrochemistry is based on current, not voltage. The anode is the electrode at which the oxidation of the electrolyte occurs and through which electrical current flows into the solution in a polarized electrochemical cell. A useful mnemonic is ACED (Anode Current Enters Device). The cathode is the electrode at which the reduction of the electrolyte solution occurs and through which electrical current passes into solution in a polarized electrochemical cell. A useful cathode mnemonic is ACDC (All Current Departs Cathode).

The polarity of the electrodes by electrochemical convention is broadly misunderstood. The misconception is that the anode polarity is always positive relative to the cathode and the cathode is always negative relative to the anode. The misunderstanding may arise from the fact that negatively charged anions move in solution toward the anode and positively charged cations move toward the cathode regardless of the type of electrochemical

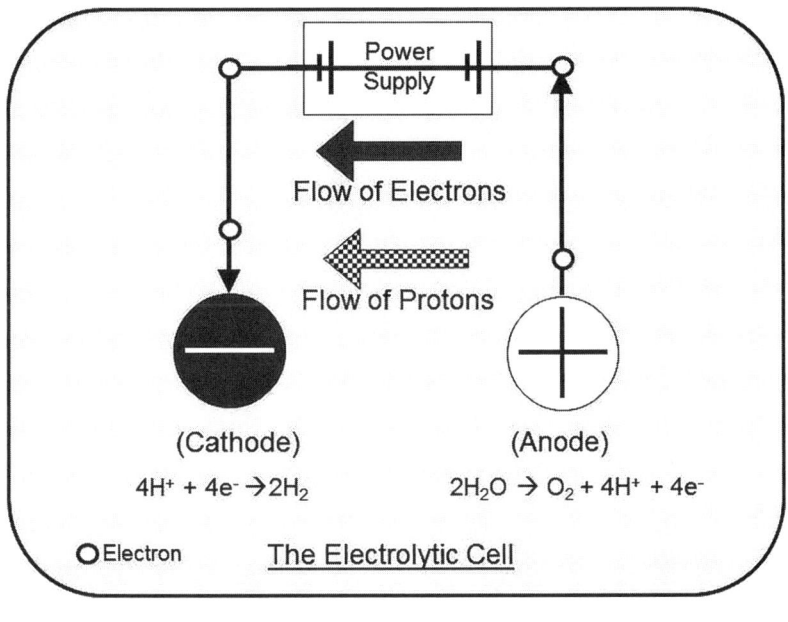

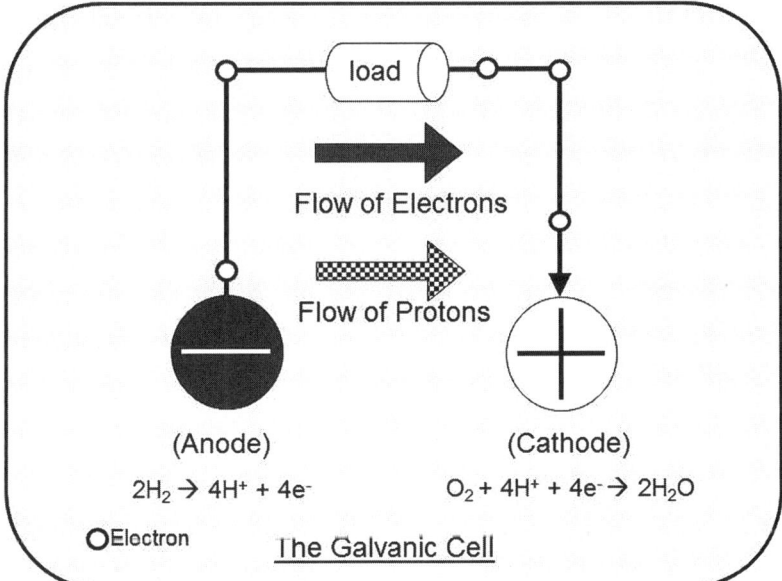

FIGURE 1 (Top) The electrolytic cell. An external electrical potential is applied between the two electrodes and a chemical reaction at the electrodes is driven against its ΔG by the external potential. Current flowing between the electrodes drives the chemical reaction, consuming electrical energy. (Bottom) The galvanic cell. Chemical energy is converted to electrical energy via an electrochemical cell. The chemical reaction at each electrode occurs spontaneously with its ΔG, releasing energy which passes in the form of electrical current between the two electrodes.

cell. However, the polarity of the anode and cathode depends on the particular device and, more importantly, whether it is consuming or providing external electrical power. In a device that consumes power (electrolytic cell, Fig. 1 [top]), by convention, the anode is positive polarity and the cathode negative. For the device that generates power (galvanic cell, Fig. 1 [bottom]) the anode is marked negative and the cathode positive. The conventions are most familiar and conceptually simpler in the galvanic cell where electrons and cations, generated at a negative polarity anode, move toward the positive polarity cathode. The movement of positively charged ions toward a positive polarity cathode may be thought to occur at the expense of energy released from the fuel. In the electrolytic cell, convention holds that the cathode has negative polarity and the anode has positive; electrons and positively charged ions move toward the negative polarity. The polarity of the cathode is most easily conceptualized in terms of where negative polarity is applied to drive the cell, e.g., regenerate the reductant of a battery (H_2 in the case of Fig. 1). In either case, it is easier to discuss the electrodes in terms of the currents that take place either into (anode) or out of (cathode) solution.

Fuel Cell Refresher

Fuel cells are galvanic electrochemical cells. Reactions at the anode and cathode occur spontaneously and release energy by direct conversion of chemical energy released from a reaction into electrical energy. Since fuel cells directly convert chemical energy to electricity, they circumvent the inefficiencies of conventional combustion in a heat engine. For example, the heat engine burns fuel to produce heat, which in turn energizes a steam engine, which drives a dynamo that produces electricity. At each energy conversion step there are limits in efficiency provided by the Second Law of Thermodynamics. The chemical energy in the fuel source is converted to heat, but not all of the heat is converted to electricity; the heat during combustion lowers the overall energy efficiency. Avoiding the energy losses due to heat is among the primary reasons that fuel cells have received so much attention.

The most basic fuel cell consists of two electrodes, an anode at which fuel oxidation occurs and a cathode at which reduction of the oxidant occurs. The electrodes are in contact with and separated by an electrolyte (Fig. 2). Ions travel through the separator to maintain charge balance of the overall chemical reaction as electrons flow from the anode to the cathode. The separator is commonly an ion-selective membrane (Barbir, 2005). In the case of a polymer electrolyte membrane (PEM) fuel cell a cation-selective membrane (such as Nafion) serves as the separator by simultaneously preventing mixing of anolyte and catholyte solutions, while selectively allowing cations to pass from the anolyte to the catholyte solution for maintenance of the charge balance. For the fuel to be oxidized as a useful source of electricity, the overall reaction must be kinetically and thermodynamically favorable.

Electricity is generated via separation and catalysis of the reaction between the fuel (electron donor) at the anode with the oxidant (electron acceptor) at the cathode in suitable electrolytes. Electrons captured from the fuel at the anode are forced through an external circuit to the cathode where they reduce the oxidant. The most common example of a fuel cell reaction in a conventional (abiotic) fuel cell is that of hydrogen and oxygen. Other fuels such as alcohols and petroleum hydrocarbons have been developed, as well as alternative oxidants such as chorine dioxide (Barbir, 2005). As electrons captured from H_2 leave the anode side and travel through the external circuit to the cathode, an imbalance in charge is created which must be equalized by the flow of cations to the catholyte solution for the electrode reactions to continue.

Anode reaction of fuel:
$$2H_2 \rightarrow 4H^+ + 4e^-$$

Cathode reaction of oxidant:
$$O_2 + 4H^+ + 4e^- \rightarrow 2H_2O$$

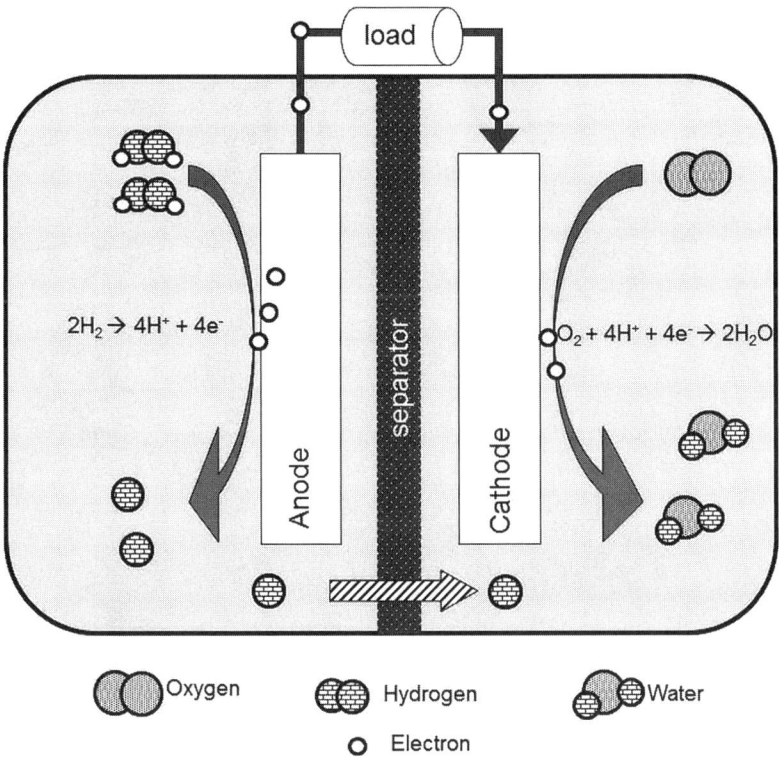

FIGURE 2 A conventional (abiotic) hydrogen fuel cell. Hydrogen is spontaneously oxidized at the anode with reduction of oxygen at the cathode. Electrons from hydrogen flow through the external circuit and reduce oxygen to water. The oxidation of hydrogen generates protons that travel across the separator, commonly an ion-selective membrane, to maintain charge balance in the overall redox reaction.

Overall cell reaction:
$$2H_2 + O_2 \rightarrow 2H_2O$$

It is noteworthy that fuel cells and batteries both release electricity as galvanic cells. However, a fuel cell is a thermodynamically open system by convention; it does not store chemical energy. Fuel and oxidant must be supplied continuously and the reaction products must exit the system. By similar convention, a battery is a thermodynamically closed system; the fuel and oxidant are contained within and therefore stored until needed. From this perspective, the battery first consumes electricity while recharging as an electrolytic cell. External power drives the chemical reactions against their ΔG to produce compounds in which chemical energy is stored until it is needed.

MICROBIAL FUEL CELLS

Overview

The microbial fuel cell is a galvanic cell that produces electricity and differs only in that the oxidation of the fuel and the transfer of electrons to the anode is the result of microbial metabolism (Fig. 3). From an electrochemical perspective, the bacteria that oxidize the fuel, liberate electrons, and transfer them to the electrode are serving as the anode catalyst while the graphite anode is the current collector. It should be noted that the bacteria at the anode are not true catalysts; they contain true catalysts, enzymes that enable their metabolism and electron transfer, but a portion of the energy liberated from the fuel is directed toward catabolic needs and cell growth and therefore is changed during

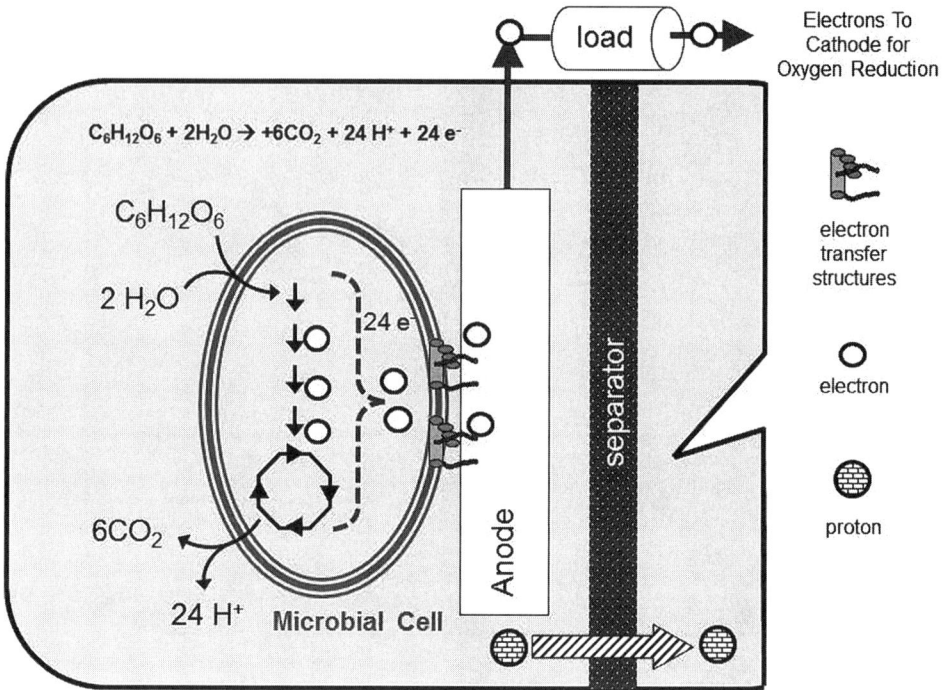

FIGURE 3 A model of a microbial fuel cell featuring direct electron transfer mechanism. Glucose is oxidized to CO_2 by the bacterium. Electrons liberated from glucose are transferred to the anode via direct contact of outer membrane and transmembrane electron carriers and transport structures.

catalysis. That said, it is reasonable to discuss microbial activity at the anode as catalysis because bacterial metabolism enables oxidation of the fuel at the anode. As a result of microbial catalysis of the oxidation at the anode, microbial fuel cells do not require the expensive metal catalysts at the anode that conventional fuel cells require. Rather, a suitable conductor, such as graphite, is commonly used for the anode (Logan, 2008). Following transfer of electrons from the microbial cell to the anode, electrons are conducted via an external connector to the cathode, where the half-cell reaction proceeds in an identical fashion to a conventional, abiotic fuel cell. Excellent state-of-the-art summaries of microbial fuel cell design, materials, measurements (Logan, 2009), and operations are available (Logan et al., 2006; Rabaey and Verstraete, 2005; Logan, 2008; Pham et al., 2009).

Although H_2 may serve as the electron donor for microbial respiration in a fuel cell, most microbial fuel cells are designed around the oxidation of biodegradable organic carbon. Bacteria oxidize organic electron donors through conventional central metabolic pathways and produce reduced electron carriers such as NADH and NAD(P)H. In order to maintain these pathways, the electron carrier coenzymes must be regenerated at the cell membrane during respiration of the terminal electron acceptor, such as O_2, NO_3^-, Fe^{3+}, etc. During anode respiration in a microbial fuel cell, the anode becomes the terminal electron acceptor from the perspective of the bacteria despite the fact that the electron eventually terminates on oxygen during the cathode reaction (Fig. 3). Most naturally occurring terminal electron acceptors are limited in concentration and are consumed upon reduction during respiration; however, because the electrons that are collected from the organisms by the anode do not alter the anode surface and are rapidly conveyed to the cathode, the anode of the microbial fuel cell provides an inexhaustible electron acceptor.

Power production from microbial fuel cells is many orders of magnitude lower than conventional fuel cells. As of this review, volumetric power densities for mixed culture fuel cells is approximately 1.6 W/m^3 (Fan et al., 2008), and for pure cultures, it is 2.2 kW/m^3 (Nevin et al., 2008). For comparison, a small-scale, single-anode hydrogen fuel cell can produce 32,500 kW/m^3 (Kamitani et al., 2008). Although over the past 10 years, power densities from microbial fuel cells have increased more than 6 orders of magnitude (Logan, 2009), the theoretical limit of power density from a bacterium (16,000 × 10^4 kW/m^3 of cell volume) has yet to be approached (Logan, 2009). The leaps in power production are the result of greatly improved fuel cell design to minimize internal resistance (Clauwaert et al., 2008; Logan, 2009) and the identification of bacteria that conserve energy for growth while respiring an anode and are physiologically well suited for respiration of the anode (Clauwaert et al., 2008; Lovley, 2006). Although other limitations exist, quantifiable disadvantages of the microbial fuel cell have been determined, such as diffusion limitations of electron donor into (Clauwaert et al., 2008; Logan, 2008) and protons out of (Biffinger et al., 2008; Clauwaert et al., 2008; Franks et al., 2009; Torres et al., 2008) the anode biofilm.

Despite the kinetic disadvantages introduced by microbial biofilms at the anode, microorganisms offer the distinct advantage of vastly greater variety of fuel sources for electricity generation. Microbial fuel cells may utilize fatty acids (Bond et al., 2002; Bond and Lovley, 2003; Freguia et al., 2010; Liu et al., 2005a; Xing et al., 2008), carbohydrates (Chaudhuri and Lovley, 2003; Rabaey et al., 2003; Rezaei et al., 2009), alcohols (Catal et al., 2008; Kim et al., 2007; Xing et al., 2008), protein (Heilmann and Logan, 2006), and acid mine drainage (Cheng et al., 2007), among many others (Logan, 2008; Logan et al., 2006; Lovley, 2006a, 2006b; Rabaey and Verstraete, 2005). It is not unreasonable to assume that, if the fuel oxidation is thermodynamically feasible, there exists an organism or consortia that may utilize the fuel as their electron donor in a microbial fuel cell. Such broad fuel capability has not been achieved in abiotic fuel cells.

Although novel catalysts are leading the way for abiotic fuel cells to utilize carbohydrate fuels (Schechner et al., 2007; Wheeler et al., 2009), abiotic fuel cells generally require greatly elevated temperatures, pressures, concentrations of reactants, and/or extremes of pH, as well as expensive rare earth metal catalysts to achieve rates of reaction suitable for electricity production from reduced organic compounds (Barbir, 2005). In contrast, the catalytic activity of microorganisms enables organic matter oxidation at the anode at relatively mild environmental conditions (Allen, 1966; Logan, 2008; Lovley, 2006a).

Complex organic matrices may also serve as electron donors in microbial fuel cells. For example, domestic, agricultural, and industrial wastewaters (Bocher et al., 2008; Feng et al., 2008; Huang and Logan, 2008; Aelterman et al., 2006a; Liu et al., 2004; Min et al., 2005a) may serve as fuels for electricity generation in microbial fuel cells. Because of the complexities of the organic constituents of these fuels, complex microbial communities arise that enable anaerobic oxidation of the fuel coupled to electricity generation (Choo et al., 2006; Kim et al., 2004; Logan and Regan, 2006; Lovley, 2006a; Lovley et al., 2004; Parameswaran et al., 2010; Rabaey et al., 2007; White et al., 2009). The rates of electricity production from such complex fuels are an order of magnitude lower than rates generated by single substrate fuels (Aelterman et al., 2006b; Clauwaert et al., 2008; Rabaey et al., 2005). However, when a complex fuel is amended with a readily biodegradable fuel, higher-power output is observed, demonstrating that rates of hydrolysis or fermentation of the complex fuel constituents may be limiting electricity production (Rabaey et al., 2005b).

Complex organic matter in biomass and waste streams such as domestic wastewater are appealing fuel sources because they are readily available and broadly distributed. Perhaps, most importantly, the reduced carbon in these fuels was ultimately derived from primary production via photosynthesis, rendering energy harvesting relatively carbon neutral compared with conventional fossil energy fuels. Therefore, utilizing waste and biomass as fuel sources for direct energy conversion via a fuel cell is a particularly attractive option because it has the potential to meet the growing public demand for both reduced carbon emissions and greater energy efficiency and independence.

The development of microbial fuel cells as a more sustainable approach toward energy production has been relatively recent; however, the capabilities of microorganisms to generate electrical potential and current are not recent observations. In fact, 2010 marks the 100th anniversary of the first report of electrical potential development and current production from bacteria (Potter, 1910). In this report, Potter evaluated several organisms for their ability to generate electrical potential at the anode of a galvanic cell including; *"Bacillus coli,"* *B. fluorescens*, *B. violaceus*, and *Sarcina lutea*. Potter reported that *B. coli* produced electric potential and a "feeble" current in the galvanic cell. Immediately after this work, Potter clearly established the ability of microbial cultures, including bacteria and yeast, to produce electrical current (Potter, 1911) and noted that the rate of electrical potential development was correlated with the concentration of the microbial cells and organic carbon.

The link between microbial activity and redox conditions in natural systems followed Potter's work in 1920 (Gillespie, 1920). Gillespie, an agricultural scientist investigating reduction potentials of soils as an indicator of soil vitality, used a standardized electrode to evaluate redox potentials of pure and mixed cultures of bacteria as well as saturated soils. He noted a progressive increase in reducing potentials in all systems following amendment of organic carbon and consumption of oxygen. Gillespie did not speculate on producing current from soil potentials as Potter had for pure cultures of microorganisms, but he nevertheless concluded that reducing conditions in saturated soils were the result of microbial activity and were impacted by the rates of oxygen supply and respiration in the soil. Measurement of electrical potential with standardized electrodes in sediments has become an important tool for assessing geochemistry in natural (Bagander and Niemisto, 1978; Reimers, 2007; Whitfield, 1969) and contaminated (Brendel and Luther, 1995; Himmelheber et al., 2008) systems and gave rise to harvesting electricity from the sediment-water interface (Reimers et al., 2001; Tender et al., 2002).

Benthic Microbial Fuel Cells and Microbial Communities

A potential gradient exists between the sediment and oxygenated water that immediately overlies it. When an electrode is placed in the sediment and connected via a suitable conductor to an electrode in the adjacent overlying water, electrons flow from the reduced sediment to the oxygen at the cathode in a manner that is analogous to a galvanic cell (Fig. 4). The potential gradient that develops is the result of microbial oxidation of organic matter coupled to the reduction of oxidants in the sediment in an order which reflects the decreasing amount of energy available per mole of organic carbon for each electron acceptor, e.g., oxygen > manganese oxides > nitrate > iron oxides > sulfate (Froelich et al., 1979). Reduction products from microbial respiration accumulate in the sediment and result in the development of a depth-dependent potential gradient as measured by a reference electrode (Bagander and Niemisto, 1978; Schindler and Honick, 1971). The open-circuit potential between the sediment-water interface and a few centimeters below may be over 0.8 V (Nielsen et al., 2007). The fuel of the benthic microbial fuel cell is the sediment organic matter in sediment that microbes consume as their electron donor. In an open, natural environment both the fuel and the oxidant, the oxygen in the overlying water, are continuously replenished.

Benthic microbial fuel cells (BMFC) harvest electricity from the naturally occurring potential gradient (Reimers et al., 2001) and

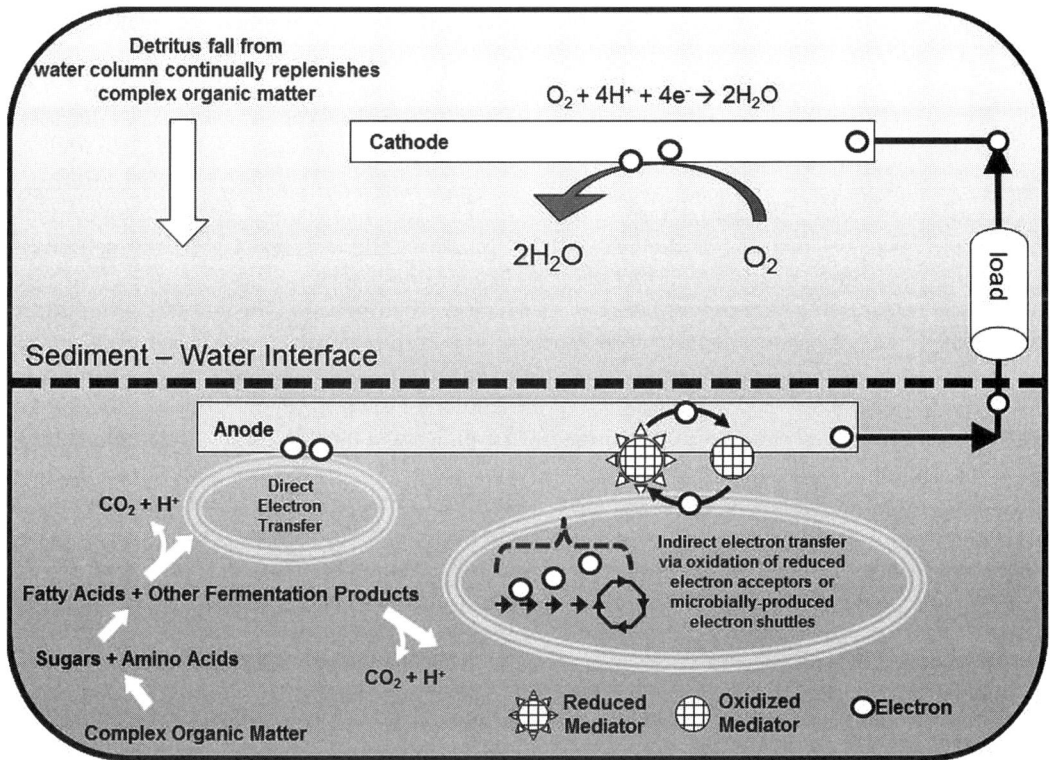

FIGURE 4 A model of a benthic microbial fuel cell (BMFC). A complex community of sediment organisms participates in hydrolysis and fermentation of detritus material to produce simple organic compounds such as acetate that are used by electrode-respiring bacteria enriched from the sediment. The cells transfer electrons to the anode of the BMFC via direct respiration of anode or through an electron mediator, such as a primary metabolite (HS^-, Fe^{2+}, or reduced humic material) or secondary metabolite electron shuttles such as flavinoid compounds.

are the first and most well-developed microbial fuel cell application (Tender et al., 2008). The electron donors at the anode were originally proposed to be microbially reduced electron acceptors that were oxidized at the anode of the fuel cell and recycled by the microbial community on or near the anode at the expense of further organic matter oxidation (Reimers et al., 2001). Examination of the microbial communities that become enriched on the anodes gave rise to alternate hypotheses, and it is now known that sediment microbes provide electrons at the anode through a variety of direct and indirect mechanisms summarized in Fig. 4.

During electricity harvesting from sediments, the microbial community attached to the anode decreases in diversity through enrichment of metal-reducing bacteria from the δ-proteobacteria. In marine and freshwater sediments, the δ-proteobacteria typically represent over 50% of the electrode-associated microbial community and of that phylotype, organisms from the family *Geobacteraceae* commonly represent over 50% (Holmes et al., 2004b; Nielsen et al., 2008, 2009; Reimers et al., 2006; Tender et al., 2002). *Geobacteraceae* are best known for the oxidation of organic matter coupled with the reduction of insoluble Fe^{3+} oxides (Lovley et al., 2004). Moreover, they are not known to produce electron shuttles and have and require attachment to Fe^{3+} oxides in order to reduce them (Lovley et al., 2004; Nevin and Lovley, 2002a, 2002b). This is significant because from the perspective of a bacterium the anode is an insoluble electron acceptor and therefore analogous to Fe^{3+} (and Mn^{4+}) oxides. These findings gave rise to the hypothesis that the anode may serve as the direct electron acceptor for microbial respiration and conservation of energy for growth in a BMFC (Fig. 3).

Several species of Geobacteraceae respire the anode of a microbial fuel cell as their terminal electron acceptor for growth, including Geobacter metallireducens, G. sulfurreducens, G. psychrophilus, Desulfuromonas acetoxidans, and Geopsychrobacter electrodiphilus (Bond et al., 2002; Bond and Lovley, 2003; Holmes et al., 2004a). In each of these studies, pure cultures of Geobacteraceae oxidized acetate and produced electricity at the anode. The growth medium was exchanged to remove planktonic cells and any electron-mediating compounds without loss or decrease in current production. These studies demonstrated that Geobacteraceae respire electrons liberated from organic matter and produce electricity in a galvanic cell with growth and attachment to the anode, the prima facie evidence that electricity production in BMFC may be the result of direct anode respiration by sediment bacteria (Bond et al., 2002; Tender et al., 2002) (Fig. 3).

In addition to Geobacteraceae, deployments of BMFC also enrich for bacteria closely related to the genera Desulfobulbus/Desulfocapsa (Holmes et al., 2004b; Reimers et al., 2006; Ryckelynck et al., 2005; Tender et al., 2002), another δ-proteobacterium, and Geothrix (Holmes et al., 2004b), an Acidobacterium. Poised-potential, pure-culture studies by Holmes and coworkers (Holmes et al., 2004a) demonstrated that Desulfobulbus propionicus oxidized S^0 to SO_4^{2-} with an electrode serving as an electron acceptor. Coupled with the findings of strong correlations between current production and porewater sulfide concentrations (Nielsen et al., 2007; Reimers et al., 2006) and the well-known electrochemical oxidation of sulfide at an anode (Ateya and Al-Kharafi, 2002) it is likely that microbe-electrode cycling of sulfur compounds as primary metabolites (Fig. 4), including sulfur disproportionation (De Schamphelaire et al., 2008), is an important component of electricity production from BMFC (Holmes et al., 2004b; Lovley, 2006a; Nielsen et al., 2007; Reimers et al., 2006; Ryckelynck et al., 2005). However, inhibition of sulfate reduction with molybdate appears not to interfere with electricity production from some BMFC deployments (Nielsen et al., 2009). Given the robustness of current production in the absence of sulfate reduction, sulfate may not be a significant component of the electricity-linked sulfur cycling in the anode biofilm. Sulfur is the best studied of the primary metabolites of microbial respiration which may serve as reductants at the anode, but others have noted that the oxidation of microbially reduced humic acids (De Schamphelaire et al., 2008; Lovley, 2006a) and ferrous iron (De Schamphelaire et al., 2008) may create alternate redox cycles in anode-associated biofilms in BMFC.

Microbial fuel cell studies with Geothrix fermentans reveal that it will respire and grow on an electrode, aided by an electron shuttle that enhances electricity production (Bond and Lovley, 2005) and permits reduction of insoluble Fe(III) oxide (Nevin and Lovley, 2002a). It does not require a mediator to respire the anode, but current increases as the mediator accumulates in the growth medium over time. G. fermentans was the first bacterium outside the Proteobacteria shown to be capable of respiration of an anode for electricity production and growth. Although a relatively small portion of the electrode-attached enrichment on the anode of BMFC (Holmes et al., 2004b), the identification of microbes related to Geothrix on sediment-deployed anodes suggests that a redox cycle with microbially produced electron shuttles that are oxidized at the anode may also contribute to electricity production in BMFC (Fig. 4). The importance of secondary metabolites as redox mediators in engineered microbial fuel cell reactors is well known and may enable bacteria that are otherwise incapable of electron transfer to an anode to participate in electricity generation in an anode biofilm (Marsili et al., 2008; Pham et al., 2008a, 2008b; Rabaey et al., 2004, 2005a;

Stams et al., 2006). Nevertheless, evidence for microbially-produced, secondary metabolites as electron shuttles in the anode community of BMFC is largely circumstantial and their relative contribution to electricity generation not known.

BMFC as a source of power is emerging as a robust (Tender et al., 2008) and evolving technology (Nielsen et al., 2007) for remote generation of electricity. The microbial communities and resulting biogeochemistry at the anodes that enable their stability are complex and still poorly understood (De Schamphelaire et al., 2008). Nevertheless, three distinct mechanisms of electron transfer to anodes of BMFC are recognized as important components of electricity production from the sediment-water interface (Fig. 4): (i) direct respiration of the anode by electrode-respiring bacteria (Bond et al., 2002; Holmes et al., 2004b; Reimers et al., 2001; Tender et al., 2002); (ii) oxidation of a microbially reduced, sediment-borne compound at the anode such as sulfide, ferrous iron, or naturally occurring electron shuttles (Lovley, 2006a; Reimers et al., 2001, 2006; Ryckelynck et al., 2005; Tender et al., 2002); or (iii) microbially produced secondary metabolite electron shuttles such as phenazine and flavinoid compounds (De Schamphelaire et al., 2008; Lovley, 2006b).

Microbial Fuel Cell Reactors and Microbial Communities

Electricity production by microorganisms occurs across a phylogenetically broad spectrum of the living world including both *Bacteria* and *Eukarya*. Even though no known selective pressure exists in nature for microbial respiration of an anode in an electrochemical cell, electricity production in a fuel cell is nevertheless commonly observed and very broadly distributed among the *Bacteria* and has been identified in α-, β-, γ-, δ-*Proteobacteria*, and phyla *Cyanobacteria*, *Firmicutes*, and *Acidobacteria*. This phylogenetic diversity among electricity-producing bacteria, and hence the mechanisms for electron transfer from the reduced electron carriers in the cell to the insoluble and external anode take on a similar physiological diversity. Knowledge of these mechanisms of anode respiration led to fuel cell design and operational parameters that are linked to the mechanism(s) of electron transfer for optimization of power outputs for a particular bacterium or culture (Chang et al., 2005; Cheng et al., 2006a, 2006b; Cheng and Logan, 2007b; Clauwaert et al., 2008; Deng et al., 2010; Fredrickson et al., 2008; Logan et al., 2006; Lovley, 2006a, 2006b; Moon et al., 2005; Oh and Logan, 2006a; Pham et al., 2005, 2009; Ringeisen et al., 2006; Zuo et al., 2007).

As described above, the closest analogues of current-harvesting electrodes in nature are probably Fe(III) and Mn(IV) oxides, all serving as insoluble, extracellular electron acceptors. Fe(III)-reducing bacteria are commonly identified in the anode-associated microbial community, most notably those of the genus *Geobacter* (Bond et al., 2002; Chae et al., 2009; Freguia et al., 2010; Ha et al., 2008; Holmes et al., 2004b; Jung and Regan, 2007; Tender et al., 2002; Xing et al., 2009), but also *Desulfobulbus* (Holmes et al., 2004b), *Desulfobacterales* (Holmes et al., 2004b), and *Geothrix* (Holmes et al., 2004b). *Shewanellaceae*, a well-known family of Fe(III)-reducing organisms, are not known to be enriched from the environment on anodes of microbial fuel cells but nevertheless oxidize fermentation products coupled to electricity generation (Baron et al., 2009; Biffinger et al., 2007, 2009; Bretschger et al., 2007; Gorby et al., 2006; Kim et al., 1999, 2002; Lanthier et al., 2008; Marsili et al., 2008; Ringeisen et al., 2006). *Rhodoferax*, another Fe(III)-reducing genus, oxidizes glucose with high coulombic efficiency in a microbial fuel cell (Chaudhuri and Lovley, 2003; Liu et al., 2007).

Lovley proposes (Lovley, 2006a, 2008a, 2008b; Lovley et al., 2004) that the complete oxidation of complex organic matter with electrodes or Fe(III) oxides as the terminal electron acceptor necessitates cooperation of a consortia of fermentative and Fe(III)-reducing microbes; Fe(III)-reducing microorganisms metabolize the fermentation products and organic compounds that fermentative microbes do not readily metabolize to CO_2 with Fe(III) oxides or current-harvesting electrodes serving as the electron acceptor. Evidence that these consortia are important for electricity production comes from the fact that most mixed microbial communities produce more power (Freguia et al., 2007; Liu et al., 2005a; Logan et al., 2006; Rabaey et al., 2004, 2005; Rabaey and Verstraete, 2005; Ren et al., 2007b; Watson and Logan, 2010) than pure cultures, with the exception of *Rhodopseudomonas palustris* (Xing et al., 2008) and *Geobacter sulfurreducens* (Nevin et al., 2008).

During oxidation of complex organic substrates in microbial fuel cells, the community that becomes enriched on the anode contains a variety of populations. Most notably are those that ferment complex substrates into simpler compounds that the electrode-reducing populations may utilize. However, identification of a particular phylotype on the anode does not necessarily imply that members of that particular group contribute directly to electricity generation (Lovley, 2008b) or that their presence is even beneficial. For instance, methanogenesis or aerobic respiration may decrease coulombic efficiency by directing electrons away from electricity production (Clauwaert et al., 2008; Lu et al., 2009). Fermentative organisms, on the other hand, are not known for electricity production yet are indispensable for electricity generation from complex assemblages of organic matter (Ren et al., 2007b, 2008). These organisms break down complex substrates into simpler compounds that are readily oxidized by the electrode-reducing community (Lovley, 2006a, 2008b). Pant and coworkers (2010) produced a comprehensive review of all organic substrates that have been explored as a fuel source in microbial fuel cells. They catalog the fuel cell architecture, operating conditions, inocula, and power outputs.

Molecular analyses of a number of microbial fuel cells inoculated with marine or freshwater sediments (Holmes et al., 2004b, 2007a; Tender et al., 2002) or other fuel sources, such as organic wastewater (Jong et al., 2006; Kim et al., 2004), activated sludge (Aelterman et al., 2006a; Kim et al., 2004; Lee et al., 2003; Wrighton et al., 2008) or oligotrophic river water (Phung et al., 2004), have indicated that fermentative microorganisms are important members of the microbial community on the anode surface. Some of the most common fermentative organisms associated with electricity production from complex organic matter include *Firmicutes* (Chae et al., 2009; Choo et al., 2006; Park et al., 2001; Tender et al., 2002; Wrighton et al., 2008), *Bacteroidetes* (Holmes et al., 2007a; Jung and Regan, 2007; Rismani-Yazdi et al., 2007), and β-proteobacteria (Chae et al., 2009).

Several microorganisms with fermentative types of metabolism have been able to grow and produce current in microbial fuel cells; however, electron transfer to the electrode carried out by the majority of these organisms was relatively inefficient (Pham et al., 2003; Park et al., 2001). For example, when *Clostridium butyricum* was grown in an anodic chamber with glucose as the electron donor, only 0.04% of the electrons available in glucose were transferred to the electrode surface (Park et al., 2001), and less than 0.02% of the electrons available from yeast extract (calculated from the chemical oxygen demand) were transferred to an electrode by *Aeromonas hydrophila* (Pham et al., 2003). In addition, only ~25% of the electrons available from the incomplete oxidation of pyruvate, lactate, and propionate were transferred to the electrode surface by *D. propionicus*, via a mixed fermentative metabolism (Holmes et al., 2004a). Rezaei and colleagues (2009) recently isolated a fermentative organism, a unique strain of *Enterobacter cloacae*, which produces electricity from cellulose in the absence of exogenous electron shuttles, but with low (14%) coulombic efficiency.

The broad diversity of organisms that are enriched on anodes from sediments in the case of BMFC, or from complex consortia such as municipal or industrial wastewaters, has led to the examination of many pure-culture bacteria in microbial fuel cells and the discovery of electricity production among a broad spectrum of the bacterial world. Although originally considered an ability that arose from the relatively unique aptitude of certain bacteria to externalize electrons for respiration of insoluble Fe(III) and Mn(IV) oxides, through the study of pure cultures that respire anodes in galvanic cells, a much more complex suite of mechanisms for electron transfer to electrodes has become apparent.

MICROBIAL ELECTRON TRANSFER TO ANODES

The anode of a galvanic cell is an insoluble electron acceptor. In order for current to be produced, electrons generated in the central metabolism inside the cell must be externalized across the cell membrane. The respiration of Fe(III) and Mn(IV) oxides may occur through direct and indirect means (Lovley et al., 2004). A similar set of mechanisms are emerging for electron transfer to anode (Lovley, 2006b; Schröder, 2007). The direct process for electron transfer occurs through attachment and a physical connection between the cell and the anode via redox-labile or -conductive structures on the outside of the cell such as cytochromes or conductive pili (Fig. 3). Indirectly, a cell may transfer electrons to the anode by producing a reduced primary metabolite such as sulfide or a secondary metabolite electron shuttle that microbially reduced and reoxidized at the anode, creating a redox cycle. In all cases,

TABLE 1 Organisms able to produce current in mediator-less fuel cells

Organism name	Donor	Reference(s)
Shewanella oneidensis	Lactate	Bretschger et al., 2007; Gorby et al., 2006; Lanthier et al., 2008
Shewanella baltica	Lactate	Liang et al., 2009
Shewanella marisflavi	Lactate	Huang et al., 2010
Aeromonas hydrophila		Pham et al., 2003
Geobacter lovleyi	Acetate	Strycharz et al., 2008
Geopsychrobacter electrodiphilus	Acetate	Holmes et al., 2004a
Geobacter metallireducens	Acetate	Gregory et al., 2004; Min et al., 2005b
Rhodoferax ferrireducens	Glucose	Chaudhuri and Lovley, 2003
Pseudomonas aeruginosa	Glucose	Luo et al., 2009; Rabaey et al., 2005a
Aeromonas sp. strain ISO2-3	Glucose	Chung and Okabe, 2009
Enterobacter cloacae	Cellulose	Rezaei et al., 2009
Geothrix fermentans	Acetate	Bond and Lovley, 2005
Comamonas denitrificans DX-4	Acetate	Xing et al., 2010
Clostridium cellulolyticum	Cellulose	Niessen et al., 2005
Thermincola sp. strain JR	Acetate	Wrighton et al., 2008
Geobacter sulfurreducens	Acetate	Bond and Lovley, 2003; Reguera et al., 2006
Clostridium butyricum	Glucose	Park et al., 2001
Aeromonas hydrophila	Yeast extract	Pham et al., 2003
Desulfobulbus propionicus	Lactate	Holmes et al., 2004a
Clostridium cellulolyticum	Carboxymethylcellulose	Ren et al., 2007a

the cell requires an electrochemical connection between the intracellular electron carriers, such as NAD, FAD, or menaquinone, and the insoluble anode of the galvanic cell. Conservation of energy and growth using Fe(III) and Mn(IV) oxides are now known to be relatively widespread among *Bacteria* and *Archaea* (Lovley et al., 2004). Similarly, the ability to transfer electrons to an anode is broadly distributed (Table 1) and arises from the need of all cells to generate energy through reduction of an electron acceptor. Not surprisingly, a great variety of specific mechanisms has emerged for bacteria to transfer electrons to an electrode; these mechanisms are comprehensively reviewed by Schröder (2007).

Indirect Electron Transfer: Metabolic Products and Exogenous Mediators

Indirect electron transfer to the electrode may be defined as any externalization of electrons to the electrode that is enabled by a soluble carrier that transports electrons from inside the cell to the electrode (Fig. 5). These soluble carriers are sometimes referred to as electron mediators or electron shuttles in the case of exogenous and endogenous carriers and, in essence, enable the organism to produce an electrochemically active product from their metabolism. Indirect electron transfer mechanisms may take the form of primary metabolic products such as H_2 or formic acid (Aston and Turner, 1984; Karube et al., 1977; Niessen et al., 2005; Schröder et al., 2003) in the case of fermentative organisms or sulfide by sulfate (Rabaey et al., 2006; Sisler, 1961, 1962; Cordas et al., 2008; Habermann and Pommer, 1991) or sulfur-reducing organisms (Holmes et al., 2004b). Alternatively, the soluble carrier may be an endogenous electron shuttle, produced by the bacterium as a secondary metabolite that aids or enables electron transfer in pure culture (Bond and Lovley, 2005; Freguia et al., 2009; Lanthier et al., 2008; Marsili et al.,

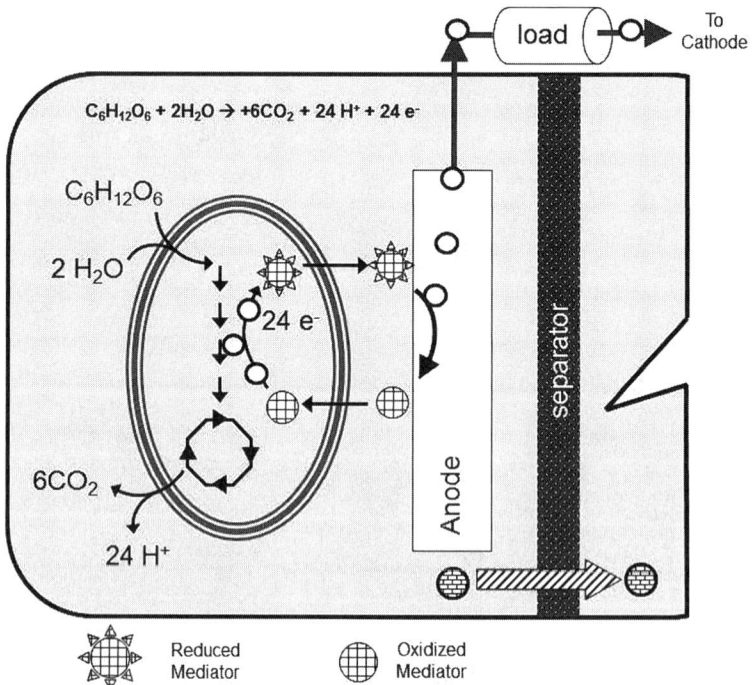

FIGURE 5 The indirect electron transfer mechanism for respiration of the anode in a microbial fuel cell. Electron transfer to the anode is enabled by a primary metabolite such as an electron shuttle (e.g., HS^-, H_2) or secondary metabolites electron shuttles such as flavins or phenazines.

2008; Rabaey et al., 2004, 2005a) or in mixed cultures (Pham et al., 2008a, 2008b; Rabaey et al., 2004, 2005a). Last, the soluble carrier may be an exogenous mediator, which is reduced by the bacteria and reoxidized at the anode, which enables or enhances current production such as methylene blue, ferricyanide, benzyl viologen, phenoxoazines, to name a few described in reviews (Allen, 1966; Allen and Bennetto, 1993; Roller et al., 1984) and in a recent survey of mediators for electricity production by *Clostridium cellulolyticum* (Sund et al., 2007).

ARTIFICIAL MEDIATORS

The earliest microbial fuel cell studies utilized commonly cultured aerobic organisms such as *Escherichia coli*, *Proteus* spp., *Pseudomonas* spp., and *Bacillus* spp. These bacteria produced very small currents that were greatly enhanced by introduction of exogenous, redox-active compounds that enabled or greatly enhanced electron transfer rates from the bacteria to the anode (Cohen, 1931; Davis, 1963; Delaney et al., 1984; Emde and Schink, 1989, 1990; Park and Zeikus, 2000; Roller et al., 1984; Tanaka et al., 1983; Thurston et al., 1985). The increase in electrical current is largely attributed to decreased overpotentials at the anode, but the soluble shuttle also enables planktonic cells to participate in electricity production and grow in the anode medium as observed by increasing cell densities (Emde and Schink, 1989, 1990; Park and Zeikus, 2000).

Neutral red ($E_0' = -0.325$ V) is the most well-studied artificial mediator of electron transfer in bacteria (Park and Zeikus, 2000). In addition to electrodes, neutral red may enable Fe(III) (McKinlay and Zeikus, 2004) and fumarate (Park and Zeikus, 1999) reduction. Neutral red may be reduced by a variety of enzymes, including hydrogenase and formate dehydrogenase, as well as receive electrons from NADH (McKinlay and Zeikus, 2004). Neutral red and many other redox-active dyes may participate in electron transfer reactions with microbial electron carriers, such as *c*-type cytochromes (Xie and Dong, 1992), quinones (Sanchez et al., 1995), and NAD (Miyawaki and Yano, 1992; Surya et al., 1994), as well as enzymes that catalyze redox reactions such as nitrite (Cristina et al., 1993) and nitrate reductase (Willner et al., 1992), fumarate reductase (Sucheta et al., 1993), and hydrogenase (Schlereth and Fernandez, 1992). In microbial fuel cells that rely on artificial mediators, the mediator enables two redox couples to occur: the first couple's reduction of the mediator to oxidative metabolism and the second couple's oxidation of the mediator to the reduction of the electrode. Ideally, the mediator has a low E_0' to readily intercept electrons from oxidative metabolism and maximize energy generation.

Artificial mediators may also enhance electron recovery (a.k.a. coulombic efficiency) for bacteria that are unable to (or in a limited way) produce electricity in their absence. *Proteus vulgaris* cultures completely oxidize sugars to CO_2 and achieve relatively high coulombic efficiency (up to 63%) using thionine as an electron mediator (Allen and Bennetto, 1993; Kim et al., 2000; Thurston et al., 1985). Similar, high coulombic efficiencies may be achieved using humic acids or the artificial analog, anthraquinone 2,6-disulfonate (AQDS) (Milliken and May, 2007). The coulombic efficiency in these cases not only depends on the mediator, but also strongly depends on the substrate fuel (Choi et al., 2001; Kim et al., 2000). Many of the bacteria that require or are enhanced by artificial mediators produce fermentation products or other soluble metabolites that may not be electrochemically reactive at the anode and reduce the coulombic efficiency (Thurston et al., 1985). In the absence of mediators these bacteria exhibit very low coulombic efficiencies. *C. butyricum* grown in an anodic chamber with glucose as the electron donor achieved only 0.04% recovery of the electrons available (Park et al., 2001). Similarly, less than 0.02% of the electrons available from yeast extract (calculated from the chemical oxygen demand) were transferred to an electrode by *Aeromonas hydrophila* (Pham et al., 2003).

In addition to the inefficiencies of electron recovery for organisms that incompletely oxidize substrates and require mediators, from an application standpoint, fuel cells are thermodynamically open systems. Therefore, artificial mediators would need to be continuously or semicontinuously fed to maintain sufficient concentrations, unless the mediator was bound to the cells (Park and Zeikus, 1999), bound to the electrode (Park and Zeikus, 2003), or remained and was cycled in the anode biofilm as proposed by Rabaey and coworkers (2004, 2005a) for biologically produced mediators.

METABOLITES AS MEDIATORS OF ELECTRON TRANSFER

Indirect electron transfer by microbially produced electron carriers includes mechanisms by which electrochemically active compounds as primary metabolites from catabolic processes such as H_2 or S^{2-} shuttle electrons from central metabolism and become oxidized at the anode (Fig. 5). Alternatively, anabolic processes that produce redox-labile, secondary metabolites may shuttle electrons to the anode. These secondary metabolites include compounds such as phenazine or flavinoid compounds that mediate electron transfer from the cell to the anode. The distinction between the primary and secondary metabolites arises from the fact that secondary metabolites do not necessitate an exogenous electron acceptor and the secondary metabolite may be regenerated at the anode for multiple electron transfer cycles. The reader is referred to a review by Schröder (2007).

Microbial fuel cells that take advantage of electrochemically reactive sulfide as a primary metabolite of microbial respiration were first demonstrated almost 50 years ago (Sisler, 1961, 1962). These fuel cells utilized *Desulfovibrio*, a dissimilatory sulfate-reducing bacterium that partially oxidizes lactate to acetate and produces sulfide as a metabolite. Similarly, a discontinuous microbial fuel cells was proposed that utilized mixed cultures of *Desulfovibrio desulfuricans* and *P. vulgaris* or *D. desulfuricans*, *P. vulgaris*, and *E. coli* to oxidize a variety of organic carbon compounds (glucose, fructose, cane sugar, starch, and crude oil) and to produce sulfide, which was stored for discontinuous energy production from sulfide oxidation at the anode similar to a battery (Habermann and Pommer, 1991). *Desulfovibrio* spp. were recently proposed to directly transfer electrons to an anode (Cordas et al., 2008); however, evaluation of the direct and indirect components of electricity generation are complicated by the reactivity of sulfide at the anode (Dutta et al., 2008; Sun et al., 2009). The application of microbial sulfate reduction in a microbial fuel cell may serve as a power source for treatment of sulfate in wastewaters (Rabaey et al., 2006). Limiting this potential application is the potential for deposition of elemental sulfur at the anode which may necessitate periodic removal (Dutta et al., 2008).

Fermentative microorganisms have been extensively studied in microbial fuel cells. As described previously, current production and coulombic efficiency are typically low but may be enhanced by the addition of artificial mediators to enable or increase rates of electron transfer to the anode. However, the primary metabolites that arise from fermentation may be electrochemically active and act to transfer electrons from the fuel to the anode. Clearly, hydrogen falls into this category and was first proposed in a microbial fuel cell by Karube and coworkers in 1977 (Karube et al., 1977). Since then, large improvements in production rates (Niessen et al., 2004, 2005) and the ability to utilize complex substrates for hydrogen production (Mohan et al., 2008; Niessen et al., 2004, 2005) have greatly improved the prospects for hydrogen fermentation and direct oxidation at the anode of a microbial fuel cell (MFC), but the accumulation of acidic fermentation products during H_2 production may shift the fermentation pathway from H_2 to alcohol production and inhibit current production (Niessen et al., 2005).

Microbial H_2 production for oxidation in a fuel cell may be more efficient and stable utilizing phototrophs (Cao et al., 2008; Cho et al., 2008; Rosenbaum et al., 2005a, 2005b). For complex substrates, a two-step process is necessitated whereby a fermentative organism produces organic acids and alcohols and *Rhodopseudomonas sphaeroides* photo-oxidized these products to produce H_2 (Rosenbaum et al., 2005b). An important drawback of H_2 oxidation as a microbially produced fuel is the requirement for expensive metal catalysts for rapid anode reaction. Likewise, it has been suggested that, with the appropriate catalyst, other microbial metabolites such as ethene (Davis, 1963) and alcohols (Lewis, 1966) may be utilized for electricity production in a microbial fuel cell. Regardless of the fermentative culture, oxidation secondary metabolites at the anode complicate analysis of the relative contribution of microbial electricity production (Cao et al., 2008) and abiotic reaction of the metabolite.

Ammonium as a microbially produced mediator for electricity generation was postulated over 40 years ago (Lewis, 1966). Ammonium, nitrogen at its lowest oxidation state, may be electrochemically reactive and directly oxidize at the anode or alternatively serve as an electron donor for bacteria that might respire the anode. He and coworkers (2009b) recently demonstrated ammonium-dependent electricity generation in a microbial fuel cell. In this study, ammonium was amended to the medium as an electron donor, rather than produced as a primary metabolite. Nevertheless, microbial nitrate reduction or ammonification produces ammonia as a primary metabolite that may be oxidized at an anode. However, the relative contribution of the abiotic reaction of ammonium at the anode and microbial oxidation with the electrode as the electron acceptor remains an open question (He et al., 2009b).

Secondary metabolites serving as electron shuttles are well known during respiration of insoluble Fe(III) and Mn(IV) oxides and have surfaced as potentially important mechanisms for bacteria to externalize electrons to anodes as well. Secondary metabolite electrons shuttles have been identified during dissimilatory Fe(III) reduction by *Shewanella oneidensis* (Marsili et al., 2008; Newman and Kolter, 2000) and *G. fermentans* (Bond and Lovley, 2005; Nevin and Lovley, 2002b). They enable *Shewanella* species to respire Fe(III) oxide at substantial distances from the cell surface (Lies et al., 2005; Nevin and Lovley, 2002b; Rosso et al., 2003) and mediate electricity generation at the anode (Marsili et al., 2008). Other species produce extracellular electron shuttles that mediate electron transfer to electrodes, including *Pseudomonas aeruginosa*, which produces phenazine electron shuttles (Rabaey et al., 2004, 2005a), and the homolactic fermenter, *Lactococcus lactis*, which produces several membrane-associated quinones that mediate electron transfer to extracellular electron acceptors like Fe(III) and current harvesting anodes (Freguia et al., 2009). These secondary metabolites, as electron transfer mediators to anodes, have been thoroughly reviewed (Schröder, 2007).

Like artificial mediators, in an open system, secondary metabolites as electron shuttles may not play a significant role where insoluble electron acceptors are available because they may be flushed or diffused away from the electrode-reducing community. When *G. fermentans* was grown in a microbial fuel cell, power was reduced by 50% when medium was replaced (Bond and Lovley, 2005). In addition, production of these electron shuttles is energetically expensive, and where diffusion is not limiting, an organism would need to continuously replace any shuttles that were removed from the environment. Therefore, it is likely that organisms that produce these electron shuttles are at a competitive disadvantage in situ. However, Rabaey and colleagues (2004, 2005a) have proposed that phenazine shuttles may remain in *P. aeruginosa* biofilms for redox cycling to the anode. They also propose that the phenazine produced by *P. aeruginosa* may also be used by other bacteria (Pham et al., 2008a, 2008b). However, regardless of the in situ relevance, studies of bacteria that produce secondary metabolite electron shuttles provide important information about the mechanisms of electron transfer to insoluble acceptors and greatly expand the potential applications of microbial fuel cells.

The issue of diffusion and loss of secondary metabolite mediators may explain why organisms that are able to directly transfer electrons without production of shuttles, such as bacteria from the family *Geobacteraceae* (Nevin and Lovley, 2000), are frequently the predominant members of communities associated with Fe(III)-reducing subsurface environments and the anode surface of many microbial fuel cells. For example, the *Geobacteraceae* can account for up to 90% of the total bacterial population in a diversity of sedimentary environments in which Fe(III) reduction is an important process (Anderson et al., 2003; Holmes et al., 2002, 2007b; Roling et al., 2001; Vrionis et al., 2005), and over 70% of the microbial communities attached to current-harvesting electrodes (Holmes et al., 2004b). One defining characteristic of *Geobacter* species is the occurrence of abundant *c*-type cytochromes that enable direct electron transfer to Fe(III) oxides; *Geobacter metallireducens* and *G. sulfurreducens* contain over 100 putative *c*-type cytochrome genes (Aklujkar et al., 2009; Methe et al., 2003).

Direct Electron Transfer: Outer Membrane Cytochromes and Conductive Pilin

ELECTRICITY PRODUCTION BY *GEOBACTERACEAE*

Microorganisms do not encounter graphite electrodes in the natural environment. Therefore, it seems likely that organisms that are able to transfer electrons to current-harvesting anodes utilize mechanisms that are similar to naturally occurring extracellular electron acceptors such as insoluble Fe(III) and Mn(IV) oxides. A substantial volume of research has been done on the mechanisms of electron transfer to these acceptors in several dissimilatory iron- and manganese-reducing bacteria, most prominently those from the families *Geobacteraceae* and *Shewanellaceae*. Figure 6 shows a comparison of the models for electron transfer to the anode by *G. sulfurreducens* electron and *S. oneidensis* developed by transcriptomic, proteomic, and genetic studies of each organism transferring electrons to insoluble Fe(III) and Mn(IV) oxides and anodes of galvanic cells.

The physiology of electron transfer to electrodes and Fe(III) and Mn(IV) oxides by *Geobacteraceae* has primarily been studied with *G. sulfurreducens*, because a genetic system (Coppi et al., 2001) and complete genome sequence (Methe et al., 2003) are available. A substantial amount of research has indicated that *G. sulfurreducens* utilizes mechanisms for electron transfer to current-harvesting anodes that are similar to those used for transfer to naturally occurring extracellular electron acceptors such as Fe(III) and Mn(IV) oxides (Fig. 6). A majority of proteins that appear to be involved in electron transfer to insoluble electron acceptors are *c*-type cytochromes (Ding et al., 2008; Holmes et al., 2006a; Lovley, 2006a; Lovley et al., 2004; Nevin et al., 2008, 2009). Transcriptomic, proteomic, and genetic studies have shown that all of the *c*-type cytochromes that appear to be most important for electron transfer to current-harvesting electrodes also appear to be involved in transfer to Fe(III) and/or Mn(IV) oxides (Table 2).

Initial microbial fuel cell studies of *G. sulfurreducens* were done with cells growing in a thin monolayer on the anode surface (Bond and Lovley, 2003). In these early studies, current densities were low (0.08 to 0.39 A/m^2) and most of the cells were growing in a layer that was rarely more than a few cells thick. More recently, *G. sulfurreducens* has been grown in flowthrough fuel cells (Nevin et al., 2008; Reguera et al., 2006). These newer fuel cell systems achieve significantly higher current densities (4.56 A/m^2) with thick biofilms on the anode surface (Nevin et al., 2008; Reguera et al., 2006). The observation of thick biofilms suggested that *G. sulfurreducens* does not require direct cell contact with the anode surface for electron transfer. Subsequent physiological studies of *G. sulfurreducens* demonstrated that they produce and use electrically conductive

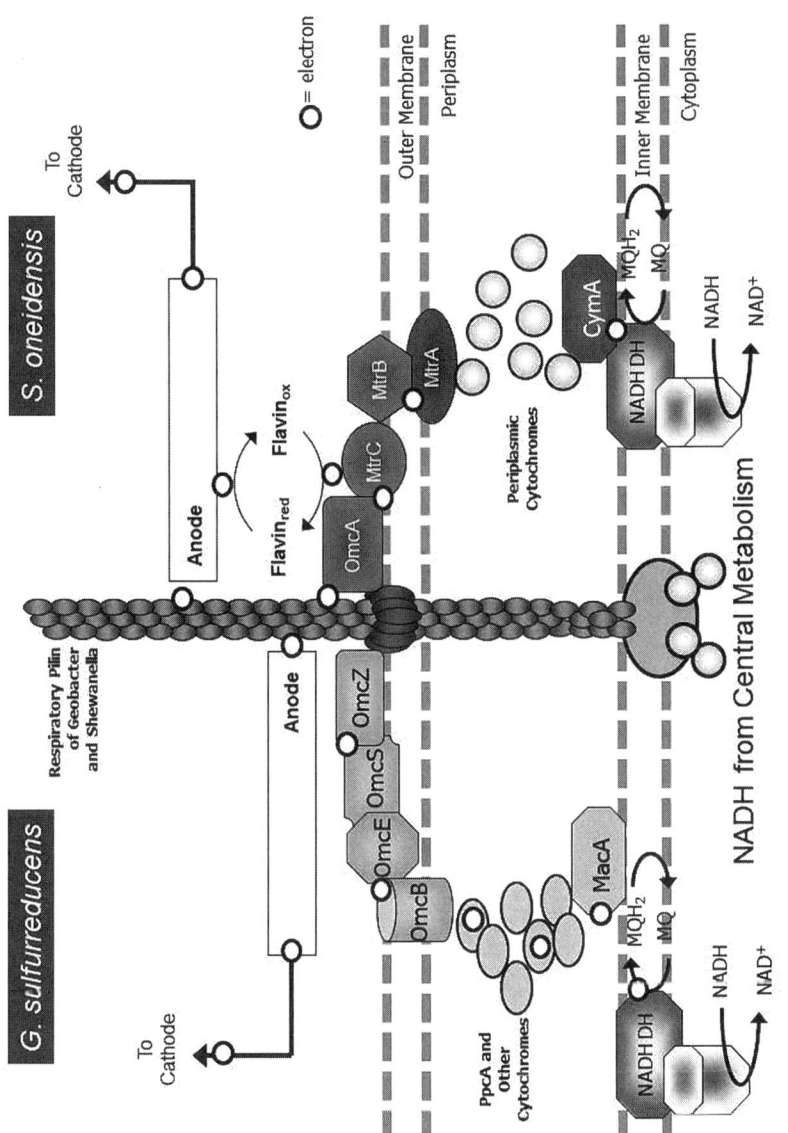

FIGURE 6 Models of the molecular basis for electron transfer to anodes by *Shewanella oneidensis* and *Geobacter metallireducens*. Oxidation of organic matter in central metabolism of both organisms produces reduced, intracellular electron carriers such as NADH. Electrons are transferred through the membranes via a series of intermediates to outer membrane cytochromes.

TABLE 2 Proteins that transcriptomic, proteomic, and genetic studies show to be involved in electron transfer by *G. sulfurreducens* to current-harvesting electrodes and/or Fe(III) and Mn(IV) oxides

Gene name	Biofilm structure	Involved in Fe(III) oxide reduction	Involved in Mn(IV) oxide reduction	References
c-Type cytochrome, OmcS (GSU2504)	Monolayer	Yes	Yes	Holmes et al., 2006; Mehta et al., 2005
c-Type cytochrome, OmcE (GSU0618)	Monolayer and thick biofilm	Yes	No	Holmes et al., 2006; Mehta et al., 2005
c-Type cytochrome (GSU2937)	Monolayer	Yes	No	Holmes et al., 2006; Mehta et al., 2005
c-Type cytochrome (GSU0594)	Monolayer	Yes	No	Holmes et al., 2006; Mehta et al., 2005
c-Type cytochrome, PpcJ (GSU2494)	Monolayer	Yes	No	Holmes et al., 2006
c-Type cytochrome (GSU2934)	Monolayer	Yes	No	Ding et al., 2008; Holmes et al., 2006
Pilin domain protein, PilA (GSU1496)	Thick biofilm	Yes	Yes	Ding et al., 2008; Nevin et al., 2009; Reguera et al., 2005
c-Type cytochrome OmcB (GSU2737)	Thick biofilm	Yes	No	Ding et al., 2008; Kim et al., 2006; Leang et al., 2003; Nevin et al., 2009
c-Type cytochrome (GSU2732)	Thick biofilm	Yes	Yes	Ding et al., 2008; Nevin et al., 2009
c-Type cytochrome, OmcZ (GSU2076)	Thick biofilm	No	Yes	Ding et al., 2008; Nevin et al., 2009

pili (a.k.a. nanowires) that may enable cells not in direct, outer membrane, contact with the electrode to respire and produce electricity (Lovley, 2006b; Nevin et al., 2009; Reguera et al., 2005, 2006).

Interestingly, molecular comparisons of the thick biofilm and monolayer fuel cell systems have suggested that several mechanisms of electron transfer are utilized by *G. sulfurreducens* to transfer electrons to the anode surface and that they have great redundancy. A number of *c*-type cytochromes are involved in electron transfer by *G. sulfurreducens* under both monolayer and biofilm growth, but the specific cytochromes vary. For example, transcriptomic and genetic studies of *G. sulfurreducens* grown in a monolayer on the anode indicated that the outer membrane *c*-type cytochrome, OmcS, is necessary for electron transfer to the anode surface (Holmes et al., 2006; Lovley, 2006b). However, OmcS does not seem to be important for growth of *G. sulfurreducens* in the anode biofilm, and the gene encoding OmcS was actually downregulated in anode-biofilm-grown cells (Nevin et al., 2009). In addition, the outer membrane *c*-type cytochromes, OmcB and OmcZ, and proteins involved in the synthesis of pilin proteins (Reguera et al., 2005, 2006) were important for growth of *G. sulfurreducens* in thick biofilms, but did not appear to be involved in growth in a monolayer (Holmes et al., 2006; Nevin et al., 2009). The only protein that appears to be involved in electron transfer in both systems is the outer membrane c-type cytochrome, OmcE (Holmes et al., 2006; Nevin et al., 2009).

ELECTRICITY PRODUCTION BY SHEWANELLACEAE

Numerous members of *Shewanellaceae* are capable of dissimilatory Fe(III) reduction (Fredrickson et al., 2008; Konstantinidis et al., 2009; Lovley et al., 2004; Nealson et al., 2002) and have been shown to produce current (Biffinger et al., 2007; Bretschger et al., 2007; Crittenden et al., 2006; Gorby et al., 2006; Kim et al., 1999, 2002; Lanthier et al., 2008; Marsili et al., 2008) and conserve energy for growth (Lanthier et al., 2008) in microbial fuel cells. The vast majority of studies on their mechanism of electron transfer to insoluble acceptors have been conducted with *S. oneidensis* MR-1 (Fredrickson et al., 2008; Konstantinidis et al., 2009). Similar to *G. sulfurreducens*, this organism has been extensively studied because a complete genome (Heidelberg et al., 2002) and genetic system (Beliaev and Saffarini, 1998; Beliaev et al., 2001; Bouhenni et al., 2005) are available. Analysis of the genome of *S. oneidensis* indicates that this organism has 42 putative c-type cytochromes, less than half of the potential c-type cytochromes found in *Geobacteraceae* genomes (Heidelberg et al., 2002; Meyer et al., 2004). As in *Geobacteraceae*, these c-type cytochromes appear to be important for electron transfer by *S. oneidensis* to anodes and other insoluble electron acceptors (Table 3). In fact, when the cytochrome *c* maturation genes *ccmC* (Bouhenni et al., 2005) and *ccmB* (Dale et al., 2007) were deleted, *S. oneidensis* was incapable of dissimilatory metal reduction. Molecular studies have also shown that the c-type cytochromes MtrA, MtrB, OmcA, MtrC, and CymA are important for current production and are involved in electron transfer to insoluble metal oxides (Beliaev et al., 2002, 2005; Beliaev and Saffarini, 1998; Bretschger et al., 2007; Fredrickson et al., 2008; Gorby et

TABLE 3 Proteins that appear to be in electron transfer by *S. oneidensis* to current-harvesting electrodes and/or Fe(III) and Mn(IV) oxides

Gene	Anode	Fe(III) oxide	Mn(IV) oxide	Reference(s)
c-Type cytochrome, MtrC (omcB) (SO1779)	Yes	Yes	No	Beliaev et al., 2001; Bretschger et al., 2007; Coursolle et al., 2010; Fredrickson et al., 2008; Myers and Myers, 2001
c-Type cytochrome, OmcA (SO1778)	Yes	Yes	No	Bretschger et al., 2007; Fredrickson et al., 2008; Myers and Myers, 1997
c-Type cytochrome, MtrA (SO1777)	Yes	Yes	Yes	Beliaev et al., 2005; Bretschger et al., 2007; Fredrickson et al., 2008; Pitts et al., 2003
c-Type cytochrome, MtrB (SO1776)	Yes	Yes	Yes	Beliaev and Saffarini, 1998; Bretschger et al., 2007; Fredrickson et al., 2008
c-Type cytochrome, CymA (SO4591)	Yes	Yes	Yes	Bretschger et al., 2007; Fredrickson et al., 2008; Myers and Myers, 1997
Type IV prepilin peptidase, PilD (SO0414)	Yes	Yes	Yes	Bretschger et al., 2007; Fredrickson et al., 2008
Type II secretion protein, GspG (SO0166)	Yes	Yes	Yes	Bretschger et al., 2007
Type II secretion protein, GspD (SO0169)	Yes	Yes	Yes	Bretschger et al., 2007; Gorby et al., 2006

al., 2006; Meyer et al., 2004; Pitts et al., 2003) (Table 3). In addition, similar to *G. sulfurreducens*, electrically conductive pilin appear to be required for electron transfer to both metal oxides and electrodes (Bretschger et al., 2007; Gorby et al., 2006) (Table 3).

While growing on the anode for electricity production, repeated replacement of the anode medium decreases current by *S. oneidensis* biofilms (Lanthier et al., 2008; Marsili et al., 2008) which is only recovered slowly (Marsili et al., 2008). Furthermore, optimal current production by *S. oneidensis* is observed when planktonic cells participate in the anode reaction (Lanthier et al., 2008). These observations are consistent with the production of a secondary metabolite as an electron shuttle that enables planktonic cells to respire the anode as previously hypothesized for reduction of insoluble Fe(III) oxides by *Shewanella* (Nevin and Lovley, 2002b; Newman and Kolter, 2000). Recent electrochemical studies of *S. oneidensis* by Marsili and colleagues (2008) identified the production of flavin compounds that enhanced electron transfer rates to Fe(III) oxides and anodes. Additionally, flavin compounds and AQDS increase rates of electron transfer to insoluble Fe(III) oxide by 18 times, but have no impact on the respiration of soluble Fe(III) citrate (von Canstein et al., 2007). MtrC, a decaheme outer membrane cytochrome, is important for extracellular transfer of electrons by *S. oneidensis* (Beliaev et al., 2001) and also appears to be important for optimal rates of flavin reduction (Coursolle et al., 2010). Interestingly, deletion mutants in MtrC are able to reduce flavin at 50% slower rates but are unable to form electricity-producing biofilms. Clearly, a complex picture is emerging for the mechanisms of electricity production by *Shewanella* (Fig. 6).

The electrical conductivity of the pilin in conjunction with the thick biofilms which form on the anodes by *G. sulfurreducens* and *S. oneidensis* suggests that the pilin promote electron transfer by cells that are not in direct contact with the anode. The biofilm thickness of *G. sulfurreducens* (Reguera et al., 2006) and *S. oneidensis* (Lanthier et al., 2008) is greater than the lengths of their pilin (Reguera et al., 2005, 2006). The question remains whether the pilin and/or cytochromes promote cell-to-cell electron transfer from the outer reaches of the biofilm to the anode for respiration.

AMBIGUOUS ANALOGIES BETWEEN ELECTRICITY PRODUCTION AND Fe(III) OXIDE REDUCTION

In most cases, mechanisms involved in transfer to electrodes are similar to those for insoluble Fe(III) oxides (Fredrickson et al., 2008; Lovley, 2006a). However, a few organisms are able to grow on insoluble Fe(III) oxides, but cannot produce current on current-harvesting electrodes. For example, the *Geobacteraceae* species *Pelobacter carbinolicus* is able to reduce Fe(III) oxide (Haveman et al., 2008; Lonergan et al., 1996; Lovley et al., 1995), but was unable to produce current in microbial fuel cells (Richter et al., 2007). Similarly, *Geobacillus* strain S2E was isolated on poorly crystalline Fe(III) oxide, but could not produce current in a microbial fuel cell without the addition of an exogenous electron shuttle (Wrighton et al., 2008). In addition, the lactic acid bacterium *Enterococcus gallinarum* was able to grow on insoluble Fe(III) oxide but could not generate current in a microbial fuel cell (Kim et al., 2005). Furthermore, the link between insoluble Fe(III) respiration and electricity production arises from the recent identification of light-independent energy generation by a phototroph in a galvanic cell (Xing et al., 2008). Xing and coworkers isolated a unique *Rhodopseudomonas palustris* strain, DX-1, which produced electricity by oxidation of a broad variety of fatty acids and alcohols with relatively high coulombic efficiencies (up to 60%). Although *Rhodopseudomonas* spp. are not known for dissimilatory iron reduction, DX-1 was isolated from a fuel cell inoculum by using Fe(III) oxide. More interestingly, an ATCC strain of *R. palustris* was unable to produce current (Xing et al., 2008). Clearly, a much greater diversity exists in strategies for extracellular electron transfer to both Fe(III) oxides and anodes than the state of the art provides.

ELECTROLYTIC CELLS AND ELECTRODES AS ELECTRON DONORS

For purposes of this discussion, a "biocathode" will refer to two different paradigms for electron transfer to bacteria. The first refers to microbial colonization of the cathode in a galvanic cell from which energy is being produced. In this case, the microbial community serves as a mediator or catalyst for electron transfer from the cathode to the catholyte and is most commonly used in microbial fuel cell applications to facilitate the aerobic cathode reaction or to enable an alternate electron acceptor for energy production. Reviews of biocathodes in microbial fuel cell applications are available by He and Angenent (2006) and Chen and coworkers (Chen et al., 2008) The second is the use of microorganisms at the cathode or anode of an electrolytic cell. At the cathode of an electrolytic cell bacteria use electrons provided by the electrode as their electron donor and in turn use the electrodes to reduce an electron acceptor such as metals or halogenated compounds. At the anode of the electrolytic cell, the most extensively explored application is hydrogen production. Detailed reviews of state-of-the-art biocathode technology and electrodes as electron donors are available by He and Angenent (2006) and Thrash and Coates (2008).

Galvanic Cell Biocathodes

The earliest findings of bacteria that could produce electricity at the anode of a galvanic cell (Cohen, 1931; Potter, 1911; Sisler, 1961) gave rise to the question in the early 1960s of whether bacteria could also participate in reactions at the cathode. Biocathode studies conducted by the U.S. Naval Research Laboratory and led by J. B. Wilson in 1963 (Wilson, 1963) were aimed at identifying a long-term, low-power energy source that was compatible with deployment in the open ocean. The model fuel cell consisted of a magnesium anode that oxidized in seawater to provide electrons and a mild steel cathode that reduced water to hydrogen and hydroxide ion. Their "biocell" experiments included mixed cultures of sulfate-reducing bacteria that were predominantly of the *Desulfovibrio* genus. Wilson reported that current and power densities of the biocell were much greater than the uncolonized control. He concluded that power from the magnesium biocell could be doubled "under favorable oceanic conditions" but that the longevity of the biocell remained an open question. Wilson and colleagues additionally questioned the role of the microorganisms in the cathode reaction, noting that it was unclear whether the microbial culture was acting as a depolarizing agent by reducing hydrogen overvoltages or whether the culture contributed to energy generation by increasing the cell potential or assisting the transport of charge into the catholyte (Wilson, 1963). Regardless, this study demonstrated that microbial colonization of the cathode imparted an advantage for power production from the fuel cell.

Traditionally, microbial fuel cells are only colonized at the anode. However, in open systems, microbial colonization of the cathode is expected. Microbial colonization of the cathode in microbial fuel cells is commonly observed in both engineered fuel cell reactors where the PEM is absent as in the single-chamber microbial fuel cell (Liu and Logan, 2004) and in BMFCs (Holmes et al., 2004b; Nielsen et al., 2008; Reimers et al., 2001, 2006, 2007; Ryckelynck et al., 2005; Tender et al., 2002). See De Schamphelaire and colleagues (2007, 2008) for a comprehensive review of biocathodes in BMFC applications.

Microbial communities were first and extensively examined on the cathodes of BMFC deployed in two marine environments (Holmes et al., 2004b). The diversity of the community on cathodes is greatly decreased at the expense of specific enrichment of certain phylotypes. However, the enriched communities were dissimilar between two deployments. α-*Proteobacteria* represented over 70% of the microbial community on cathodes deployed in a salt marsh, whereas a marine deployment had a cathode community which was over 60% γ-*Proteobacteria*. The majority of the α-proteobacterial sequences fell within the *Rhodobacter* family and the majority of the γ-proteobacterial sequences fell within the *Cycloclasticus*/type I methanotroph cluster. A 125-day deep-ocean deployment of a BMFC (Reimers et al., 2006) was also enriched with γ-*Proteobacteria,* and more specifically nearly 50% of the recovered sequences were most similar to *Pseudomonas fluorescens* and 27% most similar to *Janthinobacterium lividum*, a purple bacterium. The biogeochemical role of the cathode community in BMFC is not well understood and likely complex, but even thick cathode biofilms do not appear to immediately impact cathode potentials (Reimers et al., 2006). It has been suggested that manganese oxides or phenazines may have a role in maintaining cathode potentials (De Schamphelaire et al., 2007; Reimers et al., 2006) Another theory suggests that the cathode biofilm catalyzes and accelerates oxygen reduction rates in BMFC (De Schamphelaire et al., 2008).

Oxygen reduction at the cathode may be limited by kinetics when inexpensive catalysts are used for the electrode, as is the case for unmodified graphite where large overpotentials for oxygen reduction are observed (Zhao et al., 2006). Such limitations may be partially overcome by rare-earth metal doping of the cathode, but with greatly increased costs. A particularly attractive option is the use of graphite cathodes with bacteria in the anode chamber to catalyze the desired reaction (He and Angenent, 2006). The microbial community at the cathode of an engineered fuel cell may catalyze redox reactions that are otherwise not feasible with the goal of increasing the rate of the cathode reaction with oxygen or enabling a cathode reaction with an alternate terminal electron acceptor.

Applications of microbial fuel cells where O_2 is the desired electron acceptor but the cathode is limiting microbial colonization of the cathode could be desired (Clauwaert et al., 2007b; De Schamphelaire et al., 2008; Rabaey et al., 2008). Studies by Rabaey and coworkers (2008) demonstrate that mixed-culture microbial biocathodes increase power generation. The microbial community catalyzed the reduction of oxygen at the cathode and decreased the cathode overpotential. The microbial community on biocathodes that catalyze oxygen reduction appears to be dominated by bacteria most similar to *Bacteroidetes* (Chen et al., 2008; He et al., 2009a; Rabaey et al., 2008). Interestingly, pure-culture isolates of *Sphingobacterium* spp. and *Acinetobacter* spp. from a biocathode were unable to achieve the same power benefits as a mixed community (Rabaey et al., 2008). Power curves show that the apparent advantage for current generation using biocathodes is not immediate, rather it develops over time as the microbial community shifts and adapts to the cathode environment (Chen et al., 2008; Rabaey et al., 2008), adding to the question of whether precolonization of the cathode with an adapted microbial community would readily enhance power output from microbial fuel cells. Whether microbial communities at the cathode of a galvanic cell directly receive electrons from the electrode remains an open question.

Engineered cathode reactions with an exogenous compound that serves as an electron mediator to the microbial community for catalysis of oxygen reduction at the cathode also enhance power output from microbial fuel cells. For example, using a cathode with manganese oxide deposits, *Leptothrix discophora,* a manganese-oxidizing bacteria, enabled a manganese cycle at the cathode that greatly increased power output from a microbial fuel cell (Rhoads et al., 2005). In this cathode reaction, manganese oxide is first reduced by electrons at the cathode to Mn(II), followed by reoxidation of Mn(II) and precipitation of $MnO_{2(s)}$ by *L. discophora* according to:

$$MnO_{2(s)} + 4H^+ + 3e^- \rightarrow Mn^{2+} + 2H_2O \text{ (cathode reduction)}$$

$Mn^{2+} + 0.5O_2 + H_2O \rightarrow MnO_{2(s)}$
$+ 2H^+$ (biological reoxidation)

An analogous redox cycle at the cathode with Fe(III) in the presence of an iron-oxidizing bacterium increases the rates of electricity production from microbial fuel cells (Prasad et al., 2006; ter Heijne et al., 2006, 2007). With an acidic catholyte (~pH 2.5), Fe(III) is reduced to Fe(II) at the cathode, and *Acidithiobacillus ferrooxidans* mediates the reoxidation of Fe(II) using oxygen as the electron acceptor:

$8Fe^{3+} + 8e^- \rightarrow 8Fe^{2+}$ (cathode reduction)

$8Fe^{2+} + 8H^+ + 2O_2 \rightarrow 8Fe^{3+}$
$+ 4H_2O$ (biological reoxidation)

The long-term sustainability of this biocathode reaction may be limited by the consumption of protons by the biological reaction (ter Heijne et al., 2006), necessitating recurring adjustment of the pH to maintain acidic conditions and the solubility of Fe(III).

Microbial communities on anodes may make oxygen reduction at the cathode more efficient for power generation, but may also greatly broaden the potential electron acceptors at the cathode to enable anaerobic cathodes. Typically, depletion of oxygen at the cathode will result in a drop in cell potential and may even lead to a reversal of current flow to the anode (Oh and Logan, 2007). If anoxic conditions are expected, photosynthetic organisms may be employed for the generation of oxygen in microbial fuel cells (Cao et al., 2008; He et al., 2009a). However, a biocathode under anoxic conditions may enable alternate terminal electron acceptors such as manganese, nitrate, iron, sulfate, or carbon dioxide to serve as the oxidant at the cathode via microbial respiration. The available free energy of the overall fuel cell reaction will decrease, but that may be acceptable for certain applications where oxygen is in low concentration or where diffusion limitations of oxygen to the anode may dissipate cell potential or reduce coulombic efficiency.

Anaerobic biocathodes were first described by Lewis in 1966 (Lewis, 1966). In his review, he suggested, but did not further describe, the potential for microbial denitrification at the cathode using a culture of *Micrococcus denitrificans*. The first demonstration of a galvanic microbial fuel cell with an anaerobic biocathode produced current with complete denitrification by a mixed community at the cathode (Clauwaert et al., 2007a). Clauwaert and coworkers (2007a) determined that current was dependent on denitrification of nitrate to nitrous oxide, although some ammonium was detected. Denitrification was not the result of hydrogen production by the cathode as a potential electron donor; however, the mechanism of electron transfer from the cathode to the denitrifying community was not determined. Other biocathode studies that examined cathode-supported denitrification reported nitrite and nitrogen gas as products of microbial respiration at the cathode (Lefebvre et al., 2008). The microbial communities that colonized the cathode in these studies were not examined.

Other applications of biocathodes in galvanic cells include biological chromium reduction for removal of chromium from contaminated waste streams (Tandukar et al., 2009; Wang et al., 2008), biological treatment of perchlorate (Shea et al., 2008), and azo dyes (Liu et al., 2009). These engineered bioreactors for waste treatment in a galvanic cell offer the additional opportunity for simultaneous electricity production during contaminant removal, although at this time the power generated from such reactors is very low.

Bacteria on Electrolytic Cell Anodes

An electrolytic cell is formed when an external electrical potential is applied externally between two electrodes that is greater than the open-circuit potential of the cell. In the electrolytic cell, chemical reactions are driven by the imposed external potential and electrical energy is consumed. For example, Fig. 7 shows the electrolytic cell as an electron donor

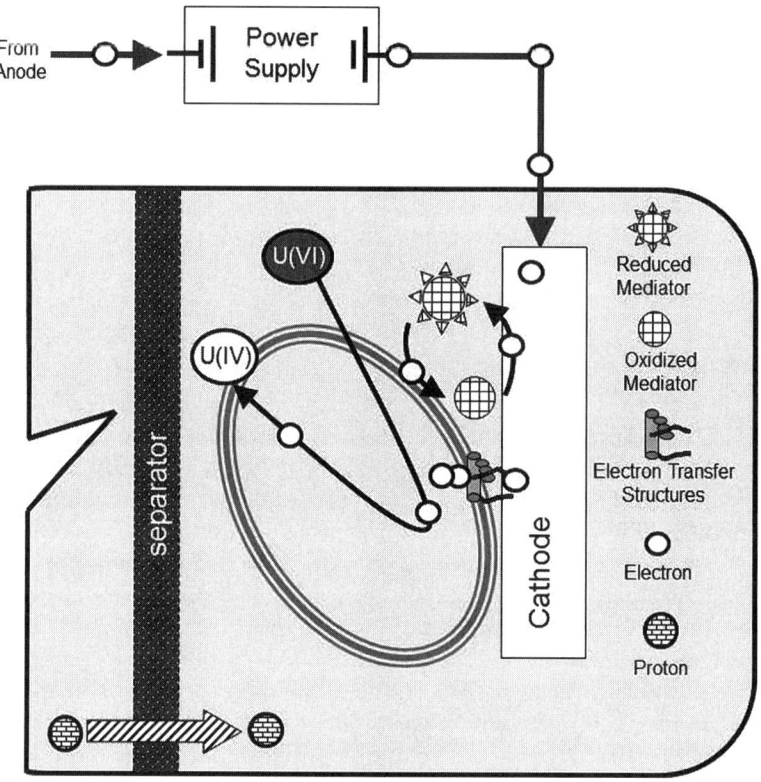

FIGURE 7 Bacteria may respire at the cathode of an electrolytic cell. Analogous to electron transfer from cells to the anode of a galvanic cell, the mechanisms of electron transfer from the cathode to bacteria in an electrolytic cell may occur through direct and indirect mechanisms.

for bacteria growing on the cathode and during respiration of U(VI) to U(IV). As described previously for anodic and cathodic reactions in galvanic cells, the rates of reaction in electrolytic cells are optimal when a suitable catalyst is employed at the electrode. Electrolytic cells with microbial communities or pure cultures at the anode have been explored extensively in recent years for the production of hydrogen. In a microbial electrolysis cell (MEC), the microbial community at the anode catalyzes the oxidation of organic substrate, coupled with abiotic hydrogen production at the cathode (Cheng and Logan, 2007a; Liu et al., 2005b; Logan et al., 2008; Rozendal et al., 2006). The principle behind the MEC is the reduction of the overall external potential necessary to promote hydrogen evolution at the cathode through the polarization and respiration of the anode by a microbial population. In this manner, the microbes at the anode assist and enable the overall oxidation of organic matter at the anode coupled with abiotic hydrogen production at the cathode using platinum-doped catalysts. Analogous MEC technology was demonstrated for abiotic transformation of nitrobenzene (Mu et al., 2009b) and azo dyes (Mu et al., 2009a) driven by microbial oxidation of organic carbon at the anode. Contaminant transformation by MEC-like process appears to be more efficient than conventional biological treatment by greatly reducing the quantity of organic electron donor necessary for contaminant reduction. The variety of mechanisms for microbial electron transfer to the anode of the MEC are likely similar to those in MFC because the process is analogous (Rozendal et al., 2006). Recently, biocathodes in MEC have been explored in single-chamber (Call and Logan, 2008; Clauwaert and Verstraete, 2009; Lee et al., 2009) and dual-chamber electrolytic cells (Jeremiasse et al., 2010; Rozendal et al., 2007). The biocathode MEC has bacterial populations on both the anode and cathode for microbial catalysis of organic matter oxidation at the anode coupled with microbial hydrogen production at the cathode. The single-chamber MEC lacks a separator (often an ion-selective membrane), and microorganisms from the inoculum are not prevented from colonizing the cathode. The single-chamber MEC appears to promote increased rates of hydrogen production (Call and Logan, 2008) but at the expense of decreased efficiency as electrons from the organic substrate may be directed toward methanogenesis (Clauwaert and Verstraete, 2009; Lee et al., 2009).

Bacteria on Electrolytic Cell Cathodes

The electrolytic cell cathode supplies electron donor via an externally applied potential for reduction of an electron acceptor in the catholyte. Microbial biofilms at the cathode in electrolytic cells enable the use of relatively inexpensive electrode materials, colonized by bacteria to increase the rate of transfer of electrons from the electrode to the electron acceptor. For the most part, electrolytic cells with microbial populations on the cathode have been explored with an overarching goal of biological reduction of oxidized contaminants. Electrodes as electron donors for biological contaminant transformation have been explored for perchlorate reduction (Shea et al., 2008; Thrash et al., 2007), uranium reduction (Gregory and Lovley, 2005), chlorinated solvent reductive dechlorination (Aulenta et al., 2009; Strycharz et al., 2008), and denitrification (Cast and Flora, 1998; Feleke et al., 1998; Gregory et al., 2004; Park et al., 2005; Sakakibara and Kuroda, 1993; Szekeres et al., 2001). With the exception of Gregory and coworkers (2004) and Park and coworkers (2005), who demonstrated a direct mechanism of electron transfer from the cathode, biological denitrification in the electrolytic cells in other studies was achieved through microbial respiration of cathodic hydrogen as an electron donor. Similarly, cathodic hydrogen may be oxidized to support chlorinated solvent-reductive dechlorination (Shimomura and Sanford, 2005; Skadberg et al., 1999).

In an electrolytic cell, stimulation of the bacteria at the cathode occurs through the provision of an electron donor. As in electron transfer from bacteria to the anode of a galvanic cell, the mechanisms of electron transfer from an anode to bacteria in an electrolytic cell occur via both direct and indirect paradigms. A superb perspective and review of electron transfer to bacteria by Thrash and Coates (2008) describes three mechanisms by which bacteria may utilize a cathode as their electron donor: (i) indirectly through oxidation of hydrogen produced by electrolysis, (ii) indirectly via oxidation of reduced electron shuttle produced by the anode, or (iii) directly via respiration of electrons supplied at the cathode surface (Fig. 7). A large variety of microorganisms are capable of hydrogen oxidation coupled to respiration (Bowien and Schlegel, 1981). For brevity, we will limit discussion to the microbial electron shuttling and direct transfer of electrons to bacteria during anaerobic respiration.

In the absence of electrolysis of water and hydrogen production at the cathode, indirect electron transfer to bacteria from the cathode occurs as a result of cathodic reduction of an artificial, exogenous, electron-shuttling compound or primary or secondary metabolite followed by reoxidation of the shuttle by the bacteria as an electron donor for respiration. As described previously for microbial respiration of an anode, the electrode and bacteria create a redox cycle with the shuttle. For example, artificial electron shuttles such as ferricyanide and neutral red are reduced at the cathode and participate in the reduction of intracellular electron carriers and microbial growth and metabolism (Emde and Schink, 1990; Park and Zeikus, 1999). Similarly, electrochemically reduced AQDS may also serve as the electron donor for stimulation of perchlorate reduction in mixed-culture communities and a perchlorate-respiring pure-culture bacterium, *Dechlorospirillum* strain VDY (Thrash et al., 2007). Electron transfer to perchlorate with VDY occurred with 100% coulombic efficiency and with the expected stoichiometry of the reduced products.

At this time, the only bacteria known to directly accept electrons from an electrode are in the *Geobacteraceae* family (Gregory et al., 2004; Gregory and Lovley, 2005; Strycharz et al., 2008). *Geobacter* spp. will directly utilize electrons from the cathode coupled to respiration of fumarate (Gregory et al., 2004; Strycharz et al., 2008), nitrate (Gregory et al., 2004), uranium(VI) (Gregory and Lovley, 2005), and tetrachloroethene (Strycharz et al., 2008). Replacement of the catholyte medium to remove shuttles did not impact current consumption or reduction rates of the terminal electron acceptor. The transfer of electrons from the cathode to the terminal electron acceptor were stoichiometric with the reduction products observed during respiration (Gregory et al., 2004; Gregory and Lovley, 2005; Strycharz et al., 2008).

The mechanism of direct electron transfer from electrodes to *Geobacteraceae* is not known. As detailed previously, *Geobacteraceae* produce outer membrane electron transfer structures that enable respiration of the anode as a terminal electron acceptor for growth (Lovley, 2006a, 2006b) and whether these same structures enable respiration of the cathode remains to be determined. As pointed out by Thrash and Coates (2008), the cells may not have been conserving energy. Instead, the electron transfer to fumarate and nitrate may have occurred directly at the fumarate and nitrate reductase, respectively.

The hypothesis of direct respiration of the cathode by *Geobacteraceae* is supported by several lines of indirect evidence. Cathodically poised electrodes inoculated with surficial sediments or aquifer solids become enriched with *Geobacteraceae* (Gregory et al., 2004; Gregory and Lovley, 2005). In addition, *Geobacter* sp. may couple ferrous iron oxidation to nitrate reduction (Finneran et al., 2002). Furthermore, a variety of other organisms outside of *Geobacteraceae* will utilize insoluble Fe(II) oxides as electron donors (Chaudhuri et al., 2001; Shelobolina et al., 2003; Weber et al., 2001, 2006). Although the smoking gun remains to be discovered, these independent findings suggest that *Geobacteraceae* may use electrons from the cathode of an electrolytic cell for respiration coupled to growth during metal respiration. Perhaps more importantly, they clearly support the hypothesis that, like microbial respiration of the anode, cathode respiration will be broadly distributed among phylotypes of *Bacteria*.

REFERENCES

Aelterman, P., K. Rabaey, P. Clauwaert, and W. Verstraete. 2006a. Microbial fuel cells for wastewater treatment. *Water Sci. Technol.* **54:**9–15.

Aelterman, P., K. Rabaey, H. T. Pham, N. Boon, and W. Verstraete. 2006b. Continuous electricity generation at high voltages and currents using stacked microbial fuel cells. *Environ. Sci. Technol.* **40:**3388–3394.

Aklujkar, M., J. Krushkal, G. DiBartolo, A. Lapidus, M. L. Land, and D. R. Lovley. 2009. The genome sequence of *Geobacter metallireducens*: features of metabolism, physiology and regulation common and dissimilar to Geobacter sulfurreducens. *BMC Microbiol.* **9:**109.

Allen, M. J. 1966. Symposium on bioelectrochemistry of microorganisms. 2. Electrochemical aspects of metabolism. *Bacteriol. Rev.* **30:**80–93.

Allen, R. M., and H. P. Bennetto. 1993. Microbial fuel cells: electricity production from carbohydrates. *Appl. Biochem. Biotechnol.* **39:**27–40.

Anderson, R. T., H. A. Vrionis, I. Ortiz-Bernad, C. T. Resch, P. E. Long, R. Dayvault, K. Karp, S. Marutzky, D. R. Metzler, A. Peacock, D. C. White, M. Lowe, and D. R. Lovley. 2003. Stimulating the in situ activity of Geobacter species to remove uranium from the groundwater of a uranium-contaminated aquifer. *Appl. Environ. Microbiol.* **69:**5884–5891.

Aston, W. J., and A. P. F. Turner. 1984. Biosensors and biofuel cells. *Biotechnol. Genet. Eng. Rev.* **1:**89–120.

Ateya, B. G., and F. M. Al-Kharafi. 2002. Anodic oxidation of sulphide ions from chloride brines. *Electrochem. Commun.* **4:**231–238.

Aulenta, F., A. Canosa, P. Reale, S. Rossetti, S. Panero, and M. Majone. 2009. Microbial reductive dechlorination of trichloroethene to ethene with electrodes serving as electron donors without the external addition of redox mediators. *Biotechnol. Bioeng.* **103**:85–91.

Bagander, L. E., and L. Niemisto. 1978. An evaluation of the use of redox measurements for characterizing recent sediments. *Estuar. Coast. Mar. Sci.* **6**:127–134.

Barbir, F. 2005. *PEM Fuel Cells: Theory and Practice.* Elsevier Academic Press, Burlington, MA.

Bard, A. J., and L. R. Faulkner. 2001. *Electrochemical Methods Fundamentals and Applications*, 2nd ed. John Wiley & Sons, New York, NY.

Baron, D., E. LaBelle, D. Coursolle, J. A. Gralnick, and D. R. Bond. 2009. Electrochemical measurement of electron transfer kinetics by *Shewanella oneidensis* MR-1. *J. Biol. Chem.* **284**:28865–28873.

Beliaev, A. S., D. M. Klingeman, J. A. Klappenbach, L. Wu, M. F. Romine, J. A. Tiedje, K. H. Nealson, J. K. Fredrickson, and J. Zhou. 2005. Global transcriptome analysis of *Shewanella oneidensis* MR-1 exposed to different terminal electron acceptors. *J. Bacteriol.* **187**:7138–7145.

Beliaev, A. S., and D. A. Saffarini. 1998. *Shewanella putrefaciens* mtrB encodes an outer membrane protein required for Fe(III) and Mn(IV) reduction. *J. Bacteriol.* **180**:6292–6297.

Beliaev, A. S., D. A. Saffarini, J. L. McLaughlin, and D. Hunnicutt. 2001. MtrC, an outer membrane decahaem c cytochrome required for metal reduction in *Shewanella putrefaciens* MR-1. *Mol. Microbiol.* **39**:722–730.

Beliaev, A. S., D. K. Thompson, M. W. Fields, L. Y. Wu, D. P. Lies, K. H. Nealson, and J. Z. Zhou. 2002. Microarray transcription profiling of a *Shewanella oneidensis* etrA mutant. *J. Bacteriol.* **184**:4612–4616.

Biffinger, J. C., J. Pietron, O. Bretschger, L. J. Nadeau, G. R. Johnson, C. C. Williams, K. H. Nealson, and B. R. Ringeisen. 2008. The influence of acidity on microbial fuel cells containing *Shewanella oneidensis. Biosens. Bioelectron.* **24**:900–905.

Biffinger, J. C., J. Pietron, R. Ray, B. Little, and B. R. Ringeisen. 2007. A biofilm enhanced miniature microbial fuel cell using *Shewanella oneidensis* DSP10 and oxygen reduction cathodes. *Biosens. Bioelectron.* **22**:1672–1679.

Biffinger, J. C., R. Ray, B. J. Little, L. A. Fitzgerald, M. Ribbens, S. E. Finkel, and B. R. Ringeisen. 2009. Simultaneous analysis of physiological and electrical output changes in an operating microbial fuel cell with *Shewanella oneidensis. Biotechnol. Bioeng.* **103**:524–531.

Bocher, B. T., M. T. Agler, M. L. Garcia, A. R. Beers, and L. T. Angenent. 2008. Anaerobic digestion of secondary residuals from an anaerobic bioreactor at a brewery to enhance bioenergy generation. *J. Ind. Microbiol. Biotechnol.* **35**:321–329.

Bond, D. R., D. E. Holmes, L. M. Tender, and D. R. Lovley. 2002. Electrode-reducing microorganisms that harvest energy from marine sediments. *Science* **295**:483–485.

Bond, D. R., and D. R. Lovley. 2003. Electricity production by *Geobacter sulfurreducens* attached to electrodes. *Appl. Environ. Microbiol.* **69**:1548–1555.

Bond, D. R., and D. R. Lovley. 2005. Evidence for involvement of an electron shuttle in electricity generation by *Geothrix fermentans. Appl. Environ. Microbiol.* **71**:2186–2189.

Bouhenni, R., A. Gehrke, and D. Saffarini. 2005. Identification of genes involved in cytochrome c biogenesis in Shewanella oneidensis, using a modified mariner transposon. *Appl. Environ. Microbiol.* **71**:4935–4937.

Bowien, B., and H. G. Schlegel. 1981. Physiology and biochemistry of aerobic hydrogen-oxidizing bacteria. *Annu. Rev. Microbiol.* **35**:405–452.

Brendel, P. J., and G. W. Luther. 1995. Development of a gold amalgam voltammetric microelectrode for the determination of dissolved Fe, Mn, O_2, and S(-ll) in porewaters of marine and freshwater sediments. *Environ. Sci. Technol.* **29**:751–761.

Bretschger, O., A. Obraztsova, C. A. Sturm, I. S. Chang, Y. A. Gorby, S. B. Reed, D. E. Culley, C. L. Reardon, S. Barua, M. F. Romine, J. Zhou, A. S. Beliaev, R. Bouhenni, D. Saffarini, F. Mansfeld, B. H. Kim, J. K. Fredrickson, and K. H. Nealson. 2007. Current production and metal oxide reduction by *Shewanella oneidensis* MR-1 wild type and mutants. *Appl. Environ. Microbiol.* **73**:7003–7012.

Call, D., and B. E. Logan. 2008. Hydrogen production in a single chamber microbial electrolysis cell lacking a membrane. *Environ. Sci. Technol.* **42**:3401–3406.

Cao, X., X. Huang, N. Boon, P. Liang, and M. Fan. 2008. Electricity generation by an enriched phototrophic consortium in a microbial fuel cell. *Electrochem. Commun.* **10**:1392–1395.

Cast, K. L., and J. R. V. Flora. 1998. An evaluation of two cathode materials and the impact of copper on bioelectrochemical denitrification. *Water Res.* **32**:63–70.

Catal, T., S. T. Xu, K. C. Li, H. Bermek, and H. Liu. 2008. Electricity generation from polyalcohols in single-chamber microbial fuel cells. *Biosens. Bioelectron.* **24**:849–854.

Chae, K.-J., M.-J. Choi, J.-W. Lee, K.-Y. Kim, and I.-S. Kim. 2009. Effect of different substrates on the performance, bacterial diversity, and bacterial viability in microbial fuel cells. *Bioresour. Technol.* **100**:3518–3525.

Chang, I. S., H. Moon, J. K. Jang, and B. H. Kim. 2005. Improvement of a microbial fuel cell performance as a BOD sensor using respiratory inhibitors. *Biosens. Bioelectron.* **20**:1856–1859.

Chaudhuri, S. K., J. G. Lack, and J. D. Coates. 2001. Biogenic magnetite formation through anaerobic biooxidation of Fe(II). *Appl. Environ. Microbiol.* **67**:2844–2848.

Chaudhuri, S. K., and D. R. Lovley. 2003. Electricity generation by direct oxidation of glucose in mediatorless microbial fuel cells. *Nat. Biotechnol.* **21**:1229–1232.

Chen, G.-W., S.-J. Choi, T.-H. Lee, G.-Y. Lee, J.-H. Cha, and C.-W. Kim. 2008. Application of biocathode in microbial fuel cells: cell performance and microbial community. *Appl. Microbiol. Biotechnol.* **79**:379–388.

Cheng, S., B. A. Dempsey, and B. E. Logan. 2007. Electricity generation from synthetic acid-mine drainage (AMD) water using fuel cell technologies. *Environ. Sci. Technol.* **41**:8149–8153.

Cheng, S., H. Liu, and B. E. Logan. 2006a. Increased power generation in a continuous flow MFC with advective flow through the porous anode and reduced electrode spacing. *Environ. Sci. Technol.* **40**:2426–2432.

Cheng, S., H. Liu, and B. E. Logan. 2006b. Power densities using different cathode catalysts (Pt and CoTMPP) and polymer binders (Nafion and PTFE) in single chamber microbial fuel cells. *Environ. Sci. Technol.* **40**:364–369.

Cheng, S., and B. E. Logan. 2007a. Sustainable and efficient biohydrogen production via electrohydrogenesis. *Proc. Natl. Acad. Sci. USA* **104**:18871–18873.

Cheng, S. A., and B. E. Logan. 2007b. Ammonia treatment of carbon cloth anodes to enhance power generation of microbial fuel cells. *Electrochem. Commun.* **9**:492–496.

Cho, Y. K., T. J. Donohue, I. Tejedor, M. A. Anderson, K. D. McMahon, and D. R. Noguera. 2008. Development of a solar-powered microbial fuel cell. *J. Appl. Microbiol.* **104**:640–650.

Choi, Y.-J., J. Song, S. Jung, and S. Kim. 2001. Optimization of the performance of microbial fuel cells containing alkalophilic Bacillus sp. *J. Microbiol. Biotechnol.* **11**:863–869.

Choo, Y.-F., J. Lee, I. S. Chang, and B. H. Kim. 2006. Bacterial communities in microbial fuel cells enriched with high concentrations of glucose and glutamate. *J. Microbiol. Biotechnol.* **16**:1481–1484.

Chung, K., and S. Okabe. 2009. Characterization of electrochemical activity of a strain ISO2-3 phylogenetically related to *Aeromonas* sp. isolated from a glucose-fed microbial fuel cell. *Biotechnol. Bioeng.* **104**:901–910.

Clauwaert, P., P. Aelterman, T. H. Pham, L. De Schamphelaire, M. Carballa, K. Rabaey, and W. Verstraete. 2008. Minimizing losses in bio-electrochemical systems: the road to applications. *Appl. Microbiol. Biotechnol.* **79**:901–913.

Clauwaert, P., K. Rabaey, P. Aelterman, L. De Schamphelaire, T. H. Ham, P. Boeckx, N. Boon, and W. Verstraete. 2007a. Biological denitrification in microbial fuel cells. *Environ. Sci. Technol.* **41**:3354–3360.

Clauwaert, P., D. Van der Ha, N. Boon, K. Verbeken, M. Verhaege, K. Rabaey, and W. Verstraete. 2007b. Open air biocathode enables effective electricity generation with microbial fuel cells. *Environ. Sci. Technol.* **41**:7564–7569.

Clauwaert, P., and W. Verstraete. 2009. Methanogenesis in membraneless microbial electrolysis cells. *Appl. Microbiol. Biotechnol.* **82**:829–836.

Cohen, B. 1931. The bacterial culture as an electrical half-cell. *J. Bacteriol.* **21**:18–19.

Coppi, M. V., C. Leang, S. J. Sandler, and D. R. Lovley. 2001. Development of a genetic system for *Geobacter sulfurreducens*. *Appl. Environ. Microbiol.* **67**:3180–3187.

Cordas, C. M., L. T. Guerra, C. Xavier, and J. J. G. Moura. 2008. Electroactive biofilms of sulphate reducing bacteria. *Electrochim. Acta* **54**:29–34.

Coursolle, D., D. B. Baron, D. R. Bond, and J. A. Gralnick. 2010. The Mtr respiratory pathway is essential for reducing flavins and electrodes in *Shewanella oneidensis*. *J. Bacteriol.* **192**:467–474.

Cristina, M., C. Cristina, M. Isabel, G. Jean, Y. L. Ming, J. P. William, D. Cees, and J. J. G. Moura. 1993. Electrochemical studies of the hexaheme nitrite reductase from *Desulfovibrio desulfuricans* ATCC 27774. *Eur. J. Biochem.* **212**:79–86.

Crittenden, S. R., C. J. Sund, and J. J. Sumner. 2006. Mediating electron transfer from bacteria to a gold electrode via a self-assembled monolayer. *Langmuir* **22**:9473–9476.

Dale, J. R., R. Wade, Jr., and T. J. Dichristina. 2007. A conserved histidine in cytochrome c maturation permease CcmB of *Shewanella putrefaciens* is required for anaerobic growth below a threshold standard redox potential. *J. Bacteriol.* **189**:1036–1043.

Davis, J. B. 1963. Generation of electricity by microbial action. *Adv. Appl. Microbiol.* **5**:51–64.

Delaney, G. M., H. P. Bennetto, J. R. Mason, S. D. Roller, J. L. Stirling, and C. F. Thurston. 1984. Electron-transfer coupling in microbial fuel cells. 2. performance of fuel cells containing selected microorganism-mediator-substrate combinations. *J. Chem. Technol. Biotechnol. B Biotechnol.* **34**:13–27.

Deng, Q., X. Y. Li, J. E. Zuo, A. Ling, and B. E. Logan. 2010. Power generation using an activated carbon fiber felt cathode in an upflow microbial fuel cell. *J. Power Sources* **195:**1130–1135.

De Schamphelaire, L., K. Rabaey, P. Boeckx, N. Boon, and W. Verstraete. 2008. Outlook for benefits of sediment microbial fuel cells with two bio-electrodes. *Microb. Biotechnol.* **1:** 446–462.

De Schamphelaire, L., K. Rabaey, N. Boon, W. Verstraete, and P. Boeckx. 2007. Minireview: the potential of enhanced manganese redox cycling for sediment oxidation. *Geomicrobiol. J.* **24:**547–558.

Ding, Y. H. R., K. K. Hixson, M. A. Aklujkar, M. S. Lipton, R. D. Smith, D. R. Lovley, and T. Mester. 2008. Proteome of *Geobacter sulfurreducens* grown with Fe(III) oxide or Fe(III) citrate as the electron acceptor. *Biochim. Biophys. Acta Proteins Proteomics* **1784:**1935–1941.

Dutta, P. K., K. Rabaey, Z. G. Yuan, and J. Keller. 2008. Spontaneous electrochemical removal of aqueous sulfide. *Water Res.* **42:**4965–4975.

Emde, R., and B. Schink. 1989. Anaerobic oxidation of glycerol by *Escherichia coli* in an amperometric poised-potential culture system. *Appl. Microbiol. Biotechnol.* **32:**170–175.

Emde, R., and B. Schink. 1990. Oxidation of glycerol, lactate, and propionate by *Propionibacterium freudenreichii* in a poised-potential amperometric culture system *Arch. Microbiol.* **153:**506–512.

Fan, Y. Z., E. Sharbrough, and H. Liu. 2008. Quantification of the internal resistance distribution of microbial fuel cells. *Environ. Sci. Technol.* **42:**8101–8107.

Feleke, Z., K. Araki, Y. Sakakibara, T. Watanabe, and M. Kuroda. 1998. Selective reduction of nitrate to nitrogen gas in a biofilm-electrode reactor. *Water Res.* **32:**2728–2734.

Feng, Y., X. Wang, B. E. Logan, and H. Lee. 2008. Brewery wastewater treatment using air-cathode microbial fuel cells. *Appl. Microbiol. Biotechnol.* **78:**873–880.

Finneran, K. T., M. E. Housewright, and D. R. Lovley. 2002. Multiple influences of nitrate on uranium solubility during bioremediation of uranium-contaminated subsurface sediments. *Environ. Microbiol.* **4:**510–516.

Franks, A. E., K. P. Nevin, H. F. Jia, M. Izallalen, T. L. Woodard, and D. R. Lovley. 2009. Novel strategy for three-dimensional real-time imaging of microbial fuel cell communities: monitoring the inhibitory effects of proton accumulation within the anode biofilm. *Energy Environ. Sci.* **2:**113–119.

Fredrickson, J. K., M. F. Romine, A. S. Beliaev, J. M. Auchtung, M. E. Driscoll, T. S. Gardner, K. H. Nealson, A. L. Osterman, G. Pinchuk, J. L. Reed, D. A. Rodionov, J. L. M. Rodrigues, D. A. Saffarini, M. H. Serres, A. M. Spormann, I. B. Zhulin, and J. M. Tiedje. 2008. Towards environmental systems biology of *Shewanella*. *Nat. Rev. Microbiol.* **6:**592–603.

Freguia, S., M. Masuda, S. Tsujimura, and K. Kano. 2009. *Lactococcus lactis* catalyses electricity generation at microbial fuel cell anodes via excretion of a soluble quinone. *Bioelectrochemistry* **76:**14–18.

Freguia, S., K. Rabaey, Z. G. Yuan, and J. Keller. 2007. Electron and carbon balances in microbial fuel cells reveal temporary bacterial storage behavior during electricity generation. *Environ. Sci. Technol.* **41:**2915–2921.

Freguia, S., E. H. Teh, N. Boon, K. M. Leung, J. Keller, and K. Rabaey. 2010. Microbial fuel cells operating on mixed fatty acids. *Bioresour. Technol.* **101:**1233–1238.

Froelich, P. N., G. P. Klinkhammer, M. L. Bender, N. A. Luedtke, G. R. Heath, D. Cullen, P. Dauphin, D. Hammond, B. Hartman, and V. Maynard. 1979. Early oxidation of organic matter in pelagic sediments of the eastern equatorial Atlantic—suboxic diagenesis. *Geochim. Cosmochim. Acta* **43:**1075–1090.

Gillespie, L. J. 1920. Reduction potentials of bacterial cultures and of water-logged soils. *Soil Sci.* **9:**199–216.

Gorby, Y. A., S. Yanina, J. S. McLean, K. M. Rosso, D. Moyles, A. Dohnalkova, T. J. Beveridge, I. S. Chang, B. H. Kim, K. S. Kim, D. E. Culley, S. B. Reed, M. F. Romine, D. A. Saffarini, E. A. Hill, L. Shi, D. A. Elias, D. W. Kennedy, G. Pinchuk, K. Watanabe, S. Ishii, B. Logan, K. H. Nealson, and J. K. Fredrickson. 2006. Electrically conductive bacterial nanowires produced by *Shewanella oneidensis* strain MR-1 and other microorganisms. *Proc. Natl. Acad. Sci. USA* **103:**11358–11363.

Gregory, K. B., D. R. Bond, and D. R. Lovley. 2004. Graphite electrodes as electron donors for anaerobic respiration. *Environ. Microbiol.* **6:**596–604.

Gregory, K. B., and D. R. Lovley. 2005. Remediation and recovery of uranium from contaminated subsurface environments with electrodes. *Environ. Sci. Technol.* **39:**8943–8947.

Ha, P. T., B. Tae, and I. S. Chang. 2008. Performance and bacterial consortium of microbial fuel cell fed with formate. *Energy Fuels* **22:**164–168.

Habermann, W., and E. H. Pommer. 1991. Biological fuel cells with sulphide storage capacity. *Appl. Microbiol. Biotechnol.* **35:**128–133.

Haveman, S. A., R. J. DiDonato, L. Villanueva, E. S. Shelobolina, B. L. Postier, B. Xu, A. Liu, and D. R. Lovley. 2008. Genome-wide gene expression patterns and growth requirements suggest that *Pelobacter carbinolicus* reduces Fe(III) indirectly via sulfide production. *Appl. Environ. Microbiol.* **74:**4277–4284.

He, Z., and L. T. Angenent. 2006. Application of bacterial biocathodes in microbial fuel cells. *Electroanalysis* **18:**2009–2015.

He, Z., J. Kan, F. Mansfeld, L. T. Angenent, and K. H. Nealson. 2009a. Self-sustained phototrophic microbial fuel cells based on the synergistic cooperation between photosynthetic microorganisms and heterotrophic bacteria. *Environ. Sci. Technol.* **43:**1648–1654.

He, Z., J. J. Kan, Y. B. Wang, Y. L. Huang, F. Mansfeld, and K. H. Nealson. 2009b. Electricity production coupled to ammonium in a microbial fuel cell. *Environ. Sci. Technol.* **43:**3391–3397.

Heidelberg, J. F., I. T. Paulsen, K. E. Nelson, E. J. Gaidos, W. C. Nelson, T. D. Read, J. A. Eisen, R. Seshadri, N. Ward, B. Methe, R. A. Clayton, T. Meyer, A. Tsapin, J. Scott, M. Beanan, L. Brinkac, S. Daugherty, R. T. DeBoy, R. J. Dodson, A. S. Durkin, D. H. Haft, J. F. Kolonay, R. Madupu, J. D. Peterson, L. A. Umayam, O. White, A. M. Wolf, J. Vamathevan, J. Weidman, M. Impraim, K. Lee, K. Berry, C. Lee, J. Mueller, H. Khouri, J. Gill, T. R. Utterback, L. A. McDonald, T. V. Feldblyum, H. O. Smith, J. C. Venter, K. H. Nealson, and C. M. Fraser. 2002. Genome sequence of the dissimilatory metal ion-reducing bacterium *Shewanella oneidensis*. *Nat. Biotechnol.* **20:**1118–1123.

Heilmann, J., and B. E. Logan. 2006. Production of electricity from proteins using a microbial fuel cell. *Water Environ. Res.* **78:**531–537.

Himmelheber, D. W., M. Taillefert, K. D. Pennell, and J. B. Hughes. 2008. Spatial and temporal evolution of biogeochemical processes following in situ capping of contaminated sediments. *Environ. Sci. Technol.* **42:**4113–4120.

Holmes, D. E., D. R. Bond, and D. R. Lovley. 2004a. Electron transfer by *Desulfobulbus propionicus* to Fe(III) and graphite electrodes. *Appl. Environ. Microbiol.* **70:**1234–1237.

Holmes, D. E., D. R. Bond, R. A. O'Neill, C. E. Reimers, L. R. Tender, and D. R. Lovley. 2004b. Microbial communities associated with electrodes harvesting electricity from a variety of aquatic sediments. *Microb. Ecol.* **48:**178–190.

Holmes, D. E., S. K. Chaudhuri, K. P. Nevin, T. Mehta, B. A. Methe, A. Liu, J. E. Ward, T. L. Woodard, J. Webster, and D. R. Lovley. 2006. Microarray and genetic analysis of electron transfer to electrodes in *Geobacter sulfurreducens*. *Environ. Microbiol.* **8:**1805–1815.

Holmes, D. E., K. T. Finneran, R. A. O'Neil, and D. R. Lovley. 2002. Enrichment of members of the family Geobacteraceae associated with stimulation of dissimilatory metal reduction in uranium-contaminated aquifer sediments. *Appl. Environ. Microbiol.* **68:**2300–2306.

Holmes, D. E., K. P. Nevin, T. L. Woodard, A. D. Peacock, and D. R. Lovley. 2007a. *Prolixibacter bellariivorans* gen nov, sp nov, a sugar-fermenting, psychrotolerant anaerobe of the phylum Bacteroidetes, isolated from a marine-sediment fuel cell. *Int. J. Syst. Evol. Microbiol.* **57:**701–707.

Holmes, D. E., J. S. Nicoll, D. R. Bond, and D. R. Lovley. 2004c. Potential role of a novel psychrotolerant member of the family Geobacteraceae, *Geopsychrobacter electrodiphilus* gen. nov., sp nov., in electricity production by a marine sediment fuel cell. *Appl. Environ. Microbiol.* **70:**6023–6030.

Holmes, D. E., R. A. O'Neil, H. A. Vrionis, L. A. N'Guessan, I. Ortiz-Bernad, M. J. Larrahondo, L. A. Adams, J. A. Ward, J. S. Nicoll, K. P. Nevin, M. A. Chavan, J. P. Johnson, P. E. Long, and D. R. Lovley. 2007b. Subsurface clade of Geobacteraceae that predominates in a diversity of Fe(III)-reducing subsurface environments. *ISME J.* **1:**663–677.

Huang, J. X., B. L. Sun, and X. B. Zhang. 2010. Electricity generation at high ionic strength in microbial fuel cell by a newly isolated *Shewanella marisflavi* EP1. *Appl. Microbiol. Biotechnol.* **85:**1141–1149.

Huang, L. P., and B. E. Logan. 2008. Electricity generation and treatment of paper recycling wastewater using a microbial fuel cell. *Appl. Microbiol. Biotechnol.* **80:**349–355.

Jeremiasse, A. W., H. V. M. Hamelers, and C. J. N. Buisman. 2010. Microbial electrolysis cell with a microbial biocathode. *Bioelectrochemistry* **78:**39–43.

Jong, B. C., B. H. Kim, I. S. Chang, P. W. Y. Liew, Y. F. Choo, and G. S. Kang. 2006. Enrichment, performance, and microbial diversity of a thermophilic mediatorless microbial fuel cell. *Environ. Sci. Technol.* **40:**6449–6454.

Jung, S., and J. M. Regan. 2007. Comparison of anode bacterial communities and performance in microbial fuel cells with different electron donors. *Appl. Microbiol. Biotechnol.* **77:**394–402.

Kamitani, A., S. Morishita, H. Kotaki, and S. Arscott. 2008. Miniaturized microDMFC using silicon microsystems techniques: performances at low fuel flow rates. *J. Micromech. Microeng.* **18:**125019–125028.

Karube, I., T. Matsunaga, S. Tsuru, and S. Suzuki. 1977. Biochemical fuel cell utilizing immobilized cells of Clostridium butyricum. *Biotechnol. Bioeng.* **19:**1727–1733.

Kim, B. C., X. L. Qian, L. A. Ching, M. V. Coppi, and D. R. Lovley. 2006. Two putative c-type multiheme cytochromes required for the expression of OmcB, an outer membrane protein essential for optimal Fe(III) reduction in *Geobacter sulfurreducens*. *J. Bacteriol.* **188**:3138–3142.

Kim, B. H., H. S. Park, H. J. Kim, G. T. Kim, I. S. Chang, J. Lee, and N. T. Phung. 2004. Enrichment of microbial community generating electricity using a fuel-cell-type electrochemical cell. *Appl. Microbiol. Biotechnol.* **63**:672–681.

Kim, G. T., M. S. Hyun, I. S. Chang, H. J. Kim, H. S. Park, B. H. Kim, S. D. Kim, J. W. T. Wimpenny, and A. J. Weightman. 2005. Dissimilatory Fe(III) reduction by an electrochemically active lactic acid bacterium phylogenetically related to *Enterococcus gallinarum* isolated from submerged soil. *J. Appl. Microbiol.* **99**:978–987.

Kim, H. J., M. S. Hyun, I. S. Chang, and B. H. Kim. 1999. A microbial fuel cell type lactate biosensor using a metal-reducing bacterium, *Shewanella putrefaciens*. *J. Microbiol. Biotechnol.* **9**:365–367.

Kim, H. J., H. S. Park, M. S. Hyun, I. S. Chang, M. Kim, and B. H. Kim. 2002. A mediator-less microbial fuel cell using a metal reducing bacterium, Shewanella putrefaciens. *Enzyme Microb. Technol.* **30**:145–152.

Kim, J. R., S. H. Jung, J. M. Regan, and B. E. Logan. 2007. Electricity generation and microbial community analysis of alcohol powered microbial fuel cells. *Bioresour. Technol.* **98**:2568–2577.

Kim, N., Y. Choi, S. Jung, and S. Kim. 2000. Effect of initial carbon sources on the performance of microbial fuel cells containing *Proteus vulgaris*. *Biotechnol. Bioeng.* **70**:109–114.

Konstantinidis, K. T., M. H. Serres, M. F. Romine, J. L. M. Rodrigues, J. Auchtung, L. A. McCue, M. S. Lipton, A. Obraztsova, C. S. Giometti, K. H. Nealson, J. K. Fredrickson, and J. M. Tiedje. 2009. Comparative systems biology across an evolutionary gradient within the *Shewanella* genus. *Proc. Natl. Acad. Sci. USA* **106**:15909–15914.

Lanthier, M., K. B. Gregory, and D. R. Lovley. 2008. Growth with high planktonic biomass in *Shewanella oneidensis* fuel cells. *FEMS Microbiol. Lett.* **278**:29–35.

Leang, C., M. V. Coppi, and D. R. Lovley. 2003. OmcB, a c-type polyheme cytochrome, involved in Fe(III) reduction in *Geobacter sulfurreducens*. *J. Bacteriol.* **185**:2096–2103.

Lee, H. S., C. I. Torres, P. Parameswaran, and B. E. Rittmann. 2009. Fate of H-2 in an upflow single-chamber microbial electrolysis cell using a metal-catalyst-free cathode. *Environ. Sci. Technol.* **43**:7971–7976.

Lee, J. Y., N. T. Phung, I. S. Chang, B. H. Kim, and H. C. Sung. 2003. Use of acetate for enrichment of electrochemically active microorganisms and their 16S rDNA analyses. *FEMS Microbiol. Lett.* **223**:185–191.

Lefebvre, O., A. Al-Mamun, and Y. H. Ng. 2008. A microbial fuel cell equipped with a biocathode for organic removal and denitrification. *Water Sci. Technol.* **58**:881–885.

Lewis, K. 1966. Symposium on bioelectrochemistry of microorganisms. 4. Biochemical fuel cells. *Bacteriol. Rev.* **30**:101–113.

Liang, P., H. Y. Wang, X. Huang, X. X. Cao, and Y. H. Mo. 2009. Influence of environmental factors on electricity production by microbial fuel cell inoculation *Shewanella baltica*. *Huan Jing Ke Xue* **30**:2148–2152.

Lies, D. P., M. E. Hernandez, A. Kappler, R. E. Mielke, J. A. Gralnick, and D. K. Newman. 2005. Shewanella oneidensis MR-1 uses overlapping pathways for iron reduction at a distance and by direct contact under conditions relevant for biofilms. *Appl. Environ. Microbiol.* **71**:4414–4426.

Liu, H., S. A. Cheng, and B. E. Logan. 2005a. Production of electricity from acetate or butyrate using a single-chamber microbial fuel cell. *Environ. Sci. Technol.* **39**:658–662.

Liu, H., S. Grot, and B. E. Logan. 2005b. Electrochemically assisted microbial production of hydrogen from acetate. *Environ. Sci. Technol.* **39**:4317–4320.

Liu, H., and B. E. Logan. 2004. Electricity generation using an air-cathode single chamber microbial fuel cell in the presence and absence of a proton exchange membrane. *Environ. Sci. Technol.* **38**:4040–4046.

Liu, H., R. Ramnarayanan, and B. E. Logan. 2004. Production of electricity during wastewater treatment using a single chamber microbial fuel cell. *Environ. Sci. Technol.* **38**:2281–2285.

Liu, L., F.-B. Li, C.-H. Feng, and X.-Z. Li. 2009. Microbial fuel cell with an azo-dye-feeding cathode. *Appl. Microbiol. Biotechnol.* **85**:175–183.

Liu, Z. D., Z. W. Du, J. Lian, X. Y. Zhu, S. H. Li, and H. R. Li. 2007. Improving energy accumulation of microbial fuel cells by metabolism regulation using *Rhodoferax ferrireducens* as biocatalyst. *Lett. Appl. Microbiol.* **44**:393–398.

Logan, B. E. 2008. *Microbial Fuel Cells*. John Wiley & Sons, Hoboken, NJ.

Logan, B. E. 2009. Exoelectrogenic bacteria that power microbial fuel cells. *Nat. Rev. Microbiol.* **7**:375–381.

Logan, B. E., D. Call, S. Cheng, H. V. M. Hamelers, T. Sleutels, A. W. Jeremiasse, and R. A. Rozendal. 2008. Microbial electrolysis cells for high yield hydrogen gas production from organic matter. *Environ. Sci. Technol.* **42**:8630–8640.

Logan, B. E., B. Hamelers, R. A. Rozendal, U. Schröder, J. Keller, S. Freguia, P. Aelterman, W. Verstraete, and K. Rabaey. 2006. Microbial fuel cells: methodology and technology. *Environ. Sci. Technol.* **40:**5181–5192.

Logan, B. E., and J. M. Regan. 2006. Electricity-producing bacterial communities in microbial fuel cells. *Trends Microbiol.* **14:**512–518.

Lonergan, D. J., H. Jenter, J. D. Coates, E. J. P. Phillips, T. Schmidt, and D. R. Lovley. 1996. Phylogenetic analysis of dissimilatory Fe(III)-reducing bacteria. *J. Bacteriol.* **178:**2402–2408.

Lovley, D. R. 2006a. Bug juice: harvesting electricity with microorganisms. *Nat. Rev. Microbiol.* **4:**497–508.

Lovley, D. R. 2006b. Microbial fuel cells: novel microbial physiologies and engineering approaches. *Curr. Opin. Biotechnol.* **17:**327–332.

Lovley, D. R. 2008a. Extracellular electron transfer: wires, capacitors, iron lungs, and more. *Geobiology* **6:**225–231.

Lovley, D. R. 2008b. The microbe electric: conversion of organic matter to electricity. *Curr. Opin. Biotechnol.* **19:**564–571.

Lovley, D. R., D. E. Holmes, and K. P. Nevin. 2004. Dissimilatory Fe(III) and Mn(IV) reduction. *Adv. Microb. Physiol.* **49:**219–286.

Lovley, D. R., E. J. Phillips, D. J. Lonergan, and P. K. Widman. 1995. Fe(III) and S^0 reduction by *Pelobacter carbinolicus*. *Appl. Environ. Microbiol.* **61:**2132–2138.

Lu, L., N. Q. Ren, D. F. Xing, and B. E. Logan. 2009. Hydrogen production with effluent from an ethanol-H-2-coproducing fermentation reactor using a single-chamber microbial electrolysis cell. *Biosens. Bioelectron.* **24:**3055–3060.

Luo, H. P., G. L. Liu, R. D. Zhang, and L. X. Cao. 2009. Isolation and characterization of electrochemical active bacterial *Pseudomonas aeruginosa* strain RE7. *Huan Jing Ke Xue* **30:**2118–2123.

Marsili, E., D. B. Baron, I. D. Shikhare, D. Coursolle, J. A. Gralnick, and D. R. Bond. 2008. *Shewanella* secretes flavins that mediate extracellular electron transfer. *Proc. Natl. Acad. Sci. USA* **105:**3968–3973.

McKinlay, J. B., and J. G. Zeikus. 2004. Extracellular iron reduction is mediated in part by neutral red and hydrogenase in *Escherichia coli*. *Appl. Environ. Microbiol.* **70:**3467–3474.

Mehta, T., M. V. Coppi, S. E. Childers, and D. R. Lovley. 2005. Outer membrane c-type cytochromes required for Fe(III) and Mn(IV) oxide reduction in *Geobacter sulfurreducens*. *Appl. Environ. Microbiol.* **71:**8634–8641.

Methe, B. A., K. E. Nelson, J. A. Eisen, I. T. Paulsen, W. Nelson, J. F. Heidelberg, D. Wu, M. Wu, N. Ward, M. J. Beanan, R. J. Dodson, R. Madupu, L. M. Brinkac, S. C. Daugherty, R. T. DeBoy, A. S. Durkin, M. Gwinn, J. F. Kolonay, S. A. Sullivan, D. H. Haft, J. Selengut, T. M. Davidsen, N. Zafar, O. White, B. Tran, C. Romero, H. A. Forberger, J. Weidman, H. Khouri, T. V. Feldblyum, T. R. Utterback, S. E. Van Aken, D. R. Lovley, and C. M. Fraser. 2003. Genome of *Geobacter sulfurreducens*: metal reduction in subsurface environments. *Science* **302:**1967–1969.

Meyer, T. E., A. I. Tsapin, I. Vandenberghe, L. De Smet, D. Frishman, K. H. Nealson, M. A. Cusanovich, and J. J. Van Beeumen. 2004. Identification of 42 possible cytochrome c genes in the *Shewanella oneidensis* genome and characterization of six soluble cytochromes. *OMICS* **8:**57–77.

Milliken, C. E., and H. D. May. 2007. Sustained generation of electricity by the spore-forming, Gram-positive, *Desulfitobacterium hafniense* strain DCB2. *Appl. Microbiol. Biotechnol.* **73:**1180–1189.

Min, B., J. R. Kim, S. E. Oh, J. M. Regan, and B. E. Logan. 2005a. Electricity generation from swine wastewater using microbial fuel cells. *Water Res.* **39:**4961–4968.

Min, B. K., S. A. Cheng, and B. E. Logan. 2005b. Electricity generation using membrane and salt bridge microbial fuel cells. *Water Res.* **39:**1675–1686.

Miyawaki, O., and T. Yano. 1992. Electrochemical bioreactor with regeneration of NAD1 by rotating graphite disk electrode with PMS absorbed. *Enzyme Microb. Technol.* **14:**474–478.

Mohan, S. V., G. Mohanakrishna, B. P. Reddy, R. Saravanan, and P. N. Sarma. 2008. Bioelectricity generation from chemical wastewater treatment in mediatorless (anode) microbial fuel cell (MFC) using selectively enriched hydrogen producing mixed culture under acidophilic microenvironment. *Biochem. Eng. J.* **39:**121–130.

Moon, H., I. S. Chang, J. K. Jang, and B. H. Kim. 2005. Residence time distribution in microbial fuel cell and its influence on COD removal with electricity generation. *Biochem. Eng. J.* **27:**59–65.

Mu, Y., K. Rabaey, R. A. Rozendal, Z. G. Yuan, and J. Keller. 2009a. Decolorization of azo dyes in bioelectrochemical systems. *Environ. Sci. Technol.* **43:**5137–5143.

Mu, Y., R. A. Rozendal, K. Rabaey, and J. Keller. 2009b. Nitrobenzene removal in bioelectrochemical systems. *Environ. Sci. Technol.* **43:**8690–8695.

Myers, C. R., and J. M. Myers. 1997. Cloning and sequencing of *cymA*, a gene encoding a tetraheme cytochrome c required for reduction of iron(III), fumarate, and nitrate by *Shewanella putrefaciens* strain MR-1. *J. Bacteriol.* **179:**1143–1152.

Myers, J. M., and C. R. Myers. 2001. Role for outer membrane cytochromes OmcA and OmcB of *Shewanella putrefaciens* MR-1 in reduction of manganese dioxide. *Appl. Environ. Microbiol.* **67:**260–269.

Nealson, K. H., A. Belz, and B. McKee. 2002. Breathing metals as a way of life: geobiology in action. *Antonie Van Leeuwenhoek* **81:**215–222.

Nevin, K. P., B. C. Kim, R. H. Glaven, J. P. Johnson, T. L. Woodard, B. A. Methe, R. J. Didonato, S. F. Covalla, A. E. Franks, A. Liu, and D. R. Lovley. 2009. Anode biofilm transcriptomics reveals outer surface components essential for high density current production in *Geobacter sulfurreducens* fuel cells. *PLoS One* **4:**e5628.

Nevin, K. P., and D. R. Lovley. 2000. Lack of production of electron-shuttling compounds or solubilization of Fe(III) during reduction of insoluble Fe(III) oxide by *Geobacter metallireducens*. *Appl. Environ. Microbiol.* **66:**2248–2251.

Nevin, K. P., and D. R. Lovley. 2002a. Mechanisms for accessing insoluble Fe(III) oxide during dissimilatory Fe(III) reduction by *Geothrix fermentans*. *Appl. Environ. Microbiol.* **68:**2294–2299.

Nevin, K. P., and D. R. Lovley. 2002b. Mechanisms for Fe(III) oxide reduction in sedimentary environments. *Geomicrobiol. J.* **19:**141–159.

Nevin, K. P., H. Richter, S. F. Covalla, J. P. Johnson, T. L. Woodard, A. L. Orloff, H. Jia, M. Zhang, and D. R. Lovley. 2008. Power output and columbic efficiencies from biofilms of *Geobacter sulfurreducens* comparable to mixed community microbial fuel cells. *Environ. Microbiol.* **10:**2505–2514.

Newman, D. K., and R. Kolter. 2000. A role for excreted quinones in extracellular electron transfer. *Nature* **405:**94–97.

Nielsen, M. E., C. E. Reimers, and H. A. Stecher. 2007. Enhanced power from chambered benthic microbial fuel cells. *Environ. Sci. Technol.* **41:**7895–7900.

Nielsen, M. E., C. E. Reimers, H. K. White, S. Sharma, and P. R. Girguis. 2008. Sustainable energy from deep ocean cold seeps. *Energy Environ. Sci.* **1:**584–593.

Nielsen, M. E., D. M. Wu, P. R. Girguis, and C. E. Reimers. 2009. Influence of substrate on electron transfer mechanisms in chambered benthic microbial fuel cells. *Environ. Sci. Technol.* **43:**8671–8677.

Niessen, J., U. Schröder, F. Harnisch, and F. Scholz. 2005. Gaining electricity from in situ oxidation of hydrogen produced by fermentative cellulose degradation. *Lett. Appl. Microbiol.* **41:**286–290.

Niessen, J., U. Schröder, and F. Scholz. 2004. Exploiting complex carbohydrates for microbial electricity generation—a bacterial fuel cell operating on starch. *Electrochem. Commun.* **6:**955–958.

Oh, S. E., and B. E. Logan. 2006. Proton exchange membrane and electrode surface areas as factors that affect power generation in microbial fuel cells. *Appl. Microbiol. Biotechnol.* **70:**162–169.

Oh, S. E., and B. E. Logan. 2007. Voltage reversal during microbial fuel cell stack operation. *J. Power Sources* **167:**11–17.

Pant, D., G. Van Bogaert, L. Diels, and K. Vanbroekhoven. 2010. A review of the substrates used in microbial fuel cells (MFCs) for sustainable energy production. *Bioresour. Technol.* **101:**1533–1543.

Parameswaran, P., H. S. Zhang, C. I. Torres, B. E. Rittmann, and R. Krajmalnik-Brown. 2010. Microbial community structure in a biofilm anode fed with a fermentable substrate: the significance of hydrogen scavengers. *Biotechnol. Bioeng.* **105:**69–78.

Park, D. H., and J. G. Zeikus. 1999. Utilization of electrically reduced neutral red by Actinobacillus succinogenes: physiological function of neutral red in membrane-driven fumarate reduction and energy conservation. *J. Bacteriol.* **181:**2403–2410.

Park, D. H., and J. G. Zeikus. 2000. Electricity generation in microbial fuel cells using neutral red as an electronophore. *Appl. Environ. Microbiol.* **66:**1292–1297.

Park, D. H., and J. G. Zeikus. 2003. Improved fuel cell and electrode designs for producing electricity from microbial degradation. *Biotechnol. Bioeng.* **81:**348–355.

Park, H.-I., D.-K. Kim, Y.-J. Choi, and D. Pak. 2005. Nitrate reduction using an electrode as direct electron donor in a biofilm-electrode reactor. *Proc. Biochem.* **40:**3383–3388.

Park, H. S., B. H. Kim, H. S. Kim, H. J. Kim, G. T. Kim, M. Kim, I. S. Chang, Y. K. Park, and H. I. Chang. 2001. A novel electrochemically active and Fe(III)-reducing bacterium phylogenetically related to *Clostridium butyricum* isolated from a microbial fuel cell. *Anaerobe* **7:**297–306.

Pham, C. A., S. J. Jung, N. T. Phung, J. Lee, I. S. Chang, B. H. Kim, H. Yi, and J. Chun. 2003. A novel electrochemically active and Fe(III)-reducing bacterium phylogenetically related to *Acromonas hydrophila*, isolated from a microbial fuel cell. *FEMS Microbiol. Lett.* **223:**129–134.

Pham, T. H., P. Aelterman, and W. Verstraete. 2009. Bioanode performance in bioelectrochemical systems: recent improvements and prospects. *Trends Biotechnol.* **27:**168–178.

Pham, T. H., N. Boon, P. Aelterman, P. Clauwaert, L. De Schamphelaire, L. Vanhaecke, K. De Maeyer, M. Hofte, W. Verstraete, and K. Rabaey. 2008a. Metabolites produced by *Pseu-*

domonas sp. enable a Gram-positive bacterium to achieve extracellular electron transfer. *Appl. Microbiol. Biotechnol.* **77:**1119–1129.

Pham, T. H., N. Boon, K. De Maeyer, M. Hofte, K. Rabaey, and W. Verstraete. 2008b. Use of *Pseudomonas* species producing phenazine-based metabolites in the anodes of microbial fuel cells to improve electricity generation. *Appl. Microbiol. Biotechnol.* **80:**985–993.

Pham, T. H., J. K. Jang, H. S. Moon, I. S. Chang, and B. H. Kim. 2005. Improved performance of microbial fuel cell using membrane-electrode assembly. *J. Microbiol. Biotechnol.* **15:**438–441.

Phung, N. T., J. Lee, K. H. Kang, I. S. Chang, G. M. Gadd, and B. H. Kim. 2004. Analysis of microbial diversity in oligotrophic microbial fuel cells using 16S rDNA sequences. *FEMS Microbiol. Lett.* **233:**77–82.

Pitts, K. E., P. S. Dobbin, F. Reyes-Ramirez, A. J. Thomson, D. J. Richardson, and H. E. Seward. 2003. Characterization of the *Shewanella oneidensis* MR-1 decaheme cytochrome MtrA: expression in Escherichia coli confers the ability to reduce soluble Fe(III) chelates. *J. Biol. Chem.* **278:**27758–27765.

Potter, M. 1910. On the difference of potential due to the vital activity of microorganisms. *Proc. Univ. Durham Phil. Soc.* **3:**245–249.

Potter, M. 1911. Electrical effects accompanying the decomposition of organic compounds. *Proc. R. Soc. Lond. B* **84:**260–276.

Prasad, D., T. K. Sivaram, S. Berchmans, and V. Yegnaraman. 2006. Microbial fuel cell constructed with a micro-organism isolated from sugar industry effluent. *J. Power Sources* **160:**991–996.

Rabaey, K., N. Boon, M. Hofte, and W. Verstraete. 2005a. Microbial phenazine production enhances electron transfer in biofuel cells. *Environ. Sci. Technol.* **39:**3401–3408.

Rabaey, K., N. Boon, S. D. Siciliano, M. Verhaege, and W. Verstraete. 2004. Biofuel cells select for microbial consortia that self-mediate electron transfer. *Appl. Environ. Microbiol.* **70:**5373–5382.

Rabaey, K., P. Clauwaert, P. Aelterman, and W. Verstraete. 2005b. Tubular microbial fuel cells for efficient electricity generation. *Environ. Sci. Technol.* **39:**8077–8082.

Rabaey, K., G. Lissens, S. D. Siciliano, and W. Verstraete. 2003. A microbial fuel cell capable of converting glucose to electricity at high rate and efficiency. *Biotechnol. Lett.* **25:**1531–1535.

Rabaey, K., S. T. Read, P. Clauwaert, S. Freguia, P. L. Bond, L. L. Blackall, and J. Keller. 2008. Cathodic oxygen reduction catalyzed by bacteria in microbial fuel cells. *ISME J.* **2:**519–527.

Rabaey, K., J. Rodriguez, L. L. Blackall, J. Keller, P. Gross, D. Batstone, W. Verstraete, and K. H. Nealson. 2007. Microbial ecology meets electrochemistry: electricity-driven and driving communities. *ISME J.* **1:**9–18.

Rabaey, K., K. Van de Sompel, L. Maignien, N. Boon, P. Aelterman, P. Clauwaert, L. De Schamphelaire, H. T. Pham, J. Vermeulen, M. Verhaege, P. Lens, and W. Verstraete. 2006. Microbial fuel cells for sulfide removal. *Environ. Sci. Technol.* **40:**5218–5224.

Rabaey, K., and W. Verstraete. 2005. Microbial fuel cells: novel biotechnology for energy generation. *Trend Biotechnol.* **23:**291–298.

Reguera, G., K. D. McCarthy, T. Mehta, J. S. Nicoll, M. T. Tuominen, and D. R. Lovley. 2005. Extracellular electron transfer via microbial nanowires. *Nature* **435:**1098–10101.

Reguera, G., K. P. Nevin, J. S. Nicoll, S. F. Covalla, T. L. Woodard, and D. R. Lovley. 2006. Biofilm and nanowire production leads to increased current in *Geobacter sulfurreducens* fuel cells. *Appl. Environ. Microbiol.* **72:**7345–7348.

Reimers, C. E. 2007. Applications of microelectrodes to problems in chemical oceanography. *Chem. Rev.* **107:**590–600.

Reimers, C. E., P. Girguis, H. A. Stecher, L. M. Tender, N. Ryckelynck, and P. Whaling. 2006. Microbial fuel cell energy from an ocean cold seep. *Geobiology* **4:**123–136.

Reimers, C. E., H. A. Stecher, J. C. Westall, Y. Alleau, K. A. Howell, L. Soule, H. K. White, and P. R. Girguis. 2007. Substrate degradation kinetics, microbial diversity, and current efficiency of microbial fuel cells supplied with marine plankton. *Appl. Environ. Microbiol.* **73:**7029–7040.

Reimers, C. E., L. M. Tender, S. Fertig, and W. Wang. 2001. Harvesting energy from the marine sediment-water interface. *Environ. Sci. Technol.* **35:**192–195.

Ren, Z., L. M. Steinberg, and J. M. Regan. 2008. Electricity production and microbial biofilm characterization in cellulose-fed microbial fuel cells. *Water Sci. Technol.* **58:**617–622.

Ren, Z., T. E. Ward, B. E. Logan, and J. M. Regan. 2007a. Characterization of the cellulolytic and hydrogen-producing activities of six mesophilic *Clostridium* species. *J. Appl. Microbiol.* **103:**2258–2266.

Ren, Z., T. E. Ward, and J. M. Regan. 2007b. Electricity production from cellulose in a microbial fuel cell using a defined binary culture. *Environ. Sci. Technol.* **41:**4781–4786.

Rezaei, F., D. F. Xing, R. Wagner, J. M. Regan, T. L. Richard, and B. E. Logan. 2009. Simultaneous cellulose degradation and electricity production by *Enterobacter cloacae* in a microbial fuel cell. *Appl. Environ. Microbiol.* **75:**3673–3678.

Rhoads, A., H. Beyenal, and Z. Lewandowski. 2005. Microbial fuel cell using anaerobic respiration as an anodic reaction and biomineralized manganese as a cathodic reactant. *Environ. Sci. Technol.* **39:**4666–4671.

Richter, H., M. Lanthier, K. P. Nevin, and D. R. Lovley. 2007. Lack of electricity production by *Pelobacter carbinolicus* indicates that the capacity for Fe(III) oxide reduction does not necessarily confer electron transfer ability to fuel cell anodes. *Appl. Environ. Microbiol.* **73:**5347–5353.

Ringeisen, B. R., E. Henderson, P. K. Wu, J. Pietron, R. Ray, B. Little, J. C. Biffinger, and J. M. Jones-Meehan. 2006. High power density from a miniature microbial fuel cell using *Shewanella oneidensis* DSP10. *Environ. Sci. Technol.* **40:**2629–2634.

Rismani-Yazdi, H., A. D. Christy, B. A. Dehority, M. Morrison, Z. Yu, and O. H. Tuovinen. 2007. Electricity generation from cellulose by rumen microorganisms in microbial fuel cells. *Biotechnol. Bioeng.* **97:**1398–1407.

Roling, W. F. M., B. M. van Breukelen, M. Braster, B. Lin, and H. W. van Verseveld. 2001. Relationships between microbial community structure and hydrochemistry in a landfill leachate-polluted aquifer. *Appl. Environ. Microbiol.* **67:**4619–4629.

Roller, S. D., H. P. Bennetto, G. M. Delaney, J. R. Mason, J. L. Stirling, and C. F. Thurston. 1984. Electron-transfer coupling in microbial fuel cells. 1. Comparison of redox-mediator reduction rates and respiratory rates of bacteria. *J. Chem. Technol. Biotechnol. B-Biotechnol.* **34:**3–12.

Rosenbaum, M., U. Schroder, and F. Scholz. 2005a. Utilizing the green alga *Chlamydomonas reinhardtii* for microbial electricity generation: a living solar cell. *Appl. Microbiol. Biotechnol.* **68:**753–756.

Rosenbaum, M., U. Schröder, and F. Scholz. 2005b. In situ electrooxidation of photobiological hydrogen in a photobioelectrochemical fuel cell based on *Rhodobacter sphaeroides*. *Environ. Sci. Technol.* **39:**6328–6333.

Rosso, K. M., J. M. Zachara, J. K. Fredrickson, Y. A. Gorby, and S. C. Smith. 2003. Nonlocal bacterial electron transfer to hematite surfaces. *Geochim. Cosmochim. Acta* **67:**1081–1087.

Rozendal, R. A., H. V. M. Hamelers, G. t. J. W. Euverink, S. J. Metz, and C. J. N. Buisman. 2006. Principle and perspectives of hydrogen production through biocatalyzed electrolysis. *Int. J. Hydrogen Energy* **31:**1632–1640.

Rozendal, R. A., A. W. Jeremiasse, H. V. M. Hamelers, and C. J. N. Buisman. 2007. Hydrogen production with a microbial biocathode. *Environ. Sci. Technol.* **42:**629–634.

Ryckelynck, N., H. A. Stecher, and C. E. Reimers. 2005. Understanding the anodic mechanism of a seafloor fuel cell: interactions between geochemistry and microbial activity. *Biogeochemistry* **76:**113–139.

Sakakibara, Y., and M. Kuroda. 1993. Electric prompting and control of denitrification. *Biotechnol. Bioeng.* **42:**535–537.

Sanchez, S., A. Arratia, R. Córdova, H. Gomez, and R. Schrebler. 1995. Electron transport in biological processes. II: Electrochemical behaviour of Q10 immersed in a phospholipidic matrix added on a pyrolitic graphite electrode. *Bioelectrochem. Bioenerg.* **36:**67–71.

Schechner, P., E. Kroll, E. Bubis, S. Chervinsky, and E. Zussman. 2007. Silver-plated electrospun fibrous anode for glucose alkaline fuel cells. *J. Electrochem. Soc.* **154:**B942–B948.

Schindler, J. E., and K. R. Honick. 1971. Oxidation-reduction determinations at the mud-water interface. *Limnol. Oceanogr.* **16:**837–840.

Schlereth, D. D., and V. M. Fernandez. 1992. Direct electron transfer between *Alcaligenes eutrophus* Z-1 hydrogenase and glassy carbon electrodes. *Bioelectrochem. Bioenerg.* **28:**473–482.

Schröder, U. 2007. Anodic electron transfer mechanisms in microbial fuel cells and their energy efficiency. *Phys. Chem. Chem. Phys.* **9:**2619–2629.

Schröder, U., J. Niessen, and F. Scholz. 2003. A generation of microbial fuel cells with current outputs boosted by more than one order of magnitude. *Angew. Chem.-Int. Ed.* **42:**2880–2883.

Shea, C., P. Clauwaert, W. Verstraete, and R. Nerenberg. 2008. Adapting a denitrifying biocathode for perchlorate reduction. *Water Sci. Technol.* **58:**1941–1946.

Shelobolina, E. S., C. G. Vanpraagh, and D. R. Lovley. 2003. Use of ferric and ferrous iron containing minerals for respiration by *Desulfitobacterium frappieri*. *Geomicrobiol. J.* **20:**143–156.

Shimomura, T., and R. A. Sanford. 2005. Reductive dechlorination of tetrachloroethene in a sand reactor using a potentiostat. *J. Environ. Qual.* **34:**1435–1438.

Sisler, F. D. 1961. Electrical energy from biochemical fuel cells. *New Sci.* **256:**110–111.

Sisler, F. D. 1962. Electrical energy from microbiological processes. *J. Wash. Acad. Sci.* **52:**181–187.

Skadberg, B., S. L. Geoly-Horn, V. Sangamalli, and J. R. V. Flora. 1999. Influence of pH, current and copper on the biological dechlorination of 2,6-dichlorophenol in an electrochemical cell. *Water Res.* **33:**1997–2010.

Stams, A. J. M., F. A. M. de Bok, C. M. Plugge, M. H. A. van Eekert, J. Dolfing, and G. Schraa. 2006. Exocellular electron transfer in anaerobic microbial communities. *Environ. Microbiol.* **8:**371–382.

Strycharz, S. M., T. L. Woodard, J. P. Johnson, K. P. Nevin, R. A. Sanford, F. E. Löffler, and D. R. Lovley. 2008. Graphite electrode as a sole electron donor for reductive dechlorination of tetrachlorethene by *Geobacter lovleyi*. *Appl. Environ. Microbiol.* **74:**5943–5947.

Sucheta, A., R. Cammack, J. Weiner, and F. A. Armstrong. 1993. Reversible electrochemistry of fumarate reductase immobilized on an electrode surface. Direct voltammetric observations of redox centers and their participation in rapid catalytic electron transport. *Biochemistry* **32:**5455–5465.

Sun, M., Z. X. Mu, Y. P. Chen, G. P. Sheng, X. W. Liu, Y. Z. Chen, Y. Zhao, H. L. Wang, H. Q. Yu, L. Wei, and F. Ma. 2009. Microbe-assisted sulfide oxidation in the anode of a microbial fuel cell. *Environ. Sci. Technol.* **43:**3372–3377.

Sund, C. J., S. McMasters, S. R. Crittenden, L. E. Harrell, and J. J. Sumner. 2007. Effect of electron mediators on current generation and fermentation in a microbial fuel cell. *Appl. Microbiol. Biotechnol.* **76:**561–568.

Surya, A., N. Murthy, and S. Anita. 1994. Tetracyanoquinodimethane (TCNQ) modified electrode for NADH oxidation. *Bioelectrochem. Bioenerg.* **33:**71–73.

Szekeres, S., I. Kiss, T. T. Bejerano, and I. M. Soares. 2001. Hydrogen-dependent denitrification in a two-reactor bio-electrochemical system. *Water Res.* **35:**715–719.

Tanaka, K., C. A. Vega, and R. Tamamushi. 1983. Thionine and ferric chelate compounds as coupled mediators in microbial fuel cells. *Bioelectrochem. Bioenerg.* **11:**289–297.

Tandukar, M., S. J. Huber, T. Onodera, and S. G. Pavlostathis. 2009. Biological chromium(VI) reduction in the cathode of a microbial fuel cell. *Environ. Sci. Technol.* **43:**8159–8165.

Tender, L. M., S. A. Gray, E. Groveman, D. A. Lowy, P. Kauffman, J. Melhado, R. C. Tyce, D. Flynn, R. Petrecca, and J. Dobarro. 2008. The first demonstration of a microbial fuel cell as a viable power supply: powering a meteorological buoy. *J. Power Sources* **179:**571–575.

Tender, L. M., C. E. Reimers, H. A. Stecher, D. E. Holmes, D. R. Bond, D. A. Lowy, K. Pilobello, S. J. Fertig, and D. R. Lovley. 2002. Harnessing microbially generated power on the seafloor. *Nat. Biotechnol.* **20:**821–825.

ter Heijne, A., H. V. M. Hamelers, and C. J. N. Buisman. 2007. Microbial fuel cell operation with continuous biological ferrous iron oxidation of the catholyte. *Environ. Sci. Technol.* **41:**4130–4134.

ter Heijne, A., H. V. M. Hamelers, V. de Wilde, R. A. Rozendal, and C. J. N. Buisman. 2006. a bipolar membrane combined with ferric iron reduction as an efficient cathode system in microbial fuel cells. *Environ. Sci. Technol.* **40:**5200–5205.

Thrash, J. C., and J. D. Coates. 2008. Review: direct and indirect electrical stimulation of microbial metabolism. *Environ. Sci. Technol.* **42:**3921–3931.

Thrash, J. C., J. I. Van Trump, K. A. Weber, E. Miller, L. A. Achenbach, and J. D. Coates. 2007. Electrochemical stimulation of microbial perchlorate reduction. *Environ. Sci. Technol.* **41:**1740–1746.

Thurston, C. F., H. P. Bennetto, G. M. Delaney, J. R. Mason, S. D. Roller, and J. L. Stirling. 1985. Glucose metabolism in a microbial fuel cell. stoichiometry of product formation in a thionine-mediated Proteus vulgaris fuel cell and its relation to coulombic yields. *J. Gen. Microbiol.* **131:**1393–1401.

Torres, C. I., H. S. Lee, and B. E. Rittmann. 2008. Carbonate species as OH- carriers for decreasing the pH gradient between cathode and anode in biological fuel cells. *Environ. Sci. Technol.* **42:**8773–8777.

von Canstein, H., J. Ogawa, S. Shimizu, and J. R. Lloyd. 2007. Secretion of flavins by Shewanella species and their role in extracellular electron transfer. *Appl. Environ. Microbiol.* **74:**615–623.

Vrionis, H. A., R. T. Anderson, I. Ortiz-Bernad, K. R. O'Neill, C. T. Resch, A. D. Peacock, R. Dayvault, D. C. White, P. E. Long, and D. R. Lovley. 2005. Microbiological and geochemical heterogeneity in an in situ uranium bioremediation field site. *Appl. Environ. Microbiol.* **71:**6308–6318.

Wang, G., L. Huang, and Y. Zhang. 2008. Cathodic reduction of hexavalent chromium [Cr(VI)] coupled with electricity generation in microbial fuel cells. *Biotechnol. Lett.* **30:**1959–1966.

Watson, V. J., and B. E. Logan. 2010. Power production in MFCs inoculated with *Shewanella oneidensis* MR-1 or mixed cultures. *Biotechnol. Bioeng.* **105:**489–498.

Weber, K. A., L. A. Achenbach, and J. D. Coates. 2006. Microorganisms pumping iron: anaerobic microbial iron oxidation and reduction. *Nat. Rev. Microbiol.* **4:**752–764.

Weber, K. A., F. W. Picardal, and E. E. Roden. 2001. Microbially catalyzed nitrate-dependent oxidation of biogenic solid-phase Fe(II) compounds. *Environ. Sci. Technol.* **35:**1644–1650.

Wheeler, D. R., J. Nichols, D. Hansen, M. Andrus, S. Choi, and G. D. Watt. 2009. Viologen catalysts for a direct carbohydrate fuel cell. *J. Electrochem. Soc.* **156:**B1201–B1207.

White, H. K., C. E. Reimers, E. E. Cordes, G. F. Dilly, and P. R. Girguis. 2009. Quantitative population dynamics of microbial communities in plankton-fed microbial fuel cells. *ISME J.* **3:**635–646.

Whitfield, M. 1969. E_h as an operational parameter in estuarine studies. *Limnol. Oceanogr.* **14:**547–558.

Willner, I., E. Katz, and N. Lapidot. 1992. Bioelectrocatalysed reduction of nitrate utilizing polythiophene bipyridium enzyme electrodes. *Bioelectrochem. Bioenerg.* **29:**29–45.

Wilson, B. J. 1963. *Experiments with a Magnesium Seawater Cell Incorporating a Bacterial Colonized Cathode.* U.S. Department of Defense, Energy Conversion Branch. U.S. Naval Research Laboratory, Alexandria, VA.

Wrighton, K. C., P. Agbo, F. Warnecke, K. A. Weber, E. L. Brodie, T. Z. DeSantis, P. Hugenholtz, G. L. Andersen, and J. D. Coates. 2008. A novel ecological role of the Firmicutes identified in thermophilic microbial fuel cells. *ISME J.* **2:**1146–1156.

Xie, Y., and S. Dong. 1992. Effect of pH on the electron transfer of cytochrome c on a gold electrode modified with bis (4-pyridyl) disulphide. *Bioelectrochem. Bioenerg.* **29:**71–79.

Xing, D., S. Cheng, B. Logan, and J. M. Regan. 2010. Isolation of the exoelectrogenic denitrifying bacterium Comamonas denitrificans based on dilution to extinction. *Appl. Microbiol. Biotechnol.* **85:**1575–1587.

Xing, D. F., S. A. Cheng, J. M. Regan, and B. E. Logan. 2009. Change in microbial communities in acetate- and glucose-fed microbial fuel cells in the presence of light. *Biosens. Bioelectron.* **25:**105–111.

Xing, D. F., Y. Zuo, S. A. Cheng, J. M. Regan, and B. E. Logan. 2008. Electricity generation by *Rhodopseudomonas palustris* DX-1. *Environ. Sci. Technol.* **42:**4146–4151.

Zhao, F., F. Harnisch, U. Schroder, F. Scholz, P. Bogdanoff, and I. Herrmann. 2006. Challenges and constraints of using oxygen cathodes in microbial fuel cells. *Environ. Sci. Technol.* **40:**5193–5199.

Zuo, Y., S. Cheng, D. Call, and B. E. Logan. 2007. Tubular membrane cathodes for scalable power generation in microbial fuel cells. *Environ. Sci. Technol.* **41:**3347–3353.

INDEX

A

Achromobacter SY8, 203
Acidithiobacillus ferrooxidans, 345
Actinobacteria, 214
   iron reducers of, 181
Aeromonas hydrophila, 332, 335
Agrobacterium, 203
Alcaligenes faecalis, 197
Alkalilimnicola ehrlichii, 199–200, 250–251, 291
Alkaliphilus metalliredigens, 219
Alkaliphilus oremlandii, 286–287, 288, 292, 293
Alteration and weathering experiments, in situ, incubation studies of, 71–73
Aluminum substitution, rate and and extent of Fe(III) reduction and, 105–106
Ammonium, electricity generation and, 337
Anaeromyxobacter, 214
Anaeromyxobacter dehalogenans, 222, 226
Anodes, and cathodes, microbial respiration of, in electrochemical cells, 321–359
   microbial electron transfer to, 333–343
Aquifers, arsenic-impacted, microbial ecology of, 83–84
   arsenic in, microbially driven mobilization of, 82–83
Aquificales, 205
Arabidopsis, 215
Arabidopsis thaliana, 261
Archaea, 314, 334
Arsenate, dissimilatory reduction of, 198
   under anaerobic conditions, 79–81
   microbial resistance to, via arsenic operon, 79
   reduction and mobilization of, by pure cultures of *G. uraniireducens*, 84–86
   respiratory reduction of, and arsenite oxidation, 198–200

Arsenate reductase activity assay, 287–290
Arsenate reductases, phylogenetic tree, based on 203 amino acid sequences, 83, 84
Arsenate respiration, regulation of, 201–202
Arsenate respiratory reductase activity, 201
   arsenate-dependent regulation of, 202
Arsenate respiratory reductases, phylogenetic analysis of, 198, 199
Arsenate-respiring bacteria, cultivation of, enrichment media and, 284–287
   methods for detection of, 283–295
Arsenic, 297–302
   aquifers impacted by, microbial ecology of, 83–84
   dissolved species of, release of, by resting incubated cells, 86, 87
   four oxidation states of, 78
   gaining energy from, 79–81
   in aquifers, microbially driven mobilization of, 82–83
   microbial cycling of, 77–78
   microbial interactions with, 299–301
   in subsurface sediments, 79–81
   microbial transformations of, biochemistry of, 79–82
   in subsurface, 77–90
   optoelectrical properties of, 298
   oxyanions, biochemical transformations of, by microbial cells, 78
   proteome and transcriptome responses to, 203–204
   toxicity of, cause of, 298
Arsenic-based metabolic pathways, 196–197
   ars regulation of, 200–201
   early microbiology studies of, 195–198
   regulation of, 200–203
Arsenic-based metabolism, 195
Arsenic cycle, closing of, 81–82
Arsenic metabolic genes, environmental expression of, 204–205

INDEX

Arsenic metabolic pathways, in prokaryotes, regulation of, 195–210
Arsenic operon, microbial resistance to AsV via, 79
Arsenic speciation, in Ohio River sediment matrix, 284–285, 286, 287
Arsenic-sulfide, biogenic precipitates of, electron micrographs of, 312, 313
 bionanoparticles of, characteristics of, 312–314
Arsenic transformation, assessment of, 283–284
Arsenicals, plasmids conferring resistance to, 197
Arsenite, 79
 oxidation of, 81–82
  arsenate respiratory reduction and, 198
  regulation of, 202–203
Arsenite oxidizers, 198
Arsenomethionine, 297

B

Bacillus arseniciselenatis, 284–285, 291
Bacillus beveridgei, 303, 311
Bacillus infernus, 181
Bacillus selenitireducens, 284–285, 287–290, 291, 292, 293, 305, 310
Bacillus subtilis, 223, 224
Bacterial cell surface reactivity, 6–7
Bacterial surfaces, geochemical reactivity of, 1–9
 two phyla of, 174
 influence on mineral precipitation in natural sediments, 4–5
 on electrolytic cell anodes, 345–347
 on electrolytic cell cathode, 347–348
 structured surface array (S-layer) of, 3
Bacteroidetes, 344
Basalts, seafloor, microbe-metal interactions on, 65–76
 microbial physiology of, 70–71
 microbiology of, 67–71
 production and alteration at mid-ocean ridges, flanks, and seamounts, 66
Batch cultures, and chemostats, 16–17
Benthic microbial fuel cells, and microbial communities, 328–331
 as source of power, 331
Beveridge, Terry J., tribute to, 1–9
Biocathodes, galvanic cell, 343–345
Biochemical probes, 289, 290
Biofilms, responses to metal-reducing bacteria, 228–230
Biogeochemistry, microbe-metal, and geomicrobiology, 53–54
Bioinformatics, and subsequent studies, 23–24
Bionanoparticles, of group 15 and 16 metalloids, applications of, 303–305

Bioremediation, of contaminated environments, cultures and microbial communities for assessing, 248
 proteomics in, 247–259
Bioremediation studies, benefit of microbial communities in, 253
Bradyrhizobium, 222
Brevibacillus, 183
Burton, E. F., 2

C

Carbonates, and silica redux, 6
Carboxydothermus ferrireducens, 188
Carboxydothermus hydrogenoformans, 188, 189
Cathodes, and anodes, microbial respiration of, in electrochemical cells, 321–359
Caulobacter, 222
Caulobacter crescentus, 224
Cellulomonas, iron reduction and, 181
CERCLA sites, 247
Chemostats, batch cultures and, 16–17
Chromate responses, to metal-reducing bacteria, 224
Chrysiogenes arsenatis, 201
Citrobacter, 314
Clostridium butyricum, 182, 335
Cobalt, assimilation into metalloenzymes, 46–48
Common oligo reference standard probe, 262–264
Community approaches, to study microbial communities, 21–22
Community genome arrays, 261
Contaminants, metal, radioactive and toxic, in subsurface aquifers, 117
 reduction of, Fe(III) reduction rates and bioavailability, 109
Contaminated environments, bioremediation of, cultures and microbial communities for assessing, 248
Crystal size, growth conditions of Fe(III) (hydr)oxides and, 103–104
Cultures, pure, isolation, characterization, and proteomic studies on, 252–253
Culturing techniques, and enrichment method, 17
Cyanobacteria, undergoing silification, 5
Cytochromes, c-type, in direct extracellular electron transfer, 188–189
 outer membrane, 338–343

D

Deltaproteobacteria, 284–285
Denaturing gradient gel electrophoresis, PCR primers and, 291–293
Desulfitobacterium, 181, 214
Desulfitobacterium hafniense, 182–183, 188, 291

Desulfitobacterium metallireducens, 188
Desulfobacterales, 331
Desulfobulbus propionicus, 330
Desulfotomaculum, 214
Desulfotomaculum reducens, 185
Desulfovibrio, 214, 215, 216, 217, 219, 224, 225, 228, 229, 231, 232, 233
Desulfovibrio desulfuricans, 223, 227, 251, 252, 336
Desulfovibrio vulgaris, 216, 221, 223, 225, 226, 227, 229–230, 232, 233, 234, 235
Desulfuromonas, 214, 219
Dethiobacter alkaliphilus, 188
Differential in gel electrophoresis analysis, 249–251
Dimethylselenide, 297
Dimethyltelluride, 297
Dissimilatory Fe(II)-reducing bacteria, gram positive, genome sequences and multiheme c-cytochromes of, 188, 189
Dissimilatory Fe(III)-reducing bacteria, 173
    gram-positive, electron transfer and, 184
    microbial fuel cell research, 182–184
Dissimilatory metal-reducing bacteria, cultured, respiration of Fe(III) (hydr)oxides and, 102–103
    gram-positive, genera of, 174, 175–180
        iron reduction by, 173–193
    in dissimilatory reduction of Fe(III), 174, 175–180
    iron-reducing, gram-positive, 174–182
    M. bellicus strain VDY, effect of Fe(II) and Fe(III) on, 166–167
        iterations of iron with, 168–169
    respiratory versatility of, 106–107
    respiring solid Fe minerals, 97
    targeted stimulation of, 93
    transfer of electrons to crystalline Fe(III) (hydr)oxides, 98
DNA-based methods, to study microbial communities, 20
Douglas, S., 5, 6

E

Electricity, generation of, in electrolytes, 324–325
    production of, and Fe(III) oxide reduction, 342–343
        by *Geobacteraceae*, 338–341
        by *Shewanellaceae*, 341–342
Electrochemical cells, classification of, 322
    current direction and polarity of, 322–324
    electrolytic, 322
    fuel cells as, 324–325
    galvanic, 322

    microbial biocatalysis of, 321–359
    microbial respiration of anodes and cathodes in, 321–359
Electrodes, and electrolytic cells as electron donors, 343–348
Electrolytic cell(s), and electrodes as electron donors, 343–348
    anodes of, bacteria on, 345–347
    cathodes of, bacteria on, 347–348
        bacteria respiring on, 346
Electron acceptors, solid-phase, mechanisms of electron transfer to, 184–186
Electron donor responses, to metal-reducing bacteria, 233–236
Electron donors, electrolytic cells and electrodes as, 343–348
Electron recovery, artificial mediators enhancing, 334–335
Electron shuttles, extracellular, 52
Electron transfer, across cell wall, 49–50
    direct, 338–343
    direct extracellular, biochemistry of, 186–188
    direct mechanism of, 186
    extracellular, direct, c-type cytochromes in, 188–189
    indirect, metabolic products of, and exogenous mediators of, 334–335
    mediated mechanisms of, 185–186
    metabolites as mediators of, 336–338
    microbial fuel cell anodes and, 184
    to insoluble metals, 49–50
Electron transport system, organization of, 216–219
Enrichment media, and cultivation of arsenate-respiring bacteria, 284–287
Enterobacter cloacae, 332
Enterococcus gallinarum, 342
Escherichia coli, 220, 262, 303, 336
    phosphate metabolism in, resistance plasmids and, 197
Ethanol, as electron donor to stimulate bioreduction of U(VI), 125–126
Eukaryotes, role in U(VI) contaminated subsurface, 129

F

Fe, reduction potential of, 96
Fe couples, redox potentials of, 96
Fe cycling, in U(IV) stability in subsurface, 151
Fe (hydr)oxides, natural, coprecipitated ions of, 95–96
Fe minerals, physicochemical properties of, 94–96
    solid, dissimilatory metal-reducing bacteria respiring, 97

Fe oxides, in abiotic U(IV) oxidation, 146–151
in U(IV) oxidation, S redox cycling and, 147–149
Fe(II), and Fe(III), use by microorganisms for respiration, 158
oxidation of, nitrate-dependent, 158–168
environmental samples and, 158–159
diversity in, 162–163
pure cultures and, 161–162
(per)chlorate-dependent, 165
solid-phase, coupled with nitrate, 163
Fe(II)-oxidizing microorganisms, anaerobic, phylogenetic diversity of, 160
Fe(III), bioavailability of, and Fe(III) reduction rates, contaminant reduction and, 109
microbial reduction of, 96–98
Michaelis-Menten kinetic dependence of, 100
mineral surface area and, 99
mineralogical and geochemical controls on, 98–107
rates of, factors influencing, 99, 100
Fe(III) (hydr)oxides, as sparingly soluble at circumneutral pH, 98–107
crystallinity of, rate of Fe(III) (hydr)oxide reduction, 101
electron transfer to, at protein level, 102, 103
enhancement of contaminant attenuation and, 108
growth conditions of, crystal size and, 103–104
intrinsic susceptibility to microbial reduction, 102
mean force adhesion upon retraction of *S. oneidensis* on, 102, 103
microbial reduction of, cascade of subsequent redox reactions and, 108–109
impact on transport of metals and radionuclides, 107–109
mineralogical controls on, 93–115
mineral aggregation state of, 104–105
nanometer- and micrometer-sized particles of, 104
physicochemical properties of, in bioreduction experiments, 95
reduction of, Fe(III) (hydr)oxide crystallinity and, 101
indirect, 98
solubility of, organic complexes influencing, 100–101
spectrum of crystallinities and bioavailabilities, 95
structural environment of, and reduction of, 101–102
structures of, 94–95
Fe(III) oxide, microbial reduction of, cation substitution and, 105

Fe(III) oxide reduction, electricity production by, 342–343
Fe(III) phases, reduction of, 93–94
Fe(III)-reducing microbes, 97
Fe(III) reduction, rate and extent of, aluminum substitution and, 105–106
rates of, reduction potential and, 106–107
Fe(III) reduction rates, and bioavailability of Fe(III), contaminant reduction and, 109
Ferrihydrite, 94–95
Firmicutes, 181, 182, 185, 188, 214, 220, 284, 285
FISH-microautoradiography, 20–21
Fluorescence in situ hybridization, 19–20
Fuel cells, as electrochemical cells, 324–325
hydrogen, 325
mediator-less, organisms producing current in, 333, 334
microbial, 325–333. *See* Microbial fuel cells
two electrodes of, 324, 325
Functional gene arrays, 262–264
Fungi, role in U(VI)-contaminated subsurface, 129
Fyfe, W. S., 1, 2, 4, 5, 6

G
Galvanic cell biocathodes, 343–345
Gel electrophoresis, two-dimensional, 250–251
Gene responses, hypothetical, to metal-reducing bacteria, 236–237
Genetics, microbial, 23
Genomes, geocycles to, 13–38
Genomics, 214
Geobacillus, 183
Geobacillus thermodenitrificans, 253
Geobacter, 214, 215, 216, 217, 219, 221, 223, 225, 228, 231, 232, 233, 253
Geobacter clade, subsurface, 128–129
Geobacter lovleyi, 222, 226
Geobacter metallireducens, 222, 223, 226, 237, 339
Geobacter species, genome sequences of, 128
Geobacter sulfurreducens, 186, 187, 220–223, 226, 230, 232, 235–237, 302, 303–305, 307, 310, 332, 338–340, 341
Geobacter uraniireducens, 222–223, 237
cultures inoculated with, dissolved arsenic speciation in, 85, 86
pure cultures of, arsenate reduction and mobilization by, 84–86
Geobacteraceae, 330, 348
electricity production by, 338–341
Geochemical reactivity, of bacterial surfaces, 1–9

GeoChip, activity of, 269–270
  application of, 266–267
  data analysis by, 266–267
  design of, 264–270
  for microbial community analysis, 270–275
  hybridization of, 265
  image analysis by, 265–266
  microarray application and, 267–270
  monitoring of microbial activity with, 261–281
  neural network analysis by, 266–267
  probe design of, 264
  quantitative capability of, 269
  sensitivity of, 268–269
  specificity of, 268
  target preparation for, 264–265
GeoChip 3.0, for global change-related studies, 274
Geocycles, to genomes, 13–38
Geomicrobial activity, in situ study of, 15–16
Geomicrobiology, microbe-metal biogeochemistry and, 53–54
Geothrix, 331
Geothrix fermentans, 330, 337
Glucose, as electron donor to stimulate bioreduction of U(VI), 125–126
Gram-negative bacteria
  distinguishing from gram-positive bacteria, 173–174
Gram-positive bacteria
  distinguishing from gram-negative bacteria, 173–174
Groundwater, effective removal of U(VI) from, 129–131

H
Halobacterium, 205
Herminiimonas arsenicoxydans, 203
Hydrogen fuel cells, 325
Hydrogenobaculum, 205
Hydrothermal environments, hyperthermophile-metal interactions, 39–63
  marine, hyperthermophiles found in, 42
Hydrothermal fluids, metal chemistry of, 40–42
Hyperthermophile metabolism, metals used in, 42–44
Hyperthermophile-metal interactions, in hydrothermal environments, 39–63
Hyperthermophilic iron reducers, description of, 49

I
Immobilization strategies, biomineralization of U(VI) with organophosphate, 129–131
Incubation studies, seafloor, 71–72
  subseafloor, 72–73
Inositol phosphates, 131

Iron. *See also* Fe(II), Fe(III)
  and biosphere, 24–26
  and oxygen, U(IV) oxidation in presence of, 147, 148
  as energy source, 27–29
  as nutrient, 26–27
  assimilation into metalloenzymes, 44–45
  assimilatory reduction of, gram-positive, 174
  at earth's surface, 24–26
  biogeochemical cycle of, microorganisms contributing to, 157
  (bio)geochemistry of, 24–29
  biomineralization of, 163–164
  biomineralization products of, metal adsorption to, 164–165
  contact with dissimilatory metal-reducing microorganisms, 51–52
  in soils and sediments, 213
  oxidation states of, 157–158
  oxidized minerals, as sink for heavy metals and metalloids, 165
  reduction of, environmental constraints on, 53
  responses of, to metal-reducing bacteria, 219
Iron chelators, 52–53
Iron formations, banded, Fe(II)-oxidizing bacteria in production of, 163
Iron oxides, oxyhydroxides, and hydroxides. *See* Fe(III) (hydr)oxides
Iron redox cycle, microbially mediated, 158–159
Iron reduction, 230–231
Iron(II), oxidation of, anaerobic respiratory, 157–172
Isotope array, 22

J
Janthinobacterium lividum, 344

K
Klebsiella pneumoniae, 262

L
Lake Matano, Indonesia, iron-rich, as case study, 29–32
Leptospirillum, 253
Leptothrix discophora, 344–345
Listeria monocytogenes, 186

M
Manganese, and oxygen, U(IV) oxidation in presence of, 152
Manganese oxides, for U(IV) oxidation, 151–152
Marine environments, GeoChip and, 273–274

Mass spectrometry, matrix-assisted laser desorption/ionization-time of flight, 250
  to drive proteomics in bioremediation studies, 250, 251–252
Matrix-assisted laser desorption/ionization-time of flight mass spectrometry, 250, 252
Metabolites, as mediators of electron transfer, 336–338
Metagenomics, metatranscriptomics, and metaproteomics, 22–23
Metal-contaminated site, Lake DePue, Illinois, 270–272
Metal-reducing bacteria, transcriptome analysis of, 213–246
Metal reduction, dissimilatory, 48–53
Metal respiration, chemistry of, 48–49
Metal-respiring bacteria, iron-containing proteins in, 214
Metalloenzymes, metal assimilation into, 44–48
Metalloids, group 15 and group 16, bionanoparticles of, applications of, 303–305
  toxic group 15 and group 16, environmental significance of, 297–302
    microbial interaction with, 302–303
    microbial oxyanion reduction of, nanoparticles formed from, 297–319
Metals, and radionuclides, transport of, microbial Fe(III) (hydr)oxide reduction and, 107–109
  contaminant, radioactive and toxic, in subsurface aquifers, 117
  fermentative reduction of, in U(VI) bioremediation, 127–128
  interactions of microbes with, 65–66
  transition, for biological cells, 213–246
Methanococcus maripaludis, 236
Methylarsonous acid, 297
Microarray technology, 261
Microbe-metal interactions, on seafloor basalts, 65–76
Microbial activity, monitoring of, with GeoChip, 261–281
Microbial communities, and benefit in bioremediation studies, 253
  and cultures, for assessing bioremediation of contaminated environments, 248
  methods available to study, 14–24
Microbial community, GeoChip for analysis of, 270–275
Microbial electron transfer, to anodes, 333–343
Microbial fuel cell anodes, electron transfer and, 184
Microbial fuel cell reactors, and microbial communities, 331–333
Microbial fuel cells, 182, 325–333
  abiotic, 327
  as sustainable energy, 328
  benthic, and microbial communities, 328–331
  direct electron transfer mechanism of, 326
  electrochemically reactive sulfide and, 336
  electron donors in, 327
Microbial life, fluid flow and, below ocean bottom, 67
Microbial oxyanion reduction, of toxic group 15 and 16 metalloids, nanoparticles formed from, 297–319
Microbial respiration, of anodes and cathodes, in electrochemical cells, 321–359
Micrococcus denitrificans, 262
Microorganisms, and processes linked to uranium reduction and immobilization, 117–138
Microscopy, in situ, 14–15
Mineral aggregation, Fe(III) (hydr)oxides and, 104–105
Mineral precipitation, sorption of, 4
Mineralogical controls, on microbial reduction of Fe(III) (hydr)oxides, 93–115
Molecular probes, 290–293
Molybdenum, assimilation into metalloenzymes, 45–46
Montastraea faveolata, 274
Murray, R. G. E., 1–4

N

N oxides, cycling with, 149–151
  for indirect biological and abiotic U(IV) oxidation, 145–146
  for oxidation of U(IV), 144–146
    direct and indirect biological pathways of, 144
Nanoparticles, formed from microbial oxyanion reduction of toxic group 15 and 16 metalloids, 297–319
Nickel, assimilation into metalloenzymes, 46–48
Nitrate, as co-contaminant with U(VI), 117
  solid-phase Fe(II) oxidation coupled with, 163
Nitrate-dependent Fe(II) oxidation, 158–168
Nitrate reductase and dimethyl sulfoxide reductase, 290
Nucleic acid quality, GeoChip analysis and, 267
Nutrient levels, at liquid-surface interface, 229

O

Ocean bottom, fluid flow and microbial life below, 67
Oil- and pesticide-contaminated sites, GeoChip and, 273

Oligotrophic groundwater, and sediments, 228
Organophosphate(s), as route to remediation of uranium-contaminated sites, 131
  biomineralization of U(VI) with, 129–131
  soluble, for subsurface precipitation and stabilization of uranium, 130–131
Oxygen, and iron, U(IV) oxidation in presence of, 147, 148
  and manganese, U(IV) oxidation in presence of, 152
  for oxidation of U(IV), 140–144
    advantages and disadvantages of, 143
Oxygen/nitrate responses, to metal-reducing bacteria, 230–233

P
PCR amplification, primers for, 284
PCR-based methods, to study microbial communities, 18–19
PCR primers, and denaturing gradient gel electrophoresis, 291
Pelobacter carbinolicus, 342
Perchlorate, reduction of, iron inhibition of, 167–168
Photoferrotrophy, 27–28
PhyloChip, 269, 272–273
Pilin, conductive, 338–343
Plasmids, conferring resistance to arsenicals, 197
Probe coverage, GeoChip analysis and, 267
Prokaryotes, arsenic metabolic pathways, 195–210
  basalt-hosted, abundance and diversity of, 67–70
Proteobacteria, 214
δ-Proteobacteria, 214, 219, 223
Proteome analysis, 248
Proteomics, applications of, 249
  environmental, 249
  gel and gel-free, 249–253
  in bioremediation, 247–259
  mass-spectrometry-driven, in bioremediation studies, 250, 251–252
Proteus vulgaris, 336
Pseudomonas, 198
Pseudomonas aeruginosa, 218, 226, 303, 337
Pseudomonas fluorescens, 344
Pseudomonas syringae, 262
Pyrobaculum, 203–204
Pyrobaculum aerophilum, 204
Pyrococcus furiosus, 262

R
Radionuclides, and metals, transport of, microbial Fe(III) (hydr)oxide reduction and, 107–109
Ralstonia metallireducens, 226

Reticulitermes flavipes, 312–314
Rhodobacter sphaeroides, 303
Rhodobacter strain SW2, 15
Rhodoferax, 331
Rhodoferax ferrireducens, 222
Rhodopseudomonas palustris, 332, 342
RNA (rRNA) gene, small-subunit ribosomal, 118
Rock, alteration at and below seafloor, 66

S
S redox cycling, U(IV) oxidation by Fe oxides and, 147–149
16S rRNA/DNA molecule, 18
Saccharomyces cerevisiae, 198
Scanning electron microscopy, 15
Seafloor incubation studies, 71–72
Selenate, dissimilatory reduction of, biochemistry of, 302
Selenium, 297–302
  bionanoparticles of, characteristics of, 311–312
  elemental, bionanoparticles of, characteristics of, 305–307
    precipitates, electron micrographs of, 305, 306
  microbial interaction with, 299–301
  optoelectrical properties of, 298
  oxyanion reduction of, 302
  toxicity of, cause of, 298
Selenocysteine, 297
Selenomethionine, 297
Shewanella, 201–202, 214–217, 219, 224, 228, 231, 233, 237, 262, 290, 291, 304, 305, 314, 337
Shewanella oneidensis, 186, 187, 218–220, 222, 224–228, 230, 231–233, 237, 252, 262, 302, 338, 339, 341, 342
Shewanella putrefaciens, 218
Shewanellaceae, 331
  electricity production by, 341–342
Siderophore system, and strontium responses, 228
Single-cell approaches, to study microbial communities, 20–21
Stable isotope probing, 21–22
Strontium responses, to metal-reducing bacteria, 228
Subseafloor incubation studies, 72–73
Subsurface, microbial transformations of arsenic in, 77–90
Subsurface *Geobacter* clade, 128–129
Sulfate reduction, 230–231
Sulfurihydrogenibium, 205
Sulfurihydrogenibium yellowstonense, 205
Sulfurospirillum, 204, 285–287, 290
Sulfurospirillum arsenophilus, 290

Sulfurospirillum barnesii, 290, 291, 292, 293
Sulfurospirillum deleyianum, 290
Surface sorbates and coprecipitates, Fe(III) (hydr)oxides and, 105–106

T
Tellurate, microbial respiration of, 302–303
Tellurium, 297–302
  biogenic, precipitate morphology of, control of, 307–311
  elemental, bionanoparticles of, 307–311
  microbial interaction with, 299–301
  optoelectrical properties of, 298
  toxicity of, cause of, 298
Tellurocysteine, 297
Telluromethionine, 297
Terrestrial sites, GeoChip and, 274
Terrestrial subsurface, ecology of U(VI)-reducing taxa in, 121–128
  soils and sediments of, environmental conditions in, 121–123
Thermincola, 186
Thermincola ferriacetica, 183
Thermincola potens, 183, 188, 189
Thermoanaerobacter, 181, 186, 214
Thiobacillus denitrificans, 144–145, 226
Transcriptome analysis, of metal-reducing bacteria, 213–246
Transcriptomics, 215–216
Transmission electron microscopy, 15
Transposon mutagenesis, 230
Tungsten, assimilation into metalloenzymes, 45–46

U
U(IV), oxidation of, abiotic, 140–143
    Fe oxides in, 146–147, 148
    Mn oxides in, 151–152
  biological, 143–144
  direct and indirect processes leading to, 139–156
  direct biological, 144–145
  environmental oxidants for, 140, 1412
  in presence of oxygen and iron, 147, 148
  in presence of oxygen and manganese, 152
  indirect biological and abiotic, 145–146
  iron oxides for, biological pathways of, 147, 148
  oxygen for, 140–144
    advantages and disadvantages of, 143
U(IV) oxides, 119
Uraninite. *See* U(IV)

Uranium, bioimmobilization of, chemistry of, 118–120
  bioreduction of, 226
  contamination from, as environmental problem, 226
  environments contaminated by, taxonomic profiling of 16S rRNA gene sequences and, 121, 124
  geochemistry of, 120
  hexavalent. *See* U(VI)
  immobilization of, "biostimulation" of, 120
  mobility in porous media, complexation and redox reactions controlling, 119
  reduced, long-term stability in subsurface, 139–140
  reduction and immobilization of, microorganisms and processes linked to, 117–138
  responses to metal-reducing bacteria, 226–228
  subsurface precipitation and stabilization of, soluble organophosphates for, 130–131
Uranium-contaminated sites, remediation of, organophosphates as route to, 131
  stimulation of native microbiota in, 139
Uranium(IV). *See* U(IV)
U(VI), biomineralization of, with organophosphate, 129–131
  bioreduction of, ethanol or glucose in stimulation of, 125–126
  bioremediation, fermentative reduction in, 127–128
    investigations of, 125
  immobilization of, biostimulation of, electron donor influencing, 127
  microbiological immobilization from water, 119
  oxidation of, nitrate-dependent, bacteria to catalyze, 150
    soluble Fe as inhibitor of, 150
  reduction methods, unifying biochemical features of, 120
  reduction of, and Fe(III) reduction, organisms capable of, 121
    to U(IV), 139, 140
  reductive immobilization of, in remediation, 117
    via biostimulation, 124–125
  removal from groundwater, 129–131
U(VI) contaminated subsurface, fungi in, role of, 129
  microbial eukaryotes in, role of, 129
U(VI)-reducing microorganisms, phylogenetic tree of, 120, 122–123
U(VI)-reducing taxa, cultivated, metabolic and phylogenetic diversity and, 120–121
  ecology of, in terrestrial subsurface, 121–128

V

Veillonella atypica, 304, 305, 307, 309, 312

W

Water, and sediments, chemistry of, in subsurface, microbial metabolism and, 121–123

Whole-genome ORF arrays, 262
Whole-genome sequence approaches (phylogenomics), 18–19
Wolinella succinogenes, 291